Periodic Table of the Elements

Main-group elements

Legend:
- Metals
- Nonmetals
- Noble gases

Transition elements

Period	1 1A	2 2A	3 3B	4 4B	5 5B	6 6B	7 7B	8 8B	9 8B	10 8B	11 1B	12 2B	13 3A	14 4A	15 5A	16 6A	17 7A	18 8A
1	1 H 1.00794																	2 He 4.00260
2	3 Li 6.941	4 Be 9.01218											5 B 10.811	6 C 12.011	7 N 14.0067	8 O 15.9994	9 F 18.9984	10 Ne 20.1797
3	11 Na 22.9898	12 Mg 24.3050											13 Al 26.9815	14 Si 28.0855	15 P 30.9738	16 S 32.066	17 Cl 35.4527	18 Ar 39.948
4	19 K 39.0983	20 Ca 40.078	21 Sc 44.9559	22 Ti 47.88	23 V 50.9415	24 Cr 51.9961	25 Mn 54.9381	26 Fe 55.847	27 Co 58.9332	28 Ni 58.693	29 Cu 63.546	30 Zn 65.39	31 Ga 69.723	32 Ge 72.61	33 As 74.9216	34 Se 78.96	35 Br 79.904	36 Kr 83.80
5	37 Rb 85.4678	38 Sr 87.62	39 Y 88.9059	40 Zr 91.224	41 Nb 92.9064	42 Mo 95.94	43 Tc (98)	44 Ru 101.07	45 Rh 102.906	46 Pd 106.42	47 Ag 107.868	48 Cd 112.411	49 In 114.818	50 Sn 118.710	51 Sb 121.76	52 Te 127.60	53 I 126.904	54 Xe 131.29
6	55 Cs 132.905	56 Ba 137.327	57 *La 138.906	72 Hf 178.49	73 Ta 180.948	74 W 183.84	75 Re 186.207	76 Os 190.23	77 Ir 192.22	78 Pt 195.08	79 Au 196.967	80 Hg 200.59	81 Tl 204.383	82 Pb 207.2	83 Bi 208.980	84 Po (209)	85 At (210)	86 Rn (222)
7	87 Fr (223)	88 Ra 226.025	89 †Ac 227.028	104 Rf (261)	105 Db (262)	106 Sg (263)	107 Bh (262)	108 Hs (265)	109 Mt (266)	110 Ds (281)	111 Rg (272)	112 ** (285)	114 ** (289)		116 ** (292)			

Lanthanide series

58 Ce 140.115	59 Pr 140.908	60 Nd 144.24	61 Pm (145)	62 Sm 150.36	63 Eu 151.965	64 Gd 157.25	65 Tb 158.925	66 Dy 162.50	67 Ho 164.930	68 Er 167.26	69 Tm 168.934	70 Yb 173.04	71 Lu 174.967

†Actinide series

90 Th 232.038	91 Pa 231.036	92 U 238.029	93 Np 237.048	94 Pu (244)	95 Am (243)	96 Cm (247)	97 Bk (247)	98 Cf (251)	99 Es (252)	100 Fm (257)	101 Md (258)	102 No (259)	103 Lr (260)

** Not yet named

Notes: (1) Values in parentheses are the mass numbers of the most common or most stable isotopes of radioactive elements. (2) Some elements adjacent to the stair-step line between the metals and nonmetals have a metallic appearance but some nonmetallic properties. These elements are often called metalloids or semimetals. Almost every list includes Si, Ge, As, Sb, and Te. Some also include B, At, and/or Po. There is no general agreement on just which elements are so designated.

The Elements

Name	Symbol	Atomic Number	Relative Atomic Mass*	Name	Symbol	Atomic Number	Relative Atomic Mass*
Actinium	Ac	89	227.028	Mendelevium	Md	101	(258)
Aluminum	Al	13	26.9815	Mercury	Hg	80	200.59
Americium	Am	95	(243)	Molybdenum	Mo	42	95.94
Antimony	Sb	51	121.757	Neodymium	Nd	60	144.24
Argon	Ar	18	39.948	Neon	Ne	10	20.1797
Arsenic	As	33	74.9216	Neptunium	Np	93	237.048
Astatine	At	85	(210)	Nickel	Ni	28	58.693
Barium	Ba	56	137.327	Niobium	Nb	41	92.9064
Berkelium	Bk	97	(247)	Nitrogen	N	7	14.0067
Beryllium	Be	4	9.01218	Nobelium	No	102	(259)
Bismuth	Bi	83	208.980	Osmium	Os	76	190.23
Bohrium	Bh	107	(262)	Oxygen	O	8	15.9994
Boron	B	5	10.811	Palladium	Pd	46	106.42
Bromine	Br	35	79.904	Phosphorus	P	15	30.9738
Cadmium	Cd	48	112.411	Platinum	Pt	78	195.08
Calcium	Ca	20	40.078	Plutonium	Pu	94	(244)
Californium	Cf	98	(251)	Polonium	Po	84	(209)
Carbon	C	6	12.011	Potassium	K	19	39.0983
Cerium	Ce	58	140.115	Praseodymium	Pr	59	140.908
Cesium	Cs	55	132.905	Promethium	Pm	61	(145)
Chlorine	Cl	17	35.4527	Protactinium	Pa	91	231.036
Chromium	Cr	24	51.9961	Radium	Ra	88	226.025
Cobalt	Co	27	58.9332	Radon	Rn	86	(222)
Copper	Cu	29	63.546	Rhenium	Re	75	186.207
Curium	Cm	96	(247)	Rhodium	Rh	45	102.906
Darmstadtium	Ds	110	(281)	Roentgenium	Rg	111	(272)
Dubnium	Db	105	(262)	Rubidium	Rb	37	85.4678
Dysprosium	Dy	66	162.50	Ruthenium	Ru	44	101.07
Einsteinium	Es	99	(252)	Rutherfordium	Rf	104	(261)
Erbium	Er	68	167.26	Samarium	Sm	62	150.36
Europium	Eu	63	151.965	Scandium	Sc	21	44.9559
Fermium	Fm	100	(257)	Seaborgium	Sg	106	(263)
Fluorine	F	9	18.9984	Selenium	Se	34	78.96
Francium	Fr	87	(223)	Silicon	Si	14	28.0855
Gadolinium	Gd	64	157.25	Silver	Ag	47	107.868
Gallium	Ga	31	69.723	Sodium	Na	11	22.9898
Germanium	Ge	32	72.61	Strontium	Sr	38	87.62
Gold	Au	79	196.967	Sulfur	S	16	32.066
Hafnium	Hf	72	178.49	Tantalum	Ta	73	180.948
Hassium	Hs	108	(265)	Technetium	Tc	43	(98)
Helium	He	2	4.00260	Tellurium	Te	52	127.60
Holmium	Ho	67	164.930	Terbium	Tb	65	158.925
Hydrogen	H	1	1.00794	Thallium	Tl	81	204.383
Indium	In	49	114.818	Thorium	Th	90	232.038
Iodine	I	53	126.904	Thulium	Tm	69	168.934
Iridium	Ir	77	192.22	Tin	Sn	50	118.710
Iron	Fe	26	55.847	Titanium	Ti	22	47.88
Krypton	Kr	36	83.80	Tungsten	W	74	183.84
Lanthanum	La	57	138.906	Uranium	U	92	238.029
Lawrencium	Lr	103	(260)	Vanadium	V	23	50.9415
Lead	Pb	82	207.2	Xenon	Xe	54	131.29
Lithium	Li	3	6.941	Ytterbium	Yb	70	173.04
Lutetium	Lu	71	174.967	Yttrium	Y	39	88.9059
Magnesium	Mg	12	24.3050	Zinc	Zn	30	65.39
Manganese	Mn	25	54.9381	Zirconium	Zr	40	91.224
Meitnerium	Mt	109	(266)				

* These atomic masses are relative to a value of exactly 12 assigned to the carbon isotope $^{12}_{6}C$. Values in parentheses are the mass numbers of the most common or most stable isotopes of radioactive elements.

The Chemistry of Everything

The Chemistry of Everything

KIMBERLEY
WALDRON

Regis University

PEARSON
Prentice
Hall

Upper Saddle River, NJ 07458

Library of Congress Cataloging-in-Publication Data

Waldron, Kimberley.
    The chemistry of everything / Kimberley Waldron.
       p. cm.
    Includes bibliographical references and index.
    ISBN 0-13-008522-7
    1. Chemistry—Textbooks. I. Title.
QD31.3.W35 2007
540—dc22

                    2005035829

Senior Editor: *Kent Porter Hamann*
Development Editor: *Irene Nunes*
Production Editor: *Donna Young*
Media Editor: *Michael J. Richards*
Project Manager: *Crissy Dudonis*
Art Direction and Cover Design: *Kenny Beck*
Editor in Chief, Science: *Daniel Kaveney*
Editor in Chief, Development: *Ray Mullaney*
Marketing Manager: *Steve Sartori*
Director of Marketing: *Patrick Lynch*
Executive Managing Editor: *Kathleen Schiaparelli*
Assistant Managing Editor: *Beth Sweeten*
Manufacturing Manager: *Alexis Heydt-Long*
Director of Creative Services: *Paul Belfanti*
Creative Director: *Juan López*
Manufacturing Buyer: *Alan Fischer*
Editorial Assistant: *Joya Carlton*
Production Assistant: *Nancy Bauer*
Marketing Assistants: *Larry Grodsky, Renee Hanenberg*
Interior Designer: *Maureen Eide*
Senior Managing Editor, Visual Assets: *Patricia Burns*
Manager, Production Technologies: *Matthew Haas*
Managing Editor, Visual Assets: *Abigail Bass*

Visual Development Editor: *Jay McElroy*
Visual Assets Production Editor: *Connie Long*
Production Manager and Illustration Art Director:
   Sean Hogan
Assistant Production Manager, Illustration: *Ronda Whitson*
Illustrations: Lead Illustrator: *Royce Copenheaver*
Manager of Electronic Composition: *Allyson Graesser*
Electronic Page Makeup: *Progressive Information Technologies*
Managing Editor, Science and Math Media: *Nicole M. Jackson*
Assistant Managing Editor, Science Supplements:
   Karen Bosch
Media Production Editors: *Dana Dunn, Ashley Booth*
Copyediting and Proofreading: *Write With, Inc.*
Director, Image Resource Center: *Melinda Patelli-Reo*
Manager, Rights and Permissions: *Zina Arabia*
Manager, Visual Research: *Beth Boyd-Brenzel*
Manager, Cover Visual Research and Permissions:
   Karen Sanatar
Image Permission Coordinator: *Robert Farrell*
Photo Researcher: *Yvonne Gerin*
Cover Photo: *Earth inside Water Droplet/David*
   Trood/Photonica/Getty Images, Inc.
Other image credits appear in the backmatter.

© 2007 Pearson Education, Inc.
Pearson Prentice Hall
Pearson Education, Inc.
Upper Saddle River, NJ 07458

Printed in the United States of America.

10 9 8 7 6 5 4 3 2 1

ISBN 0-13-008522-7

Pearson Education LTD., *London*
Pearson Education Australia PTY, Limited, *Sydney*
Pearson Education Singapore, Pte. Ltd
Pearson Education North Asia Ltd, *Hong Kong*
Pearson Education Canada, Ltd., *Toronto*
Pearson Educación de Mexico, S.A. de C.V.
Pearson Education—Japan, *Tokyo*
Pearson Education Malaysia, Pte. Ltd

About the Author

Dr. Kimberley Waldron is an Associate Professor of Chemistry at Regis University where, in addition to lower division courses for majors and nonmajors, she teaches advanced courses in the fields of inorganic and biological chemistry. She also serves as Director of the Regis Environmental Studies Program. Dr. Waldron received her bachelor's degree in chemistry from the University of Virginia and her doctoral degree in inorganic chemistry from Virginia Commonwealth University. She then worked as a postdoctoral fellow at the California Institute of Technology in the area of bioinorganic chemistry before joining the faculty at Regis University as a Clare Boothe Luce Assistant Professor of Chemistry. Dr. Waldron lives in Denver, Colorado, with her friends and family and her two dogs and one cat.

For My Dad

Brief Contents

To the Student   xv

Preface   xvii

1   Everything   AN OVERVIEW OF THE COMPOSITION OF MATTER
AND THE WAY SCIENTISTS STUDY IT   1

2   Dirt   HOW ATOMS INTERACT WITH ONE ANOTHER: AN INTRODUCTION TO
CHEMICAL BONDS AND SIMPLE REACTIONS   38

3   Diamonds   AN EXAMINATION OF CARBON ALLOTROPES, COVALENT BONDING,
AND THE STRUCTURE OF SIMPLE ORGANIC MOLECULES   82

4   Salt   THE STUDY OF THE BEHAVIOR OF IONS, INCLUDING ACIDS AND BASES,
AND THE NOTION OF EQUILIBRIUM   124

5   Film   THE STUDY OF LIGHT AND HOW ELECTRONS RESPOND TO IT   164

6   Sunshine   A STUDY OF NUCLEAR EVENTS AND THE INHERENT INSTABILITY
OF (SOME) ATOMS   208

7   Water   A LOOK AT INTERMOLECULAR INTERACTIONS, PARTICULARLY THOSE
INVOLVING THE UNIQUE PROPERTIES OF WATER   250

8   Air   A STUDY OF THE GASEOUS ATMOSPHERE IN WHICH WE LIVE AND THE LAWS
THAT DICTATE THE BEHAVIOR OF GASES   290

9   Explosives   A STUDY OF ORGANIC MOLECULES WITH SIMPLE FUNCTIONAL GROUPS
AND THE FORCES WITHIN THEM   334

10   Chains I   A STUDY OF SYNTHETIC POLYMERS AND THE WAYS THEY IMPROVE EVERYDAY LIFE   382

11   Chains II   A SURVEY OF NATURAL POLYMERS, INCLUDING PROTEINS AND NUCLEIC ACIDS   420

12   Groceries   THE CHEMISTRY OF THE FOODS WE EAT   462

13   Drugs I   HOW DRUGS ARE DESIGNED TO BENEFIT THE HUMAN BODY   508

14   Drugs II   THE DARK SIDE OF DRUG USE   550

Appendices

A   Working with Measured Numbers   A-1

B   Basic Algebra Review   A-7

C   Glossary   A-11

D   Answers to Odd-numbered Questions   A-20

Photo Credits   C-1

Index   I-1

Contents

To the Student   xv

Preface   xvii

1 Everything AN OVERVIEW OF THE COMPOSITION OF MATTER AND THE WAY SCIENTISTS STUDY IT   1

1.1   Atoms—the Basis of Everything   7

1.2   Our First Chemical Reaction   9

1.3   A Peek into Your Medicine Cabinet   14

1.4   Numbers in Science—SI Units and Conversion Factors   17

1.5   The Metric Epicurean   23

Students Often Ask   What Is the Volume of One Drop?   24

1.6   Chemistry in the Limelight   27

Summary   33   •   Key Terms   33   •   Questions   33
World Wide Web Resources   37

2 Dirt HOW ATOMS INTERACT WITH ONE ANOTHER: AN INTRODUCTION TO CHEMICAL BONDS AND SIMPLE REACTIONS   38

2.1   What Is Dirt?   40

Students Often Ask   How Much Gold Is Left in the Ground?   46

2.2   Why Do Minerals Exist?   46

2.3   Why Neutrons Matter   54

2.4   Coming to Terms with the Very Large and Very Small   58

2.5   Atmospheric Dirt: An Introduction to Redox Chemistry   64

2.6   Compass or Bacterium?   67

Students Often Ask   Why Are Gemstones Different Colors?   70

Issues in Chemistry   The Dark Side of Chemicals: Asbestos and Baby Powder   71

2.7   Dirt as Forensic Evidence   72

Summary   76   •   Key Terms   77   •   Questions   77
World Wide Web Resources   81

3 Diamonds AN EXAMINATION OF CARBON ALLOTROPES, COVALENT BONDING, AND THE STRUCTURE OF SIMPLE ORGANIC MOLECULES   82

3.1   Why Is Carbon Special?   84

3.2   Do Diamonds Really Last Forever?   86

3.3   Graphite and the Multiple Bond   90

3.4   Buckyballs and the Concept of Resonance   94

3.5   Carbon Baguettes: The Development of Nanotubes   97

Beyond the Ordinary   Nanoears   99

3.6   Organic Molecules and Electronic Bookkeeping   100

3.7   Molecules on Your Toothbrush   106

Students Often Ask   What Makes New Cars Smell So Good?   110

3.8   Molecules in Three Dimensions   110

Students Often Ask   When Drawing an Electron Dot Structure for a Tetrahedral Molecule, Does It Matter Which Atom I Show at Which Vertex of the Tetrahedron?   113

3.9   Silicon: Carbon's Big Sister   116

Students Often Ask   Is a Life-Form Based on Silicon Rather than Carbon Possible?   116

Summary   119   •   Key Terms   119   •   Questions   119
World Wide Web Resources   123

4 Salt THE STUDY OF THE BEHAVIOR OF IONS, INCLUDING ACIDS AND BASES, AND THE NOTION OF EQUILIBRIUM   124

4.1   Ionic Liquid: A Contradiction in Terms   126

4.2   Egyptian Mummies and the Polyatomic Ion   129

4.3   Keeping Your Mummy Dry   132

4.4   Gatorade and the Chemistry of Electrolytes   137

Students Often Ask   Is Fortified Water Useful for Ion Replacement or Is It Just Very Expensive Water?   140

4.5   Not Your Father's Tomato   141

Students Often Ask   How Do Home Water Purifiers Remove Ions from Tap Water?   145

4.6   The Autoionization of Water   147

Chemistry at the Crime Scene   Determining Time of Death   148

Beyond the Ordinary   Polka-dotted Airplanes   152

4.7   Bad Air, Bad Water   154

Summary   159   •   Key Terms   159   •   Questions   160
World Wide Web Resources   163

5 Film THE STUDY OF LIGHT AND HOW ELECTRONS RESPOND TO IT   164

5.1   Electron Headquarters   167

5.2   Electrons on the Move   172

5.3   Smile! Electrons Moving Through Film   174

Students Often Ask   How Do Photogray Sunglasses Turn Dark Outdoors?   177

5.4   More Rambling Electrons: Electrochemical Cells   179

5.5   Working Batteries   183

Students Often Ask   Do Electric Eels Really Have Electricity in Them?   185

5.6   Electron Movement in Organic Molecules: Free Radicals   186

Students Often Ask   Why Does Some Fruit Turn Brown When You Cut It?   191

5.7   Still More Rambling Electrons: Light-Driven Reactions   192

5.8   Electrons at the Beach and in the Darkroom   198

Students Often Ask   How Are Colors Incorporated into Color Film?   201

Summary   203   •   Key Terms   203   •   Questions   204
World Wide Web Resources   207

6 sunshine A STUDY OF NUCLEAR EVENTS AND THE INHERENT INSTABILITY OF (SOME) ATOMS 208

6.1 Weight Gain for Atoms: Fusion 210

6.2 Alchemy 214

Students Often Ask How Are New Elements Confirmed and Who Gets to Name Them? 218

6.3 Weight Loss for Atoms: Radioactive Decay 219

Issues in Chemistry Three Women 224

6.4 It's A Wonderful Half-Life 226

Beyond the Ordinary A New Old Way of "Seeing" Alpha Particles 230

Students Often Ask How Do Smoke Detectors Work? 231

6.5 Power from the Nucleus: Fission 231

Students Often Ask What Is Weapons-Grade Uranium? 236

6.6 Living Organisms and Radiation 237

Issues in Chemistry Who's Going to Take the Trash Out? 238

Issues in Chemistry The Tragedy at Chernobyl 243

Summary 245 • Key Terms 245 • Questions 245
World Wide Web Resources 249

7 Water A LOOK AT INTERMOLECULAR INTERACTIONS, PARTICULARLY THOSE INVOLVING THE UNIQUE PROPERTIES OF WATER 250

7.1 Songbirds and Hydrogen Atoms 252

7.2 Water with Water 256

Students Often Ask How Can Insects Walk on Water? 262

7.3 Solid Water 263

Students Often Ask Why Is Salt Used to Melt Ice on Wintry Roads? 266

Students Often Ask Can It Really Be Too Cold to Snow? 268

7.4 Shifting Phases 268

7.5 How to Boil Water 272

Students Often Ask Does Adding Salt to Water Make Pasta Cook More Quickly? 276

7.6 From Aerosol Cans to Scuba Diving 276

7.7 Like Dissolves Like 279

Chemistry at the Crime Scene Death by Drowning 282

Students Often Ask How Do Detergents Get Greasy Dirt Out of Clothing? 284

Summary 286 • Key Terms 286 • Questions 287
World Wide Web Resources 289

8 Air A STUDY OF THE GASEOUS ATMOSPHERE IN WHICH WE LIVE AND THE LAWS THAT DICTATE THE BEHAVIOR OF GASES 290

8.1 Trouble in Tokyo 293

8.2 "Better Killing Through Chemistry" 296

Beyond the Ordinary Mind Your Own Bees-ness 298

8.3   Keep a Lid on It   299

Students Often Ask   Why Are Tires Filled with Air Instead of Being Solid?   302

8.4   Under Pressure   305

8.5   Turn Up the Thermostat   308

Students Often Ask   How Does a Thermos Keep Hot Things Hot and Cold Things Cold?   309

Students Often Ask   Why Is It More Difficult for an Airplane to Take Off at High Altitude or on a Hot Day?   312

8.6   Designer Gas Laws   316

Issues in Chemistry   Do Human Pheromones Really Exist?   322

8.7   Stay Cool: Ozone and Global Warming   323

Issues in Chemistry   The Tuvalu Blues   327

Summary   328   •   Key Terms   329   •   Questions   329
World Wide Web Resources   333

9 Explosives A STUDY OF ORGANIC MOLECULES WITH SIMPLE FUNCTIONAL GROUPS AND THE FORCES WITHIN THEM   334

9.1   The Smoking Gun   336

Beyond the Ordinary   Disappearing Ships   341

9.2   Keeping Track of Hydrocarbons   341

Students Often Ask   What Is the Meaning of the Octane Number Reported for Gasoline?   349

9.3   Cocktails and Anesthetics   350

Issues in Chemistry   MTBE in Gasoline   356

9.4   Bigger Bangs   359

Students Often Ask   Why Do People with Heart Conditions Take Nitroglycerin?   362

9.5   High Explosives, Low Explosives   363

Students Often Ask   What's the Most Explosive Compound Known?   366

9.6   Chaos   367

9.7   Arson and the Analysis of Explosives   370

Chemistry at the Crime Scene   The Dog Nose Knows 375

Summary   376   •   Key Terms   376   •   Questions   377
World Wide Web Resources   381

10 Chains I A STUDY OF SYNTHETIC POLYMERS AND THE WAYS THEY IMPROVE EVERYDAY LIFE   382

10.1   Dubble Bubble   384

10.2   Monomers to Polymers   388

Students Often Ask   If Super Glue Adheres Immediately to Everything It Touches, Why Doesn't It Adhere to the Inside Walls of the Super-glue Tube?   391

10.3   Groovy, Baby: The Science Behind Polyester   392

10.4   Design-Your-Own Plastics   397

10.5   Get Out of My Way!   403

Chemistry at the Crime Scene   Synthetic Fibers   405

10.6   The Weakest Link   407

Students Often Ask   If Esters Are Hydrolyzed by Water, Why Can
We Wash Polyester Clothes?   412

Issues in Chemistry   University of Arizona Garbage Project   413

Summary   414   •   Key Terms   415   •   Questions   415
World Wide Web Resources   419

11 Chains II   A SURVEY OF NATURAL POLYMERS, INCLUDING PROTEINS AND NUCLEIC ACIDS   420

11.1   Tapping Mother Nature   422

11.2   Amino Acids: Nature's Building Blocks   427

Students Often Ask   What Is the Difference Between a Peptide and a Protein?   435

Issues in Chemistry   Transgenic Plants—Friend or Foe?   436

11.3   Protein: Nature's Jack of All Trades   437

11.4   The Secret Language of Chains   445

11.5   Genetic Engineering: The DNA Shuffle   452

Chemistry at the Crime Scene   DNA—True to Life   455

Summary   456   •   Key Terms   456   •   Questions   457
World Wide Web Resources   461

12 Groceries   THE CHEMISTRY OF THE FOODS WE EAT   462

12.1   The Fat Tax   465

12.2   Organized Fats   472

Students Often Ask   What Are Trans Fats, and Why Should I Worry About Them?   473

12.3   Bite, Chew, and Swallow   477

Beyond the Ordinary   The Most Amazing Enzymes Known   481

12.4   Sugar, Sugar   483

Students Often Ask   Are Artificial Sweeteners Really Okay to Eat?   486

Chemistry at the Crime Scene   A Practical Guide to Dining Out
Before Being Murdered   489

12.5   Weight Watching   491

12.6   Fake Food   496

Issues in Chemistry   Honey-Mustard Glazed Irradiated Pork Tenderloin
with Grilled Vegetables   499

Summary   501   •   Key Terms   501   •   Questions   502
World Wide Web Resources   507

13 Drugs I   HOW DRUGS ARE DESIGNED TO BENEFIT THE HUMAN BODY   508

13.1   Eat Your Broccoli!   510

13.2   The Ideal Drug   518

Students Often Ask   Why Do Patients Undergoing Chemotherapy for Cancer
So Often Get Sick from the Medicine?   520

13.3   Gumming Up the Works   523

13.4   On the Other Hand . . .   528

13.5   Left or Right, Right or Wrong?   536

Students Often Ask   Why Are There So Many Names for the Same Drug?   538

13.6   Brand Name or Generic?   539

Students Often Ask   How Do Time-Release Medications Work?   542

Summary   545   •   Key Terms   545   •   Questions   545
World Wide Web Resources   549

14 Drugs II THE DARK SIDE OF DRUG USE   550

14.1   That Ubiquitous Nitrogen Atom   552

Students Often Ask   What Causes Bad Breath?   556

14.2   "Speed"   557

Students Often Ask   What Is the Difference between Ecstasy and Speed?   562

14.3   Who's on Drugs?   564

14.4   The State of the Art   568

Chemistry at the Crime Scene   Drug Testing at Sporting Events   572

14.5   Cocaine, Coca-Cola, Crack   575

Students Often Ask   Does Coca-Cola Really Contain Cocaine?   581

14.6   Smoke and Mirrors   581

Postscript   586   •   Summary   588   •   Key Terms   588
Questions   581   •   World Wide Web Resources   593

Appendices

Appendix A
Working with Measured Numbers   A-1

Appendix B
Basic Algebra Review   A-7

Appendix C
Glossary   A-11

Appendix D
Answers to Odd-numbered Questions   A-20

Photo Credits   C-1

Index   I-1

Application Features

Beyond the Ordinary

A New Old Way of "Seeing" Alpha Particles, p. 230
Disappearing Ships, p. 341
Mind Your Own Bees-ness, p. 298
Nanoears, p. 99
Polka-dotted Airplanes, p. 152
The Most Amazing Enzymes Known, p. 481

Chemistry at the Crime Scene

A Practical Guide to Dining Out Before Being Murdered, p. 489
Death by Drowning, p. 282
Determining Time of Death, p. 148
DNA—True to Life, p. 445
Drug Testing at Sporting Events, p. 572
Synthetic Fibers, p. 405
The Dog Nose Knows, p. 375

Issues in Chemistry

Do Human Pheromones Really Exist? p. 322
What Is Weapons-Grade Uranium? p. 236
Honey-Mustard Glazed Irradiated Pork Tenderloin
   with Grilled Vegetables, p. 499
MBTE in Gasoline, p. 356
The Dark Side of Chemicals: Asbestos and Baby Powder, p. 71
The Tragedy at Chernobyl, p. 243
The Tuvalu Blues, p. 327
Three Women, p. 224
Transgenic Plants—Friend or Foe? p. 436
University of Arizona Garbage Project, p. 413
Who's Going to Take the Trash Out? p. 238

Students Often Ask

Are Artificial Sweeteners Really Okay to Eat? p. 486
Can It Really Be Too Cold to Snow? p. 268
Do Electric Eels Really Have Electricity In Them? p. 185
Does Adding Salt to Water Make Pasta Cook
   More Quickly? p. 276
Does Coca-Cola Really Contain Cocaine? p. 581
How Are Colors Incorporated into Color Film? p. 201
How Are New Elements Confirmed and Who Gets
   to Name Them? p. 218

How Can Insects Walk on Water? p. 262
How Do Detergents Get Greasy Dirt Out of Clothing? p. 284
How Do Home Water Purifiers Remove Ions? p. 145
How Do Photogray Sunglasses Turn Dark Outdoors? p. 177
How Do Smoke Detectors Work? p. 231
How Do Time-Release Medications Work? p. 542
How Does a Thermos Keep Hot Things Hot
   and Cold Things Cold? p. 309
How Much Gold Is Left in the Ground? p. 46
If Esters Are Hydrolyzed by Water, Why Can We Wash
   Polyester Clothes? p. 412
If Super Glue Adheres Immediately to Everything It Touches,
   Why Doesn't It Adhere to the Inside Walls of the Super-glue
   Tube? p. 391
Is a Life Form Based on Silicon Rather than Carbon
   Possible? p. 116
Is Fortified Water Useful for Ion Replacement or Is It Just Very
   Expensive Water? p. 140
What Are Trans-Fats, and Why Should I Worry
   about Them? p. 473
What Causes Bad Breath? p. 556
What Is the Difference between a Peptide and a Protein? p. 435
What Is the Difference between Ecstasy and Speed? p. 562
What Is the Meaning of the Octane Number Reported
   for Gasoline? p. 349
What Is the Volume of One Drop? p. 24
What Is Weapons–Grade Uranium, p. 236
What Makes New Cars Smell So Good? p. 110
What's the Most Explosive Compound Known? p. 366
When Drawing an Electron Dot Structure for a Tetrahedral
   Molecule, Does It Matter Which Atom I Show at Which
   Vertex of the Tetrahedron? p. 113
Why Are Gemstones Different Colors? p. 70
Why Are There So Many Names for the Same Drug? p. 538
Why Are Tires Filled with Air Instead of Being Solid? p. 302
Why Do Patients Undergoing Chemotherapy for Cancer So
   Often Get Sick from the Medicine? p. 520
Why Do People with Heart Conditions Take
   Nitroglycerin? p. 362
Why Does Some Fruit Turn Brown When You Cut It? p. 191
Why Is It More Difficult for an Airplane to Take Off at High
   Altitude or on a Hot Day? p. 312
Why Is Salt Used to Melt Ice on Wintry Roads? p. 266

To the Student

For many of you, this course will be the only science course you will take during your four years of college. And while you may not be required to study the sciences in-depth while you are in college, you will encounter scientific issues again and again after you graduate. After all, the sciences are becoming more and more intertwined in the everyday experiences of most twenty-first century Americans. Pick up a daily newspaper or read one online and look at every article in it. How often do you read about energy consumption, the cloning debate, or issues related to drugs, including those found on the street or those obtained with a prescription? Did you encounter articles about weapons of mass destruction, stem cell research, global warming, the spread of disease, or the debate over genetically modified organisms?

Chances are that you *did* encounter articles with some link to the sciences. For this reason, the goal of this book is to make you more cognizant of the sciences in general and, perhaps, to spark within you an interest in scientific issues. My wish for students who have read this book is this: when you encounter science-related headlines in the newspaper, maybe your curiosity will be piqued, and maybe you will want to invest a few minutes of your morning to reading that article. And, after reading this text, hopefully your understanding of that article will be deeper and more satisfying than before.

It is clear that we cannot cover the entire realm of chemistry, and all of its related disciplines, in one book. No book can do that. We also cannot claim that you will have a comprehensive knowledge of the sciences, chemistry in particular, after reading this book. Instead, *The Chemistry of Everything* contains the fundamentals, the need-to-know facts of chemistry that are a prerequisite to understanding a broad palette of issues in the sciences. We cover several disciplines within chemistry in this book and show you the highlights of each. In many cases, you must learn some concepts that require light mathematical skills, but in these cases we walk you through the material in a straightforward and painless way. We try to make the effort of learning the chemistry worth your while.

Once you have mastered some fundamentals, you are then equipped to understand how those concepts are applied in everyday circumstances. For example, we teach you about the architecture of organic molecules, which is simpler than many people expect. Armed with this information, we can explore the structure of asbestos fibers and the related mineral talc. By understanding what these substances look like at the molecular level, we can see why they pose risks for human lung tissue. In another chapter, we use our understanding of the structures of organic molecules to understand the properties of rubber (like its water resistance). We then tell you about the first civilization to use rubber, a story that demonstrates that clever inventions do not have to have practical, or even serious, uses.

In many ways, a book for nonmajors is a delight to write because there is no prescribed list of topics that must be covered, as there is for more advanced texts. However, this freedom of choice also brings with it the responsibility for making the right choices. What subjects should be included? I have put considerable thought into the collection of topics included in this text, with the intention of providing you with the highest level of scientific literacy in the short time that you spend in this course.

When you have checked your science requirement off your list, we urge you to consider keeping this book when your class is finished. Once you have graduated and have moved away from campus into your first house or apartment, consider putting *The Chemistry of Everything* on your bookshelf. We hope that this book will be one that you will turn to when you encounter the sciences in your everyday life.

Sincerely,
Kim Waldron
kwaldron@regis.edu

Preface

It is daunting when we have only one or two semesters to teach the topic of chemistry to nonmajors students. If this course is the only formal exposure these students will have to the sciences, how can we possibly select the short list of topics to cover with them? How can we grab and hold their attention? What information will be most useful for them when they are productive citizens trying to make a contribution to society? In other words, what does a person really need to know to be scientifically literate?

One way to approach this course is to consider the types of scientific issues that citizens encounter from day to day. Certainly, there are the big news headline topics that we hear about frequently: global warming, stem cell research, future energy sources, nuclear waste, weapons of mass destruction. Without a doubt, the student who has completed this course should have developed a small but significant scientific vocabulary that includes words that we hear regularly in the news.

News from the newspaper and television is not the only place where science intersects our lives. In fact, many scientific issues are just part of our mundane daily routines. Right now, someone might be wondering whether she can recycle a plastic container that might have been thrown in the trash. Someone else might be weighing the pros and cons of buying premium gasoline. Yet another person might be thinking about whether or not the bread he is making will rise if his apartment is too cold. A student who has completed a nonmajors course in chemistry should be able to grapple with such questions by applying their new knowledge to these situations.

All of this is what we think of as scientific literacy.

Every nonmajors textbook is unique, and authors of such books have a certain amount of freedom to select topics and to teach students in the way they think is best. For this reason, it seems appropriate to start with the basic principles on which this book is based. The following short statements encapsulate the essence of this book and give you a sense of its character, its quality and, hopefully, its charm.

*A nonmajors text should somehow blend chemistry content with compelling applications.

The style in which this is done is what separates one nonmajors book from another. In *The Chemistry of Everything*, all of the chemistry fundamentals are there, and they are divided as follows: eight chapters on physical/inorganic chemistry (Chapters 1 through 8), two chapters on organic chemistry (Chapters 9 and 10), and four chapters on biochemistry (Chapters 11 through 14). Each chapter has a unique one-word theme that is woven throughout, and that theme is linked to applications that show off the chemistry that is taught. The one-word titles are designed to arouse curiosity. However, for the convenience of instructors looking for exactly where a given chemical concept is presented, there is a comprehensive list of specific chemistry topics covered in each chapter adjacent to the theme title.

For example, Chapter 4, "Salt," explores the chemistry of salts and their applications as electrolytes. The chapter begins with the science of ionic liquids, a topic that is not found (yet) in many chemistry textbooks. This topic was chosen as an up-to-date and conceptually interesting way to teach the packing of ions into crystal lattices. We then move on to the subject of Egyptian

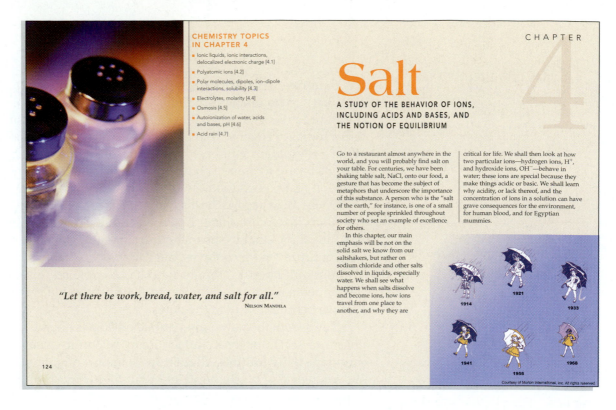

mummies, which teaches the idea of ion hydration spheres and the use of salts to absorb moisture. Later in the chapter, the story of the invention of Gatorade® is told in Section 4.4, p. 137, and the ions in Gatorade are used to teach the concept of molarity. In the remaining sections of the chapter, more salt chemistry is taught along with attention-grabbing applications. For example, the genetic engineering of tomatoes to grow in salty soil illustrates fundamental concepts of salt chemistry such as osmosis and concentration gradients. Home water purifiers are used to explain how ion-exchange resins work. Boxed sections in this chapter cover related topics such as the determination of time of death by analysis of ocular fluids and the use of pH indicators to spot airplane fuselage defects.

This is but one example of how content and application are entwined into a coherent story with an overarching theme, the title of the chapter.

*Nonmajors courses should include content that a majors course does not have.

Sometimes we think of a nonmajors course as a watered down version of a majors course. Certainly, nonmajors need enough fundamental chemistry to understand the scientific issues that they encounter in everyday life. However, a well-taught nonmajors course can and should include many things that a majors course lacks. If we had the time in our general chemistry courses, it would be a luxury to include discussions about some of the topics mentioned below. In a nonmajors course, we can take the time to explore these relevant, but tangential, topics and you will find them sprinkled throughout this text:

1. A nonmajors course is an ideal venue for discussing **ethical issues** in scientific research, which carry with them the burden of responsible use. Can a scientist work on the design of a lethal chemical weapon without considering its effects on humankind? Should all questions in science be pursued regardless of the consequences that the answers bring?

2. **Environmental issues** are pervasive in our society, and every citizen is called upon to help with efforts to preserve the nature around us. Thus, we devote space to the issues of global warming and the greenhouse effect. In addition, Chapter 10, "Chains I", is focused on artificial polymers, but we also take time to discuss how sustainable agriculture can be used to produce natural, Earth-friendly materials.

3. In recent years, **forensic science** has moved to the forefront of the media, and it offers a fascinating vehicle for teaching chemistry. Perhaps it is the natural fascination that we humans have with the macabre, but it seems to be a rare student nowadays who has not seen an episode of a forensic detective show. You will find boxed sections entitled "Chemistry at the Crime Scene" peppered throughout *The Chemistry of Everything*. These forensic stories are related to the chemistry being taught, and they range from the analysis of soil samples in a homicide investigation, to the determination of the time of death, to the use of rug fibers to identify a serial killer.

*The work that many scientists do is actually fascinating.

It is not unusual to see scientists depicted in film and on TV as lonely, hard-working geeks who spend every free minute working hard in the lab, often alone. Thankfully, the rise in popularity of forensic shows on TV has begun to dispel the nerdy scientist myth. In *The Chemistry of Everything*, students are introduced to scientists who do really interesting work. For example, in Chapter 7, "Water", we meet Dustin Rubenstein, a promising young ornithologist who uses variation in isotope levels in water to track bird migration (p. 254). Hopefully, this book will convince some students that doing science can actually be enjoyable, and even exhilarating.

*Nonmajors do not need to follow rigorously the subdivisions within chemistry that dictate many of our majors courses.

Over the past several decades, the boundaries between the traditional subdiciplines of chemistry have become more and more blurred. For this reason, *The Chemistry of Everything* is not strictly broken down into physical, inorganic, organic, analytical, and biological specializations. As we mentioned earlier, there is a broad division of chapters into these categories, but within those chapters, there is some mixing of disciplines. For example, in Chapter 3, "Diamonds," the theme centers around diamonds and the newly discovered allotropes of carbon. In this chapter, though, there is also an introduction to covalent bonding and a tutorial on how carbon frameworks and simple organic molecules are depicted. The last four chapters of the book are devoted to biochemistry. You will find, though, that these chapters include examples and boxed sections that fall into other chemistry subdisciplines, yet are relevant to biology. For example, in Chapter 12, "Groceries," there is a discussion of the use of radioactive isotopes to irradiate food. In Chapter 13, "Drugs I," there is a box focused on the design of time-release capsules made from artificial polymers.

In an attempt to interweave basic chemical principles throughout all chapters, we have added a feature called "Recurring Themes in

254 CHAPTER 7 • WATER

▶ FIGURE 7.4 For the birds. Their knowledge of isotope chemistry led Dustin Rubenstein (shown here) and advisor Richard Holmes of Dartmouth University to a study of the migration patterns of the black-throated warbler.

The Dartmouth biologists reasoned that it may be possible to use the distribution of isotopes in water to follow the migration patterns of the black-throated warbler, an insect-eating songbird that travels between the Caribbean and the eastern United States (Figure 7.4). The legwork had already been done: careful measurements of the ratios of hydrogen isotopes in water samples revealed that this ratio differs from one place to another. What does isotope distribution have to do with birds, though? Simple—birds eat insects, insects eat plants, and plants take up water from soil that has a distinctive isotope ratio. Thus, it should be possible to analyze the water in these birds and, from that analysis, determine where the birds have eaten and therefore their migration path. In a flock of birds migrating together, you will find traces of water from different places, and a map of these locations should provide a record of the migration path for the flock.

The Dartmouth biologists used measurements of deuterium to track the migration patterns of the black-throated warbler across the United States and the Caribbean, analyzing hydrogen-to-deuterium ratios in the water contained in the birds' feathers. Thankfully, this project did not involve a group of scientists chasing birds and pulling out their feathers! After mating, the birds molt, and the discarded feathers were collected and analyzed.

What Holmes and Rubenstein found was big news in the ornithology community. Birds that migrate to the western Caribbean come from the northern United States and southern Canada, whereas those that migrate to the eastern Caribbean come from farther south in the United States (Figure 7.5). In this study, the migration path of the black-throated warbler was established using measurements of hydrogen isotopes. This information will allow ornithologists to track these birds in the future and to assess the risks, such as those from environmental pollutants, that may threaten the black warbler population. Thus, we have now seen a third example of how the measurement of isotopes can be very valuable. (The other two, in case you have forgotten, were the Elgin marble study and carbon-dating of the prehistoric "iceman," both in Chapter 2.)

Chemistry." These Recurring Themes are placed as notes in the margins and trace the thread of fundamental chemical principles through the text. A list of the ten Recurring Themes appears inside the back cover, with page references for each theme.

*Some topics within chemistry are naturally more interesting to students than others.

We are all aware of those mind-numbing topics that make students' eyes glaze over. But, as any seasoned chemistry instructor knows, there are ways to make even metric units and VSEPR theory interesting. In *The Chemistry of Everything*, we have tried to spice up drier topics with some creative applications. For example, when metric units are introduced in Chapter 1, they are taught in the context of a recipe for chocolate brownies that has been "translated" from English units into metric units (p. 23). When we teach electron bookkeeping, we use the example of free radicals, how they cause food rancidity, and how preservatives prevent free-radical formation and preserve food (p. 172). When VSEPR theory and organic structures are introduced, we use "molecules on your toothbrush" as sample structures.

*Four types of enrichment features add flavor to the chemistry content.

The *Students Often Ask* feature answers everyday questions that students wonder about. For example, in Chapter 4 there is a *Students Often Ask* feature entitled "Is Fortified Water Useful for Ion Replacement, or Is It Just Very Expensive Water?" (p. 140). In this boxed section, we use fresh knowledge of salt chemistry to explore the contents of several designer waters. In *Beyond the Ordinary*, we look at unusual natural phenomena and curious scientific discoveries. For example, the feature entitled "Polka-dotted Airplanes" (p. 152) talks about the use of pH indicator dyes to spot small areas of corrosion on airplane fuselage. *Issues in Chemistry* focus on the development of the field, applications, and discoveries. For example, in Chapter 6, an *Issues in Chemistry* feature entitled "Three Women" focuses on the role that chemists Ida Noddack, Irene Curie, and Lise Meitner played during the 1930s in research that led to the discovery of nuclear fission. And because many students love anything related to forensic science, we have included *Chemistry at the Crime Scene*. The feature "Determining Time of Death" (p. 148) uses the student's knowledge of pH to discuss the use of ocular fluids to accurately pinpoint time of death. Throughout this book, we have tried to present topics in a way that has direct relevance for the student.

*Reading a nonmajors text should not be stressful.

Many nonmajors are apprehensive about chemistry before they ever buy a chemistry textbook or attend a college chemistry course. With this book, we hope to make the process of learning chemistry a pleasant and entertaining experience. Sure, there are sections of this book that are pure chemistry.

However, we have tried to follow up the chemistry with something interesting, so that the effort invested in learning the chemistry is worth it.

*A nonmajors course should include mathematics when needed to explain a fundamental concept.

We must teach nonmajors enough mathematics to help them understand basic chemical principles such as pH and unit conversions. In *The Chemistry of Everything*, mathematics is taught when necessary, and two appendices—Appendix A: Working with Measured Numbers and Appendix B: Basic Algebra Review—provide supplemental mathematics help if needed. In instances where students often stumble over a calculation, a "Problem Solving Hint" is provided to walk the student step-by-step through the math. For example, when pH is discussed, there is a Problem Solving Hint on how to enter logarithms into a standard calculator. In general, equations are written out completely and calculations are performed in sequential steps so that students who are not math wizards can follow along easily.

*Twenty-first century students are conceptualizers and visualizers.

This is especially true for nonmajors who may not be accustomed to thinking analytically or quantitatively. In *The Chemistry of Everything*, the art program and text together are designed to provide a conceptualized approach to understanding chemistry. For example, in Chapter 8, "Air," the gas laws are presented using an easy-to-understand, common-sense approach. We do not emphasize the importance of memorizing each gas law and who discovered it. Rather, the emphasis is on how gas laws are evident all around us. This discussion is complemented with visually stimulating artwork. In fact, the art throughout the book shows students what is perceivable to the human eye, and then shows how this behavior or phenomenon is a direct result of what is happening at the molecular level.

*In majors and nonmajors courses alike, practice is required for mastery of chemistry.

Nowhere is the phrase "practice makes perfect" more appropriate than in the chemistry classroom. For this reason, we have provided not only within-chapter questions, which are designed to test the students' comprehension as they read along, but also an extensive battery of end-of-chapter questions. These questions are divided into three groups: "The Painless Questions," "More Challenging Questions," and "The Toughest Questions." They are designed to suit any level of nonmajors students. In addition, a number of questions that require web research are provided at the end of each chapter. Some of them might ask students to read a web article, for example, and write a short essay on the biases perceived in it. Another might instruct the student to look up data on the watershed in their home county. When combined with the test bank provided on the Waldron website, the available questions should suit the needs of instructors with diverse class requirements.

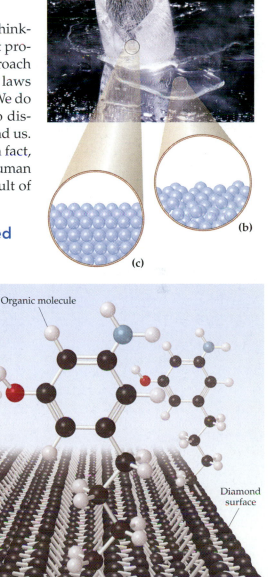

(a)

(b)

(c)

Organic molecule

Diamond surface

Print and Media Resources

For the Instructor

Instructor's Resource Manual & Full Solutions Manual (0-13-039177-8) by Kimberley Waldron, Regis University; and Joseph Laurino, University of Tampa. This manual features lecture outlines with presentation suggestions, teaching tips, suggested in-class demonstrations, and topics for classroom discussion. It also contains full solutions to all the end-of-chapter problems from the text.

Test Item File (0-13-039179-4) by Barbara Mowery, York College of Pennsylvania. This printed test bank includes more than 850 questions.

Instructor Resource Center on CD/DVD (0-13-039176-X) Brian Polk, Rollins College; Mary Sisak, Slippery Rock University; Richard Jarmon, College of DuPage; and Mark E. Ott, Jackson Community College. This lecture resource provides a fully searchable and integrated collection of resources to help you make efficient and effective use of your lecture preparation time, as well as to enhance your classroom presentations and assessment efforts. This resource features almost all the art from the text, including tables, in .JPEG, .PDF and PowerPoint™ formats; two pre-built PowerPoint™ presentations including (1) a lecture-based presentation for each chapter and (2) CRS (Classroom Response System) questions; and all interactive and dynamic media objects from the Companion Website. This CD/DVD also features a search-engine tool that enables you to find relevant resources via key terms, learning objectives, figure numbers, and resource type. This CD/DVD also contains TestGen test-generation software and a TestGen version of the Test Item File that enables professors to create and tailor exams to their needs, or to create online quizzes for delivery in WebCT, Blackboard or CourseCompass. Other features of this CD/DVD are Word™ files of the Instructor's Resource Manual and the Test Item File.

ONEKEY COURSE MANAGEMENT. *With contributions by Mark E. Ott, Jackson Community College; Rill Ann Reuter, Winona State University; Bette Kreuz, University of Michigan–Dearborn; and Christopher L. Truitt, Texas Tech University.* OneKey offers the best teaching and learning resources all in one place. All of the content in OneKey is available in **WebCT, Blackboard**, or Pearson Education's **CourseCompass**. Visit http://cms.prenhall.com or contact your Prentice Hall sales representative for details.

- OneKey for *The Chemistry of Everything* is all *your students* need for access to your course materials anytime, anywhere.
- OneKey is all *you* need to plan and administer your course. Conveniently organized by textbook chapter, these compiled resources help you save time and help your students reinforce and apply what they have learned in class.

OneKey content—Along with all the material from the Companion Website (**www.prenhall.com/waldron**), OneKey includes:

- Resources from the **Instructor's Resource Center on CD/DVD.**
- Test bank questions, converted from our **TestGen** test item file.

More basic courses, loaded with just the Test Item file, are also available.

Instructor's Manual to Laboratory Manual (0-13-039178-6) by Lawrence McGahey, The College of St. Scholastica. This Annotated Instructor's Manual combines the full student laboratory manual with front and back appendices covering the proper disposal of chemical waste, safety instructions for the lab, descriptions of standard lab equipment and materials, answers to questions, and more.

Transparency Pack (0-13-045512-1) by Rill Ann Reuter, Winona State University. This set contains 175 full-color transparencies from the text. Chosen specifically to put principles into visual perspective, it can save you time while you are preparing your lectures.

For the Student

Study Guide & Selected Solutions Manual (0-13-187537-X) by Joseph Laurino, University of Tampa. This book assists students through the text material with chapter overviews and practice problems for each major concept in the text, followed by two or three self-tests with answers at the end of each chapter. This book also provides solutions only to those problems that have a short answer in the text's Answers section (problems numbered in blue in the text).

Companion Website with GradeTracker (www.prenhall.com/waldron). Now with GradeTracker to provide gradebook functionality for instructors who wish to use it. This robust online resource is designed to specifically support and enhance *The Chemistry of Everything*. Key features:

- Live Art—Several key principles and reactions from the text are illustrated in a clean and compelling animated format.
- ABC News Videos—National news stories with relevance to the field of chemistry.
- E-Questions and World Wide Web Resources—These end-of-chapter questions from the textbook can be answered by using the links supplied for each chapter.
- Molecule Gallery—A sizeable selection of Chime 3D molecular models that students can manipulate to view various perspectives.
- Practice Quiz, Chapter Quiz, Conceptual Quiz—Multiple-choice quizzes, featuring hints and instant feedback.
- Math Tutorial—A math tutorial covering all relevant math skills needed to succeed in an introductory chemistry course.

Student Accelerator CD-ROM. This student CD, which contains animations, videos, and molecules from the Companion Website, accelerates the performance of the website when students download high-bandwidth media to prevent restriction by slow connections. It can also be used apart from the Companion Website if a student doesn't have a live Internet connection. *The Chemistry of Everything* Accelerator CD comes packaged with every new student textbook.

Laboratory Manual (0-13-187536-1) by Lawrence McGahey, The College of St. Scholastica. This manual contains 32 finely tuned experiments chosen to introduce students to basic lab techniques and to illustrate core chemical principles. Pre-lab and post-lab questions are included with detachable report sheets. Includes safety, disposal, and waste management information.

Acknowledgments

When I began this project, I had no idea how many people could be involved in the production of a single textbook. Without any one of them, some aspect of this book would be incomplete. While literally dozens of people have invested their time and energy into this book, there are a handful of individuals who have contributed a significant part of themselves to it, and who have literally grown grey hairs and lost sleep during its writing. First and foremost among them is Irene Nunes, my Developmental Editor. She was the one person that I was permitted to call at any time of day or night, and often did. She is the one who read every single word many times over, the one who worried about every punctuation mark, and the one who knows me best. When you spend years working closely with another person, as I have with Irene, you become inextricably linked. Irene seems like a big sister to me, only one I have never met in person. The quality of this book is largely a result of her hours of hard work, and for that I am very grateful.

There are two editors at Prentice Hall who also belong on the list of significant contributors to this text. They are Ray Mullaney, Editor in Chief for book development, and Kent Porter Hamann, Senior Editor. Ray and Kent have given a final polish to this book and have orchestrated the compilation of all of the pieces—photos, artwork, and text. They each toiled over every word of the final drafts, and their combined years of experience with scientific publishing have proved invaluable for this greenhorn author. I thank them for believing that this book would be something worth publishing.

Donna Young, the Production Editor, is owed a debt of gratitude for her determination combined with unlimited patience during the production phase. I would also like to thank Yvonne Gerin, my photo researcher, for her diligence and for being one of the friendliest people I know. I would also like to thank Joya Carlton, Editorial Assistant; Kathleen Schiaparelli and Beth Sweeten in Production; Connie Long and the cast of talented artists at Artworks; and Steve Sartori, Executive Marketing Manager, for working his magic in the marketing campaign. Finally, I'd like to thank Karen Karlin for volunteering to help with the book at the eleventh hour when her expertise and attention to detail were greatly appreciated.

Literally dozens of college chemistry professors across the country worked as reviewers for this text. One accuracy reviewer, Prof. Rill Ann Reuter of Winona State, read every chapter at least twice. I thank her for her long hours of meticulous, painstaking research on my behalf. Other reviewers, including Prof. Joe Laurino (University of Tampa) and Prof. Stephen Summers (Seminole Community College), worked on a significant portion of the book and provided extraordinarily useful feedback. A special thanks is owed to Prof. Nora Stackpole (University of South Florida), who first suggested the name for this text.

There is insufficient room here to personally thank each and every reviewer who made an important contribution to this book. I am profoundly grateful to these individuals, who are listed below.

Reviewers

Christopher A. Baumann, *University of Scranton*
Paul H. Benoit, *University of Arkansas*
David E. Bergbreiter, *Texas A&M*
Sharmaine S. Cady, *East Stroudsburg University*
Paul Chamberlain, *George Fox University*
Pamela Coffin, *University of Michigan*
Kathleen Cornely, *Providence College*
Tom D. Corso, *Canisius College*
Veronica Curtis-Palmer, *Northeastern Illinois University*
Dwaine Davis, *Forsyth Tech*
S. Todd Deal, *Georgia Southern University*
John P. DiVencenzo, *Middle Tennessee State University*
William Donovan, *University of Akron*
Daniel Domin, *Tennessee State University*
Paul F. Endres, *Bowling Green State University*
Crystal Gambino, *Manatee Community College*
Jennifer Garlitz, *Bowling Green State*
Marcia L. Gillette, *Indiana University*
Susan J. Glenn, *University of South Carolina, Aiken*
Cliff Gottleib, *Shasta College*
Robert Hammond, *East Carolina University*
Sally Harms, *Wayne State College*
Eric Helms, *SUNY Geneseo*
Theresa S. Hill, *Rochester Community and Technical College*
Linda Hobart, *Finger Lakes Community College*
John C. Hogan, *Louisiana State University*
Donna K. Howell, *Angelo State University*
Michael W. Jones, *Texas Tech University*
Liliane Khouri, *Hudson Valley Community College*
Angela King, *Wake Forest University*

Cindy Klevickis, *James Madison University*
Kathleen Knierim, *Univerity of Lousiana at Lafayette*
Rebecca Krystyniak, *St. Cloud State University*
Martha Kurtz, *Central Washington University*
Joseph P. Laurino, *University of Tampa*
Alan Levine, *University of Louisiana*
Christine Martey-Ochola, *Shippensburg University*
Kristen Murphy, *University of Wisconsin, Milwaukee*
David S. Newman, *Bowling Green State University*
Maria Pacheco, *Buffalo State University*
Kim Pamplin, *Abilene Christian University*
David Peitz, *Wayne State College*
Cindy Phelps, *California State University, Chico*
Brian Polk, *Rollins College*
Allison Hurley Predecki, *Shippensburg University*
Rill Ann Reuter, *Winona State University*
Elsa C. Santos, *Colorado State University*
Svein Saebo, *Mississippi State University*
Jeffrey A. Smiley, *Youngstown University*
Duane Smith, *Nicholls State*
Stephen P. Summers, *Seminole Community College*
Nora Stackpole, *University of Southern Florida*
Erach R. Talaty, *Wichita State University*
Joseph C. Tausta, *SUNY Oneonta*
Christopher L. Truitt, *Texas Tech University*
Paul E. Vorndam, *Colorado Community College*
Robert W. Wallace, *Bentley College*
Robert W. Widing, *University of Illinois*
Thomas A. Zona, *Illinois State University*

Sincerely,
Kim Waldron
kwaldron@regis.edu

CHEMISTRY TOPICS IN CHAPTER 1

- The periodic table, classes of subatomic particles, electron density [1.1]

- Elements, compounds, molecules, chemical reactions, balanced chemical equations [1.2]

- Separation of substances in a mixture, homogeneous and heterogeneous mixtures [1.3]

- Sizes of atoms, SI units, conversion factors, dimensional analysis [1.4]

- Common units of measure, measurement of temperature [1.5]

- The scientific method, testing hypotheses, scientific models [1.6]

"They say that everything in the world is good for something."

JOHN DRYDEN

Everything

AN OVERVIEW OF THE COMPOSITION OF MATTER AND THE WAY SCIENTISTS STUDY IT

During your years in college, you may decide to take a course in music or art appreciation. These courses will teach you to listen to a sonata, for example, in order to determine its structure and form, or to go to an art gallery and recognize the style used to create a particular painting or sculpture. You will also learn some of the history of each discipline so that you can understand the evolution of a piece of music or a school of art.

This book aims to do something similar with chemistry. It describes all the basics you need in order to appreciate what chemistry is and the impact it has on so many aspects of our daily lives. We also would like to give you some perspective on how science is done. How can simple observations turn into momentous scientific discoveries? What kind of people make revolutionary breakthroughs in science?

Let's begin to answer these questions with a story....

Modern Americans might wince at the idea of turning back the calendar to the end of the nineteenth century. The average person living during the 1890s in urban America would read by gaslight, get around by horse and buggy, and communicate by sending letters that would typically take days or weeks to reach their destina-

▲ **FIGURE 1.1   Everything was slower then.**

tion (Figure 1.1). Contrast that with today, when we wait impatiently for fractions of a second for our computers to respond to our keystrokes and agonize through the minutes it takes our microwave ovens to heat our food. The intervening century brought with it technological advances of which no other century can boast. The 1900s saw the extensive use of telephones, automobiles, electricity, and computers, as well as a dramatic improvement in life expectancy. It was also a century of radical change in the way scientists view the world. This new world view was responsible, in part, for the new ideas and inventions that make twenty-first-century life easier on the one hand, but much more difficult and complex on the other.

Some of the major scientific advances of the twentieth century were born of discoveries made at the tail end of the nineteenth century. As is often true in the sciences, the discovery of X-rays in 1895 was serendipitous. This form of radiation was discovered by Wilhelm Roentgen of Germany during an experiment he had designed to study something altogether different. X-rays turned out to be the newest, hottest find in decades, and within one year, X-ray machines were being widely used to peek inside the human body (Figure 1.2).

While the hullabaloo about X-rays rocked 1890s civilization, an equally important discovery was being made, again quite by accident. While studying how light affects minerals, a French scientist named Antoine Henri Becquerel found that certain rocks emit some kind of powerful radiation. What made this discovery so astonishing was that, whereas the creation of X-rays requires an input of electricity, Becquerel's radiation was being emitted with no input of anything. All on their own, without being heated or

▶ **FIGURE 1.2   Seeing inside the body.**   Wilhelm Roentgen used his new X-ray technology to take an X-ray photograph of his wife's left hand. The X-rays can pass through flesh, but not through bone or metal. For this reason, the image shows only the bones inside her hand and her ring.

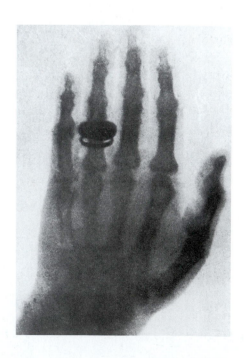

subject to an electric current, the rocks glowed in the dark. Becquerel published his peculiar findings about glow-in-the-dark rocks, but most people in scientific circles took little notice.

One person who did notice was a young Polish woman studying for her doctoral degree in Paris, one of only two women working on a Ph.D. in Europe at the time. She was looking for something completely new and different for her dissertation research, and Becquerel's findings fit the bill perfectly. In a matter of months, Marie Sklodowska Curie (Figure 1.3) had characterized and measured these mysterious emissions, which she called radioactivity. Working with her husband, Pierre Curie (Figure 1.4), a respected French scientist, Marie began the first study of radioactive materials, discovering several new chemical substances along the way.

▲ FIGURE 1.3   Marie Curie.

The discoveries made at the end of the nineteenth and beginning of the twentieth century set what was then "modern" science on its ear. Until that time, matter was thought to be inert and made up of minuscule, rigid particles that were indestructible and unchangeable. The suggestion that matter could absorb energy in one form (electricity) and give it off in another form (X-rays) was heady stuff. What was more unbelievable was the notion that matter could give off energy on its own in the form of radiation. Everything known at the time about how substances behaved was brought into question. As is often the case in scientific research, the discoveries made by Roentgen, Becquerel, and the Curies raised more questions than they answered.

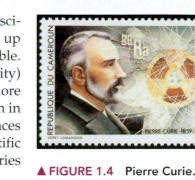

▲ FIGURE 1.4   Pierre Curie.   Pierre Curie shown on a stamp (from the Republic of Cameroon) that highlights his work with radioactive elements.

Because scientists are driven, in part, by the desire to understand mysteries in the natural world, the next several decades found physicists and chemists thinking hard about how matter can behave in these new and peculiar ways. One such scientist was a German named Max von Laue. Just as X-rays could be used to photograph the human skeleton, von Laue reasoned, they could also be used to photograph atoms arranged in a crystal. He bounced the idea off of some scientists while on a ski trip one day and was bombarded with heavy skepticism. Von Laue persevered, though, and eventually pioneered the first means of probing atoms. His X-ray "pictures" were only one of the pieces that chemists fit into a growing puzzle that would be used to understand the atom.

In the 1940s, a young British chemist named Dorothy Hodgkin (Figure 1.5) decided to use X-ray

British Winner Is a Grandmother

▲ FIGURE 1.5   Dorothy Hodgkin. Dorothy Hodgkin is commemorated on a British postage stamp. This headline from one newspaper announcing Hodgkin's being awarded the 1964 Nobel Prize in Chemistry has an interesting way of characterizing her. (Headline: *The New York Times*)

MOLECULE
Penicillin

▼ **FIGURE 1.6   August 6, 1945.**
Explosion of the atomic bomb over
Hiroshima, Japan, near the end of World
War II.

technology to do something no one had yet attempted: figure out how atoms are bound together in small assemblages called molecules. She used X-rays to show how atoms are linked together in the molecules cholesterol and penicillin. These structures were important because they provided a "blueprint" of a molecule. That blueprint could then serve as the basis for the creation of large quantities of that molecule. This was particularly important in the case of penicillin, a substance desperately needed to treat soldiers during World War II. At the same time they were being used to create medicines to treat ailing soldiers, however, the new discoveries about the atom were also being used to understand how the atom could be split. This knowledge would eventually end the war as the United States developed the first atomic bombs and used them on the Japanese cities of Hiroshima and Nagasaki in the summer of 1945 (Figure 1.6).

Shortly after the war, another young British chemist, Rosalind Franklin, shown in Figure 1.7, used the groundwork laid by Hodgkin and others to take the first X-ray pictures of *deoxyribonucleic acid* (DNA), the molecule that holds an organism's genetic information. Franklin's work demonstrated that there is more than one form of DNA, and her photographs were purportedly used by two scientists, American James Watson and Englishman Francis Crick, who were trying to determine DNA's structure. In his account of their discovery, Watson writes,

> The instant I saw the picture my mouth fell open and my pulse began to race . . . . [T]he black cross of reflections which dominated the picture could arise only from a helical structure.

Once the structure of DNA was known, its method of self-replication became obvious. The two strands of the double helix separate from each other, and then each strand acts as a template for building two new double helixes (Figure 1.8). Thus the X-ray picture of DNA revealed much more than just the three-dimensional structure of a molecule. It helped scientists understand how cells pass on genetic information from one generation to the next. In the 1980s and 1990s, this knowledge led to a revolution in the manipulation of the DNA double helix. The capability to "cut and paste" selected pieces of a DNA

▼ **FIGURE 1.7   Unraveling the secrets of DNA.   (a)** The X-ray image of DNA produced by Rosalind Franklin and used in the determination of DNA's three-dimensional structure. **(b)** Franklin died from cancer before the Nobel Prize in Physiology or Medicine was awarded for this joint discovery. The prize went to James Watson, Maurice Wilkins, and Francis Crick in 1962. (Nobel prizes are not awarded posthumously.)

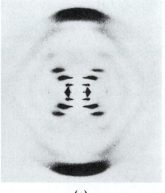

(a)

(b)

Old   Old

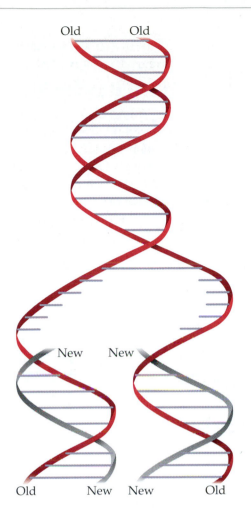

New   New

Old   New   New   Old

◀ **FIGURE 1.8** **The famous double helix.** A schematic drawing of the structure of the double helix of DNA. The two strands of the double helix split apart, allowing copies of the DNA sequence to be produced.

 **MOLECULE**
DNA

molecule has led to the discovery and manufacture of hundreds of new drugs and also to the production of genetically engineered food crops.

All of the scientists mentioned in the previous paragraphs (except Franklin) received Nobel prizes for work that was very much their own, but was a continuation of earlier work done by other scientists. This is how science happens. One person contributes a piece of information to the scientific community, which then scrutinizes and tests that information. Then someone else adds the next piece to the puzzle. Before long, the accumulated information might be applied to a practical problem that serves the public. For example, our newfound knowledge of the atom has brought us everything from smoke detectors to CAT and MRI scans to laser-based checkout machines at the grocery store.

Along with the incredible potential for improving our lives, however, these scientific advances have also brought the heavy weight of responsibility for their proper use. Scientists working to create an atomic bomb during World War II used their brand-new knowledge of the atom to develop a weapon of mass destruction. In the 1990s, molecular biologists cloned the first mammal using new techniques for manipulating genes (Figure 1.9). Were these acceptable uses of scientific knowledge? This question remains a hot topic of debate.

In your lifetime, there will be issues like these that will force policy makers and the public to decide how far our knowledge should be allowed to take us. That is part of what this book is about: providing you with

▼**FIGURE 1.9** The March 10, 1997, *Time Magazine* special report on cloning. One article was titled "The Age of Cloning: A line has been crossed, and reproductive biology will never be the same for people or for sheep."

Money for Science: Slim Pickings Ahead?

Wind farm gets initial state OK, but data sought

Commercial Risks For Science

In Fish vs. Farmer Cases, The Fish Loses Its Edge

Face it: Safety of cosmetics is in doubt

Endangered Species Act Reconsidered

House panel warns ballplayers, league to testify on steroids

Stem Cell Researchers Move Closer To Cloning Us

Himalayan Glaciers Melting Fast

EPA Criticized On Mercury Emissions Rule

Vote on drilling looms

Talks Push For Alternate Fuels

Mental illness and poverty: Does one cause the other?

Diet and Lose Weight? Scientists Say 'Prove It!'

Brain science v. death penalty

New Strains Of Mad Cow Materialize

Women continue to demand hormones despite research

▲ FIGURE 1.10   **Understanding the stories behind the headlines.**   As a responsible citizen, you should be able to understand the general ideas underlying controversial science issues.

some of the fundamentals necessary to understand scientific issues (Figure 1.10). Can you understand the concerns raised in a newspaper article discussing a topic in chemistry? Can you inform yourself about issues concerning the environment? Can you spot a ruse designed to influence you to buy or invest in some dubious scientific product or venture? We hope that, after reading this book, your answers to these questions will be "yes" and that taking this course will help you develop a healthy sense of skepticism. We also hope you will learn to evaluate issues and products with a skeptic's eye and come to conclusions that fit in with your beliefs and your way of life. In short, we want you to be able to evaluate the reliability of the scientific information you hear from the scientific community, from the media, and from your political representatives.

In this book, the chemistry concepts are introduced in the framework of a story from the news, an everyday experience, or a modern technique used to achieve something. For example, in Chapter 2, we talk about how dirt at a crime scene can be used to pinpoint suspects in a homicide case. First, though, we teach you what chemical substances are contained in dirt and how those substances are different from one another. Most chapters in this book have stories related to forensic science, medicine, or the environment. We hope that our approach, whereby we *teach* you chemistry and *show* you chemistry, will provide you with a perspective on the sciences that will enrich every day of your life.

1.1 | Atoms—the Basis of Everything

Chemistry is the study of all forms of **matter**, the "stuff" of which all physical material is composed. All matter is made up of atoms. An **atom** is the smallest possible unit of matter that cannot be broken down by chemical or physical means. At the time of this writing, there are 114 different types of atoms known. Each type is called an **element** and has its own name and its own one-, two-, or three-letter symbol. The arrangement of elements shown in Figure 1.11 is called the **periodic table of the elements**. Each row in this table is called a **period**, and each column is called either a **group** or a **family**.

The atoms of the elements tend to get successively more massive as you move from left to right across a period and as you move from top to bottom down a group. The reason for the increase in mass is found in the center of each atom, termed the **nucleus**. The nucleus contains particles known as **protons**, which are positively charged, and **neutrons**, which carry no

Atoms can be broken down by *nuclear* means, a topic we discuss in Chapter 6.

▼ FIGURE 1.11 The periodic table of the elements.

1 1A																	18 8A
1 **H** 1.00794	2 2A											13 3A	14 4A	15 5A	16 6A	17 7A	2 **He** 4.00260
3 **Li** 6.941	4 **Be** 9.01218											5 **B** 10.811	6 **C** 12.011	7 **N** 14.0067	8 **O** 15.9994	9 **F** 18.9984	10 **Ne** 20.1797
11 **Na** 22.9898	12 **Mg** 24.3050	3 3B	4 4B	5 5B	6 6B	7 7B	8	9 8B	10	11 1B	12 2B	13 **Al** 26.9815	14 **Si** 28.0855	15 **P** 30.9738	16 **S** 32.066	17 **Cl** 35.4527	18 **Ar** 39.948
19 **K** 39.0983	20 **Ca** 40.078	21 **Sc** 44.9559	22 **Ti** 47.88	23 **V** 50.9415	24 **Cr** 51.9961	25 **Mn** 54.9381	26 **Fe** 55.847	27 **Co** 58.9332	28 **Ni** 58.693	29 **Cu** 63.546	30 **Zn** 65.39	31 **Ga** 69.723	32 **Ge** 72.61	33 **As** 74.9216	34 **Se** 78.96	35 **Br** 79.904	36 **Kr** 83.80
37 **Rb** 85.4678	38 **Sr** 87.62	39 **Y** 88.9059	40 **Zr** 91.224	41 **Nb** 92.9064	42 **Mo** 95.94	43 **Tc** (98)	44 **Ru** 101.07	45 **Rh** 102.906	46 **Pd** 106.42	47 **Ag** 107.868	48 **Cd** 112.411	49 **In** 114.818	50 **Sn** 118.710	51 **Sb** 121.76	52 **Te** 127.60	53 **I** 126.904	54 **Xe** 131.29
55 **Cs** 132.905	56 **Ba** 137.327	57 ***La** 138.906	72 **Hf** 178.49	73 **Ta** 180.948	74 **W** 183.84	75 **Re** 186.207	76 **Os** 190.23	77 **Ir** 192.22	78 **Pt** 195.08	79 **Au** 196.967	80 **Hg** 200.59	81 **Tl** 204.383	82 **Pb** 207.2	83 **Bi** 208.980	84 **Po** (209)	85 **At** (210)	86 **Rn** (222)
87 **Fr** (223)	88 **Ra** 226.025	89 **†Ac** 227.028	104 **Rf** (261)	105 **Db** (262)	106 **Sg** (263)	107 **Bh** (262)	108 **Hs** (265)	109 **Mt** (266)	110 **Ds** (281)	111 ** (272)	112 ** (285)		114 ** (289)		116 ** (292)		

Period (left margin label)

		58 **Ce** 140.115	59 **Pr** 140.908	60 **Nd** 144.24	61 **Pm** (145)	62 **Sm** 150.36	63 **Eu** 151.965	64 **Gd** 157.25	65 **Tb** 158.925	66 **Dy** 162.50	67 **Ho** 164.930	68 **Er** 167.26	69 **Tm** 168.934	70 **Yb** 173.04	71 **Lu** 174.967
*Lanthanide series		90 **Th** 232.038	91 **Pa** 231.036	92 **U** 238.029	93 **Np** 237.048	94 **Pu** (244)	95 **Am** (243)	96 **Cm** (247)	97 **Bk** (247)	98 **Cf** (251)	99 **Es** (252)	100 **Fm** (257)	101 **Md** (258)	102 **No** (259)	103 **Lr** (260)

†Actinide series

** Not yet named

electrical charge. All the atoms of a given element contain more protons and generally more neutrons than do the atoms of any element coming earlier in the periodic table, and protons and neutrons have mass.

The protons and neutrons in the nucleus are packed together very tightly, and consequently the nucleus takes up only a very, very small portion of the atom. To get an idea of scale, imagine a map of North America on which someone has drawn a circle passing through Washington, D.C., and San Francisco (Figure 1.12). If the diameter of this circle represented the diameter of an atom, the nucleus would be about the size of a department store and located somewhere in southern Nebraska. As we shall see shortly, the atom is extraordinarily small, and so nuclei are that much smaller.

All atoms of a given element contain the same number of protons. This number of protons is called the element's **atomic number** and defines the element. It is this atomic number that appears above each elemental symbol in the periodic table shown in Figure 1.11. The table tells you that the atomic number of hydrogen, H, is 1, meaning every hydrogen atom contains one proton in its nucleus. Therefore, every time you have an atom containing only one proton, that atom *must* be a hydrogen atom, by definition. Likewise, every time you have an atom containing two protons, that atom

▼ **FIGURE 1.12 Getting a feel for the atom's size.**   The circle drawn on this map of North America represents the circumference of an atom. With the diameter of the circle equal to the distance from San Francisco to Washington, D.C., a department store located in Nebraska would represent the relative size of the atom's nucleus.

must be a helium atom, He. If an atom has eight protons, it *must* be oxygen, O, and if an atom has 53 protons, it *must* be iodine, I. (Two atoms of the same element can contain different numbers of neutrons, an idea we explore in Chapter 2.)

In addition to protons and neutrons, atoms also contain particles called **electrons**, which are located outside the nucleus and carry a negative electrical charge. Because they are negatively charged, the electrons are attracted to the positively charged protons in the nucleus. The numbers of electrons and protons in an atom must balance each other so that the atom has a net electrical charge of zero. Thus, the number of electrons in atoms of hydrogen, helium, oxygen, and iodine must be one, two, eight, and 53, respectively.

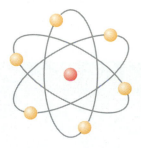

▲ **FIGURE 1.13   A wrong way of imagining electrons.**   This outdated way of representing the atom shows electrons moving in elliptical orbits around the nucleus.

QUESTION 1.1   How many protons and electrons are there in an atom of (*a*) indium, In; (*b*) magnesium, Mg; (*c*) aluminum, Al; (*d*) arsenic, As; (*e*) zirconium, Zr?

ANSWER 1.1   (*a*) 49 protons, 49 electrons; (*b*) 12 protons, 12 electrons; (*c*) 13 protons, 13 electrons; (*d*) 33 protons, 33 electrons; (*e*) 40 protons, 40 electrons.

In the introduction to this chapter, we talked about how at the beginning of the twentieth century there was a radical change in the way we think of matter. In fact, much of what was learned during that time was about the nature of electrons. The original understanding of electrons was that they whirled around the atomic nucleus in elliptical orbits, as shown in Figure 1.13. Experiments showed, however, that electrons do not orbit the nucleus the way planets orbit the sun. In fact, these experiments showed that electrons cannot be pinned down to one location at all. The best we can do is to think of electrons as a cloud around the nucleus. Sometimes the electrons are in one part of the cloud, and at other times they are in another part of the cloud. Because we can never know exactly where the electrons in an atom are, it is often more practical to think not of discrete electrons, but rather of the **electron density** in the atom, a "smearing" of the average positions of all electrons (Figure 1.14).

▼ **FIGURE 1.14   The correct way of imagining electrons today.**   The modern view of the atom depicts electron-density regions rather than discrete electrons around the atomic nucleus.

The notion of electron density allows us to understand how atoms interact with one another. A **chemical reaction** is an interaction between substances. In a chemical reaction, atoms in the reacting substances can give up electrons to other atoms, take away electrons from other atoms, or share electrons with other atoms. To see how this happens, let's consider our first chemical reaction: the detonation of nitroglycerin.

RECURRING THEME IN CHEMISTRY   Chemical reactions are *nothing more than exchanges or rearrangements of electrons,* which is why we pay particular attention to electrons throughout this book.

1.2 | Our First Chemical Reaction

We begin our foray into chemical reactions with a dramatic example: the detonation of nitroglycerin. To understand how this or any other chemical reaction can occur, however, we must first have a basic understanding of the substances involved. Nitroglycerin is a highly unstable chemical **compound**, which is any substance made up of atoms of two or more elements. An element is any substance composed of only one type of atom. In other words, elements contain atoms that all have the same number of protons. In an element, the atoms can exist either singly or joined together in some

(a)   Neon (Ne)   Gold (Au)   Fluorine gas (F$_2$)

(b)   Dry ice, CO$_2$ Water, H$_2$O   Ammonia, NH$_3$

▲ **FIGURE 1.15 Elements versus compounds. (a)** Elements contain only atoms of the same element. However, they can be individual atoms, as in neon, or atoms that are joined together, as in solid gold and fluorine gas. **(b)** Compounds contain atoms of more than one element.

way. The element neon, for example, is made up of individual neon atoms unattached to one another, as shown in Figure 1.15*a*. The element gold is a collection of gold atoms all joined together, and the element fluorine is made up of individual F_2 units. (Each unit is composed of a pair of fluorine atoms.) As we'll see throughout this book, though, most substances contain more than one element and therefore are compounds (Figure 1.15*b*). Nitroglycerin, for example, is a highly unstable chemical compound containing four elements: carbon, hydrogen, nitrogen, and oxygen.

In every substance in which atoms are joined together, what connects the atoms is some type of chemical bond. A **chemical bond** is an interaction between the electrons of joined atoms. As we'll see in future chapters, there are several varieties of chemical bonds. In all the substances we look at in this chapter, however, the bonds holding the atoms together all involve a *sharing* of electrons. When electrons are shared between atoms in a substance, we refer to the smallest unit that makes up this substance as a **molecule**. The compound nitroglycerin, for instance, is made up of molecules because its atoms are held together by shared electrons. Molecules can also be the basic units of substances that are classified as elements, as just mentioned for fluorine, F_2. As another example, oxygen gas, O_2, is an element because it contains only oxygen. Because the two oxygen atoms in each O_2 unit are held together by shared electrons, each O_2 unit is a molecule. Thus, on the one hand, the distinction between element and compound is in the number of *different elements* a substance contains. On the other hand, the term *molecule* is concerned with the *type of bond* holding individual atoms together.

QUESTION 1.2   Which of these substances are elements and which are compounds: (*a*) silver chloride, AgCl; (*b*) bromine, Br_2; (*c*) boron triiodide, BI_3; (*d*) molybdenum trioxide, MoO_3; (*e*) uranium, U; (*f*) fluorine, F_2?

ANSWER 1.2   An element contains atoms of one kind only. Thus (*b*), (*e*), and (*f*) are elements. Because (*a*), (*c*), and (*d*) all contain more than one type of atom, they are compounds.

The nitroglycerin molecule is unstable and readily undergoes a reaction—so readily that the simultaneous reaction of a large number of nitroglycerin molecules makes a powerful explosion (Figure 1.16). In the 1860s, Alfred Nobel mixed nitroglycerin with a chalklike material, a process that stabilized the nitroglycerin so that it could be transported safely and used in controlled explosions. This new mixture, called *dynamite*, was a moneymaker for Nobel, and he used some of those funds to initiate the Nobel prizes.

A **chemical equation** is a notation chemists use to represent a chemical reaction. It consists of two sets of symbols separated by an arrow. The chemical equation that depicts the explosion of nitroglycerin is shown in Figure 1.17. The starting materials in any chemical reaction are called **reactants** (they *react*), and the substances present after the reaction has been carried out are called **products** (they are *produced*). By convention, all reactants are shown to the left of the arrow in a chemical equation and all products are shown to the right of the arrow.

In any chemical reaction, some or all of the chemical bonds in the reactants are broken and then new bonds are formed as the products are created. All the atoms present before a chemical reaction begins are still there after the reaction is finished. What has changed is the way these atoms are connected

▼ **FIGURE 1.16   Barnyard antics.**
Dynamite comes in handy.
(www.CartoonStock.com)

Old McDonald had a farm

to one another. During the reaction, the atoms in the reacting substances are rearranged relative to one another.

The symbols written in a chemical equation are called **chemical formulas**. For instance, the combination of letters and subscript numbers $C_3H_5N_3O_9$ shown in Figure 1.17 is the chemical formula for nitroglycerin. The letters tell us that the compound nitroglycerin contains the elements carbon, hydrogen, nitrogen, and oxygen. The subscripts indicate how many atoms of each element are present in *one molecule* of nitroglycerin. Thus, the formula $C_3H_5N_3O_9$ tells us there are three carbon atoms, five hydrogen atoms, three nitrogen atoms, and nine oxygen atoms in one nitroglycerin molecule.

The (*l*) that follows the chemical formula stands for "liquid" and tells us that the nitroglycerin is in the liquid state in this reaction. The (*g*) shown next to the chemical formula for each product molecule in Figure 1.17 designates a gas. A third symbol of this type, not needed in Figure 1.17, is (*s*) for any reactant or product that is a solid. These symbols are often not shown in a chemical equation. They are important only when the *phase* of a reactant or product—whether it is a liquid, solid, or gas—is important for some reason. In our nitroglycerin reaction, it is useful to know that the nitroglycerin is a liquid and that the four products are all gases. This information about phase is important because, in any explosive reaction, the products are gases, and it is the high-speed expansion of these gases that causes the explosion.

QUESTION 1.3   Trinitrotoluene (TNT), $C_7H_5N_3O_6$, is another explosive material that detonates easily. (*a*) How many atoms does one molecule of TNT contain? (*b*) How many atoms do three molecules of TNT contain? (*c*) How many atoms of nitrogen are contained in seven molecules of TNT?

ANSWER 1.3   (*a*) 21 atoms. (*b*) 21 atoms per molecule × 3 molecules = 63 atoms. (*c*) 3 N atoms per molecule × 7 molecules = 21 N atoms.

▶ **FIGURE 1.17   The components of a chemical equation.**   This equation represents the reaction in which nitroglycerin explodes.

LIVE ART
Components of a Chemical Equation

MOLECULE
Nitroglycerin

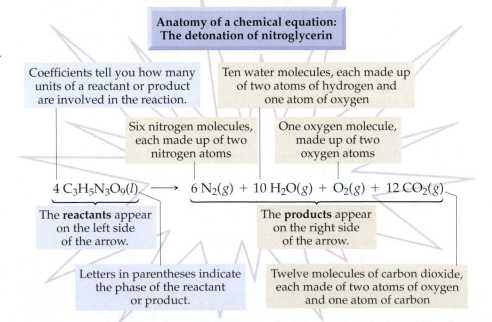

Anatomy of a chemical equation:
The detonation of nitroglycerin

Coefficients tell you how many units of a reactant or product are involved in the reaction.

Ten water molecules, each made up of two atoms of hydrogen and one atom of oxygen

Six nitrogen molecules, each made up of two nitrogen atoms

One oxygen molecule, made up of two oxygen atoms

$$4\,C_3H_5N_3O_9(l) \longrightarrow 6\,N_2(g) + 10\,H_2O(g) + O_2(g) + 12\,CO_2(g)$$

The **reactants** appear on the left side of the arrow.

The **products** appear on the right side of the arrow.

Letters in parentheses indicate the phase of the reactant or product.

Twelve molecules of carbon dioxide, each made of two atoms of oxygen and one atom of carbon

There are four products formed in the nitroglycerin reaction: nitrogen gas, N_2; water gas (more commonly known as *water vapor*), H_2O; oxygen gas, O_2; and carbon dioxide gas, CO_2. Notice that in the chemical equation shown in Figure 1.17, there is a number, or **coefficient**, before each reactant and product except O_2. When no coefficient is given, it is understood that the coefficient is the number 1. These coefficients indicate the numbers of molecules present before and after the reaction.

This brings us to a fundamental tenet of chemistry: *in any chemical reaction, we can neither make matter nor destroy it*. Thus, we must have equal numbers of each atom on the two sides of any correctly written chemical equation. On the left side of the equation in Figure 1.17, there are four nitroglycerin molecules indicated. This means

$$4 \times 3 = 12 \text{ carbon atoms}$$
$$4 \times 5 = 20 \text{ hydrogen atoms}$$
$$4 \times 3 = 12 \text{ nitrogen atoms}$$
$$4 \times 9 = 36 \text{ oxygen atoms}$$

All these atoms must also appear on the right side of the arrow. Let's count the atoms of each element on the right side:

$$\text{Carbon (only in } CO_2) \ 12 \times 1 = 12 \text{ atoms}$$

$$\text{Hydrogen (only in } H_2O) \ 10 \times 2 = 20 \text{ atoms}$$

$$\text{Nitrogen (only in } N_2) \ 6 \times 2 = 12 \text{ atoms}$$

$$\text{Oxygen (in } CO_2, H_2O, \text{ and } O_2)(12 \times 2) + (10 \times 1) + (1 \times 2) = 36 \text{ atoms}$$

The number of each atom on the left side of the equation matches the number of each atom on the right side. Thus, we say that this is a **balanced equation**.

Notice that the numbers of *molecules* on the two sides of a chemical equation do not have to be balanced. In the nitroglycerin equation, for instance, there are only four reactant molecules on the left, but 29 product molecules on the right. Instead, it is the *numbers of each kind of atom* that must be balanced.

It is important that chemical equations be balanced not only because matter can neither be created nor destroyed, but also because the coefficients tell us how many units of each reactant take part in the reaction and how many units of each product are formed. We know from our balanced nitroglycerin equation that if we start with four molecules of nitroglycerin, we end up with six molecules of nitrogen, ten molecules of water, one molecule of oxygen, and 12 molecules of carbon dioxide. We begin with four molecules of a liquid on the left and end up with $(6 + 10 + 1 + 12) = 29$ molecules of gas on the right. This is typical of explosions: they usually produce lots of gaseous molecules.

MOLECULE
Carbon Dioxide

MOLECULE
Water

RECURRING THEME IN CHEMISTRY   Chemical reactions do not create or destroy matter. Therefore, the types and numbers of atoms must be the same on the two sides of a chemical equation.

QUESTION 1.4   The equation showing the reaction between isooctane, C_8H_{18}, a component of gasoline, and oxygen, O_2, is

$$2\,C_8H_{18} + 25\,O_2 \longrightarrow 16\,CO_2 + 18\,H_2O$$

(*a*) Is this equation balanced? (*b*) If you begin the reaction with two molecules of isooctane, how many molecules of water, H_2O, will you end up with? (*c*) If you begin the reaction with ten molecules of isooctane, how many molecules of water will you end up with?

ANSWER 1.4 (*a*) Yes. There are 16 carbon atoms, 36 hydrogen atoms, and 50 oxygen atoms on each side of the equation. (*b*) For every two molecules of isooctane used, 18 molecules of water are formed. (*c*) There are five times more molecules of isooctane (10) than what is shown in the balanced equation. Thus, you must produce five times more water molecules: $5 \times 18 = 90$ molecules of water.

Chemical reactions are taking place constantly all around us and within us. Except for atoms that enter and leave Earth's atmosphere, the atoms that exist today on our planet are the same ones that have been exchanging partners in chemical reactions since Earth was formed. Thus, the atoms that make up our bodies and everything else in the world are very old. They have been recycled through different chemical compounds many, many times and now have ended up with us for a while. Knowing this, we begin our study of chemistry with this simple notion: everything that exists is made up of atoms found on the periodic table, and those atoms are continually swapped in chemical reactions.

Substances frequently exist in a form that includes water molecules. This is true for Epsom salt, which includes seven water molecules in addition to $MgSO_4$. Thus, we see the word *heptahydrate* (hepta-, "seven"; -hydrate, "water") in the name.

1.3 | A Peek into Your Medicine Cabinet

With the staggering number of substances on Earth, how can we possibly get a handle on the diversity offered by our 114 elements? As a start, let's peek inside a medicine cabinet to see what sorts of substances might be lurking there. We can begin with what we already know: everything must be composed of atoms of the elements listed on the periodic table.

Matter can be divided into two large categories: pure substances and mixtures. A **pure substance** is matter that has a definite composition. That is, a pure substance either will be an element in its pure form or, if it is a compound, will have a fixed ratio of one atom to another. Thus, chemical formulas can always be written for pure substances. A **mixture** is defined as two or more pure substances combined together. The composition of a mixture may vary from one part of the sample to another. Consequently, the ratio of substances in one part of the sample can be very different from the ratio of substances in another part of the sample. Let's now go to the medicine cabinet to see examples of both pure substances and mixtures.

The first thing we see is a carton of Epsom salt (Figure 1.18). Epsom salt has the chemical formula $MgSO_4 \cdot 7H_2O$, and its formal name is magnesium sulfate heptahydrate. Epsom salt is a pure substance. We refer to it as a compound because it always has a fixed ratio of atoms of different elements: always one magnesium atom for every atom of sulfur, every 11 atoms of oxygen, and every 14 atoms of hydrogen. Thus, a pure sample of the compound Epsom salt contains only $MgSO_4 \cdot 7H_2O$, in the same way that a pure sample of the compound nitroglycerin contains only $C_3H_5N_3O_9$.

You should note that the term *pure substance* is a little bit misleading because it suggests that something can be truly pure. In the real world, however, an absolutely pure substance is impossible to achieve. The meaning of the term *purity* depends on context. Take water, for example. If you say that the lake you are about to jump into has very clean water, you are probably comparing the purity of that water to the water in a lake that is more polluted. If you are talking to a salesperson selling water purifiers for the home, then *pure* means that the water run through the purifier has fewer impurities than your tap water. If you are talking to a chemical company selling ultrapure water, then purity means the water may be the purest water found anywhere, but it is still not entirely free of impurities. The fact is, no matter how many times the water is purified using the most sophisticated chemical instruments, it will never be absolutely pure. Consequently, we must use the term *pure substance* loosely, meaning that, impurities aside, the substance is composed of only one element or one compound.

Pure substances are rare on Earth because nature tends toward disorder, and disorder means mixing. Thus, most things in our world are mixtures. How can you distinguish a mixture from a pure substance? To find out, let's use two items from our medicine cabinet, acetone and tincture of iodine. Acetone (nail polish remover), C_3H_6O, is a pure substance. Tincture of iodine is a mixture of the element iodine and the compound isopropyl (rubbing) alcohol. If you analyze ten samples of acetone, you will find that every sample contains a ratio of three carbon atoms to six hydrogen atoms to one oxygen atom. This must be the case because the atoms are connected together by chemical bonds that establish the atoms in a fixed ratio. In contrast, the tincture of iodine is a mixture because it was made by dissolving some mass of iodine crystals, a pure substance, into isopropyl alcohol, another pure substance. If the most meticulous chemist tried to prepare this mixture the same way ten times, the result will be ten slightly different tinctures because it is impossible to prepare *exactly* the same mixture every time.

Another mixture you might find in your medicine cabinet is a solution of hydrogen peroxide, H_2O_2. Your pharmacist can prepare 2 percent, 5 percent, or 10 percent solutions of hydrogen peroxide in water. These solutions are made by mixing the pure substance hydrogen peroxide with the pure substance water in various ratios.

A mixture can have a smooth texture that makes it impossible to see boundaries between the various components. We refer to this as a **homogeneous mixture** (also called a **solution**) because it is completely uniform in appearance. The two mixtures previously mentioned—tincture of iodine and hydrogen peroxide solution—are both homogeneous mixtures because they look the same throughout and the components are thoroughly mixed in each case. As odd as it might seem, the nail clippers you might find in a medicine cabinet are another example of a homogeneous mixture. Clippers are usually made from stainless steel, a homogeneous mixture of iron, chromium, nickel, manganese, and other elements. Even though it is a solid, stainless steel is a homogeneous mixture because all the components are completely and uniformly mixed. The air inside the medicine cabinet, composed mainly of nitrogen gas and oxygen gas, is another homogeneous mixture.

Heterogeneous mixtures have boundaries where you can see where one component stops and another begins because the composition changes from one place to another in a sample. Think about the items in your medicine cabinet that require shaking before use (Figure 1.19). If you have not used these items lately, you may have some fine examples of heterogeneous mixtures. Your bottle of Pepto-Bismol®, for example, may contain separate layers where settling has occurred. The heterogeneous mixture you know as nail polish usually includes a small metal ball that helps mix the components when you shake the bottle. Before they are mixed, these are both examples of heterogeneous mixtures.

Some mixtures are heterogeneous regardless of how much you shake them. For example, toothpaste containing little flecks or stripes of color is certainly a heterogeneous mixture that cannot be shaken.

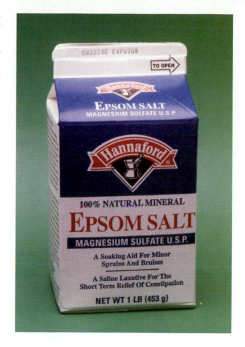

▲ FIGURE 1.18 **A pure substance.** The product we know as Epsom salt is a pure substance, containing only magnesium sulfate.

MOLECULE
Hydrogen Peroxide

▼ FIGURE 1.19 **Shake well before using.** Some of the bottles in your medicine cabinet may contain heterogeneous mixtures. These products usually require shaking before use.

It is possible to separate the components of a mixture by physical means. To demonstrate this, let's consider another mixture found in many medicine cabinets: saline solution. Saline solution is a mixture because it is a combination of table salt (NaCl) and water, both pure substances. Because this is a mixture, it is possible to separate the two components by physical means, but how? One way is to let the solution sit open until all the water has evaporated. What is left in the container is pure NaCl.

If you are eager to do the separation immediately, you can speed up the process by boiling off the water. In either case, you are effecting a **physical change**, a process that alters a substance without changing it into some other substance. In this case, water has undergone the physical change of being converted from the liquid phase to the gas phase. It is H_2O both before and after the change, however.

QUESTION 1.5   (*a*) Describe two ways you could separate a mixture of wood shavings and iron filings. (*b*) When the separation is complete, what do you have? Identify each component as element, compound, homogeneous mixture, or heterogeneous mixture.

ANSWER 1.5   (*a*) You could add water and allow the wood to float to the surface and the iron filings to sink. Because iron is magnetic, you could use a magnet to pull the iron away from the wood. (*b*) When the separation is complete, you have iron filings (element) and wood shavings (heterogeneous mixture). The shavings must be a mixture, because trees are complex mixtures of compounds and elements.

In addition to physical changes, substances can also undergo *chemical changes*. A **chemical change** is one that results in the breakdown of a pure substance into one or more other pure substances. To see what is involved in a chemical change, let's think further about the saline solution we have separated into NaCl and water. We know that these compounds are both pure substances, and so they cannot be broken down further by physical change. What would happen, though, if we subjected the water to a chemical change? Each water molecule is formed by the sharing of electrons between two hydrogen atoms and one oxygen atom. A chemical change can break apart the strong bonds between these atoms. For example, water can be converted to oxygen gas and hydrogen gas via a big shock of electric current in a process known as *electrolysis*:

$$2\,H_2O(\ell) \xrightarrow{\text{Electricity}} 2\,H_2(g) + O_2(g)$$

The electrolysis of water is a chemical change because it changes a pure substance, water, into something else, hydrogen gas and oxygen gas.

QUESTION 1.6   Is the foregoing electrolysis equation balanced?

ANSWER 1.6   Yes; there are four hydrogen atoms on each side and two oxygen atoms on each side.

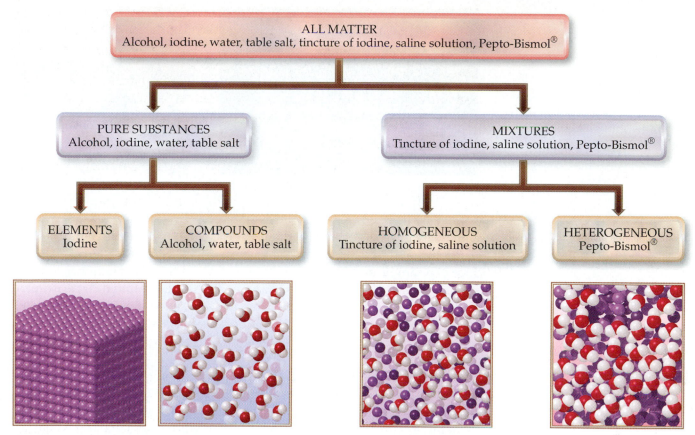

ALL MATTER
Alcohol, iodine, water, table salt, tincture of iodine, saline solution, Pepto-Bismol®

PURE SUBSTANCES
Alcohol, iodine, water, table salt

MIXTURES
Tincture of iodine, saline solution, Pepto-Bismol®

ELEMENTS
Iodine

COMPOUNDS
Alcohol, water, table salt

HOMOGENEOUS
Tincture of iodine, saline solution

HETEROGENEOUS
Pepto-Bismol®

▲ **FIGURE 1.20   A classification scheme for matter.** All matter can be divided into pure substances and mixtures. Pure substances can be further broken down into elements and compounds. Mixtures can be further broken down into homogeneous mixtures and heterogeneous mixtures.

QUESTION 1.7   Identify each of the following as a physical change or a chemical change: (*a*) octane burns in a car engine, (*b*) water freezes at 0 °C, (*c*) lake water evaporates on a summer day.

ANSWER 1.7   (*a*) Octane burns by combining with oxygen in the air to produce carbon monoxide, carbon dioxide, and other compounds. Thus the burning of octane is a chemical change. (*b*) When water freezes, it forms ice, which is still H_2O. Thus freezing is a physical change. (*c*) The liquid water from the lake enters the air in the form of a gas. The change from liquid water to gaseous water is another example of a physical change.

Figure 1.20 summarizes the matter-classification scheme we just discussed.

1.4 | Numbers in Science—SI Units and Conversion Factors

In the 1980s, a breakthrough in scientists' ability to probe very small things took a gigantic leap. A type of microscope called a *scanning tunneling microscope* (STM) was used to view single atoms, and researchers at IBM figured

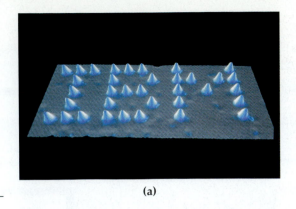
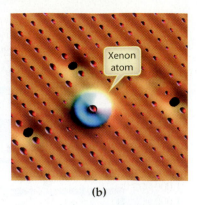

Xenon
atom

(a)   (b)

► **FIGURE 1.21 Seeing individual atoms.** (a) The letters IBM spelled out using individual atoms. (b) An image of a xenon atom made by a scanning tunneling microscope. The xenon atom is sitting on a nickel surface.

out a way to arrange atoms to spell out the company's name (Figure 1.21*a*). The heart of the STM is a very, very tiny probe that bounces along the surface of a sample and "senses" its contours. Once this information is fed into a computer, an image of the surface, at the level of atomic detail, is produced. Figure 1.21*b* shows an STM image of a xenon atom on a nickel surface. The outline of the atom is blurry because, as noted earlier, the electrons exist not at specific locations in the atom, but rather as a smear of electron density around the nucleus.

The fact that it has taken extraordinary technological advances to view single atoms attests to their very small size, but just how small are they? To get a feel for just how small they are, imagine a piece of paper. The paper is a mere 0.010 centimeter thick, but it would take 1,400,000 atoms of hydrogen stacked on top of one another to span this distance! As another example, consider the silver coin shown in Figure 1.22. This 1-ounce coin contains 160,000,000,000,000,000,000,000 atoms of silver, Ag.

Chemists routinely work with enormous numbers, like the number of atoms in our silver coin, as well as with extremely small numbers, such as the diameter of a silver atom, which is 0.000 000 000 153 meter. Thus, they need some convenient way to express the exceedingly large and small values that often must be counted or measured. Scientists worldwide use an adaptation of the *metric system*. The metric system has been around since the 1790s, but in 1960 it was standardized to the *Systéme International* (SI). In this system, each type of measurement, such as length or volume, has a *base unit*, and other units are multiples of the base unit. The names of these multiple units are formed by adding a prefix to the name of the base unit. These prefixes and the letter used to designate each one in an abbreviation are listed in Table 1.1. For example, the SI base unit for length is the *meter*, and the abbreviation is m. One length unit that is a multiple of the meter is the *millimeter*. Table 1.1 tells us that the letter representing *milli-* is m, and so the abbreviation for millimeter is mm. To cite just two more examples, other length units based on the meter are the *kilometer* and the *centimeter*. We see from Table 1.1 that the abbreviation for kilometer must be km and that for centimeter must be cm.

Table 1.2 lists the base units we use in this book. They are a mixture of SI units and metric units. In cases when the SI and metric unit are not the same, we shall sometimes use the traditional metric unit if it is the one more commonly used by chemists. For example, the SI base unit for volume is the cubic meter (m^3), but chemists most often express volumes in either liters (L) or milliliters (mL), both metric units. Additionally, chemists sometimes use the metric temperature unit, the Celsius degree, rather than the SI temperature unit, the kelvin.

▼ **FIGURE 1.22 Too many to count.** A 1-ounce silver coin contains 160,000,000,000,000,000,000,000 atoms of silver.

TABLE 1.1 Metric Prefixes		
Multiply the number of base units by ...	**To get the number given by the prefix ...**	**Symbol or letter used for designation**
1/1,000,000	mega-	M
1/1000	kilo-	k
10	deci-	d
100	centi-	c
1000	milli-	m
1,000,000	micro-	μ
1,000,000,000	nano-	n
1,000,000,000,000	pico-	p
1,000,000,000,000,000	femto-	f

You will often see the terms *SI* and *metric* used interchangeably. As you've just learned, though, they are similar, but not exactly alike. For simplicity, in this book, we use the term *metric system* to refer to all metric *and* SI units, with the understanding that the SI system is a modification of the metric system.

Table 1.1 shows that there are 1000 mm in 1 m and 1,000,000,000 nanometers (nm) in 1 m:

$$1 \text{ m} = 1000 \text{ mm}$$
$$1 \text{ m} = 1,000,000,000 \text{ nm}$$

Instead of writing these values next to each other separated by an equals sign, we can write them in fraction form. The value on one side of the equals sign becomes the numerator, and the value on the other side of the equals sign becomes the denominator. It does not matter which value you put in the numerator and which in the denominator. Therefore, we can write

$$\frac{1 \text{ m}}{1000 \text{ mm}} \quad \text{or} \quad \frac{1000 \text{ mm}}{1 \text{ m}}$$

$$\frac{1 \text{ m}}{1,000,000,000 \text{ nm}} \quad \text{or} \quad \frac{1,000,000,000 \text{ nm}}{1 \text{ m}}$$

We call these fractions **conversion factors**. Because any fraction is equal to 1 when the numerator is equal to the denominator, each conversion factor is equal to 1:

$$1 = \frac{1 \text{ m}}{1000 \text{ mm}} = \frac{1000 \text{ mm}}{1 \text{ m}}$$

$$1 = \frac{1 \text{ m}}{1,000,000,000 \text{ nm}} = \frac{1,000,000,000 \text{ nm}}{1 \text{ m}}$$

TABLE 1.2 Base Units Used in This Book		
Physical quantity	**Name of base unit**	**Abbreviation**
Amount of substance	Mole	mol
Length	Meter	m
Mass	Gram	g
Volume	Liter	L
Temperature	Kelvin or degree Celsius	K or °C
Time	Second	s

Let's look at how conversion factors are used. Figure 1.23 shows the radii of some atoms expressed in picometers. Suppose we want to express the radius of a krypton atom, 110 pm, in meters. (The chemical symbol for the element krypton is Kr.) We know from Table 1.1 that there are 1,000,000,000,000 pm in 1 m, and we get our two possible conversion factors from this information:

$$\frac{1 \text{ m}}{1,000,000,000,000 \text{ pm}} \quad \text{or} \quad \frac{1,000,000,000,000 \text{ pm}}{1 \text{ m}}$$

To decide which factor to use to convert picometers to meters, we write the given value—110 pm in this example—and then multiply it by the conversion factor that allows us to cancel the unit we want to get rid of (pm) and leaves us with the unit we are seeking (m):

$$110 \text{ pm} \times \frac{1 \text{ m}}{1,000,000,000,000 \text{ pm}} = 0.000\,000\,000\,110 \text{ m}$$

In this book we will use rules for significant figures when we express numbers that result from mathematical operations. For example, this equation correctly shows 10 zeroes to the right of the decimal place. The rules for significant figures are explained in Appendix A.

If your math is rusty, a review of basic algebra can be found in Appendix B.

Our answer, 0.000 000 000 110 m, is unwieldy and awkward to write out. Now you can see the convenience of metric prefixes. They allow us to convert values containing long strings of zeros to more convenient values. Thus, scientists most often use picometers to express atomic radii. Meters are too large a unit to use when expressing a measurement for something as small as an atomic radius!

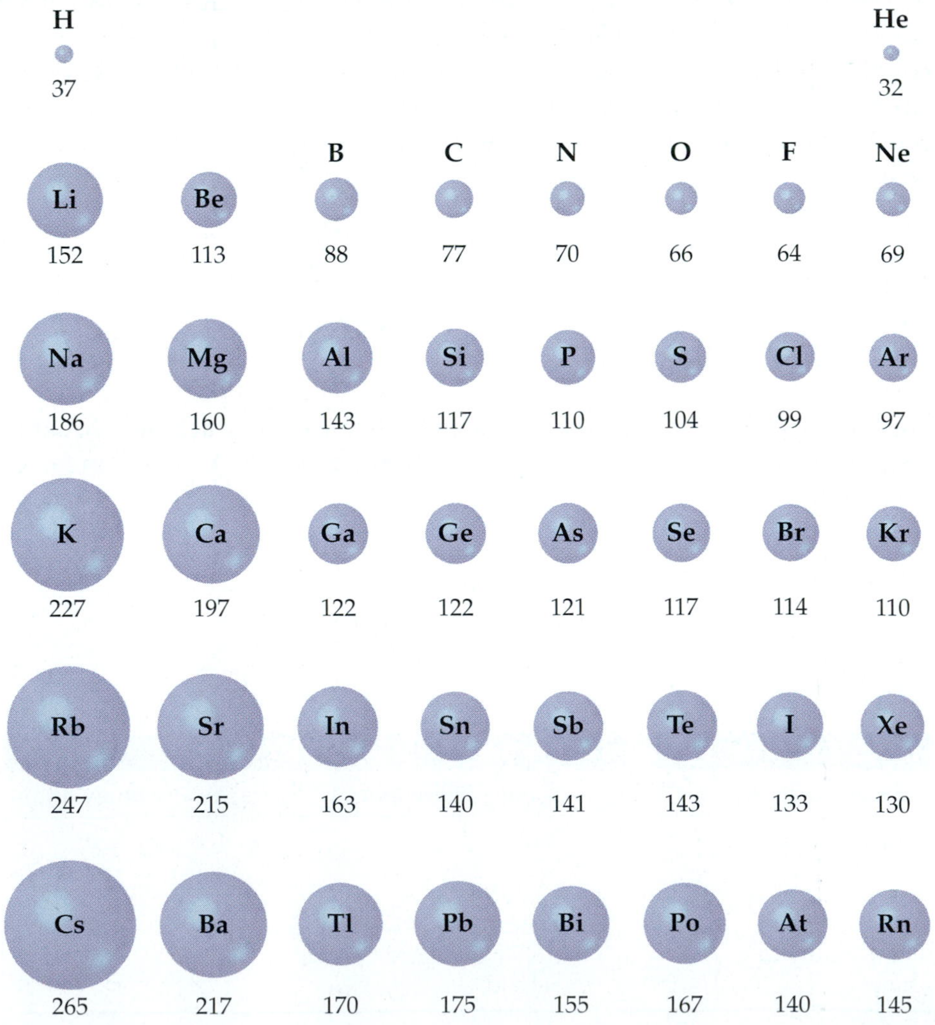

H							He
37							32

		B	C	N	O	F	Ne
Li	Be						
152	113	88	77	70	66	64	69

Na	Mg	Al	Si	P	S	Cl	Ar
186	160	143	117	110	104	99	97

K	Ca	Ga	Ge	As	Se	Br	Kr
227	197	122	122	121	117	114	110

Rb	Sr	In	Sn	Sb	Te	I	Xe
247	215	163	140	141	143	133	130

Cs	Ba	Tl	Pb	Bi	Po	At	Rn
265	217	170	175	155	167	140	145

▶ FIGURE 1.23 **Atomic radii.** The radii in picometers of atoms of some of the elements in the periodic table.

The metric system was designed to be easy to use; everything is based on multiples of ten. If you are a math wizard, you can probably convert from one metric unit to another in your head. If so, you might be wondering why it is necessary to use this system of cancellation and conversion factors, called **dimensional analysis**. The reason for using dimensional analysis is that you may not always be working with metric units when converting one thing to another. Also, even when you are using only metric units, conversions can become complex, as the example that follows demonstrates.

In the STM image shown in Figure 1.21*a*, the distance from the top to the bottom of each letter is 0.000 000 000 001 30 km. Let us use this information to determine the radius of one of the atoms, and let us express that radius in a unit that does not require us to write so many zeros. We begin by noting that, in each letter, there are five atoms from top to bottom. Also, let's ignore the small space between adjacent atoms, as indicated in Figure 1.24. Ignoring this space, we can say that the radius of one atom is approximately one-tenth of the total letter height (remember, the radius of a circle is half the diameter):

$$\text{Radius of one atom} = \frac{0.000\ 000\ 000\ 001\ 30\ \text{km}}{10} = 0.000\ 000\ 000\ 000\ 130\ \text{km}$$

We already know that atomic radii can be expressed conveniently in picometers. A look at Table 1.1 shows that we might also use nanometers because this unit is close to the picometer. Because we are starting with kilometers, it is best to return to the base unit, meters, and then change meters to the desired unit. The two conversion factors that relate kilometers to meters are

$$\frac{1000\ \text{m}}{1\ \text{km}} \quad \text{or} \quad \frac{1\ \text{km}}{1000\ \text{m}}$$

Which should we use? The answer is that we use the factor on the left because our starting number is in kilometers and only this factor lets us cancel kilometers and end up with meters:

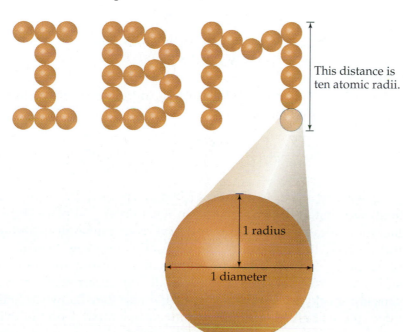

This distance is ten atomic radii.

1 radius

1 diameter

◀ FIGURE 1.24   **Adjusting the IBM atoms to help with the math.** The atoms of Figure 1.21*a* shown with no space between adjacent atoms. When we make this simplifying assumption, the letter height is equivalent to ten times the radius of each atom.

$$0.000\ 000\ 000\ 000\ 130\ \cancel{km} \times \frac{1000\ m}{1\ \cancel{km}} = 0.000\ 000\ 000\ 130\ m$$

Now we can convert to nanometers,

$$0.000\ 000\ 000\ 130\ \cancel{m} \times \frac{1,000,000,000\ nm}{1\ \cancel{m}} = 0.130\ nm$$

or to picometers,

$$0.000\ 000\ 000\ 130\ \cancel{m} \times \frac{1,000,000,000,000\ pm}{1\ \cancel{m}} = 130\ pm$$

So we can express the radius of one atom in the IBM image as 0.130 nm, 130 pm, or 130,000 fm. All of these are correct, but picometers are most convenient (only one zero and no decimal point) and thus the one most often used for atomic radii.

Now that we know the length of the radii of the atoms shown in the IBM image, can we figure out which element the IBM people used for this remarkable achievement? Figure 1.23 tells us these must be xenon atoms, Xe, which have a radius of 130 pm. Now you can appreciate the scale of this IBM experiment. Each xenon atom is only 130 pm in radius, which means even billions of xenon atoms clustered together are much too small to see with the naked eye.

QUESTION 1.8 Express the radius of an atom of rubidium, Rb, in nanometers.

ANSWER 1.8 Figure 1.23 tells us that the radius of a rubidium atom is 247 pm. We can convert first to meters and then to nanometers in one step:

$$247\ \cancel{pm} \times \frac{1\ \cancel{m}}{1,000,000,000,000\ \cancel{pm}} \times \frac{1,000,000,000\ nm}{1\ \cancel{m}} = 0.247\ nm$$

QUESTION 1.9 A photographer taking a photograph at night sets her camera's shutter speed to 4 seconds (s). Express this length of time in microseconds (μs).

ANSWER 1.9 There are 1,000,000 μs in 1 s, a fact that can be used to create a conversion factor:

$$4\ \cancel{s} \times \frac{1,000,000\ \mu s}{1\ \cancel{s}} = 4,000,000\ \mu s$$

There are a couple of things to be aware of when using dimensional analysis. First, it does not matter in what order you place the conversion factors in a calculation. Once you have your numerators and denominators correct, the units cancel and the math works out the same regardless of the order. Second, if you are faced with a question that requires dimensional analysis, you may not need all the information given. Beware! This is a sneaky trick professors play because they know students try to fit every piece of given information into a calculation. Don't fall victim to it!

1.5 | The Metric Epicurean

Along with American industry, American and English kitchens may be the last nonmetric strongholds. Such measuring devices as cups, teaspoons, and tablespoons are old friends, and many cooks would balk at the idea of trading them in for metric measures. You may find the metric system in your refrigerator—all those 2-L bottles of soda—but the old standbys *quart* and *gallon* are still the most frequently seen volume units in the United States. Figure 1.25 shows a recipe for "metric" brownies. Except for the *drop*, everything in the recipe is expressed in metric units. It looks more like a chemistry laboratory manual than a recipe for brownies!

The volume unit *milliliter* used to measure out the sugar, flour, and walnuts is based on the *liter*, the long-established metric unit for volume. Volume is three-dimensional, and so it can also be expressed in units of length. In fact, the milliliter is defined as 1 cubic centimeter (cm^3), a volume about the size of a thimble (Figure 1.26). An alternative abbreviation for *cubic centimeter*, one you might see more frequently in nonscientific writing, is cc. Thus, $1\ mL = 1\ cm^3 = 1\ cc$.

In addition to *milli-*, all the other metric prefixes can be placed in front of the word *liter*. Thus nanoliter (nL), picoliter (pL), and femtoliter (fL) are all legitimate volume units.

If you get a hankering for brownies and would like to try these, here are the conversion factors you will need: 454 g = 16 ounces; 1 L = 1.0567 quarts; 1 cm = 0.394 in. Also, the conversion factor 1 kg = 2.2046 lb will help you to calculate your weight gain once you've eaten all 28 brownies. (Hint: One brownie has a mass of roughly 50 g.)

▼ FIGURE 1.25 **May I have 250 mL of milk with that?** A recipe for brownies using metric units of measurement.

Metric Brownies

226 g butter
226 g unsweetened chocolate squares
5 eggs
714 g sugar
15 mL vanilla extract
354 mL flour
472 g chopped walnuts

1. Preheat oven to 450 K. Grease and flour a 23 cm × 33 cm baking pan.

2. Melt the first two ingredients in a saucepan and heat gently until melted with constant stirring.

3. Thoroughly mix eggs, vanilla extract, and sugar together. Mix in the chocolate mixture, the nuts, and the flour.

4. Pour brownie mixture into the prepared pan and bake at 450 K for 30 to 40 minutes.

5. Cool brownies.

Makes 25 large brownies.

▼ FIGURE 1.26 **A milliliter is a cubic centimeter.** A sewing thimble has a volume of approximately 1 mL, which is the same volume as 1 cm^3, which is the same volume as 1 cc.

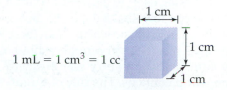

$1\ mL = 1\ cm^3 = 1\ cc$

QUESTION 1.10 A plastic bottle contains 2.0 L of soda. Express this volume in milliliters and in cubic centimeters.

ANSWER 1.10 We know that 1000 mL = 1 L and that 1 mL = 1 cm³. Therefore,

$$2.0 \text{ L} \times \frac{1000 \text{ mL}}{1 \text{ L}} = 2000 \text{ mL}$$

$$2000 \text{ mL} \times \frac{1 \text{ cm}^3}{1 \text{ mL}} = 2000 \text{ cm}^3$$

QUESTION 1.11 Express the length 5,600,000,000 nm in a metric length unit containing no zeros.

ANSWER 1.11 There are nine decimal places after the "5" in our number of nanometers, and we know that 1,000,000,000 nm = 1 m. Setting up our conversion factor so that all units but meters disappear, we get

$$5,600,000,000 \text{ nm} \times \frac{1 \text{ m}}{1,000,000,000 \text{ nm}} = 5.6 \text{ m}$$

In our recipe for metric brownies, the unit used for the oven temperature is K, which stands for *kelvins*. The **kelvin** is the base unit of the **Kelvin temperature scale**, as noted earlier when we talked about Table 1.2. The Kelvin temperature scale replaced the **Celsius temperature scale** (°C) in scientific work back when scientists decided that the Celsius scale did not define temperature well enough to be used for certain applications.

STUDENTS OFTEN ASK

What Is the Volume of One Drop?

If you look back at our recipe for metric brownies, you will see that the vanilla extract is measured out in drops. Drops are tricky because the size of a drop depends on many things. For example, the eye dropper you use can have a wide mouth or small mouth. Also, the drop size will depend on whether the dropper is made of glass or plastic.

This variation is manifested in the confusing variety of drop definitions. There are the U.S. drop (0.0616 mL), the U.K. drop (0.0592 mL), the metric drop (0.05 mL), and the Winchester drop (0.0616 mL). With all these drops, it is not surprising that people do not use them as a strict standard of measurement. Your best bet is to think of a drop as merely a very rough way to measure out a small volume and not worry about exactly how much liquid one drop contains. The rule of thumb many chemists use is 20 drops per milliliter, and so one drop is equal to about 0.05 mL.

The Celsius scale is based on assigning 0 °C as the melting point of water and 100 °C as its boiling point. This might seem like a good universal standard. After all, almost everyone has access to water, and all water has the same composition, H_2O. Thus, water should serve as a workable world-wide standard for temperature, and in fact the Celsius scale serves us well for most purposes. However, although the melting point of water is fairly consistent from one location to another, the boiling point varies with altitude. Consequently, even though countries that use the metric system invariably use the Celsius scale for temperature, the fact that one of the reference points on this scale is variable makes it difficult to use the scale for scientific purposes, which frequently call for the temperature to be known very accurately.

The Kelvin scale is based on a fundamental reference point that is intrinsic to everything: the temperature at which all motion stops. Suppose you have molecules in the gas phase. The molecules are moving all the time, and how fast they move depends on their temperature. The higher the temperature, the faster the molecules move. If you lower the temperature, the molecules slow down and eventually adhere to one another, turning the gas to a liquid. The molecules are still moving, but not as fast as when the temperature was higher. Continue to lower the temperature, and the molecules slow down even more, until at some point the liquid becomes a solid. The molecules in the solid are still moving (although relatively slowly). If you continue to lower the temperature, however, sooner or later you will reach a temperature at which all molecular motion ceases. This fundamental temperature at which all motion stops is the same for everything and is taken as the zero reference point on the Kelvin temperature scale. We write it 0 K and call it *absolute zero*, and the Kelvin temperature scale is also called the *absolute temperature scale*.

The Kelvin scale is related to the Celsius scale by these simple relationships:

$$K = °C + 273.15$$
$$°C = K - 273.15$$

Thus, you can report the temperature of boiling water as 100.00 °C or as 373.15 K. The temperature at which we want to bake our brownies is 450 K − 273 = 177 °C.

In the United States, we frequently use the *Fahrenheit temperature scale* (°F). The conversion from Fahrenheit to Celsius is

$$°C = \frac{5}{9}(°F - 32)$$

and the conversion from Celsius to Fahrenheit is

$$°F = \frac{9}{5}(°C) + 32$$

So if our oven temperature should be 177 °C and our stove dial is calibrated in degrees Fahrenheit, we should set the dial to

$$\frac{9}{5}(177 \, °C) + 32 = 351 \, °F$$

Note that when expressing a temperature using the Kelvin scale, you do not include the degrees symbol (°). Instead, you simply write a number followed by a space and then the letter K. Therefore, you would write 30 °C, but 303 K.

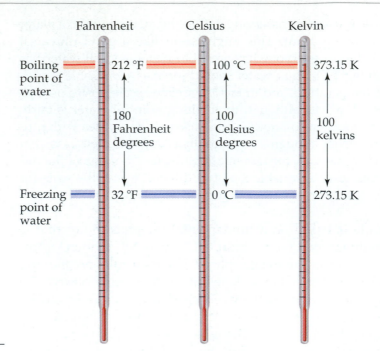

▶ FIGURE 1.27 The three most commonly used temperature scales.

You will rarely see temperatures expressed in kelvins unless you are reading scientific material. In this book, we shall use the Celsius scale. Figure 1.27 summarizes the relationships among the three temperature scales.

QUESTION 1.12 Express the temperature 100.00 K in both degrees Celsius and degrees Fahrenheit. Is this temperature much hotter or much colder than ambient temperatures on Earth?

ANSWER 1.12

$$°C = K - 273.15$$

$$= 100.00\ K - 273.15 = -173.15\ °C$$

$$\frac{9}{5}(-173.15\ °C) + 32 = -280\ °F$$

This is pretty cold! The lowest temperature recorded on Earth was −128.6 °F in Vostock, Antarctica, on July 21, 1983.

Let's look once more at our metric brownie recipe. The amounts of butter and chocolate are expressed in *grams*, the unit for mass given in Table 1.2 on p. 19. As with all other metric units, this one is easy to manipulate using prefixes. For example, if you decided to make 10 recipes of brownies, you would need 2270 g of butter. We can express this mass in kilograms by using the conversion factor 1 kg = 1000 g:

$$2270\ \cancel{g}\ \text{butter} \times \frac{1\ kg}{1000\ \cancel{g}} = 2.27\ kg\ \text{butter}$$

Likewise, if you decided to make one brownie (which no one would ever want to do), you would need

$$1\ \cancel{\text{brownie}} \times \frac{227\ g\ \text{butter}}{28\ \cancel{\text{brownies}}} = 8.1\ g\ \text{butter}$$

This amount of butter could also be expressed in milligrams, where $1\text{ g} = 1000\text{ mg}$:

$$8.1\text{ g butter} \times \frac{1000\text{ mg}}{1\text{ g}} = 8100\text{ mg}$$

QUESTION 1.13 Before you leave for college, you get some news about your new roommate from France, a country that uses the metric system. His height is 198 cm, and his weight is 79 kg. What does your new roommate look like? (Hint: 1 cm = 0.394 in. and 1 kg = 2.20 lb.)

ANSWER 1.13 First express these conversion factors in fraction form:

$$\frac{0.394\text{ in.}}{1\text{ cm}} \quad \text{or} \quad \frac{1\text{ cm}}{0.394\text{ in.}} \quad \text{and}$$

$$\frac{2.20\text{ lb}}{1\text{ kg}} \quad \text{or} \quad \frac{1\text{ kg}}{2.20\text{ lb}} \quad \text{and} \quad \frac{12\text{ in.}}{1\text{ ft}} \quad \text{or} \quad \frac{1\text{ ft}}{12\text{ in.}}$$

Now choose one from each pair of conversion factors, and use dimensional analysis to convert from metric to English:

$$198\text{ cm} \times \frac{0.394\text{ in.}}{1\text{ cm}} \times \frac{1\text{ ft}}{12\text{ in.}} = 6.5\text{ ft}$$

and

$$79\text{ kg} \times \frac{2.20\text{ lb}}{1\text{ kg}} = 170\text{ lb}$$

Your new roommate is very tall and very skinny!

The only unit in our recipe we have not discussed yet is the time unit. Luckily, most of the world uses similar systems for expressing time. The official metric unit is the *second*, and larger and smaller units can be made from it. For example, ultrafast lasers currently being developed flash more than 1,000,000,000,000,000 times per second. We call them *femtosecond* (fs) lasers because there are 15 zeros after the 1. Likewise, units like the *millisecond* (ms), *microsecond* (μs), and *nanosecond* (ns) are all legitimate ways to express time.

For longer lengths of time, the units minutes, hours, and years are commonly used.

1.6 | Chemistry in the Limelight

If you have taken a course in world history, you may have noticed something about the impact of science on society: major changes in the way we think about nature can alter the course of history, as we saw in the chapter introduction. In the early twentieth century, scientists found that it is possible to split the atom and to tap the energy within to produce forces unheard of before. As we know, the upshot was the first nuclear weapons, which brought an end to World War II. Since then, our understanding of the nucleus has allowed us to tap its energy for peaceful uses, such as the generation of electrical energy at nuclear power plants. Because these plants produce radioactive waste products, however, the exploitation of nuclear energy for power has been, and will continue to be, one of the most divisive areas of public debate. We must decide whether the cheap power we get from nuclear plants is worth the cost of the radioactive waste produced. Do our power plants put people at risk from terrorists? From power-plant accidents?

You, the reader of this book, will someday be making your wishes known to the politicians making decisions like these, and that is why you need to know how science works. We hope that the critical-thinking skills developed in taking this chemistry course will serve you well for the rest of your life.

One question often heard in public debate is, should scientists be permitted to pursue any particular avenue of research? This question has been the focal point of research on cloning organisms. Should governments take control of the natural progression of science? What *is* the natural progression of science? How do scientists decide what is important to study, which questions should be answered and which ignored?

Well, the questions scientists choose to work on usually depend largely on two things. First, the natural momentum of scientific research rolls along, and it is often tempting to try to answer the next logical question. For example, if energy research progresses to a point at which a new hydrocarbon fuel source is discovered, a logical next step is to find out as much as possible about this fuel. A second criterion a scientist often uses for selecting a research project is how important the topic is to the well-being of the human race. For instance, we know that chemical weapons have been researched and synthesized by countries around the world, including the United States. Parallel studies that focus on the development of antidotes to chemical weapons, therefore, are a possible focus of chemical research. This is certainly a topic worthy of study.

A scientist most often addresses only a very small part of a larger problem. For example, a chemist may be trying to figure out why a certain compound in aerosol cans causes depletion of the ozone layer. This chemist may decide to change the compound slightly to see if the adverse effects still occur. Another chemist working independently on the same problem may try to figure out what other, more benign compounds can be substituted in the aerosol formulation. Yet another chemist may study how the compound affects the chemistry of natural waters. As experiments are completed, each scientist submits his or her findings to a scientific journal. There, the data are extensively *peer reviewed*—that is, sent out to other scientists for evaluation.

Peer reviewers scrutinize the data to be sure the experiments were done properly and the most logical conclusions reached. Sometimes the reviewers may repeat the experiment described to be sure the results are reproducible in another laboratory. If the work is found to be acceptable, it is published and thereby becomes available to the wider scientific community. The three scientists working on the compound that causes ozone depletion can then read one another's findings in the literature and update their own experimental designs accordingly. In this way, little gains are made toward a broader understanding of the original question. Breakthroughs are built upon and make way for newer breakthroughs (Figure 1.28). This whole approach to doing science—including verification of reproducibility and peer review—is how science gets done, and notable findings eventually filter out to the public.

How are experiments carried out in a way that will withstand the scrutiny of peer review? The general approach to experimentation is referred to as the **scientific method**. Let's look at an example of how it works. Suppose you are an amateur scientist having coffee at a bookstore. You overhear a debate about whether a piece of toast accidentally knocked off a kitchen table lands more frequently buttered side up or buttered side down (Figure 1.29). The prevailing consensus is that toast always lands buttered side down. Later that day, after giving this problem considerable thought, you decide to perform a series of exhaustive bread and toast experiments in your kitchen. You begin by formulating a **hypothesis**, which is an initial best guess about some question concerning nature:

> **Hypothesis:** The buttered side of a slice of bread or toast knocked off a kitchen table always lands facing the floor.

Note that a *scientific* hypothesis is different from other kinds of hypotheses because it can be tested by experiment. If you ever take a class in philosophy, you may be asked to discuss some hypothesis about the meaning of life, say, or the existence of an afterlife. This type of hypothesis is not a scientific hypothesis because it cannot be tested by experiment.

Armed with the knowledge that answers await in your kitchen, you stock up on bread and butter and set to work. Done correctly, your experiments should convince even the most stubborn skeptic, and so it is important to anticipate every objection that might arise when you unveil your results. You therefore design an experiment to include every combination of buttered and unbuttered bread and toast. The obvious place to begin is pushing slices off a typical kitchen table. Because table height might be an important factor, you measure the height of yours. Having decided to start with buttered toast, you then slide a slice, buttered side up, off the table. However, how do you know what will happen the next time, and the time after that? To find out, you repeat the experiment 19 more times and record your results. To be as consistent as possible, you try to use the same force and wrist motion each time.

You're not done yet, however. You repeat your experiment with unbuttered toast, buttered bread, and unbuttered bread, doing 20 trials each time and recording the results. With the buttered bread, you again have the buttered side up at the beginning of each run. With the unbuttered bread and unbuttered toast, you put a tiny spot of red ink on one side and make this the side initially facing up in each trial.

One objection to your procedure could be the question of whether or not it matters if the same slice of toast or bread is used for all 20 trials. In other words, does a slice of toast/bread that has fallen repeatedly begin to fall differently after a while? You suspect that this question is unimportant, but you check it anyway by running all the experiments over, but starting with a fresh slice of toast or bread every time. Five hours later, you've got all your data.

VOLUME 37 ISSUE 17   AMERICA'S FINEST NEWS SOURCE™   10–16 MAY 2001

SCIENCE

New Technological Breakthrough To Fix Problems Of Previous Breakthrough

Above: A Texas A&M chemist works on the breakthrough.

▲ **FIGURE 1.28 How science is done.** This headline from *The Onion*, September 12, 2002, pokes fun at the way science works. (Reprinted with permission of *THE ONION*. Copyright 2001, by ONION, INC. www.theonion.com/Getty Images, Inc.)

▼ **FIGURE 1.29 Toast falling to the floor.** Toast does unpredictable things at breakfast time.

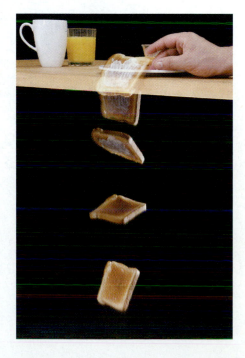

QUESTION 1.14   Which are scientific hypotheses: (*a*) Water boils at different temperatures depending on altitude, (*b*) a soul exists in every human being, (*c*) apples fall from trees at a constant speed?

ANSWER 1.14   (*a*) and (*c*) are scientific hypotheses because they can be tested. Choice (*b*) is a hypothesis, but not a *scientific* one, because it cannot be tested.

After looking at the data, you attempt to come up with a **theory**, which is an explanation of the observed phenomenon based on your collected data. Your data tell you that, with the buttered side initially facing up, both buttered toast and buttered bread land buttered side down 90 percent of the time. Your findings also show the same result when the toast and bread are unbuttered: with the side marked with red ink initially facing up, both bread and toast land red side down 90 percent of the time. These results show that the presence or absence of butter and whether or not the bread is toasted are irrelevant to the observed phenomenon. Buttered or not, toasted or not, 90 percent of the slices land on the side that was facing up when the slice left the table.

You don't know *why* this happens, though, and so you cannot formulate a theory. You decide to try tables of different heights and repeat your original experiments at each new table height. Aha! Table height does make a difference! Tables only a little taller than a kitchen table cause most slices to land buttered/red side up, and extremely tall tables cause most slices to land buttered/red side down.

Now it is time to formulate a theory. Having observed that each slice rotates as it falls, you theorize that table height determines how many rotations can be completed before the slice hits the floor. Your expert recommendation to the clumsy, buttered-toast-eating public is this: Eat your toast at a slightly taller table.

This experiment was actually performed by Robert Matthews of Aston University in Great Britain. His findings show that toast usually does land buttered side down when dropped from a table that is 1.5 m high. Taking the analysis a step further, he wrote,

> ... So why are tables the height they are? Simple: to be convenient for humans. So why are humans the height they are? [Because] ... there is a limit to the safe height for bipedal, essentially cylindrical creatures like humans. The limit is around 3 metres—above that height, a simple fall results in gravity accelerating the skull to such a high kinetic energy that the chemical bonds in the skull are ruptured, causing severe fracturing. This limit, in turn, sets a maximum height on tables suitable for creatures with human articulation of about 1.5 metres—which is still not high enough to prevent toast from landing butter-side down.

While this buttered-toast example may be tongue in cheek, it does demonstrate how a theory can develop from a testable hypothesis. Table 1.3 summarizes the way the scientific method was used to get to the bottom of the buttered-toast controversy.

One interesting characteristic of any theory is that it can never be *proved*. Did Matthews prove anything with his toast experiments? Can a scientist

TABLE 1.3	Summary of Scientific Method Applied to Buttered-Toast Question
The question	When a slice of bread or toast falls from a table, does the face that was initially up always land facing the floor?
Initial hypothesis	The buttered side of a slice of bread or toast knocked off a kitchen table always lands facing the floor.
Experimental design	Re-create all possible scenarios for a falling slice of bread, including buttered/unbuttered, toasted/untoasted, and various table heights.
Experimental results	For a table 1.5 m high, the side of toast/bread facing up almost always lands face down. Which side lands face down is a function of table height.
Theory proposed	Each slice rotates as it falls. The table height determines the number of rotations that can be completed before impact.
Conclusion	If you are going to drop buttered toast and want to keep the floor from getting greasy, it is best to drop the toast from a table that is either a bit taller or a bit shorter than 1.5 m.

ever prove something by experimentation? The answer is no. A theory provides the latest and best interpretation of experimental results, but any theory is inherently unprovable because the next set of experiments can show that the theory is wrong. Thus, we can say that evidence *supports* a theory, and we can say that evidence *disproves* a theory, but we cannot say that evidence *proves* a theory.

Throughout this book, you will see the scientific method at work. For example, the periodic table has undergone constant changes and updates as new information about the elements has emerged. Arguments go back and forth in the scientific literature about new elements as they are discovered, and the rest of us wait patiently for the scientific community to reach a consensus before we know whether or not to pencil in the new element on our periodic tables. It is important to remember that the periodic table is a **model** invented by scientists to classify and organize matter. It arranges the elements in a manner that is logical and pleasing in its regularity and symmetry.

We also use models to depict atoms and molecules because models are our means for understanding nature. After you read this book, you may find yourself picturing molecules as balls (representing the atoms making up the molecules) connected by sticks (representing the chemical bonds between atoms). It is important to remember, however, that no molecule really looks like this and that this depiction is just one way we can represent our understanding of molecular structure. Advances in computer technology have brought models from balls with holes in them for sticks to beautiful three-dimensional graphics.

Chemistry textbooks try to illustrate molecules in three dimensions, and there are tricks you will learn in Chapter 3 for depicting three-dimensional molecules on a two-dimensional sheet of paper. None of these methods can show you more than one view of a molecule at a time, however. What if you could take an actual molecule, make it larger, and flip it this way and that? This is in essence what computer technology has brought to the field of molecular modeling. Ten years ago, the file for a small three-dimensional molecule was too large for most computers to handle. With the

▲ FIGURE 1.30 **Satellite tobacco mosaic virus.**   The diameter of this virus, perhaps the smallest known, is only 17 nm. This structure was determined using a modern-day adaptation of Max von Laue's original X-ray technology. (Courtesy of Steven Larson and Alex McPherson. Larson, S.B., Day J., Greenwood, A. and McPherson, A. (1998). Refined structured of satellite tobacco mosaic virus at 1.8A resolution. J. Molecular Biology 277, 37–59.)

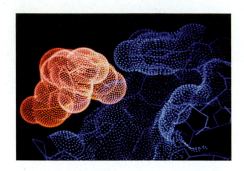

▲ FIGURE 1.31 **Computer model of interacting molecules.**   Computer-generated model of two large biological molecules recognizing and then binding to each other. The red is a drug molecule attaching itself to a protein in the body.

gigabyte storage space and high-speed memory available today, however, a reasonably priced computer can display molecules containing thousands of atoms (Figure 1.30), and the display can be rotated so that it can be viewed from the top or front or bottom or side.

Today's computer workstations, which are more powerful than a standard desktop system, can be used to watch molecules move and vibrate. You can tell your workstation to depict a molecule at a temperature of 0 K, where it is motionless, then ask for a virtual warming experiment and watch the molecule begin to move. Every year, more and more experiments like this are performed on computers, and this capability can save months of work in the laboratory. For example, medical drugs are often designed to physically "dock" into the surface of a protein (Figure 1.31). Before the computer technology was available, chemists would build models by hand, and those models would be only rough approximations of the actual molecule. Today, a molecule's structure can be precisely represented via computer graphics, with the correct length between every pair of atoms, as well as the correct topology of the molecule's surface.

The compact disc that accompanies this book makes extensive use of molecular models, and we invite you to play with them.

The purpose of this book is to put chemistry in the limelight, but what is limelight anyway? Where did that term come from? We're glad you asked. Limelight was produced by heating lime, a mixture of calcium-containing compounds, in a hydrogen–oxygen flame. The bright light created in this chemical reaction was used for theatrical productions before electric lights were available (Figure 1.32). This is just one example of the everyday applications of chemistry we shall introduce to you in this book. In addition to showing you the ways in which chemistry is ever present in our lives, we shall also show you that science is meant to be criticized and mistrusted. After all, that is the way science happens. We question everything, and then when we are finally satisfied, we know that good work has been done.

▼ FIGURE 1.32 **The production of "limelight" is a chemical reaction.**   In this vintage limelight projector, light was produced by heating solid calcium oxide (CaO, also known as lime). A lens helped to produce a spotlight effect, so before electric lights were available performing artists were always "in the limelight."

Chapter 1

SUMMARY

Chapter 1 introduces us to everything—all matter—and how chemists study it. The periodic table arranges the 114 known elements that are the components of all matter. Every atom of every element contains protons and electrons, and these particles characterize the element and give it its properties. Individual atoms can make chemical bonds with one another through chemical reactions. Reactions are represented by chemical equations in which the reactants are written to the left of an arrow and the products are written to the right.

A pure substance may contain atoms of only one element (in which case the pure substance is called an element) or atoms of more than one element (in which case the pure substance is call a compound). The ratio of atoms of different elements in a compound is fixed and invariable from sample to sample, and we can write chemical formulas for them, such as $C_6H_{12}O_6$ (glucose), $C_9H_8O_4$ (aspirin), or NaCl (table salt). There are many varieties of compounds. Molecules, for example, are compounds in which the atoms are held together by a sharing of electrons.

When pure substances, whether they are elements or compounds, are mixed together, they can form either smooth, homogeneous mixtures or poorly mixed, heterogeneous mixtures. Mixtures can be separated into pure substances through physical change, which often involves changes in phase, such as boiling or melting.

The SI system is an internationally agreed-upon set of units used for measuring time, length, mass, and other quantities. It is an adaptation of the traditional metric system. Prefixes may be combined with a base unit name to change the unit by various orders of magnitude. Examples include the microgram, μg, which is 1/1,000,000 of a gram, and the millisecond, ms, which is 1/1000 of a second. These units, and units from other systems of measurement, can be manipulated using a method called dimensional analysis, whereby conversion factors are arranged to allow for cancellation of unwanted units.

Careful experimentation is the hallmark of the scientific method, which is a series of steps that take a scientific question from hypothesis to theory. When scientific results are submitted for publication, scientists reviewing the work look for the logical sequence of steps inherent in the scientific method to ensure that the results are legitimate and reproducible.

KEY TERMS

atom	chemical reaction	group	matter	product
atomic number	coefficient	heterogeneous	mixture	proton
balanced equation	compound	mixture	model	pure substance
Celsius temperature	conversion factor	homogeneous	molecule	reactant
scale	dimensional analysis	mixture	neutron	scientific method
chemical bond	electron	hypothesis	nucleus	solution
chemical change	electron density	kelvin	period	theory
chemical equation	element	Kelvin temperature	periodic table	
chemical formula	family	scale	physical change	

QUESTIONS

The Painless Questions

1. Why were the discoveries made by Roentgen, Becquerel, and the Curies revolutionary for their time? Name two discoveries from the twentieth century that relied on the work of these pioneering scientists.

2. Comment on this statement:

    Hypotheses and theories are really the same thing. Both tell you the expected outcome of a series of scientific experiments.

Marie and Pierre Curie

3. Give the total number of atoms and the number of atoms of each element contained in one molecule of (a) benzene, C_6H_6; (b) bromine, Br_2; (c) nitrous oxide, N_2O.

4. Which length is longest: 1,350,000,000,000 pm, 13,500,000,000 nm, or 0.135 m? Use dimensional analysis to defend your choice.

5. A homogeneous mixture contains only aspirin, $C_9H_8O_4$, and vitamin C, $C_6H_8O_6$. (a) Is aspirin a compound or an element? (b) Is aspirin a pure substance or a mixture? (c) Is vitamin C a compound or an element? (d) Is vitamin C a pure substance or a mixture?

6. Atoms containing 33 protons are found in a murder victim's blood. (a) These are atoms of which element? (b) How many electrons does each of these atoms contain?

7. Isoflurane, $C_3H_2ClF_5O$, is an inhaled anesthetic used in hospital operating rooms. Which term(s) describes isoflurane: mixture, pure substance, element, compound?

8. Explain why it is incorrect to refer to a theory as being "proved."

9. Which are elements and which are compounds: (a) BF_3, (b) H_2, (c) S_8, (d) C_6H_{12}?

10. Classify each item as homogeneous mixture, heterogeneous mixture, or pure substance: (a) orange juice with pulp, (b) milk, (c) bottled water, (d) cottage cheese, (e) iced tea.

11. It takes the average Londoner about 5,000, 000,000 ns to dunk a cookie into tea. How many seconds is this?

12. Identify each of these changes as chemical or physical: (a) gasoline burning in a car engine, (b) snow melting, (c) alcohol boiling, (d) dynamite exploding.

13. Which mass is largest: 4.5 g, 0.000 045 kg, or 450,000 μg? Use dimensional analysis to defend your choice.

14. (a) Is it possible to break down a diatomic element such as chlorine, Cl_2, using a physical process? (b) Is this breakdown possible using a chemical reaction? (c) What products result from the breakdown of one molecule of Cl_2?

15. Use coefficients to balance this equation:

$$C_3H_8 + O_2 \longrightarrow CO_2 + H_2O$$

16. Describe how a scanning electron microscope works and how it can be used to visualize individual atoms on a surface.

More Challenging Questions

17. Using the information in Tables 1.1 and 1.2, give the full name for each of these abbreviations: (a) μmol, (b) fm, (c) Mm, (d) ds.

18. This reaction takes place when hydrogen sulfide gas, H_2S, is released during a volcanic eruption:

$$2\,H_2S + 3\,O_2 \longrightarrow 2\,SO_2 + 2\,H_2O$$

(a) Is this equation balanced? (b) How many atoms of each element make up the reactants? (c) How many atoms of each element make up the products? (d) If you begin the reaction with six molecules of hydrogen sulfide and nine molecules of oxygen, how many molecules of water are produced?

19. Comment on this statement:

An element consists of individual atoms that are separate from one another. All of the atoms in a sample of an element must have the same number of protons and neutrons.

20. (a) Describe two ways a mixture of water and sand (made up primarily of silicon dioxide, SiO_2) can be separated into its two components. (b) Is sand an element or a compound? (c) Can physical or chemical change (or both) be used to break the water and silicon dioxide down further?

21. Consider this chemical equation representing the reaction between rust, Fe_2O_3, and acid:

$$Fe_2O_3 + 4\,HCl \longrightarrow 2\,FeCl_3 + 3\,H_2O$$

One of the coefficients in this equation is incorrect, and thus the equation is not balanced. (a) Which coefficient is it? (b) What should this number be changed to? (c) After you make this coefficient change, check that the equation is balanced by counting atoms.

22. There is an old myth that cold water boils more rapidly than hot water. Using Table 1.3 as a guide, develop a strategy for testing this myth, beginning with a hypothesis and outlining a series of experiments you could perform to test it. What are the possible outcomes of your experiment?

23. Comment on this statement about the use of computers to view molecules:

> You can inject a sample of a molecule into one of today's sophisticated new computers, and within seconds a full three-dimensional structure of the molecule will appear on the monitor.

24. Here is a hypothesis:

> Microwave ovens heat all foods to the same temperature in the same amount of time, regardless of water content.

Formulate a series of experiments you could perform to test this hypothesis. Describe the possible outcomes of your experiments.

25. You are calling the telephone company to discuss your bill. Your expected wait time on hold is 20,000,000,000,000,000,000 fs. (a) Convert this time to seconds and decide if you would be willing to wait this long. (b) What is this length of time in minutes? (c) in years?

26. In nature, zinc is sometimes found in the form of zinc sulfide, ZnS. One step in converting this compound to pure zinc metal is to convert it first to zinc oxide, ZnO:

$$2\ ZnS(s) + O_2(g) \longrightarrow 2\ ZnO(s) + 2\ SO_2(g)$$

(a) What coefficient must be added in front of the oxygen molecule to balance this equation? (b) What is the phase of each reactant and product? (c) If you begin the reaction with 1 unit of zinc sulfide, how many units of zinc oxide will be produced?

27. The candela (cd) is the metric base unit for luminous intensity, a measure of the brightness of light. Express a luminous intensity of 34.5 kcd in millicandelas.

28. Picometers are the most convenient metric unit for expressing atomic radii. However, the length unit angstrom (Å) is also sometimes used, where 1 Å = 0.000 000 000 1 m. Using information from Figure 1.23 and dimensional analysis, express the radius of each of the following atoms in angstroms: (a) indium, In; (b) calcium, Ca; (c) lithium, Li.

29. The medical drug RU-486, formally known as *mifepristone*, is the hormone in the "morning-after pill," used to prevent implantation of a fertilized egg in a human uterus. The chemical formula for mifepristone is $C_{29}H_{35}NO_2$. (a) How many atoms are there in one molecule of mifepristone? (b) How many nitrogen atoms are there in five molecules of mifepristone? (c) How many carbon atoms are there in two molecules of mifepristone?

30. A recipe written by a scientist calls for baking muffins at 425 K. (a) Convert this temperature to degrees Fahrenheit. (b) Is this a reasonable temperature for baking?

The Toughest Questions

31. The density of any substance is the amount of mass of the substance contained in a given volume of space. (a) What unit could be used to express the numerator of a density unit? (b) What unit could be used to express the denominator? (c) If you know the density of some substance and also know the mass of a sample of that substance, what else do you know about the sample?

32. Oxygen makes up about 61 percent of the mass of the human body, and hydrogen makes up about 10 percent. If the body is mostly water, explain why the percentage of oxygen is higher than that of hydrogen.

33. The mass of an electron is estimated to be 0.000 000 000 000 91 fg, and the mass of a proton is estimated to be 0.000 000 000 000 000 000 001 7 mg. (a) Use dimensional analysis to convert both masses to grams. (b) Which particle has the larger mass? (c) By what factor do these two masses differ?

34. The metric prefix *giga-* (G) represents 1,000,000,000 times any base unit used with it. The distance from Earth to the sun is 150 Gm, and light travels at a speed of 300,000,000 m/s. How many minutes does it take light from the sun to reach Earth?

35. If you line up atoms of tellurium, Te, so that each atom is touching its two neighbors and the diameters form a straight line 2 nm long, how many atoms do you have? (Hint: You will need data from Figure 1.23.)

36. In the laboratory, you come across an unlabeled bottle containing some chemical substance. Because it was purchased at a chemical company, you suspect that what is in the bottle is a pure substance and not a mixture. (*a*) Could you determine whether this substance is an element or a compound? (*b*) Is it possible that this substance is absolutely pure with no impurities? Why or why not?

37. Caffeine is a compound that has the chemical formula $C_8H_{10}N_4O_2$. Samples of caffeine are collected from sources in Quebec, Morocco, Turkey, Norway, and Chile. The samples from Turkey and Chile both contain five hydrogen atoms for every oxygen atom. In the other three samples, the hydrogen–oxygen ratio is found to vary widely from this value. What can you conclude about these three samples?

38. The speed of light is 300,000,000 m/s. If light leaves one point and travels to a point 4,000,000 km away, how long will it take to get there?

39. In the movie *Total Recall*, humans have colonized Mars and are living there. The villain, who controls an enormous machine that creates breathable air,

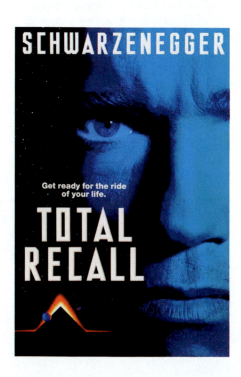

charges people to buy this air. The machine, as it turns out, makes oxygen gas from H_2O by melting an underground glacier. Is this scheme viable? Why or why not?

40. Atoms of element A have three times more protons than atoms of element B, and A and B are in the same column of the periodic table. Subtract the number of protons in B from the number in A, and your answer is equal to the number of protons in atoms of a third element in the same column as A and B. Identify A and B.

41. On a certain September day in Utah, the maximum and minimum temperatures differed by 30 °F. What is this temperature difference expressed in kelvins?

42. We learned that 1 mL is the volume of a cube that is 1 cm on each side. (This volume can also be expressed as $1 cm^3 = 1$ cc.) How many milliliters would fit into a cube that is 1 m on each side? (The volume of this cube is $1 m^3$.)

43. If a farmer walks at a pace of 2 miles per hour across his 12-furlong field, how many minutes does it take him? (Hint: There are 8 furlongs in 1 mile.)

E-Questions

Go to **www.prenhall.com/waldron** to find these questions in electronic form, complete with hyperlinks directly to the various websites cited in the questions.

44. Read the article "Why Purified Water Is Bad For You" at **IonizedWater**. This article contains information that could be alarming to people who drink purified water. How can you collect more information about this topic to evaluate this article? Take a look at the references cited. Do you trust these references, or would you require other sources of information before you made your decision? What is your overall opinion of this article? Does it convince you that purified water is dangerous? Why or why not?

45. Go to **Numericana**, which has a comprehensive list of units of measurement. (*a*) Go to the links for the following units and write a conversion factor that relates each unit to the appropriate SI base unit: diamond mark, U.S. bushel, slug. (*b*) Sometimes we need prefixes for numbers that go beyond femto- and giga-. What are the definition and abbreviation for yottasecond, zeptogram, exaliter, and petameter?

46. The term *karat* (K) is used to describe the purity of gold. Go to **ScientificAmerican** and answer the following questions about the purity of gold: (*a*) What is the definition of karat? (*b*) What is an alloy? (*c*) What is the fineness of gold? (*d*) Is 24K gold pure?

47. Read the article "Pseudoscience—Too Good to Be True?" at **ChemMatters**. Several examples of

"pseudoscience" are given, ranging from claims that the moon landings were staged to stories about extrasensory perception. After reading this article, answer these questions: (*a*) What do the authors mean by their statement

It's easily argued that the simplified "steps of the scientific method" that appear in many of our texts convey a naïve and oversimplified picture of the way science works in practice.

(*b*) What is the difference between *junk science* and *pseudoscience*? (*c*) Look through the article, and choose your favorite example of pseudoscience. Research this topic using other sources, and write an essay that supports or refutes the claim made.

48. Colored inks are made by mixing various pigments, a process described at **WhatStuff**. After reading this short article, answer these questions: (*a*) How is *ink* formally defined? (*b*) What are the differences between printing inks and writing inks? (*c*) Why does newspaper ink rub off onto your fingers when you touch it? (*d*) According to the site, what does the future hold for the ink industry?

WORLD WIDE WEB RESOURCES

As with the E-Questions, go to **www.prenhall.com/waldron** to find these questions in electronic form, complete with hyperlinks directly to the various websites cited. Some of links that follow contain research articles related to the topics in this chapter and could be used as the basis for a writing assignment in your course. Other sites are of general interest.

Some of following links contain research articles related to the topics in this chapter and could be used as the basis for a writing assignment in your course. Other sites are of general interest.

- Ever wondered how outside air temperature is measured precisely? Find out at **MadScientist**. Using these guidelines from the National Weather Service, you can find a suitable location for an accurate temperature measurement and information on what type of thermometer is needed.

- To learn about blunders made in scientific work, go to **Whoops**. Here you will find stories ranging from a technical error made on the British two-pound coin to a mistake on the cover of Pink Floyd's "The Dark Side of the Moon."

- Read the article "Marie and Pierre Curie and the Discovery of Polonium and Radium" at **NobeleMuseum**.

- Two articles posted at **NobeleMuseum** discuss the impact that science has had on various facets of society and the responsibility of scientists to perform ethical research: "On Being a Scientist: A Personal View" by John C. Polanyi (Nobel Prize in Chemistry 1986) and "Science and Humanity in the Twenty-First Century" by Sir Joseph Rotblat (Nobel Peace Prize 1995).

CHEMISTRY TOPICS IN CHAPTER 2

- Survey of the periodic table, the diversity of elements, covalent bonding [2.1]

- Ionic bonding, metallic bonding, salts, anions, cations, electronegativity, bond polarity [2.2]

- Mass number, neutrons, isotopes [2.3]

- Scientific notation, moles, Avogadro's number, molar mass [2.4]

- An introduction to oxidation and reduction reactions [2.5]

- Magnetism [2.6]

- Control experiments [2.7]

"The most mysterious place on Earth is right beneath our feet."

W. B. LOGAN

Dirt

HOW ATOMS INTERACT WITH ONE ANOTHER: AN INTRODUCTION TO CHEMICAL BONDS AND SIMPLE REACTIONS

The word *dirt* leads your mind in different directions, depending on context. It is the derogatory term for gossip ("I've got some dirt about so-and-so"), a way to describe something grimy or filthy, and a synonym for the soil that covers most of Earth's landmasses. From the point of view of this last definition, we think of *dirt* as the stuff under our feet—the part of the ground that blows around, lands in the wrong places, and must be wiped away. As we shall see, however, dirt is more than just a nuisance and a reason to clean. Therefore let's begin this chapter by getting the dirt on dirt.

The terms *dirt* and *soil* have similar meanings, with both terms referring to the matter on Earth's surface. However, *dirt* often has a colloquial meaning, while *soil* is the more scientific term. We shall use these terms interchangeably in this chapter.

2.1 | What Is Dirt?

The photographs on the first page of this chapter are images of soil, or dirt, in different landscapes. The striking variations in color tell you that these dirt samples are composed of different things—in other words, the natural components of dirt vary from one location to another. The elements making up any given soil sample determine its capacity for supporting living plants. We know, for instance, that certain crops thrive in some areas, but not in others. If you examine the color and dampness of the dirt in the desert versus that in a rain forest, you will find that the former is typically pale and dry and the latter is rich, dark, and moist. Differences in soils result from variations not only in the amount of water in the soils, but also from variations in elemental composition.

Figure 2.1 shows the periodic table of the elements drawn with the height of each element's box proportional to that element's abundance in an average soil sample. Clearly, soil is dominated by the elements oxygen and silicon. These elements come together to make up a **mineral**, which is a naturally occurring crystalline solid. Besides oxygen and silicon, there are dozens of other elements present in soil, all of them important to some process occurring there. Carbon is the sixteenth most abundant element in dirt. Most of the carbon present in dirt comes from decomposed animal and vegetable matter. These carbon-based, or **organic**, compounds were once part of living things that are now being recycled by the soil. In this chapter, we focus on the **inorganic** (that is, nonorganic) components of dirt, in particular the substances that form from interactions between the elements typically found in dirt.

Even an element present in dirt in only trace quantities can be significant in some way. For example, when manganese, which is very scarce in dirt, is concentrated into crystals, it can impart a pink hue to the crystals. When this happens, the result is the precious stone pink topaz. Thus, small variations in the amount and type of elements in dirt samples can account for marked changes in appearance from one sample to the next.

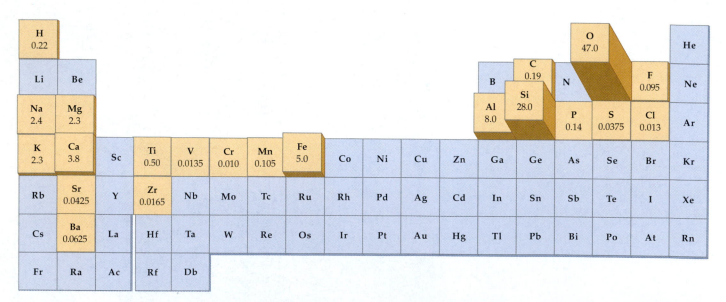

▲ **FIGURE 2.1   The distribution of elements in soil.**   In this periodic table, the height of each element box is proportional to the element's percentage in a typical soil sample.

Roughly 50 percent of a typical sample of dirt is composed of decaying matter (the organic part) and minerals (the inorganic part). The other half is water and air pockets, as shown in Figure 2.2. Soil is similar to many living things in that it makes use of gases and water. The water in dirt is infused with dissolved chemical compounds that are mobile within the matrix and serve as a transport system for nutrients.

Figure 2.3a shows a version of the periodic table that has a thick stepped line dividing the table into two parts. Except for hydrogen, all the elements to the left of the line are **metals**, which are typically shiny, lustrous solids. On the right side of the line are **nonmetals**, typically either gases or brittle, dull solids. The metals and nonmetals are further divided into groups according to their properties. Figure 2.3b shows where each group is located on the periodic table. **Halogens** and **noble gases** are found in columns 17 and 18, respectively. **Alkali metals** are all the elements in column 1 except hydrogen, and **alkaline earth metals** are all the elements in column 2. The **transition metals** occupy the center block on the periodic table. They are so called because this block forms a transition between the left and right sides of the periodic table.

▲ **FIGURE 2.2** **A worm's-eye view of soil.** In this schematic drawing, each dark region represents a soil particle surrounded by a film of water. The open spaces are air pockets where the exchange of gases can take place.

QUESTION 2.1 Table 2.1 shows the breakdown of elemental abundances in the human body and Earth's crust. (a) Which one element makes up a significant percentage of both the crust and the human body? (b) Some, but not all, metals are present both in the crust and in the human body. According to the table, which metals are present at levels of at least 0.19 percent in the crust, but less than 0.01 percent in the body? (c) The percentage of carbon is significant in the body and in Earth's crust. Speculate on where this element might be found in each.

ANSWER 2.1 (a) Oxygen. In the crust, oxygen exists primarily in the form of silicon dioxide. In the body, it is an essential part of many biomolecules required for life. (b) Silicon, aluminum, iron, and titanium. (c) In the crust, carbon is found mostly in degrading organic matter. In the human body, carbon is a primary constituent of all biomolecules.

As Figure 2.1 indicates, many metals are found in soil, but each is present only in very small quantities. Analysis of the metals in a given sample

TABLE 2.1	The Abundance of Elements in Earth's Crust and in the Human Body		
Earth's Crust		**Human Body**	
Element	Percent	Element	Percent
O	47	H	63
Si	28	O	25.5
Al	7.9	C	9.5
Fe	4.5	N	1.4
Ca	3.5	Ca	0.31
Na	2.5	P	0.22
K	2.5	Cl	0.08
Mg	2.2	K	0.06
Ti	0.46	S	0.05
H	0.22	Na	0.03
C	0.19	Mg	0.01

(a)

1 / 1A	2 / 2A	3 / 3B	4 / 4B	5 / 5B	6 / 6B	7 / 7B	8 / 8B	9 / 8B	10	11 / 1B	12 / 2B	13 / 3A	14 / 4A	15 / 5A	16 / 6A	17 / 7A	18 / 8A
1 H 1.00794																	2 He 4.00260
3 Li 6.941	4 Be 9.01218											5 B 10.811	6 C 12.011	7 N 14.0067	8 O 15.9994	9 F 18.9984	10 Ne 20.1797
11 Na 22.9898	12 Mg 24.3050											13 Al 26.9815	14 Si 28.0855	15 P 30.9738	16 S 32.066	17 Cl 35.4527	18 Ar 39.948
19 K 39.0983	20 Ca 40.078	21 Sc 44.9559	22 Ti 47.88	23 V 50.9415	24 Cr 51.9961	25 Mn 54.9381	26 Fe 55.847	27 Co 58.9332	28 Ni 58.693	29 Cu 63.546	30 Zn 65.39	31 Ga 69.723	32 Ge 72.61	33 As 74.9216	34 Se 78.96	35 Br 79.904	36 Kr 83.80
37 Rb 85.4678	38 Sr 87.62	39 Y 88.9059	40 Zr 91.224	41 Nb 92.9064	42 Mo 95.94	43 Tc (98)	44 Ru 101.07	45 Rh 102.906	46 Pd 106.42	47 Ag 107.868	48 Cd 112.411	49 In 114.818	50 Sn 118.710	51 Sb 121.76	52 Te 127.60	53 I 126.904	54 Xe 131.29
55 Cs 132.905	56 Ba 137.327	57 *La 138.906	72 Hf 178.49	73 Ta 180.948	74 W 183.84	75 Re 186.207	76 Os 190.23	77 Ir 192.22	78 Pt 195.08	79 Au 196.967	80 Hg 200.59	81 Tl 204.383	82 Pb 207.2	83 Bi 208.980	84 Po (209)	85 At (210)	86 Rn (222)
87 Fr (223)	88 Ra 226.025	89 †Ac 227.028	104 Rf (261)	105 Db (262)	106 Sg (263)	107 Bh (262)	108 Hs (265)	109 Mt (266)	110 Ds (281)	111 ** (272)	112 ** (285)		114 ** (289)		116 ** (292)		

(b)

1 / 1A	2 / 2A	3 / 3B	4 / 4B	5 / 5B	6 / 6B	7 / 7B	8 / 8B	9 / 8B	10	11 / 1B	12 / 2B	13 / 3A	14 / 4A	15 / 5A	16 / 6A	17 / 7A	18 / 8A
1 H 1.00794																	2 He 4.00260
3 Li 6.941	4 Be 9.01218											5 B 10.811	6 C 12.011	7 N 14.0067	8 O 15.9994	9 F 18.9984	10 Ne 20.1797
11 Na 22.9898	12 Mg 24.3050											13 Al 26.9815	14 Si 28.0855	15 P 30.9738	16 S 32.066	17 Cl 35.4527	18 Ar 39.948
19 K 39.0983	20 Ca 40.078	21 Sc 44.9559	22 Ti 47.88	23 V 50.9415	24 Cr 51.9961	25 Mn 54.9381	26 Fe 55.847	27 Co 58.9332	28 Ni 58.693	29 Cu 63.546	30 Zn 65.39	31 Ga 69.723	32 Ge 72.61	33 As 74.9216	34 Se 78.96	35 Br 79.904	36 Kr 83.80
37 Rb 85.4678	38 Sr 87.62	39 Y 88.9059	40 Zr 91.224	41 Nb 92.9064	42 Mo 95.94	43 Tc (98)	44 Ru 101.07	45 Rh 102.906	46 Pd 106.42	47 Ag 107.868	48 Cd 112.411	49 In 114.818	50 Sn 118.710	51 Sb 121.76	52 Te 127.60	53 I 126.904	54 Xe 131.29
55 Cs 132.905	56 Ba 137.327	57 *La 138.906	72 Hf 178.49	73 Ta 180.948	74 W 183.84	75 Re 186.207	76 Os 190.23	77 Ir 192.22	78 Pt 195.08	79 Au 196.967	80 Hg 200.59	81 Tl 204.383	82 Pb 207.2	83 Bi 208.980	84 Po (209)	85 At (210)	86 Rn (222)
87 Fr (223)	88 Ra 226.025	89 †Ac 227.028	104 Rf (261)	105 Db (262)	106 Sg (263)	107 Bh (262)	108 Hs (265)	109 Mt (266)	110 Ds (281)	111 ** (272)	112 ** (285)		114 ** (289)		116 ** (292)		

▲ **FIGURE 2.3 Group names for the elements.** (a) The stepped line divides this periodic table into two regions. Elements on the pink side of the line are metals, and elements on the blue side are nonmetals. (b) The class of elements known as metals is further divided into three groups: the alkali metals (column 1), the alkaline earth metals (column 2), and the transition metals (columns 3 through 12). Columns 17 and 18 of the nonmetals have familiar names: the halogens (column 17) and the noble gases (column 18).

of soil can reveal the level and type of human activity taking place at that location. A recent study of one area of the American South, for instance, examined the levels of lead and zinc in river and reservoir sediments across the region shown in Figure 2.4. Samples were taken from river and reservoir beds along the waterways running from northern Georgia to the Gulf of Mexico off the Florida panhandle. Levels of lead, which exists naturally in soil, are elevated when cars and trucks burn fossil fuels containing lead, an "antiknock" agent. Zinc also exists naturally in soils, but levels are

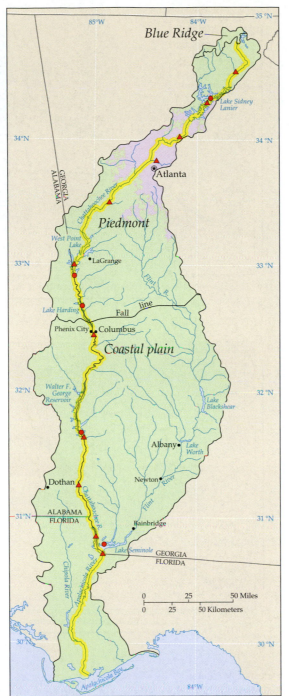

◀ **FIGURE 2.4** **Metal levels in water sediments.** This study on zinc and lead levels in riverbed and reservoir sediments was conducted in the waterways of Alabama, Georgia, and Florida. The main route of the study is highlighted in yellow. The circles and triangles show the locations where core samples were collected.

elevated by the burning of fossil fuels and by the wear of automobile tires, which are manufactured from rubber that contains zinc.

In this study, geochemists from the U.S. Geological Survey (USGS) sought to correlate human activities over the past three decades with levels of lead and zinc in the soil. Figure 2.5 consists of two graphs, part *a* showing lead levels on the vertical axis and part *b* showing zinc levels on the vertical axis. On both graphs, the horizontal axis is the distance from Apalachicola Bay, Florida, located at the southernmost point on the map of Figure 2.4. The point farthest to the right on the horizontal axis represents Lake Sidney Lanier, located near the top of Figure 2.4. The data in Figure 2.5 include not only lead and zinc levels as a function of distance from Apalachicola Bay, a *spatial* relationship, but *temporal* changes as well. In river and reservoir sediments, there is constant

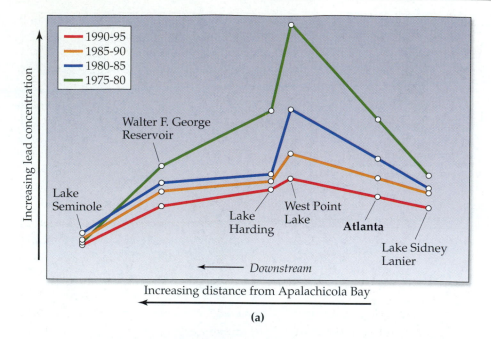

(a)

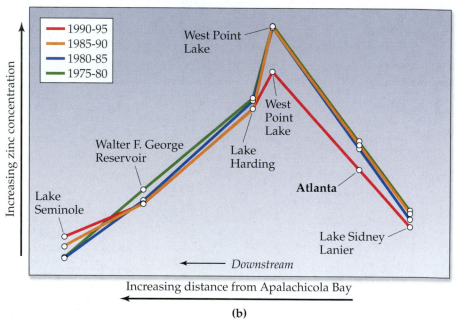

(b)

▶ FIGURE 2.5   **Data from the lead–zinc study.**   **(a)** Lead levels for four time intervals. **(b)** Zinc levels for the same four time intervals. In each graph, the horizontal axis gives the distance (in kilometers) north from Apalachicola Bay.

layering of new deposits of silt. Therefore, deeper and deeper samples of soil, represented by the four plots on each graph, reveal changes in elemental composition over the past three decades, beginning in 1975.

Look closely at these graphs. Is there any difference between what has happened to lead levels over the years and what has happened to zinc levels? The answer is yes. Figure 2.5*b* shows that, at a given location, levels of zinc have been relatively constant since 1975, whereas Figure 2.5*a* shows that, at a given location, levels of lead have dropped significantly over this time period. Why might lead and zinc levels change differently over time? The answer lies in the removal of lead from gasoline in the mid-1970s. As you can see in Figure 2.5*a*, lead levels were quite high from 1975 to 1980 (represented by the green line), as much as 140 μg of lead per gram of soil. The latest sampling (represented by the red line) shows that lead levels dropped by more than 50 percent over the intervening 20 years.

The situation is much different for zinc. A look at the data shows that the levels of both lead and zinc are highest downstream from Atlanta, a busy urban area. Whereas the data for lead downstream from Atlanta show a steady decrease in levels over time, the data for zinc show relatively little variation over time. The scientists who performed the study suggest that, unlike the case with lead, a well-known and feared toxin, levels of zinc have not been an issue with the American public. Therefore, there have been no efforts to remove zinc from fossil fuels or from tire rubber.

An average soil sample contains 0.0012 percent lead, 28 percent silicon, and 47 percent oxygen. In the late 1970s, however, the highest lead level found in the American South was 0.014 percent (about ten times higher than the national average) in West Point Lake, which is just south of Atlanta. These facts demonstrate that even minute levels of lead are cause for concern, and although the levels of other elements, such as oxygen and silicon, are much higher, they do not pose a risk to the public.

QUESTION 2.2 (*a*) How many protons and electrons does a lead atom contain? (*b*) How many protons and electrons does a zinc atom contain? (*c*) Are these elements metals or nonmetals? (*d*) Are these elements transition metals?

ANSWER 2.2 (*a*) 82 protons, 82 electrons. (*b*) 30 protons, 30 electrons. (*c*) Both elements are metals because they both appear to the left of the stepped line in Figure 2.3*a*. (*d*) Zinc is a transition metal. Lead is not.

QUESTION 2.3 In Figure 2.5, there is a gradual increase in both lead and zinc as you move from Atlanta to West Point Lake, south of Atlanta. To what do you attribute this increase, given that Atlanta is by far the largest population center in this study?

ANSWER 2.3 The huge number of cars and trucks driven in Atlanta and the coal-burning plants needed to generate power for the city were both sources of lead and zinc pollution. The contaminants originating in Atlanta were washed into local rivers and streams, then migrated through the waterways to West Point Lake and other places south of the city.

Elements have an enormous range of characteristics, and yet they are distinguished from one another by nothing more than a straightforward tally of protons. For example, in terms of effects on the human body, lead is one of the more dangerous elements known. It is a potent neurotoxin (nerve poison) and causes developmental defects in growing fetuses and in children. Thus, we now have laws restricting the use of lead in residential paint, food cans, gasoline, and ceramics. Right next to lead on the periodic table is the element bismuth, which has 83 protons and 83 electrons in each atom, compared with lead's 82 of each. That extra proton and electron make a world of difference. Humans can ingest copious quantities of bismuth without worrying about birth defects. In fact, millions of us do just that every day because bismuth subsalicylate is the active ingredient in Pepto-Bismol. Unlike its neighbor lead, bismuth is not found on the EPA's lists of toxic substances, soil contaminants, or air pollutants.

So what makes one substance potentially dangerous to humans and another substance harmless? What makes some things explosive and others

How Much Gold Is Left in the Ground?

Gold is very rare in Earth's crust. Indeed, this metal makes up only about 0.000 000 4 percent of the crust, and its rarity is one of the reasons it is sought after so fervently. Nonetheless, because of the huge mass of the crust, this percentage turns out to be a substantial amount. It is estimated that there were 36,000 tons of minable gold in the ground before we humans began taking it out. Of that total, roughly 22,000 tons, or about 60 percent, has been mined. Obviously, the gold that was easiest to find has already been found. It is unlikely that you could find much gold the old-fashioned way, which means either by panning or by discovering veins that are easily accessible. In order to get to the 40 percent that remains, you would have to invest first in exploration and development and then in extraction and refining.

It is not surprising that stories about individuals finding gold on their own are getting rarer and rarer. If you do get the itch to go looking for gold, though, where is most of that remaining 40 percent going to be? The best place to find gold in the United States is the Great Basin area, which comprises all of Nevada and part of western Utah. Other good places to look include the northern Rocky Mountains, the Pacific coast, southern California, and southern Arizona.

inflammable? Why don't we worry that our car tires will dissolve when it rains? Why is chocolate more fattening than lettuce? The answer to all these questions lies in the way atoms of different elements come together to make compounds and in the strength of the forces that hold atoms together. In Chapter 1, we saw several examples of chemical bonds in which electrons are shared between atoms; these are called **covalent bonds**. There are many other ways that atoms can interact, however, and we shall survey them in the next section. The nature of the chemical bonds holding a substance together is what dictates whether that substance will explode, dissolve in water, be fattening, or be harmful to humans.

RECURRING THEME IN CHEMISTRY The individual bonds between atoms dictate the properties and characteristics of the substances that contain them.

2.2 | Why Do Minerals Exist?

Our discussion of atoms in minerals raises an interesting question: why do atoms form minerals in the first place? Or, to put it another way, why does an atom interact with other atoms? In order to address this question, we must digress and look at interactions between atoms in simple systems. We shall return to how these interactions take place in minerals shortly.

The answer to our question lies in an atom's electron distribution, which determines the atom's place on the periodic table. Column 18 of the table contains the noble gases. The term *noble* derives from the reluctance of these elements to interact with other atoms, a reference to the nobility, known for their aloofness toward commoners. Each noble gas has an ideal number of electrons. But why are certain numbers of electrons ideal and others not?

Let's answer this question by looking at neon as an example. Neon has ten electrons surrounding its nucleus. This number is ideal for neon because, as it happens, ten electrons and ten protons have a maximum attraction to one another. It is virtually impossible to take an electron away from a neon atom, because all ten are held so tightly by the nucleus. Nor will a neon atom accept any more electrons. With 11 or 12 electrons shoe-horned in around a nucleus containing ten protons, things start to get crowded and an ideal situation no longer exists. Thus, a special stability is attained with ten electrons, the number that strikes an ideal balance between two counteracting forces: pull from the nucleus versus crowding around it. That is why we say that neon is noble: it does not interact with anything else.

This special stability is also attained by the five other noble gases, each with its own special number of electrons. It is attained with two electrons in helium, He; with 18 in argon, Ar; with 36 in krypton, Kr; with 54 in xenon, Xe; and with 86 in radon, Rn. Like neon, all these other noble gases also are unreactive, and we say that they have a **noble gas configuration** of electrons.

An element in any column except column 18 might benefit from accepting or losing some electrons to achieve a noble gas configuration. Consider sulfur, S, two columns to the left of the noble gases. A sulfur atom would benefit from the addition of two electrons to achieve a complement of electrons like that of argon, the nearest noble gas. Likewise, magnesium, Mg, a column 2 element, would benefit from losing two electrons in order to have an electron configuration like that of neon, the noble gas closest to magnesium in the table. Only the outermost electrons, called **valence electrons**, are added or taken away as an atom tries to achieve a noble gas configuration.

You can determine how many valence electrons an atom of a given element contains by beginning at the left end of the period in which the element belongs and counting to the right until you reach the element, adding one valence electron for each box you count. Consider the fluorine atom, F, which has a total of nine electrons. How many of these are valence electrons? To find out, we begin at the left end of period 2 (because fluorine is a period 2 element) and count from lithium across to fluorine, realizing that each successive element has one more valence electron: lithium has one, beryllium has two, ..., oxygen has six, and fluorine has seven.

The electrons in an atom that are not valence electrons are called **core electrons** and do not participate in chemical bonds. Because fluorine has a total of nine electrons, seven of which are valence electrons, the remaining two must be core electrons.

Fluorine's position on the periodic table one box to the left of the noble gas neon tells us that fluorine requires one additional electron in order to have an electron configuration like that of neon. It would be beneficial, therefore, for an atom of fluorine to acquire one electron somehow. If two fluorine atoms come together in a chemical bond, they are able to share electrons with each other in a way that gives each atom eight outermost (valence) electrons, which is the number of valence electrons in neon. We can depict this type of bond by using dots to symbolize the electrons. Each pair of dots drawn between two atoms represents a chemical bond:

$$\ddot{\underset{\cdot\cdot}{F}}\cdot \; + \; \cdot\ddot{\underset{\cdot\cdot}{F}}\!: \;\longrightarrow\; \ddot{\underset{\cdot\cdot}{F}}\!:\!\ddot{\underset{\cdot\cdot}{F}}\!:$$

In a fluorine molecule, each fluorine atom shares one of its valence electrons with its partner, giving each atom a total of eight valence electrons. Therefore, it is energetically favorable for two individual fluorine atoms, F, to exist as one fluorine molecule, F_2. We refer to F_2 as a **diatomic** molecule, because it contains two like atoms. Because the chemical bond in the molecule involves shared electrons, it is a covalent bond.

If the process of changing the number of electrons around an atom can give that atom an electron configuration like that of the nearest noble gas, then it is likely that a chemical bond will form. The single covalent bond between the atoms of fluorine in F_2 is depicted with a line that indicates that two electrons are being shared:

$$F—F$$

Because elements in the same column of the periodic table have the same number of valence electrons, they generally have similar properties. It makes sense then that other column 17 elements—in other words, other halogens—exist in nature as diatomic molecules, just as fluorine does. The diatomic molecules include the halogens fluorine, F_2, chlorine, Cl_2, bromine, Br_2, and iodine, I_2, as well as hydrogen, H_2, nitrogen, N_2, and oxygen, O_2. Now you know why halogens are on this list of diatomic elements: they are more stable as molecules than as individual atoms because, by being part of a molecule, each atom achieves a noble gas configuration of electrons.

Covalent bonds can form between atoms of the same element, as in the case of fluorine, or between atoms of different elements. As a result, there is an enormous assortment of covalent bonds in nature, each one forming in a way that gives each atom a noble gas configuration, if possible. The nature of a given covalent bond depends on the elements involved. In fluorine, a bond can form as a result of equal sharing of electrons between atoms. This is called a *nonpolar covalent bond* because there is no lopsidedness to the electron sharing. Each fluorine atom pulls on the shared electrons with the same force. Some covalent bonds, however, are very **polar**. That is, there is unequal sharing of the electrons in the bond, and the result of this unequal sharing is that one side of the molecule is electron rich and the other side is electron poor.

Water is a classic example of a molecule containing *polar covalent bonds:*

Each arrow with a plus sign at its tail points along a polar covalent bond in the water molecule. The plus end of the arrow indicates the atom that pulls *less* electron density toward itself, and the point of the arrow indicates the atom that pulls *more* electron density toward itself. In this case, the oxygen in each O–H bond pulls more electron density from the covalent bond toward itself, and this bond is therefore very polar. Fluorine (completely nonpolar) and water (very polar) are extreme examples of covalent bond polarity, and there are bonds with every degree of intermediate polarity between them.

A chemical bond results both when electrons are shared by atoms—as in the case of fluorine molecules and water molecules—and when electrons are *transferred* from one atom to another. To see how this electron transfer takes place, let's begin by looking at what can happen to the electrons in calcium atoms. Each calcium atom has two electrons more than the nearest noble gas, which is argon. A calcium atom can lose its outermost two electrons to form an **ion**, the term used to describe any electrically charged atom, as shown in Figure 2.6a. We depict the calcium ion as Ca^{2+} to show that it is missing two of its electrons. (Remember, electrons have a negative

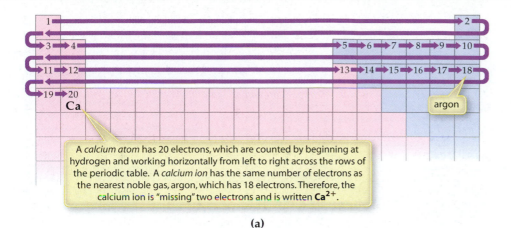

A *calcium atom* has 20 electrons, which are counted by beginning at hydrogen and working horizontally from left to right across the rows of the periodic table. A *calcium ion* has the same number of electrons as the nearest noble gas, argon, which has 18 electrons. Therefore, the calcium ion is "missing" two electrons and is written **Ca²⁺**.

(a)

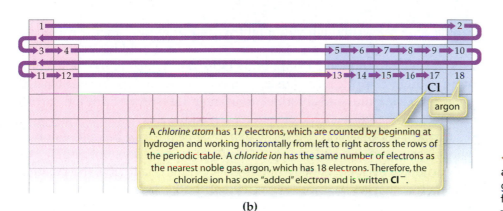

A *chlorine atom* has 17 electrons, which are counted by beginning at hydrogen and working horizontally from left to right across the rows of the periodic table. A *chloride ion* has the same number of electrons as the nearest noble gas, argon, which has 18 electrons. Therefore, the chloride ion has one "added" electron and is written **Cl⁻**.

(b)

◄ **FIGURE 2.6 Creating ions from atoms.** **(a)** Calcium atoms form ions by giving up two electrons. **(b)** Chloride ions form by accepting one electron.

charge. If they are removed from an atom, the atom becomes a positively charged ion because it now has more protons than electrons.)

Let's also consider chlorine, an element from the other side of the periodic table. A chlorine atom has seven valence electrons, just like the atoms of its sister element in column 17, fluorine. Therefore a chlorine atom requires one more electron to achieve a full complement of electrons like the nearest noble gas, argon. The atom can get that extra electron by forming a covalent bond with another chlorine atom in the same way that fluorine does (and, as we know, the element chlorine exists as the diatomic molecule Cl_2). Or, chlorine can add an extra electron to make an ion, Cl^-, as shown in Figure 2.6b.

Both Ca^{2+} and Cl^- are ions, but they are often distinguished with the more specific terms *cation* (pronounced CAT-eye-un) and *anion* (pronounced AN-eye-un). A **cation** is a positively charged ion, and an **anion** is a negatively charged ion.

Where do the electrons lost by cations go when they leave the atom, and where do anions get the extra electrons they need? The answer is that ions do not exist in isolation; they must be paired with oppositely charged ions. Thus, the electrons lost by any cation must be transferred to the anion with which the cation is paired. The result is a neutral compound called a **salt**. Salts are formed from the complete transfer of one or more electrons from one atom to another to create a cation and an anion. The strong electrical attraction between the positively charged cation and the negatively charged anion is called an **ionic bond**.

There is no sharing of electrons in ionic bonds; these are *not* covalent interactions. Therefore, we do not draw lines between ions in the chemical

▶ **FIGURE 2.7** Ions from groups 1, 2, 16, and 17. This partial periodic table shows the noble gases (column 18) plus the four columns flanking the noble gases (columns 16 and 17 and columns 1 and 2). The elements in columns 16 and 17 gain electrons to make anions having the same number of electrons as does the nearest noble gas. The elements in columns 1 and 2 lose electrons to make cations having the same number of electrons as the nearest noble gas.

1 1A	2 2A				16 6A	17 7A	18 8A
H^+ (no electrons)							He
Li^+ (2 electrons, like He)	Be^{2+} (2 electrons, like He)				O^{2-} (10 electrons, like Ne)	F^- (10 electrons, like Ne)	Ne
Na^+ (10 electrons, like Ne)	Mg^{2+} (10 electrons, like Ne)				S^{2-} (18 electrons, like Ar)	Cl^- (18 electrons, like Ar)	Ar
K^+ (18 electrons, like Ar)	Ca^{2+} (18 electrons, like Ar)				Se^{2-} (36 electrons, like Kr)	Br^- (36 electrons, like Kr)	Kr
Rb^+ (36 electrons, like Kr)	Sr^{2+} (36 electrons, like Kr)				Te^{2-} (54 electrons, like Xe)	I^- (54 electrons, like Xe)	Xe

formula for a salt. Although you might write F–F to indicate the molecule F_2, with the line representing a covalent bond, you should never use a line in the formula for an ionic compound. Table salt, which is the ionic compound sodium chloride, is a classic example. We write NaCl, *never* Na–Cl! In NaCl, both sodium and chlorine are in their ionic form. Each ion carries one electrical charge: a charge of +1 on each sodium cation and a charge of −1 on each chloride anion. We know this must be the case because chlorine always forms a −1 ion and sodium always forms a +1 ion, both noble gas configurations. Because these two ions always pair one to one, the positive and negative charges balance so that the net charge in NaCl is zero.

We now know that elements located in columns just before and just after the noble gases always form ions that have the same number of electrons as the nearest noble gas. This idea is summarized in Figure 2.7. For example, all elements in column 16 require two electrons to attain a noble gas configuration of electrons. Thus, oxygen makes the anion O^{2-}, and sulfur makes the anion S^{2-}. This means that the anions O^{3-} and S^{3-} cannot exist because they do not have a noble gas configuration of electrons. Likewise, all elements in column 2 must shed two electrons to achieve a noble gas configuration of electrons. Thus, calcium and magnesium make the cations Ca^{2+} and Mg^{2+}, respectively, but *never* Ca^+ or Mg^{3+}.

In a solid crystal of any salt, the ions alternate in a regular, repeating, three-dimensional pattern, as Figure 2.8 shows. When we write the name or chemical formula of a salt we simply say, for instance, sodium chloride (NaCl) or calcium fluoride (CaF_2). It is not necessary to indicate that solid salts are actually arrays of billions of cations and anions; this fact is understood. The term **formula unit** is used to indicate the smallest repeating unit in a salt, such as KI or MgI_2. Molecules are different from salts in that molecules most often exist as discrete entities and not in extended arrays. Liquid water, for example, is composed of individual molecules, each containing exactly two hydrogen atoms and exactly one oxygen atom. The atoms are linked by covalent bonds, and the molecular formula, H_2O, tells you exactly how many of each atom are present.

QUESTION 2.4 What salt forms from the elements calcium and chlorine?

ANSWER 2.4 $CaCl_2$. Calcium makes the cation Ca^{2+}, and chlorine makes the anion Cl^-. Because salts must have zero net charge, we must combine one ion of calcium with two chloride ions. The resulting salt is calcium chloride.

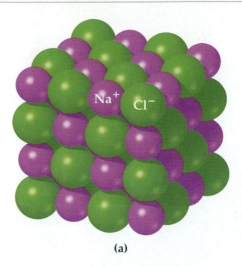

(a)

(b)

MOLECULE
Sodium Chloride

◄ **FIGURE 2.8** **The ionic compound sodium chloride.** **(a)** A schematic drawing of a portion of a sodium chloride crystal, showing the regular array of sodium ions and chloride ions. **(b)** Sodium chloride crystals magnified several hundred times.

QUESTION 2.5 Identify and count the cations and anions in these ionic compounds: (a) K_3N, (b) K_2O, (c) CaF_2, (d) $BaCl_2$.

ANSWER 2.5 (a) Three K^+ cations and one N^{3-} anion, (b) two K^+ cations and one O^{2-} anion, (c) one Ca^{2+} cation and two F^- anions, (d) one Ba^{2+} cation and two Cl^- anions.

Salts have the most polar bonds imaginable. The electrons are not simply concentrated near one or the other atom in the bond. Rather, they are *transferred fully* from one atom to the other. Thus, in the continuum of bond polarities, ionic bonds fall at the most polar end. As we shall see in later chapters, the polarity of the bonds in a given substance is a reliable predictor of many of its chemical properties. We shall look at examples of covalent bonds in Chapter 3 and ionic bonds in Chapter 4.

What determines whether two atoms form a covalent bond or an ionic bond with each other? It boils down to a simple rule of thumb: the closer two nonmetal atoms are to each other on the periodic table, the more likely that a bond between them will be covalent. A covalent bond forms, for instance, between the elements silicon and oxygen, which are neighbors on the periodic table. The farther apart two atoms are on the table, the more likely the bond between them will be ionic. For example, rubidium, Rb, which is in column 1, and oxygen, from column 16, form the salt Rb_2O, rubidium oxide.

How, though, can we know what is meant by the terms *close* and *far* on the periodic table? It would be convenient to be able to quantify this concept. The term **electronegativity**, the tendency of an atom to draw electrons toward itself, is our gauge for the level of polarity a bond will have. As Figure 2.9a shows, the most electronegative elements are in the upper right corner of the periodic table, and the least electronegative elements are in the lower left corner. In general, if the electronegativity difference between two atoms is 2.0 or greater, the bond between them will be ionic. If the electronegativity difference between two atoms is less than 2.0, the bond between them will be covalent. Electronegativity differences just slightly less than 2.0 lead to polar covalent bonds, and as the difference gets closer and closer to zero, the bond between the two atoms becomes more nonpolar. This trend is summarized in Figure 2.9b.

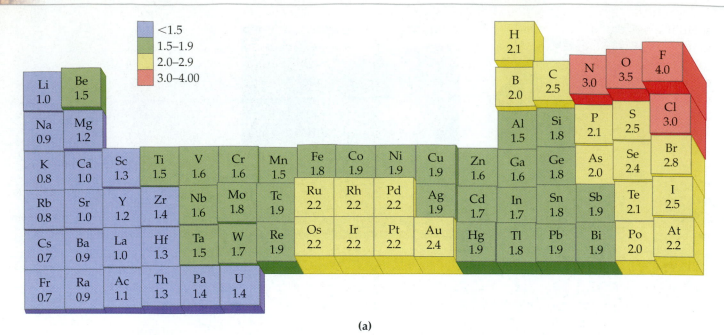

(a)

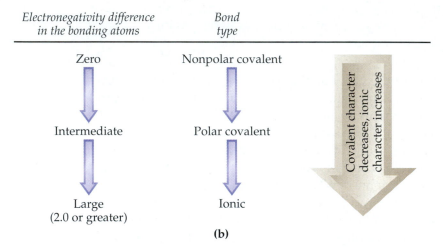

(b)

▲ **FIGURE 2.9 Electronegativity values.** **(a)** Fluorine, in the top right corner of the periodic table, has the largest electronegativity value of all the elements. Cesium and francium, in the lower left corner, have the smallest value. **(b)** Ionic bonds are formed when the electronegativity difference between two atoms is 2.0 or greater. Covalent bonds are formed when the electronegativity difference between two atoms is less than 2.0. As the difference between the two values gets smaller, the covalent bond moves from polar covalent to nonpolar covalent.

QUESTION 2.6 Is the bond between each of these pairs of atoms ionic or covalent: (*a*) K bonding with Cl, (*b*) C bonding with Cl, (*c*) C bonding with O, (*d*) Rb bonding with F?

ANSWER 2.6 In each pair, you must find the difference in the electronegativity values, which are given in Figure 2.9*a*. A difference of 2.0 or higher means an ionic bond; a difference of less than 2.0 means a covalent bond. (*a*) $3.0 - 0.8 = 2.2$, ionic; (*b*) $3.0 - 2.5 = 0.5$, covalent; (*c*) $3.5 - 2.5 = 1.0$, covalent; (*d*) $4.0 - 0.8 = 3.2$, ionic.

The preceding examples show that ionic bonds tend to form between metal and nonmetal and that covalent bonds tend to form between nonmetal and nonmetal. To name just two of the combinations we have seen, the metal sodium forms an ionic bond with the nonmetal chlorine to give the salt NaCl, and the nonmetal oxygen forms a polar covalent bond to the nonmetal hydrogen to give the molecule H_2O. We have one more combination to consider. What type of interaction takes place between metal and metal? In the pure, solid phase, metals have their own style of bonding and do not follow our rules based on electronegativity.

Consider a pure, unadulterated chunk of magnesium, an alkaline earth metal. By losing two electrons, each Mg atom can form the ion Mg^{2+} and achieve an electron configuration like that of neon. Because magnesium is the only element in solid magnesium, however, there is no nonmetal atom to accept those electrons and form an anion. Thus, the electrons produced from the ionization of the magnesium atoms form a *sea of charge* that surrounds and neutralizes the positive charges of the magnesium cations, as shown in Figure 2.10. A **metallic bond**, therefore, is similar to an ionic bond in that the atoms become ionized. It is also similar to covalent bonding in that electrons are shared, but metallic bonds are different because the sea of electrons is fluid. As a result, solid metals are malleable and can be shaped into jewelry and coins. Because electrons in a solid metal are mobile, these materials readily conduct heat and electricity.

Now it is appropriate to turn back to the subject of minerals and to reexamine the question of why they form. We can guess the answer: they form so that the atoms they contain can optimize their numbers of electrons. What kind of bond exists in a mineral? Is it ionic, covalent, or metallic? In fact, any of the bond types can exist in a mineral, depending on the element(s) it contains. For example, carbon exists in nature in the crystalline form of the mineral we know as diamond. In diamond, covalent bonds between carbon atoms hold the crystal together. The mineral fluorite, CaF_2, is used in the manufacture of steel. It is also a crystalline solid, but its bonds are very different from those in diamond. Calcium forms cations, Ca^{2+}, that pair with anions of fluorine, F^-, in a one-to-two ratio to give the salt CaF_2. These bonds are ionic. A piece of solid copper contains only Cu ions, and we know that a sea of electrons, each donated from the atoms of copper in the solid, holds the piece of copper together. What all these examples have in common is that in each case the atoms are arranged in a regular, repeating, three-dimensional arrangement—in other words, they are all crystalline solids. However, the types of bonds in these crystalline solids are very different from one another.

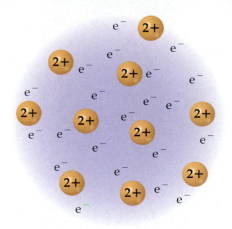

▲ FIGURE 2.10 Chemical bonding in metals. The sea-of-electrons model of metallic bonding, represented schematically for an alkaline earth element.

QUESTION 2.7 Chalcopyrite is an iridescent greenish-black mineral that has the chemical formula $CuFeS_2$. (*a*) What is the total number of atoms in each formula unit of chalcopyrite? (*b*) How many atoms of each element are present in one formula unit of chalcopyrite?

The mineral chalcopyrite.

ANSWER 2.7 (*a*) Four. (*b*) One copper atom, one iron atom, and two sulfur atoms.

2.3 | Why Neutrons Matter

Dirt is made up, ultimately, of atoms or combinations of atoms. These atoms come together to form chemical bonds of different types: covalent, ionic, metallic. As Figure 2.1 shows, silicon and oxygen are the most ubiquitous elements in soil. Thus, we would expect soil to be rich in compounds containing these elements. In fact, the most abundant compound in soil is silicon dioxide, SiO_2, in which the silicon and oxygen atoms are connected in a network of covalent bonds.

Up to now, we have focused on the dramatic differences that exist between elements: metal versus nonmetal, lead versus bismuth, halogen versus noble gas. Next, we are going to look at the differences that can exist *within one element*. We have learned that the number of protons gives an element its essential character and that the electrons do the chemistry in the atom. They are the outermost, most easily accessed particles in the atom, and they are added, removed, and shared during chemical reactions. What of the neutrons in the nucleus? What do they do? Why should we pay attention to them?

> **RECURRING THEME IN CHEMISTRY** Chemical reactions are *nothing more than exchanges or rearrangements of electrons*, which is why we pay particular attention to electrons throughout this book.

The number of neutrons in one atom of a given element can be different from the number of neutrons in another atom of the same element. (Remember, it is the number of protons that determines the identity of an element, not the number of neutrons.) The total number of neutrons plus protons in the nucleus of an atom is referred to as the atom's **mass number**. This mass number is sometimes shown either as a superscript before the symbol for an element or as a number following the element name:

$$^{48}Ti \text{ or titanium-48} \qquad ^{50}Ti \text{ or titanium-50}$$

Both titanium-48 and titanium-50 represent the element titanium, and the two forms are *isotopes* of the element. **Isotopes** are atoms having the same number of electrons, the same number of protons, but different numbers of neutrons. The periodic table tells us that titanium has 22 protons, and therefore both titanium-48 and titanium-50, because they are both titanium, *must* contain 22 protons. There is a difference in the numbers of *neutrons* in the two symbols, however, and this means the names represent two isotopes of titanium. For the isotope that has a mass number of 48, the number of neutrons must be

> Isotopes also can be shown with the atomic number as a subscript below the mass number, like this:
>
> $$^{48}_{22}Ti$$

$$\text{Mass number} = \text{number of protons} + \text{number of neutrons}$$
$$\text{Number of neutrons} = \text{mass number} - \text{number of protons} =$$
$$48 - 22 = 26$$

In most cases in chemistry, the number of neutrons in an atom is unimportant; elements and compounds usually react in the same way regardless of the isotopes involved. However, there are many instances when isotopes are critically important. For example, most carbon atoms are carbon-12, a few are carbon-13, and very few are carbon-14. The ratio of carbon-12 to carbon-14 remains constant in living organisms, but the concentration of carbon-14 decreases when an organism dies. (We shall discuss why this happens in Chapter 6.) Therefore, the ratio of carbon-12 to carbon-14 measured by **carbon dating** is an indicator of the current age of something that was once alive. Carbon dating was used to determine the age of Ötzi, the prehistoric Austrian "iceman" discovered by hikers in 1991. The dating showed that he was buried in ice about 5300 years ago.

QUESTION 2.8 The isotope titanium-50 has a mass number of 50. (*a*) How many protons, neutrons, and electrons does an atom of this isotope contain? (*b*) Of the two titanium isotopes titanium-48 and titanium-50, which would you expect to have the greater mass?

ANSWER 2.8 (*a*) Any isotope of titanium must contain 22 protons. Because the isotope is electrically neutral, the number of electrons must be equal to the number of protons, and so an atom of titanium-50 contains 22 electrons. The number of neutrons is the mass number minus the number of protons: $50 - 22 = 28$ neutrons. (*b*) Titanium-50 has two more neutrons than does titanium-48. You know that neutrons are tiny, but they do have mass, and therefore titanium-50, with 28 neutrons, must be more massive than titanium-48, which contains only 26 neutrons.

To see another reason neutrons can be important, take a look at Figure 2.11. These pie charts show, for five elements, the percentages of each isotope found in a typical naturally occurring sample of each element. As you can see, a given element can have a small number of isotopes, as is the case for chlorine (two isotopes) and for carbon and oxygen (three isotopes each), or there can be a large number of isotopes, as with calcium and tin. The percentages shown in Figure 2.11 are average values calculated from large numbers of samples taken from around the world, and the isotope percentages in any one sample taken from a specific location may differ from those shown in the figure. A sample of dirt taken from a particular farm in Nebraska, for instance, may contain 75.75

◄ **FIGURE 2.11** The natural abundances of isotopes of chlorine, carbon, oxygen, calcium, and tin.

LIVE ART
Isotopic Abundances

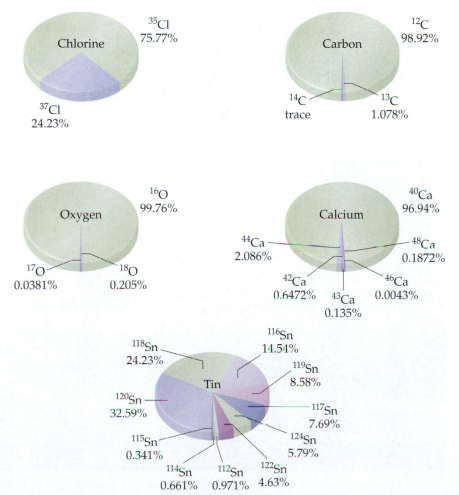

percent chlorine-35 and 24.25 percent chlorine-37. Likewise, a sample of dirt collected in the Mekong Delta of Vietnam may contain 74.31 percent chlorine-35 and 25.69 percent chlorine-37. Such variation in isotope ratios can be used to determine where a particular sample of dirt or piece of rock (or any other natural object) came from, as we shall see in the example that follows.

As a graduate student at the University of Georgia, Scott Pike wanted to answer a question that has been on the minds of art historians and archeologists for years: where did the marble used by the builders of the Greek Parthenon (shown in Figure 2.12*a*) come from? Marble is a mineral made up

▶ FIGURE 2.12 **Isotopes in ancient Greece. (a)** The Parthenon in Athens, built about 450 B.C. (The Parthenon, Athens, Greece. Blaine Harrington III/Stock Market.) **(b)** A portion of the Elgin marbles, which once adorned the temple.

(a)

(b)

primarily of calcite, or calcium carbonate, $CaCO_3$. One formula unit of calcium carbonate contains calcium, carbon, and oxygen atoms that come together in a specific way to form the solid substance we know as marble. However, the relative amounts of the different isotopes of these three elements may differ from one sample of marble to another, depending on where the sample was mined.

The marble used in the pieces of the Parthenon today known as the Elgin marbles (Figure 2.12*b*) were known to contain a very high percentage of oxygen-18, more than the 0.205 percent average value shown in Figure 2.11. Acting on this information, Pike analyzed the marble in various quarries located on the southern slope of Mount Pendelikon, all likely candidates for where the marble could have originated. Figure 2.13 shows a map of Europe

◄ **FIGURE 2.13** **Map of Europe with the Attic peninsula of Greece enlarged.** The location of Mount Pendelikon, where the marble was mined, and the city of Athens, where the marble was used in the Parthenon, are indicated.

and the location of the quarries in relation to the city of Athens, home to the Parthenon and the Elgin marbles. The isotope data Pike collected showed that the highest percentages of the oxygen-18 isotope are found in the quarries near the top of Mount Pendelikon. Thus, it is likely that the Elgin marbles originated in this area, a location that would be difficult to pinpoint without the use of oxygen isotope data.

QUESTION 2.9 Marble contains the elements carbon and calcium, in addition to oxygen. (*a*) Using Figure 2.11 as a reference, which isotope of carbon would most samples of marble contain? (*b*) Which isotope of calcium would most samples of marble contain?

ANSWER 2.9 (*a*) Carbon-12 and (*b*) calcium-40.

QUESTION 2.10 Another graduate student has decided to use a different isotope to study the Elgin marbles. She wants to use the element tin, Sn, rather than oxygen because there are many isotopes of tin, as Figure 2.11 shows. Because there is a greater natural variety of tin isotopes, she believes it should be possible to pinpoint the origin of the Elgin marbles more accurately with this isotope. As her research advisor, why will you steer her away from this idea?

ANSWER 2.10 Her basic idea is sound. It would be convincing to have isotopic data for several isotopes of tin so that she could do a more exact comparison between the marble in the Elgin marbles and the marble in the quarries. The only problem is that marble is made mostly of calcium carbonate, which does not contain tin. Even if the marble in the quarries and in the Elgin pieces contains trace quantities of tin, these levels are likely to be too small to measure, especially for the rarer tin isotopes.

2.4 | Coming to Terms with the Very Large and Very Small

We humans sometimes have difficulty relating to very large and very small numbers, and so we often see large and small numbers explained in a familiar context (Figure 2.14). For example, a recent news article reported that the United States spent 1.157 billion dollars on a new stealth bomber, a sum that would feed the entire country of Haiti for 14 years. This enormous sum of money is put into perspective, in this case showing how far it could go toward addressing the problem of hunger in a third-world country. The media uses analogies like this all the time, often in a way that biases the information, as in this stealth bomber–Haiti example.

We do the same thing with the extraordinarily small. Here is an example from chemistry: hydrogen atoms are so small that 800,000 of them will fit side by side across the diameter of a blond human hair. (Blond hair is thinner than black hair.) This comparison helps us to imagine the size of a hydrogen atom. Here is another example that illustrates how small atoms and molecules are: an 8-oz glass of water contains 7.9×10^{24} H_2O molecules.

Clearly, this glass contains oodles of water molecules, but what is the meaning of the exponential term 7.9×10^{24}? Very often, when a number is

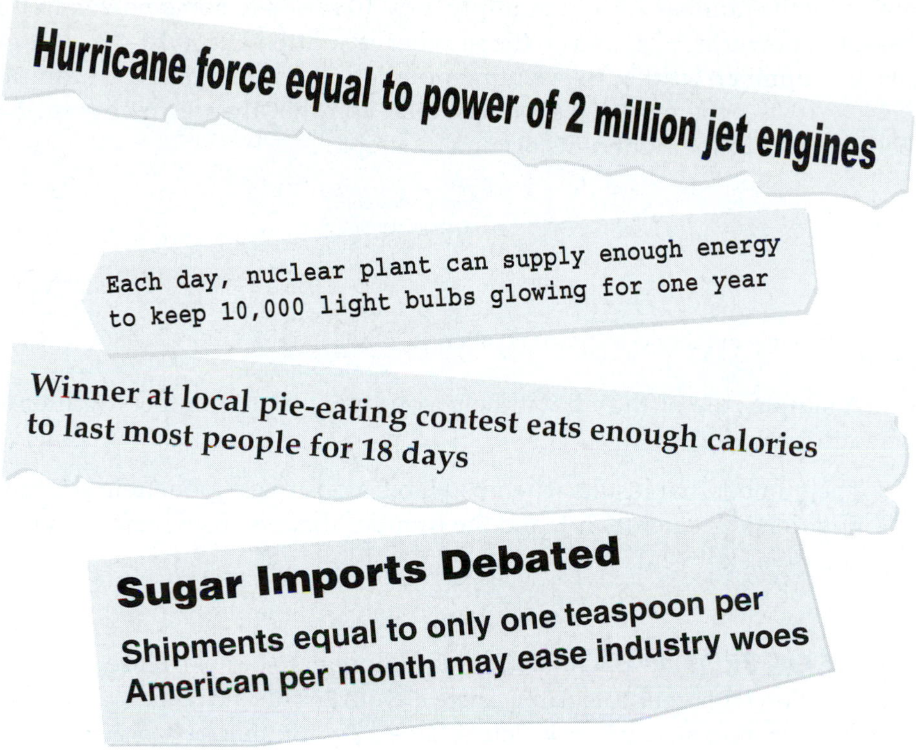

Hurricane force equal to power of 2 million jet engines

Each day, nuclear plant can supply enough energy to keep 10,000 light bulbs glowing for one year

Winner at local pie-eating contest eats enough calories to last most people for 18 days

Sugar Imports Debated

Shipments equal to only one teaspoon per American per month may ease industry woes

◀ FIGURE 2.14   **Making sense of the news.**   In the media, large and small numbers are often put into a context to which readers can relate.

exceedingly small or large, scientists use **scientific notation** for convenience; it is their way of dealing with unwieldy numbers. To see how this notation works, let's take a number that is not that large, but will serve as a good first example. To express the number 457,000 in scientific notation, you move its decimal point to the location that results in a number between 1 and 10, which in this case means 4.57. Next, you express the number of places you moved the decimal point as 10 raised to that power: 4.57×10^5. When you have moved the decimal point of the original number to the left, as you did here, the exponent is positive. When you are converting a number less than 1 to scientific notation, you must move the decimal point to the right to obtain a number between 1 and 10, and in this case the exponent is negative. For example, 0.000 004 57 becomes 4.57, and because you had to move the decimal point six places to the right to get 4.57, the exponent is negative: 4.57×10^{-6}.

Thus, the notation 7.9×10^{24} represents a very large number, one in which the decimal point of the original number has been moved 24 places to the left of its original location:

Original position of decimal point

A very large number   7,900,000,000,000,000,000,000,000

The same number in scientific notation 7.9×10^{24}

Note that in scientific notation the number before the multiplication sign *must* fall between 1 and 10. The decimal point must be placed right after the first digit—which is 7 in this case—and not one place to the left of the first digit (giving 0.79) or one place to the right (giving 79). In scientific notation, we write the number 7.9 and multiply it by 10 raised to some power. In this case, the power is +24, and so the number is written 7.9×10^{24}.

If a number written in scientific notation is very small—for example, 1.24×10^{-26}—the negative exponent tells you that the decimal point has been moved to the right of its original position:

Original position of decimal point

A very small number   0.000 000 000 000 000 000 000 000 012 4

The same number in scientific notation 1.24×10^{-26}

A simple rule of thumb will help you manipulate numbers written in scientific notation:

The number that results when you have a positive exponent in scientific notation is greater than 1, and the number that results when you have a negative exponent is less than 1.

HINT ON PROBLEM SOLVING: It is important to be able to use scientific notation with your calculator. There are dozens of calculator brands out there, but most scientific calculators have a button that says either EE or EXP. To enter a number written in scientific notation, enter the decimal number, press the EE/EXP key, and then enter the exponent. For example, for 1.24×10^{-26}, you would enter 1.24, then EE, then −26.

Most often when students arrive at a number in a calculation that is incorrect by a factor of 10, it is because they entered the 10 part of 10^{-26} instead of just the −26 part. (Doing so multiplies the decimal number by an extra 10.) *It is not necessary to enter the number 10 when you use the EE/EXP key.*

You will sometimes see numbers in scientific notation where the first number is 1. In writing such a number, the 1 is often omitted for convenience:

1×10^{-19} is the same as writing 10^{-19}

To enter the number 10^{-19} into your calculator, however, you must include the 1: enter 1, then EE/EXP, then −19.

QUESTION 2.11 Write these numbers in scientific notation: (a) 0.000 034 5, (b) 9,080,000, (c) 0.2.

ANSWER 2.11 (a) 3.45×10^{-5}, (b) 9.08×10^{6}, (c) 2×10^{-1}. Note that in this last example, it is uncommon to use scientific notation for numbers close to 1. It is easier to write 0.2 than to write 2×10^{-1}. Scientific notation is usually reserved for very large and very small numbers.

QUESTION 2.12 What is odd about the number 88.6×10^{5}?

ANSWER 2.12 The decimal number in scientific notation should be between 1 and 10. Here, the number is 88.6, which is greater than 10. The correct way to write this number is 8.86×10^{6}. Try entering 88.6×10^{5} into your calculator. Your calculator may correct it for you.

In the same way that bankers work with dollars and hair stylists work with strands of hair and gardeners work with blades of grass, the working medium for chemists is atoms. Over the years chemists have devised ways of handling their small size. We learned earlier that 7.9×10^{24} water molecules is the number that would fill an 8-oz glass. This trick of thinking of many molecules amassed into a familiar volume allows us to imagine the astonishing tininess of a single atom. Because they most often work with quantities of atoms that are visible to the human eye, chemists have defined 6.02×10^{23} atoms as a unit called the **mole** (abbreviated mol). This number arises from an internationally agreed-upon standard: one mole is the number of carbon atoms in 12.000 g of the isotope carbon-12. This happens to be the number 6.02×10^{23}, also known as **Avogadro's number**, named for the chemist who first measured it. The mole is a convenient unit because the quantities of matter that we deal with every day are expressed easily in moles. For example, the tennis racket shown in Figure 2.15 has a frame made of 100 percent graphite, a form of elemental carbon. This frame contains about 40 mol of carbon atoms.

It is important to remember that the mole is nothing more than a counting device. You can have a mole of anything—tables, chickens, atoms, toes. Thus, chemists use the word to count not only numbers of atoms, but also numbers of molecules, ions, or any other small particles. If you have 1 mol of water, you have 6.02×10^{23} *molecules* of water because molecules are discrete entities held together by covalent bonds. If you have 1 mol of table salt, you have 6.02×10^{23} *formula units* of NaCl because solid salts exist in a repeating crystalline lattice defined by formula units. If you have 0.5 mol of copper, you have $0.5 \times 6.02 \times 10^{23} = 3.01 \times 10^{23}$ *atoms* of copper because copper is an elemental substance made up of individual atoms. Figure 2.16 shows 1-mol quantities of some familiar substances.

QUESTION 2.13 If there is 0.5 mol of dogs in your backyard (your *very big* backyard!), (a) how many moles of dog tails are there in the yard? (b) how many moles of dog ears? (c) how many moles of dog legs?

▲ FIGURE 2.15 **Carbon atoms in a graphite tennis racket.** This tennis racket, made of 100 percent graphite, contains about 40 mol of carbon atoms, which means 2.4×10^{25} atoms.

► FIGURE 2.16 **One-mole quantities of everyday substances.** Clockwise from top: table sugar, copper metal, carbon, and table salt.

ANSWER 2.13 (*a*) 0.5 mol of tails because there is one tail per dog:

$$0.5 \; \text{mol dogs} \times \frac{1 \; \text{mol tails}}{1 \; \text{mol dogs}} = 0.5 \; \text{mol tails}$$

(*b*) 1 mol of ears because there are two ears per dog:

$$0.5 \; \text{mol dogs} \times \frac{2 \; \text{mol ears}}{1 \; \text{mol dogs}} = 1 \; \text{mol ears}$$

(*c*) 2 mol of legs because there are four legs per dog:

$$0.5 \; \text{mol dogs} \times \frac{4 \; \text{mol legs}}{1 \; \text{mol dogs}} = 2 \; \text{mol legs}$$

QUESTION 2.14 If you have 1 mol of water, H_2O, how many moles of hydrogen atoms do you have? how many moles of oxygen atoms?

ANSWER 2.14 Each molecule of water contains one oxygen atom and two hydrogen atoms. Therefore, each mole of water molecules contains 1 mol of oxygen atoms and 2 mol of hydrogen atoms.

If you look at a typical periodic table, you will see a number written under each element symbol. This number is the element's **molar mass**, defined as the mass in grams of 1 mol of that element. But wait a minute. Why is the molar mass of carbon shown as 12.011 rather than 12.000? The answer lies with mother nature. Recall that we specified that 12.000 g is the mass of 1 mol of carbon-12. Now, if all carbon in nature were carbon-12, the periodic table would list the molar mass as 12.000. However, each molar mass given on the periodic table is the *weighted average* of one mole of each of the naturally occurring isotopes of a given element. As we know from Figure 2.11, carbon has three common isotopes, and this is the reason the molar mass value given for carbon is greater than 12.000. Most of the carbon in any naturally occurring sample is carbon-12, but some is carbon-13 and some is carbon-14.

Let's consider another example: chlorine. Chlorine exists naturally as two isotopes, chlorine-35 and chlorine-37. Figure 2.11 tells us that the more abundant isotope is the former, representing 75.77 percent of all chlorine atoms. Thus, if you were to analyze 100 atoms of chlorine, on average, about 76 of them would be the lighter isotope. The numeric value for the molar mass of chlorine given on the periodic table is 35.453. This is the *weighted average* of the molar masses of these two isotopes, and it takes into account the average abundance of each chlorine isotope in nature.

The molar mass of chlorine listed on the periodic table is 35.453, but 35.453 *what?* Whenever we express a number that is measured, there should always be a unit that goes with the number; otherwise, we do not know what the number means. Recall that molar mass tells us the mass in grams of 1 mol of an element. Another way of saying this is *the mass per 1 mol*, and the word *per* usually indicates a fraction. Therefore, the unit for molar mass should have two parts: mass in the numerator and moles in the denominator. Thus, the molar mass for chlorine is 35.453 g/mol, read "35.453 grams per mole."

The molar masses of all the elements are expressed in this way. In fact, it is possible to find the mass of any substance for which you know the chemical formula simply by summing the molar masses of all the atoms in the formula. The result is the molar mass of that substance. For example, to find the molar mass of ammonia, NH_3, you use the molar masses for nitrogen and hydrogen from the periodic table. In the calculation, you take into account that each mole of NH_3 contains 1 mol of nitrogen atoms and 3 mol of hydrogen atoms:

Molar mass of NH_3

$$= \left(\frac{1 \text{ mol N}}{1 \text{ mol NH}_3} \times \frac{14.01 \text{ g}}{1 \text{ mol N}}\right) + \left(\frac{3 \text{ mol H}}{1 \text{ mol NH}_3} \times \frac{1.01 \text{ g}}{1 \text{ mol H}}\right)$$

$$= \frac{17.04 \text{ g}}{1 \text{ mol NH}_3}$$

or 17.04 g/mol

Thus, 1 mol of ammonia has a mass of 17.04 g.

In this book we will use rules for significant figures when we express numbers that result from mathematical operations. For example, this equation correctly shows 2 digits to the right of the decimal place. The rules for significant figures are explained in Appendix A.

If your math is rusty, a review of basic algebra can be found in Appendix B.

QUESTION 2.15 What is the molar mass of potassium superoxide, KO_2, a substance found in the breathing apparatus used by firefighters?
ANSWER 2.15

Molar mass of KO_2

$$= \left(\frac{1 \text{ mol K}}{1 \text{ mol KO}_2} \times \frac{39.10 \text{ g}}{1 \text{ mol K}}\right) + \left(\frac{2 \text{ mol O}}{1 \text{ mol KO}_2} \times \frac{16.00 \text{ g}}{1 \text{ mol O}}\right)$$

$$= \frac{71.10 \text{ g}}{1 \text{ mol KO}_2}$$

or 71.10 g/mol

QUESTION 2.16 The diatomic chlorine molecule, Cl_2, is used as a disinfectant for drinking water. (*a*) What is its molar mass? (*b*) How many moles of chlorine atoms make up 18 mol of chlorine molecules?
ANSWER 2.16 (*a*) The molar mass is:

$$2 \times \frac{35.45 \text{ g}}{1 \text{ mol}} = \frac{70.90 \text{ g}}{\text{mol}}$$

(*b*) Each mole of chlorine molecules contains 2 mol of chlorine atoms. Therefore, 18 mol of chlorine molecules contain 36 mol of chlorine atoms.

2.5 | Atmospheric Dirt: An Introduction to Redox Chemistry

Evidently it is not uncommon to have a goanna (which is a large lizard) living in your attic if you reside in certain areas of Australia. Because they can be very large (more than 5 ft in length), it is not unheard of to have a goanna break through the ceiling of your living room for an unexpected visit. Paul Hancox, a resident of New South Wales, Australia, remembered this when he heard something crash through his roof on the evening of December 14, 1999. To his surprise, his nocturnal visitor turned out to be a meteorite, not a four-legged reptile.

Meteoroids are, in effect, atmospheric dirt. They form when small pieces of *asteroids*, large orbiting masses originating from a planet, break off and venture out on their own. Occasionally, a meteoroid ends up passing into Earth's atmosphere and landing on the ground, at which point it is called a *meteorite* (Figure 2.17). Why might a chemist be interested in meteorites? One reason is that the chemistry of meteorites can give us clues about the goings-on on the meteorite's mother planet, as we shall see shortly. Another reason is that meteorites contain metals in different forms, and inorganic chemists often think about how metals combine with other elements, particularly oxygen.

Some meteorites are made up of a mixture of elemental iron and the mineral magnetite (Figure 2.18). Recall that mixtures can have varied compositions, which means one meteorite might be 25 percent iron and 75 percent magnetite, while another might have the opposite composition. Magnetite itself contains iron and has the chemical formula Fe_3O_4. It is natural to wonder why iron is sometimes in the form of elemental iron and other times in the form of magnetite or some other iron-containing compound. Why doesn't iron simply remain iron? The answer lies in the reactions that occur when iron is in the presence of oxygen. As we know, oxygen requires two electrons to achieve a noble gas configuration like that of neon. Thus, when something reacts with oxygen, it gives the oxygen electrons and ends up with fewer electrons than it started with. We call this process, the loss of electrons, an **oxidation reaction**, and the thing that steals electrons, oxygen in this case, is called an **oxidizing agent**. (Anything that steals electrons away from something else is referred to as an oxidizing agent, even if it does not contain oxygen.)

Elemental iron is very susceptible to oxidation and readily gives up its electrons to oxygen from the air:

$$4\,Fe + 3\,O_2 \longrightarrow 2\,Fe_2O_3$$

▲ FIGURE 2.17 A photo of a meteorite crater located near Winslow, Arizona.

RECURRING THEME IN CHEMISTRY Chemical reactions are *nothing more than exchanges or rearrangements of electrons,* which is why we pay particular attention to electrons throughout this book.

▶ FIGURE 2.18 **Two views of magnetite. (a)** A crystal of the mineral magnetite. **(b)** A thin section of a meteorite sample viewed under a light microscope.

(a)

(b)

This balanced equation actually involves two separate processes. The oxidation of iron is one of the two:

$$4\,Fe \longrightarrow 4\,Fe^{3+} + 12\text{ electrons} \qquad \text{(oxidation of iron)}$$

The four Fe^{3+} ions produced are missing a total of $4 \times 3 = 12$ electrons, which means 12 electrons must also be produced. However, this is not a valid chemical equation as written because you cannot produce electrons that remain unused. Thus, there must be a second process that uses up those electrons.

Oxygen is a diatomic molecule in which the two oxygen atoms are held together by two bonds. When the molecule accepts electrons, these bonds break and O^{2-} anions are formed. This process, whereby electrons are gained, is known as a **reduction reaction**:

$$3\,O_2 + 12\text{ electrons} \longrightarrow 6\,O^{2-} \qquad \text{(reduction of oxygen)}$$

Because iron is the thing that gives electrons to oxygen, we call the iron a **reducing agent**. In this case, six oxygen anions are formed, each carrying two negative charges. This requires twelve electrons on the left side of the equation.

QUESTION 2.17   Identify each of these reactions as being either oxidation or reduction:

(a) $Cu^+ \rightarrow Cu^{2+} + 1$ electron

(b) $Co^{2+} + 1$ electron $\rightarrow Co^-$

(c) $S + 2$ electrons $\rightarrow S^{2-}$

ANSWER 2.17   When electrons are lost, as in part (a), they appear on the right side of the equation. This loss of electrons is an oxidation reaction. When electrons are gained, as in parts (b) and (c), they appear on the left side of the equation. Both (b) and (c) are reduction reactions.

We previously mentioned that it is "illegal" in chemistry terms to write equations that show electrons on the left or right side. The equations we wrote before, which show electrons as either reactants or products, are equations for what are called **half reactions**. To create a complete and balanced chemical equation, we must combine the equation for our oxidation half reaction and the equation for our reduction half reaction in a way that cancels out the electrons. The reaction represented by the resulting equation is called a **redox reaction** because it is a combination of **red**uction and **ox**idation:

Start with the equation for the reduction half reaction:

$$3\,O_2 + 12\text{ electrons} \longrightarrow 6\,O^{2-} \qquad (1)$$

Now write the equation for the oxidation half reaction:

$$4\,Fe \longrightarrow 4\,Fe^{3+} + 12\text{ electrons} \qquad (2)$$

Add them together:

$$3\,O_2 + 12\text{ electrons} + 4\,Fe \longrightarrow 6\,O^{2-} + 4\,Fe^{3+} + 12\text{ electrons} \qquad (3)$$

Cancel out the electrons on both sides:

$$3\,O_2 + 4\,Fe \longrightarrow 6\,O^{2-} + 4\,Fe^{3+} \qquad (4)$$

This process is complex, and it is not critical that you know how to manipulate and add equations for oxidation and reduction half reactions. What is important is that each step make sense to you. In Equations 1 and 2, we have simply copied the equations for the oxidation half reaction and the reduction half reaction. In Equation 3, we add together Equations 1 and 2 to get the equation for a redox reaction, with 12 electrons on the left and 12 electrons on the right. Whenever you have the same thing on both sides of a chemical equation, those things can be removed by cancellation. This gives Equation 4, a balanced chemical equation. Like all other chemical equations, the equation describing a redox reaction must have a balanced number of atoms of each element on the two sides. The important lesson from this example is that electrons produced by an oxidation reaction must somehow be consumed by a reduction reaction. Likewise, whenever a reduction reaction consumes electrons, you must account for the source of those electrons.

> **RECURRING THEME IN CHEMISTRY** Chemical reactions do not create or destroy matter. Therefore, the types and numbers of atoms must be the same on the two sides of a chemical equation.

The redox reaction shown in Equation 4 has two products: four Fe^{3+} ions and six O^{2-} ions. Because iron oxide is a solid salt (just like NaCl or MgS), we know that its anions and cations are stacked together in a solid crystalline lattice and that its positive and negative charges must balance each other. We also know that salts are written as formula units, which give the smallest ratio of one ion to another. To figure out the correct chemical formula for iron oxide, *we must be certain that the formula contains equal numbers of negative and positive charges.* Using trial and error, we can write some possible chemical formulas for iron oxide:

FeO_2 is not electrically neutral because one Fe^{3+} and two O^{2-} ions give a net charge of -1.

Fe_2O_2 is not electrically neutral because two Fe^{3+} and two O^{2-} ions give a net charge of $+2$.

Fe_3O_2 is not electrically neutral because three Fe^{3+} and two O^{2-} ions give a net charge of $+5$.

Fe_2O_3 is electrically neutral because two Fe^{3+} and three O^{2-} ions give a net charge of zero.

The chemical formula Fe_2O_3 is correct because the two Fe^{3+} cations give six positive charges that balance the six negative charges on the three O^{2-} anions. Thus, we can rewrite Equation 4 with the correct chemical formula for iron oxide:

$$3\,O_2 + 4\,Fe \longrightarrow 2\,Fe_2O_3$$

▼ **FIGURE 2.19 Oxidation.** This truck is coated with a layer of iron oxide that formed from the oxidation of the iron metal used to make the truck body.

The product of this oxidation of iron, Fe_2O_3, is more commonly known as *rust* (Figure 2.19), but iron may also oxidize to other iron compounds. One example is Fe_3O_4, called ferrosoferric oxide. In this compound, the iron is present in two forms, Fe^{2+} and Fe^{3+}. Ferrosoferric oxide is the form of iron that makes up magnetite, one of the compounds found in meteorites. Now the composition of our meteorite makes more sense. It contains some elemental iron, Fe, but it also contains some oxidized iron in the form of magnetite because, at some point, the elemental iron came into contact with oxygen. Furthermore, as we shall see in the next section, magnetite can form in different ways. The way it formed in some Martian meteorites has been the subject of shocking news reports about Mars and its history, as we shall see in Section 2.6.

QUESTION 2.18 The elemental substances copper, Cu, and oxygen, O_2, combine to form the compound copper oxide, CuO, which contains Cu^{2+} cations and O^{2-} anions:

$$2\,Cu + O_2 \longrightarrow 2\,CuO$$

(a) In this redox reaction, which substance is being oxidized and which is being reduced? (b) Which reactant is the oxidizing agent? (c) Which reactant is the reducing agent?

ANSWER 2.18 (a) The copper is oxidized, and the oxygen is reduced. The reaction begins with neutral Cu atoms and ends up with Cu^{2+} ions. (You know that the ions must be Cu^{2+} because CuO contains one O^{2-} anion, which must have two positive charges balancing it.) Thus, the copper loses electrons, which means it is oxidized. The oxygen gains these electrons, which means it is reduced. (b) Oxygen is the oxidizing agent because it takes electrons away from the copper. (c) Copper is the reducing agent because it gives away two of its electrons.

QUESTION 2.19 As noted above, the iron cations in ferrosoferric oxide, Fe_3O_4, are a mixture of Fe^{2+} and Fe^{3+}. Of the three iron ions, how many are Fe^{2+} and how many are Fe^{3+}?

ANSWER 2.19 In this chemical formula, the oxygens each contribute two negative charges, for a total of $2 \times 4 = 8$. Therefore, the number of positive charges on the iron ions must add up to 8. Two Fe^{3+} ions plus one Fe^{2+} ion give a total of eight positive charges.

2.6 | Compass or Bacterium?

Meteorites containing magnetite are easy to identify because they are strongly magnetic. Magnetite is not found just in meteorites, however; it is an abundant mineral here on Earth, too. If you have a videotape around the house, then you have a tiny sample of magnetite. In fact, Greek navigators devised the first compass by fixing a small magnetite rod onto a cork and floating it in a bowl of water.

It is clear that magnetite's magnetism makes it a handy substance, but what makes it magnetic? Why are some substances magnetic and other substances not? Electrons hold the answer. Every electron is like a tiny magnet, and in most substances electrons are paired. The two electrons in a pair orient parallel to each other with the "north" pole of one electron next to the "south" pole of the other, in the same way that two bar magnets align next to each other (Figure 2.20a). Substances with all electrons paired are not magnetic because the oppositely oriented electrons in each pair cancel each other's magnetism. In substances that exhibit **magnetism**, most electrons in each atom are paired up, but a few are not. In a magnetic field, the unpaired electrons all point in the same direction, as shown in Figure 2.20b. A chunk of magnetite has billions and billions of iron atoms, each one containing electrons that are mostly paired up. However, a few electrons in each atom of iron are not paired up. It is these electrons that are lined up parallel to

▼ FIGURE 2.20 **Magnetism in matter.** (a) Two bar magnets align so that opposite poles are next to each other. (b) Each circle represents one unpaired electron in an atom of a magnetic substance.

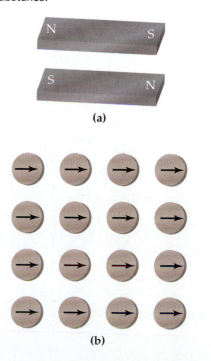

(a)

(b)

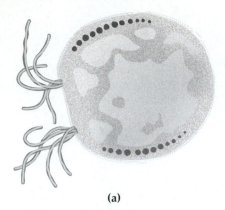

(a)

(b)

▲ **FIGURE 2.21** **Magnets as a navigation tool.** (a) A schematic rendering of a simple magnetotactic bacterium, showing strands of magnetite beads on each side. (b) A more elaborate magnetotactic bacterium.

one another and contribute to the magnetism of magnetite. The result is an accumulation of little magnets that make one larger, stronger magnet.

Lately, magnetite has cropped up in some surprising places. In an effort to understand the inexplicable navigating capabilities of some animal species, scientists looking for the answer have analyzed the brain tissue of these animals. What they have found is that members of several species have magnetite crystals in their cells, and those crystals are thought to lead the organisms from one place to another, perhaps in search of food, oxygen, or the perfect mate. Magnetite has been identified in the membranous outer layers of bacteria and the brains of whales, platypuses, moles, and anteaters. Magnetite in the brains of racing and homing pigeons may be responsible for the ability these birds have to find their target. In fact, there is a hot debate about the presence of magnetite in the human brain. A study performed on brain tissue in 1995 strongly suggests that *there is* magnetite present, and these little magnets might explain some people's uncanny sense of direction or ability to sense the presence of another person nearby.

Magnetite was first found in certain bacteria back in the 1970s when a researcher observed that the bacteria tend to swim south to north. Since then, these *magnetotactic* bacteria have been studied extensively, and the type of magnetite they produce has been characterized. Figure 2.21*a* shows a schematic drawing of a simple magnetotactic bacterium. The dark gray spots along the sides of the bacterium are tiny crystals of magnetite connected together so that they resemble beads on a necklace. More important is that the two chains of magnetite "beads" are oriented parallel to each other, with the magnets in each bead all pointing in the same direction. These little magnets align naturally with any external magnetic field, including the natural magnetic field of Earth, and this alignment allows the bacteria to move along a north–south axis in search of what they need. Their ability to move puts them a step ahead of most other bacteria, which must tumble randomly in search of food. Some magnetotactic bacteria have evolved highly sophisticated navigation mechanisms. Figure 2.21*b* shows a schematic drawing of the magnetite crystal chains in one bacterium that has five strands spanning its length.

We refer to the formation of inorganic crystals, like magnetite, in living organisms as *biomineralization*, literally the formation of minerals in biological contexts. Until now, we have been using the terms *crystal* and *mineral* interchangeably, but there is a difference between them. What does the term *mineral* imply? The answer is that all minerals are crystals, but not all crystals are minerals. There are several criteria that a substance must meet before it can be classified as a mineral. First, it must be inorganic. Thus, a crystal made of organic molecules, such as glucose ($C_6H_{12}O_6$) or benzene (C_6H_6), cannot be classified as a mineral. Second, the substance must be natural. Hence, a crystal grown in the laboratory cannot be a mineral. Finally, it must be a "pure" substance (although limited impurities are allowed). That is, it must have the same chemical composition throughout, and every sample of a given mineral, whether it is found in Tanzania or Tibet or Texas, must have that same composition. Consequently, we can write chemical formulas for minerals because they are pure substances. Two examples of mineral formulas are those for quartz, SiO_2, and one type of feldspar, $KAlSi_3O_8$.

Sometimes a mineral's appearance can be slightly different depending on where it was mined or where it was formed. For example, magnetite that formed geologically looks different from magnetite formed in a bacterium.

When formed by geological processes, magnetite crystals are large, round, and riddled with defects in the crystalline structure. When formed by bacteria, however, magnetite crystals are small, oblong, and defect free. This fact has caused an uproar at NASA, where researchers are studying a Martian meteorite that landed in the Allan Hills region of Antarctica about 13,000 years ago. It seems that some of the magnetite crystals found in this meteorite resemble magnetite crystals formed by bacteria here on Earth. To some people, this similarity suggests that Mars may have harbored living bacteria at some point in its history. Many scientists are not convinced by the data, though, and believe that the harsh atmosphere of Mars is much too unfriendly to support any kind of life-form. Figure 2.22 compares magnetite from a magnetotactic bacterium found here on Earth with a magnetite sample from the Allen Hills meteorite. You be the judge.

QUESTION 2.20 You are a notorious skeptic. Even after reading about the Allan Hills meteorite, you have doubts about the existence of life on Mars. Name one objection you would raise with anyone arguing that the meteorite crystals indicate life on Mars.

ANSWER 2.20 There are many possible answers. One objection is that the meteorite is 13,000 years old and could have been contaminated by magnetotactic bacteria that originated on Earth. Another objection is that the conclusion that there at one time was life on Mars is based on the assumption that small, defect-free magnetite crystals like the ones formed by bacteria on Earth can be formed only in the same way on Mars.

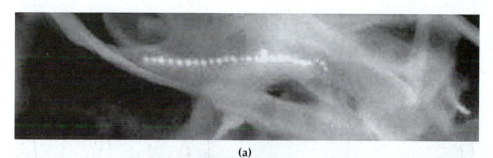

(a)

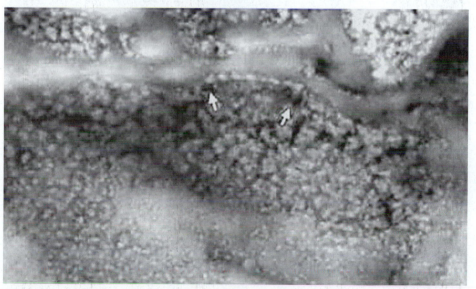

(b)

◀ **FIGURE 2.22** **Two more views of magnetite.** Microscope photographs of **(a)** magnetite crystals inside a magnetotactic bacterium found on Earth and **(b)** magnetite crystals found in the Allan Hills meteorite, which is presumed to be a Martian rock that landed on Earth 13,000 years ago.

STUDENTS OFTEN ASK

Why Are Gemstones Different Colors?

We just saw that the same mineral can have a different appearance depending on where it was formed. It is also true that the same mineral can look dramatically different depending on the impurities it contains. Let's look at two examples. *Beryl*, which has the chemical formula $Be_3O_{18}Al_2Si_6$, and *corundum*, Al_2O_3, are both clear, completely colorless crystals when absolutely pure. In nature, however, impurities become incorporated into crystals as they grow, and these impurities replace a few of the ions that belong in the crystalline lattice.

When the impurities are transition metals, which are located in the center block of the periodic table, brilliantly colored crystals can result (Figure 2.23a). For example, beryl containing tiny amounts of Cr^{3+} ions takes on a green color and becomes the gemstone *emerald*. The same chromium impurity in a corundum crystal gives a red color, and the result is known to us as *ruby*. *Yellow topaz* is formed from corundum containing Fe^{3+} ions, and *sapphire* results when the corundum also contains Fe^{2+} and Ti^{4+} impurities. *Quartz*, another colorless crystal, takes on many different colors when transition-metal impurities are present. Several examples of colored quartz are shown in Figure 2.23b.

▼ **FIGURE 2.23   Impurities make gemstones what they are. (a)** At a mine near Franklin, North Carolina, you can buy a bucket of dirt and keep any stones it contains. This handful of stones—some valuable, some not—was collected by sifting through one bucket of dirt. **(b)** Samples of quartz containing different transition metal impurities. Clockwise from top left: colorless (pure) quartz, yellow quartz (citrine), smoky quartz, and purple quartz (amethyst).

(a)

(b)

QUESTION 2.21   Which are transition metals: (*a*) calcium, Ca; (*b*) tungsten, W; (*c*) iodine, I; (*d*) strontium, Sr; (*e*) iridium; Ir, (*f*) gallium; Ga?

ANSWER 2.21   Tungsten and iridium. (See Figure 2.3*b*.)

QUESTION 2.22   What is the molar mass of (*a*) the mineral beryl and (*b*) the mineral corundum?

ANSWER 2.22   (*a*) 542.90 g/mol, (*b*) 101.96 g/mol.

The Dark Side of Minerals: Asbestos and Baby Powder

So far we have seen examples of many minerals, some sought for their color and brilliance. The strong crystalline lattices in minerals might seem to confer inertness and stability, making these substances incapable of harming living organisms. We certainly have nothing to fear from a sapphire!

Such is not the case with all minerals, though. In recent years, Americans have reacted to alarms set off by the health community about the mineral *asbestos* and its connection with serious lung disease. In the 1950s and 1960s, our view of asbestos was quite different. The soft, fibrous mineral was lauded for its benefits as a fireproof material and thermal insulator. As a result, asbestos was used in buildings and in fireproof clothing, like the type shown in Figure 2.24*a*. Because asbestos is light and stringy, it is *friable*; that is, it is easily reduced to a powdery form that can produce breathable dust. Now we are scrambling to ensure that asbestos is removed from any building where it might be present in the air.

There are several types of asbestos, the most common of which are the blue (crocidolite) variety and the white (crysotile) variety. (The latter is shown in Figure 2.24*b*.) Both varieties have similar chemical formulas. However, asbestos crystals form differently depending on the conditions under which they are grown. Crocidolite grows as long, straight fibers, and crysotile grows as curly threads.

Many studies have attempted to quantify the risks of exposure to these two forms of asbestos. It was found that, in places where crocidolite is predominant, the health risks are significant. Individuals who work in areas of western Australia and South Africa (both

(a)

(b)

◄ **FIGURE 2.24** **Asbestos.** **(a)** Even though we know about the dangers of asbestos, protective clothing made of this fireproof material is still used by industrial workers today. **(b)** A sample of crysotile asbestos, the white variety.

locations where crocidolite is mined) have significantly higher incidences of lung and chest cancers than does the general population. What is surprising, however, is that crysotile, which accounts for about 95 percent of the asbestos found in the United States, is more benign because of its curly fibers. In addition to not being trapped in the lungs as readily as are crocidolite fibers, crysotile fibers also break down more easily when they are incorporated into lung tissue. Consequently, the health risks associated with inhaling crysotile asbestos are thought to be minimal. One study showed that a person has a higher chance of dying as a result of a high-school football casualty than as a result of exposure to crysotile asbestos.

Part of the concern about the enormous quantities of dust produced in the World Trade Center (WTC) collapse on September 11, 2001, was that workers at the site and the citizens of Manhattan would be exposed to asbestos dust (Figure 2.25). More than 75 percent of the WTC dust samples analyzed contained crysotile asbestos. As a result, cleanup at the site has been performed with extraordinary care, reminding everyone that asbestos continues to scare us, despite what we know about how minimal the dangers of the crysotile form are.

Talc is quite another story, although we hear little of its danger to our health. Many body powders contain primarily talc, and efforts to remove these products from stores have been met with skepticism. Why worry about talcum powder? The answer to this question is that talc has the same chemical formula as does asbestos and, like asbestos, can grow in long, stringy strands that are very pliable and even more friable. A recent medical study published in the *American Journal of Epidemiology* found that women who use talcum powder had a higher-than-normal risk of developing ovarian cancer. However, a committee of the National Toxicology Program, the people charged with placing substances on a list of known cancer-causing agents, was not convinced by the data. Consequently, talcum powder has not been placed on the list.

▼ FIGURE 2.25   **Asbestos exposure on September 11.** People running through the dust created by the collapsing buildings of the World Trade Center.

2.7 | Dirt as Forensic Evidence

Let's end the chapter with the topic we started with—dirt. Realizing that dirt contains many minerals allowed us to learn some interesting and basic chemical concepts: covalent, ionic, and metallic bonding; Avogadro's number; molar mass; oxidation and reduction; and so on. Let's see now how dirt can make itself useful in the detection of criminal activity.

Any sample of dirt may contain dozens of different minerals, and these compounds can be separated from one another and classified using techniques available to the forensic scientist. In addition, dirt in Nevada is different from dirt in Newfoundland, which is different from dirt in Norway or Nigeria. In fact, a given sample of dirt is almost like a fingerprint in that it is often possible to rule out many locations from which the sample could *not* have originated. A telltale sample can pinpoint the scene of a crime, in some cases down to a few acres or even a few square yards.

Special Agent Ronald Rawalt has been a geological expert for the FBI for over 30 years. If you think that geologists are specialists who research and study the structure and surface of the Earth, then you may be surprised to learn that geologists can also be forensic experts. In this highly focused subdiscipline, Rawalt is among the best. One of his most famous cases began when a Pennsylvania police officer was late for work one day. When his car was found with his blood in it, the case became a homicide investigation. And when some interesting dirt was found in the car's wheel wells, Agent Rawalt was on the case. By carefully analyzing this dirt, he was able to determine several things about the path the car took before it was found by police. First, he was able to identify a specific mixture of fluorescent paint chips that is found on highways that slope uphill and turn at the same time. Second, he was able to identify a small fossil that is found in a very specific kind of limestone. He researched this particular limestone, and learned it is found only in a few very narrow veins, one of which was located in the area from which the officer was missing. Rawalt was able to pinpoint the region where this thin vein crosses a highway. Within that region, there was one area where the highway turned and sloped upward. The police officer's body was found at that exact location, a few feet down the hill from the highway.

Another of Rawalt's cases took him to a completely different location: the soils of central Mexico. The case involved the 1985 murder of U.S. Drug Enforcement Administration employee Enrique Camarena-Salazar, whose body was found in a region to the west of Mexico City where the soil contains distinctive chunks of a dark-colored glass. The body itself was covered with a completely different type of soil, however, which suggested that the body had initially been buried in one location, and then moved. Rawalt examined the dirt found on the body and determined that it was volcanic in origin. In Mexico, a belt of volcanic activity runs west to east in a belt that passes about 100 km to the north of the site where Camarena's body was found. Rawalt's further investigation of the volcanic soil found on the body revealed three clues: it must come from an area with a specific thickness of ash; it must contain certain trace minerals; and it must be located in a region where there is burned vegetation. One such area was located within the volcanic belt. In this area, the original site of the burial was located by a team of specially trained scent dogs, a finding that eventually led to the identification of the people who murdered Agent Camarena.

A legendary pioneer in forensic evidence, Edmond Locard, insisted that, *no matter what*, whenever a person comes in contact with any other thing, an exchange of material evidence takes place. Clearly, this is true for dirt, as the preceding excerpt aptly demonstrates. It is interesting that limestone is the mineral that led the police to the victim's body when you consider that limestone is one of the most common minerals known. Most limestone forms from the shells and skeletons of marine animals. In "new" limestone, these fossil shells and skeletons may still be visible, as shown in Figure 2.26a. Most older limestone has *lithified*, which means that it has compacted down into solid rock. An example can be found in the famous White Cliffs of Dover, shown in Figure 2.26b. Lucky for Special Agent Rawalt, the limestone he found was "new," and the fossil architecture was still recognizable and analyzable.

There have been many high-profile cases that used dirt as physical evidence. In 1932, the infant son of the famous aviator Charles Lindbergh was kidnapped. Dirt left on a windowsill at the Lindbergh home was one of many clues to the kidnapper's methods. In another case, information obtained from dirt was the pivotal evidence in the 1960 murder of Adolph Coors III in Morrison, Colorado (Figure 2.27). The soil in this region is very distinctive and was one of three soil layers found in the wheel wells of a yellow automobile

▶ **FIGURE 2.26   Limestone old and new. (a)** Newer limestone rocks often show evidence of fossil remains. The rock in this photograph contains pieces of shell that were once part of a living organism. **(b)** Older limestone often appears homogeneous and chalky. The White Cliffs of Dover, located on the southeastern coast of England, are composed almost entirely of limestone, formed as small marine organisms were deposited at this location over centuries.

(a)

(b)

that had been seen near a bridge on the Coors property. It was under this bridge that bloodstains were found on the day Coors went missing. Another layer in the wheel wells was equally distinctive—pink feldspars found only in the Pikes Peak area at high altitude. It was in this location that Coors's body was discovered eight months after his disappearance.

The car, burned nearly beyond recognition, was found in Atlantic City, New Jersey. The top layer in the wheel well, diagrammed in Figure 2.28, contained smoothed grains of sand typical of beach soils. Forensic geologists identified the top layer as coming from the New Jersey dump where the car was found. The next layer down matched the unique feldspars composition of the high-altitude dump where Coors's body was found, and the third layer down was an exact match of the mixture of minerals that characterizes the soil around the Coors ranch. The owner of the car, a man named Joseph Corbett, Jr., was identified as a hunter who frequented the area of the Coors ranch in his yellow car. All this evidence was combined into a strong case against Corbett, and he was convicted of the Coors murder.

▶ **FIGURE 2.27   Headline news.** This headline appeared during the search for the missing Adolph Coors III. A team of investigators gathered at the site of the Coors kidnapping in Morrison, Colorado (left). A photo of Adolph Coors III (right).

ol. 68, No. 192   Wednesday, Feb. 10, 1960   5 Cents, 52 Pages

Adolph Coors III Disappears; FBI Enters Search

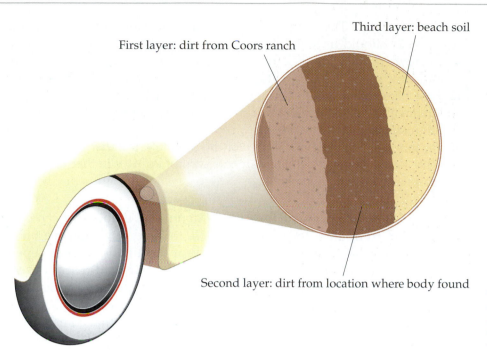

First layer: dirt from Coors ranch

Third layer: beach soil

Second layer: dirt from location where body found

◄ FIGURE 2.28 **Dirty clues.** This schematic rendering represents the three layers of dirt found in the wheel wells of Corbett's car.

The Coors case is certainly a remarkable example of the use of dirt as physical evidence, but can scientists ever make an exact match of one sample with another? The answer is no. The goal, instead, is to state that the evidence *supports* a given claim, and the accuracy of a piece of data must be taken into account whenever real numbers are going to be scrutinized by a jury. In most cases, it is necessary to collect *control samples* from the area of interest. In a **control experiment**, these control samples can be compared with the samples linked to the crime, in this case the dirt found under the wheel wells of the yellow car. For the Coors case, 421 control samples were collected. Investigators took dozens of samples from the Coors ranch, from the location where the body was found, and from the location where the car was finally found. Without control samples, it would have been impossible to connect an unknown sample with a specific location.

RECURRING THEME IN CHEMISTRY Control experiments are included in every respectable scientific analysis. For example, chemists studying a molecule that is dissolved in a watery medium must run the same tests on the watery medium that they run on the molecule. Doing so establishes that the medium is not playing a role in the results obtained on the molecule of interest.

QUESTION 2.23 A scientist analyzing pink topaz finds high levels of chromium in the sample. It is tempting to say that chromium produces the pink color because it is the color-producing agent in many other colorful minerals. However, what critical control experiment must be done in order to pinpoint the role of chromium in the crystal?

ANSWER 2.23 The colorless version of the crystal must be tested for chromium. (In fact, the presence of manganese, not chromium, is the reason for the pink color.)

QUESTION 2.24 You have been told that shellfish caught just off the coast of your hometown contain high levels of mercury caused by industrial waste from nearby factories. In addition to measuring the mercury content in shellfish in the area, what control experiments should have been run by the researchers collecting this information?

ANSWER 2.24   There are several possible answers. One is that the researchers should have measured mercury levels in shellfish caught in an area where the water is known to be free of mercury; this experiment would have given them the natural level of mercury in shellfish. Another control experiment would be to test the mercury levels in the water just upstream and just downstream from the nearby factories. If the water is clean upstream of the factories, but contaminated just downstream from them, they are probably the source of the mercury. If there is contamination upstream, however, the researchers cannot know whether or not the factories are discharging mercury.

On one hand, the complexity of dirt is daunting because dirt can contain anything from glass to paint to crystals to dead bugs. On the other hand, the things that contribute to the complexity of dirt also make it a perfect resource for the forensic scientist. It has been estimated that it is very unlikely that two indistinguishable samples of dirt could have come from locations 1000 feet apart. Dirt will not often reveal why a crime was perpetrated, but it can finger a criminal who has shoes or tires that come in contact with the ground. So if you are planning a crime, keep this in mind: dirt is all around you, and it will almost always betray where you were and when you were there.

SUMMARY

There are literally thousands of things, both artificial and natural, that could be found in a sample of dirt. Among the most conspicuous substances in the latter category are minerals. Minerals are naturally occurring, pure substances that contain one or more elements in a regular crystalline array. Natural soils contain widely varying quantities of many of the elements on the periodic table. In this chapter, we have seen examples of some elements in soil, such as lead, that can cause health risks even in minute quantities. All the atoms of a given element have the same number of protons in the nucleus. The number of neutrons in the nucleus determines which isotope of a given element we are dealing with.

A mineral sample that can be seen with the naked eye contains an enormous number of atoms. When dealing with these unwieldy numbers, scientists use scientific notation for simplicity. The masses of atoms are so small that their numeric values cannot be easily manipulated using conventional metric units. Furthermore, the extremely small masses of atoms make it difficult to discuss the number of atoms present in a sample of matter visible to the naked eye. This colossal number, on the order of 10^{23}, is handled with a convenient unit: the mole. The mole is a counting device that enumerates 6.02×10^{23} of anything, and it is especially useful for keeping track of numbers of both atoms and molecules.

The molar mass of an element is the number provided on the periodic table; the units of molar mass are grams per mole. Molar masses can be calculated for substances made up of more than one atom by summing the molar masses of all the elements present.

Meteorites are examples of natural objects that contain metals in different forms. In this chapter, we discussed a meteorite in which iron exists both as elemental iron and iron in its oxidized state. Oxidation, the loss of electrons, must take place with reduction, the gain of electrons. These two half reactions, when combined, make up a redox reaction. Elements like iron, which are susceptible to oxidation by air, undergo an oxidation reaction to form an oxide. The meteorite we studied contained magnetite, a magnetic substance. Magnetite owes its magnetism to the iron atoms in it, and these atoms have all their electrons aligned in the same direction.

The extraordinary diversity of minerals makes them ideal as physical evidence in criminal investigations. As we saw in both the Pennsylvania murder case and the Coors case, dirt and the minerals it contains are often key pieces of physical evidence. Whenever physical evidence is collected, however, control samples must be collected at the same time. This is true, too, for most scientific experiments: The accuracy of the data depends on the validity of the experimental design.

KEY TERMS

alkali metal	covalent bond	ionic bond	mole	polar bond
alkaline earth metal	diatomic	isotope	noble gas	redox reaction
anion	electronegativity	magnetism	noble gas	reducing agent
Avogadro's number	formula unit	mass number	configuration	reduction reaction
carbon dating	half reaction	metal	nonmetal	salt
cation	halogen	metallic bond	organic	scientific notation
control experiment	inorganic	mineral	oxidation reaction	transition metal
core electron	ion	molar mass	oxidizing agent	valence electron

QUESTIONS

The Painless Questions

1. Referring to Figure 2.1, rank these elements in order of abundance in Earth's crust: vanadium, barium, hydrogen, sodium, iron, phosphorus. Indicate whether each element is a metal or nonmetal.

2. Take another look at Figures 2.4 and 2.5. You have already established that zinc concentrations rise in the area of West Point Lake (Figure 2.5b) as a result of automobile usage and burning of fossil fuels in the Atlanta metropolitan area. To what area south of Atlanta must you travel before the level of zinc contamination decreases to the level at Lake Sidney Lanier north of Atlanta?

3. The typical atom is about 0.000 000 000 1 m in diameter. (a) Write this number in scientific notation. (b) Use a metric prefix before *meter* to make the unit a more reasonable one. That is, choose a metric prefix that makes the use of scientific notation unnecessary.

4. (a) If an element has 56 protons, 82 neutrons, and 56 electrons, what are its symbol and mass number? (b) How is an atom with 56 protons, 78 neutrons, and 56 electrons related to the atom in part (a)?

5. Write these measured numbers in scientific notation: 5,600,000,000 nm, 0.000 047 6 g, 4500 mg. For each value, change the metric prefix so that scientific notation is unnecessary.

6. Which metal would be most likely to contribute to the color of a gemstone: cesium, Cs; beryllium, Be; cobalt, Co; or oxygen, O? Explain your choice.

7. Brown asbestos has properties similar to those of blue asbestos. Which would you consider to be a greater health hazard, brown asbestos or white asbestos?

8. Write out the complete number for (a) 6.7×10^4 kg of dirt, (b) 9.777×10^{-5} ng of cobalt metal, (c) 6.02×10^{23} atoms of silver.

9. Comment on this statement:

   An oxidizing agent is something that accepts electrons from something else, whereas a reducing agent donates electrons to something else.

10. Calcium forms calcium oxide, CaO, as well as calcium bromide, $CaBr_2$. (a) Identify the cation and anions in these formulas. (b) Explain why the ions in calcium oxide have a one-to-one ratio, but the ions in calcium bromide have a one-to-two ratio.

11. A certain piece of copper pipe is composed of (nearly) pure copper. (a) Describe the bonding in this piece of metal. (b) How is it possible that there are no anions in the metal?

12. (a) Give an example of two atoms that form a nonpolar bond with each other. (b) Is it possible for two atoms of *different* elements to form a nonpolar bond? How?

13. Name two criteria that a crystal must meet in order to be classified as a mineral.

14. Comment on this statement:

    Electronegativity is a measure of an atom's ability to pull electron density toward itself. Atoms with the highest electronegativity tend to be located in the lower right corner of the periodic table.

15. The equation

$$Cu^{2+} + 2\,\text{electrons} \longrightarrow Cu^+$$

does not make sense as written. (a) How can it be corrected? (b) If you write the corrected equation in reverse (so that products and reactants are switched), what type of reaction does this become?

More Challenging Questions

16. Keeping in mind the steps of the scientific method (Section 1.6), plan a series of experiments designed to determine whether several unknown pure elemental substances are metals or nonmetals. (a) Describe the experiments you would perform. (b) Describe the control experiments you would run. (c) What properties would you examine in these substances?

17. Write the chemical formula for the ionic compound that forms between (a) barium and oxygen, (b) rubidium and nitrogen, (c) magnesium and fluorine.

18. Figure 2.11 shows that the rarest isotope of tin is tin-115. How many neutrons does this isotope have?

19. A particular geographic location can have an unusually high concentration of a given element. As a result, unusual minerals are often found in such places. One example is smithsonite, a brilliant blue-green stone that is also found in pink and white varieties. The chemical formula for smithsonite is $ZnCO_3$. What is the molar mass of this mineral?

Smithsonite.

20. The masses of the proton, neutron, and electron have been determined to be 1.67×10^{-27} kg, 1.67×10^{-27} kg, and 9.11×10^{-31} kg, respectively. (a) Using these values, determine the total mass of one atom of ^{16}O. (b) Does your calculated value match the value given for the molar mass of oxygen on the periodic table? Explain any discrepancy between the two values.

21. The mineral galena has the chemical formula PbS. What is the mass of 1 mol of galena?

Galena.

22. El Capitan Peak in California's Yosemite Valley is a solid mass of limestone. Look up the term *limestone* in a dictionary. (a) What is this mineral made from? (b) What is the chemical formula for limestone?

El Capitan.

23. Students are sometimes confused by the differences among the terms *salt, crystal, mineral,* and *formula unit*. Pretend you are explaining these terms to a friend (or find a friend who is willing to listen). In what contexts are these terms used, and how are they similar or different from one another?

24. This photograph shows the mineral malachite, $Cu_2(CO_3)(OH)_2$, mixed with the mineral azurite, $Cu_3(CO_3)_2(OH)_2$. Determine the molar mass of each of these two substances.

Malachite (green) and azurite (blue).

25. Using the electronegativity values given in Figure 2.9a, rank the bonds between these pairs of atoms, from most polar to least polar. For each pair, indicate whether you expect the bond to be covalent or ionic. (a) Cs bonding with F, (b) Si bonding with P, (c) Cl bonding with Cl, (d) Mg bonding with O, (e) O bonding with Pb.

26. Artificial rubies can be produced by the Verneuilli process, in which aluminum oxide (the chemical name for corundum) is subjected to very high temperatures and infused with trace impurities. The resulting crystals are absolutely flawless and reach 100 percent purity. (a) How might artificial rubies be less desirable than real ones? (b) How would you distinguish between the two?

27. (a) Explain why the mole is a convenient way to count things like atoms and molecules. (b) Why is Avogadro's number so large?

28. (a) Explain how some substances can be magnetic. (b) On the atomic level, how do magnetic substances differ from substances that are not magnetic?

29. (a) What can be learned from unnaturally elevated levels of trace metals in a dirt sample? (b) How is it possible to gather both spatial (based on location) and temporal (based on time elapsed) data from dirt?

30. Comment on this statement:

> As a result of the recent findings on the dangers of talcum powder, cosmetics manufacturers are considering making body powder from ground asbestos rather than talc. Because these two materials have similar structures and consistencies, one should be able to substitute readily for the other.

The Toughest Questions

31. Imagine you are a professional scientist who has created 300 atoms of a new element. You determine that the element has 116 protons, and you decide to give it the symbol Oh, for *ohgreatium*. Of the 300 atoms, 24 exist as the ohgreatium-232 isotope, 266 as the ohgreatium-234 isotope, and 10 as the ohgreatium-235 isotope. How many neutrons does each isotope contain?

32. Some anions and cations have more than one atom. When the chemical formulas for these *polyatomic ions* are written, the atomic symbols are enclosed in parentheses and the charge is written as a superscript outside the parentheses. We shall look closely at several examples of polyatomic ions in later chapters. For now, let's make salts from them by applying the rule we already know: salts must have a balance of positive and negative charges. Write the chemical formula for the salt that forms from each of these ion pairs: (a) Na^+ and $(PO_4)^{3-}$, (b) $(NH_4)^+$ and Se^{2-}, (c) Mg^{2+} and $(CO_3)^{2-}$.

33. Summarize the evidence presented in Section 2.6 in support of there once having been life-forms on Mars.

34. The popular cholesterol-lowering drug Lipitor® has the molecular formula $C_{33}FH_{35}N_2O_5$. (a) What is its molar mass? (b) What is the mass, in grams, of one molecule of Lipitor? (c) If you take a pill that contains 15.0 mg of Lipitor, how many molecules have you ingested?

35. Give examples to support or refute this statement:

> The periodic table is arranged so that elements which are close to one another have similar properties.

36. A certain diamond contains 1.81×10^{24} atoms of carbon. How many moles of carbon does it contain?

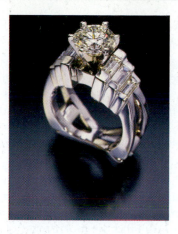

Diamond.

37. Because of its prismatic shape, the mineral scapolite is named for the Greek word for "shaft." Its chemical formula is quite complex: $Na_2Ca_2Al_3Si_9O_{24}Cl$. What is the molar mass of the mineral?

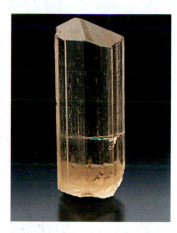

Scapolite.

38. A homicide detective is working on a case in which an unusual brand of talcum powder was found in a suspect's car. A fine white powder was also found on the body of the victim. (a) How might the forensic scientists assigned to the case confirm that the powder on the victim's body is talcum powder? (b) What control experiments should be performed if this is to be a thorough investigation?

39. A bank thief drives from Mississippi to New York to rob the Federal Reserve Bank. The thief leaves faint footprints in the bank lobby, and the investigators on the case collect samples of the dirt in the prints in the hope of discovering its origin. What control samples should be collected to ensure that the results are valid?

40. (a) Describe what is meant by a valence electron. (b) Why are valence electrons important for the chemistry of a given element? (c) Is it possible for an atom or ion to have no valence electrons? If so, provide an example.

41. (a) If a meteorite lands in your living room, how can you quickly determine if it *does not* contain magnetite? (b) Why do we think of meteorites as mixtures rather than compounds or elemental substances? (c) What kind of magnetite would you expect to find in a meteorite made in part by the action of bacteria?

42. The gemstone aquamarine is composed chiefly of the mineral beryl, $Be_3Al_2Si_6O_{18}$. The brilliant blue color of aquamarine comes from iron impurities. (a) From the chemical formula, can you guess where beryl gets its name? (b) Find the molar mass of this mineral.

Aquamarine.

43. Identify each of the following as an oxidation half reaction, a reduction half reaction, or a redox reaction (in each case, identify each reactant as an oxidizing agent or a reducing agent):

(a) $Fe^{2+} + Cu^+ \rightarrow Cu + Fe^{3+}$

(b) $Mn^{5+} + 4$ electrons $\rightarrow Mn^+$

(c) $Cl^- \rightarrow Cl + 1$ electron

44. An anion is formed when a neutral atom takes on additional electrons. For example, sulfur forms S^{2-}, chlorine forms Cl^-, and tellurium forms Te^{2-}. (a) Explain why each of these elements has a specific number of negative charges in its anionic form. (b) Why do you think Ne^- or Ar^{2-} would be unlikely to form? What factors would preclude the addition of electrons to these atoms?

E-Questions

Go to **www.prenhall.com/waldron** to find these questions in electronic form, complete with hyperlinks directly to the various websites cited in the questions.

45. Go to **Webmineral**, and choose the link Chemical Composition. You will see a periodic table where you can select an element and find out which minerals contain it. Choose your favorite element. From the list that comes up, choose five minerals that contain this element. Calculate the molar mass of each of your five minerals, and check your answer against the information on the webpage (given in the column headed MW for the out-of-date term *molecular weight*). Look at detailed information for each mineral. Under Locality, the mining location is provided in many cases. Is information provided about the mining location of your minerals?

46. Go to the U.S. Geological Survey at **USGS**. Beginning with "USGS Information by State," follow the links: Your State → Mineral Information → Map of Principal Mineral-Producing Localities. What minerals are mined in your county and in surrounding counties? Now follow the links USGS Information by State → Your State → Factsheet. Look through the geology news for your state. In four or five sentences, summarize the story you find most interesting.

47. The article "Mars Is Dead" summarizes evidence refuting claims of life on Mars. Read this article at **PopularMechanics**. Why is the earlier evidence suggesting that life once existed on Mars being questioned today? Who sponsored the new experiments? Does the new study definitively show that life never existed on Mars?

48. Read a story about the discovery of the first magnetotactic bacteria at **StrangeHorizon**. What was Blakemore looking for when he found these bacteria? Did Blakemore use the scientific method when trying to decipher the behavior of these bacteria? Is magnetite the only mineral found in magnetotactic bacteria?

49. Check out **Fidelitrade**. The daily prices for precious metals are given in U.S. dollars per troy ounce (32 troy oz is approximately equal to 1 kg). What is the price of gold today? What would the price of 1 mol of gold be in today's market? Which precious metal is worth the most today: Au, Ag, Pd, or Pt?

WORLD WIDE WEB RESOURCES

As with the E-Questions, go to **www.prenhall.com/waldron** to find these questions in electronic form, complete with hyperlinks directly to the various websites cited.

NOTE: Some of links that follow contain research articles related to the topics in this chapter and could be used as the basis for a writing assignment in your course. Other sites are of general interest.

- You can send any earth-science questions to **Ask-A-Geologist**, and a real live geologist will respond in a few days. The site has a few rules about what to send. First, be sure to put a return address in your e-mail. Second, they will not answer exam questions or homework problems! Third, they will not answer financial questions or questions concerning products or companies.

- If you are interested in purchasing a chunk of meteorite, go to **NwaSale**, where pieces of a meteorite called Northwest Africa 482 are for sale. What is the price of a 1.4-g piece of this meteorite? What do you get with your purchase? What features do the most desirable meteorite fragments have?

- Go to **GoldInstitute** and read the featured article on environmental issues. Write a paper that explores the biases you perceive in the article, if any. Does the article extol the mining industry? Can you find alternative sources that offer another point of view on this issue?

- The Mineral Collector's Page can be found at **Mineralant**. There are many interesting things at this website, but best among them is the collection of photographs of fluorescent minerals taken under ultraviolet light by a professional photographer. They are worth a look.

- The U.S. Geological Survey website at **USGS** has an interesting paper on the use of aerial photography to map geological features on Earth's surface entitled "Remote Sensing in the USGS Mineral Resource Surveys Program in the Eastern United States."

- If you would like to go looking for some precious metals on your own, take a look at the National Mineral Resource Assessment the **USGS**.

CHEMISTRY TOPICS IN CHAPTER 3

- Period 2 elements, the uniqueness principle, network solids [3.1]
- Single, double, and triple bonds between carbon atoms [3.2]
- Allotropes, bond energy, bond length [3.3]
- Delocalized electrons, resonance [3.4]
- Chemistry on a nanometer scale [3.5]
- Electron dot structures, the octet rule, nonbonding electrons [3.6]
- Drawing molecules: full structures, condensed structures, line structures [3.7]
- VSEPR theory, three-dimensional shape in molecules [3.8]
- Silicon chemistry [3.9]

"O Diamond! Diamond! thou little knowest the mischief done!"

ISAAC NEWTON

Diamonds

AN EXAMINATION OF CARBON ALLOTROPES, COVALENT BONDING, AND THE STRUCTURE OF SIMPLE ORGANIC MOLECULES

Nothing says prosperity like a flawless five-carat diamond. In the film *Roman Holiday*, Audrey Hepburn, as a young princess, is bejeweled in diamonds, the stone preferred by royalty and the social elite. Why are diamonds considered by some people to be a lifelong investment? You might guess rarity, but there are many gemstones that are less abundant than diamonds, but still far less expensive. Or you might guess that diamonds are special because they are clear, colorless, and shimmering. But today it is possible to manufacture beautiful diamond fakes that can fool even a veteran jeweler. The true test of a genuine diamond is its toughness. Nothing natural is harder or more durable. If you have a bona fide diamond, it will far outlast any other gemstone.

In this chapter, we shall look at the special properties of diamond. In particular, we shall explore the nature of covalent bonds, which can be very weak or very strong (as in the case of diamond). After looking at covalent bonds in other natural

forms of carbon besides diamond, we shall look at how these bonds are arranged in various molecules. What holds a molecule together? Why does every molecule have a distinctive three-dimensional shape? How can we predict what that shape will be? We shall answer these questions as we survey the chemistry of carbon and the nature of the covalent bonds formed by it and by other elements.

3.1 | Why Is Carbon Special?

The arrangement of the periodic table helps us predict the properties of elements. For example, we expect all the halogens (column 17) to undergo similar chemical and physical changes, and they do. They are all very reactive and pull electrons away from other atoms. Trends predicted by the periodic table are typically reliable, but anomalies occur in period 2, as shown in Figure 3.1, which plots atomic size versus atomic number for some of the group 14 elements. As you can see, carbon is smaller than you would expect based on the size trend for silicon, germanium, and tin. In fact, this is also true for all the elements from boron to neon (period 2): their size is smaller than predicted. What causes this anomaly?

An atom is very small when it has only a few electrons and those electrons are held close to the nucleus, which is the case for atoms of the period 2 elements. In atoms farther down in a group, there are additional electron layers farther from the nucleus, and these outer electrons feel less pull from the positive charge of the nucleus. As a result, an element *other than* the top element in a group defines the size trend of the atoms in the group. Chemists refer to this phenomenon as the **uniqueness principle**: the period 2 elements are unique because of their paucity of electrons and because of the proximity of those electrons to the nucleus.

QUESTION 3.1 On the vertical axis of Figure 3.1, atomic size is expressed in picometers. Atomic size is also sometimes expressed in angstroms ($1 \text{ Å} = 10^{-10}$ m) and nanometers. Express the atomic size of carbon in these units.

ANSWER 3.1 There are 10^{12} pm in 1 m and 10^{10} Å in 1 m. Therefore, a picometer is a smaller length than an angstrom (100 pm fit into 1Å). According to Figure 3.1, the atomic size of carbon is about 77 pm, and so the conversions are

$$77 \text{ pm} \times \frac{1 \text{ Å}}{100 \text{ pm}} = 0.77 \text{ Å}$$

$$77 \text{ pm} \times \frac{1 \text{ m}}{10^{12} \text{ pm}} \times \frac{1 \text{ nm}}{10^{-9} \text{ m}} = 0.077 \text{ nm}$$

Because it is small, a carbon atom is able to approach other atoms closely, sharing electrons in a covalent bond. When a carbon atom bonds to other carbon atoms, the atoms are especially close, and the bond is especially strong. As a result, carbon atoms tend to form long chains, and the chains may form rings. When a carbon atom makes bonds to four other atoms, the four bonds point to the four corners of a **tetrahedron**—a pyramid made up of four faces, with each face an equilateral triangle, as shown in Figure 3.2. As we shall see throughout this text, the tetrahedron is a very common structural motif in carbon-containing molecules.

RECURRING THEME IN CHEMISTRY The structure of carbon-containing substances is predictable because each carbon atom almost always makes four bonds to other atoms.

QUESTION 3.2 In any chemistry course, it is important to know how to draw tetrahedral structures in three-dimensional perspective. Describe what makes the tetrahedron on the right in Figure 3.2 look three dimensional, and practice drawing this perspective. What would the bonding geometry around the central carbon look like if depth were not added to the drawing?

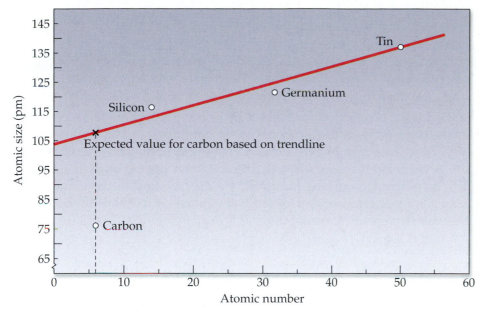

Atomic size of four group 14 elements as a function of atomic number. The trendline was determined from the sizes of the elements silicon, germanium, and tin.

ANSWER 3.2 Two of the four bonds in a tetrahedral structure are drawn in the plane of the paper. No three-dimensional perspective is required for them, and so they are shown as two regular lines, one running from C to vertex 1 and the other running from C to vertex 4. The other two bonds are drawn in the third dimension. One goes back into the page, away from the viewer; this is indicated by a dashed line. The other comes out of the paper, toward the viewer; this is indicated by a solid wedge that gets larger at the end closer to the viewer.

The bond angles for a tetrahedron are all 109.5 degrees. If, instead of using this wedge-and-dash method to add depth, you used regular lines for all four bonds, the molecule would look flat, with the four bonds all in the plane of the page and all at 90 degrees to one another.

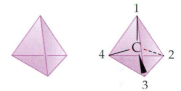

▲ FIGURE 3.2 **A tetrahedron.** Left: Equilateral triangles form the four faces of a tetrahedron. Right: A tetrahedron with a carbon atom inside. When this atom bonds to four others, each of the others is located at a vertex of the tetrahedron.

MOLECULE
Diamond

▼ FIGURE 3.3 **The three-dimensional structure of diamond.** Each carbon atom is bonded to four others.

Diamond's structure, depicted in Figure 3.3, is an extraordinary example of covalent bonding. Notice that, except for those atoms around the perimeter of the drawing, each carbon atom makes a covalent bond with four other carbon atoms. This drawing shows only a small part of the whole. In a real diamond, there are millions of other carbon atoms extending out from this structure in all directions. A solid that has an extended system of repeating covalent bonds, like diamond, is referred to as a **network solid**. Network solids are generally very robust because, as the pattern of covalent bonding repeats, the strength of the overall structure increases. Its three-dimensional network of covalent bonds is what makes diamond the hardest natural substance known.

The extraordinary toughness of diamond makes it reluctant to undergo any type of chemical reaction. All of its carbon atoms are "satisfied" in the sense that they all have a full complement of electrons around the nucleus. To see how this is so, consider one carbon atom. As Figure 3.4 shows, it has four valence electrons and so needs to share four more in order to achieve the ideal: an octet like that in the nearest noble gas, neon. When a carbon atom is surrounded by four other carbon atoms, each of the four valence electrons in the central atom pairs up with one valence

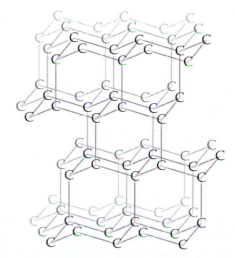

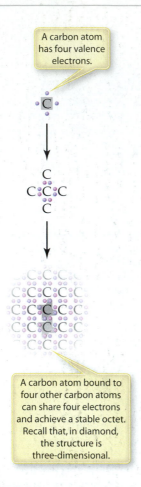

A carbon atom has four valence electrons.

A carbon atom bound to four other carbon atoms can share four electrons and achieve a stable octet. Recall that, in diamond, the structure is three-dimensional.

▲ FIGURE 3.4 **Covalent bonds in carbon atoms.**   Carbon atoms have four valence electrons. The electron dot structure of a carbon atom has one dot placed at the midpoint of each side of an imaginary square surrounding the atomic symbol. In order to have an octet of electrons, a carbon atom may share one electron with each of four other carbon atoms. In this rendering, each covalent bond is represented by a blue dot and a purple dot. The blue dot represents one valence electron from one of the atoms making the bond, and the purple dot represents one valence electron from the other atom in the bond.

electron from one of the surrounding atoms, forming a covalent bond made up of two shared electrons. This sharing allows each carbon in the extended network of diamond to have eight valence electrons, the ideal number. Of the eight valence electrons around each carbon, four "belong" to that carbon and four come from the atoms on the other end of the covalent bonds.

The octet of electrons achieved by every carbon atom of diamond makes diamond inert like a noble gas. In both, each atom has an octet of valence electrons and therefore has no reason to undergo a chemical reaction. A chemical reaction, after all, occurs when the circumstances of the atom or molecule could be improved by the reaction. In the case of diamond, its ideal arrangement of electrons makes it resistant to acids and other reactive chemical compounds and explains why, at least in theory, a diamond is forever.

QUESTION 3.3   The covalent bonding in silicon, found just below carbon on the periodic table, is similar, but not identical, to the bonding in carbon. Discuss one similarity and one difference you might expect.

ANSWER 3.3   Because silicon is just below carbon on the periodic table, you know it has the same number of valence electrons: four. Therefore, you can expect silicon to make four bonds to other atoms, just as carbon does. (This similarity can be extended down the periodic table. Every element in a group should share some properties with every other element in the group because they all have the same number of valence electrons.) You know carbon, being a period 2 atom, is anomalously small relative to the other elements in column 14. Therefore, you should expect to see differences between carbon and silicon as a result of this size disparity. Silicon cannot approach other atoms as closely as carbon can, and so bonds involving silicon atoms are longer and often weaker than those involving carbon atoms.

3.2 | Do Diamonds Really Last Forever?

Diamond's resistance to virtually every kind of physical or chemical stress may make it seem like a boring substance with nothing but looks to recommend it. Over the last two decades, however, diamond has been a trendy research subject for engineers and materials scientists. Besides being resistant to both chemicals and physical abuse, diamond is one of the best-known conductors of heat and has very low "stiction" (a coined word indicating a tendency to *stick* and cause *friction*). These properties make diamond ideal as a coating for artificial human joints. The low stiction means the surfaces of diamond-coated joints slide across each other easily. Diamond's hardness means that the coating wears down very, very slowly, and diamond's resistance to chemical reaction means that its presence in the body does not elicit an allergic response in patients.

Diamond is also the material of choice for computer components that must be good thermal conductors. The design of computers and other electronic devices is often limited by the spacing between components, because the heat produced must somehow be dissipated. Because a diamond chip is a much more efficient heat sink than a silicon chip, we can expect to see electronic and mechanical devices becoming smaller and smaller as silicon components are replaced with diamond. Figure 3.5 shows a measuring device known as a vernier caliper built at Argonne National Laboratory. It was the first microelectronic machine made entirely of diamond.

Its inertness makes diamond a desirable engineering material with nearly unlimited uses. Why, though, would anyone want to force a diamond to undergo a chemical reaction? Ask this question of Stacey Bent (Figure 3.6), and she will tell you that diamond would be even more useful if its surface could be modified. Bent, a chemical engineer at Stanford University, has used extremely low pressure to "trick" diamond into adding a small organic molecule to its surface (Figure 3.7). This is the first time diamond has been coerced into reaction, and this first demonstrated reaction paves the way for further manipulation of diamond's surface. The appended organic molecules might be made to act as tethers for attaching diamond to other molecules. Or diamond might become a newfangled solid support for organic reactions. The opportunities are endless.

If diamond is one of the most inert substances around, how is it possible to make it undergo a chemical reaction? In order to answer this question, consider the surface of diamond for a moment. What happens when the extended network of tetrahedral covalent bonds ends at the surface? Do the bonds stick out into space with nothing attached? The answer is no. You already know every carbon atom must have four bonds, and this restriction applies at the diamond surface. Each atom cannot continue in a tetrahedral bonding arrangement, though, because that would extend the diamond infinitely. The answer to this conundrum lies in the fact that, although carbon must make four bonds, those bonds do not have to be to four other atoms. Carbon, and many other atoms, can make not only **single bonds** (in which two electrons are shared by two atoms), but **multiple bonds** as well.

Recall that the uniqueness principle says that period 2 elements are very small. Thus, they can get very close to one another, and this is what makes multiple bonding between carbon atoms possible. The multiple bond known as a **double bond** (represented by two lines between atoms) is composed of a pair of bonds, each containing two electrons, for a total of four shared electrons. The multiple bond known as a **triple bond** (represented by three lines between atoms) is composed of a triplet of bonds, each containing two electrons, for a total of six shared electrons. Figure 3.8 shows examples of carbon-containing molecules containing single and multiple bonds. Notice that, in both molecules shown, every carbon atom makes a total of four bonds.

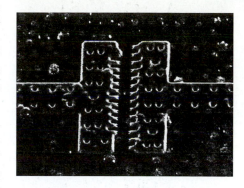

▲ FIGURE 3.5 A vernier caliper fashioned out of pure diamond.

▲ FIGURE 3.6 Diamond researcher Stacey Bent. Professor Stacey Bent, a chemical engineer at Stanford University, is working on the modification of diamond surfaces.

Heavier elements—molybdenum (Mo, element 42) is one example—are known to form quadruple bonds!

QUESTION 3.4 The structure of the precursor to vitamin A is shown in Figure 3.8a. (a) How many single bonds between carbon atoms does it contain? (b) How many double bonds between carbon atoms? (c) How many triple bonds between carbon atoms? (d) What is the chemical formula for this molecule? (e) What is its molar mass to two decimal places?

ANSWER 3.4 (a) 15, (b) 4, (c) 1, (d) $C_{20}H_{28}O$, (e) 284.44 g/mol.

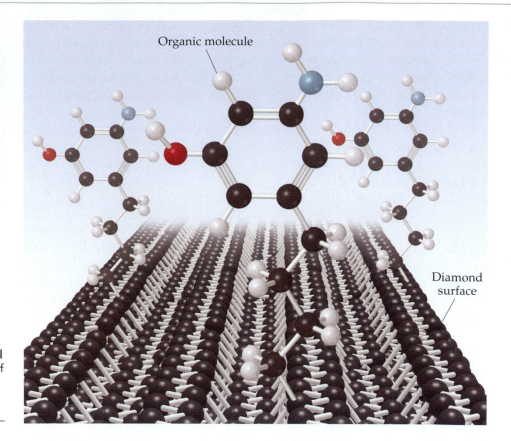

▶ **FIGURE 3.7** **Diamond in a chemical reaction.** A schematic representation of several molecules of the same organic compound attaching to the surface of a diamond.

▶ **FIGURE 3.8** **Multiple bonds in carbon compounds.** Notice that no matter whether the bond is single, double, or triple, each carbon always makes four bonds in organic compounds.

LIVE ART
Multiple Bonds

(a)

(b)

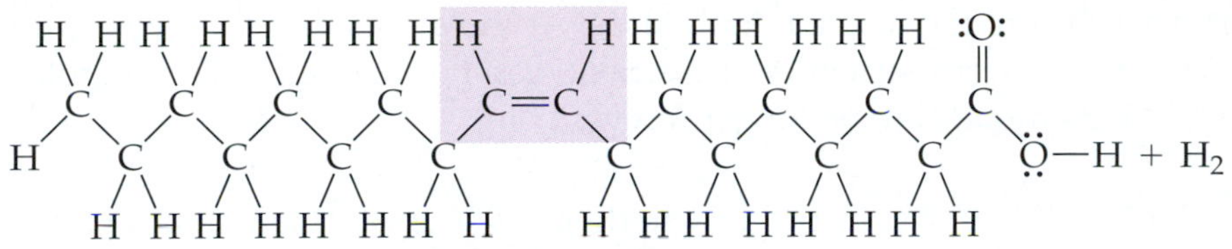

Oleic acid (an unsaturated fat)

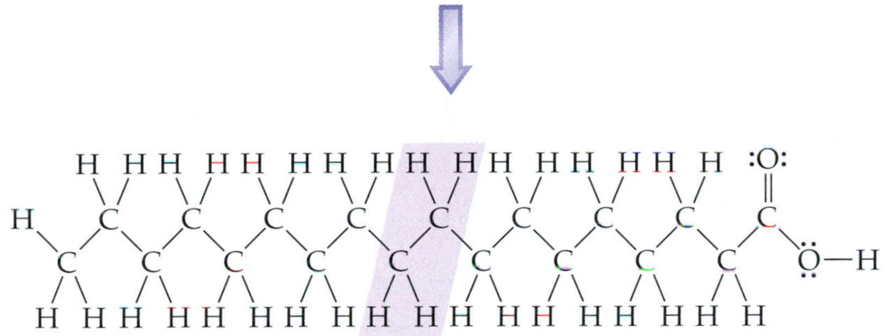

Stearic acid (a saturated fat)

MOLECULE
Oleic Acid

◀ **FIGURE 3.9 An addition reaction in a fat.** Oleic acid, an unsaturated fat, reacts with hydrogen, H_2, which places one hydrogen on each of the atoms of double bond. The result is stearic acid, a saturated fat.

QUESTION 3.5 Acrylonitrile is the precursor to Orlon, a synthetic material used to make carpet fibers, among other things. The structure of the acrylonitrile molecule is

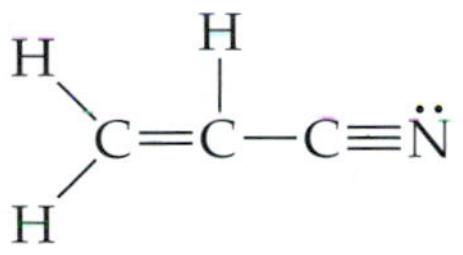

(a) How many single bonds between carbon atoms does it contain? (b) How many double bonds between carbon atoms? (c) How many triple bonds between carbon atoms? (d) What is the chemical formula for this molecule? (e) What is its molar mass to two decimal places?

ANSWER 3.5 (a) 1, (b) 1, (c) 0, (d) C_3H_3N, (e) 53.06 g/mol.

Bent's diamond research gives us an opportunity to introduce a new chemical reaction: the **addition reaction**. In an addition reaction, double bonds are changed to single bonds. Let's consider a simple example. Oleic acid, the primary constituent of peanut oil, contains one double bond. It can be forced to undergo an addition reaction with hydrogen, H_2, as shown in Figure 3.9. We refer to a molecule containing one or more multiple bonds as **unsaturated**, because hydrogen atoms can be added to the multiple bonds to make them into single bonds. In other words, because it can accept additional H atoms, the molecule containing the multiple bond is *unsaturated* with respect to H atoms. A molecule containing only single bonds is called a **saturated** molecule because it will not accept any more H atoms. Thus, this addition reaction begins with oleic acid, an unsaturated molecule, and the product is a saturated molecule, stearic acid.

Bent did the same thing with diamond in a process represented in Figure 3.10. At the surface of diamond, double-bonded carbon atoms "cap off" the ends of tetrahedra and cover the diamond surface. Because the surface has double-bonded carbon atoms, Bent was able to use an addition reaction to break open these bonds and attach new atoms to the surface, proving that diamonds are not as inert as we once thought they were. To the naked eye, this new diamond looks the same

LIVE ART
Addition Reaction

▼ **FIGURE 3.10 An addition reaction on the surface of diamond.** The molecule to be added, represented by the symbol R–H, is an organic molecule containing a C–H bond. The result is a saturated product with R and H attached to it.

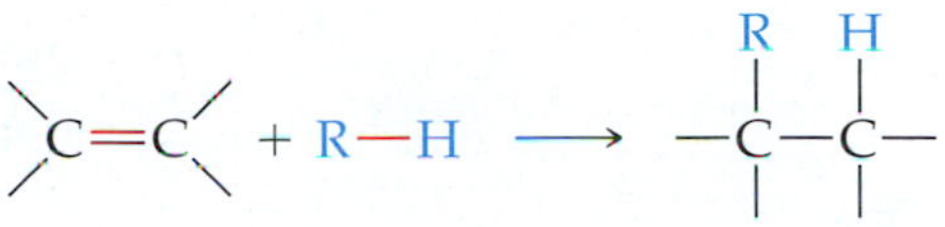

| Double bond on diamond surface | Organic molecule to be attached | Single bond between carbons on diamond surface with new organic molecule R attached |

as it did before the reaction took place. At the atomic level, however, it is a different story: the structure of diamond can be changed and manipulated. Indeed, diamonds are no longer forever.

QUESTION 3.6   Indicate whether each molecule is saturated or unsaturated:

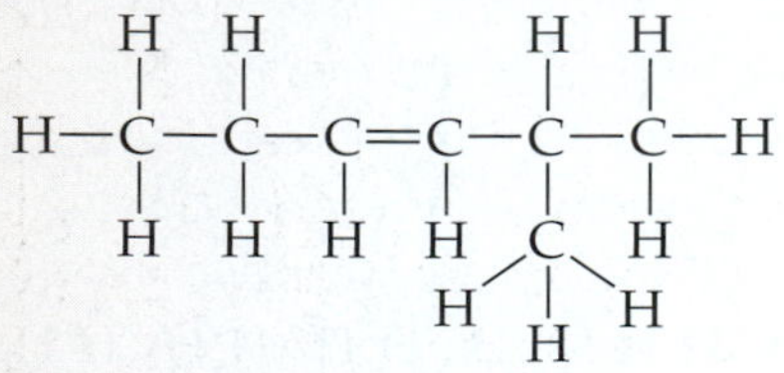

$$H-\underset{\underset{H}{|}}{\overset{\overset{H}{|}}{C}}-\underset{\underset{H}{|}}{\overset{\overset{H}{|}}{C}}-C=C-\underset{\underset{\underset{H}{|}}{\overset{H}{C}}}{\overset{\overset{H}{|}}{C}}-\underset{\underset{H}{|}}{\overset{\overset{H}{|}}{C}}-H$$

(a)

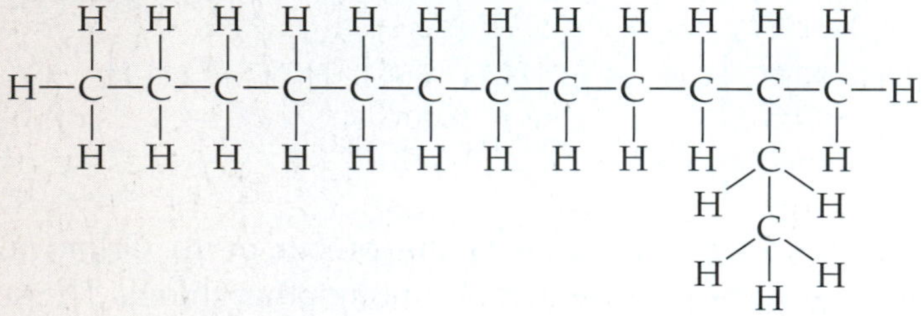

$$H-\overset{H}{\underset{H}{C}}-\overset{H}{\underset{H}{C}}-\overset{H}{\underset{H}{C}}-\overset{H}{\underset{H}{C}}-\overset{H}{\underset{H}{C}}-\overset{H}{\underset{H}{C}}-\overset{H}{\underset{H}{C}}-\overset{H}{\underset{H}{C}}-\overset{H}{\underset{H}{C}}-\overset{H}{\underset{H}{C}}-\overset{H}{\underset{C}{C}}-\overset{H}{\underset{H}{C}}-H$$

(b)

$$H-\overset{H}{\underset{H}{C}}-\overset{H}{\underset{H}{C}}-\overset{H}{\underset{H}{C}}-\overset{H}{\underset{H}{C}}-C\equiv C-H$$

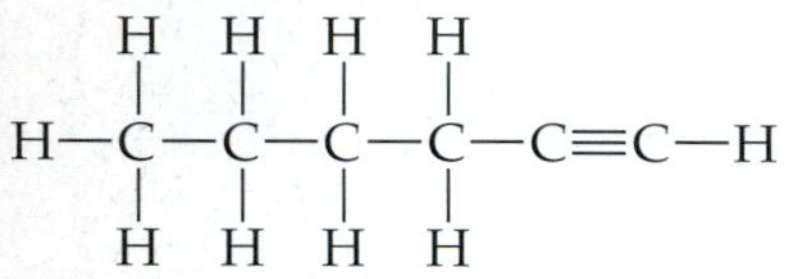

(c)

ANSWER 3.6   (a) unsaturated, (b) saturated, (c) unsaturated.

3.3 | Graphite and the Multiple Bond

So far, we have considered only one form of carbon: the hard, rigid substance we know as diamond. You need look no farther than the pencil on your desk, however, to spot another form: the graphite in the pencil's center—the black material commonly referred to as pencil "lead." Like diamond, graphite contains only carbon atoms. This substance is very different from diamond, however. It is a greasy, black solid so slippery that it makes an excellent industrial lubricant. We say that diamond and graphite are allotropes of carbon, and an **allotrope** is defined as any one of two or more different forms of an element.

If, like diamond, graphite is pure carbon, how can it have such a radically different look and feel? The answer lies in the covalent bonds that hold graphite together. Figure 3.11 shows a schematic drawing of graphite. What is the most striking difference between diamond's structure (Figure 3.3) and that of graphite? The most noticeable difference, besides the presence of double bonds, is that all the bonds in graphite are in the same plane (the plane of the page). In contrast, the bonds in diamond are in many planes. In other words, graphite's structure is a flat, two-dimensional sheet, and diamond's structure is three dimensional. This is a natural result of the chemistry of carbon, which makes four bonds. In graphite, each carbon is bound to only three other car-

MOLECULE
Graphite

▼ **FIGURE 3.11   The structure of graphite.**   All the carbon atoms in this sheet of graphite are in the same plane. Each carbon atom makes a total of four bonds, two single and one double.

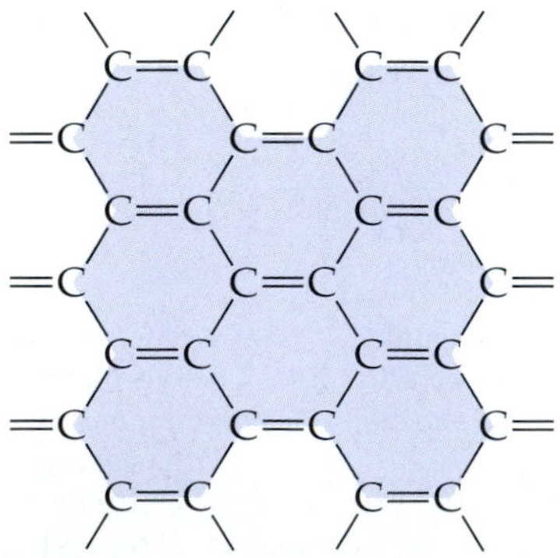

TABLE 3.1   Bond Energies and Lengths in Carbon–Carbon Bonds

Bond type	Bond energy (kJ/mol)	Bond length (pm)
Single C–C	376	154
Double C=C	611	133
Triple C≡C	835	120

bons via one double and two single bonds. Therefore, each carbon in graphite does not have a fourth carbon to which it can bond in the third dimension.

From what we know about diamond and how its structure dictates its physical properties, what should we be able to predict from the structure of graphite? Do the double bonds in graphite make it stronger or weaker than diamond? Consider Table 3.1. The strength of a bond, or **bond energy**, is measured in energy units, and the SI unit for energy is the **joule** (J). The table shows the bond energies for single, double, and triple bonds between carbon atoms, and the distance between atoms in each bond, referred to as **bond length**. What do the data say about the relative strengths of single and double bonds? Clearly, a double bond is stronger, which means that the amount of energy needed to break a double bond is more than the amount needed to break a single bond. What happens to the distance between carbon atoms as we go from a single bond to a double bond? The data tell us that the bond shortens. These trends continue when you consider double versus triple bonds: triple bonds are stronger and shorter.

You can think of the joule as a unit similar to the Calorie, which is a unit used to measure the energy in food. The Calorie we are familiar with (capital C) is actually a kilocalorie (kcal), which is 1000 calories (small c): 1 Cal = 1000 cal = 1 kcal. Because 1 cal = 4.184 J, the energy in a 800-Cal (800,000-cal!) bowl of ice cream is equivalent to about 3×10^6 J = 3000 kJ.

QUESTION 3.7 The energy values in Table 3.1 are kilojoules of energy contained in 1 mol of bonds. Calculate the energy of one carbon–carbon triple bond, reporting your answer in joules.

ANSWER 3.7 You must use Avogadro's number to convert from the per-mole value given in the table to a per-bond value. There are 6.02×10^{23} individual C≡C bonds in 1 mol of C≡C bonds:

$$\frac{835 \text{ kJ}}{1 \text{ mol bonds}} \times \frac{1 \text{ mol bonds}}{6.02 \times 10^{23} \text{ bonds}} \times \frac{1000 \text{ J}}{1 \text{ kJ}} = \frac{1.39 \times 10^{-18} \text{ J}}{1 \text{ bond}}$$

Let's look more closely at the differences between single, double, and triple bonds. We know that each carbon contributes four electrons to all bonds it makes, and it must share electrons to get the other four it needs. Figure 3.12a shows one carbon bound to four other atoms (or groups of atoms), which are given the symbols R_1, R_2, R_3, and R_4. Figures 3.12b and 3.12c show one carbon bound to three other atoms (R_1, R_2, and R_3) and one carbon bound to two other atoms (R_1 and R_2), respectively. In order to accommodate the four-bonds-to-carbon rule, we must have double and triple bonds in b and c.

We can draw several conclusions from Figure 3.12. First of all, carbon can satisfy the four-bonds-to-carbon rule by bonding to two, three, or four other atoms. When it bonds to four other atoms, all the bonds must be single. When it bonds to three other atoms, one of the bonds must be double and the others single. When it bonds to two other atoms, one of the bonds must be triple and the other single. Second, we can see by looking at the numbers of electrons in each bond in Figure 3.12 and the bond strengths in Table 3.1 that the number of electrons in a bond is directly proportional to the strength of the bond. In fact, electrons are the "glue" that holds two atoms together in

▼ FIGURE 3.12 **Bonding patterns in carbon-containing molecules.** Carbon can bond **(a)** to four other groups by making single bonds to each, **(b)** to three other groups by making two single and one double bond, and **(c)** to two other groups by making one single and one triple bond. The bottom drawings show the valence electrons from the two atoms participating in each bond. These drawings remind you that each line in a chemical bond stands for two electrons.

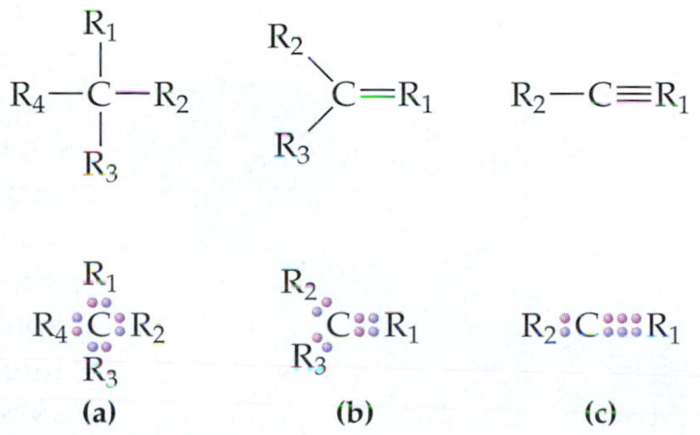

▶ **FIGURE 3.13** **Multiple sheets of graphite stacked on one another.** The individual sheets are held together by very weak noncovalent forces that allow the sheets to slide across one another easily.

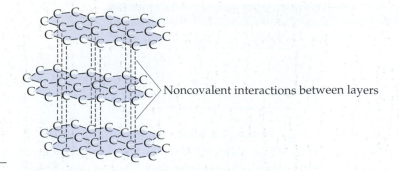

Noncovalent interactions between layers

Astute readers will wonder why carbon cannot make two double bonds instead of a triple and a single. The answer is that carbon does do this, and the resulting compound is called an *allene*. Allenes do exist, but they are very rare and usually unstable. Alkynes—compounds with triple bonds—are more common and more stable than allenes.

a chemical bond: the more glue, the stronger the bond. As the number of electrons in a bond increases, the bond gets stronger, and a stronger bond means that the electrons are pulling the two atoms closer together. Therefore, more bonds between two atoms mean shorter bond length. This is true for all sorts of chemical bonds, not just those between two carbon atoms.

Now, let us return to the structure of graphite—a soft solid that is both flaky and greasy—and diamond—a superhard, transparent crystal. Do these differences in physical properties make sense in light of the structures of these two carbon allotropes (Figures 3.3 and 3.11)? The easy answer is no. Graphite contains double bonds, but diamond does not (except on its surface). Therefore, we might expect graphite to be the tougher material. Consider the three-dimensional structure of graphite, however. The single layer shown in Figure 3.11 is only one flat sheet of many. These sheets are stacked on top of one another to build up a three-dimensional solid, as shown in Figure 3.13.

What holds the sheets of graphite together? Apparently very little. Every carbon-to-carbon covalent bond in graphite lies *within* an individual sheet. There is no way for a carbon atom in a given sheet to make covalent bonds to atoms in the sheets above or below. Therefore, the sheets are held together only by very weak, intermolecular forces (which we shall discuss in Chapters 4 and 7). Graphite certainly has strength *within* each flat sheet, but these sheets cannot bind covalently to one another. Therefore, the sheets slide across one another easily, providing graphite with its most useful attribute—lubricating ability. (The tube of lubricant shown in Figure 3.14 lists graphite as its first ingredient.) Diamond, on the other hand, has bonds in three dimensions rather than two, and the result is an extremely tough lattice that can withstand tremendous physical stress.

▼ **FIGURE 3.14** **Slippery stuff.** Lubricants often contain graphite because of its slipperiness.

QUESTION 3.8 Imagine a new allotrope of carbon has been found, one having alternating single and triple bonds. What can you say about its structure? Will it be as strong as diamond? How do you know?

ANSWER 3.8 Alternating single and triple bonds mean the structure is

$$\ldots \equiv C - C \equiv C - C \equiv C - C \equiv C - C \equiv C - C \equiv C -$$
$$C \equiv C - C \equiv C - \ldots$$

The carbon atoms in this chain are unable to bond to anything else because each one already has four bonds. Therefore, the bonds in this molecule form a line (as opposed to the bonds in graphite, which form a sheet). Knowing that triple bonds are stronger than single bonds, we can speculate that, along its long axis only, this new molecule should be stronger than diamond. However, because it does not have strong bonds in three directions, as diamond does, it will not be as hard.

QUESTION 3.9 From what you know about the structure of graphite, do you think it would be easy or difficult to append organic molecules to it the way Stacey Bent did with diamond?

ANSWER 3.9 Easy, because its double bonds can undergo addition reactions.

We now know that a double bond between two carbon atoms is stronger than a single bond between the two atoms, and that a triple bond between carbons atoms is stronger than either a single or a double bond. In fact, bonds also differ in strength as a function of the identity of the two atoms involved, which can be either two atoms of the same element or atoms of two different elements. For example, the N–H single bond of ammonia is stronger than the C–H bond in methane. And C–H bonds generally are stronger than C–C single bonds, which are stronger than C–N single bonds. The fact that organic molecules contain bonds of different strengths has been exploited over and over in chemical reactions. If you hit an organic molecule with enough energy (in the form of heat or light), its bonds break, and the weaker ones break first.

Chemists have been doing this for years—breaking the weakest bonds in a molecule so that a piece of the molecule can be used for something else. For example, in Figure 3.15, the reactant on the left side of the arrow breaks into two smaller molecules in the presence of light. The bond that breaks is indicated in red. For reasons that go beyond the scope of this book, this is the weakest bond in the molecule and therefore the first to break when light is shone on it. This is the way chemists do chemistry—knowing they can probably break the weakest bond, they use that information as the basis for a strategy to create a new molecule—until recently, that is.

Scientists at Germany's Wurzberg University have come up with a way to break bonds that *are not* the weakest ones in a molecule. To do this, they use a femtosecond source of light whose flash lasts 10^{-15} second. The light, which is tuned to a very specific energy, breaks bonds in a target molecule, and the products are analyzed by computer. If the computer analysis says the desired products were not formed (in other words, if the desired bond was not broken), the light is adjusted to a very slightly different energy and the process is repeated. The products are again analyzed, and if they are still not what the researchers want, the light is adjusted again and again by tiny increments until the desired compounds are produced. In this way, the scientists are able to converge on breaking one specific bond in the target molecule.

▼ **FIGURE 3.15** **Breaking bonds in a molecule.** The bond shown in red is weaker than the other bonds in the reactant molecule. When energy is added in the form of light, this bond breaks, and the molecule is split into two pieces.

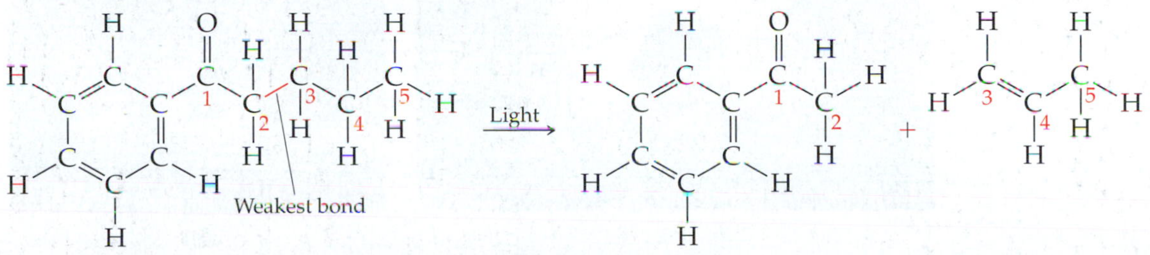

H
|
H   C   H
\ ‖ /
C   C
‖   ‖
C   C   Strongest bond
/   \   /   CO
H   C   Fe
‖   ⋮   One of these bonds broken
H   Cl   CO
|
H

Weakest bond

▲ **FIGURE 3.16 Bypassing the weakest link.** Even though the weakest bond in the molecule is the one between Fe and Cl, researchers at Wurzberg University were able to break one of the bonds that connects Fe with CO while keeping the Fe—Cl bond intact.

The molecule shown in Figure 3.16 was used in a demonstration of this technique. In an ordinary chemical reaction, you would expect added energy to break the bond between iron and chlorine because it is the weakest bond in the molecule. However, the demonstration showed that it is possible to break the bond between iron and either one of the carbon monoxide, CO, groups. With this new technology, chemists will be able to break stronger bonds in a molecule while leaving weaker ones intact, giving researchers an entirely new approach in designing new therapeutic drugs.

QUESTION 3.10 Would the technique developed by the Wurzberg researchers be useful for cleaving the bonds in diamond?

ANSWER 3.10 No, because all the bonds in diamond are approximately the same strength. This technique is useful only for molecules containing bonds of different strengths.

3.4 | Buckyballs and the Concept of Resonance

There are some ideas about nature that scientists take for granted. Look in any chemistry textbook published before 1985, and you will find this indisputable fact: carbon has only two allotropes, diamond and graphite. This was a fact of life for chemists for decades. After all, carbon is one of the best understood elements. What more can we learn about it? Periodically, though, what scientists know as fact is called into question, and the paradigm shifts. This occurred in 1985 when three scientists—Robert Curl of the United States, Harry Kroto of the United Kingdom, and Richard Smalley of the United States—found that natural carbon has other forms besides graphite and diamond. More amazing was the fact that the most abundant newly discovered form of carbon, called C_{60}, was highly symmetrical. This form became popular as much for its aesthetic appeal as for its chemical importance.

Curl, Kroto, and Smalley knew that this allotrope contained 60 carbon atoms and was extraordinarily stable, but they did not yet know its structure. It occurred to them that the number 60 is found in some well-known manufactured products, such as soccer balls and geodesic domes (Figure 3.17). Both

▶ **FIGURE 3.17 Buckminsterfullerene large and small.** **(a)** A soccer ball is constructed from alternating pentagons and hexagons. **(b)** The geodesic dome, designed by Buckminster Fuller, has the same structure as a soccer ball.

(a)

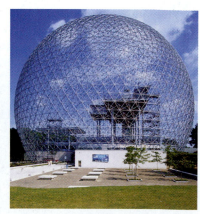

(b)

of these familiar shapes have the same structure: pentagonal rings surrounded by hexagonal rings to make a closed-cage arrangement. This is a motif found both in nature and in manufactured products, because it is the only way to arrange 60 vertices in a closed symmetrical structure. Knowing this, Curl, Kroto, and Smalley postulated such a structure for their carbon allotrope (Figure 3.18). The structure was confirmed, and the three men took home the 1996 Nobel Prize in Chemistry for their work. They named C_{60} *buckminsterfullerene* after Buckminster Fuller, the architect who patented the geodesic dome. This molecule is now affectionately known as the "buckyball."

Studies show that the C_{60} structure is much more stable than it should be. Other molecules having comparable numbers of carbon atoms are more easily broken down and more susceptible to chemical attack. Why is the buckyball so stable? As usual, we must turn to the structure of the molecule to understand its properties. Let's begin with Figure 3.19*a*, which shows a complex molecule with one carbon atom at each vertex. Except for some around the perimeter, each carbon is bonded to three others, and so each must have two single bonds and one double bond. (The perimeter atoms bonded only to two others are not really breaking the four-bonds rule. Remember, this is just a partial structure, with more hexagons and pentagons extending out in all directions.) If you can imagine holding this flat structure in your hands and gently bending the edges up into a bowl shape, you will have roughly one-half of a buckyball, as shown in Figure 3.19*b*.

The stability of the buckyball structure is a result of the network of alternating single and double bonds connecting the pentagons and hexagons. This alternating pattern is common in carbon-containing molecules because it strengthens the molecular structure. When double and single bonds alternate, the electrons in these bonds are able to spread out and diffuse over all the bonds in the region of alternation. This phenomenon, known as the **delocalization** of electrons, provides extra stability for the molecule. This extra stability is referred to as **resonance stabilization**. Resonance stabilization makes a contribution to the special stability of buckminsterfullerene, which has this alternating system of double and single bonds.

The molecule *benzene* provides a second simple example of resonance stabilization. This small organic molecule is often drawn with alternating double and single bonds that can be shown in either of two equivalent ways (Figure 3.20*a*). However, evidence reveals that the bonds are not truly double and single, but rather a hybrid of the two—more like a bond and a half, sometimes referred to as 1.5 bonds. In fact, the benzene molecule contains *identical* 1.5 bonds between each pair of adjacent carbon atoms in the ring. This even distribution of 1.5 bonds all around the benzene ring can be represented as shown in Figure 3.20*b*, with a dotted line between each pair of carbon atoms. A more frequently seen representation of benzene is the circle inside a hexagon shown in Figure 3.20*c*. The circle suggests the equality of the six carbon–carbon bonds. We shall use the circle-in-a-hexagon method throughout this book.

You may be wondering what has happened to the letters representing carbon and hydrogen atoms in the benzene structure shown in Figure 3.20*c*.

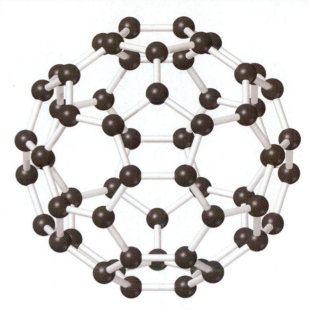

▲ **FIGURE 3.18   Buckyball, anyone?**   The structure of C_{60}, buckminsterfullerene—better known as the buckyball—is like that of the soccer ball and geodesic dome of Figure 3.17.

MOLECULE
Buckminsterfullerene

▼ **FIGURE 3.19   The bonding in buckminsterfullerene.   (a)** Hexagons surround a pentagon in this fragment of a buckyball structure. The structure is flat and lies in the plane of the page. There is a carbon atom at each vertex, and except for some of those around the perimeter, each carbon atom makes one double and two single bonds. **(b)** If the flattened structure of **(a)** is bent into a bowl shape, you can see one curved surface of the buckyball.

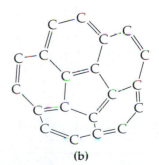

(a)

(b)

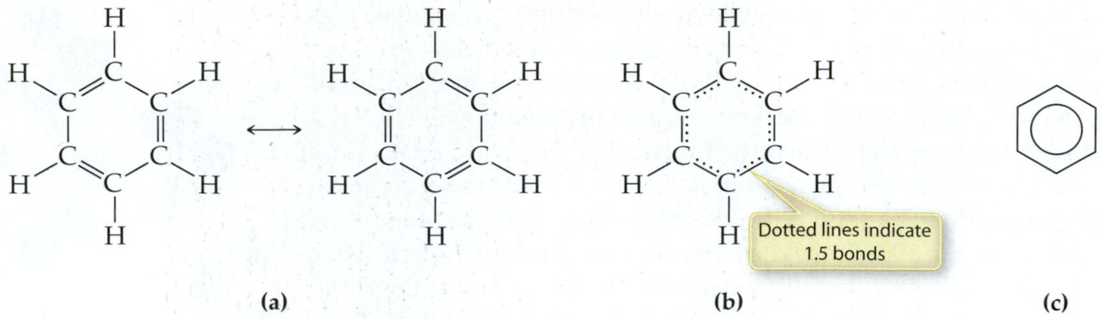

MOLECULE
Benzene

▲ **FIGURE 3.20 Representing benzene. (a)** The structure of benzene, C_6H_6, is represented by either of these two drawings. Even though the alternating double bonds are shown in different positions in the two versions, they both represent the same molecule. **(b)** Each bond between carbon atoms in benzene is not a double bond, not a single bond, but rather a "1.5 bond." This is sometimes represented by using dotted lines for the bonds. **(c)** In this text, we shall always represent the benzene molecule by a circle within a hexagon. The circle indicates that the three double bonds are "smeared" around the whole ring and equally shared by all six carbon atoms.

Dotted lines indicate 1.5 bonds

(a)   (b)   (c)

If so, for now just note that each vertex of the hexagon represents one carbon atom and as many hydrogen atoms as needed to obey the four-bonds rule. We shall discuss this shorthand way of drawing structures in Section 3.7.

QUESTION 3.11   Which of these molecules are resonance stabilized?

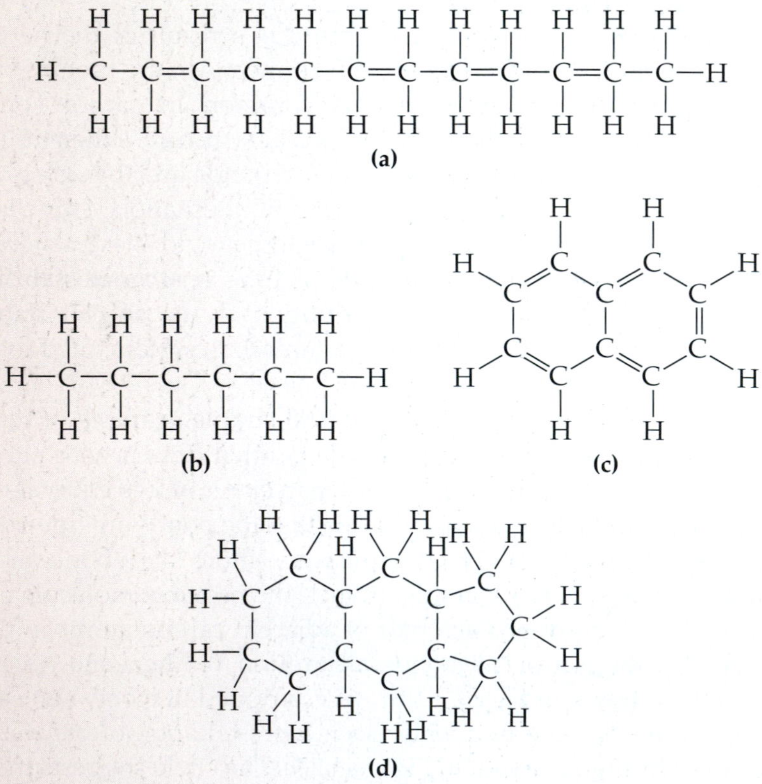

(a)

(b)   (c)

(d)

ANSWER 3.11   Parts *a* and *c* because they contain alternating double and single bonds.

QUESTION 3.12   In each molecule, identify the bonds that have delocalized electrons:

(a)

(b)

ANSWER 3.12   Regions of electron delocalization occur wherever double bonds alternate with single bonds:

(a)

(b)

3.5 | Carbon Baguettes: The Development of Nanotubes

The discovery of buckyballs in 1985 spurred the growth of a new field linking materials engineering and chemistry. Researchers began jumping on the buckyball bandwagon by the hundreds, trying to tap the limitless possibilities these caged structures offer. In 1991, scientists in Japan made the first cigar-shaped carbon cage by changing the conditions under which the carbon is deposited as the cage is created. These little hollow cages, called **nanotubes**, have been compared to a baguette loaf of bread wrapped in chicken wire and capped off at the ends. Figure 3.21 shows Alex Zettl, a nanotube researcher at the University of California at Berkeley, holding a nanotube model. Of course, nanotubes are not this big. The width of one nanotube is roughly one ten-thousandth the width of a human hair! This value is an estimate, because nanotubes form in all shapes and sizes: some bent, some

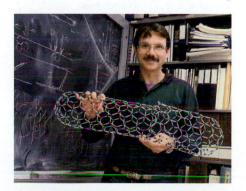

▼ FIGURE 3.21   **Nanotubes.**   Alex Zettl, a researcher at the University of California at Berkeley, holds a model of a carbon nanotube.

▲ FIGURE 3.22 **Holiday greeting from the cutting edge of science.** Aligned carbon nanotubes depicted in a cruciform shape on a holiday greeting card. Each short, curved, reddish-orange "feather" is a separate nanotube.

long, some curly. In fact, the main obstacle to making nanotubes do remarkable things has been getting them to behave themselves.

Scientists in Australia came up with a way to make nanotubes grow in an ordered fashion. They found that, when lined up end to end, nanotubes have properties that are desirable in the electronics industry. These properties are due to the long carbon skeleton of alternating double and single bonds. A notable discovery like this would ordinarily be shipped off immediately for publication in a prestigious scientific journal. Instead, the Australian scientists decided to send out the news on a holiday greeting card. Figure 3.22 shows the card, which said simply, "Aligned Carbon Nanotubes." As a result of this breakthrough, carbon nanotubes are being developed for use in flat-panel displays, as tiny diodes, as material for VELCRO®, and in miniature computer memory chips.

One of the most promising nanotube applications has come from the National Renewable Energy Laboratories (NREL) in Golden, Colorado, where scientists are using nanotubes to make car engines that run on fuel cells. Fuel cells, which were used as the energy source for the Apollo space missions, are based on a very simple chemical reaction:

$$2\,H_2(g) + O_2(g) \longrightarrow 2\,H_2O(l)$$

There is one obvious advantage to using fuel cells in cars: they produce water waste instead of the carbon dioxide and carbon monoxide produced by a gasoline engine. The problem with fuel cells is that they require large volumes of hydrogen gas, a requirement that presents a difficult engineering problem for automobile engines. The Apollo spacecraft employed enormous fuel cells weighing 500 pounds, a scale impractical for consumer use. In order for any significant amount of hydrogen to be stored in a small space (an automobile gas tank, for instance), the gas must be cooled to –253 °C, the temperature at which it liquefies. For an automobile, this would require unwieldy, high-priced refrigeration equipment.

Enter the nanotube, with its network of double and single bonds. The scientists at NREL found that carbon nanotubes, because of their double bonds, bind hydrogen gas via an addition reaction, shown as reaction 1 in Figure 3.23. When the reaction reverses, the hydrogen gas is released for use in a fuel cell (reaction 2 in Figure 3.23). The researchers were able to store enough hydrogen at room temperature to make fuel cells a practical option for the car of the future. They call their nanotube-filled reservoirs "nanotanks" and hope to coax these little carbon molecules into storing enough hydrogen to run a fuel cell car as far as a gasoline-burning car.

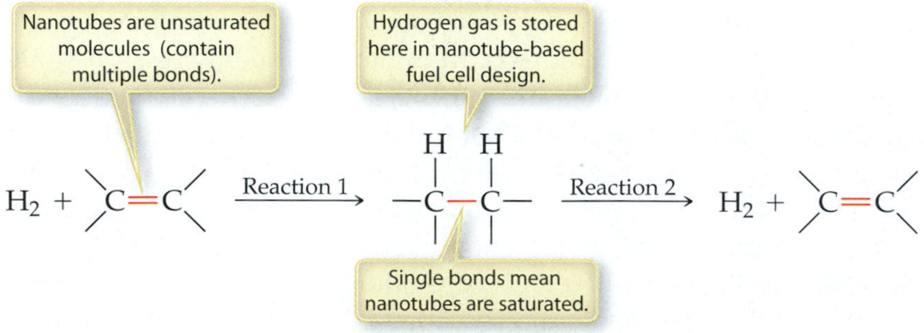

▶ FIGURE 3.23 **Nanotank technology.** Reaction 1: Carbon nanotubes, which contain double bonds and are therefore unsaturated molecules, undergo an addition reaction with hydrogen gas. One hydrogen atom is placed on each carbon atom of the C=C double bonds to convert unsaturated nanotubes to saturated ones. Reaction 2: The hydrogen is stored this way only temporarily. When the nanotank fuel is burned in an automobile engine, the H_2 is released and unsaturated nanotube molecules are regenerated.

Nanotubes are unsaturated molecules (contain multiple bonds).

Hydrogen gas is stored here in nanotube-based fuel cell design.

Single bonds mean nanotubes are saturated.

QUESTION 3.13   In a fuel cell reaction, how many molecules of water are formed from the reaction of ten molecules of hydrogen and five molecules of oxygen?

ANSWER 3.13   According to the balanced equation, you get one molecule of water for every one molecule of hydrogen. Therefore, you get ten molecules of water.

QUESTION 3.14   Comment on this statement:

> Nanotubes and diamonds are very similar structurally because both have an alternating system of double and single bonds.

ANSWER 3.14   It is true that both diamond and nanotubes have double and single bonds. However, diamond has double bonds only on its surface, with the rest of the diamond lattice containing all single bonds. Like buckyballs, nanotubes have double and single bonds alternating in a single layer of carbon atoms that enclose a hollow center.

Nanoears

BEYOND THE ORDINARY

Ears are ingenious devices. They employ projections on hair cells called stereocilia that bend as a function of sound frequency. These hair movements are translated into electrical signals that are passed along the auditory nerve to the brain, where they are recognized as sound, loudness, and pitch. Figure 3.24*a* shows a guinea pig's stereocilia sticking up, healthy and attentive. Figure 3.24*b* shows the same stereocilia after the pig has listened (without earplugs) to very loud rock music.

Hearing researchers are looking for ways to regrow hair cells that have been ruined by disease or noise. However, scientists at the Jet Propulsion Laboratory (JPL) in Pasadena, California, are taking a completely new approach to hearing research. These scientists reasoned that nanotubes are a lot like human hairs, but much stronger and much thinner. They also reasoned that nanotubes are a lot like stereocilia in that both are very good at transmitting electrical impulses. Thus, they are putting nanotubes to work as "nanoears" that pick up noises so faint that no one has heard them before—sounds in the human bloodstream.

The researchers are not even certain such sounds exist, and they admit the idea is off the wall. However, they reason that if everyday human-scale activities make sound, why not molecular-scale activities? Interactions between macromolecules in the body—for example, the docking of two large proteins—might make some kind of perceivable noise. In the words of the JPL team, "Acoustics at the molecular level is uncharted territory. There's a whole world buzzing down there, and many questions are looking for answers. Our stereocilia will begin to provide them."

The JPL workers postulate that a healthy person's biochemistry may sound completely different from that of a sick person. They plan to use nanoears to eavesdrop on cell cultures and listen in on the mechanical processes taking place inside the cell. This technology might someday give new meaning to the phrase "You don't sound so good."

▼ **FIGURE 3.24   Ear damage caused by noise.   (a)** Hair cells from a healthy guinea pig ear. **(b)** The same hair cells after the pig has listened to loud music.

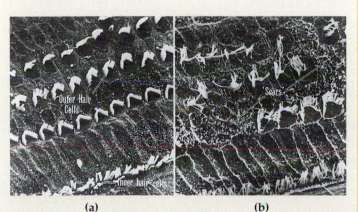

(a)   (b)

3.6 | Organic Molecules and Electron Bookkeeping

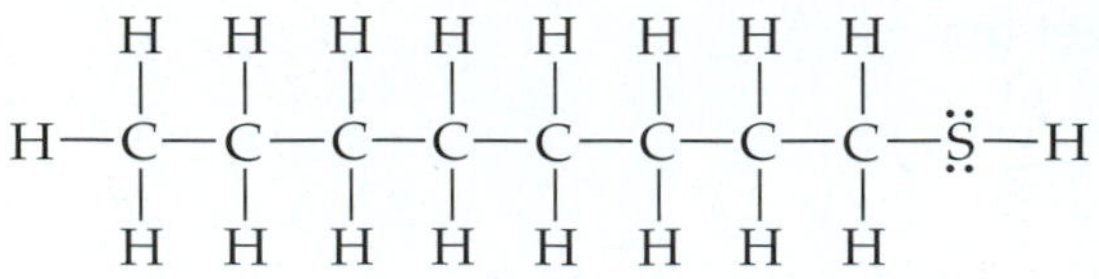

Thiol
(a)

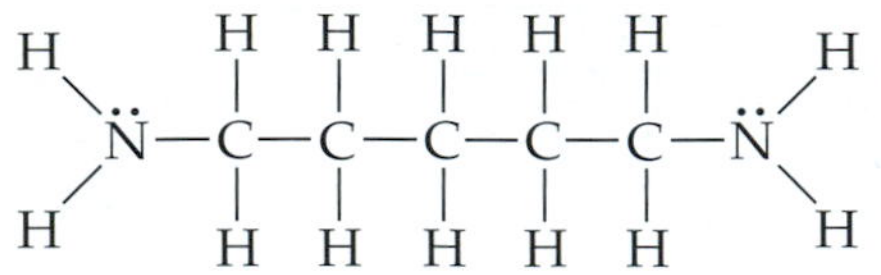

Amine
(b)

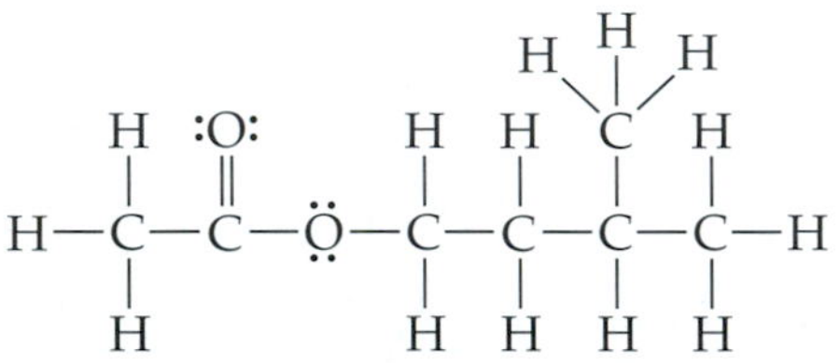

Oxygen-containing compound
(c)

▲ **FIGURE 3.25** **Some typical organic molecules.** Organic molecules can contain elements in addition to carbon and hydrogen. **(a)** A sulfur-containing organic molecule is called a *thiol*. Thiols, known for their disagreeable odor, are added to natural gas so that gas leaks can be detected by the human nose. **(b)** A nitrogen-containing organic molecule is called an *amine*. Amines are also smelly; the name of the amine shown here is cadaverine. **(c)** An oxygen-containing organic molecule. Molecules like this one are sometimes used to manufacture perfumes. This molecule, isopentyl acetate, gives bananas their characteristic scent.

RECURRING THEME IN CHEMISTRY Chemical reactions are *nothing more than exchanges or rearrangements of electrons*, which is why we pay particular attention to electrons throughout this book.

Electron dot structures are also known as Lewis structures.

So far we have seen two examples of network solids: diamond and graphite. These substances are officially classified as molecules because they contain atoms connected by covalent bonds. In the more traditional sense of the word, however, a molecule is discrete and does not continue indefinitely. For example, a buckyball is a discrete molecule because it is held together by covalent bonds, and you can draw it without indicating that it goes on continuously. The same is true for nanotubes.

The term *molecule* usually conjures up something smaller than a buckyball or a nanotube, and we have already seen several examples of smaller organic molecules, such as isooctane and butane. These two compounds are called *hydrocarbons* because they contain only hydrogen and carbon. But organic molecules can contain a limited number of other elements, most commonly sulfur, nitrogen, and oxygen. Notice in each of the molecules in Figure 3.25 that the fundamental structure is a chain of carbon atoms. This is true of all organic molecules. Much of this book is devoted to such molecules, and we shall discuss many that play a role in our lives. For now, though, let us focus on how small organic molecules are constructed.

The trick to understanding the structure and behavior of organic molecules lies in knowing the distribution of electrons in every bond, which means we must have a system for electron bookkeeping. Because chemical reactions are, in the most fundamental sense, nothing more than rearrangements of electrons, it is important to understand where the electrons are in the first place. Once we know where the electrons are in a molecule, we can predict its three-dimensional structure.

We begin with methane, CH_4, and draw its **electron dot structure**, a sketch indicating the position of all valence electrons in the molecule. The first step is to count the total number of valence electrons on all atoms in the molecule. We allot four for each carbon atom and one for each hydrogen atom. Therefore, the total number of valence electrons in a methane molecule is

$$1 \text{ carbon atom} \times \frac{4 \text{ valence electrons}}{1 \text{ carbon atom}} = 4 \text{ valence electrons from carbon}$$

$$4 \text{ hydrogen atoms} \times \frac{1 \text{ valence electron}}{1 \text{ hydrogen atom}}$$
$$= 4 \text{ valence electrons from hydrogen}$$
$$\text{Valence electron total} = 4 + 4 = 8$$

These eight electrons must be distributed among the five atoms in methane.

The next step can be tricky because you must know in advance the way in which the atoms are connected. The molecular formula will usually give you this information. For CH_4 and many other molecules, the central atom is listed first in the molecular formula. It is understood that the hydrogen atoms are connected to it like so, where each line represents a single covalent bond formed by two electrons, one electron from H and one electron from C:

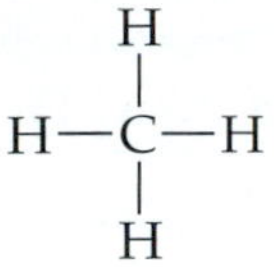

$$
\begin{array}{c}
\text{H} \\
| \\
\text{H}-\text{C}-\text{H} \\
| \\
\text{H}
\end{array}
$$

Such a drawing is called the *skeletal structure* of the molecule.

Very often, the atoms in a molecule are connected in a straight line, and the order given in the molecular formula is the order of connectivity in the molecule. For example, the formula HCN, for hydrocyanic acid, tells you that the skeletal structure for this molecule is

$$\text{H}-\text{C}-\text{N}$$

MOLECULE
Hydrogen Cyanide

QUESTION 3.15 Show how the atoms are connected in (*a*) BCl_3, (*b*) NO_2, (*c*) HSCN.

ANSWER 3.15

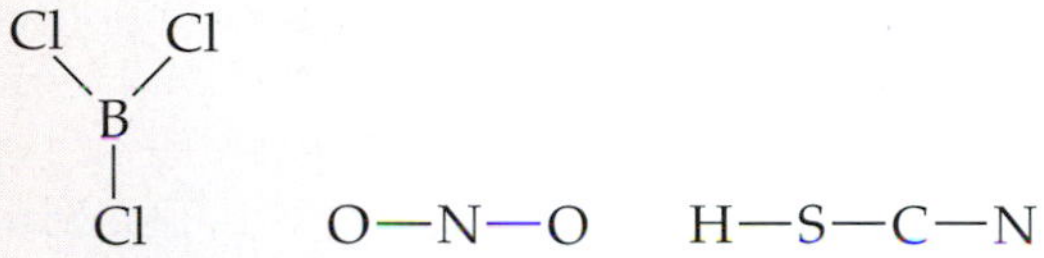

$$
\begin{array}{c}
\text{Cl} \quad \text{Cl} \\
\backslash \quad / \\
\text{B} \\
| \\
\text{Cl}
\end{array}
\qquad
\text{O}-\text{N}-\text{O}
\qquad
\text{H}-\text{S}-\text{C}-\text{N}
$$

(a)   (b)   (c)

As mentioned, each line in these skeletal structures represents two valence electrons (forming one covalent bond).

In the next step, you add up all the electrons represented by the lines in your skeletal structure and subtract this number from the total number of valence electrons. The thing to remember when drawing electron dot structures is that, except for H, every atom in a molecule must end up with eight valence electrons. (Each H must end up with two valence electrons.) As we previously calculated, methane has eight valence electrons, and the four lines in our methane skeletal structure mean all eight of them are used up in the four bonds. Therefore, you have no more electrons to place, and all you need to do is check that every atom has eight valence electrons (or a duet for H), which indeed is the case. The atoms in most organic molecules have a total of eight valence electrons when you include the electrons they share with other atoms. This concept is referred to as the **octet rule**.

In Figure 3.26, the electron dot structure for methane is drawn with purple dots symbolizing electrons from carbon and blue dots symbolizing electrons from hydrogen. Remember, though, that carbon and hydrogen do not have different types of electrons. An electron from hydrogen is the same kind of particle as an electron from carbon, or from potassium or bromine or germanium. The colored dots are just an aid as we carry out our electron-bookkeeping task. The electrons in a covalent bond do not have a memory: they are shared, and their origin is not important.

▼ FIGURE 3.26 **Electron dot structure for methane.** A carbon atom has four valence electrons, indicated by purple dots. When bound to four atoms of hydrogen, the carbon atom shares one electron from each hydrogen atom (indicated by blue dots). This gives the carbon atom an octet of valence electrons.

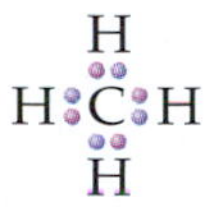

$$
\begin{array}{c}
\text{H} \\
\text{H}\overset{\cdot\cdot}{\underset{\cdot\cdot}{\text{C}}}\text{H} \\
\text{H}
\end{array}
$$

MOLECULE
Methane

QUESTION 3.16 Is this skeletal structure plausible for CH_4?

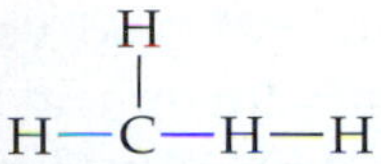

$$
\begin{array}{c}
\text{H} \\
| \\
\text{H}-\text{C}-\text{H}-\text{H}
\end{array}
$$

ANSWER 3.16 No. Carbon atoms make four bonds. Hydrogen has only one valence electron and can make only one bond to other atoms.

TABLE 3.2	Bonding Rules for Elements in Uncharged Organic Molecules
Element	**Number of bonds it always forms**
H	One
O	Two
S	Two
N	Three
C	Four

So far we have learned two things that will make drawing organic molecules easier: (1) carbon always makes four bonds, because it has four valence electrons and must share four more to form an octet; and (2) hydrogen always makes one bond, because it has one valence electron and needs only one more to form a duet. The same type of conclusions can be applied to other elements, as Table 3.2 shows. For example, oxygen and nitrogen, when present in uncharged organic molecules, always make two and three bonds, respectively. A look at the distribution of valence electrons shows you why:

$$:\ddot{\text{O}}: \quad \cdot \ddot{\text{N}} \cdot$$

You can see from these two drawings that, when drawing valence electrons for an atom, you must distribute them in a particular way around the element symbol. As an example, let us consider the period 2 elements from boron to neon. Because boron has three valence electrons, picture the symbol B as being enclosed inside a square and place the three valence electrons on three sides of the square (it does not matter which three sides you choose):

$$\cdot \dot{\text{B}} \cdot$$

Carbon is next in period 2; its fourth valence electron goes on the fourth side. Nitrogen has five valence electrons, and so we pair up the fifth one with one of the other four to make a pair. We pair oxygen's sixth valence electron with one of the other single electrons (it does not matter which one). When you follow this pattern, fluorine has three pairs of valence electrons and one single one, and the noble gas neon has four pairs:

Now it is clear why nitrogen atoms form three bonds in organic molecules and oxygen atoms form two. Nitrogen has three single electrons that must pair up with something to give nitrogen an octet. Consider ammonia, a molecule made up of one nitrogen surrounded by hydrogen. How many hydrogen atoms must there be? The answer is three because N needs three electrons and each H supplies one of them. The molecular formula for ammonia is therefore NH_3. Oxygen, as we know, makes two bonds to other atoms in uncharged molecules. This makes sense because it has two "vacancies" (in other words, two unpaired valence electrons). To fill them, it must share electrons to make an octet.

Take a look back at Figure 3.25, and verify that the rules for carbon, hydrogen, oxygen, and nitrogen are upheld for the structures shown. Note places in which **nonbonding pairs** of valence electrons (pairs that do not take part in the bonding) are indicated. They are important and are usually drawn on atoms that have them.

Let's try another electron dot structure, adding additional rules as we need them for more complex molecules. From the formula HCOOH, formic acid, the connectivity of the atoms is not obvious. In such cases, this book will always give you the skeletal structure of the molecule. For HCOOH, this structure is

$$
\begin{array}{c}
\text{O} \\
| \\
\text{C} \\
\diagup \quad \diagdown \\
\text{H} \qquad \text{O—H}
\end{array}
$$

Formic acid skeletal
structure

Does this skeletal structure represent the correct electron dot structure? To find out, we must know the total number of valence electrons:

$$1 \text{ carbon atoms} \times \frac{4 \text{ valence electrons}}{1 \text{ carbon atom}} = 4 \text{ valence electrons from carbon}$$

$$2 \text{ hydrogen atoms} \times \frac{1 \text{ valence electron}}{1 \text{ hydrogen atom}}$$

$$= 2 \text{ valence electrons from hydrogen}$$

$$2 \text{ oxygen atoms} \times \frac{6 \text{ valence electrons}}{1 \text{ oxygen atom}} = 12 \text{ valence electrons from oxygen}$$

Valence electron total $= 4 + 2 + 12 = 18$

We need 18 valence electrons in a correct electron dot structure for formic acid. The four bonds in our skeletal structure contain eight electrons, and now we must place the other $(18 - 8 =)$ 10. The rule is that, when assigning these "leftover" electrons in an electron dot structure, we begin with one of the atoms connected to the central atom and add electrons until this atom has its octet. For formic acid, we can add leftover electrons to either O atom:

Six of ten "leftover"   Four of ten "leftover"
electrons used   electrons used

(The H connected to the central C is not a choice because that H already has its duet.) If we have not used up all the electrons that need assigning, our next step is to choose another atom bonded to the central atom and continue placing the electrons:

Final four electrons used   Final six electrons used

Either way, we end up with the structure:

One line represents
two valence electrons.

Formic acid incorrect
electron dot structure

Now we have our 18 valence electrons assigned in the electron dot structure of formic acid, which is good. But there's a problem. What is wrong with the structure shown? The answer is that carbon has only six electrons around it, not the octet it must have. (This is not good.) Another problem is that the top O, although it has an octet, has only one bond to other atoms, and we know from Table 3.2 that it must make two bonds. This must also be remedied.

The solution to these two problems brings us to our next rule. If your central atom comes up short on bonds (and therefore electrons), move a nonbonding pair from some atom bonded to the central atom to make a double bond between the two. With formic acid, we have three choices. We can try to make a double bond to the hydrogen bonded to the carbon or to either oxygen. Table 3.2 tells us that H always makes one bond and O always makes two. The H bonded to C already has its one bond, and the lower O already has its two, which leaves the top O:

$$:O:$$
$$\parallel$$
$$C$$

H   Ö—H

Formic acid correct
electron dot structure

Now, each oxygen has two nonbonding pairs of electrons and makes two bonds. This is the way it should be.

We should make a final check of our formic acid structure. Does it have its full complement of 18 electrons? Yes. (Remember, when we are doing electron bookkeeping, each line of a multiple bond represents two electrons. Thus, there are four electrons in the $C{=}O$ bond.) Does our structure have the correct number of bonds to each atom? Yes: carbon has four, each oxygen has two, and each hydrogen has one, all matching the numbers given in Table 3.2.

QUESTION 3.17 Draw an electron dot structure for (a) methylene chloride, CH_2Cl_2; (b) chloroform, $CHCl_3$; (c) carbon tetrachloride, CCl_4.

ANSWER 3.17

H	:Cl:	:Cl:
:Cl—C—H	:Cl—C—H	:Cl—C—Cl:
:Cl:	:Cl:	:Cl:
Methylene chloride	Chloroform	Carbon tetrachloride
(a)	**(b)**	**(c)**

QUESTION 3.18 Draw the electron dot structure for acetic acid. The skeletal structure is

H   O
| ||
H—C—C—O—H
|
H

Acetic acid skeletal structure

MOLECULE
Formic Acid

LIVE ART
The Building of an Electron Dot Structure

ANSWER 3.18   The electron dot structure must have the correct number of bonds to each atom (Table 3.2), an octet (or duet) of valence electrons around each atom, and nonbonding pairs in the proper places:

$$\begin{array}{ccc} H & :\!\ddot{O}: & \\ | & \| & \\ H\!-\!C\!-\!C\!-\!\ddot{O}\!-\!H \\ | & & \\ H & & \end{array}$$

Acetic acid electron dot structure

If you compare the electron dot structure for acetic acid with that of formic acid, you can see that the right side of the molecule is the same in the two drawings. In fact, identical pieces of molecules appear over and over in organic chemistry. As you see more of them, you will begin to recognize these pieces.

Let's do one more example and use all our rules. As its name suggests, cyclohexene, C_6H_{10}, has six (*hex-*) carbons in a ring (*cyclo-*) shape:

$$\begin{array}{c} C\!-\!C \\ C \qquad C \\ C\!-\!C \end{array}$$

Cyclohexene skeletal structure

We know each carbon in cyclohexene must make four bonds to other atoms and each hydrogen must make one. Because each carbon in the skeleton makes two bonds, we know from Table 3.2 that each must make two more, and these additional bonds must be with hydrogen atoms. *In general, hydrogen atoms are present in organic molecules in any positions at which carbon atoms have vacancies.*

The formula C_6H_{10} tells us our cyclohexene molecule requires ten hydrogens. We can imagine distributing them in either one of two ways:

Two carbons with one hydrogen each

One carbon with no hydrogens

At this stage in our drawing, either way is okay because we have more work to do. Having assigned all ten H, we have run out of H before the two carbons shown in red can make four bonds each. Therefore, we must adjust our structure. Let us count valence electrons first. Each carbon donates four

$(4 \times 6 = 24)$, and each hydrogen donates one $(1 \times 10 = 10)$, giving a total of 34. Each of the preceding structures contains 16 single bonds, accounting for 32 electrons. Therefore, each structure is short two electrons. In the structure on the right, one of the red carbons has its four requisite bonds, leaving us with only one red carbon to work with. The solution to this problem is to continue with the structure on the left—the one in which two of the carbons have one hydrogen each—and use the two electrons to form a $C=C$ double bond:

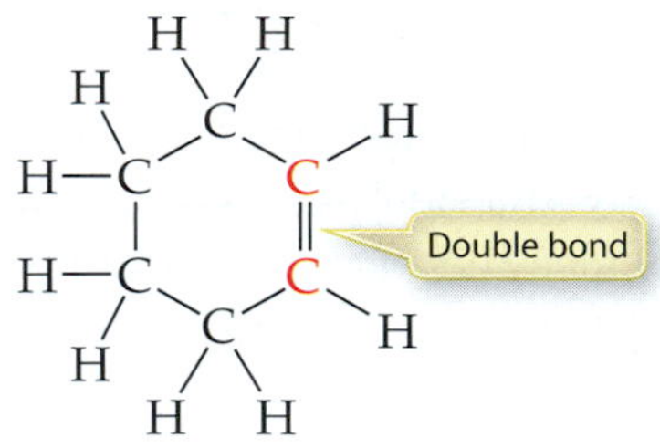

Cyclohexene

Is this structure correct? Let's check it out. Does it have its full complement of 34 electrons? Yes. (Remember, there are four electrons in the $C=C$ bond.) Does it have the correct number of bonds to each atom? Yes: four to each C and one to each H. Have we accounted for the six carbons and ten hydrogens in our molecular formula? Yes. This is the correct electron dot structure. Notice that there are no nonbonding pairs in this molecule. This is because the molecule is a hydrocarbon, and neither carbon nor hydrogen has any nonbonding pairs.

3.7 | Molecules on Your Toothbrush

Think for a moment about your toothpaste. What sorts of things does it contain? We hear a lot about the fluoride in toothpaste. Its job is to put fluoride ions into the solid core of each tooth, making the tooth stronger and healthier. Some toothpastes also contain hydrogen peroxide, touted as a gum-disease fighter, but what else is in your toothpaste? What is the bulk of the viscous goop you squirt onto your toothbrush every day?

A list of ingredients for a typical toothpaste is shown in Table 3.3. There must be an abrasive that scours the surface of your teeth. This is typically a hard solid that is ground up and suspended in the paste. Calcium

TABLE 3.3 Ingredients in a Typical Toothpaste	
Ingredient	Purpose
Active ingredient: sodium monofluorophosphate	Decay prevention
Glycerin	Moistener
Water	Consistency
Calcium carbonate	Mild abrasive
Hydrated silica	Mild abrasive
Xylitol	Flavor
Carrageenan	Thickener
Sodium lauryl sulfate	Detergent
Oil of wintergreen	Flavor

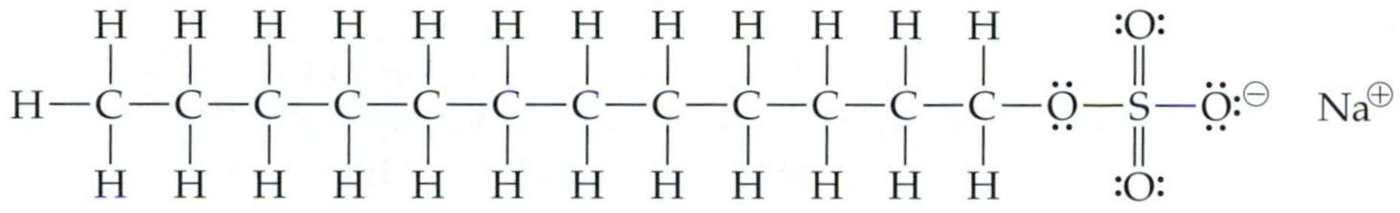

Sodium lauryl sulfate

(a)

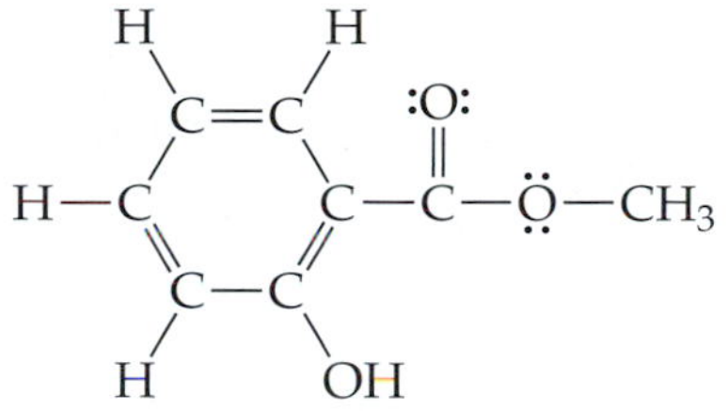

Oil of wintergreen

(b)

FIGURE 3.27 **Full structures.**
(a) The full structure for sodium lauryl sulfate, an ingredient commonly found in toothpaste. **(b)** The full structure for oil of wintergreen, a flavoring agent often found in toothpaste.

carbonate, the primary component in marble (discussed in Chapter 2), is often used as this abrasive. Also needed is some soap or detergent to pull offensive matter off your teeth. Sodium lauryl sulfate (Figure 3.27a), a long organic molecule, is usually used. The paste also needs binders for consistency and an antibacterial preservative to keep the paste from spoiling. Because all of this does not taste very good, there is bound to be a large quantity of flavoring, usually something minty, such as oil of wintergreen (Figure 3.27b). Mix these things together in the right proportions and you have toothpaste.

Look for a moment at the drawings in Figure 3.27. Try copying the two structures three times. How long did it take you? These molecules are depicted in a way that shows every single atom. Imagine how tedious it would be to draw each molecule 100 times! At some unknown point in the nineteenth century, this thought must have occurred to a chemist who, for some reason, had to draw dozens of molecules. He may have thought to himself (there were not many women chemists back then) that writing out these **full structures**, as they are called, is unnecessarily time consuming. Thus, abbreviated ways of drawing molecules were developed.

One time-saver is the **condensed structure**. Typically, organic molecules end in a **methyl group**, which is made up of one carbon and three hydrogens:

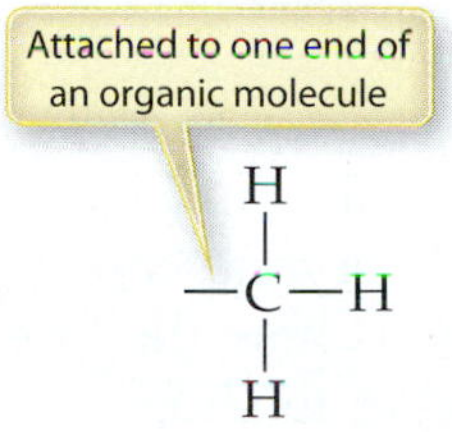

Attached to one end of an organic molecule

Attached to the end of an organic molecule

$$-\overset{\displaystyle H}{\underset{\displaystyle H}{C}}-H$$

$$-CH_3$$

Methyl group full structure

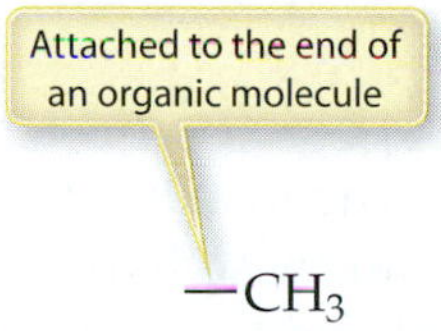

Methyl group condensed structure

The condensed form $-CH_3$ is a much more convenient way to designate a methyl group, especially if you are using a computer. A carbon in the middle of a chain in an organic molecule, known as a **methylene group**, can likewise be drawn as:

Triacontanyl hexadecanoate. See Question 3.19.

Attached to groups on both sides

$$-\overset{\displaystyle H}{\underset{\displaystyle H}{C}}-$$

Methylene group full structure

Attached to groups on both sides

$$-CH_2-$$

Methylene group condensed structure

Thus, the condensed structure of our toothpaste detergent, sodium lauryl sulfate, looks like this:

$$CH_3CH_2CH_2CH_2CH_2CH_2CH_2CH_2CH_2CH_2CH_2CH_2-\overset{..}{\underset{..}{O}}-\overset{:O:}{\underset{:O:}{S}}-\overset{..}{\underset{..}{O}}{}^{\ominus} \quad Na^{\oplus}$$

Sodium lauryl sulfate condensed structure

Even writing condensed structures can be tiresome, though, as some molecules have more than 20 carbon atoms in their chains. Therefore, we simplify things further by using a **shortened condensed structure**, in which all methylene groups are represented by one methylene unit plus a subscript indicating how many of these units there are in the molecule:

$$CH_3(CH_2)_{11}-\overset{..}{\underset{..}{O}}-\overset{:O:}{\underset{:O:}{S}}-\overset{..}{\underset{..}{O}}{}^{\ominus} \quad Na^{\oplus}$$

Sodium lauryl sulfate shortened condensed structure

QUESTION 3.19 Triacontanyl hexadecanoate is one of the main ingredients in beeswax. Its full structure is shown in the left margin (with hydrogen atoms left off to avoid clutter). Draw the shortened condensed structure, and in that structure show all nonbonding pairs of electrons.

ANSWER 3.19

$$CH_3(CH_2)_{29}-\overset{..}{\underset{..}{O}}-\overset{:O:}{C}-(CH_2)_{14}CH_3$$

Our weary chemist, trying to save more time, may have thought, "Why am I drawing all these hydrogen atoms? Everyone knows they are there. Why should I have to include them?" This is how the **line structure** was born, in which, in addition to not bothering to show any H, we do not show any C either. No H need be shown, because all chemists (and students of chemistry) know that a set number of them must be attached to each C, and no C need be drawn, because everyone agrees on this convention: in a line structure, each angle indicates a carbon atom and, unless there is some other atom written there, the end of each line also indicates a carbon atom:

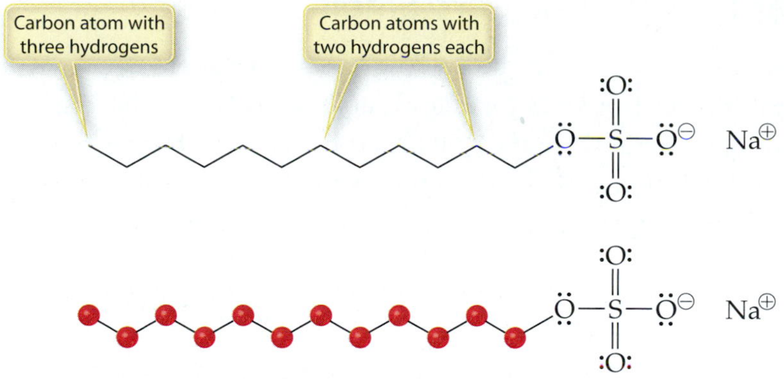

Carbon atom with three hydrogens

Carbon atoms with two hydrogens each

$$
\ddot{O}-\overset{\displaystyle :O:}{\underset{\displaystyle :O:}{\overset{\displaystyle \|}{\underset{\displaystyle \|}{S}}}}-\ddot{O}^{\ominus} \quad Na^{\oplus}
$$

$$
\ddot{O}-\overset{\displaystyle :O:}{\underset{\displaystyle :O:}{\overset{\displaystyle \|}{\underset{\displaystyle \|}{S}}}}-\ddot{O}^{\ominus} \quad Na^{\oplus}
$$

Sodium lauryl sulfate line structure

The zigzag drawing at the top is the line structure. The bottom drawing is included to show you how the line structure works: the red circles show that each zig and each zag represents a carbon atom. It is understood that there are three hydrogens on the leftmost carbon and two hydrogens on every other carbon, so that every carbon always makes four bonds. All atoms other than carbon and hydrogen must be drawn, and if those atoms have nonbonding pairs, the pairs must be included.

Let's take another example from our recipe for toothpaste. Figure 3.27*b* shows the full structure for oil of wintergreen, and the line structure for this molecule is

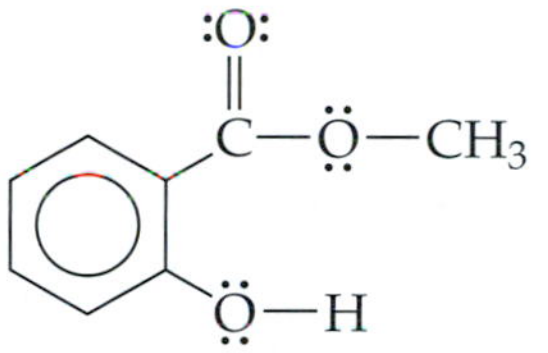

$$
\overset{\displaystyle :O:}{\underset{\displaystyle \|}{C}}-\ddot{O}-CH_3
$$

$\ddot{O}-H$

Oil of wintergreen line structure

As you can see, rings are easily drawn using line structures.

QUESTION 3.20 The skeletal structure for the cholesterol molecule is

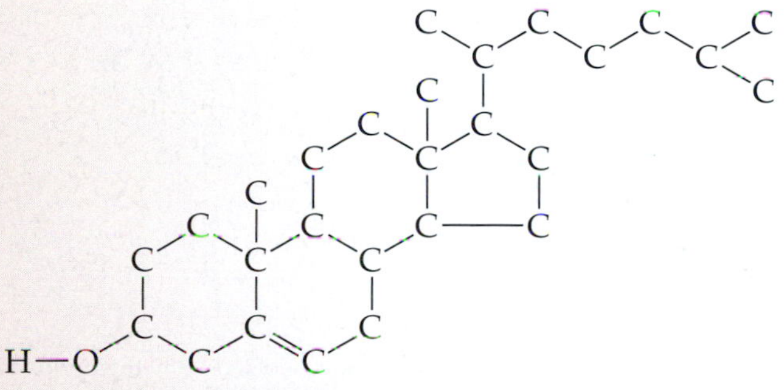

Cholesterol skeletal structure

Draw a line structure for this molecule.

ANSWER 3.20

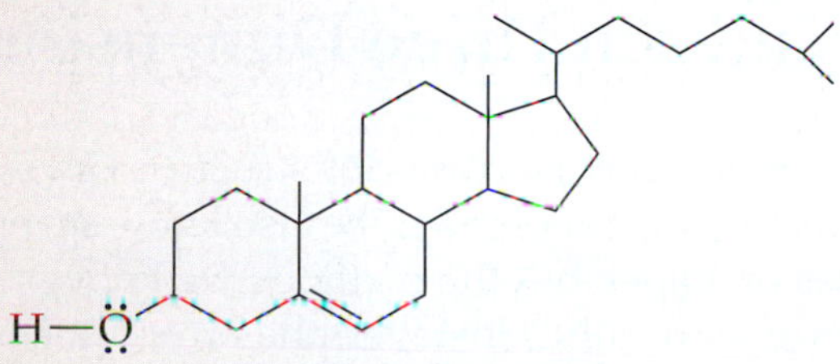

Cholesterol line structure

STUDENTS OFTEN ASK

What Makes New Cars Smell So Good?

Because companies that make air fresheners have been trying for years to duplicate the smell of a new car interior, quite a bit of research has gone into the analysis of the air in new cars. As it turns out, the aroma comes from the adhesives and solvents used in the upholstery and other plastic components. Molecules from the adhesives and solvents escape easily into the air in the car,

and we say such molecules are **volatile**. The molecules most abundant in the air of a new car are *siloxanes*, which contain silicon, oxygen, and small hydrocarbon groups. Figure 3.28 shows line structures for the three siloxanes that are most concentrated in new car interiors. In addition to these, there are 25 or more other volatile organic compounds present. (Volatile organic compounds are the VOCs you see listed on the labels for paints and other solvent-containing products.)

QUESTION 3.21   Connect each part of each name in Figure 3.28 to some feature of the molecule that goes with the name. For example, what does *hexamethyl-* tell you about compound 1; what does *cyclopenta-* tell you about compound 3, and so on?

ANSWER   3.21   *Hexamethyl-*, *octamethyl-*, and *decamethyl-* mean that the molecules contain six (*hexa-*), eight (*octa-*), and ten (*deca-*) methyl groups, respectively. Each name contains *cyclo-* because each molecule is in the shape of a ring. *Trisiloxane* in the name for compound 1 means that there are three silicon atoms and that they alternate with oxygen atoms. Likewise, *tetrasiloxane* means four silicon

atoms in compound 2, and *pentasiloxane* means five silicon atoms in compound 3.

You may be wondering about whether the mix of VOCs found in the air inside a new car is safe. The answer is that we do not know. The government does not have restrictions on VOC levels in new cars, partly because the odor—and therefore the molecules—dissipate after a month or two. The rule of thumb is this: if the concentration of molecules in the new car fogs up the windows, then the interior should be aired out before the car is sold.

In other environments—new buildings, for example—the allowable level of VOCs is about 10 mg per cubic meter of air. Levels higher than this can cause headaches and irritate your eyes and throat. Average measurements of VOCs in new cars are between 30 and 40 mg per cubic meter. Should we worry about inhaling this heady mixture of volatile molecules? Possibly yes. There is evidence that harmful levels of undesirable molecules are what make up the aroma of a new car. So, if you are making a new car purchase, you may want to consider a convertible.

▼ **FIGURE 3.28**   Three siloxane molecules found in new-car scent.

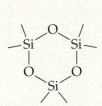

Compound 1: hexamethylcyclotrisiloxane

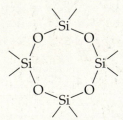

Compound 2: octamethylcyclotetrasiloxane

Compound 3: decamethylcyclopentasiloxane

3.8 | Molecules in Three Dimensions

Now we know how to draw electron dot structures showing the distribution of valence electrons in a molecule. We also know several ways to depict organic molecules on paper, but the methods we know so far depict molecules in two dimensions only. However, the *three*-dimensional shape of a molecule is one of the primary factors that determine how the molecule will behave. Throughout this book, we shall see dozens of examples—an insecticide, a drug used to treat AIDS, a building block for plastic, a food molecule—in which the shape of the molecule is paramount to its function. Because few molecules are flat, we must figure out how to draw them with a three-dimensional perspective. To do this, we use something called **valence shell electron pair repulsion (VSEPR) theory**. This is a fancy name for a simple concept: all the atoms surrounding a central atom in a molecule try to get as far away from one another as possible. This is intuitively sensible because atoms and bonds contain electrons, and electrons repel one another if they get close enough together.

Let's look at a simple example. Ammonia, NH_3, has a central nitrogen surrounded by three hydrogens. Therefore, you might assume (incorrectly) that the three hydrogens take positions at the corner of a triangle with nitrogen in the center. This geometry would place the hydrogen atoms as far apart as possible:

$$
\begin{array}{ccc}
H & & H \\
 & \diagdown \; \diagup & \\
 & N & \\
 & | & \\
 & H & \\
\end{array}
$$

Incorrect structure of ammonia

This structure is incorrect, however, because we have forgotten one important group that is part of the nitrogen: the nonbonding pair of electrons. Nitrogen has five valence electrons, and the three bonds to hydrogen account for only three of them. The fourth and fifth make up one nonbonding pair, and this nonbonding pair also wants to be as far away as possible from the hydrogens (and from the bonding electron pairs):

$$
H \longrightarrow \cdot \ddot{N} \cdot \longleftarrow H
$$
$$
\cdot H
$$

Thus, the ammonia molecule has not three but four electron pairs around the nitrogen. (Remember, each single covalent bond contains two electrons.) Three of the four pairs are in covalent bonds, and the fourth is a nonbonding pair. Now, in order to put four electron pairs as far from one another as possible, they should be positioned not on the four sides of a flat square as depicted, but pointing to the four corners of a tetrahedron, as shown in Figure 3.29a. This tetrahedral arrangement makes the angular distance between any two vertices 109.5 degrees, the largest possible angle. The slight difference in this molecule, as Figure 3.29b shows, is that the angles are not 109.5 degrees, but only 106.7 degrees. This new bond angle results from repulsion between the nonbonding pair and the electrons in the three N–H bonds. The representation in Figure 3.29b is what we call the **electron-pair geometry** for the molecule.

Electron-pair geometry does not quite show us what the molecule looks like in three dimensions. To see this, you have to grasp the important point that, even though it is depicted in the drawing showing the electron-pair geometry, *the nonbonding pair is not considered part of the molecule's three-dimensional shape*. The three-dimensional shape is called the **molecular geometry**, and a drawing of it shows the three-dimensional arrangement of *only the atoms* in the molecule and the bonds between them. Even though they take up space, nonbonding pairs do not occupy a vertex in the molecular geometry for a molecule. To depict a molecular geometry, we make a drawing of the electron-pair geometry for the molecule, then move each

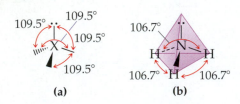

▲ **FIGURE 3.29 Electron-pair geometry of ammonia.** (a) All angles in a tetrahedron without nonbonding pairs are 109.5 degrees. (b) Both the bonds formed by the central N atom and its nonbonding pair of electrons are shown. When the electron pairs in the three bonds and the nonbonding pair get as far from one another as possible, they form a tetrahedral shape around the central atom.

Move nonbonding electrons to be next to their atom

Electron-pair geometry for ammonia

Molecular geometry for ammonia

▲ **FIGURE 3.30 Molecular geometry of ammonia.** To get from an electron-pair geometry to a molecular geometry, begin with the electron-pair geometry, move nonbonding pairs in toward the central atom, and redraw the molecules without vertices shown.

What is the difference between a tetrahedron and a three-sided pyramid, you ask? That's easy. A tetrahedron is a special kind of pyramid, one in which all four faces are identical equilateral triangles. In a three-sided pyramid, the four triangles forming the faces need not be identical to one another.

MOLECULE
Ammonia

▼ **FIGURE 3.31 A ball-and-stick model of an ammonia molecule.** The blue ball represents a nitrogen atom, and each silver ball represents a hydrogen atom. The shape shown here is the same as the shape shown on the right in Figure 3.30: pyramidal.

nonbonding pair directly adjacent to the atom to which it belongs (without moving any of the atoms in the drawing).

Let us again use ammonia as an example. The ammonia electron-pair geometry in Figure 3.29*b* shows a tetrahedron formed by the three H atoms and the nonbonding pair. If you start with this drawing of the electron-pair geometry and move the nonbonding pair toward the center atom (as indicated in the central portion of Figure 3.30), you end up with a structure showing the molecular geometry for ammonia. Notice that, in the drawing of the molecular geometry, each atom in the molecule sits in exactly the same position as in the drawing of the electron-pair geometry. The nitrogen is still in the center of the tetrahedron, and the hydrogen atoms are at three of its vertices. The difference is that the nonbonding pair no longer occupies the fourth vertex. Instead, it sits right next to the nitrogen.

Remember, *you do not change the positions of the atoms after moving the nonbonding pair.* Once you have sketched the electron-pair geometry with all bonded atoms and all nonbonding pairs as far apart as possible, *the atoms stay in those positions* when you move from a drawing of electron-pair geometry to a drawing of molecular geometry.

If you look at ammonia's molecular geometry without the tetrahedron drawn in (as shown in Figure 3.30), you see that the molecule no longer looks tetrahedral. Rather, it is *pyramidal*, and a nitrogen atom sits at the top of the pyramid. If you were to build a model of ammonia with balls and sticks, this is what it would look like (Figure 3.31). When you are asked to draw a molecule, you are being asked to draw a molecular geometry. The molecular geometry is what we have been shooting for.

That being so, why must we bother to draw the electron-pair geometry? The connection between these two drawings is that *you need to know the electron-pair geometry in order to determine the molecular geometry.* A drawing of a molecule's electron-pair geometry shows all bonding electron pairs plus all nonbonding pairs, all positioned as far apart as possible. A drawing of the molecule's molecular geometry shows bonds and atoms positioned exactly as in the drawing of the electron-pair geometry, plus the nonbonding pairs sitting directly adjacent to their respective atoms.

This process of drawing molecules in three dimensions using VSEPR theory may seem a bit complex at first. However, here is some good news about this method: in organic chemistry, the same structural motifs appear again and again. For example, we can be sure that, whenever nitrogen is bound to three other atoms in an uncharged organic molecule (which is often), the molecule has the pyramidal molecular geometry seen in ammonia.

Let's consider the structure of a simple molecule that your toothbrush knows well: water. First, we must know the electron dot structure. We always begin with the total number of valence electrons. In the water molecule, we have one O with six valence electrons plus two H with one valence electron each, for a total of eight valence electrons. We know the water molecule has the oxygen attached to the two hydrogens because each hydrogen can bind to only one other atom, and the oxygen must bind to two:

$$H\text{—}O\text{—}H$$

With this skeletal structure, we have used up four valence electrons in two bonds. This leaves four valence electrons to be assigned. We know oxygen

When Drawing an Electron Dot Structure for a Tetrahedral Molecule, Does It Matter Which Atom I Show at Which Vertex of the Tetrahedron?

STUDENTS OFTEN ASK

When drawing organic molecules, you will often come up with two electron dot structures that seem equally valid. For example, for the molecule methylene chloride, CH_2Cl_2, which is correct?

$$H \qquad\qquad H$$
$$| \qquad\qquad\quad |$$
$$:\ddot{C}l-C-H \quad or \quad :\ddot{C}l-C-\ddot{C}l: \ ?$$
$$| \qquad\qquad\qquad |$$
$$:\ddot{C}l: \qquad\qquad\quad H$$

Methylene chloride

The answer is that both are correct, as you can see by considering the molecular geometry:

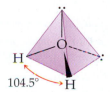

Molecular geometry for methylene chloride

All these depictions of the molecular geometry are equivalent because, in each of them, the atoms are connected to the central carbon in the same fashion. These drawings simply show the same molecule in different orientations. Thus, the two representations of methylene chloride shown here are equivalent, and you can correctly draw the electron dot structure either way.

When *four different groups* are attached to a central carbon atom, it does matter where the groups are placed, as we shall see in Chapter 13.

has two nonbonding pairs plus two places to make a bond. Therefore, the remaining four electrons must reside on oxygen:

$$H-\ddot{O}-H$$

Now we can predict the molecular geometry, and there is a general rule that helps us here: in any molecule containing only two or three atoms, all the atoms must lie in the same plane. In other words, the molecule is flat. Therefore, we know that the two H and the O all lie in the same plane, but is the line connecting them straight or bent? If it is bent, what are the angles? The answer comes from the electron-pair geometry and VSEPR theory, which requires that we count the number of electron pairs around the central atom. This means counting both bonding pairs *and* nonbonding pairs. Because the oxygen has two nonbonding pairs and two bonds, the electron-pair geometry for the molecule must be tetrahedral (just as with the ammonia molecule), since this geometry puts the four electron pairs as far apart as possible. The observed bond angle is smaller than the theoretical value of 109.5 degrees for a tetraheadron. As we saw for the ammonia molecule, the bond angle between atoms is compressed by the presence of nonbonding pairs.

Electron-pair geometry for water

This is the electron-pair geometry, but our goal is to determine the molecular geometry. Therefore, we push the two nonbonding pairs right up against the O and see what geometry we are left with—three atoms in a plane with the line connecting the three bent at an angle of 104.5 degrees:

Molecular geometry for water

Thus, the water molecule has a tetrahedral electron-pair geometry but a *bent* molecular geometry.

> **QUESTION 3.22**   Sketch the electron-pair geometry and the molecular geometry for the molecule H_2S.
>
> **ANSWER 3.22**   This molecule is similar to the water molecule with the one difference that sulfur replaces oxygen. Because S and O are in the same column of the periodic table, we know hydrogen sulfide has a similar electron pair and molecular geometry to water:

Hydrogen sulfide, tetrahedral   Hydrogen sulfide,
electron-pair geometry   bent molecular geometry

Once you have considered a few dozen organic structures, you will begin to see patterns that make predicting geometries easier. For example, you know that in an uncharged organic molecule, each kind of atom makes a characteristic number of bonds and has a characteristic number of nonbonding pairs of electrons. This information is summarized in Table 3.4 for the atoms most frequently found in organic molecules. You also know that VSEPR theory gives the same answer again and again when you consider the same type of atom in the same environment, and so you will become familiar with the various shapes and bond angles associated with the atoms of Table 3.4.

For example, you know a carbon atom always makes four bonds, but those four bonds may be single, double, or triple. Further, the geometry around a carbon atom depends on this information. When a carbon atom makes four single bonds, the geometry always has the carbon atom floating at the center of a tetrahedron and the bond angles are always 109.5 degrees:

Methane: four bonds, zero nonbonding pairs
Tetrahedral electron-pair geometry, tetrahedral molecular geometry

When a carbon atom is bound to three other atoms, one of the bonds must be a double bond. The modification in determining molecular geometry when multiple bonds are involved is that *we make no distinction between single and multiple bonds*; instead, we count double and triple bonds around a central atom as *one bond*. In order to spread these three bonds as far apart as possible, we put them in one plane with each bond pointing toward one corner of a triangle:

TABLE 3.4   Rules for Neutral Atoms in Organic Molecules Acting as Central Atoms (Summary of VSEPR Rules)

Element	Electron dot structure	Number of nonbonding pairs around neutral atom	If the *total* number of electron pairs around atom is …	The electron-pair geometry must be …	And the molecular geometry must be …
Oxygen	$:\overset{..}{\underset{..}{O}}:$	Always two	Four	Tetrahedral	Bent
Sulfur	$:\overset{..}{\underset{..}{S}}:$	Always two	Four	Tetrahedral	Bent
Nitrogen	$\cdot\overset{.}{\underset{.}{N}}\cdot$	Always one	Four	Tetrahedral	Pyramidal
Carbon	$\cdot\overset{.}{\underset{.}{C}}\cdot$	Always zero	Two	Linear	Linear
			Three	Flat triangular	Flat triangular
			Four	Tetrahedral	Tetrahedral

Because the three bonds from carbon all lie in the same plane, we expect the bond angles to be 120 degrees. (In fact, the measured value is close to this predicted value.) We call this geometry *flat triangular*, and the name applies both to the electron-pair geometry around the carbon atom and to the molecular geometry around it. The geometry around the carbon atom in the organic compound known as formaldehyde is flat triangular:

Formaldehyde: three bonds, zero nonbonding pairs
Flat triangular electron-pair geometry, flat triangular molecular geometry

(You may have noticed something about VSEPR structures: when you have no nonbonding pairs around a central atom, the electron-pair geometry and the molecular geometry are identical. This is because the difference between electron-pair geometry and molecular geometry for a given molecule involves only the position of nonbonding pairs. When there are no nonbonding pairs, which is always true for carbon, the two geometries are the same.)

When a carbon atom is bound to two other atoms via one single bond and one triple bond, all the atoms are along a straight line, which means both bond angles are 180 degrees:

$$H-C\equiv C-H$$

Acetylene: two bonds, no nonbonding pairs
Linear electron pair-geometry, linear molecular geometry

MOLECULE
Acetylene

Refer to Table 3.4 whenever you are asked to sketch either the electron-pair geometry or the molecular geometry of an organic molecule.

QUESTION 3.23 Draw the three-dimensional structure of carbon tetrachloride, CCl_4, and describe the molecular geometry around the central atom.

ANSWER 3.23

Carbon tetrachloride

The central atom is carbon, and the molecular geometry around it is tetrahedral.

TABLE 3.5 Bond Lengths for C–C, Si–Si, and Ge–Ge Bonds	
Bond type	**Bond length (pm)**
Carbon–carbon	154
Silicon–silicon	235
Germanium–germanium	245

3.9 | Silicon: Carbon's Big Sister

One step down from carbon on the periodic table you will find silicon, which shares some of carbon's properties. Like carbon, silicon has four valence electrons and makes four bonds to other atoms. Silicon makes a network solid similar to diamond, with each silicon atom in the lattice bound to four other silicon atoms in a tetrahedral arrangement. The similarities end there, however.

As noted in Section 3.1, carbon, because of its small size, is able to approach other atoms very closely and form multiple bonds with them. Table 3.5 shows the bond lengths for C–C, Si–Si, and Ge–Ge bonds. Silicon, being larger, cannot get close enough to other silicon atoms to make more than one bond. As we move down any column of the periodic table, the number of electrons around each atom increases, and, with few exceptions, the size of the atom increases, too. As a result, in general, larger elements make longer bonds with other elements than do smaller elements. Longer bonds are weaker bonds because the electrons on each atom do not overlap and interact as effectively. Both the network solid made up solely of silicon atoms and the one made solely of germanium atoms have a diamondlike structure, but you can bet that the bonds within the silicon network solid

STUDENTS OFTEN ASK

Is a Life-Form Based on Silicon Rather than Carbon Possible?

There are dozens of science fiction stories about alien life-forms whose chemistry is silicon based. The reason writers choose silicon more than any other element for their aliens is simple: silicon is directly below carbon on the periodic table and is therefore the element most likely to replace carbon in other

forms of life. But could silicon really replace carbon? The two are similar in many ways: they both have four valence electrons, and as a result they are both capable of forming long chains. There is one crucial difference, however: carbon has extraordinary versatility because of its small size. It can make double and triple bonds, a property that contributes to the enormous molecular diversity in organic compounds.

Another similarity between silicon and carbon is that both react with oxygen to form an oxide. However, because carbon dioxide is a gas at the temperatures needed to support life, it can be exchanged between life-forms easily

(humans, animals, and plants, to be specific). Silicon dioxide, on the other hand, is a rigid solid at biological temperatures, the stuff of sand and glass. Because it is not a gas, silicon cannot be exchanged between organisms easily.

Finally, the strengths of the bonds in carbon compounds are in the correct range for being broken and re-formed at the ambient temperatures of Earth. Because the Si atom is larger than the C atom, silicon generally makes weaker bonds. This means that, on Earth, most bonds formed by silicon atoms in biological molecules would fall apart. Thus, silicon is not suitable for the biochemistry of living organisms.

and those in the germanium network solid are not nearly as strong as are those in diamond.

As is true for carbon, silicon's size and its position in the periodic table make it suited to very specific functions. When silicon atoms form Si–Si bonds, those bonds extend in a network solid. Some of the valence electrons in the silicon atoms in the network are occasionally able to move from a lower energy level to a higher one (Figure 3.32). This jump to a higher energy level requires an outside source of energy, which can be heat or light. Thus, when a silicon sample is either heated or exposed to high-energy light, some of the electrons absorb the energy and are promoted to the higher energy level.

An electron moving to the higher-energy level leaves behind a "hole" in the network (Figure 3.33). Because this hole represents a missing negative charge, we say the hole is positively charged. The positively charged hole attracts an electron from someplace else in the network, and that electron, as it migrates to the hole, leaves a fresh hole behind at its original location. Thus, there is a great deal of "traffic" within the silicon network. Because moving electrons constitute an electrical charge, silicon can conduct an electric current when the required energy is available in the form of heat or high-energy light. We refer to silicon as a *semiconductor* because it is conductive only under specific conditions.

The ability of silicon to conduct electricity under specific conditions can be enhanced by adding an impurity to the silicon in a process known as **doping**. One often-used dopant is aluminum, an element with three valence electrons. Because each silicon atom in a silicon crystal lattice has four valence electrons, the presence of an occasional aluminum atom with its three valence electrons means you have, in effect, added positive holes to the crystal, as shown in Figure 3.34a. For this reason, dopants that are electron-deficient (as aluminum is) are referred to as *p-type dopants* (*p* for *positive*). Any hole created by the presence of an Al atom in the lattice kicks off conduction by luring an electron from a neighboring silicon atom to the hole. The electron movement creates a new hole at that silicon atom, an electron from somewhere else moves into *that* hole, and so on. An electric current is created.

Another type of dopant is arsenic, which has five valence electrons. If you add small amounts of arsenic to a growing silicon crystal, the arsenic is

▲ FIGURE 3.32   Electron movement in a silicon network solid. When solid silicon is either heated or exposed to high-energy light, electrons in the atoms can absorb some of the energy and jump to a higher energy level.

Higher energy level

Electron can jump this gap

Lower energy level

Semiconductor

In contrast, pure metals, in which a sea of electrons bonds the atoms together, are *conductors* because they conduct an electric current under all conditions.

▼ FIGURE 3.33   Creating a hole in a silicon network. As the electron of Figure 3.32 moves to the higher-energy level, it leaves behind a hole in the network lattice. Because this space is "missing" a negatively charged electron, it acts like a positively charged particle in that it can attract other electrons to itself. Thus, the hole is considered to be positively charged.

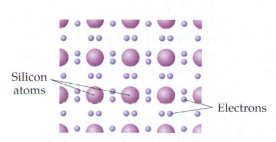

Silicon atoms

Electrons

All electrons in lower energy level

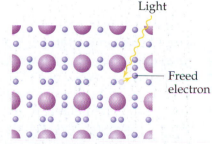

Light

Freed electron

In presence of light, one electron promoted to higher energy level

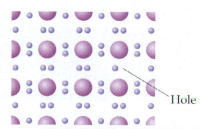

Hole

Positively charged hole exists where electron used to be

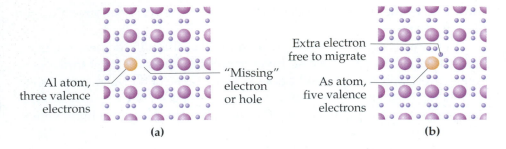

▶ **FIGURE 3.34 Semiconductor dopants.** **(a)** Because it has only three valence electrons instead of the four of each silicon atom, an aluminum atom doped into a silicon crystal provides a positive hole that helps initiate electron movement in the crystal. **(b)** With its five valence electrons, an arsenic atom added to a silicon crystal provides an extra electron that can move around in the crystal lattice.

taken up into the crystal, as shown in Figure 3.34*b*. The "extra" electron provided by each arsenic atom can move around in the network. This type of dopant is called an *n-type dopant* because it supplies excess negative charges (that is, electrons) to the crystal. The arsenic electrons increase the number of electrons free to move around in the network and thereby enhance the conductivity of the silicon crystal.

QUESTION 3.24 (*a*) What other elements besides aluminum would make suitable p-type dopants for silicon semiconductors? (*b*) What other elements besides arsenic would make suitable n-type dopants for silicon semiconductors?

ANSWER 3.24 (*a*) Any element in the same family as aluminum, which means boron, gallium, indium, and thallium. (*b*) Any element in the same family as arsenic, which means nitrogen, phosphorus, antimony, and bismuth.

▲ **FIGURE 3.35 Creating a diode.** A diode is made by placing a p-type semiconductor adjacent to an n-type semiconductor. The movement of electrons from the n-type side to the p-type side causes a decrease in the energy of the system, and the lost energy can be given off as light.

You may be asking yourself: why not use a metal to conduct electricity? Aren't metals the best electrical conductors? The answer is yes. Metals are superior electrical conductors. However, they conduct under a wide range of conditions and are therefore conductive most of the time. It is advantageous to have a conductor that can be controlled, the way a semiconductor can be. A semiconductor will conduct *only when you give it the proper energy*, which means you can dictate when it is conductive and when it is not.

You can also control *how* the electricity is conducted. The semiconductor device known as a *diode* is made by placing a p-type silicon crystal next to an n-type silicon crystal, as shown in Figure 3.35. The juxtaposition of oppositely doped crystals allows the holes in the p-type crystal to be filled by electrons from the n-type crystal. Having n-type electrons fill p-type holes eases the tension of having these charges separated: the separation of positive and negative charges on the two sides of a *p–n junction* is a high-energy situation. After all, positive and negative charges seek one another out. When electrons from the n-type crystal travel over to fill in the holes in the p-type crystal, the energy of the system drops: now the electrons are where they wanted to be. The energy released when the system energy drops can be in the form of light. In this case, the diode is called a *light-emitting diode* (LED). If you have a flat-screen monitor in your home, you might be taking advantage of LED technology. LEDs can be found in everything from light bulbs to solar power stations. Just one of many, many applications is shown in Figure 3.36.

▼ **FIGURE 3.36 Semiconductors at work.** The light given off by this dog collar comes from a light-emitting diode.

SUMMARY

We started this chapter with a look at carbon and what makes it different from other elements. The uniqueness principle tells us that carbon's small size accounts for its incorporation into tens of thousands of molecules. Diamond is a simple, elegant example of carbon's ability to make strong bonds with other carbon atoms. Diamond also demonstrates the most common structural motif for carbon: the tetrahedron. The extraordinarily strong bonds between carbon atoms in diamond make it the hardest natural substance known. Another allotrope of carbon, graphite, has double as well as single bonds in its structure. Its two-dimensional framework makes it strong within that plane, but the bonds between graphite sheets are necessarily weak, making the look and feel of graphite entirely different from the look and feel of diamond. Recently, a new collection of carbon allotropes was discovered, including the buckyball and the carbon nanotube. These caged structures, which contain double and single bonds, present opportunities for such innovative applications as microelectronic devices and drug-delivery vehicles.

We learned how to simplify the tedious process of drawing organic molecules. Full and condensed structures are valuable because they depict every atom and the connectivity between atoms is clear. Experienced organic chemists know how to draw line structures and how to interpret them; these structures are the simplest way to draw organic molecules, and you should master the task of drawing them.

The electron dot structures of atoms and the molecules that contain them provide us with a sense of how electrons are distributed in molecules. This electron-bookkeeping system allows us to draw a molecule in two dimensions and then determine the number of attached atoms or groups around any central atom. When this number is known, we can sketch the electron-pair geometry of the molecule, and from the electron-pair geometry, we can sketch the molecular geometry. The molecular geometry is the structure we normally see depicted in three-dimensional drawings of molecules. Single and multiple bonds between carbon atoms differ in bond strength and length. We talked about a new technique that can be used to break bonds other than the weakest in a molecule.

Semiconductors are based on the concept that certain elements, most notably silicon, have a gap between a lower energy level and a higher energy level. The distance between these levels in silicon is ideal for allowing electrons to jump from one level to another. Silicon is a semiconductor because this "jumping" process can be controlled by the input of energy.

KEY TERMS

addition reaction	double bond	methyl group	octet rule	single bond
allotrope	electron dot structure	methylene group	resonance	tetrahedron
bond energy	electron-pair	molecular geometry	stabilization	triple bond
bond length	geometry	multiple bond	saturated	uniqueness principle
condensed structure	full structure	nanotube	semiconductor	unsaturated
delocalization	joule	network solid	shortened condensed	volatile
doping	line structure	nonbonding pair	structure	VSEPR theory

QUESTIONS

The Painless Questions

1. Name the three allotropes of carbon we discussed in this chapter. Briefly describe the physical appearance of each one, and comment on how their appearance is related to their structure.

2. The elements of group 16 are called the *chalcogens*. What trend do you expect to see in the sizes of atoms in this family? Do you expect oxygen to have properties representative of the other elements in this family? Why or why not?

3. Are calcium carbonate and hydrated silica, two of the toothpaste ingredients listed in Table 3.3, organic or inorganic compounds? How do you know?

4. Refer to Table 3.1 to answer these questions: Which is/are the weakest bond(s) in this molecule? Which is/are the strongest bond(s)?

$$H_3C-C\equiv C-CH_3$$

5. Which of these molecules are organic: (*a*) SO_2, (*b*) $CH_3(CH_2)_8CH_3$, (*c*) HCl, (*d*) CH_3COOH, (*e*) C_{60}, (*f*) $CH_3COCH_2CH_3$?

6. Which of these elements is the best choice for a p-type dopant in a semiconductor: (*a*) carbon, (*b*) silicon, (*c*) germanium, (*d*) phosphorus, (*e*) indium?

7. Leucocyanidin, an organic molecule isolated from the petals of Asian cotton flowers, is used to treat vascular disorders. Draw its full structure, given that its line structure is

Leucocyanidin

8. Safflower oil is composed primarily of a polyunsaturated oil. What is meant by *polyunsaturated*? Draw the full structure of any polyunsaturated carbon chain containing ten carbons.

9. The length of the carbon–carbon bonds in benzene is 139 pm. Express this value in (*a*) angstroms, (*b*) nanometers, (*c*) micrometers.

10. Describe the difference in structure between a buckyball and a nanotube. Why do nanotubes come in different lengths?

11. Write a balanced chemical equation for the reaction that takes place in a fuel cell. If you want the reaction to produce 7 mol of water, how many moles of oxygen and hydrogen must react?

12. The dibromomethane molecule has a tetrahedral carbon surrounded by two bromine atoms and two hydrogen atoms. Indicate whether each of these three structures is a correct way to draw this molecule:

Three versions of dibromomethane molecule

13. The electron-pair geometry of water is tetrahedral. Why then is the H–O–H angle in water 104.5 degrees and not 109.5 degrees?

14. Count the total number of valence electrons in each molecule (Hint: Count seven valence electrons for iodine):

Methyl iodide   Ethylmethyl ketone

15. (*a*) Draw the electron dot structure for phosphorus. (*b*) How many covalent bonds does phosphorus usually make? (*c*) What other element makes the same number of bonds in organic molecules?

More Challenging Questions

16. Silicon is the preferred element for semiconducting devices. What property of silicon makes it ideally suited to this role?

17. Bond energies are most often expressed in either kilojoules or kilocalories (1 kcal = 4.184 kJ). How many kilojoules of energy are contained in a piece of cherry pie that contains 750 kilocalories?

18. Acetylene is a fuel used in torches. When it burns, it produces large quantities of energy. Acetylene's chemical formula is C_2H_2, and the atoms are attached in the order H–C–C–H. (*a*) Draw an electron dot structure for the acetylene molecule. (*b*) What is the three-dimensional shape of the molecule? (*c*) What is its molar mass?

19. Imagine that a full nanotank in your car can bind enough hydrogen gas to take you 300 miles. Each 1 mL of nanotubes has a mass of 3.5 g, and the combined mass of all the nanotubes in the full tank is 12.8 kg. (*a*) What is the volume of your nanotank in liters? (*b*) How many grams of nanotubes do you need to take you 1 mile?

20. (*a*) Why was it difficult for Stacey Bent to modify the surface of a diamond? (*b*) What feature in the structure of diamond made it possible for her to succeed? (*c*) Would it be possible to attach molecules to a buckminsterfullerene molecule?

21. A diode is designed to have a high-energy state that results from the juxtaposition of two types of semiconductors. What two types of material are placed side by side? How does the p–n junction function?

22. Consider palmitic acid, an organic molecule that composes about 25 percent of butter and lard:

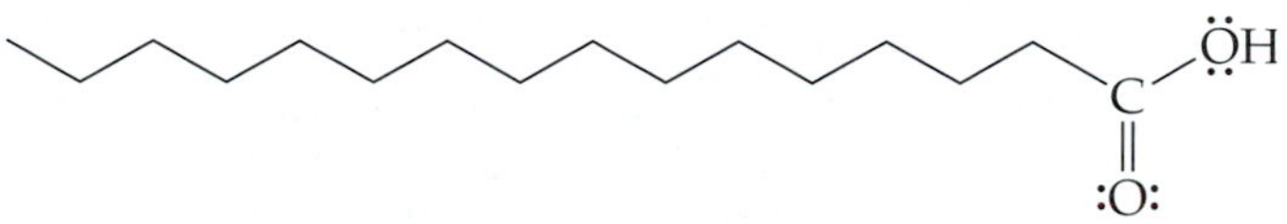

Palmitic acid

(a) Is this molecule saturated or unsaturated? (b) How many methyl groups does it contain? (c) How many methylene groups? (d) Can this molecule undergo an addition reaction?

23. The halogens, found in column 17 of the periodic table, are commonly found in organic molecules. In general, how many bonds does a halogen make to other atoms?

24. Consider this molecule:

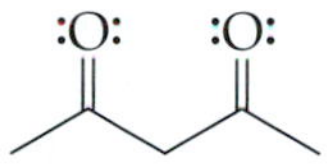

(a) What is the molecular formula of this molecule? (b) How many atoms does it contain? (c) Of this total number of atoms, what is the maximum number that are in the same two-dimensional plane?

25. Like graphite, the buckyball is known to have lubricating properties. From an understanding of the structures of these two molecules, explain how buckyballs are able to act as lubricants.

26. Predict the product of each addition reaction:

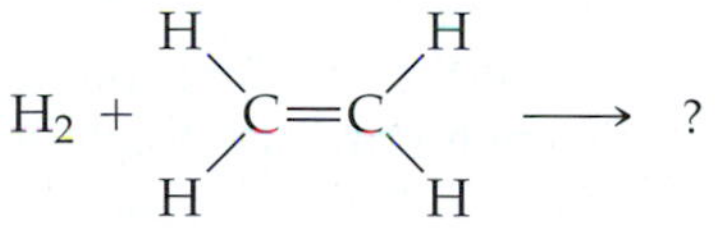

(a)

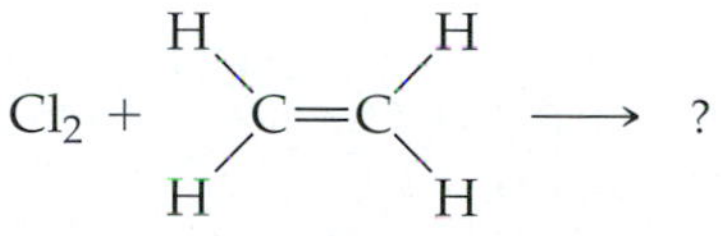

(b)

27. Draw a line structure for each molecule in Figure 3.8.

28. Phosphorus atoms are occasionally found in organic molecules. For example, the biomolecules DNA and RNA both have a backbone that contains phosphorus atoms. Based on your knowledge of electron dot structures, predict which one of the following compounds is most likely to exist: (a) PO_4, (b) PBr_7, (c) PCl_3, (d) PS_2.

29. Draw the line structure for each molecule:

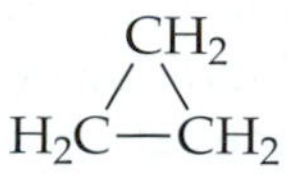

Cyclopropane 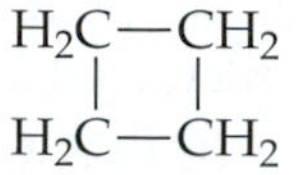   Cyclobutane

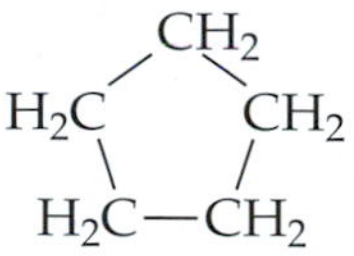

Cyclopentane 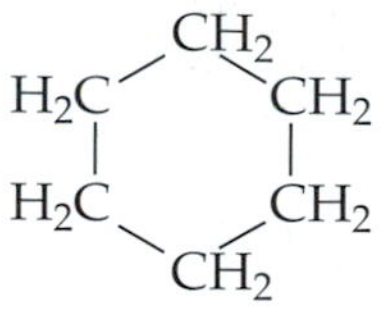   Cyclohexane

30. Cinnamaldehyde and anethole are both sweet-smelling organic compounds used in flavorings and perfumes. For each, (a) draw a line structure, (b) write the chemical formula, (c) calculate the molar mass.

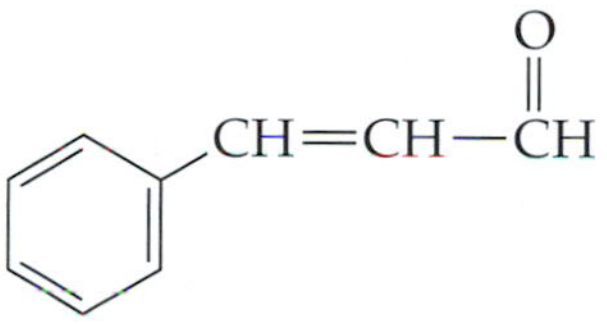

Cinnamaldehyde 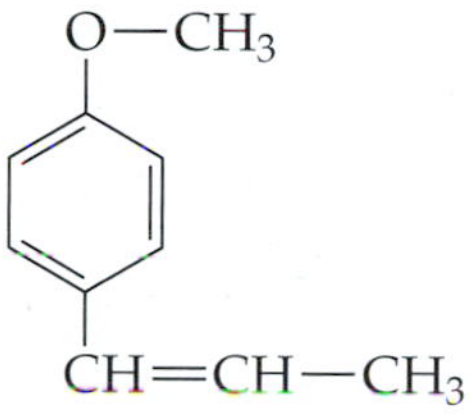   Anethole

The Toughest Questions

31. For Prozac and cocaine, (a) write the chemical formula, (b) calculate the molar mass, (c) draw the line structure.

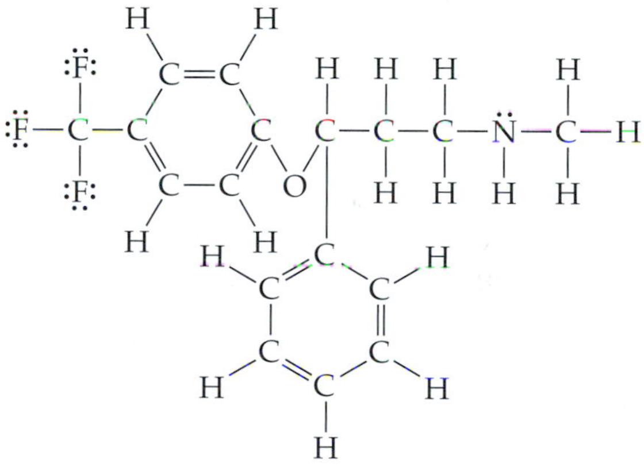

Prozac

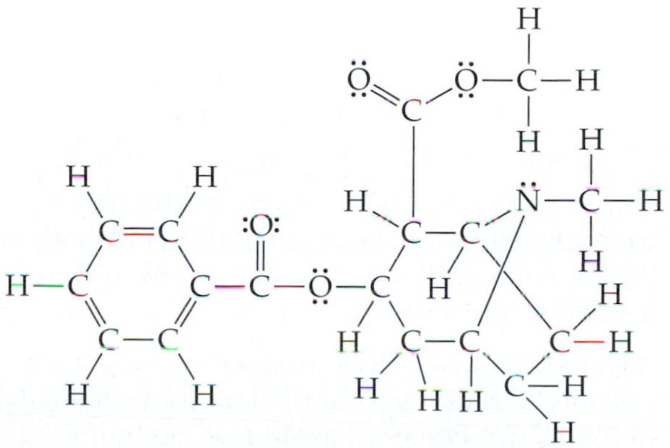

Cocaine

32. (*a*) What is the molecular geometry surrounding the red atom in each of these molecules? (*b*) What is missing from structures 1 and 2? (*c*) Correct these two line structures, and then draw the full structure for each of them.

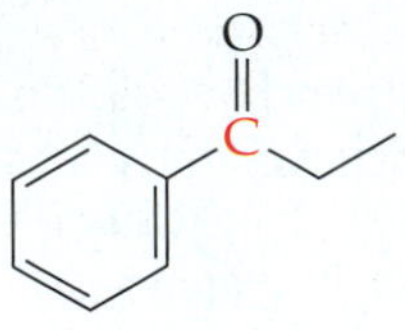

Structure 1 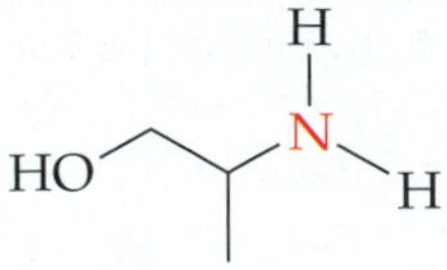   Structure 2

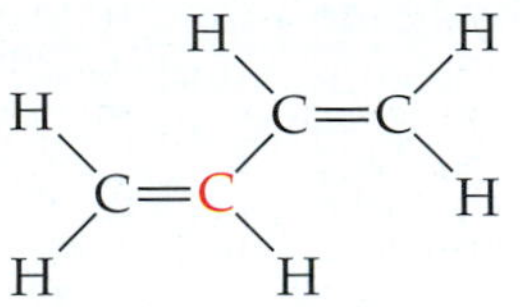

Structure 3

33. VSEPR theory can extend beyond central atoms surrounded by four electron pairs. Consider a central atom having six electron pairs around it. Try to sketch a structure that would position these six pairs as far apart from one another as possible.

34. The length of each carbon–carbon bond in benzene is 139 pm. Referring to Table 3.1, comment on this information. Does this value make sense based on what you know about the structure of benzene? Is the carbon–carbon bond length in benzene more like the length of a typical single bond or a typical double bond? Why are all the carbon–carbon bond lengths in benzene the same?

35. These two structures contain mistakes:

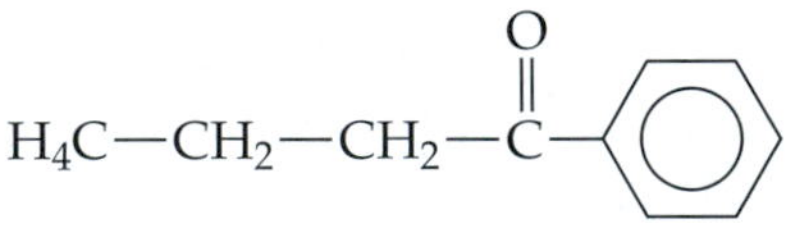

(a)

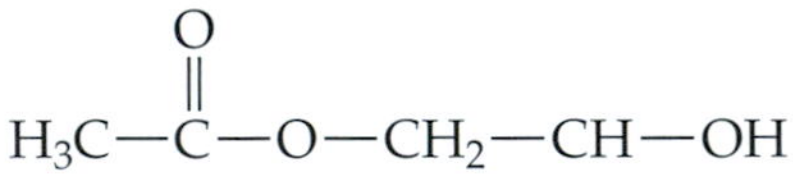

(b)

Redraw them correctly as full structures, and then draw them as line structures.

36. (*a*) Is the way the carbon atoms are connected in graphite (Figure 3.11) the same as the way they are connected in a buckyball (Figures 3.18 and 3.19)? (*b*) Is a buckyball no more than a sheet of graphite fashioned into a sphere? Explain.

37. Ascorbic acid, which is more commonly known as vitamin C, is an important vitamin for maintenance of connective tissues in the human body. (*a*) Draw its

line structure. (*b*) Write its chemical formula, and then calculate its molar mass.

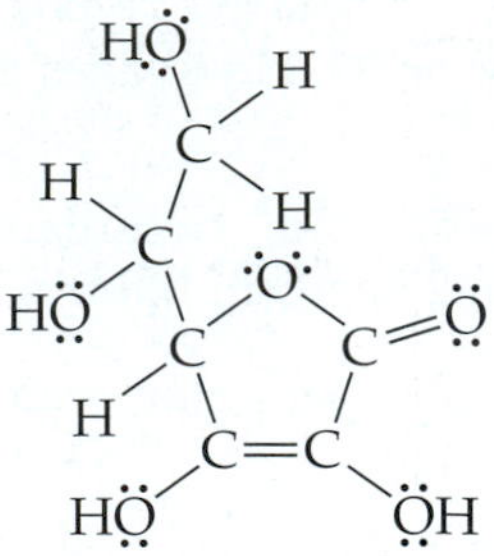

Ascorbic acid, aka vitamin C

38. Glycerol is sometimes used in toothpaste as a moistening agent. (*a*) Draw the full structure. (*b*) Determine the chemical formula and molar mass.

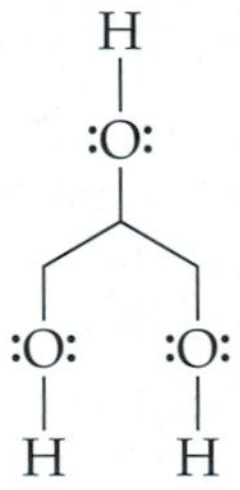

Glycerol

39. Retinol is important for vision in the human eye. (*a*) Draw its full structure. (*b*) Determine its chemical formula and molar mass. (*c*) The retinol molecule is highly stable. What feature of the molecule is responsible for this stability?

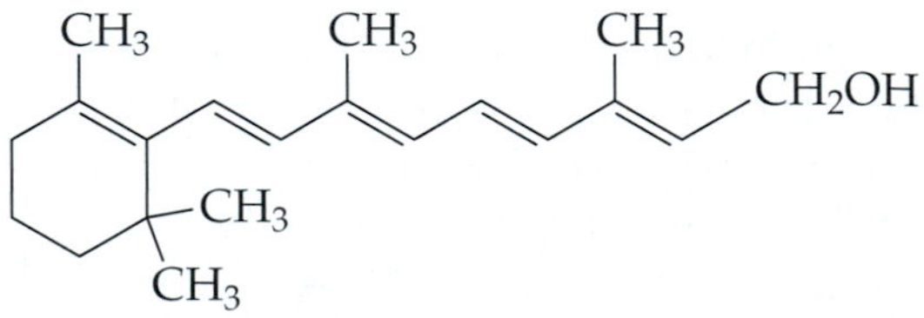

Retinol

40. Based on what you know about bond energies and lengths, predict the following about phenylacetylene: (*a*) Which bond is shortest? (*b*) Which bond will break most easily when energy is added? (*c*) How do you know?

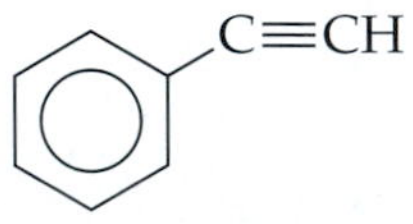

Phenylacetylene

41. For each of these molecular formulas, there is more than one skeletal structure:

(*a*) $C_5H_{10}Cl_2$, (*b*) $C_6H_{10}O$, (*c*) C_6H_7N, (*d*) $C_{10}H_{16}O$.

Create as many different molecules as possible for each formula, showing both the line structure and the full structure for each molecule you come up with.

E-Questions

Go to **www.prenhall.com/waldron** to find these questions in electronic form, complete with hyperlinks directly to the various websites cited in the questions.

42. Check car manufacturers' websites to find out the advertised mileage for at least two hybrid cars. Convert each miles-per-gallon value to the SI units kilometers per liter.

43. Read more about silicon and semiconductor devices at **HowStuffWorks** and answer these questions: (a) In the schematic drawing of a diode, identify each part and explain its purpose. (b) How do transistors differ from diodes? What do transistors do? (c) Why are LEDs superior to incandescent lights?

44. Chemists have discovered that the molecule tetracyanoethylene has bonds twice as long as those in diamond. Read about this finding at **NewScientist** and answer these questions: (a) How long is the bond measured in tetracyanoethylene? (b) Compare this value with other carbon–carbon bond lengths shown in Table 3.1. (c) Briefly explain why this bond is shorter than expected.

45. Scientists at the Skaggs Institute for Chemical Biology in San Diego claim to have created a new nanotube smart drug. Their invention heralds a new class of drugs that may be used to treat bacterial infections. Read about this finding at **ScienceDaily** and answer these questions: (a) What is the mechanism of action of these new drugs? (b) According to the article, why are new antibacterial drugs needed? (c) Why are these drugs called nanotubes? Are they similar in structure to the nanotubes discussed in this chapter?

46. Diamond films have been used as a supertough coating for everything from prosthetic devices to airplane wings. Read more about this at **NewScientist** and answer these questions: (a) Why are diamond films useful to engineers? (b) Why is the new method for making diamond films better than the traditional method? (c) What is the role of hydrogen in the creation of diamond films?

WORLD WIDE WEB RESOURCES

As with the E-Questions, go to **www.prenhall.com/waldron** to find these questions in electronic form, complete with hyperlinks directly to the various websites cited.

Some of the links that follow contain research articles related to the topics in this chapter and could be used as the basis for a writing assignment in your course. Other sites are of general interest.

- At **EnchantedLearning**, you will learn how to make your own polyhedron from paper. The website includes complete instructions and a template for cutting out the basic form.
- The article at **ScientificAmerican** answers the question, "Have buckyballs been put to any practical uses?"

- Type the word *nanotube* into your web browser. Look through the links provided, and record at least three new applications of nanotube technology. Which application is the most recent? Which applications use nanotubes in products that are currently available to the public?
- A layperson's review of femtosecond lasers can be found at **ScienceMagazine** in the article "Just a Light Squeeze," by David Voss.
- A story about how femtosecond lasers can be used to create advanced fiber optic communication materials can be found in the article "Laser Makes History's Fastest Holes" at **HarvardGazette.**

CHEMISTRY TOPICS IN CHAPTER 4

- Ionic liquids, ionic interactions, delocalized electronic charge [4.1]

- Polyatomic ions [4.2]

- Polar molecules, dipoles, ion–dipole interactions, solubility [4.3]

- Electrolytes, molarity [4.4]

- Osmosis [4.5]

- Autoionization of water, acids and bases, pH [4.6]

- Acid rain [4.7]

"Let there be work, bread, water, and salt for all."

NELSON MANDELA

Salt

A STUDY OF THE BEHAVIOR OF IONS, INCLUDING ACIDS AND BASES, AND THE NOTION OF EQUILIBRIUM

Go to a restaurant almost anywhere in the world, and you will probably find salt on your table. For centuries, we have been shaking table salt, NaCl, onto our food, a gesture that has become the subject of metaphors that underscore the importance of this substance. A person who is the "salt of the earth," for instance, is one of a small number of people sprinkled throughout society who set an example of excellence for others.

In this chapter, our main emphasis will be not on the solid salt we know from our saltshakers, but rather on sodium chloride and other salts dissolved in liquids, especially water. We shall see what happens when salts dissolve and become ions, how ions travel from one place to another, and why they are

critical for life. We shall then look at how two particular ions—hydrogen ions, H^+, and hydroxide ions, OH^-—behave in water; these ions are special because they make things acidic or basic. We shall learn why acidity, or lack thereof, and the concentration of ions in a solution can have grave consequences for the environment, for human blood, and for Egyptian mummies.

1914

1921

1933

1941

1956

1968

4.1 | Ionic Liquid: A Contradiction in Terms

Chemists define the term *salt* more broadly than laypersons do. When the average person thinks *salt*, what usually comes to mind is table salt, sodium chloride. For a chemist, however, a salt is *any* ionic solid, be it sodium chloride, magnesium bromide, potassium sulfate, or calcium iodide. Recall from Chapter 2 that salts typically are formed from a cation from the left side of the periodic table combined with an anion from the right side of the table. For example, a potassium atom gives up one electron in order to have the same number of valence electrons as argon (Figure 4.1). In giving up the electron, the potassium can form a salt with, say, a chlorine atom, which accepts one electron in order to have the same number of valence electrons as argon. These two species come together in an **ionic interaction**, an attraction between positive and negative charges. The interaction is so strong that the resulting salt is rock solid and robust. One result of this ultrastrong ionic interaction is that salts typically have extremely high *melting points*, the temperature at which the solid becomes a liquid. (Sodium chloride, for instance, melts at 801 °C.)

QUESTION 4.1   How many electrons are there in (*a*) a potassium atom, (*b*) a chlorine atom, (*c*) a calcium ion, (*d*) a fluoride anion, (*e*) a potassium ion, (*f*) an iodine atom, (*g*) a chloride anion, (*h*) a sulfide anion?

ANSWER 4.1   (*a*) 19, (*b*) 17, (*c*) 18, (*d*) 10, (*e*) 18, (*f*) 53, (*g*) 18, (*h*) 18.

Recently, however, chemists have had to modify the way they think about salts. What they once took for fact—that all salts are solids at room temperature—has been toppled by the rediscovery of an old idea. In the 1940s, salts that melt at low temperatures, called *ionic liquids*, were discovered. They were largely ignored until recently, when scientists began to realize how useful they could be. A salt that is a liquid at room temperature has some very useful properties. Used as a solvent for other substances, for example, it could replace solvents typically used for organic synthesis—nasty compounds like benzene and methylene chloride. An ionic liquid could be used without damage to the environment because salts are, for the most part, benign. Every year, pharmaceutical and chemical companies spend billions of dollars to dispose of and recycle waste organic solvents safely. However, an ionic solvent could be recycled and reused inexpensively. Now hundreds of scientists have decided that ionic liquids deserve further scrutiny, and whole institutes have been founded to study them.

What exactly is an ionic liquid? To a scientist, the term seems like an oxymoron. After all, ionic compounds are high-melting-point solids, not liquids. Clearly, ionic liquids must be different from run-of-the-mill salts like sodium chloride. In fact, they *are* different in one very important way: the ions in ionic liquids do not pack tightly into a crystalline lattice the way the ions of a traditional ionic substance do. The main reasons for this difference are ion size and shape. Sodium ions and chloride ions are both essentially spherical and have similar radii—116 pm and 167 pm, respectively. For this reason, the ions easily stack into a pattern of alternating cations and anions, as shown in Figure 4.2a. The cations and anions of ionic liquids, however, are not as evenly

▼ FIGURE 4.1 **Forming an ionic solid.** A potassium ion is formed when a potassium atom loses one electron. A chloride ion is formed when a chlorine atom gains one electron. Both ions have the same number of electrons as argon (18). These two ions combine in a one-to-one ratio to form an ionic solid, the salt potassium chloride.

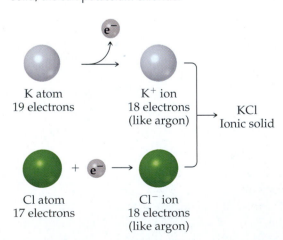

K atom
19 electrons

K⁺ ion
18 electrons
(like argon)

Cl atom
17 electrons

Cl⁻ ion
18 electrons
(like argon)

KCl
Ionic solid

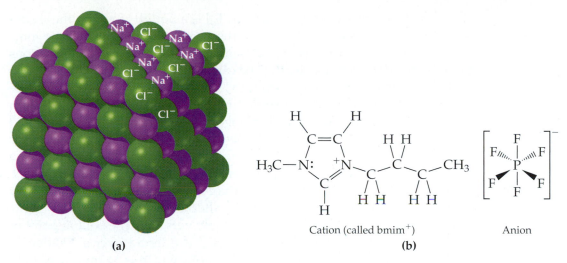

(a)   Cation (called bmim⁺)   Anion
                                                        (b)

▲ **FIGURE 4.2   Solid and liquid-ionic compounds.   (a)** The structure of a NaCl crystal, a "traditional" ionic substance, a solid at room temperature. **(b)** The structure of the bmim⁺ cation and PF_6^- anion, which combine to form a substance known as an ionic liquid. One of the most useful properties of ionic liquids is that they are liquids at room temperature.

matched either shapewise or sizewise, and this difference makes it difficult for the ions to maneuver into a crystalline lattice.

Figure 4.2*b* shows the cation and anion in one particular ionic liquid. The cation is called bmim⁺ for simplicity. It consists of a flat ring of carbon and nitrogen atoms plus two organic "arms" containing singly bonded carbons. Because of the single bonds in the four-carbon arm, the geometry around the carbons in the arm is tetrahedral and, as a result, the arm has a flexible zigzag shape. Thus the bmim⁺ cation resembles a pancake with flailing arms, which means it looks nothing like a sodium cation.

The bmim⁺ cation is much bigger than a sodium ion, but they both have the same charge: +1. For bmim⁺, that single charge is spread out, or *delocalized*, over the five atoms of the ring. (See Section 3.4 if you need to review delocalization.) The sodium cation, in contrast, has its one positive charge confined to a very small area. When an ion has its charge concentrated into a small volume, the ion attracts an ion of opposite charge forcefully. However, when the same charge is delocalized over many atoms, the strength of that charge is diminished and attraction between oppositely charged ions is weaker.

The anion in Figure 4.2*b*, PF_6^-, is spherical and much smaller that the bmim⁺ cation. Like the cation, however, it has multiple atoms, and the single negative charge is delocalized over all of them. This situation is very different from what we find with the chloride anion, which is very compact and has its one negative charge packed into a small volume.

QUESTION 4.2   What is the molar mass of the bmim⁺ cation shown in Figure 4.2*b*?

ANSWER 4.2   The molecular formula for the cation is $[C_8H_{15}N_2]^+$. Its molar mass is 139.22 g/mol.

The differences between the sizes and shapes of the ions in the ionic liquid [bmim][PF₆] versus those in NaCl make for striking variations in melting points. Sodium chloride has a melting point of 801 °C, but the ionic liquid has a melting point of 0 °C! Thus, at room temperature (about 25 °C), sodium chloride is a solid and [bmim][PF$_6$] is a liquid. How can these substances be so unlike each other when they are both held together by ionic bonds? The answer is that the small, same-shape sodium and chloride ions can pack together very efficiently. The cations and anions in the NaCl crystal lattice attract one another strongly, and this makes sodium chloride extremely difficult to melt. The similarity in ion size and shape in a sodium chloride crystal makes the crystal analogous to a crate full of neatly stacked oranges. The ionic liquid, though, is like a crate filled with lamps and hair dryers: there is no way to stack them neatly and close to one another. As a result, the ionic bonds in [bmim][PF$_6$] are much weaker than those in NaCl.

Ionic liquids are becoming popular very quickly, not only because they offer a more environmentally friendly alternative to traditional organic solvents, but also because they are extremely versatile. The boiling point of [bmim][PF$_6$] is about 200 °C. This means it is a liquid over a 200-degree temperature range—unlike water, for example, which is a liquid over only a 100-degree range. If you look at the cation structure in Figure 4.2*b*, you will see there is one arm containing one carbon and one arm containing four. This ion could just as easily have five carbons on one arm and two on the other, or any number of alternative configurations. Each variant will have slightly different properties, including melting point. In the words of Kenneth Seddon, an ionic-liquid researcher at Queen's College in Belfast, "You've got this incredible flexibility for getting the system right." He estimates that more than one trillion different ionic liquids could, in theory, be made.

Figure 4.3 illustrates one ionic-liquid application. Suppose you have water contaminated with heavy metal ions from industrial pollution. These ions dissolve better in an ionic liquid than in water. Water does not mix with the ionic liquid, however. Rather, the two liquids form two layers like the oil and vinegar layers in a bottle of salad dressing. When you put the dirty water and the ionic liquid in a separating funnel like the one shown

RECURRING THEME IN CHEMISTRY The individual bonds between atoms dictate the properties and characteristics of the substances that contain them.

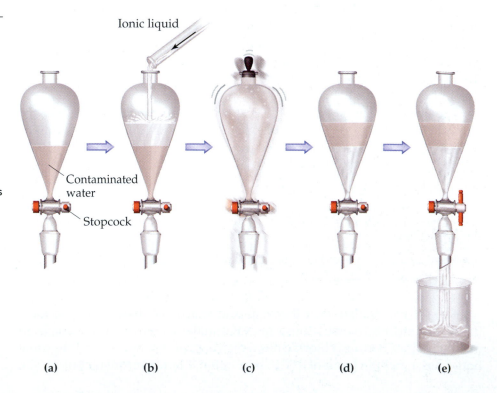

▶ FIGURE 4.3 **Cleaning water: an ionic-liquid application.** **(a)** Water contaminated with heavy metal ions is poured into a separating funnel. **(b)** An ionic liquid is added to the funnel, and two layers form, with the water on the bottom and the ionic liquid on top. **(c)** The two liquids are mixed thoroughly. **(d)** After sitting for a while, the liquids again separate into two layers. The contaminating ions are now in the ionic-liquid top layer, leaving only clean water in the bottom layer. **(e)** The water is collected through the stopcock at the bottom of the funnel.

Ionic liquid

Contaminated water

Stopcock

(a)   (b)   (c)   (d)   (e)

in Figure 4.3, the ionic-liquid layer will sit on top of the water layer. Shake the funnel a bit to allow the metal ions to come into contact with the ionic liquid, then open the stopcock and let the clean water flow out. The contaminating ions remain behind in the ionic liquid, and the water is purified. There is no doubt that ionic liquids will have hundreds of uses very soon. James Davis, an ionic-liquids researcher at the University of California, Davis, put it best: "That looks like an orchard I'm never going to finish picking from."

QUESTION 4.3 Ionic liquids are not your typical salts. How is the structure of an ordinary salt different from that of an ionic liquid?

ANSWER 4.3 A typical salt is composed of small positively and negatively charged ions that fit snugly next to one another in a crystalline lattice. Ionic liquids contain large, bulky ions that are unable to fit into a regular crystalline lattice.

4.2 | Egyptian Mummies and the Polyatomic Ion

Being an early Egyptian was not easy. The Egyptians believed that your body as well as your spirit would continue in the afterlife, but only if very specific rules were followed. In particular, it was important that the body be preserved so that it would be intact in the afterlife. This preservation was accomplished through mummification (Figure 4.4), a ritual that has been studied for centuries. There are still conflicting ideas, though, about exactly how mummification was done. For example, Egyptologists are fairly certain that *natron*, a mixture of salts, was used in mummification, but its purpose is still a topic of controversy among scholars. Was the natron used to pickle the body, in much the same way NaCl pickles a cucumber? Was it used as a drying agent? The remnants of early Egyptian civilization support the latter theory. To date, no one has found any of the large tubs that would have been necessary for pickling a body in water. What today's researchers *have* found are large, wide tables suitable for laying out a body to dry. These tables may have been used for mummification, but they also may have had some other unknown purpose. Clearly, there are still many unanswered questions about the ritual of mummification.

In 1994, scientists at the University of Maryland School of Medicine decided to try mummification for themselves, reasoning that a modern mummy would provide a replica of an ancient specimen. Because their modern mummy would involve a modern-day person with a lifetime medical history, it would provide a known standard with which other mummies could be compared. First the Maryland scientists went in search of someone to mummify. Their criteria were strict: the person must have died of natural causes, never have had surgery, and have filled out a donor card. The body of a man from Baltimore who had donated his body to science fit the bill. Wanting to mimic the mummification ritual as closely as possible, the researchers worked on every detail. They considered everything from the vessels that held the embalming solutions (the vessels were handmade and painted with authentic hieroglyphics) to the embalming table, which was copied from ancient tables found in Egypt. They even

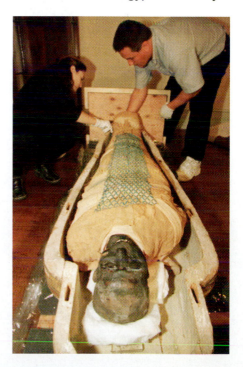

▼ FIGURE 4.4   An Egyptian mummy.

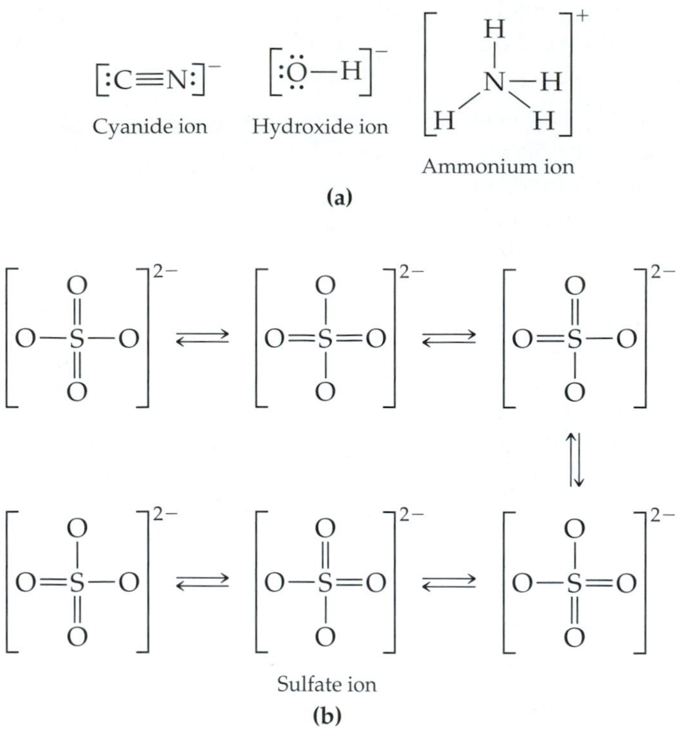

(a)

Sulfate ion

(b)

▲ **FIGURE 4.5 Polyatomic ions.** **(a)** The cyanide ion, ammonium ion, and hydroxide ion are three examples of polyatomic ions. **(b)** Six resonance stabilized forms of the sulfate polyatomic ion.

went to Egypt to procure spices, oils, and salts to make natron from a local source.

Natron is composed of four salts: sodium carbonate, Na_2CO_3, sodium bicarbonate, $NaHCO_3$, sodium chloride, NaCl, and sodium sulfate, Na_2SO_4. From this list, only sodium chloride is familiar to us so far in this book. We have seen that the simplest ions are single atoms that have been ionized to form a cation or anion. Simple salts, such as sodium chloride, are composed of this type of ion, but the other three salts listed contain ions made up of not one atom, but several. Take, for example, sodium carbonate. Sodium is familiar to you, but the carbonate ion, CO_3^{2-}, probably is not. Likewise, the ions in ionic liquids are complex. Bmim$^+$, for example, contains 33 atoms! Ions of this type are referred to as **polyatomic ions** because they contain more than one atom. The electron dot structures for three polyatomic ions are depicted in Figure 4.5a.

Think back to our procedure for drawing electron dot structures from Chapter 3, the six steps of which are reproduced in Table 4.1. The six steps can be condensed into two rules for electron dot structures: (a) a correctly drawn structure must have the correct number of valence electrons, and (b) in a correctly drawn structure, each atom must form the number of bonds listed in Table 3.2.

Do the electron dot structures for the polyatomic ions in Figure 4.5a follow these two rules? Let us look at rule 1 first. We expect $4 + 5 = 9$ valence electrons from the two atoms in the cyanide ion, but because the ion has a -1 charge, we must add one electron: $9 + 1 = 10$ valence electrons. A count of the valence electrons in cyanide indeed gives this total: six in the triple bond plus four in the two nonbonding pairs. Ammonium ion and hydroxide ion should have eight valence electrons each, and indeed they do. Thus, rule 1 holds for ions and molecules alike.

A glance back at Table 3.2, where you will see that carbon must form four bonds, nitrogen three, and oxygen two, tells you that rule 2 goes right

---

TABLE 4.1   Procedure for Drawing a Molecule's Electron Dot Structure

1. Count the total number of valence electrons in the molecule.

2. Draw the molecule's skeletal structure from its molecular formula.

3. Calculate the number of electrons used to show the skeletal structure, and subtract this number from the total calculated in step 1.

4. Assign the remaining electrons to atoms connected to the central atom, following the octet rule for all atoms except H (which should have a duet of electrons).

5. If you have any electrons left after completing step 4, place them on the central atom. If you have run out of electrons before the central atom has an octet, move one or more nonbonding electron pairs from an adjoining atom to the central atom to form a multiple bond.

6. Check that every atom has an octet of electrons (duet for H) and that all the valence electrons calculated in step 1 have been placed.[a]

---

[a]There are some exceptions to the octet rule, but you will not need to know them for the electron dot structures in this book.

out the window for the examples in Figure 4.5a. There are two reasons for this. First, we are dealing with ions, not neutral molecules. Recall from Chapter 3 that the rules governing electron dot structures apply only to *uncharged molecules*. (Now you know why.) For example, nitrogen makes four bonds (not three) when it has a charge of +1, as it does in the ammonium ion of Figure 4.5a. Second, most simple polyatomic ions are not organic molecules because they are not based on a carbon-chain framework. Rather, they are *inorganic* compounds that happen sometimes to contain carbon. Thus, when you leave the realm of organic chemistry, rule 2 no longer applies.

Inorganic compounds have a wider range of structures available to them because they are made up of a wider variety of elements than organic compounds are. For example, the permanganate ion, MnO_4^-, contains manganese, an element not found in most organic molecules. Additionally, some atoms in inorganic compounds can make more than four bonds to other atoms and often do so. In our study of organic bonding in Chapter 3, the maximum number of bonds formed was the four bonds of carbon. But in the inorganic sulfate ion, for instance, the sulfur makes *six* bonds—double bonds with two of the oxygens and single bonds with the other two oxygens (Figure 4.5b). This atom can form all these bonds because it is larger than carbon and can fit more atoms around it. Thus polyatomic ions follow rule 1 for valence electrons, but not rule 2 for numbers of bonds to each atom.

QUESTION 4.4 Why are six versions of the sulfate ion shown in Figure 4.5b? What do we call this condition?

ANSWER 4.4 The bonds in the sulfate ion alternate between double and single, and so we can show the bonds in different (but equivalent) positions. Each bond is in actuality a bond and a half, and we call the condition *resonance stabilized* (Section 3.4).

We know from Chapter 2 that simple salts are electrically neutral. For example, magnesium chloride contains Mg^{2+} ions and Cl^- ions (note that in the name of a salt, the cation name is followed by the anion name). In order to make an electrically neutral salt from these ions, we must have two chloride ions for each magnesium ion: $MgCl_2$. The same rule applies to salts containing polyatomic ions. For example, ammonium carbonate, which contains ammonium ion, NH_4^+, and carbonate ion, CO_3^{2-}, must have zero net charge. Therefore, there must be two ammonium ions for every carbonate ion: $(NH_4)_2CO_3$. We enclose polyatomic ions in parentheses when we add subscripts to them so that the total number of atoms is perfectly clear. For example, the subscript 2 in the formula for ammonium carbonate tells us that each subscript inside the parentheses is to be multiplied by 2. Thus, one formula unit of this salt contains two atoms of nitrogen, eight atoms of hydrogen, one atom of carbon, and three atoms of oxygen. Sodium phosphate, which contains Na^+ ions and polyatomic PO_4^{3-} ions, has the formula Na_3PO_4. Because sodium ion is monatomic, it is not enclosed in parentheses, and we understand that there are three of them in this formula. As we shall see, polyatomic ions are as important physiologically as are simpler ions. For example, carbonate ions are responsible for maintaining certain ion concentrations in blood, and peroxide ions (O_2^{2-}) are thought to be partly responsible for aging. Table 4.2 lists common polyatomic ions.

TABLE 4.2   Common polyatomic ions

Cations

Hg_2^{2+}	Mercury(I)	NH_4^+	Ammonium

Anions

NO_3^-	Nitrate	CO_3^{2-}	Carbonate
SO_4^{2-}	Sulfate	HCO_3^-	Hydrogen carbonate (bicarbonate)
OH^-	Hydroxide		
CN^-	Cyanide	ClO_4^-	Perchlorate
PO_4^{3-}	Phosphate	MnO_4^-	Permanganate
HPO_4^{2-}	Hydrogen phosphate	CrO_4^{2-}	Chromate
$H_2PO_4^-$	Dihydrogen phosphate	O_2^{2-}	Peroxide

QUESTION 4.5 Using either a polyatomic ion from Table 4.2 or any monatomic ion you like, form a salt from each polyatomic ion in Figure 4.5. Write the name and chemical formula for each salt you form.

ANSWER 4.5 Because the choice of ion is yours, there are many correct answers, and only one is given here for each ion in Figure 4.5: sodium cyanide, $NaCN$; ammonium iodide, NH_4I; magnesium hydroxide, $Mg(OH)_2$; calcium sulfate, $CaSO_4$.

QUESTION 4.6 Using the information given in Table 4.2, determine whether or not each chemical formula given here is electrically neutral. If it is not, change the number of one ion in the formula to make the formula have zero charge. Then give the name corresponding to each correct formula: (a) CaO_2, (b) KCO_3, (c) $Mg_3(PO_4)_2$, (d) $Li(CN)_2$, (e) $KMnO_4$, (f) $(NH_4)_2CrO_4$.

ANSWER 4.6 (a) Correct because the ions are Ca^{2+} and O_2^{2-}; calcium peroxide. (b) Incorrect because the ions are K^+ and CO_3^{2-}; correct formula K_2CO_3; potassium carbonate. (c) Correct; Mg^{2+} and PO_4^{3-}; magnesium phosphate. (d) Incorrect; Li^+ and CN^-; correct formula $LiCN$; lithium cyanide. (e) Correct; K^+ and MnO_4^-; potassium permanganate. (f) Correct; NH_4^+ and CrO_4^{2-}; ammonium chromate.

4.3 | Keeping Your Mummy Dry

A **desiccant** is a drying agent, something that removes water from a substance.

The natron used in mummification removes the water present in the body tissue, but how does a salt act as a desiccant? What properties of salts allow them to attract water? In order to answer these questions, we must go back to the geometry of the water molecule:

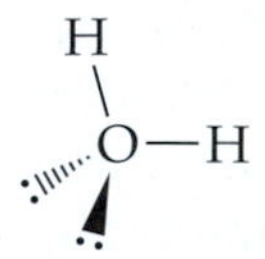

Tetrahedral electron-pair geometry 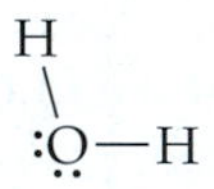   Bent molecular geometry

Recall that the water molecule has a bent geometry, with a 104.5° bond angle between its three atoms. As described in Section 3.8, the electrons in a water molecule are concentrated on one side of the molecule, leaving the other side electron-deficient. How do we know, though, where electron density will build up and where it will be deficient? In Section 2.2, we introduced the concept of *electronegativity*, the tendency for an atom to draw electrons toward itself. Figure 4.6 shows a partial table of electronegativity values. (Refer back to Figure 2.9 for the complete table.) The value for oxygen is 3.5, meaning that it has a strong propensity for attracting electrons, and the value for hydrogen is 2.1. Thus, in any bond between oxygen and hydrogen, the electrons are pulled toward the oxygen atom and away from the hydrogen atom, as described in Chapter 2 for water.

Whenever a molecule has one end that is relatively more positive and another end that is relatively more negative, we say that the molecule is **polar**, and we refer to this imbalance as a **dipole**. The water molecule's dipole can be thought of as an imaginary arrow running through the oxygen atom:

The head of the arrow points to the end of the molecule having the greater electron density (the more negative end because electrons carry a negative charge). A crosshatch resembling a plus sign on the tail of the arrow points to the electron-deficient end (the more positive end). Another way to depict a molecular dipole is with plus and minus signs on the Greek lowercase delta, δ:

$$\overset{\delta-}{\ddot{\text{O}}}\text{—H} \atop \text{H} \ \delta+$$

A $\delta-$ sign is placed at the negative end of the molecule, and a $\delta+$ sign is placed at the positive end. These **partial charges** suggest there is part of a charge in one place and part of a charge in another. These are not quite full charges like the ones found on ions—but almost.

QUESTION 4.7   The ammonia molecule has a dipole. Which way does the imaginary arrow representing the dipole point? In other words, which part of the molecule is relatively more negative and which part is relatively more positive?

ANSWER 4.7   The electron density is highest on the nitrogen atom because its electronegativity (3.0) is greater than that of the hydrogen atoms (2.1). The hydrogen atoms are electron-deficient because the one electron each hydrogen atom contributes to an N–H covalent bond is pulled toward the nitrogen. Thus, each N–H bond has a dipole, with the negative end at the nitrogen and the positive end at the hydrogen. The average of these three dipoles gives one global dipole for the molecule. This global dipole begins at a point in the middle of the three hydrogens ($\delta+$) and extends through the center of the nitrogen ($\delta-$):

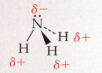

Dipoles depicted with partial charges

The three individual N–H dipoles

The average of the three individual dipoles gives one global dipole for the molecule.

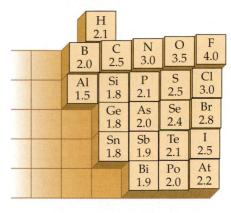

▲ **FIGURE 4.6   Electronegativity values.**   A partial table of the electronegativity values for selected elements.

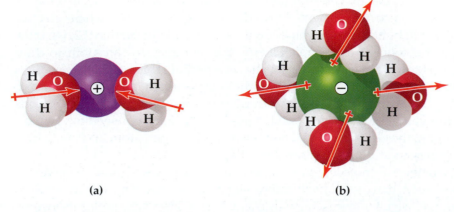

► FIGURE 4.7 **Hydrated ions.**
(a) Hydration of a cation by water molecules. **(b)** Hydration of an anion by water molecules. Notice that the water molecules are oriented differently in the two cases. In **(a)**, the highly electronegative oxygen atoms are closest to the positively charged cation. In **(b)**, the oxygen atoms are farther from the negatively charged anion.

(a)　　　　　　　　　　　　(b)

The partial charges on the water molecule allow us to predict which end of the molecule will interact either with another polar molecule or with an ion. For example, a sodium ion is attracted to water molecules because, being positively charged, the ion is attracted to the negative end of each water molecule. Thus, water molecules pack in around the sodium ion as closely as possible. We say the ion is **hydrated**, or surrounded by water, and the process is called **hydration**. Hydration also happens with anions because the positive end of a water molecule is attracted to the negative charge on an anion. These interactions are shown schematically in Figure 4.7. Because water, which has a dipole but is not an ion, is interacting with ions, we think of hydration as an **ion–dipole interaction**. The ion–ion (that is, ionic) interaction in sodium chloride arises because of the attraction between charges. Collectively, these two types of interaction are referred to as **noncovalent interactions** because no electrons are shared.

Here is a rule of thumb we shall see over and over whenever we are dealing with substances in solution: *like dissolves like*. This means that substances with similar polarities tend to dissolve easily in one another. Therefore, if something is ionic, like a salt, it will dissolve in a solvent that is either ionic or polar, such as an ionic liquid or water. Likewise, nonpolar substances tend to dissolve readily only in one another. Thus, a nonpolar compound, like benzene, should easily dissolve in another nonpolar substance, such as octane.

The hydration of ions is so favorable that most salts dissolve readily in water. Substances that carry an electrical charge, whether a partial charge (dipoles) or a full charge (ions), interact with one another through attraction between positive and negative. If you drop a crystal of table salt into a glass of water, the salt quickly disappears as it dissolves in the water. The uniform crystalline lattice of the sodium chloride crystal is gradually broken down as water molecules pluck ions away from it and pull them into the surrounding solution, as shown in Figure 4.8.

RECURRING THEME IN CHEMISTRY Like dissolves like.

QUESTION 4.8 Which of these pairs interact via an ion–dipole interaction: (*a*) water molecule and magnesium ion, (*b*) ammonia molecule and iodide ion, (*c*) lithium ion and bromide ion, (*d*) water molecule and ammonia molecule?

ANSWER 4.8 (*a*) and (*b*). In (*a*), the negative end of the water molecule interacts with the positively charged magnesium ion. In (*b*), the positive end of the ammonia molecule interacts with the negatively charged iodide ion. There are no dipoles in (*c*) and no ions in (*d*), meaning that neither case can be an ion–dipole interaction. (Ammonia and water do interact with each other, though, as you will see in Chapter 7.)

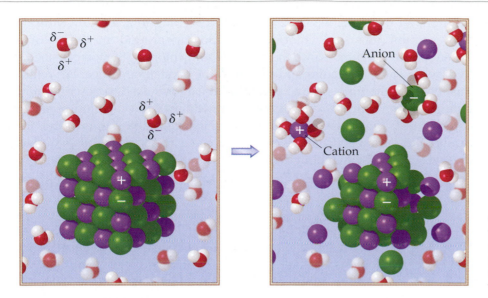

◄ **FIGURE 4.8** **Dissolution of a salt in water.** In this ion–dipole interaction, the cations and anions in solid NaCl are plucked away from the crystal by water molecules.

LIVE ART
Dissolving NaCl

QUESTION 4.9 Sketch the interaction between water molecules and (*a*) an ammonium ion, (*b*) a sulfate ion.

ANSWER 4.9 Be certain you draw the negative side of the water molecules pointing toward the positively charged ammonium ion, NH_4^+, and the positive side of the water molecules pointing toward the negatively charged sulfate ion, SO_4^{2-}:

Alert readers are probably wondering if there is a limit to the amount of salt you can dissolve in water. After all, if you add a tablespoon of water to a bucketful of sodium chloride, common sense tells you the salt will not dissolve. There simply is not enough water.

It is true that most salts dissolve readily in water. A chemist would say either that such salts are *soluble in water* or that their **solubility in water** is high. But there is always a limit to the solubility of any solid. If you have a lot of sodium chloride, but only a small amount of water, you will run out of water before the process shown in Figure 4.8 is complete. The available water will hydrate ions from the NaCl crystal one by one. Eventually, if enough NaCl is present, all the water molecules will be used up in the process of hydration. When this happens, you end up with a pile of solid NaCl in the watery solution. In Figure 4.9, sodium chloride has been added to a volume of water, and the solubility of the salt has been exceeded. The water has absorbed as many ions as it can. We say this solution is **saturated**—no more NaCl can dissolve.

Once a salt's solubility limit has been reached and you have a pile of salt sitting at the bottom of a solution, it may seem like nothing is happening in the solution. Nothing could be further from the truth. In our saturated NaCl

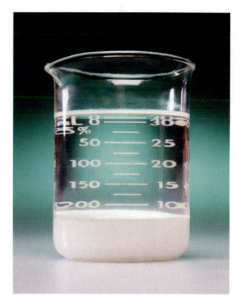

▲ **FIGURE 4.9** A saturated solution of sodium chloride in water.

Recall from Chapter 3 that the term *saturated* is also used to describe a long carbon chain containing all single bonds. Although this double usage might seem confusing, it rarely causes problems because the term is used in the context of organic hydrocarbons in one instance and in the context of molecules or ions in solution in the other.

solution, for example, we have a pile of salt sitting in a solution saturated with sodium ions and chloride ions moving around in the solution. Whenever an ion bumps into the solid sodium chloride, the ion can shed its hydrating water molecules and return to the crystalline lattice. We refer to this return to the solid state as **precipitation**, a term used to describe things falling, like rain. In this case, the ions are "falling" out of solution and re-entering the solid phase. As ions return to the solid phase, some other ions can be plucked away from the crystal by water molecules. Thus, you have these reactions going in both directions, indicated by a double arrow:

Solid NaCl crystalline lattice $\rightleftharpoons$ Na$^+$ ions and Cl$^-$ ions in solution

Solid phase   Liquid phase

When a saturated solution has been sitting long enough, there is no net change in the size of the pile, and we say the system is at **equilibrium**. Nevertheless, the ions are still moving back and forth between liquid and solid. In a solution at equilibrium, the rate at which ions leave the solid and

▶ FIGURE 4.10 **Systems at equilibrium.** **(a)** In a saturated solution, ions of the crystalline lattice are in equilibrium with hydrated ions in solution. **(b)** In a glass of water, the solid ice is in equilibrium with the water around it. Water molecules in humid air are in equilibrium with water on the lake's surface.

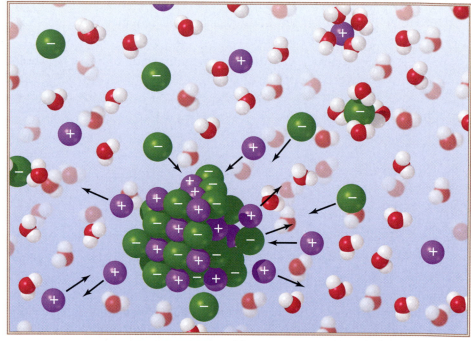

(a)

(b)

go into solution equals the rate at which ions return to the solid lattice:

$$\frac{\text{Rate at which ions in a}}{\text{salt dissolve in liquid}} = \frac{\text{Rate at which ions precipitate}}{\text{into solid}}$$

Equilibria exist in many other situations, a few of which are shown in Figure 4.10. The surface of a liquid can be at equilibrium with gaseous molecules in the air above the liquid. The surface of a liquid can be at equilibrium with another liquid surface, as happens with oil and water. A solid can become a gas directly, and the gas and solid can reach a point of equilibrium. We shall see examples of systems at equilibrium throughout this book.

Now that we understand the concept of equilibrium, we can better appreciate the University of Maryland mummification project. The process used to mummify a body is designed to remove water, because where there is water, there are likely to be bacteria. Bacteria decompose human flesh, and you do not want to be partially decomposed in the afterlife. Thus, natron is packed into the body cavity. The surrounding tissue, which is primarily water, comes into contact with the natron. The ions that compose the solid salts in the natron become surrounded by molecules of water. Because there is much more salt than water, all the water gets absorbed by the salt, and equilibrium is never reached. There is simply too much salt. Consequently, the tissues are completely dehydrated. The ability of salts to suck water from whatever they touch makes them useful as preservatives for things other than mummies. Meat is cured, for example, by coating it with a heavy layer of sodium chloride. This process removes water from the meat and thereby prevents spoilage by bacterial growth.

The same properties of salts that make them useful can also make them harmful, however. Scientists at the University of Missouri–Rolla have been studying why some of Egypt's antiquities are crumbling. Many sandstone structures, such as the Temple of Luxor (Figure 4.11), are deteriorating at an alarming rate. The Missouri–Rolla workers have determined that a salty residue resulting from Egypt's dry climate becomes wedged in the nooks and crannies of the sandstone monuments. The salts in the residue absorb water from the air and swell, pushing on the sandstone until finally the monument crumbles. In the words of Richard Stephenson of Missouri–Rolla, "These salt precipitates may be expanding and possibly speeding the degradation by exploding the sandstone from the inside out."

▲ FIGURE 4.11 **The damage salt can do.** The Temple of Luxor in Egypt, as well as other historic monuments in that country, are threatened by the presence of salt.

4.4 | Gatorade and the Chemistry of Electrolytes

The average human sweats out about 1.5 L of water every day, but a person exercising vigorously can lose up to 16 L of water in a day, about 20 percent of her or his body mass. The liquid lost by sweating must be replaced by drinking either water or a reasonable substitute. Sweat is not pure water, however. It contains salts because body fluids are salty.

When salts dissolve in water, they produce ions, and the presence of these ions means the solution can conduct an electric current. Figure 4.12 shows that pure water will not light up a light bulb, but a solution containing salt will. Because dissolution of a salt in water yields a solution capable of conducting electricity, both salts and the ions forming them are frequently referred to as **electrolytes**. This term is often used in reference

▲ FIGURE 4.12 **Ions can create an electric current.** The beaker on the left contains pure water. The light bulb is dark because pure water cannot conduct electricity, and so this setup does not constitute a complete electric circuit. The beaker on the right contains a solution of some electrolyte dissolved in water. The bulb is lit because the ions in the solution conduct electricity and so the circuit is complete.

to ions dissolved in body fluids, and so we often hear about maintaining *electrolyte balance*. The dissolved salts in the body are critical for maintaining blood pressure, neural function, and healthy cells. Therefore an athlete losing 16 L of sweat is also losing valuable electrolytes that must somehow be replaced.

Because sports drinks have a chemical makeup similar to that of sweat, they replace the electrolytes and water lost during physical exertion. Of course, even though these drinks are sweat impersonators, manufacturers wisely leave the word *sweat* out of drink names (no one wants to drink a product called "Sweat Substitute"). Instead, we have Gatorade®, the granddaddy of sports drinks.

Imagine it is the year 1964 and you are the coach of the struggling University of Florida Gators football team. Your players are dropping like flies during practice in the sizzling Florida heat. You try the usual dehydration remedy—salt tablets—but they do not help. Then you try orange juice, but it causes stomach cramps. Finally you go to the team doctor, Dana Shires, for help. He turns to a University of Florida medical researcher by the name of Robert Cade, and together they work out a solution (literally).

What Cade and Shires realized was that salt tablets replace the salts lost by sweat, but not the water. Without water, salt tablets just make dehydration worse. They also found that the reason orange juice, which is about 10 percent sugar, causes stomach cramps is that high sugar concentrations in the stomach keep water from passing through the stomach wall and into the body's cells, where it is needed. Armed with this knowledge, Cade and Shires formulated a sweatlike concoction. First they determined that 1 L of sweat contains about 100 mg each of sodium and potassium ions and about 25 mg of magnesium ions (balanced by an appropriate concentration of anions—in this case, chloride ions). They mixed these salts in water as a first step, but the solution had a most objectionable flavor. They then made the solution about 6 percent sugar—enough sugar to mask the bad taste, but not enough to cause stomach cramps. The drink was a godsend for the Florida Gators. No more stomach cramps. No more dehydration. In 1966, they went to the Orange Bowl. The score was Georgia Tech, 12, Gators, 27. The *Jacksonville Journal* described the impact Gatorade had on the team this way:

> The Gators went 7–4 that year [1965], then 9–2 in '66. Time after time, the Gators outplayed their opponent late in the game, even under conditions that sportscaster Keith Jackson said "would make a salamander sweat."

Twenty-first-century Gatorade is not the same drink it was in 1965. Nowadays, no Mg^{2+} ions are used, but sodium chloride and potassium chloride are still added in the concentrations given in Table 4.3. When these salts hit water, they form hydrated ions, and this transformation can be written in the form of a chemical equation:

$$NaCl(s) \xrightarrow{H_2O} Na^+(aq) + Cl^-(aq)$$

$$KCl(s) \xrightarrow{H_2O} K^+(aq) + Cl^-(aq)$$

TABLE 4.3   Concentration of Electrolytes in Some Liquids

Electrolyte	Human blood plasma	Human sweat	Gatorade	Orange juice	Coconut milk	Carrot juice
Na^+	140 mM	20 mM	20 mM	0.43 mM	5.7 mM	3.9 mM
Mg^{2+}	5.2 mM	8.0 mM	0	4.5 mM	19 mM	5.8 mM
K^+	4.3 mM	10 mM	6.6 mM	100 mM	116 mM	150 mM
Cl^-	100 mM	38 mM	27 mM	110 mM	160 mM	170 mM
Sugar			6%	10%	3%	9%

In each of these equations, the salt on the left side is a solid, indicated by the notation (*s*). After the salt is added to water, the ions separate and commingle, and the (*aq*) notation on the right means the salts are dissolved in a watery, or **aqueous**, solution.

Table 4.3 compares the ion concentrations in several substances, including human blood. The unit mM indicates **millimolar**, a concentration unit that tells you the number of millimoles of some substance per liter of solution. (There are 1000 millimoles in one mole.) A related unit, M, stands for **molarity** (also called *molar concentration*), which indicates the number of moles of some substance per liter of solution:.

$$\text{Molarity} = \frac{\text{moles of substance}}{\text{liter of solution}}$$

Concentration units are independent of volume. That is, you can have a 1.0-M NaCl solution in a swimming pool or in a soda can. The concentrations are the same, and every liter of solution in the pool and in the can will contain 1 mol of salt.

Let us consider a straightforward example. If you have a 6.00-L bucket filled with an aqueous sodium chloride solution that contains 0.520 mol of sodium chloride, the concentration of the solution is

$$\text{Molar concentration} = \frac{0.520 \text{ mol NaCl}}{6.00 \text{ L solution}} = \frac{0.0867 \text{ mol}}{1.00 \text{ L solution}} = 0.0867 \text{ M}$$

QUESTION 4.10 An intravenous saline solution typically contains 0.750 g of sodium chloride per 100.0 mL of solution. What is the molar concentration of NaCl in this solution?

ANSWER 4.10

$$\text{Molar concentration} = \frac{0.750 \text{ g NaCl}}{100.0 \text{ mL solution}} \times \frac{1000 \text{ mL}}{1.00 \text{ L}} \times \frac{1 \text{ mol NaCl}}{58.50 \text{ g NaCl}}$$

$$= \frac{0.128 \text{ mol NaCl}}{1 \text{ L solution}} = 0.128 \text{ M NaCl}$$

QUESTION 4.11   A certain lake has a volume of 56,500 L. It contains 452 mol of magnesium chloride. What is the millimolar concentration of magnesium chloride in the lake water?

ANSWER 4.11   The lake contains 452 mol of magnesium chloride in 56,500 L of water. Thus,

$$\frac{452 \text{ mol}}{56,500 \text{ L solution}} \times \frac{1000 \text{ mmol}}{1 \text{ mol}} = \frac{8.00 \text{ mmol}}{1 \text{ L solution}} = 8.00 \text{ mM}$$

STUDENTS OFTEN ASK

Is Fortified Water Useful for Ion Replacement, or Is It Just Very Expensive Water?

Go to a health food store's beverage aisle, and you will find an entirely new sort of drink: fortified water. This new trend in "performance drinks" follows on the heels of the bottled-water craze of the 1990s, when sales of bottled water in the United States exceeded seven billion dollars. Now the next generation of water is on the market, and only time will tell whether the American consumer is going to accept this latest product in drinkable nutrition.

Fortified waters claim to be much more than just good water. "Fortified" with vitamins, herbs, electrolytes, or sugars, these drinks are being marketed as a twenty-first-century alternative to Gatorade. After all, Gatorade has been around for decades, and it does not contain any New Age herbal pick-me-ups, like ginkgo biloba or St. John's wort. However, do herbal additives and vitamin supplements really improve athletic performance, or is Gatorade still the best thing for a thirsty athlete?

Table 4.4 shows the ingredients in some fortified waters. None of these drinks contains as much sodium ion as Gatorade, and for good reason: the sodium ion levels in Gatorade (20 mM) make the water very salty, and this taste must be masked with a lot of sugar. TRINITY Natural Mineral Dietary Supplement has about 14 percent of the sodium ion in Gatorade and human sweat. GLACÉAU vitaminwater has one-third the potassium in human sweat and no sodium. Snapple Rain has the same amount of sugar found in orange juice and therefore probably causes stomach cramps—meaning you may not want to take this drink with you to a marathon. GLACÉAU smartwater contains calcium, magnesium, potassium, chloride, and bicarbonate ions, only three of which are found in sweat. It is missing sodium, though—the cation lost most in sweat.

Part of the appeal of these drinks seems to be their *psychological* effect (in your head) rather than their *physiological* effect (in your body). The bottles are very smart looking, and the price tags make you think these products must be good for you. However, even though GLACÉAU vitaminwater and Snapple Rain do contain impressive amounts of vitamins and herbs, many of these products are mostly water. GLACÉAU smartwater claims to be "nutrient enhanced," but contains no vitamins or sugars. The GLACÉAU people claim *nutrient* means the drink is nourishing, not that it gives you energy. They also claim their water will make you smarter. The label reads:

> Side effects may include being called nerd, dork, geek, dweeb, brainiac, know-it-all, smarty-pants, smart-aleck, bookworm, egghead, four-eyes, Einstein … May include sudden and inexplicable aversion to all less intelligent forms of water, apply liberally and frequently to dry people.

So if you still feel smart after spending $1.39 for 33.8 oz of smartwater, then this is probably the drink for you.

TABLE 4.4	Electrolyte Concentrations and Ingredients in Selected Fortified Water Products			
Product	Contains vitamins or herbs?	Potassium concentration (mM)	Sodium concentration (mM)	Sugar (%)
GLACÉAU vitaminwater, revive flavor	Vitamins	3.24	0	4
TRINITY Natural Mineral Dietary Supplement	No	0	2.76	0
GLACÉAU smartwater, Nutrient Enhanced Water	No	Some	0	0
Snapple® Rain	Herbs	0	1.84	10
Alcer's Miracle Sports Water™	No	0.010	0	0

QUESTION 4.12   Some athletes swear by diluted orange juice to keep stomach cramps at bay. Suppose you start with 1 L of orange juice and add to it just enough water to form 2 L of solution. Will this dilution decrease the molar concentrations listed in Table 4.3? Will this diluted orange juice make a decent sweat-replacement drink?

ANSWER 4.12   The defining equation for molar concentration, molarity = number of moles/volume of solution, tells you that molar concentration goes down when you dilute a solution. When you dilute orange juice, the numerator on the right side of this equation stays the same because you are not adding any more orange juice. However, the denominator doubles because your final volume after adding the water is 2 L. Therefore, the molar concentration is one-half its original value. This makes sense because if you dilute something, it becomes less concentrated and its molar concentration gets smaller.

Once you have diluted the juice, each concentration in the orange-juice column of Table 4.3 is reduced to half the value shown in the table. The sugar concentration drops to 5 percent, into the range found in Gatorade. However, the potassium ion concentration in the diluted solution, 50 mM, will be much higher than that in sweat, and the diluted sodium ion concentration, 0.21 mM, will be much too low. Therefore, even though dilution keeps orange juice from causing stomach cramps, it does not provide the proper electrolyte concentrations to replace human sweat.

4.5 | Not Your Father's Tomato

In the Department of Pomology at the University of California, Davis, researchers are interested in only one thing: tomatoes. A tomato plant has very strict growing requirements: plenty of sunshine, plenty of fertilizer, plenty of water, and most of all, the right soil. Tomato plants are especially finicky when it comes to soil salt concentration: too high a concentration means no tomatoes. The reason is that salts in the soil hold on to water (as we know), which means the water stays in the soil rather than flowing into the plants. When tomatoes are planted in extremely salty soil, water is actually sucked *out* of the plants, so they become dehydrated and die.

What determines the direction in which water flows in a tomato plant in salty soil? To answer this question, we have to return to the notion of equilibrium. Take a look at Figure 4.13*a*, which shows a U-shaped tube with two compartments separated by a membrane. A *membrane* is a barrier that limits the passage of molecules and ions from one side of the membrane to the other. Membranes can be manufactured, as is the one in Figure 4.13*a*, or they can be natural, as in the case of cell membranes. As we shall see shortly, cell membranes regulate flow between compartments in living organisms. The membrane shown in Figure 4.13*a* is called a **semipermeable membrane**, meaning it allows only certain molecules

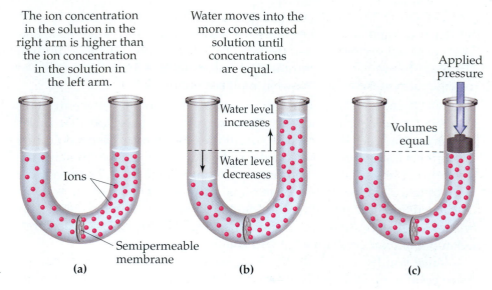

The ion concentration in the solution in the right arm is higher than the ion concentration in the solution in the left arm.

Water moves into the more concentrated solution until concentrations are equal.

Applied pressure

Water level increases

Water level decreases

Volumes equal

Ions

Semipermeable membrane

(a)

(b)

(c)

► **FIGURE 4.13 Osmosis across a semipermeable membrane.   (a)** The solution in the right arm has a higher concentration of ions (represented by the pink dots) than the solution in the left arm. Note that the water levels are equal. **(b)** Because of the difference in ion concentrations, water molecules move from the left arm into the right arm through the semipermeable membrane. After equilibrium has been reached, the ion concentrations are equal, but the water levels are unequal.

LIVE ART
Osmosis

(in this case, water) to pass through. Ions, which are bulky because they are surrounded by water molecules, are too large to pass through this semipermeable membrane. Initially, conditions in the tube are unbalanced, with the left arm containing water plus only a few ions and the right arm containing water plus many ions.

What does the system look like after a while? Well, over time, the system moves away from the initial, unbalanced condition and toward equilibrium, which is reached when the ion concentration is the same in the two arms. Because only water molecules can cross the membrane, equilibrium is attained by water molecules moving across the membrane from left to right, thereby diluting the solution in the right arm until the ion concentration is the same as in the left arm. This movement of water molecules across a semipermeable membrane from a region of low ion concentration to a region of higher ion concentration is the definition of **osmosis**.

But wait a minute. If air is pressing down on both water surfaces equally (and this is a safe assumption), how can the water levels change? The answer is that the upward force exerted by the right-arm water surface as water enters that arm is stronger than the downward force exerted by the air above the water surface. Although it seems peculiar, the water levels are uneven at equilibrium for the case shown in Figure 4.13b, but the *ion concentrations* are the same.

QUESTION 4.13 Comment on this statement: A tube like the one shown in Figure 4.13, equipped with a membrane through which only water can pass, is filled on both sides with equal amounts of the same salt solution. For this system, no water molecules pass through the membrane because both sides have exactly the same concentration of ions.

ANSWER 4.13   This statement is not true. The membrane is permeable to water, and thus water is always moving back and forth across it. Because the two sides are in equilibrium, however, equal numbers of water molecules move across the membrane in each direction, and consequently there is no net movement of water.

It is possible to have the same water level in the two arms of the tube in Figure 4.13, but to achieve this you must add energy to the system. After all, the system as represented in Figure 4.13b is at equilibrium and would prefer to stay there. To move it back to the initial condition, where the water levels are equal and the salt concentrations are different, you must apply pressure to the right side and push down on the water surface. If your luck holds and the membrane does not break, you can make the water levels equal again (Figure 4.13c).

The same sort of phenomenon takes place across cell membranes, which are semipermeable. They allow water through, but most other things require some sort of energy-consuming transport mechanism in order to pass through a cell membrane. Imagine a cell containing 140 mM K^+ and sitting in a solution containing 4 mM K^+ (Figure 4.14). We say this cell has a potassium **concentration gradient** because the concentrations of potassium are unequal. Moving *down* the gradient means moving from the region of high K^+ concentration (inside) to the region of low K^+ concentration (outside). This is the natural direction of movement, just as water flows naturally downhill. Moving *up* the concentration gradient means moving from the low-K^+ side (outside) to the high-K^+ side (inside). Movement in this direction—sometimes called movement *against* the gradient—requires an input of energy, just as you have to input energy in order to pump water uphill.

If our cell membrane is permeable only to water, then over time water will flow, via osmosis, from outside (low K^+ concentration) to inside (high K^+ concentration). The water crosses the membrane in an attempt to equalize the concentration of K^+ ions on the two sides of the membrane. Because there is a marked difference in ion concentration across the membrane, the cell will take in as much water as it can. Eventually, however, the membrane gives out under the pressure exerted by the incoming water, and then the cell bursts. Figure 4.15a shows a cell under these conditions: initially, the concentration of ions inside is greater than the concentration of ions outside. In this case, water travels from outside to inside, and the cell swells. It is possible for a cell to burst under these conditions. Figure 4.15b shows the opposite situation. The initial concentration of ions inside is lower than the concentration of ions outside. In this case, water travels from inside to outside, and a shrunken, dehydrated cell is left behind. Figure 4.15c shows a cell with equal ion concentrations on each side of the cell membrane for comparison.

Figure 4.15b is the key to seeing why a tomato plant is unhappy in salty soil. The cells in a typical tomato plant are adapted to do best when the total salt concentration in the soil is about 4 mM. At this concentration, the cells are plump and healthy. When the soil has a higher salt concentration, the cells shrink because water exits and goes into the soil, just as in Figure 4.15b. Studies on how high-salt soils affect crop yields estimate that about 40 percent

RECURRING THEME IN CHEMISTRY   Systems naturally seek out equilibrium, and so energy is required to move a system away from equilibrium.

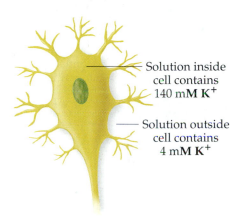

▲ **FIGURE 4.14   Ion concentrations inside and outside a cell.**   In a normal cell, the concentration of potassium ions inside the cell is much higher (140 mM) than the K^+ ion concentration outside the cell (4 mM). The cell is *not* in equilibrium with the solution in which it sits, and it is the cell membrane that maintains this nonequilibrium condition.

Solution inside cell contains **140 mM K^+**

Solution outside cell contains **4 mM K^+**

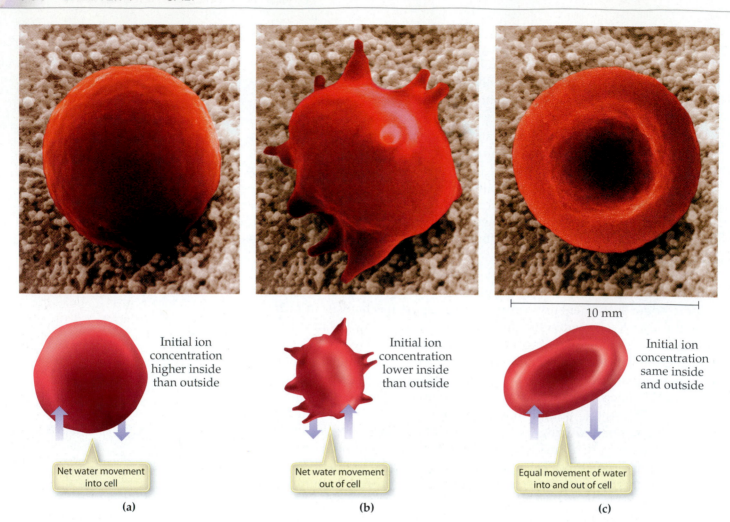

Initial ion concentration higher inside than outside

Net water movement into cell

(a)

Initial ion concentration lower inside than outside

Net water movement out of cell

(b)

Initial ion concentration same inside and outside

Equal movement of water into and out of cell

(c)

10 mm

▲ FIGURE 4.15 **Osmosis in living cells.** **(a)** A cell that has a higher ion concentration inside than outside swells as water moves into the cell. **(b)** A cell that has a lower ion concentration inside than outside shrivels as water leaves the cell. **(c)** A cell in which the ion concentration inside is the same as the ion concentration outside maintains its original size.

▼ FIGURE 4.16 **Tomatoes for the whole world?** Genetically engineered tomatoes can grow in salty soil because they accumulate salt in their leaves rather than in the fruit.

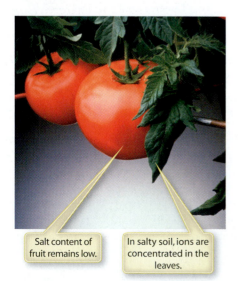

Salt content of fruit remains low.

In salty soil, ions are concentrated in the leaves.

of the world's irrigated land is unusable for tomato production because of high salt levels.

The pomologists at UC, Davis may have a solution to this problem. They have genetically engineered a tomato plant that can live in soil in which the total salt concentration is 200 mM—fifty times the normal concentration (Figure 4.16). How did they do it? They inserted into some ordinary tomato plants a gene from a cabbage plant that tells cells to produce miniature "*ion pumps.*" These pumps insert themselves into cell membranes and transport certain ions into cell compartments called *vacuoles.* The result is a tomato plant that pumps excess ions against the concentration gradient into the vacuoles in the plant, which are all located in cells in the plant's leaves. The fruit stays the same and, according to the pomologists, tastes great. They hope this technology will be applied to other plants and that newly developed salt-tolerant crops may increase the world's food supply.

How Do Home Water Purifiers Remove Ions from Tap Water?

If you have ever taken a bath or shower in hard water, you know why it is a problem. Soap scum forms in your hair, on your body, and in the tub, and this scum does not rinse away easily the way soap should. The result is a bad hair day (Figure 4.17), sticky skin, and a dirty bathtub.

Hard ions—Ca^{2+}, Mg^{2+}, and Fe^{2+} being three examples—are responsible for making water hard. What these ions have in common is two charges squeezed into a small volume. In general, the rule of thumb is that the smaller the volume available to contain the charges on an ion, the harder the ion. Why do multiple charges cause problems in your bathtub? Because they interact with soap to form the precipitate we know as soap scum.

▲ **FIGURE 4.17   The perils of hard water.**   Someone is having a very bad hair day.

QUESTION 4.14   Is the Mg^{2+} ion harder than the Sr^{2+} ion ? Why?

ANSWER 4.14   Yes, it is. We know that magnesium ions are smaller than strontium ions because Mg is located above Sr on the periodic table. Thus, the volume in which the two charges on a magnesium ion reside is smaller than the volume in which the two charges on a strontium ion reside.

Luckily, an **ion exchanger** can transform hard water to soft water by removing hard ions. Water-softening devices contain a resin made up of tiny balls stacked in a tube. If you could look closely at each molecule on the surface of one of the resin balls (which you cannot do), you would find organic tethers covalently (and permanently) attached to the ball surface, as represented in Figure 4.18. At the other end of each tether is an anionic group, such as SO_3^-. Because there is an anion on each tether, there must be cations around to balance the anions. These cations are Na^+, and there is one Na^+ for every SO_3^-. The resin-packed tube is connected to a water source, such as the pipe that brings hard water into your house. Water coming into the house flows through the ion exchanger and into the bathtub.

Figure 4.19 illustrates how water is softened by ion exchange. When a hard ion in the water, such as Mg^{2+}, encounters one of the tethers, the 2+ cation is attracted to the anion end of the tether more strongly than the 1+ sodium ion is. The interaction between Na^+ and the anion is disrupted as the Mg^{2+} takes the place of Na^+ on the resin surface. This can happen because one ion can easily replace another. Because it is doubly charged, one Mg^{2+} ion displaces two Na^+ ions on the resin. The evicted Na^+ ions travel down and out of the tube in the running water. The same thing happens with hard Ca^{2+} ions. They displace two sodium ions and remain attached to the resin.

When the resin gets filled with hard ions, there is no room for any more of them to bind. At this point, you will be glad you bought the deluxe ion exchanger because it has a mechanism to *regenerate* the resin automatically. In this process, the resin is rinsed with sodium chloride over and over again until the Mg^{2+} and Ca^{2+} ions eventually get rinsed away and the resin has Na^+ bound to it once again. If you bought the economy model, you may have to do this last step manually. Either way, your hair looks better when it is rinsed with soft water, and you can thank your ion exchanger for that.

QUESTION 4.15 Would it be possible to design an ion-exchange resin that exchanges anions rather than cations? If so, how would it differ from a cation-exchange resin?

ANSWER 4.15 Yes. Just replace the SO_3^- anions on the tether molecules with cations and use a soft anion, such as chloride ion, as the balancing ion. When water containing hard anions flows through the resin, the hard anions originally in the water displace the soft ones on the resin, and the water ends up containing only soft chloride ions.

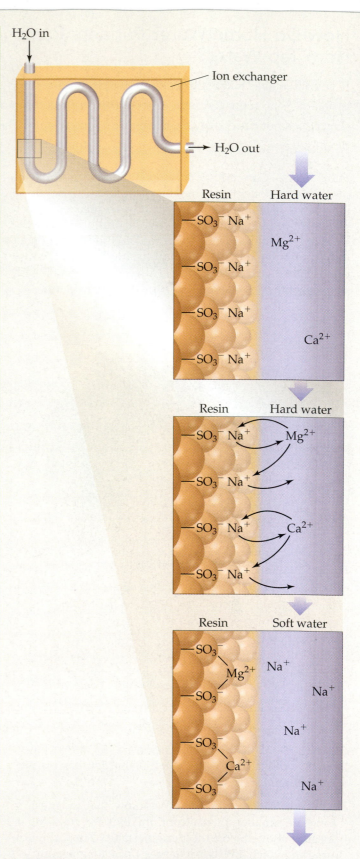

▲ **FIGURE 4.19   Ion exchanger hard at work.** Hard water going through the resin contains Mg^{2+} and Ca^{2+} ions, which are removed from the water and replaced by Na^+ cations pulled off the resin. The "hard" Mg^{2+} and Ca^{2+} ions end up on the tethers attached to the resin surface.

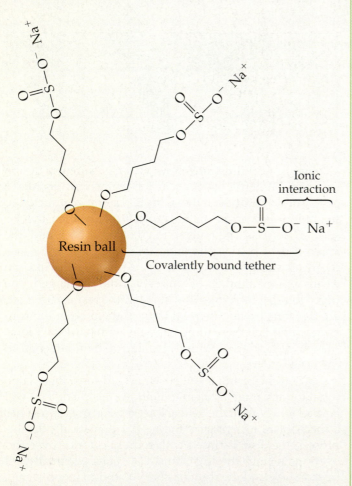

▲ **FIGURE 4.18   An ion-exchange resin.** The surface of each resin ball is covered with covalently attached tethers. The tether has a negative charge that allows it to bind to cations via ionic bonds.

4.6 | The Autoionization of Water

Water undergoes a reaction called **autoionization** that, as the name implies, produces ions:

$$H_2O \rightleftharpoons OH^- + H^+$$

To take a closer look at what happens when water autoionizes, we need to start with the electron dot structure for water:

$$H-\ddot{O}-H$$

There are eight valence electrons in this molecule divided equally between bonds and nonbonding pairs. When water autoionizes, one of the two H–O bonds breaks, and both electrons in the bond go to the oxygen. This produces one hydroxide ion, OH^-, which keeps all eight electrons, and one hydrogen ion, H^+, which is left with none:

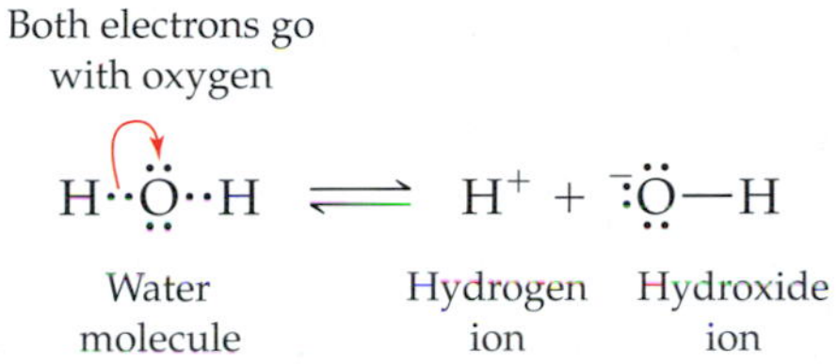

Both electrons go
with oxygen

$$H\cdot\cdot\ddot{O}\cdot\cdot H \rightleftharpoons H^+ + \ddot{\cdot\cdot}\ddot{O}-H$$

Water   Hydrogen   Hydroxide
molecule   ion   ion

> **RECURRING THEME IN CHEMISTRY**   Chemical reactions are *nothing more than exchanges or rearrangements of electrons*, which is why we pay particular attention to electrons throughout this book.

In high concentrations, both hydroxide ions and hydrogen ions can cause big problems, from acid rain to metabolic alkalosis. There is no need to worry about either ion in good drinking water, though, because the amount of each is very small.

Notice that the autoionization reaction shown above has a double arrow. This double arrow means that the reaction goes in both directions at once. This is similar to the pile of salt sitting in a saturated salt solution discussed in Section 4.3: opposing reactions are taking place at the same time. In both instances, the system is at equilibrium when the rates of the two opposing reactions are equal. For the autoionization of water, equilibrium means that water is ionizing to form H^+ and OH^- at the same rate these two ions are coming together to form water:

Forward reaction $H_2O \longrightarrow OH^- + H^+$

Reverse reaction $OH^- + H^+ \longrightarrow H_2O$

When forward rate = reverse rate, the system is at equilibrium.

What really matters for creatures that are mostly water, like us, is how much of the water ionizes. As it happens, the extent of ionization has been measured, and scientists can say with assurance that, in pure water at 25°C, the concentration is 1.00×10^{-7} M for both H^+ and OH^-. Now, 1.00 L of pure water contains 55.5 mol of water molecules. This means that only about one in every 555,000,000 molecules of water ionizes to form a hydroxide ion and a hydrogen ion. Thus, pure water is mostly water molecules—but there are a few ions in there as well—and pure water must contain equal numbers of hydrogen ions and hydroxide ions. Why? Because when one water molecule autoionizes, it forms exactly one H^+ and one OH^-.

The concentration of hydrogen ions and hydroxide ions in water is so important that chemists many years ago devised a way to simplify how the numbers are reported. Rather than saying that pure water has a hydrogen ion concentration of 1.00×10^{-7} M, we say that pure water has a pH of 7.0, where **pH** is defined as the negative logarithm of the hydrogen ion molar concentration:

$$pH = -\log[H^+]$$

The brackets around the H^+ indicate that we are talking about the molar concentration of this ion. Therefore, if you have a hydrogen ion concentration of 1.00×10^{-7} M in pure water, the pH of the water is

$$pH = -\log(1.0 \times 10^{-7}\,M) = 7.00$$

To determine a pH value using a scientific calculator, enter the number that comes after the word *log,* and then press the "log" key. For the pH value we are working with here, for instance, enter "1.0," press the "E" or "exp" key, enter "−7," then press the "log" key. The pH is the resulting number multiplied by −1.

In chemistry, the lowercase letter *p* before something always indicates we must take the negative logarithm of that something. For example, **pOH** is a mathematical operation performed on the concentration of hydroxide ion:

$$pOH = -\log[OH^-]$$

QUESTION 4.16   What is the pOH of pure water at 25 °C?

ANSWER 4.16   As just noted, in pure water $[OH^-] = 1.00 \times 10^{-7}$ M. Therefore,

$$pOH = -\log(1.0 \times 10^{-7}\,M) = -(-7.00) = 7.00$$

Do not confuse the term *chemically neutral* with the term *electrically neutral.* The former refers to the concentration of H^+ ions in a solution, with a solution being neutral when $[H+] = 1.00 \times 10^{-7}$, which means pH = 7.00. The latter refers to a balance between the number of positively charged species in a sample (cations usually) and the number of negatively charged species (either anions or electrons).

You are probably already familiar with the pH scale. When the pH of a substance is 7.00, the substance is said to be **chemically neutral**, and you probably know that when we deviate far away from neutrality, substances are potentially dangerous. Now you know that the word *neutral* refers to the situation in which hydrogen ion concentrations and hydroxide ion

Chemistry
CHEMISTRY AT THE CRIME SCENE

Determining Time of Death

In a homicide investigation, it is standard procedure to measure the temperature of the corpse as soon as possible. After death, a body cools off at the rate of 1 to 1.5 degrees per hour; so investigators can work back from the corpse's temperature to establish the approximate time of death. This estimating method is skewed, however, by clothing on the body and by the ambient temperature. As a result, forensic chemists have turned to other methods, one involving pH and another involving ion concentrations.

The blood pH in a living individual is about 7.5. About two hours after death, however, the blood pH begins to rise, causing the muscles to become taut. This phenomenon, called *rigor mortis*, can be used to estimate time of death because the stiffness becomes more pronounced over the first 12 hours. One to two days later, the pH drops again, and the body loosens up. An experienced investigator can use this change to make a rough determination of time of death: if a cold body with no rigor mortis is found, death probably occurred more than one day before.

concentrations are equal. The only way to change the pH of pure water from 7.00 is to add something to it, which means it is no longer pure water. When an **acid**, a source of H^+ ions, is added to pure water, the concentration of H^+ ions increases because the acid releases them into the water. When a **base**, a source of OH^- ions, is added to water, the concentration of OH^- ions increases.

Imagine a generic acid, HA, where A stands for *anion*, the species formed from the ionization of the acid:

$$\text{Ionization of an acid in water} \qquad HA \xrightarrow{H_2O} H^-(aq) + A^-(aq)$$

This acid HA behaves the same way a soluble salt does in water, in the sense that both ionize completely in water:

$$\text{Ionization of a salt in water} \qquad NaCl \xrightarrow{H_2O} Na^+(aq) + Cl^-(aq)$$

It is the same kind of reaction. Both the salt and the acid ionize completely to form positive and negative ions surrounded by water molecules, as shown in Figure 4.20. The cations have the negative side of the polar water molecules pointing toward them (the oxygen side), and the anions have the positive side of the polar water molecules pointing toward them (either H). The only difference is that the acid produces the cation H^+ and a salt produces such cations as Mg^{2+} or Na^+.

As we have seen, when we add acid to water, the acid forms H^+ ions. (The acid also forms an equal number of anions, but we are not interested in them here.) These H^+ ions join the H^+ ions already present in water from autoionization. Consider a solution that is prepared by adding a strong acid to water to make 1.00 L of a solution that contains 1.00×10^{-3} mol of hydrogen ions. Relative to this hydrogen ion concentration, the number of hydrogen ions contributed to the 1.00 L of solution by the autoionization of water, 1×10^{-7} mol, is negligibly small, and therefore we can ignore these hydrogen ions in our calculation. (This will be the case for all the examples in this book.) What is the pH of the solution after the acid has been added? In order to answer this question, we must remember that we need to find the hydrogen ion molar concentration before we can use the pH equation. Because we have 1.00 L of solution, the molar concentration is

$$[H^+] = \frac{1.00 \times 10^{-3} \text{ mol}}{1.00 \text{ L}} = 1.00 \times 10^{-3} \text{ M}$$

$$\log(1.00 \times 10^{-3} \text{ M}) = -3.00$$

$$pH = -\log(1.00 \times 10^{-3} \text{ M}) = -(-3.00) = 3.00$$

The pH changed from 7.00 (in pure water) to 3.00 when the acid was added. *Acids, by definition, decrease pH by increasing* $[H^+]$. According to our pH equation, as $[H^+]$ increases, pH decreases.

How is it possible to make the pH of a solution greater than 7.00? Recall that, because of autoionization, pure water contains hydroxide ions at a concentration of 1.00×10^{-7}. If we add *additional* hydroxide ions, the pH increases. Imagine we add 5.50×10^{-2} mol of hydroxide ions to a 1.00-L flask and

In this chapter, we deal with *strong acids* only, which are those that ionize completely. The ionization equation for a strong acid is drawn with a single arrow pointing to the right. Some acids are *weak*, which means they ionize only partially. These reactions are drawn with a double arrow to indicate that ionization is not complete.

LIVE ART
Ion Hydration

▼ **FIGURE 4.20 Ion hydration.** When the salt NaCl is dissolved in water, the Na^+ ions and Cl^- ions are both surrounded by water molecules, or hydrated, in aqueous solution. The same process occurs when an acid, here represented by the general formula HA, is mixed with water. The acid molecules ionize, and each ion is hydrated by the surrounding water molecules.

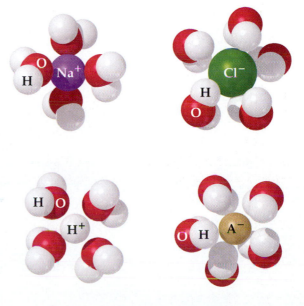

then fill the flask with water. (As we did for the acid, we shall ignore the OH^- ions that result from the autoionization of water.) Even though it is the pH of this basic solution we want to find, we need to start with the OH^- concentration rather than with the H^+ concentration and calculate the pOH:

$$[OH^-] = \frac{5.50 \times 10^{-2} \text{ mol}}{1.00 \text{ L}} = 5.50 \times 10^{-2} \text{ M}$$

$$\log(5.50 \times 10^{-2} \text{ M}) = -1.26$$

$$pOH = -\log(5.50 \times 10^{-2} \text{ M}) = -(-1.26) = 1.26$$

Because chemists are used to thinking in terms of pH rather than pOH, they use a simple relationship to convert back and forth between pOH and pH:

$$pOH + pH = 14.00$$

Knowing this relationship, we can determine the pH of our solution:

$$pH = 14.00 - pOH = 14.00 - 1.26 = 12.74$$

The pH changed from 7.00 (in pure water) to 12.74 when the base was added. *Bases, by definition, increase pH by increasing $[OH^-]$.*

QUESTION 4.17 You cannot buy a bottle containing only hydroxide ions in solution. The solution must also contain cations to balance the negative charge on the OH^- ions. Therefore, common forms that bases take include KOH, NaOH, and LiOH. If you add 3.00×10^{-3} mol of KOH to a 0.500-L flask and then fill the flask with water, what is the pH of the solution?

ANSWER 4.17 KOH dissociates just like a salt. It ionizes completely in water:

$$KOH(s) \longrightarrow K^+(aq) + OH^-(aq)$$

Therefore, if you have 3.00×10^{-3} mol of KOH in 0.500 L of solution, the molar concentration of OH^- is

$$[OH^-] = \frac{3.00 \times 10^{-3} \text{ mol}}{0.500 \text{ L}} = 6.00 \times 10^{-3} \text{ M}$$

The pOH of the solution is

$$pOH = -\log(6.00 \times 10^{-3} \text{ M}) = -(-2.22) = 2.22$$

and the pH is

$$pH = 14.00 - pOH = 14.00 - 2.22 = 11.78$$

We know that a pH less than 7.0 indicates an acidic solution and a pH greater than 7.0 indicates a basic solution. A pH scale is shown in Figure 4.21. Notice that many substances that are slippery or caustic, such as a solution of baking soda or ammonia, are bases and have pH values greater than 7.0. Many substances that are tart, such as fruit juices, are acidic and therefore have pH values less than 7.0.

You can see from Figure 4.21 that orange juice is more acidic than tomato juice, but how much more? We can answer this question by first reversing

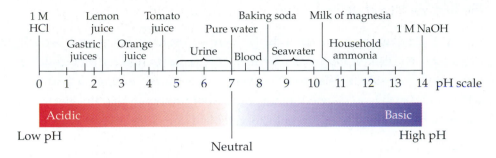

our pH equation from the logarithm form to what is called the *inverse logarithm* form:

$$\text{Logarithm form} \qquad pH = -\log[H^+]$$

$$\text{Inverse logarithm form} \quad [H+] = 10^{-pH}$$

Then we plug pH values from Figure 4.21 into the inverse equation:

$$\text{Tomato juice} \quad pH = 4.5 \quad [H^+] = 10^{-4.5} = 3.2 \times 10^{-5}\,M$$

$$\text{Orange juice} \quad pH = 3.5 \quad [H^+] = 10^{-3.5} = 3.2 \times 10^{-4}\,M$$

This inverse logarithm operation is available on *every* scientific calculator. For the orange-juice calculation, for instance, on some calculators you type in "−3.5" and then press the "10^x" key to get an answer of 3.2×10^{-4}. On other calculators, you type in "−3.5" and then press the "inv" key followed by the "log" key.

The result we obtained tells us that 1 L of tomato juice contains 0.000032 mol of H^+ ions and that 1 L of orange juice contains 0.00032 mol of H^+ ions. Thus, the acidity of orange juice is ten times the acidity of tomato juice. The beauty of the pH scale is illustrated here: if you change the pH by one unit, you change $[H^+]$ by a factor of 10. In our example, the pH changes by one unit, from 4.5 to 3.5, as we move from tomato juice to orange juice, and we found that the H^+ concentration in orange juice is ten times higher than the H^+ concentration in tomato juice.

Let us look at one more example. Because the pH values of blood (pH = 7.5) and household ammonia (pH = 11.5) differ by four pH units, we can infer that blood contains $10 \times 10 \times 10 \times 10 = 10{,}000$ times more H^+ ions than ammonia does. This makes things simple. The pH scale shrinks a wide range of H^+ concentrations down to a convenient scale, typically ranging from 0 to 14.

QUESTION 4.18 Suppose you add 5.000×10^{-3} mol of H^+ to 500.0 mL of water. (a) What is the pH of the solution? (b) What other substance has a pH value in this range (use Figure 4.21)?

ANSWER 4.18 (a) The molarity of the hydrogen ions is

$$\frac{5.000 \times 10^{-3}\,\text{mol}}{500.0\,\text{mL}} \times \frac{1000\,\text{mL}}{1.000\,\text{L}} = 1.000 \times 10^{-2}\frac{\text{mol}}{\text{L}} = 1.000 \times 10^{-2}\,M$$

$$pH = -\log(1.000 \times 10^{-2}\,M) = 2.000$$

(b) This solution has approximately the same pH as lemon juice (given as about 2.3 in Figure 4.21) and as gastric juice (about 1.8).

Beyond
BEYOND THE ORDINARY

Polka-dotted Airplanes

Figure 4.22 shows a pH meter, a device used to measure H^+ concentrations. The meter consists of electronic circuitry connected to a glass tube called a *glass electrode.* Inside the electrode is a very concentrated solution of acid, and when the electrode is inserted into a solution of unknown pH, the electronic components of the meter are able to calculate the difference between the pH inside the electrode and the pH of the solution being measured.

Before portable meters were widely available, pH was usually measured using indicator solutions. An **indicator** is typically a brightly colored organic molecule that changes color abruptly as the pH is changed. For example, the indicator phenolphthalein changes from colorless at pH values below 8.3 to bright pink at pH values greater than 8.3, as shown in Figure 4.23. Sometimes old-fashioned methods, like the use of indicators, are better than expensive electronic gadgets like sophisticated pH meters. Some engineers who study airplane corrosion have taken advantage of indicator color changes to help locate places on an airplane fuselage that are beginning to break down. In the past, corrosion on airplanes was spotted with a good pair of eyes and a powerful flashlight. More recently, fancy electronic corrosion detectors have been developed, but they have limited sensitivity.

Engineers faced with this problem knew that fuselage corrosion produces hydroxide ions, and so, they reasoned, why not use a pH test to look for corroding spots? One major component of airplane fuselages is aluminum, and when aluminum corrodes in the presence of water and salts, the following reaction occurs:

$$4\,Al(s) \longrightarrow 4\,Al^{3+}(aq) + 12\text{ electrons}$$

The electrons then react with water and oxygen:

$$3\,O_2 + 6\,H_2O + 12\text{ electrons} \longrightarrow 12\,OH^-$$

Hydroxide ions increase pH, and it is these ions the engineers hoped to detect. Thus they painted their planes with a colorless paint to which phenolphthalein had been added and waited for corrosion to begin. What they found was that, wherever corrosion was just beginning, a pink spot would turn up as the phenolphthalein changed from colorless to pink in the presence of hydroxide ions.

QUESTION 4.19 A solution that has a volume of 250 mL contains 0.19 mol of sodium hydroxide. If a drop of phenolphthalein indicator is added, what color will the solution be?

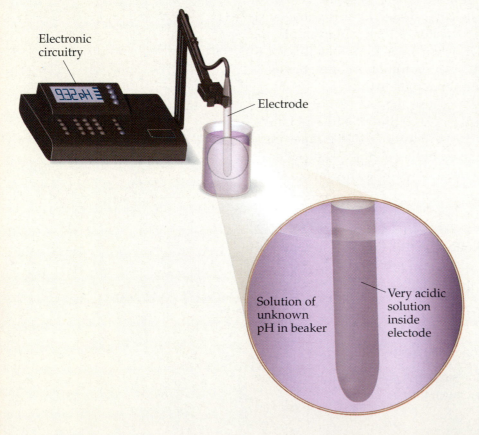

Electronic circuitry

Electrode

Solution of unknown pH in beaker

Very acidic solution inside electrode

◀ **FIGURE 4.22 A pH meter.** This device can tell us the pH of any solution when the glass electrode attached to the meter is immersed in the solution. Inside the glass electrode is a very acidic reference solution, and the meter's electronic circuitry is able to determine by what amount the hydrogen ion concentration inside the electrode differs from the hydrogen ion concentration in the solution outside the electrode.

ANSWER 4.19 NaOH ionizes to Na$^+$ and OH$^-$, and [OH$^-$] in this solution is

$$\frac{0.19\ \text{mol}}{250\ \text{mL}} \times \frac{1000\ \text{mL}}{1\ \text{L}} = 0.76\ \text{M}$$

$$\text{pOH} = -\log(0.76\ \text{M}) = 0.12$$

$$\text{pH} = 14.00 - 0.12 = 13.88$$

The solution will be bright pink because that is the color this indicator turns at pH values above 8.3.

This way of tracking corrosion has impressed a lot of people because it is much more sensitive than any of the high-priced instruments developed for the same purpose (and it is dirt cheap to boot!). The engineers who developed the new paint found that pink spots appear whenever there is a corroded spot only 15 μm deep, a crevice much too small to be seen with the naked eye. In their own words, "The new method would be especially useful for detecting the most troublesome corrosion, which is concealed around rivets and in the joints where sheets of metal overlap." These researchers think it will take several years to perfect the paint, which they believe will find uses in commercial and military aircraft. Thus, the plane of the future might be white with a coating of this new paint so that pink spots could easily be seen on the white background.

There are indicators available for almost any pH range. In fact, a product called a *universal indicator* is a mixture of several different organic molecules, each one

pH below 8.3   pH above 8.3

▲ **FIGURE 4.23** **Phenolphthalein as an indicator of pH.** Phenolphthalein changes color abruptly with changes in pH. At pH values below 8.3, a solution containing phenolphthalein is colorless. Once the pH is raised to above 8.3, the solution turns bright pink.

changing color at a different pH. This type of indicator is sensitive enough to allow solutions that differ by one pH unit to be distinguished, as shown in Figure 4.24.

▼ **FIGURE 4.24** **A pH indicator for all occasions.** **(a)** A universal indicator is a mixture of indicators. The combination is chosen so that, over a pH range from 1 to 12, the universal solution changes color with every one pH unit increment. In the solution shown here, one of the 12 indicators is orange at pH 1, another is pale green at pH 7, and another is deep purple at pH 11. **(b)** Different household products mixed with a universal indicator give solutions of different colors. You should be able to see from this photograph and then referring to the chart in part **(a)** that vinegar has a pH of about 3, club soda about 4, and ammonia about 10.

Most acidic   Most basic

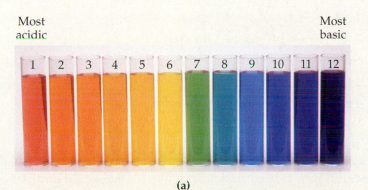

(a)

(b)

TABLE 4.5 Common Strong Acids and Bases	
Name	**Chemical formula**
Acids	
Nitric acid	HNO_3
Sulfuric acid	H_2SO_4
Hydrochloric acid	HCl
Perchloric acid	$HClO_4$
Hydrobromic acid	HBr
Hydriodic acid	HI
Bases	
Sodium hydroxide	NaOH
Potassium hydroxide	KOH
Calcium hydroxide	$Ca(OH)_2$
Barium hydroxide	$Ba(OH)_2$
Lithium hydroxide	LiOH

4.7 | Bad Air, Bad Water

Most aqueous solutions are not exactly at pH 7.0, and so there must have been acid or base added to move the pH away from chemical neutrality. A list of strong acids and strong bases is shown in Table 4.5. All these substances completely ionize in water, forming either hydrogen ions (H^+) or hydroxide ions (OH^-):

$$\text{Acids in water} \quad HClO_4 \xrightarrow{H_2O} H^+ + ClO_4^-$$

$$HI \xrightarrow{H_2O} H^+ + I^-$$

$$\text{Bases in water} \quad KOH \xrightarrow{H_2O} K^+ + OH^-$$

$$NaOH \xrightarrow{H_2O} Na^+ + OH^-$$

QUESTION 4.20 State whether each substance is an acid, base, or salt, and write an equation that shows its ionization in water: (a) KCl, (b) HCl, (c) LiOH, (d) $Sr(OH)_2$, (e) HI, (f) $HClO_4$.

ANSWER 4.20 (a) is the only salt in the list; (b), (e), and (f) are acids; and (c) and (d) are bases.

$$KCl \longrightarrow K^+(aq) + Cl^-(aq)$$
$$HCl \longrightarrow H^+(aq) + Cl^-(aq)$$
$$LiOH \longrightarrow Li^+(aq) + OH^-(aq)$$
$$Sr(OH)_2 \longrightarrow Sr^{2+}(aq) + 2OH^-(aq)$$
$$HI \longrightarrow H^+(aq) + I^-(aq)$$
$$HClO_4 \longrightarrow H^+(aq) + ClO_4^-(aq)$$

You can go to a chemical catalog and find any number of acids and bases for sale. They all undergo reactions similar to those shown here. How is it, though, that natural waters become acidic or basic? Clearly, no one has added acid or base to them from a bottle. The answer to how natural waters

TABLE 4.6 Common Pollutants Found in Air		
Pollutant	Concentration that exceeds National Air Quality Standards (NAAQS)[a]	Time interval for measurement
Ozone, O_3	125 ppb	1 hour
Carbon monoxide, CO	35.5 ppm	1 hour
Sulfur dioxide, SO_2	35 ppb	1 year
Nitrogen dioxide, NO_2	54 ppb	1 year

[a] Values are obtained by averaging the measured concentrations of the pollutant over the time interval indicated. In the monitored region, a measurement cannot exceed this value more than once in a given year.

become acidic or basic is exceedingly complex because many things can affect pH. Minerals on the ocean floor are in equilibrium with the water above, for example, and aquatic life-forms remove nutrient molecules from the water and leave other molecules behind. These and dozens of other factors affect the pH of the water.

One of the most significant factors that determine the pH of a body of water is the air above the water. Clean air is mostly nitrogen and oxygen, in addition to small amounts of other naturally occurring gases. However, human activities have added several gases to the mix, albeit in small amounts. Table 4.6 lists some common industrial air pollutants and their allowable maximum levels. These levels are set forth by the Environmental Protection Agency in the National Air Quality Standards (NAAQS). This list uses two units of concentration that are frequently used to express a very small amount of one thing mixed in with something else: **parts per million** (ppm) and **parts per billion** (ppb). For example, the concentration given for ozone in Table 4.6 is 125 ppb. This means that, in an air sample that contains this concentration of ozone, for every 1,000,000,000 atoms and molecules of various gases in the sample, 125 of them, on average, are ozone molecules. Ozone, an important natural part of Earth's upper atmosphere, is produced when sunlight and smog react with oxygen molecules. Although ozone in the upper atmosphere helps protect us from damaging ultraviolet rays, this same substance near Earth's surface is toxic and can cause lung disorders in people who breathe it regularly.

A second example from Table 4.6 is carbon monoxide, CO, another air pollutant produced by human activities. It is a gas that causes asphyxiation by binding to the hemoglobin in blood. According to Table 4.6, the NAAQS value for CO is 35.5 ppm. This means that, in an air sample containing this concentration of CO, about 36 of every 1,000,000 gas molecules and atoms will be carbon monoxide molecules.

QUESTION 4.21 Imagine that an air sample obtained in your city in 1985 contained 5 ppm ozone and that a sample collected in 2005 contained 2 ppm ozone. (*a*) What is the difference between the number of ozone molecules in 1,000,000 air atoms and molecules in 1985 and the number in 2005? (*b*) What percentage of molecules and atoms in air were ozone in 1985? (*c*) In a sample containing 1 mol of air atoms and molecules (that is, a total of 6.02×10^{23} atoms and molecules), about how many were ozone molecules in 2005? (*d*) Is either the 1985 level or the 2005 level a cause for concern?

ANSWER 4.21 (*a*) In 1985, there were five ozone molecules per 1,000,000 air atoms/molecules. In 2005, there were two ozone molecules per 1,000,000 air atoms/molecules. The difference is therefore $5 - 2 = 3$ ozone molecules. (*b*)

$$\frac{5 \text{ molecules of ozone}}{1,000,000 \text{ atoms and molecules of air}} \times 100\% = 0.0005\%$$

(*c*) You can use the percentage calculated in part (*b*) to answer this question. Remember that you always divide a percentage by 100 before you multiply by it. Thus, for 0.0005 percent, you multiply by 0.000005:

$$(6.02 \times 10^{23} \text{ atoms and molecules of air}) \times 0.000005$$
$$= 3.01 \times 10^{18} \text{ ozone molecules}$$

(*d*) Both levels are cause for concern because they both exceed the value given in Table 4.6. Note that the value in Table 4.6 is in units of parts per *billion*, not parts per *million*. A concentration of 125 ppb is the same as a concentration of 0.125 ppm, as you can see by doing dimensional analysis:

$$\frac{125 \text{ ozone molecules}}{1 \text{ billion air atoms/molecules}} \times \frac{1 \text{ billion air atoms/molecules}}{1000 \text{ million air atoms/molecules}}$$
$$= \frac{0.125 \text{ ozone molecules}}{1 \text{ million air atoms/molecules}}$$

The other two gases listed in Table 4.6 are the real culprits when it comes to the pollution of natural waters. Sulfur dioxide reacts with oxygen molecules in the air to form sulfur trioxide, which easily dissolves in water to form sulfuric acid, H_2SO_4:

MOLECULE
Sulfur Dioxide

$$2\,SO_2(g) + O_2(g) \longrightarrow 2\,SO_3(g)$$
$$SO_3(g) + H_2O(\ell) \longrightarrow H_2SO_4(aq)$$

MOLECULE
Sulfur Trioxide

Sulfuric acid has two hydrogen ions (also referred to as *protons*) that can be released. The first H^+ comes off easily. The HSO_4^- anion can then lose its H^+, but this one is more difficult to remove. Therefore, the second reaction is an equilibrium, indicated by the double arrow:

MOLECULE
Sulfuric Acid

First H^+ comes off easily $H_2SO_4 \longrightarrow H^+ + HSO_4^-$
Second H^+ comes off reluctantly $HSO_4^- \rightleftharpoons H^+ + SO_4^{2-}$

In natural waters, sulfuric acid is a problem because it releases protons, as shown by these two chemical equations. The protons, as we know, reduce the pH of the water to abnormally low levels.

Sulfur dioxide is produced mainly from the burning of coal, and in the United States most sulfur dioxide pollution comes from power plants in the Midwest. The plants burn coal mined in eastern states, and this coal contains more sulfur than coal mined in the West. The sulfur dioxide produced by the plants drifts eastward and falls into natural waters in the form of **acid rain**.

QUESTION 4.22 Which of the following are present in an aqueous solution of sulfuric acid: H_2SO_4, HSO_4^-, H^+, OH^-, SO_4^{2-}, H_2O?

ANSWER 4.22 Everything except H_2SO_4, which ionizes completely. The HSO_4^- ionization reaction is an equilibrium, meaning it does not go to completion, and therefore there is some HSO_4^- in the solution. The fact that hydroxide ion is present in the solution is tricky because this ion is not shown in either of the preceding equations. However, because there is water present, hydroxide ions are also present (but in very negligible amounts, as noted in Section 4.6).

Nitrogen dioxide is the fourth pollutant in Table 4.6. Like sulfur dioxide, nitrogen dioxide is formed from the burning of fossil fuels. However, whereas sulfur dioxide is produced from the burning of coal, nitrogen dioxide is produced from the burning of hydrocarbon fuels, such as gasoline. When dissolved in an aqueous medium such as snow or rain, nitrogen dioxide forms nitric acid, which then ionizes to produce hydrogen ions and nitrate ions:

$$3\,NO_2(g) + H_2O(\ell) \longrightarrow 2\,HNO_3(aq) + NO(g)$$
$$HNO_3(aq) \longrightarrow H^+(aq) + NO_3^-(aq)$$

Thanks to the Clean Air Act of 1970, emission of nitrogen dioxide from automobile exhaust has decreased. The guidelines in this law require that cars emit less than 0.40 g of nitrogen dioxide per mile. In the 1970s, automobile manufacturers insisted that the Clean Air Act was untenable and unrealistic, but they were forced to reduce emission of nitrogen dioxide as well as carbon monoxide. Today, catalytic converters remove a significant percentage of toxic gases from car exhaust. By 2001, for instance, the environmental levels of nitrogen dioxide had dropped by 11 percent.

Not all acid rain is the result of human activities. For example, rainwater is naturally acidic (pH 5.0 to 6.0) because carbon dioxide from the air dissolves in it. The water in lakes and streams is also acidic as a result of natural emissions of sulfur dioxide and nitrogen dioxide. Sulfur dioxide is produced in large quantities during volcanic eruptions, and nitrogen dioxide is produced naturally by plants and by lightning. A lake that is unaffected by the activities of humans should have a pH between 6.0 and 7.0. When excessive amounts of acid rain produced by air pollutants affect the pH of natural water, the lake environment begins to change. As the pH drops, aquatic plants begin to die, and this adversely affects waterfowl that depend on those plants for food. As the pH continues to drop, the bacteria that decompose organic matter begin to die, leaving a buildup of organic matter. Below pH 4.5, no fish can live.

One of the most long-lasting, comprehensive acid rain studies ever done was carried out by scientists at the University of Wisconsin–Madison. They obtained permission to acidify one half of Trout Lake in northern Wisconsin (Figure 4.25), and then they compared that half with the other half of the lake, which was left untouched at normal pH levels as the control experiment. From 1984 to 1990, the pH in the half being acidified was gradually reduced from 6.1 to 4.7. Certain fish survived at the beginning of the acidification period, but their offspring did not as the pH got lower and lower. This half of the lake became very clear, and as a result ultraviolet light from the Sun was able to penetrate deeper than before, prompting the growth of algae called "elephant snot" on the lake bottom. The ever-increasing acidity caused mercury from the soil forming the lake floor to be leached into the water. The zooplankton species that existed before 1984 died and were replaced by a more acid-tolerant strain.

MOLECULE
Nitric Acid

▼ **FIGURE 4.25** Satellite view of Trout Lake in northern Wisconsin. The two halves of the lake can be seen in the center separated by a narrow opening.

Trout Lake

RECURRING THEME IN CHEMISTRY Control experiments are included in every respectable scientific analysis. For example, chemists studying a molecule that is dissolved in a watery medium must run the same tests on the watery medium that they run on the molecule. Doing so establishes that the medium is not playing a role in the results obtained on the molecule of interest.

After acidification was complete, the researchers studied the recovery of Trout Lake for the next decade. They found that the pH and the concentrations of dissolved compounds returned to normal levels in only a few years. The biological recovery took longer, however, as restoration of aquatic plants and fish took several additional years. Nevertheless, this research should give us hope, as it tells us that the damage done to natural waters by acid rain is not irreversible.

QUESTION 4.23 Trout Lake contains 1.00×10^9 L of water. How many moles of protons (hydrogen ions) did it contain before it was acidified?

ANSWER 4.23 The pH was 6.1 before acidification. You must first convert this value to $[H^+]$ by taking an inverse logarithm:

$$[H^+] = 10^{-pH} = 10^{-6.1} = 7.9 \times 10^{-7} \text{ M} = 7.9 \times 10^{-7} \text{ mol/L}$$

$$\frac{7.9 \times 10^{-7} \text{ mol H}^+}{1 \text{ L}} \times 1.00 \times 10^9 \text{ L} = 790 \text{ mol H}^+$$

Acid rain continues to be a stress on our planet's ecosystems, but there are indications that the hard work that has gone into restricting gaseous emissions from power plants, factories, and automobiles seems to be paying off. Some parts of the country once had rain with a pH of 4.0. Now those areas have levels nearly one pH unit higher. For every optimistic report, however, there are several reports of the damage acid rain continues to inflict. Our continued vigilance is needed.

SUMMARY

Ionic liquids are salts that have low melting points and therefore are liquids at room temperature. Because they are large and unwieldy, the ions in ionic liquids do not fit into a regular crystalline lattice easily. The salts that make up natron, the substance used in mummification, include polyatomic ions, such as sulfate (SO_4^{2-}) and carbonate (CO_3^{2-}). Most inorganic salts, like those found in natron, have high melting points because their ions pack tightly into a crystalline lattice.

As a salt dissolves in water, the ions making up the salt become surrounded by water molecules. When this happens, the negative and positive ends of the polar water molecules point toward the salt cations and anions, respectively, forming a hydration shell around each ion. This ion–dipole interaction is why salts dissolve readily in water: ions are taken, one by one, from the crystalline lattice of a solid salt into aqueous solution. If there is enough salt, the water will dissolve the crystalline lattice until the solution is saturated with ions, and the leftover salt remains in the solid phase. The ions in a saturated solution are not static, however. They move out of the crystal into the water and back into the crystal at equal rates, and we say that the solution is in equilibrium with the solid.

Ionic compounds are often referred to as electrolytes because when they dissolve in water, the resulting solution can conduct an electric current. The concentration of an electrolyte (or any other pure substance) in water is commonly expressed as molarity, M. A solution's molarity is its molar concentration—the number of moles of a substance dissolved in 1 L of the solution.

Semipermeable membranes allow some substances to pass through and restrict others. The movement of water molecules across a semipermeable membrane from a region of low ion concentration to a region of high ion concentration is referred to as osmosis. The water molecules continue to move across the membrane until equilibrium is reached, at which point the ion concentration is the same on the two sides of the membrane.

Pure water forms ions naturally in a process known as autoionization. The products of this reaction are hydrogen ions, H^+, which make a solution acidic, and hydroxide ions, OH^-, which make a solution basic. When these ions exist in solution in equal amounts, as they do in pure water, the solution is neither acidic nor basic. It is chemically neutral. If you add an acid, a source of protons, to water, the H^+/OH^- ratio increases and the pH decreases. If you add a base, a source of hydroxide ions, to water, the H^+/OH^- ratio decreases and the pH goes up. If the amount of added acid or base is known, the pH of the solution can be determined using the relationship

$$pH = -\log[H^+]$$

where $[H^+]$ is the molar concentration of hydrogen ions in the solution. The pOH is calculated analogously—by taking the negative logarithm of the hydroxide ion concentration. The pOH and pH are related by the equation

$$pH + pOH = 14$$

The pH of a solution can be determined with an indicator, which is a chemical that changes color as the pH changes, or with a pH meter.

As a result of automobile and factory emissions, gaseous pollutants are released into the air. The concentrations of these pollutants in air are typically expressed in units of parts per million or parts per billion. The most serious gaseous pollutants are ozone, carbon monoxide, sulfur dioxide, and nitrogen dioxide. The latter two gases react with water to form sulfuric acid and nitric acid, respectively. When these acids dissolve in natural waters and in water vapor, they produce protons, thereby lowering the pH of the water or water vapor. Sulfur dioxide and nitrogen dioxide therefore contribute to acid rain, which disrupts the natural pH balance of lakes and rivers.

KEY TERMS

acid	dipole	ion–dipole	partial charge	saturated solution
acid rain	electrolyte	interaction	parts per billion	semipermeable
aqueous solution	equilibrium	ionic interaction	parts per million	membrane
autoionization	hard ion	millimolar	pH	solubility in water
base	hydrated ion	molarity	pOH	
chemically neutral	hydration	noncovalent	polar molecule	
concentration	indicator	interaction	polyatomic ion	
gradient	ion exchanger	osmosis	precipitation	

QUESTIONS

The Painless Questions

1. The radius of a sodium ion is 116 pm, and that of a chloride ion is 176 pm. Express these lengths in (*a*) angstroms, (*b*) nanometers.

2. Which pairs interact via an ion–dipole interaction: (*a*) a potassium ion and hydrogen sulfide, (*b*) two water molecules, (*c*) an ammonia molecule and a calcium ion, (*d*) a calcium ion and a fluoride ion?

3. Sketch the interaction between (*a*) a calcium ion and surrounding water molecules, (*b*) a carbonate ion and surrounding water molecules.

4. On a humid day, you may find that the salt in a salt shaker is clumpy and difficult to shake out. What is the reason for this?

5. Write the chemical formula and name for the salt formed from (*a*) lithium and sulfur, (*b*) sodium and oxygen, (*c*) magnesium and bromine, (*d*) calcium and fluorine, (*e*) strontium and iodine.

6. (*a*) If you dissolve some NaOH in water, do you expect the pOH to be greater or less than 7? (*b*) Do you expect the pH to be greater or less than 7?

7. The indicator methyl orange is red below pH 4.0 and yellow above pH 4.5. One of the solutions shown here contains KOH and one contains $HClO_4$. Which solution is which?

8. Write a chemical equation for the ionization in water of (*a*) HCl, (*b*) HI, (*c*) H_2SO_4, (*d*) $HClO_4$.

9. Indicate whether each of these pH values is acidic or basic: (*a*) 4.3, (*b*) 9.0, (*c*) 11.2, (*d*) 5.3, (*e*) 7.2.

10. How is it possible for a salt-tolerant tomato plant to live in high-salt conditions? How is the salt in the soil taken up by the plant in a way that leaves the tomatoes edible?

11. To treat hard water, you installed an inexpensive ion exchanger in your incoming water line. It worked well at first, but now the water is hard again. What makes water hard? Why is the exchanger no longer working as well as it was initially? What can you do to remedy the problem?

12. Over 1 year, the highest ozone concentration measured in a particular urban center is 0.22 ppm. Referring to Table 4.6, indicate whether this concentration meets the ozone standard set forth in the NAAQS.

13. Refer to Table 4.3 (page 139) and the data below to answer these questions. (*a*) Tomato juice is supposed to be good for you, but would it make a good electrolyte-replacement drink? According to the data shown here, which would be a better electrolyte-replacement drink, the juice containing salt or the no-salt juice? Why? (*b*) How do these tomato-juice numbers compare with those for Gatorade given in Table 4.3? (*c*) Which makes the better electrolyte replacer—orange juice, coconut milk, or tomato juice?

Ingredient	Tomato juice, no salt	Tomato juice, with salt
Sugar (%)	4.2	4.2
Magnesium ions (mM)	4.5	4.5
Potassium ions (mM)	56	56
Sodium ions (mM)	4.4	160

14. Sketch the interaction between (*a*) a proton and several water molecules, (*b*) a hydroxide ion and several water molecules.

More Challenging Questions

15. A concentration of 1 ppm is equivalent to a concentration of 1 mg/L. The concentration of sodium ions in tomato juice is 0.160 M. What is this concentration in parts per million?

16. The chemistry of corrosion had to be considered in the design of phenolphthalein paint for airplanes. (*a*) How do water and oxygen contribute to the corrosion of metals? (*b*) Why does phenolphthalein paint turn pink when corrosion is occurring?

17. For each of the following, state whether the substance is an acid, base, or salt, and write an equation that shows its ionization in water: (a) $KClO_4$, (b) $RbOH$, (c) H_2CrO_4, (d) $Ca(CN)_2$, (e) HNO_3, (f) $MgBr_2$.

18. Speculate on why heavy metal ions tend to move from water into an ionic liquid.

19. The photograph on the left was taken in the 1920s, and the one on the right shows the same statue 70 years later. Acid rain is the primary cause of the deterioration. (a) What two atmospheric pollutants contribute most to acid rain? (b) What strong acid is formed by each of these pollutants? (c) Write chemical equations showing how each of these strong acids ionizes in water to produce corrosive protons.

| July 1935 | June 1994 |

20. Write the chemical formula for (a) potassium phosphate, (b) potassium permanganate, (c) calcium hydrogen carbonate, (d) ammonium sulfide, (e) barium hydroxide, (f) rubidium perchlorate.

21. Hydrogen sulfide, H_2S, is structurally analogous to water. Sketch the electron-pair geometry and the molecular geometry for hydrogen sulfide. Sketch an arrow through your drawing that shows the positive and negative ends of the molecule's dipole.

22. Calculate the pH of a 1.25×10^{-2} M solution of HCl in water.

23. The old-fashioned way to treat dehydration called for ingestion of salt tablets to replace electrolytes lost in sweat. Why is this an ineffective treatment? Speculate on what happens when a salt tablet is swallowed.

24. A 591-mL bottle of one brand of fortified water contains 30 mg of potassium ions. (a) What is the molarity of the potassium ions? (b) How many potassium ions are contained in this bottle?

25. State whether each of the following electrolytes is more concentrated, less concentrated, or at the same concentration relative to human sweat: (a) magnesium ions in coconut milk, (b) chloride ions in Gatorade, (c) sodium ions in Gatorade, (d) sugar in orange juice, (e) magnesium ions in carrot juice.

26. An ion-exchange resin is designed with this organic tether attached to the resin:

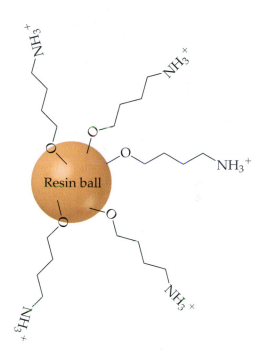

(a) Which type of ion exchanger is this? (b) Which type of ions will it remove? (c) With this tether, what soft ion could be used to replace the hard ions in incoming water?

27. A sample of air is analyzed to determine its concentration of nitrogen dioxide. It is found that, for every 6,000,000 air atoms and molecules, 420 are nitrogen dioxide. What is the concentration of NO_2 in this air sample (a) in parts per billion? (b) in parts per million? (c) Which of these two units is more appropriate for expressing this concentration?

28. Consider a tube like the one shown in Figure 4.13. There is an aqueous solution of glucose in each arm. The membrane is impermeable to glucose, but permeable to water. True or false: (a) If the water in the left arm is at the same level as the water in the right arm, the system must be at equilibrium. (b) At equilibrium, the glucose concentration in the left arm is equal to the glucose concentration in the right arm.

29. When 87.5 g of magnesium chloride is added to a 5.0-L container and the container is then filled with water, what are $[Mg^{2+}]$ and $[Cl^-]$, assuming all the salt dissolves?

The Toughest Questions

30. In the Trout Lake study, the pH was adjusted from 6.1 to 4.7 over a 7-year period. (a) Calculate $[H^+]$ before and after acidification. (b) If the number of hydroxide ions in the lake after acidification was 6.02×10^{23}, what is the volume of the lake?

31. Draw a version of $bmim^+$ that contains five carbon atoms in each arm. Calculate the molar mass of this new cation to two decimal places.

32. Draw an electron dot structure for the carbonate ion, CO_3^{2-}. (Hint: Be sure to account for the two negative charges.)

33. The salt $Mg(OH)_2$ has limited solubility in pure water. (a) If the maximum concentration of magnesium hydroxide in water is 1.4×10^{-4} M, how many moles of this salt can dissolve in 250 mL of water? (b) Write a chemical equation that shows the ionization of magnesium hydroxide in water. (c) What is the pH of 1.0 L of a saturated solution of magnesium hydroxide? (d) If you place 65.0 g of magnesium hydroxide in 2.0 L of pure water, will all of it dissolve or will some of it remain as a solid in equilibrium with the solution?

34. A 1.0-L bottle of a certain dietary supplement contains 18.75 mg of hydrogen carbonate ions, 0.9 mg of fluoride ions, and 15 mg of sodium ions. (a) What salts were added to water to create this distribution of electrolytes? (b) What is the molarity of each electrolyte? (c) Are there equal numbers of moles of cations and anions in this formulation?

35. The concentration of magnesium ion in human blood plasma is 5.2 mM. (a) If you have 4 pints of blood, how many moles of magnesium ions does the sample contain? (b) If magnesium were the only cation in blood, what molar concentration of chloride ions would be required to balance the charge on the cations? [Hint: 1 pint = 0.5 quarts and 1.0567 quarts = 1.00 L.]

36. The genetically altered tomato plants designed by pomologists at the University of California Davis can live in a 200-mM sodium chloride solution. How many grams of this salt would you have to add to a 1.0-L flask in order to make a 200-mM solution of sodium chloride in water?

37. (a) Which of the body fluids listed in the table at the top of the next column has the highest pOH? (b) Which shows the greatest variation in pH? (c) Which contains the highest concentration of hydroxide ions? (d) Imagine that you have 100 mL of each body fluid

listed in the table. Choose the fluid that, when mixed with pancreatic juice in small amounts, will bring the pH to neutral.

Body fluid	pH
Blood plasma	7.4
Liquid outside living cells	7.4
Liquid inside living cells	7.0
Saliva	5.8–7.1
Gastric juice	1.6–1.8
Pancreatic juice	7.5–8.8
Intestinal juice	6.3–8.0
Urine	4.6–8.0
Sweat	4.0–6.8

38. Write the chemical formula and calculate the molar mass of (a) sodium chromate, (b) potassium cyanide, (c) sodium nitrate, (d) magnesium carbonate, (e) ammonium peroxide, (f) calcium hydroxide. For each substance, write a chemical equation showing its complete dissociation into ions.

39. Nitric acid is a strong acid used in the manufacture of nitroglycerin, a powerful explosive. Write a chemical equation that shows the reaction of nitric acid in water. What is the pH of a 4.5×10^{-4} M solution of nitric acid?

40. (a) If you have two solutions, one made by adding 0.25 mol of NaOH to 1.0 L of water and one made by adding 0.25 mol of $Mg(OH)_2$ to 1.0 L of water, which will have the higher pOH? (b) the higher pH? (c) Why are the pH values of these solutions different? (Hint: Refer to Question 33.)

41. A 1.5-L sample of water taken from a lake in an area where acid rain is prevalent contains 9.0×10^{19} H^+ ions. What is the pH of the sample?

42. Nitrogen dioxide is converted to nitric acid in air and then dissolves in water to form ions. If you begin with 4.5 mol of NO_2 gas, how many moles of protons result from these reactions?

43. As Figure 4.12 illustrates, electrolytes dissolved in water produce ions that can conduct electricity. True or false: (a) A strong acid in the beaker of a setup such as that shown in Figure 4.12 will produce a bright light. (b) If the beaker contains a solution of polyatomic ions, the bulb will not light up. (c) The lower the electrolyte concentration in the beaker, the more weakly the bulb glows. Explain each of your answers.

E-Questions

Go to **www.prenhall.com/waldron** to find these questions in electronic form, complete with hyperlinks directly to the various websites cited in the questions.

44. Go to the website for the **EPA**. Starting with Local Drinking Water Information, follow the link to your state, and answer the following questions: (*a*) Where does your water come from? (*b*) What is the name of your water system? (*c*) What is contained in your local water-quality report?

45. Go to the **USDA**. In the Searchable Database, look up the composition of a fruit juice or other drink you like. Try to obtain information based on 100-g quantities of that drink. Because drinks are mostly water, you can assume that this mass of sample is equivalent to 100 mL of the drink. From the data given, calculate the molar concentrations of all electrolytes listed.

46. Go to **Bevnet** for a review of Life O_2 SuperOxygenated Water. (*a*) Is this article biased? (*b*) Is it written by someone representing the drink company or by an impartial observer? (*c*) What claims does the company make about its "superoxygenated" water? (*d*) How does the company support claims of enhanced endurance in consumers who use this product?

47. Read about a product called Skinny Water at **SkinnyWater**. After perusing this website, answer the following questions: (*a*) What is the slimming ingredient that the manufacturers add to their water? (*b*) Do the manufacturers provide reputable data to support the claim that their water makes people skinny? (*c*) Where does the water in Skinny Water come from? Now perform a search on your web browser for Skinny Water. Look at some of the reviews that others have posted about this product. After reading these reviews and the information on the website, what is your conclusion about Skinny Water?

48. Go to **ACS** to the article "Scrubbing Task for Ionic Liquid," which describes the use of an ionic liquid in removing carbon dioxide gas from a mixture of gases and incorporating it in the form of a small organic compound. (*a*) What is the small organic molecule that is the incorporated form of carbon dioxide in this experiment? (*b*) Why is it desirable to remove carbon dioxide from natural gas? (*c*) How does an ionic liquid used in this study differ from the one described in this chapter?

WORLD WIDE WEB RESOURCES

As with the E-Questions, go to **www.prenhall.com/waldron** to find these questions in electronic form, complete with hyperlinks directly to the various websites cited.

Some of links that follow contain research articles related to the topics in this chapter and could be used as the basis for a writing assignment in your course. Other sites are of general interest.

- Read more about the acid in your stomach at **Scientific American** and answer the question, "Why don't digestive acids corrode the stomach lining?" The article goes into considerable detail on the chemistry taking place in your stomach.

- Read more about salts at **Antarctica**. This site explains how fish can survive in the salty water of Antarctica and provides instruction for a simple experiment on how the sea freezes. Also take a look at **SoupOrg**, which has a brief history of table salt, including a Chinese story about how it was discovered.

- Go to the **EPA**. Beginning with where you live, follow these links: Search Your Community → Surf Your Watershed → Choose a Code for a Local Water Source → Look for Environmental Websites Involving This Watershed. This last link will tell you which organizations in your community are interested in preserving or monitoring your watershed. Look at the websites for these organizations. What information do they offer to the public? Is every organization interested in environmental preservation? Specifically, what do these organizations do to improve the state of your local watershed or raise awareness about this issue?

CHEMISTRY TOPICS
IN CHAPTER 5

- Energy levels and sublevels, ground and excited states [5.1]

- Electron configurations [5.2]

- Oxidation and reduction half reactions, reversible reactions [5.3]

- Batteries, electrochemical cells, reduction potentials [5.4]

- Dry cells, anodes, cathodes [5.5]

- Free radicals, antioxidants [5.6]

- The electromagnetic spectrum, line spectra, conjugated double-bond systems [5.7]

"I hate cameras. They are so much more sure than I am about everything."

JOHN STEINBECK

Film

THE STUDY OF LIGHT AND HOW ELECTRONS RESPOND TO IT

Imagine you are a present-day photographer who has been taken back in time to the early 1800s, before photography and photographic film existed. Stuck in the nineteenth century, you decide to continue your career as a capturer of still images, a visual historian in a prephotographic world. To begin, you must somehow get some film because, after all, a camera is no good without it. How can film be created, though, and what goes on when light reflected from an object strikes it? What kind of substance changes when exposed to light, and how can an image somehow be fixed onto such a substance?

In 1824, Nicophore Niépce (pronounced KNEE-eps) was pondering thoughts like these when he came up with a revolutionary idea. He reasoned that asphalt solidifies in the presence of light and that more light makes the asphalt solidify faster. Based on this observation, he coated a stone plate with a layer of asphalt and put it into a *camera obscura*, a box with a single hole through which an image *outside* the box is projected onto an interior wall of the box (Figure 5.1*a*). Eight hours later, he rinsed off the portion of the asphalt that had not yet hardened and revealed an image of his front yard. This first photograph, which Niépce called a *heliograph* (*helio-*, "sun"; *graph*, "write"), is shown in Figure 5.1*b*. You can see this heliograph on display in the Gernsheim Collection at the University of Texas at Austin.

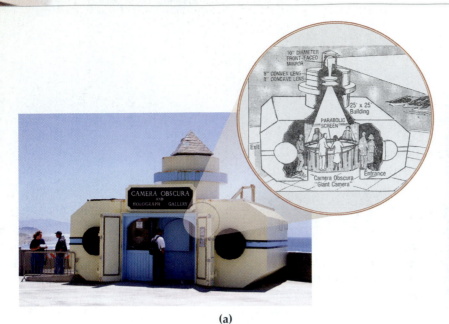

(a)

(b)

▲ **FIGURE 5.1** **Photography in prephotography times.** **(a)** This building on the shore of the Pacific Ocean near San Francisco is a life-sized version of a camera obscura. Light that enters the camera at the top is directed onto a screen inside the building using mirrors and lenses. On this screen inside the camera, people can view the scene of the ocean and the beach outside the camera. **(b)** The heliograph taken by Niépce in 1824 using asphalt on a stone plate.

▲ **FIGURE 5.2** **An early daguerreotype.** The image was created by coating a plate with a slurry of silver salts, exposing the coated plate to sunlight, and then covering the exposed surface with mercury vapor.

RECURRING THEME IN CHEMISTRY Chemical reactions are *nothing more than exchanges or rearrangements of electrons*, which is why we pay particular attention to electrons throughout this book.

We do not, of course, use asphalt-coated plates to make modern photographs, and for that we can thank Louis Daguerre, a colleague of Niépce's. Working collaboratively, these two men developed an improved photographic plate that consisted of a copper surface coated with silver, a light-sensitive metal. Working alone after Niépce's death, Daguerre had the idea of coating the exposed silvered plate with mercury vapor to "fix" the permanent positive image. Early photographs made in this way, which he called *daguerreotypes* (Figure 5.2), were incredibly detailed. Soon photographic portraits and other images were being produced in both Europe and the United States. Like Niépce's heliograph, the daguerreotype was produced by the action of light on a surface—in this case a copper plate coated with silver—the precursor of contemporary films.

Clearly, the daguerreotype was much more complex than Niépce's heliograph and involved more than merely drying asphalt. Daguerre discovered that chemical reactions occur when light hits a layer of silver salts, and we explore some of these reactions in this chapter. As we know from previous chapters, a chemical reaction is nothing more than a redistribution of electrons. In this chapter, our theme will be the different ways electrons can move and what prompts them to do so. In the chemical reactions that are the heart of photography, electron movement is triggered by light, and we shall see several examples of light-initiated reactions. Electrons can be coaxed into movement by sources of energy besides sunlight, however. Electricity and heat, for example, also animate electrons and get them moving. Therefore, these types of energy can also prompt chemical reactions.

Before we can dive into the things that make electrons move, we must first take a look at exactly where electrons should be in the first place. Then we shall look at where they can go from there.

5.1 | Electron Headquarters

Let us begin by looking at where the electrons in an atom spend most of their time. The location of electrons in an atom must be dictated, in part, by the amount of space around the nucleus. You cannot, for instance, cram 20 electrons in the space immediately adjacent to an atomic nucleus. Instead, we know that electrons are distributed around the nucleus in a way that puts each electron *as close to the nucleus as possible while allowing it its own region of uncrowded space.* To this end, electrons are arranged in a series of **energy levels** that get progressively farther from the nucleus. The level closest to the nucleus is the level of lowest energy, and the energy of each successive level increases as you move away from the nucleus. The level closest to the nucleus is called the *first energy level,* the next one out from the nucleus is the *second energy level,* and so forth.

The hydrogen atom provides the simplest example. The energy levels in a hydrogen atom are depicted in Figure 5.3*a.* The single electron in the hydrogen atom is usually located as close as possible to the nucleus, which means in the first energy level (Figure 5.3*b*). When electrons are located in the lowest available energy level in an atom, we say that the atom is in its

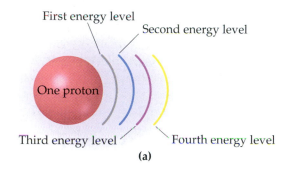

First energy level
Second energy level
One proton
Third energy level
Fourth energy level
(a)

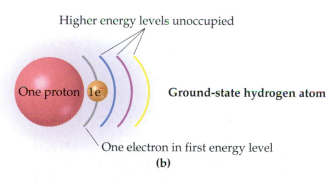

Higher energy levels unoccupied
One proton 1e⁻
Ground-state hydrogen atom
One electron in first energy level
(b)

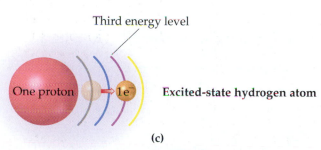

Third energy level
One proton 1e⁻
Excited-state hydrogen atom
(c)

◀ **FIGURE 5.3 Energy levels in a hydrogen atom. (a)** A hydrogen nucleus (a single proton) and the first four energy levels surrounding the nucleus. **(b)** In a hydrogen atom in the ground state, the electron is in the first energy level. **(c)** An example of an excited-state hydrogen atom, with the electron occupying the third energy level.

LIVE ART
Energy Levels in a Hydrogen Atom

ground state. It is possible to promote that electron from the ground state to one of the higher energy levels, a process that happens in photographic film. If we were to move the single electron in hydrogen to, say, the third energy level, we would say the atom is in an **excited state** because its electron is not in the lowest energy state available (Figure 5.3c).

As we shall see, an excited-state atom is formed whenever energy is added to the atom, and the added energy causes an electron to move from the ground-state energy level to some higher energy level. (This process can go in reverse, too: when an excited electron returns to the ground state, energy is released.) In the case of photography, added energy in the form of light makes a ground-state electron jump to an excited state. When this happens, the atoms in photographic film undergo a series of reactions that results in the formation of a negative image.

Now let us consider the other atoms in the periodic table. The energy levels in multi-electron atoms are more complex than the energy levels in the hydrogen atom. This is true because the second electron added is repelled by the first electron. The third electron is repelled by the other two, and so on. These complex electron–electron repulsions in a multi-electron atom cause the energy levels to split into **energy sublevels**, as shown in Figure 5.4. Each sublevel is designated with a number and a letter. An example is the 1s sublevel, which is the only sublevel in energy level 1. The sublevels of energy level 2 are designated 2s and 2p; those of energy level 3 are 3s, 3p, 3d; and those of energy level 4 are 4s, 4p, 4d, 4f.

When we must distribute multiple electrons in a ground-state atom, we begin with the lowest energy sublevel given in Figure 5.4. This is the 1s sublevel, which can accommodate a maximum of two electrons. Therefore, we place two electrons in this sublevel. The next two electrons go into the 2s sublevel, which also can hold only two electrons. (As we shall see, *any s* sublevel can hold a maximum of two electrons.) The fifth electron will be placed in the 2p sublevel, which is the next lowest in energy. The 2p sublevel can accommodate a maximum of six electrons, as do all the other *p*

▶ FIGURE 5.4 **Energy levels and sublevels.** The first four energy levels for the hydrogen atom are shown on the left. When an atom contains more than one electron, energy levels split into sublevels, shown on the right. The energy of the sublevels increases as we move from the bottom to the top of the diagram.

LIVE ART
Energy Levels and Sublevels

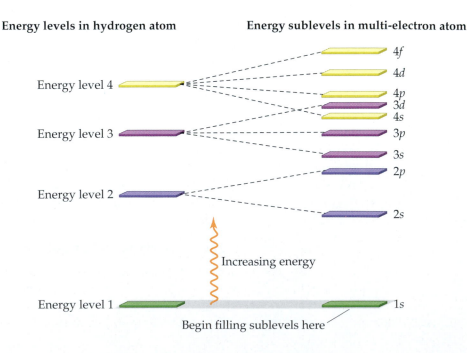

sublevels. Thus, electrons 5 through 10 are placed in this sublevel. The d sublevel accommodates a maximum of ten electrons, and the f sublevel a maximum of 14 electrons. We therefore continue placing electrons one by one in the various sublevels until we have assigned all of the electrons in the atom (or ion) we are working with.

Let us look at some examples. An atom of the element sulfur contains 16 electrons. Referring to Figure 5.4, we can see how to distribute these electrons, starting with the lowest energy sublevel. We put two electrons in the $1s$ sublevel, two in the $2s$ sublevel, and six in the $2p$ sublevel. So far, we have assigned ten electrons, and we have six more to place. The next energy sublevel shown in Figure 5.4 is $3s$, which can accommodate two of our six remaining electrons. The last four go into the $3p$ energy sublevel (which is able to hold a maximum of six electrons).

We can represent electron placement in a multi-electron atom with an **electron configuration**, a notation that lists the sublevels of increasing energy in a horizontal series. Each energy sublevel is written in full-size type, with a superscript number to the right of each sublevel indicating the number of electrons in the sublevel. Our sulfur atom, for instance, has the electron configuration

$$1s^2 2s^2 2p^6 3s^2 3p^4$$

A correct electron configuration will have the sum of the superscripts equal to the total number of electrons in the atom. In this case, the superscripts total 16, the total number of electrons in a sulfur atom. Using the same logic, we find that the electron configuration for a sodium atom, Na, is $1s^2 2s^2 2p^6 3s^1$, and the electron configuration for a fluorine atom, F, is $1s^2 2s^2 2p^5$.

For our next example, let us consider arsenic, As, element 33 on the periodic table. In this example, we shall learn how to determine an electron configuration without the help of Figure 5.4. After all, you may not have a copy of Figure 5.4 handy when you suddenly start wondering about the electron configuration of some atom. Lucky for you, the periodic table—which can be found almost anywhere—serves as a guide to electron configurations, making the use of Figure 5.4 unnecessary. In order to assign an electron configuration using the periodic table, though, you must know which parts of the table correspond to the different energy sublevels.

In Figure 5.5, the periodic table is divided into four blocks shown in different colors. Each color represents a different energy sublevel letter, and we can use this information to determine any electron configuration. The s block, shown in yellow, begins with the $1s$ sublevel in the top row and continues down through the $7s$ sublevel. Each horizontal row in the s block contains two elements, telling us that a maximum of two electrons can be placed in each s sublevel. The p block (pink) begins with the $2p$ sublevel in the top row, followed by $3p$, $4p$, $5p$, and $6p$. Six elements span the p block, meaning that each p sublevel can hold a maximum of six electrons. The d block (green) begins with the $3d$ sublevel, and each row spans ten elements. Thus, each d sublevel can hold a maximum of ten electrons. Finally, the f block (tan) begins with the $4f$ sublevel, and each row spans 14 elements. Thus, each f sublevel can hold a maximum of 14 electrons. Notice that you must leave the main table at element 57 (lanthanum) and detour to the $4f$ elements placed below the main table. After element 71 (lutetium), you return to the main table to element 72 (hafnium) and continue. When you reach element 89 (actinium), the elements take another detour to the

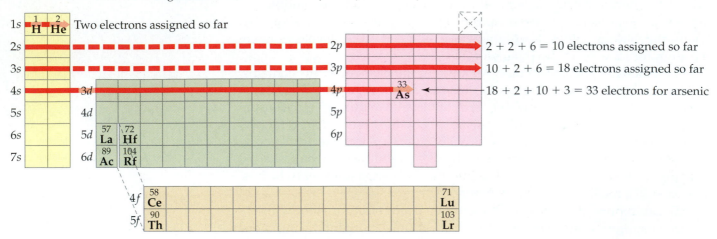

▲ **FIGURE 5.5** **Using the periodic table to determine electron configurations.** This version of the periodic table is split into blocks that match the energy sublevel letter designations *s*, *p*, *d*, and *f*. Following the red arrows across the periods allows us to determine the electron configuration of any element. The example here is arsenic, with 33 electrons.

f block below the main table, this time to the second row, which includes all of the 5*f* elements. After element 103 (lawrencium), you return to the main table at element 104 (rutherfordium).

To use the periodic table to write an electron configuration, we read from left to right beginning at the top left, as shown by the red horizontal arrows in Figure 5.5. If you follow these arrows across the periods, you will see that the order in which the energy sublevels are intersected by the arrows matches the sublevel order shown in Figure 5.4. Because most periodic tables do not show you energy sublevels (the way Figure 5.5 does), you must memorize the letters associated with the four blocks. You also must know that the *s* block begins with 1*s*, the *p* block begins with 2*p*, the *d* block begins with 3*d*, and the *f* block begins with 4*f*. Once you have mastered these things, writing electron configurations becomes a piece of cake. Let us now illustrate this idea by returning to the element arsenic.

Each arsenic atom has a total of 33 electrons, and writing an electron configuration for an arsenic atom means assigning each electron to an energy sublevel. Beginning at the upper left corner of the periodic table in Figure 5.5, reading left to right, we begin with the 1*s* sublevel and we put two electrons in it. Thus, our first entry in the electron configuration of arsenic is $1s^2$. There being no more squares in period 1 in Figure 5.5, we move down one row to the 2*s* sublevel, which we fill with two electrons. We then write $2s^2$ next in the electron configuration. Following the red arrow in Figure 5.5, we move across the dashed portion of the arrow to the first row of the *p* block. We place six electrons in the 2*p* sublevel, one electron for each of the six squares in this row, and then write $2p^6$ in our configuration. Moving down to period 3, we next put two electrons in the 3*s* sublevel and six electrons in the 3*p* sublevel, giving us $3s^2 3p^6$.

Now things get tricky. The next row begins with the 4*s* sublevel (which, as usual, holds two electrons). Notice in Figure 5.5, though, that following the red arrow which starts in the 4*s* sublevel means that the next block we come to is not a *p* sublevel, but rather the 3*d* sublevel (which holds ten electrons). Therefore, we place our next ten electrons in the ten squares that span the $3d$ sublevel and add $4s^2 3d^{10}$ to our electron configuration. So far, we have assigned $2 + 2 + 6 + 2 + 6 + 2 + 10 = 30$ electrons. Because arsenic has 33 electrons, we must continue to the next sublevel and put three electrons in it. The red arrow in Figure 5.5 tells us that the next sublevel is $4p$, which holds up to six electrons, and so we place our three remaining electrons here. The electron configuration for an arsenic atom is therefore

$$1s^2 2s^2 2p^6 3s^2 3p^6 4s^2 3d^{10} 4p^3$$

The marvelous thing about the periodic table is this: when you have read from left to right across the periods to get to the appropriate number of electrons, you end up at the element you were considering in the first place. For this example, 33 electrons bring you to the $4p$ energy sublevel. Figure 5.5 shows you that the third element across the $4p$ energy sublevel is arsenic.

QUESTION 5.1   Write the electron configuration for (a) a lithium atom, and (b) an argon atom.

ANSWER 5.1   (a) $1s^2 2s^1$, (b) $1s^2 2s^2 2p^6 3s^2 3p^6$.

QUESTION 5.2   Identify the element represented by each electron configuration: (a) $1s^2 2s^2 2p^6 3s^2 3p^6 4s^2$,   (b) $1s^2 2s^2 2p^6 3s^2 3p^6 4s^2 3d^6$, (c) $1s^2 2s^2 2p^6 3s^2 3p^6 4s^2 3d^{10} 4p^6 5s^2 4d^{10} 5p^5$.

ANSWER 5.2   (a) calcium, (b) iron, (c) iodine.

It is important to remember two things about this system for assigning electrons to energy sublevels. First of all, each electron in an atom is exactly the same as every other electron in the atom. Each has the same electrical charge $(1-)$ and the same mass. However, they differ from one another by virtue of their location in the atom. An electron's location with respect to the nucleus determines its attraction to protons in the nucleus. For example, an electron in an energy sublevel close to the nucleus is held very tightly by the pull of the protons. An electron in the outermost occupied sublevel of an atom will be held least tightly by the nucleus because it is farthest away. It makes sense, then, that the outermost electrons are the easiest to remove from the atom, and this is in fact the case.

QUESTION 5.3   For each of the following, give the number and letter of the energy sublevel described: (a) the sublevel closest to the nucleus in an aluminum atom, Al; (b) the sublevel of the outermost electrons in a germanium atom, Ge; (c) the sublevel of outermost electrons in a tellurium atom, Te.

ANSWER 5.3   (a) $1s$, (b) $4p$, (c) $5p$.

The second thing to remember when you are thinking about the layered arrangement of electrons in an atom is that this is merely a model scientists have developed so they can visualize the atom more easily. Electrons do not actually stay at fixed distances from the nucleus the way Figure 5.4 seems to suggest. However, this model can help you understand that electrons exist in different environments around the atom, even though they are all the same kind of particle. Knowing this helps you see why some electrons are more easily pulled away from the nucleus than others. Also, chemical reactions always involve a shuffling of electrons, so all of this information should help you understand why chemical reactions can and do occur. If you decide to imagine the atom as an onion with a center that represents the nucleus surrounded by layers of electrons, it is important to remember that this image comes courtesy of your imagination and that real electrons are elusive and not completely understood.

5.2 | Electrons on the Move

The preceding discussion gives us a picture of the electrons in an atom arranged in an orderly fashion. Each electron is held in a particular region by the pull of the nucleus, with this pull greatest for the innermost electrons and least for the outermost ones. Let us explore this idea using the magnesium atom. A magnesium atom has 12 electrons around the nucleus, and the two in the $1s$ sublevel are held very tightly. Clearly, those electrons are not going to participate in any chemical reactions. The next eight electrons (in the $2s$ and $2p$ sublevels) are also held snugly to the nucleus. The two outermost electrons, called the atom's valence electrons (Section 2.2), which are in the $3s$ energy sublevel, experience less pull from the nucleus than the other electrons in the atom. Thus, valence electrons are the ones that are most easily pulled away from the atom, and therefore *these are the electrons the atom uses to make chemical bonds*. Knowing this gives a deeper meaning to the electron configuration for magnesium:

Core electrons, held tightly by nucleus

Valence electrons, held less tightly by nucleus

$$1s^2 \, 2s^2 \, 2p^6 \, 3s^2$$
Magnesium

As noted in the previous paragraph, one reason valence electrons feel less nuclear pull is their greater distance from the nucleus. A second reason is that, in the case of magnesium, there are ten core electrons (Section 2.2) located between the nucleus and these outermost electrons. These core electrons shield the valence electrons from the pull of the nucleus. Thus, it should not surprise you that chemical reactions almost always involve valence electrons because they are the easiest to coax away.

You can think of electrons like the bottles in a refrigerator full of one brand of bottled water (Figure 5.6). Just as electrons in an atom are all the same, those bottles are the same: they all contain equal amounts of the same brand of water. However, the fate of those bottles depends on their location in the refrigerator. Those located closest to the door, analogous to valence electrons, are most likely to be chosen by a thirsty person making a trip to the refrigerator. Those in the back, analogous to core electrons, are harder to reach and are less likely to be chosen.

What might pull a valence electron away from its position in the outermost occupied energy level of an atom? Two of the most common influences that make electrons move are light and heat. In fact, if you ever run chemical reactions in a laboratory, you will find that many are jump-started either by shining light on the flask containing the reactants or, more commonly, by heating the flask. How do light and heat, both sources of energy, cause electrons to move? Let us consider heat first. As we

▼ FIGURE 5.6 **Cool electrons.** A refrigerator containing water bottles can be likened to an atom containing electrons. With the deep interior of the refrigerator representing the atom's nucleus, the bottles in front are analogous to the outer, valence electrons in an atom, and the bottles behind the front row are analogous to the inner, core electrons.

learned in Section 1.5, absolute zero is the zero point on the Kelvin temperature scale (0 K), and it represents the temperature at which the motion of atoms and molecules stops. At 0 K, all the electrons in any atom are in their ground state and do not have the energy required to leave the atom. As you warm up a substance from 0 K, wiggling and shaking start to occur at the atomic level. As the temperature increases, electrons, especially valence electrons, gradually become poised for movement. As the temperature continues to increase, the valence electrons acquire energy that can permit them to do one of two things: jump from the valence energy sublevel to a higher sublevel or, if enough energy is available, leave the atom altogether.

When *exactly the right amount* of energy is available, any one of the electrons in an atom can jump from its ground-state sublevel to a higher sublevel. Where the electron ends up depends on the amount of energy supplied. Light can also cause an electron to jump from the ground state to an excited state, and we shall look more closely at this concept at the end of the chapter. We shall also see some remarkable applications of this kind of electron movement, such as the colors of leaves in autumn and the action of sunscreens.

It is possible to add so much energy to an atom that the added energy promotes a valence electron right up and out of the atom. In fact, you have already seen examples in which one or more valence electrons are kicked out of an atom to make a positive ion. For example, a sodium atom, which has one valence electron, loses that electron to form a Na^+ ion. Recall that ions often form when a noble gas electron configuration can be achieved (Section 2.2). We can see now that when a noble gas configuration is achieved, *the outermost occupied energy level is filled to capacity with electrons.* This gives an atom or ion a special stability that comes from having just the right number of electrons pulled toward the nucleus.

Let us consider calcium as an example. Its electron configuration is

$$1s^2 2s^2 2p^6 3s^2 3p^6 4s^2$$

The noble gas that precedes calcium in the periodic table is argon. Its electron configuration is

$$1s^2 2s^2 2p^6 3s^2 3p^6$$

Comparing the two configurations, you can see that the only difference between them is that calcium has two $4s$ valence electrons not present in argon. Thus, to achieve a noble gas configuration of electrons, calcium must give up those two electrons. If it gives them up, it has a total of $20 - 2 = 18$ electrons remaining, each of which carries a negative charge. The element calcium always has 20 protons, each of which carries a positive charge. This is true regardless of the number of electrons it loses. Thus, the charges in a calcium ion are no longer balanced the way they are in a neutral calcium atom (20 electrons and 20 protons). There are now more protons than electrons, and calcium carries a positive charge of 2+.

Following this logic, we can see that sulfur requires two extra electrons to achieve the electron configuration of the noble gas argon. Thus, we expect sulfur to take on two electrons to become the ion S^{2-}. This ion can form a salt with the calcium ion in a one-to-one ratio: CaS, calcium sulfide, as shown in Figure 5.7. Whenever ions form, there must be a change in the location and number of electrons. Here, for instance, the formation of a

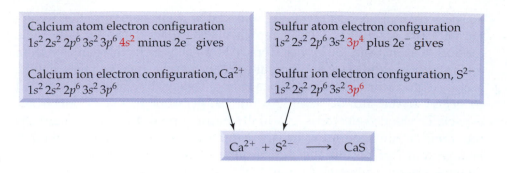

▶ **FIGURE 5.7 Ionization of sulfur and calcium atoms.** When a sulfur atom gains two electrons and a calcium atom loses two electrons, the oppositely charged ions can come together to form the salt CaS, calcium sulfide.

sulfur ion means that a sulfur atom must grab two electrons from something else. Likewise, the formation of a calcium ion means that calcium must donate two electrons to something else. In calcium sulfide, calcium donates two of its electrons to sulfur, and as a result, both ions achieve a noble gas configuration of electrons.

QUESTION 5.4 Because the salts formed by these pairs of ions are excited when light shines on them, the salts can be used in photographic films. Write the chemical formula for the salt formed in each case: (a) Cl^- and Ag^+, (b) Ag^+ and I^-, (c) Br^- and Ag^+.

ANSWER 5.4 (a) AgCl (silver chloride), (b) AgI (silver iodide), (c) AgBr (silver bromide).

You may be wondering why we are talking about electrons and salts in a chapter about film. There are two reasons: (1) the chemistry of photography is all about electron movement, and (2) the substances in film involved in the shuffling of electrons are salts. We can see now why it is important to understand energy sublevels. By knowing where each electron in an atom is located with respect to the nucleus, we can predict which electrons will move and which ones will stay put. Further, we now know that there are two things that can happen to an electron in an atom hit with some outside source of energy. That electron can be promoted to a higher energy level. Or, if a lot of energy is added, that electron can leave the atom entirely and leave behind a positive ion. That wayward electron can move to an atom that needs an extra electron in order to achieve a noble gas configuration, and by accepting the electron, that atom becomes an anion. This is exactly what happens with the atoms embedded in a piece of photographic film, as you are about to see.

5.3 | Smile! Electrons Moving Through Film

The fundamental idea behind photographic film has not changed much since Daguerre was making daguerreotypes back in the nineteenth century. In order to make any kind of photographic film, you must start with some kind of substance that changes when exposed to light. Black-and-white film has two essential components: a silver salt and a solid surface to stick it to. Recall that salts (such as CaS, for instance) are composed of positive and negative ions. The salts most often used in modern films are silver bromide, AgBr, and sometimes silver iodide, AgI.

Silver bromide is made up of silver cations, Ag^+, and bromide anions, Br^-, but it is the silver cations that do the important work in photography. Light striking a piece of film that contains silver bromide provides energy used by the bromide ions to promote an electron up and out of the ion, leaving an electrically neutral bromine atom:

$$Br^- + \text{light energy} \longrightarrow Br + 1 \text{ electron}$$

The presence of a stray electron is not permitted in a typical chemical equation, and we refer to the reaction represented by an equation that shows electrons as a *half reaction* (Section 2.5). The electrons produced by the action of light on bromide ions are absorbed by silver ions to make electrically neutral metallic silver:

$$Ag^+ + 1 \text{ electron} \longrightarrow Ag$$

When we talk about *metallic* silver or *metallic* copper or *metallic* iron, we mean nothing more than the metal atoms in their electrically neutral state, with no positive or negative charges. This is the form of metals with which we are most familiar. For example, silver jewelry and copper coins contain neutral silver atoms and neutral copper atoms, respectively, rather than ions of silver or copper.

Recall from Section 2.5 that half reactions in which electrons are exchanged are called *oxidations* and *reductions*. When we combine an oxidation half reaction with a reduction half reaction, the result is a *redox reaction* in which electrons are released by oxidation reactions and then used in reduction reactions. In fact, most of the complex reactions of photography—including image formation and development—are redox reactions of some kind.

QUESTION 5.5 In the initial reaction of photography between silver ions and bromide ions, which reactant is reduced (gains electrons) and which is oxidized (loses electrons)?

ANSWER 5.5 Because bromine begins as a bromide anion and loses an electron to become a neutral bromine atom, it is oxidized. The silver ion is reduced because it accepts an electron from the bromide ion.

QUESTION 5.6 Write the missing reactant or product for each half reaction. For each, indicate whether an oxidation or reduction is taking place.

(a) ____ $\longrightarrow$ Br + 1 electron
(b) K^+ + 1 electron $\longrightarrow$ ____
(c) Mg $\longrightarrow$ 2 electrons + ____
(d) O + 2 electrons $\longrightarrow$ ____

ANSWER 5.6 (a) Br^-, oxidation; (b) K, reduction; (c) Mg^{2+}, oxidation; (d) O^{2-}, reduction.

How does a half reaction that produces metallic silver lead to a photographic image on a piece of film? To answer this question, think for a moment about the piece of film before any light shines on it. Silver bromide, AgBr, is made up of Ag^+ ions and Br^- ions that exist in transparent crystals (Figure 5.8a). These tiny crystals of silver bromide, referred to in

▼ FIGURE 5.8 **Ionic and metallic silver.** **(a)** In solid silver bromide crystals, all of the silver is in the ionic form, Ag^+. **(b)** In objects made of metallic silver, all of the silver is in the metallic form, Ag, which is the shiny silver used in jewelry, silverware, and coins.

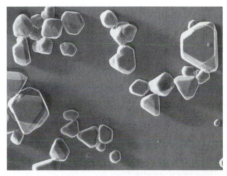

(a)

(b)

photographic jargon as *grains*, are attached to a solid backing. When light strikes a piece of film, some of the silver ions are converted to metallic silver, Ag, which is opaque and has the metallic sheen characteristic of metals used in jewelry and coins (Figure 5.8*b*). Thus, the chemical reactions of photography convert silver from its salt form (transparent) to its metallic form (opaque).

After exposure to light, some of the Ag^+ ions in the film grains are converted to Ag atoms. In areas exposed to the most light, the grains contain more Ag than Ag^+, whereas in areas exposed to the least light, the grains contain more Ag^+ that Ag. If a given grain contains four or more silver atoms after exposure to light, developing the film converts the remaining Ag^+ ions in that grain to silver atoms. Thus, areas exposed to lots of light will have most grains containing only silver atoms after the film has been developed. These areas will be opaque. Likewise, areas exposed to low light will remain transparent because in these areas most of the grains do not contain a minimum of four silver atoms. Areas that are exposed to medium light will have a mixture of opaque and transparent grains and will appear gray.

The result is a negative image, called *negative* because the areas exposed to the most light become the darkest. Because these areas are laden with silver atoms, you cannot see through them any more than you can see through a silver dinner knife. The negative image is then used to create a *positive*, which is the familiar "picture" you pick up at the store. In the negative-to-positive conversion, light passes through the negative onto a sheet of photographic paper. The lightest places on the negative let the most light through and therefore become the darkest places on the positive (as shown in Figure 5.9), and vice versa: the darkest places on the negative let the least light through and become the lightest places on the positive.

Alert readers may have a question about the photographic process just described above: What keeps the silver bromide grains attached to the film? What keeps them from sliding off a piece of plastic the same way table salt slides off a dinner plate? The answer is that some substance is required to keep the grains from moving. The substance used is gelatin, the ingredient in gelatin dessert that makes it into a gel.

▶ FIGURE 5.9 **Creating a photographic positive.**   A negative image (left) is converted to a positive image (right) when light passes through the negative onto a piece of photographic paper.

How Do Photogray Sunglasses Turn Dark Outdoors?

The chemistry of photogray sunglasses is very similar to the chemistry of black-and-white film, but with a few important differences. Photogray glasses have very finely divided silver chloride crystals imbedded in the glass. When sun hits the silver ions in these crystals, the ions are reduced to silver atoms, just as in black-and-white film. The silver salts are so sparsely scattered through the

glass that the result is a darkening of the glass. The glass does not become completely opaque the way film does, however. Glasses that suddenly become completely obscured when you walk outside would not sell very well!

There is another significant difference. What would happen when you came back inside if the chemistry used for black-and-white film were used in photogray glasses? The answer is that the glasses would remain dark because the reactions of black-and-white film are not **reversible chemical reactions**. That is, they cannot move in the direction that produces silver ions once again. Because photogray glasses contain silver chloride rather than silver bromide, the reactions that take place in the glass *are* reversible, so that the darkened glass can become transparent again.

Photogray glasses also have copper ions imbedded in the glass, which allow the chlorine atoms formed in the redox reaction that reduced Ag^+ to Ag to be reduced back to chloride ions via the reaction

$$Cu^+ + Cl \longrightarrow Cu^{2+} + Cl^-$$

The Cu^{2+} ions created in this reaction then react with silver atoms in a redox reaction that oxidizes them back to silver ions:

$$Cu^{2+} + Ag \longrightarrow Cu^+ + Ag^+$$

In this way, the redox process that changes silver chloride to silver atoms and chlorine atoms in the presence of sunlight can be reversed when the wearer moves indoors.

QUESTION 5.7 In the reversible redox reaction that takes place in photogray sunglasses, do the copper ions act as a reducing agent or an oxidizing agent?

ANSWER 5.7 Both. The first reaction shows that Cu^+ acts as a reducing agent by donating an electron to a chlorine atom to make a chloride ion. The second reaction shows that Cu^{2+} acts as an oxidizing agent by accepting an electron from the silver atom to produce a silver ion.

A highly purified form of gelatin prepared from cow bones is used for modern film, and the grains cannot move around in the gelatin any more than a person can swim through a pool full of Jell-O®. Thus, black-and-white film is little more than microcrystalline silver bromide mixed into gelatin and spread over a plastic backing. We shall look briefly at color film later in this chapter.

The basic idea behind film is simple—a silver halide in gelatin on plastic. There is actually a great deal of complicated chemistry going on, however, and that chemistry has been dramatically improved in recent years. One advance in photographic technology involves the type of silver halide used. As you may recall from our discussion of crystals in Chapter 2, there are many shapes a crystal can adopt. As it happens, a silver halide grain tends to be multifaceted, as shown on the left in Figure 5.10a. When light strikes a piece of film with grains of this type imbedded in a gelatin matrix, the light tends to reflect off some of the crystal faces in the same way it reflects off the faceted surface of a diamond. This means that not all the light that hits the film enters the grain and

Halide is the name given to the anion of any halogen, which is any element in column 17 of the periodic table (Section 2.1). The halides include fluoride ions, chloride ions, bromide ions, and iodide ions.

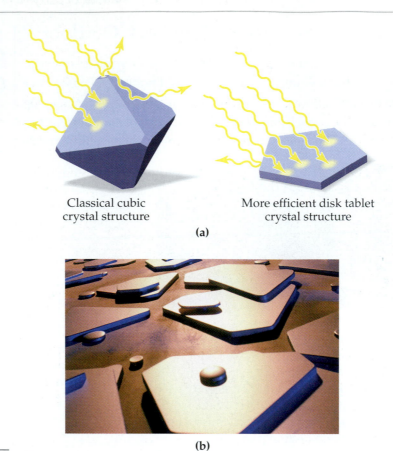

Classical cubic
crystal structure

More efficient disk tablet
crystal structure

(a)

(b)

▶ **FIGURE 5.10 Crystal shapes in photographic film. (a)** Left: The classical shape of a cubic crystal; right: the disk shape of the crystals used in photographic films. **(b)** A photomicrograph of the crystals in a piece of photographic film.

contributes to the photographic image; some of the light is lost by reflection. To remedy this problem, photographic scientists have created much flatter, disk-shaped crystals of silver halide, shown on the right in Figure 5.10*a*. When these disks are used for film, their two largest surfaces are parallel to the film surface. The result is that incoming light hits a single relatively large, flat surface (Figure 5.10b). This means that less of the light hitting the film is reflected and more is available to produce the photographic image. In practical terms, this means that if you are taking pictures in low-light situations, the small amount of light hitting the film will be used much more efficiently.

When it comes to film design, Polaroid film is a triumph of ingenuity. Have you ever noticed how thick a piece of this type of film is? Given the complexity of the photographic process, it is amazing there is film available that can produce a fully developed image right before your eyes. In order to do this, Polaroid film must have all the chemicals involved both in the film–light interaction and in the development and fixing of the final photograph. This is done by stacking more than a dozen chemical layers onto a plastic support, a design that makes the film thick and heavy. (It also gives the film a chemical odor.) Once a photograph is taken with a Polaroid camera, the film is forced through tight rollers as it leaves the camera. This squeezing initiates the mixing of various chemicals in different layers of the film, and development begins. Color Polaroid film is even more remarkable because the faithful reproduction of color requires that many more layers of chemicals be added to the film, as we shall see later in this chapter.

5.4 | More Rambling Electrons: Electrochemical Cells

In the 1963 film *From Russia With Love*, James Bond uses state-of-the-art gadgetry to track down the bad guys. His arsenal of gizmos includes a car telephone (Figure 5.11a) and a small electronic device used to receive short messages (what today we would call a pager) from MI6. The pager and car phone wowed 1963 audiences, and each subsequent new Bond film featured ever more outrageous contraptions. The secret spy camera used by Bond in the 1967 film *On Her Majesty's Secret Service* was considered revolutionary at the time (Figure 5.11b), but most of the miniature cameras on the market today are much smaller.

Perhaps the smallest camera yet is part of a device called a *video pill* (Figure 5.12), a capsule just slightly larger than a normal pill and designed to be swallowed. The capsule contains a tiny camera that sends a video feed to a receiver the swallower wears around the waist. This new invention shows extraordinary promise for people who suffer from gastrointestinal disease. Instead of shoving a long camera probe into a patient's gut, the physician can simply prescribe a video pill. The receiver records up to eight hours of digital video from the patient's gastrointestinal tract, data that the doctor downloads into a computer. Two to three days later, the camera reappears at the other end of the patient. Needless to say, patients who have tried the technology favor it over the traditional method—no hospital gowns, no probes down the throat. They can go about their daily life while the video pill helps to diagnose their problem.

The video pill is not used to study the large intestine because the life of its battery is only eight hours, enough time to view just the stomach and part of the small intestine. In fact, the limiting factor for the miniaturization of many battery-operated gadgets, including cameras, is the requirement for smaller and smaller batteries. The smallest film camera must house not only the optics and the film, but also the battery power source. When you consider the components that must go into a functioning battery, you see that it is amazing that batteries are as small as they are.

To understand batteries, we return to the recurring idea in this chapter: electrons in motion. A **battery** works by moving electrons around, and so it should not surprise you to learn that half reactions are what allow batteries to run. Because electrons must move from one part of a battery to another, the components that make up a battery must be able to conduct an electric current. Recall from Section 2.2 that metals are good electrical conductors because they are held together by a network of metallic bonds in which electrons flow throughout the metal. Because this "sea of electrons" makes it easy for electrons to move, metal components are a primary part of any battery.

Beginning here, let us continue our investigation of batteries with a simple device called an **electrochemical cell**, which is an up-market term for a device that uses chemical reactions to push electrons from one place to another. As we shall see, batteries are simply a more sophisticated version of the electrochemical cell shown in Figure 5.13. This particular cell contains copper and zinc, and these solid pieces of metal are called **electrodes**. The solid copper electrode (Cu) is immersed in a solution of

(a)

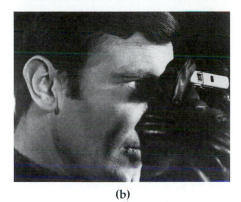

(b)

▲ **FIGURE 5.11 Bond, James Bond.** **(a)** James Bond speaking on a vintage car phone. **(b)** Bond's camera, an example of the miniaturization of photographic equipment.

▼ **FIGURE 5.12 A video pill.** This little battery-operated device will allow your doctor see what is going on in your digestive tract.

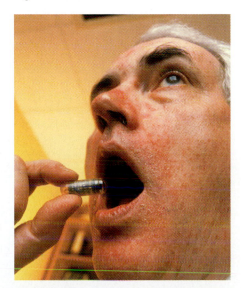

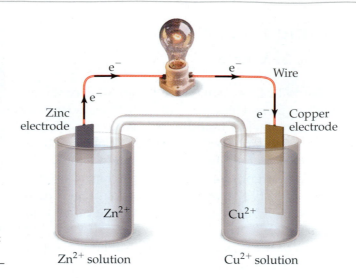

▶ **FIGURE 5.13 Diagram of an electrochemical cell.** Electrons move from the zinc electrode, through the wire and the light bulb, and into the copper electrode. At the copper electrode, the electrons reduce Cu^{2+} ions to metallic copper.

copper ions (Cu^{2+}), and the solid zinc electrode (Zn) is immersed in a solution of zinc ions (Zn^{2+}). Between the electrodes is a wire that is able to conduct electricity because it is also made of metal. In the middle of the wire is a light bulb that lights up when electricity is flowing through the wire. Be aware that when we say electricity is flowing through a wire, what we really mean is that *electrons are moving through the wire in one direction.* Another way to express this is to say that the wire *carries an electric current.*

In the case of our electrochemical cell, we know that electrons are moving through the wire in one direction, but how do we know which direction? Are electrons moving from the copper electrode to the zinc electrode, or are they moving in the opposite direction? The answer lies in the half reactions taking place in the cell, and so let us consider them for a moment. Our cell uses two solid metals, Zn and Cu, as well as their cations, Zn^{2+} and Cu^{2+}. The half reactions that interconvert the metals with their respective ions are

$$Cu(s) \rightleftharpoons Cu^{2+}(aq) + 2 \text{ electrons}$$
$$Zn(s) \rightleftharpoons Zn^{2+}(aq) + 2 \text{ electrons}$$

Take note of two things in these equations. First, the (*aq*) alongside each ion reminds us that the ions are in aqueous solution. We know that the ions, which are charged, dissolve readily in water because it is highly polar. Water that contains dissolved ions is able to conduct an electric current through the electrochemical cell.

RECURRING THEME IN CHEMISTRY Like dissolves like.

The second thing to notice about these half reactions is that they have double arrows, which tells you they can go in either direction. The most important thing to remember about any pair of oxidation–reduction half reactions is that one cannot occur without the other. If electrons are produced in one half reaction, they must be used up in another. This way, no stray electrons are ever allowed to remain unused. If one of the half reactions shown goes in the forward direction and produces electrons, the other half reaction must go in the reverse direction and use up those electrons.

Now let us get back to our original question about which way electrons move in an electrochemical cell. For our cell, the half reactions can be paired in one of the two following ways:

Redox reaction 1

$$Cu(s) \longrightarrow Cu^{2+}(aq) + 2 \text{ electrons} \qquad \text{oxidation of Cu atoms paired with}\ldots$$

$$Zn^{2+}(aq) + 2 \text{ electrons} \longrightarrow Zn(s) \qquad \ldots \text{reduction of } Zn^{2+} \text{ ions}$$

Redox reaction 2

$$Zn(s) \longrightarrow Zn^{2+}(aq) + 2 \text{ electrons} \qquad \text{oxidation of Zn atoms paired with}\ldots$$

$$Cu^{2+}(aq) + 2 \text{ electrons} \longrightarrow Cu(s) \qquad \ldots \text{reduction of } Cu^{2+} \text{ ions}$$

In redox reaction 1, the electrons leave the copper and are used by the zinc. Thus, in the setup of Figure 5.13, the electrons flow from right to left (from the copper electrode to the zinc electrode). In redox reaction 2, the electrons move from the zinc to the copper, which means from left to right in Figure 5.13. Which is correct?

In order to figure out the direction of electron flow in an electrochemical cell, we must define a new term. Consider the reduction of each ion for a moment:

$$Zn^{2+}(aq) + 2 \text{ electrons} \longrightarrow Zn(s) \qquad \text{reduction of zinc ion to zinc metal}$$

$$Cu^{2+}(aq) + 2 \text{ electrons} \longrightarrow Cu(s) \qquad \text{reduction of copper ion to copper metal}$$

Batteries work because different ions have different affinities for electrons. For the two ions we are considering, the copper ion's affinity for electrons is greater than the zinc's. Or, put another way, copper ions have a greater potential for reduction than do zinc ions. This **reduction potential** is measurable for metal ions, and you can consult a list of measured values if you want to know whether one half reaction has a larger reduction potential than another. Because the reduction potential of copper ions is larger than the reduction potential of zinc ions, it is the copper ions that gain electrons and are reduced in the copper–zinc redox reaction. This means that the zinc must undergo oxidation. Thus, of the two redox reactions shown above, the second reaction must be correct.

The two half reactions can be added to give the balanced redox reaction, in which no electrons are shown because the total number to the left of the arrows is equal to the total number to the right of the arrows:

RECURRING THEME IN CHEMISTRY Chemical reactions do not create or destroy matter. Therefore, the types and numbers of atoms must be the same on the two sides of a chemical equation.

$$Zn(s) \longrightarrow Zn^{2+}(aq) + \cancel{2 \text{ electrons}} \qquad \text{oxidation half reaction}$$
$$\underline{Cu^{2+}(aq) + \cancel{2 \text{ electrons}} \longrightarrow Cu(s) \qquad \text{reduction half reaction}}$$

$$Zn(s) + Cu^{2+}(aq) \longrightarrow Zn^{2+}(aq) + Cu(s) \qquad \text{balanced redox reaction}$$

QUESTION 5.8 (a) For the half reactions

$$Mg(s) \longrightarrow Mg^{2+}(aq) + 2 e^-$$
$$2 e^- + Ni^{2+}(aq) \longrightarrow Ni(s)$$

write a balanced equation showing the redox reaction using the half reactions in the direction in which they are written. (b) Nickel ions have a larger reduction potential than magnesium ions. In which direction will this redox reaction go?

ANSWER 5.8 (a) $Mg(s) + Ni^{2+}(aq) \rightarrow Mg^{2+}(aq) + Ni(s)$. (b) In the direction shown in part (a), because that equation shows the Ni^{2+} ions being reduced.

QUESTION 5.9 (*a*) For the half reactions

$$Au(s) \longrightarrow Au^{3+}(aq) + 3\,e^-$$
$$3\,e^- + Cr^{3+}(aq) \longrightarrow Cr(s)$$

write a balanced equation showing the redox reaction using the half reactions in the direction they are written. (*b*) Gold ions have a larger reduction potential than chromium ions. In which direction will this redox reaction go?

ANSWER 5.9 (*a*) $Au(s) + Cr^{3+}(aq) \rightarrow Au^{3+}(aq) + Cr(s)$. (*b*) In the direction opposite the direction shown in part (*a*) because that equation shows the Cr^{3+} ions being reduced. Because the Au^{3+} ions are more easily reduced than the Cr^{3+} ions, the proper equation for the reaction is $Au^{3+}(aq) + Cr(s) \rightarrow Au(s) + Cr^{3+}(aq)$.

The half reactions taking place in our electrochemical cell are illustrated in more detail in Figure 5.14. On the left, we can see that the zinc metal undergoes oxidation, which means that Zn atoms in the metal become Zn^{2+} ions and float away into the solution. This process produces two electrons for each Zn atom ionized. These electrons travel up through the zinc metal, a good electrical conductor, and into the wire, which also conducts electricity easily. The electrons flow into the light bulb, where they light the bulb. From there, they travel down into the solid copper, where they become available to the copper ions in the aqueous solution surrounding the copper electrode. As we already know, these copper ions readily undergo reduction, and as they accept electrons, they become copper atoms that take their place in the lattice of the solid copper electrode. The direction of electron flow is driven by this final half reaction: the copper ions require electrons in order to become copper atoms.

▶ **FIGURE 5.14 Details in an electrochemical cell.** On the left, zinc metal atoms leave the solid zinc electrode as they are oxidized to Zn^{2+}. On the right, incoming electrons reduce Cu^{2+} ions in solution to metallic copper, which can then join the solid lattice of the copper electrode.

LIVE ART
Electrochemical Cell

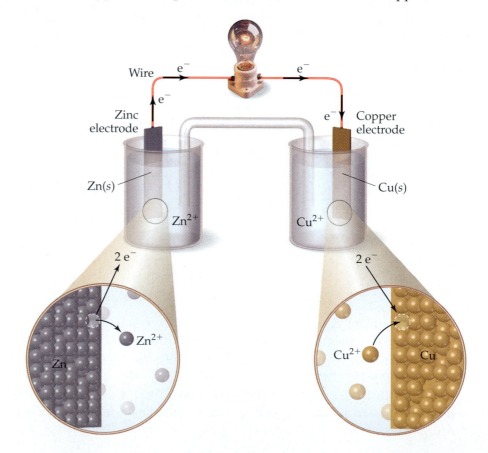

This demand pulls electrons from the zinc electrode up through the wire, through the light bulb, and down into the copper electrode.

The current that exists as electrons travel from the zinc compartment to the copper compartment in Figure 5.14 cannot continue indefinitely. Eventually, the supply of copper ions in the beaker containing the copper electrode will be depleted, and the redox reaction will peter out. If you have ever had a dead flashlight battery, then you already have experience with the limitations of batteries.

QUESTION 5.10   An electrochemical cell has electrodes made from nickel and iron and solutions containing Ni^{2+} ions and Fe^{2+} ions. (*a*) If the reduction potential for nickel ions is greater than that for iron ions, which of the two ions will be reduced? (*b*) Write the reduction half reaction for this electrochemical cell. (*c*) Write the oxidation half reaction for this electrochemical cell. (*d*) Write a balanced equation for the net redox reaction. (*e*) Do electrons flow from the nickel electrode to the iron electrode or vice versa?

ANSWER 5.10   (*a*) Nickel ions. (*b*) $Ni^{2+}(aq) + 2\,\text{electrons} \rightarrow Ni(s)$. (*c*) $Fe(s) \rightarrow Fe^{2+}(aq) + 2\,\text{electrons}$. (*d*) $Ni^{2+}(aq) + Fe(s) \rightarrow Ni(s) + Fe^{2+}(aq)$. (*e*) Vice versa, from iron electrode to nickel electrode.

QUESTION 5.11   An electrochemical cell has electrodes made from lead and manganese sitting in solutions of Pb^{2+} ions and Mn^{2+} ions, respectively. (*a*) If the reduction potential for manganese ions is less than that for lead ions, which of the two ions will be reduced? (*b*) Write the reduction half reaction for this electrochemical cell. (*c*) Write the oxidation half reaction. (*d*) Write a balanced equation for the net redox reaction. (*e*) Do electrons flow from the manganese electrode to the lead electrode or vice versa?

ANSWER 5.11   (*a*) Lead ions. (*b*) $Pb^{2+}(aq) + 2\,\text{electrons} \rightarrow Pb(s)$. (*c*) $Mn(s) \rightarrow Mn^{2+}(aq) + 2\,\text{electrons}$. (*d*) $Pb^{2+}(aq) + Mn(s) \rightarrow Pb(s) + Mn^{2+}(aq)$. (*e*) Vice versa, from manganese electrode to lead electrode.

5.5 | Working Batteries

Now, we are equipped to consider a real battery. Figure 5.15 illustrates a typical flashlight battery. This type of battery is called a **dry cell** because it contains no aqueous solutions. Rather than the solutions shown in Figures 5.13 and 5.14, the ionic species in a dry cell exist in an electrolyte paste, which prevents the battery from leaking. In many cases, the electrolyte paste contains both ions required for the battery to work.

A typical reaction in a flashlight dry cell consists of these half reactions:

$$Mn^{4+} + 1\,\text{electron} \longrightarrow Mn^{3+} \quad \text{reduction}$$
$$Zn(s) \longrightarrow Zn^{2+} + 2\,\text{electrons} \quad \text{oxidation}$$

We can assign these two half reactions to different parts of the battery shown in Figure 5.15. One electrode is solid zinc, and the other is graphite coated with Mn^{4+} ion. (Graphite, a form of carbon we met in Chapter 3, conducts electricity.) The two electrodes are distinguished according to the half reaction in which they participate. The **anode** is the electrode where oxidation takes place, meaning that the zinc electrode is the anode here. The **cathode** is the

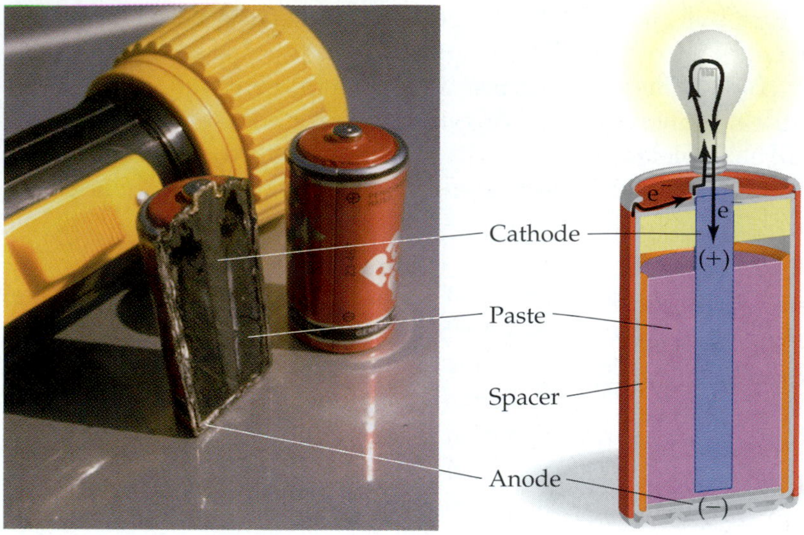

▶ **FIGURE 5.15** **A typical dry cell.** The cathode and anode are separated by a conductive electrolyte paste and an inert material called a spacer. The cathode is the positive terminal of the battery, and the anode is the negative terminal.

Cathode

Paste

Spacer

Anode

electrode where reduction takes place; thus, the Mn^{4+}-coated graphite is our cathode in this case. The anode, which in this instance is the casing of the battery, is called the *negative terminal* of the battery because it is where electrons are produced by the oxidation half reaction. Those electrons leave the battery to light up the flashlight bulb and then return to the battery via the cathode, which is called the *positive terminal*—where reduction takes place.

There are dozens of configurations for batteries, and over the years batteries have been optimized to do particular jobs. For example, car batteries must be able to provide a burst of electrical energy powerful enough to start a car. These batteries are heavy because the electrodes are made from lead (which has a high atomic mass). If you were designing a wristwatch, you would want to use a smaller, less powerful, and much lighter battery. If you were designing a pocket-sized camera, you would also want a battery that is small and light. However, you would not want to use a watch battery for a camera because camera batteries must be able to produce a small burst of energy to power the flash mechanism. This is just something a watch battery—designed to produce a tiny steady current over a long period of time—cannot do.

You have probably seen the ultrasmall batteries found in watches, hearing aids, and cameras. These are often disk shaped, like the one illustrated in Figure 5.16. In this arrangement, all the essential components of a battery must be present, and they must all fit into a tiny space. Figure 5.16 shows that a battery casing can contain both anode and cathode. In this instance, the top of the disk is the anode and the casing around the disk periphery is the cathode. The interior contains two kinds of electrolyte paste separated by an inert spacer material, with each paste suited for one of the electrodes. This configuration is used in the smallest photographic batteries, which are usually *lithium batteries*. In a lithium battery, lithium metal is oxidized and titanium ions are reduced:

$$Li \longrightarrow Li^+ + 1 \text{ electron}$$
$$Ti^{4+} + 1 \text{ electron} \longrightarrow Ti^{3+}$$

▼ **FIGURE 5.16** **Smaller and smaller.** A miniaturized battery must contain all the components found in full-size batteries.

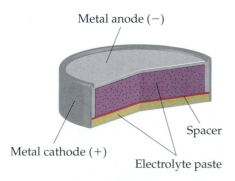

Metal anode (−)

Spacer

Metal cathode (+)

Electrolyte paste

Lithium batteries are widely used because they are extremely small and very, very light. A look at the periodic table shows you that lithium, atomic number 3, is the lightest metal, and titanium, atomic number 22, is the second lightest transition metal (after scandium, atomic number 21). A traditional

Do Electric Eels Really Have Electricity in Them?

Electric eels and batteries have a great deal in common, but to understand why, we have to reconsider the notion of reduction potential for a moment. Recall that the reduction potentials of the two half reactions occurring in a battery tell us which way electrons flow. Thus, if we know which reduction potential is greater, we can predict which way electrons travel through the battery. However, the magnitude of the difference between the two reduction potentials also tells us something.

If the difference is large, electrons are pushed through the battery more vigorously than if the difference is small. That is, a large *potential difference* tells us the oxidation half reaction readily produces electrons that are pulled *forcefully* toward the cathode so that reduction can take place. If the potential difference is smaller, electrons still travel in the same direction, but the force with which they move through the battery is weaker. We refer to this potential difference, or force of electron flow, with the familiar term *voltage*. The unit used to measure voltages is called, not surprisingly, the volt (V).

The body of an electric eel might remind you of a flashlight, which has two or more batteries stacked on top of one another *in series*. When batteries are aligned positive to negative in this way, the voltages produced by the individual batteries add together to produce one overall voltage that is used to power the bulb. In Figure 5.17, for example, each battery can produce 1.5 V, and so when you

▲ FIGURE 5.18   **South American electric eel.**

stack two together, the total voltage is 3.0 V. When these two batteries are used in a flashlight, this is the voltage available to the bulb.

Electric eels use the same kind of arrangement. *Electrophorus electricus*, found in South America, is one of the most powerful electric eels known (Figure 5.18). Its tail, which makes up about 80 percent of the length of its body, contains *electrocytes*—small, disk-shaped cells that stack up like batteries in a flashlight. (Note that this word is *electroCyte*, with a c, and not *electroLyte*.) These stacked cells generate a powerful electric voltage that is cumulative as a result of their stacked configuration. Eels can produce low voltages of about 5 to 10 V, which they use to navigate through muddy water and for communication. Some can also produce strong jolts, up to 600 V (about five times the voltage in a typical household wall socket), used to stun prey. There are reports of electric eels toppling humans and horses walking through water, even when the eels are more than 3 m away!

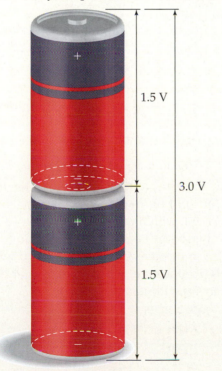

1.5 V

3.0 V

1.5 V

▲ FIGURE 5.17   **Batteries connected in series.**   When batteries are connected in series, which means the positive terminal of one abuts the negative terminal of its neighbor, the voltages are cumulative. Thus, two 1.5-V batteries combine to produce a potential difference of 3.0 V.

QUESTION 5.12   Studies on electric eels have shown that the longest ones, up to 1 m in length, produce the strongest jolts of electricity. From what you know about batteries, why do you think that longer eels can produce more electricity?

ANSWER 5.12   Batteries stacked in series have additive voltages. The electrocytes in eels are stacked in series. Thus, the longer the eel, the greater the number of electrocytes in series and therefore the greater the voltage.

AA dry-cell battery has a mass of roughly 23 g. The same-size AA battery that uses lithium chemistry has a mass of only 15 g. If cameras are going to continue to shrink in size and mass, so must their batteries, and lithium-based batteries are likely to be an attractive choice for camera manufacturers.

QUESTION 5.13   (a) Identify each of the lithium-battery half reactions as either oxidation or reduction. (b) Write a balanced equation for the redox reaction for this battery. (c) Which half reaction occurs at the anode, and which half reaction occurs at the cathode?

ANSWER   5.13   (a) $Li \rightarrow Li^+ + 1$ electron   is   oxidation; $Ti^{4+} + 1$ electron $\rightarrow Ti^{3+}$ is reduction. (b) $Li + Ti^{4+} \rightarrow Li^+ + Ti^{3+}$. (c) Lithium oxidation at anode, Ti^{4+} reduction at cathode.

5.6 | Electron Movement in Organic Molecules: Free Radicals

Batteries are assembled primarily from metal components because the electrons produced in the redox reactions must be able to travel from one part of the battery to another, and metals are good conductors of electricity. Our discussion of electron movement thus far has not included any mention of organic molecules because these molecules are generally not good electrical conductors. However, organic molecules do take part in chemical reactions that move electrons around. In the world of organic molecules, though, chemical reactions are a bit different, and different terms are used to describe them.

Very often when an organic molecule undergoes a chemical reaction, a bond breaks. Because organic molecules are built around carbon-linked frames, the bond that breaks is sometimes a carbon–carbon bond. Recall that the bonds between atoms in organic molecules are covalent, which means the two atoms forming each bond share electrons so that each atom can have an octet of electrons (or duet in the case of H):

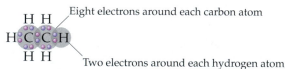

If a single covalent bond in an organic molecule breaks, the two electrons making up that bond can split up in one of two ways. First, the bond can undergo **heterolytic cleavage**, which means both electrons go with one atom of the bond and the other atom takes none. This is by far the most common way bonds in organic molecules break. The second way this bond can break is through **homolytic cleavage**, which means that each atom of the bond takes one electron away with it:

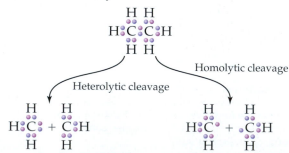

Each of the molecular fragments produced in any homolytic cleavage has an electron that is not paired up. Chemists refer to molecules containing unpaired electrons as **free radicals**. Even though they are much rarer than heterolytic reactions, homolytic reactions play an important role in thousands of industrial processes and in the human body, because they create free radicals.

A note on notation. When writing the structure for a free radical, organic chemists usually place a dot at the atom containing the unpaired electron, as in

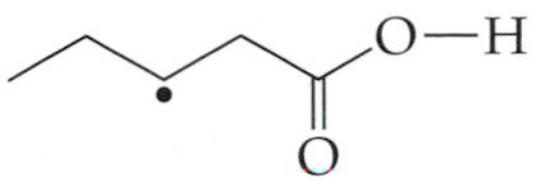

Chemical structures are often full of dots! Be careful not to confuse the single dot associated with a free radical with the duo of dots associated with a nonbonding pair of electrons.

The chemistry of free radicals is closely linked to the chemistry of oxygen and the process of oxidation. As you might suspect, oxygen is a powerful oxidizing agent and reacts with many substances. That is why there are hundreds of chemical compounds that contain oxygen plus one other element: oxygen is ubiquitous in the Earth's atmosphere, and many elements are readily oxidized by it. For example, iron oxidizes to rust, and because our atmosphere is about 21 percent oxygen, rusting is unavoidable. In fact, we humans spend considerable time working on preventing or reversing the effects of oxygen: we store fruit in the refrigerator to prevent it from oxidizing (a process that turns it brown), we use rustproofers on metal windows and on car chrome, and we eat antioxidants to prevent unwanted oxidative reactions in our bodies. Clearly, we require oxygen to live, but what makes it so hard to live with?

To answer this question, let us look at an example that illustrates the damaging effects of oxygen and the relationship between this damage and free radicals. Many foods contain *lipids* (more commonly known as *fats*), which are organic molecules that contain a long hydrocarbon chain. In the presence of molecular oxygen (from the air, say), lipids give up an electron to the oxygen in an oxidation reaction. A carbon–hydrogen bond in the lipid breaks homolytically, and a free radical is formed:

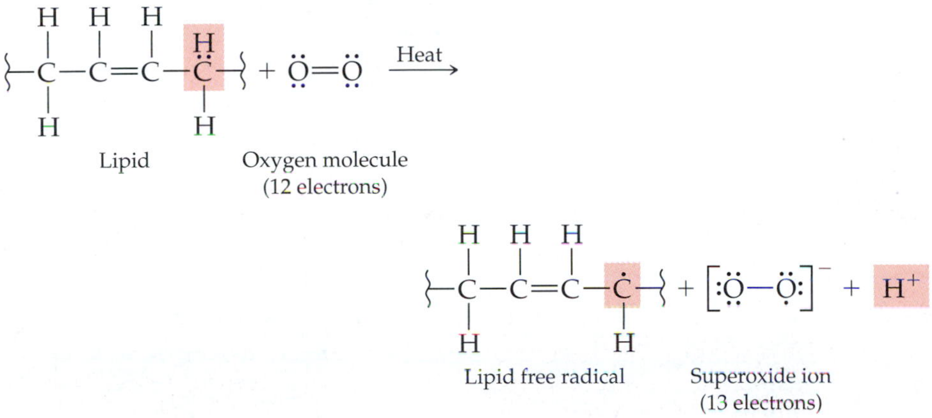

The lipid free radical goes on to form unsavory molecules that cause a rancid, unpleasant taste in the lipid-containing food. (One example is the characteristic taste of stale potato chips.) Because this reaction occurs over time, the longer the product sits on the grocery shelf, the more rancid it gets. This is where antioxidants come into the picture. An **antioxidant** is a molecule that can donate an electron to a free radical and thereby return the free radical to its preoxidized state:

$$\left\{ \begin{array}{c} H \\ | \\ C \\ | \\ H \end{array} - \begin{array}{c} H \\ | \\ C \\ \end{array} = \begin{array}{c} H \\ | \\ C \\ \end{array} - \begin{array}{c} \cdot \\ C \\ | \\ H \end{array} \right\} + \text{antioxidant} + \boxed{H^+} \longrightarrow$$

$$\left\{ \begin{array}{c} H \\ | \\ C \\ | \\ H \end{array} - \begin{array}{c} H \\ | \\ C \\ \end{array} = \begin{array}{c} H \\ | \\ C \\ \end{array} - \begin{array}{c} H \\ | \\ \ddot{C} \\ | \\ H \end{array} \right\} +$$

Stable free-radical form of antioxidant after donating one electron to lipid

Thus, if you add an antioxidant to a lipid-containing food, the food can stay on the shelf much longer because the lipids stay intact.

Figure 5.19 shows the ingredient labels on two boxes of cereal, with the antioxidant included in each cereal highlighted. Three of the most common antioxidants used to preserve food are all in the family of organic molecules known as *tocopherols*. Figure 5.20 shows the structures of three of the most widely used tocopherols: BHT, BHA, and vitamin E. If you look closely at these structures, you will notice two similarities—each molecule contains a benzene ring, and each contains an —OH group attached to that ring.

▶ **FIGURE 5.19 Antioxidants on the job.** The ingredient lists for these two breakfast cereals name two common food antioxidants: BHT, which stands for *butylated hydroxytoluene*, and vitamin E. Both substances help keep the product fresh and preserve flavor.

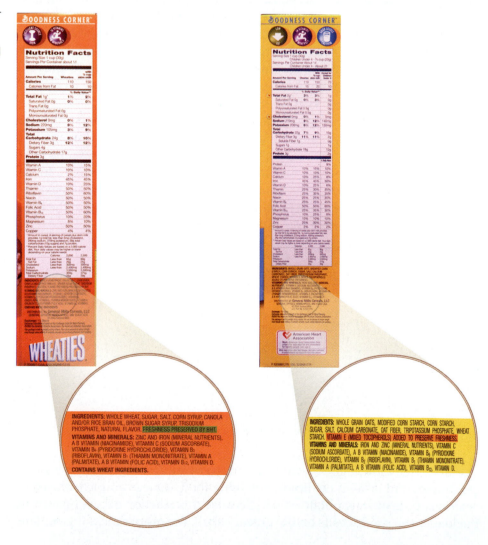

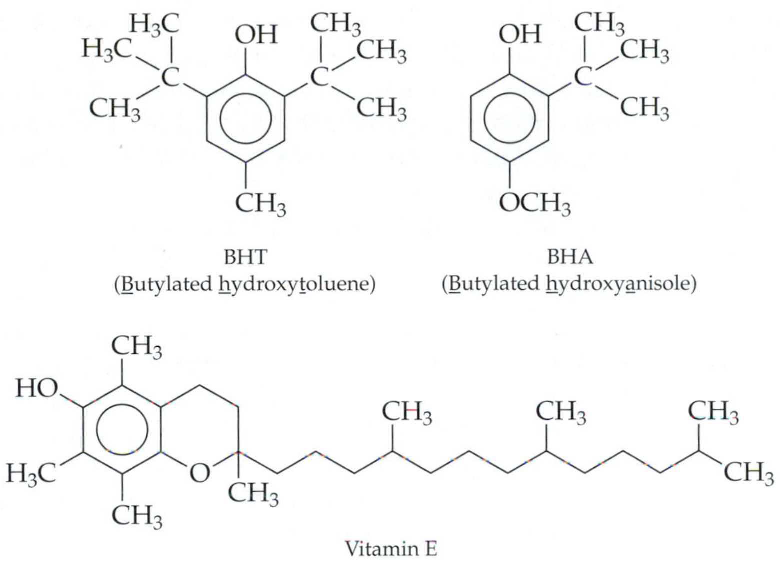

◀ FIGURE 5.20 The structures of three common antioxidants.

Organic molecules that contain an —OH group are referred to as **alcohols**. A special name, **phenol**, is given to alcohols in which the —OH group is attached to a benzene ring. The combination of benzene ring and —OH group allows a phenol to behave as an antioxidant because, when an electron is given up by the oxygen of the —OH group, the effect of the missing electron is not just localized on the —OH group. Rather, the effect is spread out all around the benzene ring because of delocalization (Section 3.4). The benzene ring exerts a stabilizing influence on the site of the missing electron, and this gives phenols a way to cope with unpaired electrons. Thus, the shelf life of a product containing a lipid (such as potato chips or cereal) is extended when BHA or some other antioxidant is present, because any lipid free radicals formed can be converted back to (nonrancid) lipid molecules:

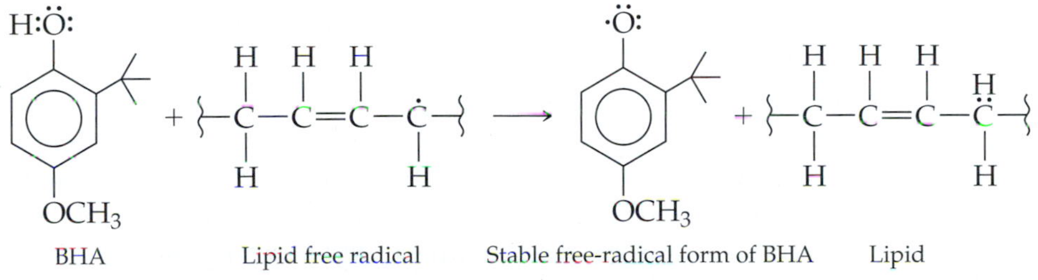

Previously, we learned that oxygen can change a lipid to a free radical, as with lipid-containing products in the grocery store. In our bodies, oxygen can also wreak havoc—but it usually does so only after it has been converted to one of several highly reactive species. Let us consider one type. In living cells, oxygen itself can form unwanted free radicals, such as the *superoxide ion*:

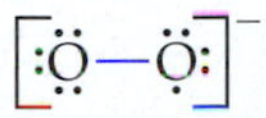

Superoxide ion, with one unpaired electron

The formation of oxygen-containing free radicals can be prompted by such things as toxic pollutant molecules and exposure to excessive levels of ultraviolet light from the sun. Once superoxide ions form, they can react with lipids in the body and convert them to free radicals, which then break down just as lipids do in a bag of potato chips. Because cell membranes are composed of lipid-based molecules, irreparable cell damage can be caused by superoxide ions and other highly reactive oxygen free radicals. Further, these free radicals attack DNA and protein molecules and can cause significant and permanent changes in their structures. For these reasons, our bodies have developed defense mechanisms that fight free radicals, especially those formed from oxygen molecules.

Our bodies' defense strategy against free radicals includes several kinds of antioxidants. As we have just seen, antioxidants that are organic molecules, like the tocopherols, typically have a means for delocalizing electrons. These molecules often include the vitamins, as we saw when we noted that vitamin E is an antioxidant. Antioxidants can also be *antioxidant enzymes*, which are enormous protein molecules whose job it is to defuse unwanted free radicals. The antioxidant enzyme *superoxide dismutase*, for example, catalyzes a reaction that converts four unwanted superoxide free radicals to two molecules of water and three molecules of oxygen. Notice that, in this balanced chemical equation, the number of electrons on the left side is equal to the number of electrons on the right side, as it must be:

> **RECURRING THEME IN CHEMISTRY** Chemical reactions do not create or destroy matter. Therefore, the types and numbers of atoms must be the same on the two sides of a chemical equation.

$$4\left[\ddot{O}-\ddot{O}\right]^{-} + 4\,H^{+} \xrightarrow{\text{Superoxide dismutase}} 2\ H\underset{}{\overset{\ddot{O}}{\diagup}}_{H} + 3\,\ddot{O}=\ddot{O}$$

Superoxide ions	Protons	Water molecules	Oxygen molecules
$4 \times 13 = 52$ electrons	0 electrons	$2 \times 8 = 16$ electrons	$3 \times 12 = 36$ electrons

Before any unwanted reaction can occur between superoxide ions and lipids or DNA, this enzyme is able to react with the damaging superoxide ions and convert them to water and oxygen molecules.

Many of the foods we eat contain natural antioxidants, but you will also find dozens of antioxidant formulations in the health food store or in the vitamin aisle of your local grocery store. With all the hype over the link between antioxidants, aging, and cancer, it is no wonder that antioxidants have been put into capsules to supplement our daily intake from food. Many health professionals, however, insist that a healthy diet provides plenty of antioxidants.

The enzyme superoxide dismutase, which is made in our bodies, must contain zinc and copper in order to function properly. These two metals ions must be obtained from the food we eat. For this reason, these elements are often included on lists of antioxidants, such as the one shown in Table 5.1. Likewise, the element selenium, Se, is required for proper function of the antioxidant enzyme *glutathione peroxidase*. Thus, we see selenium on the list of antioxidant elements, too.

The rightmost column in Table 5.1 shows that there is a limit to the RDA values for some antioxidants (vitamins A and E, Cu, and Se). These substances can cause health problems when taken in excess. Excessive

TABLE 5.1	Recommended Daily Allowances (RDAs) of Specific Antioxidants	
Antioxidant	RDA	Amount consumed each day not to exceed . . .
Vitamin A	2700 to 3300 mg	17,000 mg
Vitamin C	75 to 90 mg	—
Vitamin E	8 to 10 mg	14,000 mg
Cu	0.9 to 3 mg	10 mg
Se	1.5 to 3.0 μg	600 μg
Zn	12 to 15 mg	—

TABLE 5.2   Antioxidant Content of Some Foods[a,b]

	White rice	Chicken breast	Canned tuna	Oranges	Spinach	Carrots	Mangoes	Pecans	Seaweed	Lettuce	Walnuts	Shredded wheat
Vitamin A (mg)	0	0	13	150	6251	8024	510	37.3	77.3	4937	13	0
Vitamin C (mg)	0	0	0	53.2	28.1	5.9	27.7	1.1	3.0	18.0	1.3	20.1
Vitamin E (mg)	0.04	0.07	—	0.18	2.03	0.66	1.12	1.40	0.87	0.29	0.70	0.66
Cu (mg)	0.069	0.069	0.039	0.045	0.13	0.045	0.110	1.20	0.130	0.029	1.59	0.384
Se (μg)	7.5	11.7	65.7	0.5	0.53	0.1	0.04	4.53	0.7	0.6	4.9	4.5
Zn (mg)	0.49	0.65	0.48	0.07	1.0	0.24	0.6	3.8	1.23	0.18	3.09	3.10

[a] All portions are 100 g.
[b] Whenever a food provides more than 50 percent of the recommended daily allowance listed in Table 5.1, the value is highlighted in yellow.

levels of vitamin A in pregnant women, for example, have been shown to cause birth defects. Thus, while it may be a good idea to supplement your diet with antioxidant capsules, it is possible to get too much of a good thing.

If we want to optimize the levels of antioxidants in our diets, which foods provide the highest amounts? Conventional wisdom says that foods with color are the best antioxidant sources. Table 5.2 compares the antioxidant content of 12 foods that may be on your grocery list. Notice that while colorful fruits and vegetables are excellent sources of vitamins A, C, and E,

Why Does Some Fruit Turn Brown When You Cut It?

STUDENTS OFTEN ASK

When you cut into such fruits as apples, pears, and bananas, the newly exposed tissues—because they are no longer protected by the skin of the fruit—come into contact with oxygen in the air. Once air hits the cut surface, a series of complex biochemical reactions produces molecules with a brown pigment.

Albert Szent-Györgyi, the Hungarian chemist who discovered vitamin C, noticed that the juice of fruits that *do not* turn brown when cut, such as tomatoes and citrus fruits, can prevent browning when sprinkled onto fruits that *do* turn brown. He found that the molecule that prevents browning is *ascorbic acid*, the chemical name for vitamin C:

CH₂OH

HO—C

O

O

H

HO   OH

Vitamin C (ascorbic acid)

This is why soaking cut apple slices in lemon juice keeps them from turning brown. Being an antioxidant (Tables 5.1 and 5.2), vitamin C is readily oxidized and can react with the oxygen in air, thereby preventing that oxygen from turning the apple slices brown. (We shall discuss an entirely different type of food browning—the browning that takes place with the intense heat of a grill—in Chapter 12.)

QUESTION 5.14 How does the structure of vitamin C differ from the structure of the antioxidants shown in Figure 5.20?

ANSWER 5.14 The vitamin C molecule does not contain a benzene ring. Thus, the presence of a benzene ring is not an absolute requirement in the structure of antioxidant compounds, although it is found in them quite often.

they are not necessarily the best source of the antioxidants Se, Cu, and Zn. For example, some selenium is found in spinach and carrots, but you can get much more from tuna and chicken. While seaweed and spinach are good sources of zinc, they do not contain as much as zinc as is found in pecans, walnuts, and shredded wheat. This is one reason a balanced diet is important: to fulfill your need for all these antioxidants, you must consume foods from different food groups.

QUESTION 5.15   If you want to get enough of all the antioxidants listed in Table 5.2, which two antioxidants would it be best to obtain from antioxidant supplements? You will need to refer to the RDA values listed in Table 5.1 to answer this question.

ANSWER 5.15   The absence of any yellow highlighting in the vitamin E and Zn rows of the table shows that none of the foods listed supply even 50 percent of the recommended daily allowance of these two antioxidants. Thus, it is reasonable to supplement your diet with these antioxidants.

5.7 | Still More Rambling Electrons: Light-Driven Reactions

You may have noticed that light can change the appearance of everyday materials: newspapers turn yellow-brown, for instance, and laundry hung on a clothes line gets bleached. These changes are the result of the action of sunlight, and they may be caused either by ionization reactions or by the type of free-radical reactions we have seen in this chapter. In this section, we shall look at how light causes electrons to move from the ground-state energy level to some higher energy level, and how it causes chemical reactions to take place.

Most people think of light as the brightness we see from a light bulb or from the sun, but scientists define this term much more broadly. In a scientific context, light is a form of energy, and there are several categories of light energy. The **electromagnetic spectrum**, shown in Figure 5.21, is a continuum of light energies, and the different regions of the spectrum are given different names, many of which may be familiar to you. In Figure 5.21, light, or **electromagnetic radiation**, of the highest energy is shown on the right, and light of the lowest energy is shown on the left. The kinds of light we usually classify as dangerous are at the high-energy end of the spectrum. For example, we know that **ultraviolet light** (UV) can cause severe sunburn and that we should protect ourselves from X-rays with heavy metal shields. On the left are the more benign varieties of electromagnetic

▶ FIGURE 5.21   **The electromagnetic spectrum.**   Electromagnetic radiation that is low in energy (longer wavelength) is shown on the left. Energy increases and wavelength decreases from left to right. The visible portion of the spectrum runs from approximately 700 nm at the red end to 400 nm at the violet end.

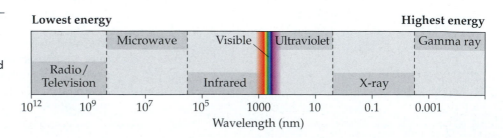

radiation: radio and television waves, microwaves, and infrared light. We certainly do not worry about radio waves having any adverse effect on our health (except when bad music is transmitted by them)!

Infrared light is most often associated with heat, and cameras loaded with film sensitive to infrared light give us a picture of hot and cold areas in a photograph. An infrared photograph of a cat's face, for example, might show the warmer areas in red and the cooler areas in bluish gray. Which parts of the face would you expect to have which color? See Figure 5.22 to find out the answer. **Visible light**, the only light to which our eyes are sensitive, is located in the middle of the electromagnetic spectrum.

▲ **FIGURE 5.22   Infrared photography shows warm and cool regions.**   In this infrared photograph of a cat's face, the bluish-gray regions are cool and the red regions are warm.

QUESTION 5.16   Rank these forms of electromagnetic radiation in order of increasing energy, lowest first: microwave, radio, gamma, infrared, visible.

ANSWER 5.16   Radio < microwave < infrared < visible < gamma.

Why are some kinds of light gentler than others? Another look at Figure 5.21 shows you that light is classified according to its wavelength—the distance from one peak to the next in a wave of light, as shown in Figure 5.23. The closer together the peaks are, the higher the energy of the light. (Or, to say it another way, the shorter the wavelength, the higher the energy.) The more energy light has, the more capable it is of penetrating solid matter, including the human body.

We talked about energy and light earlier in the chapter in the context of moving electrons in photographic film and the light that activates this process. We also know that heat energy and light energy can cause an electron to move out of an atom altogether, leaving behind a positively charged ion. If the amount of energy is not enough to cause the electron to leave the atom, a ground-state electron may instead be promoted to an excited state. We know that there are discrete energy levels and sublevels present in every atom (Figure 5.4). A ground-state atom of gallium, Ga, for example, has the electron configuration $1s^2 2s^2 2p^6 3s^2 3p^6 4s^2 3d^{10} 4p^1$. A very important thing to remember when we are discussing energy sublevels is that

> any two energy sublevels in an atom are separated by a specific amount of energy.

Thus, as you move from one sublevel to any other sublevel, there is a specific change in energy, a change that can be measured easily.

Let us look at how electron promotion to excited states can happen. When a single atom is excited by a specific amount of energy, one of its electrons can *absorb* the energy and move from one energy sublevel to a

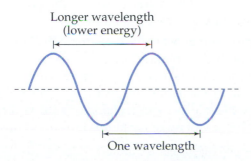

Longer wavelength
(lower energy)

One wavelength

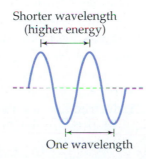

Shorter wavelength
(higher energy)

One wavelength

◀ **FIGURE 5.23   The definition of wavelength.**   One wavelength is the distance between any two adjacent peaks in a wave. As these two waves show, the shorter the wavelength, the higher is the energy of the wave. (A memory aid: L/L, the *l*onger the wavelength, the *l*ower the energy.)

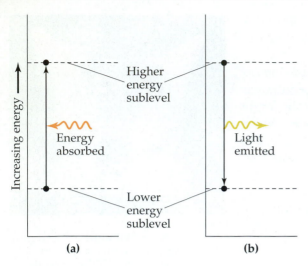

▲ **FIGURE 5.24** **Energy absorption and emission by an atom.** **(a)** When energy is supplied, most often in the form of light or heat, an electron can absorb the energy and move from a lower-energy sublevel to a higher-energy sublevel. **(b)** When the electron returns to its original sublevel, energy is given off (emitted), in this case in the form of light.

higher-energy sublevel (Figure 5.24a). In order for this to occur, the added energy must be equivalent to the difference in energy between the sublevel the electron leaves and the sublevel the electron enters. All the energy absorbed is later *emitted* when the electron returns back to its original sublevel, as shown in Figure 5.24b.

Now consider what happens when we excite a collection of hydrogen atoms with energy from a source that is made up of different energies of light. Because different energies of light interact with the electrons in our collection of hydrogen atoms, each electron can be promoted to one of several different excited states. An electron in an excited state does not stay there indefinitely, though. At some point, that electron returns to a lower energy level, and the energy of the light emitted as a given electron drops back to a lower level corresponds to the energy difference between the two levels. (Recall from earlier in the chapter that when we are discussing hydrogen, because it is not a multi-electron atom, we speak in terms of energy levels rather than energy sublevels.) Thus, an excited electron might move from energy level 5 down to energy level 2, and the energy difference between these levels corresponds to a specific wavelength of light emitted. Another excited electron may move from level 3 to level 1 and emit a different wavelength of light.

If you passed the light given off by our collection of hydrogen atoms through a prism, the light would separate into these many different discrete wavelengths, and we refer to this series of specific wavelengths of light as a **line spectrum**. For hydrogen, four of the wavelengths emitted are in the visible region of the electromagnetic spectrum (Figure 5.25a). These lines form as electrons move from an excited state to the second energy level, as shown in Figure 5.25b. For example, an electron that moves from level 5 to level 2 emits blue light, and an electron that moves from level 3 to level 2 emits yellowish-orange light. Notice that the greater the distance an electron moves, the higher the energy of the light it emits. Thus, the longest line in Figure 5.25b represents light having the shortest wavelength (and the highest energy).

Many elements produce a characteristic line spectrum in the visible region. For example, excited sodium atoms produce a brilliant yellow color as electrons fall back to lower-energy levels. Figure 5.26 shows the yellow flame of sodium along with the brightly colored flames produced by several other elements. This phenomenon is used in the manufacture of fireworks, in which different elements are used to produce all the various colored flashes of light.

QUESTION 5.17 You may have noticed that industrial lighting can have different hues. Figure 5.27 on page 196 shows two examples. The lighting on the left has a bluish glow, whereas the lighting on the right has a yellowish-orange glow. Why are the colors different?

ANSWER 5.17 Different elements are used to create the bulbs. Mercury was used for the lights in the left photograph, and sodium was used for the lights in the right photograph. These lamps are referred to as mercury vapor lamps and sodium vapor lamps, respectively. Notice the color of industrial lighting the next time you get a chance. See if you can distinguish between the bluish glow of mercury lamps and the yellowish glow of sodium lamps.

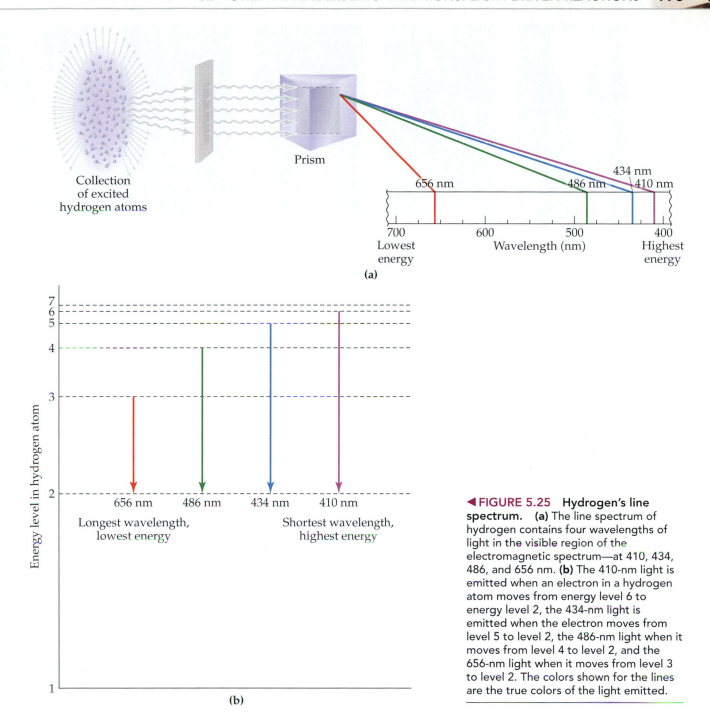

◄ FIGURE 5.25   Hydrogen's line spectrum.   (a) The line spectrum of hydrogen contains four wavelengths of light in the visible region of the electromagnetic spectrum—at 410, 434, 486, and 656 nm. **(b)** The 410-nm light is emitted when an electron in a hydrogen atom moves from energy level 6 to energy level 2, the 434-nm light is emitted when the electron moves from level 5 to level 2, the 486-nm light when it moves from level 4 to level 2, and the 656-nm light when it moves from level 3 to level 2. The colors shown for the lines are the true colors of the light emitted.

QUESTION 5.18   To perform a flame test, you dip a wire made of some inert material into a solution containing ions of an unknown element and then hold the wire in a flame. Why can elements be distinguished in this way?

ANSWER 5.18   Each element has a unique distribution of spacings between its energy sublevels. Therefore, the colors of light that correspond to those energy differences are characteristic of that element.

The region of the electromagnetic spectrum from red to violet is referred to as *white light*, and sunlight is mainly white light (along with some ultraviolet and other radiation). The entire visible portion of the

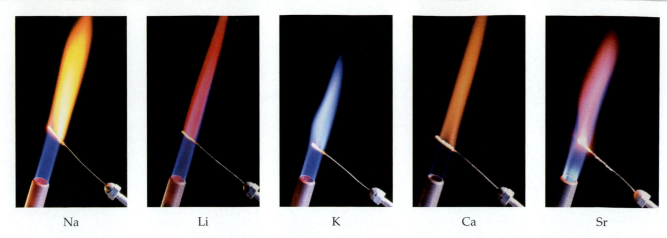

Na   Li   K   Ca   Sr

▲ FIGURE 5.26 **Flame tests for identifying elements.** After a wire is dipped into a solution containing ions of some element, the wet end of the wire is put into the flame of a burner. The color to which the flame changes depends on the type of ions in the solution. A sodium solution causes the flame color to change to bright yellow, for instance, lithium deep pink, calcium orange, and so on.

spectrum can be seen when either sunlight or white light from some other source passes through a prism, as illustrated in Figure 5.28. This assortment of colors should look familiar to you because the colors you see in a rainbow are those of the visible portion of the electromagnetic spectrum. The rainbow is created as these colors are separated one from the other by water droplets, which act as tiny prisms in the air after a rainstorm. Thus, white light gives off all the wavelengths of visible light. In contrast, excited atoms of *one specific element* give off *only very specific wavelengths of light* because these wavelengths correspond to the spacing between the energy levels (or sublevels) of that given element.

Thus far, we have seen that electrons can move up and down from the ground state to an excited state and back. Our example of the hydrogen atom showed that a source of energy can excite electrons up to various excited states, and then a line spectrum is produced as those atoms relax back to the ground state by *emitting* light of specific wavelengths. It

▼ FIGURE 5.27 **Not all light bulbs are created equal.** Industrial lighting produced by the emission of light by different elements. A mercury vapor lamp is shown at left, and a sodium vapor lamp is shown at right.

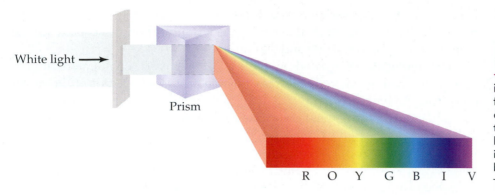

is also possible for some organic molecules to interact with visible light. For the remainder of this discussion, however, we will be concerned only with light they *absorb* rather than light they emit. Organic molecules that are brightly colored usually contain an alternating pattern of double and single bonds. These molecules are said to be **conjugated**, and in conjugated molecules, the electrons do not remain between the atoms of one bond. Rather, conjugation means that the electrons can spread out and become delocalized over all the areas of the molecule in which alternating double and single bonds exist. To see how such a molecule can have color, let us look at a few hydrocarbons and see how they respond to light.

Consider the two conjugated hydrocarbon molecules shown in Figure 5.29. Hydrocarbons with a conjugated carbon chain absorb light from the visible region of the electromagnetic spectrum if the length of the conjugated segment is sufficiently long. In general, if the number of bonds in the conjugated system is roughly ten or more, the molecule absorbs visible light. Take a look at the molecule shown in Figure 5.29a. Because this molecule is conjugated over only five bonds, we expect it to appear colorless to our eyes because it does not absorb light in the visible region. In general, conjugated organic molecules absorb light of increasing wavelength as the amount of conjugation increases. Therefore, a conjugated organic molecule with between two and ten conjugated double bonds

Absorbs light of shorter wavelengths (higher energy)

(a)

◀ FIGURE 5.29 Conjugated double bonds. **(a)** A hydrocarbon molecule with five double bonds in its conjugated system absorbs high-energy light. **(b)** The structure of beta-carotene, a molecule that contains 11 double bonds in its conjugated system. The greater number of double bonds in the conjugation sequence means this molecule absorbs low-energy light.

Absorbs light of longer wavelengths (lower energy)

(b)

absorbs in the ultraviolet region (between 200 and 400 nm), whereas a molecule with more than ten conjugated double bonds absorbs in the visible region (between 400 and 700 nm).

The molecule illustrated in Figure 5.29b has 11 conjugated double bonds, and so it is colored. You may be able to guess the color of the molecule from its name: *beta-carotene*. (It is yellow-orange like a carrot.) How, though, can we know what color this highly conjugated organic molecule will be? Experimental evidence shows that beta-carotene absorbs most of the wavelengths at the blue-violet end of the visible region of the electromagnetic spectrum. When white light, which contains all colors of the visible spectrum, is absorbed by beta-carotene, the molecule absorbs violets and blues and reflects all of the remaining colors back to our eyes. Thus, this molecule reflects back light with yellow, red, and orange wavelengths and has an overall hue that is a combination of these colors (in other words, orange). This molecule, which is abundant in carrots, is essential for healthy vision in humans.

The brightly colored tie-dyed jacket and jeans seen in Figure 5.30 were made using the vivid colors of various conjugated organic molecules.

▲ **FIGURE 5.30 Colored dyes are the result of conjugated double bonds.** The dyes used to create these colorful pieces of clothing owe their colors to highly conjugated organic molecules.

5.8 | Electrons at the Beach and in the Darkroom

Pretend you have just been hired by a chemical company to design a sunscreen that will protect people from dangerous ultraviolet light. You consider a few alternatives. The simplest approach would be to prepare an opaque cream that looks white on the skin and prevents sunlight from getting through. This is the basis for zinc oxide cream, the white stuff people used to see on the nose of many a lifeguard (Figure 5.31). Most of today's sunscreens are transparent and colorless, however, which means that they contain neither ZnO (which would make them opaque) nor organic molecules that absorb visible light (which would give them color). After all, visible light is not damaging to skin, and so there is no reason a sunscreen product needs to absorb that light and keep it from reaching the skin. Ultraviolet light is damaging, however, and so if you are charged with designing a sunscreen, you want a conjugated organic molecule that absorbs ultraviolet light. The molecule shown in Figure 5.29a absorbs ultraviolet light and has five double bonds in its conjugated system. Therefore, a molecule similar to this one would absorb ultraviolet light and thereby lessen skin damage.

▼ **FIGURE 5.31 Sun protection.** Sunblock creams containing zinc oxide (ZnO) work because their opaqueness prevents sunlight from reaching the skin.

The ultraviolet portion of the electromagnetic spectrum is divided into two regions: UVA light (lower energy) and UVB light (higher energy), as shown in Figure 5.32. Traditionally, sunscreens have been designed to block UVB light because its higher energy makes it more damaging to skin. UVB light has been blamed not only for sunburn, but also for the majority of skin cancers. Recently, however, debate has centered around protection from UVA light as well as UVB. New studies show that UVA light may contribute to both premature aging and skin cancer.

Figure 5.33 shows two organic molecules currently used in sunscreen formulations. One protects the wearer from UVB light, and the other protects from UVA light. Can you tell which is which? The two molecules are

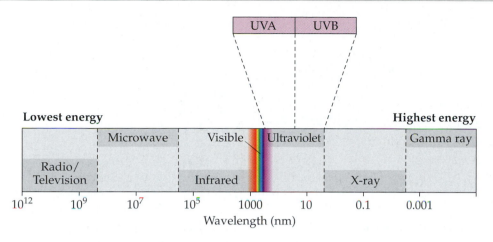

◄ **FIGURE 5.32   Two kinds of ultraviolet light.** The ultraviolet portion of the electromagnetic spectrum is divided into two regions: the UVA region (lower energy) and the UVB region (higher energy).

octyl methoxycinnamate, which contains five double bonds in its conjugated system, and benzophenone, which contains seven. The regions of conjugation are highlighted in Figure 5.33. (Recall from Section 3.4 that the circle notation inside the benzene ring represents three double bonds.) Because octyl methoxycinnamate has a less extensively conjugated system of double bonds, you would expect the ultraviolet light it absorbs to have higher energy (a shorter wavelength) than that of the ultraviolet light absorbed by benzophenone. This is indeed the case: octyl methoxycinnamate is a commonly used sunscreen ingredient for UVB protection, and benzophenone protects against UVA. Because there is risk associated with exposure to both UVA and UVB, you may see sunscreens that absorb both types of ultraviolet light. Most of these formulations contain two types of molecules, one to absorb UVA light and one to absorb UVB light.

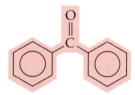

Octyl methoxycinnamate, five conjugated double bonds

Benzophenone, seven conjugated double bonds

▲ **FIGURE 5.33   Summertime helpers.** The chemical structures of two popular sunscreen molecules: octyl methoxycinnamate and benzophenone.

QUESTION 5.19 Many new sunscreen formulations contain sunscreen molecules that are less polar than their predecessors. What advantage does this provide for a sunbather at the beach?

ANSWER 5.19 Because water is a polar molecule, the polar molecules in older sunscreens are attracted to the water molecules and therefore wash off easily. The use of molecules that are less polar makes the sunscreen "waterproof."

RECURRING THEME IN CHEMISTRY Like dissolves like.

The dangers associated with ultraviolet light have been attributed to light-driven reactions that cause bonds in organic molecules to break homolytically. The products of homolytic reactions are free radicals, and as you now know, free radicals can lead to the breakdown of important biological molecules. Excessive exposure to sunlight can cause free-radical damage that can eventually lead to a wide variety of skin cancers.

Free-radical damage by ultraviolet light is not limited to the organic molecules found in human cells. For example, scientists at the University of North Carolina (UNC) have been researching the effects of sunlight on beer—a beverage that gets much of its flavor from organic molecules. Beer manufacturers have known for years that the action of sunlight gives beer a "skunky" taste, and so many manufacturers package their beer in dark glass containers to reduce its exposure to sunlight. Until recently, however, the substances responsible for the offensive flavor were unknown.

The reaction that the UNC chemists believe is behind the skunky taste is

+ source of sulfur $\xrightarrow{\text{UV light}}$ + Gives rotten-egg taste to skunky beer

Bond broken homolytically

The reactant molecule comes from the hops used to make the beer. Because hops are a primary contributor to the taste of beer, they cannot be eliminated from the recipe. When the beer is exposed to ultraviolet light, this hop-derived molecule undergoes a free-radical reaction at the site indicated. This bond breaks homolytically and produces two products, one of which contains a sulfur atom. Some sulfur-containing molecules are notoriously stinky and have a rotten-egg stench. The sulfur-containing product of the light-driven reaction is such a molecule and is what gives beer a skunky flavor. In fact, a structural cousin of this molecule is actually the offensive sulfur-containing molecule stored in the glands of skunks.

Another example of how free radicals can be produced by light involves the changing of leaf color in deciduous trees (Figure 5.34). As we have seen in many examples, highly conjugated molecules absorb light readily. They also undergo free-radical reactions and can absorb the effects of an otherwise dangerous free electron. That is, the highly conjugated structures of pigments and dyes make them well suited to spreading electron density over many atoms. Molecules that are not extensively conjugated have no way to cope with free radicals, and so these molecules are highly reactive because they are eager to return to a form in which all of their electrons are paired.

The leaves of deciduous trees contain many *organic pigments* that give them a wide spectrum of colors. The molecular structures for two leaf pigments are shown in Figure 5.35. Notice that these molecules are extensively conjugated. The slight differences in structure and in extent of conjugation give these pigments different colors. The traditional thinking about fall colors has been that trees manufacture all their color pigment in the spring and then the green pigments slowly fade away in the fall, leaving yellows and reds behind as weather turns cooler. New research has shown, however, that trees may actually produce red and orange pigments as fall nears. One hypothesis says these pigments act as a sunscreen to absorb sunlight that can damage the deteriorating leaf and cause it to die prematurely. As we have already seen in this chapter, extensively conjugated organic molecules are able to become stable free radicals. Thus, the red and orange pigments we see in autumn leaves can accept a lone electron and prevent it from causing damage to the rest of the leaf.

We began this chapter with Nicophore Niépce, a pioneer of photography, who used light and a camera obscura to create an image on an asphalt-covered plate. This introduced the notion of saving an image in a permanent way. Today, we take for granted that we can make a color photo-copy of an image or take a color photograph. With modern films, light coming from an image can be captured by light-sensitive salts mounted on a solid support.

▲ FIGURE 5.34 **Autumn leaves.** Recent research suggests that trees may produce red and orange pigments as fall nears and that these pigments may act as a sort of sunscreen which keep the leaves from dying prematurely.

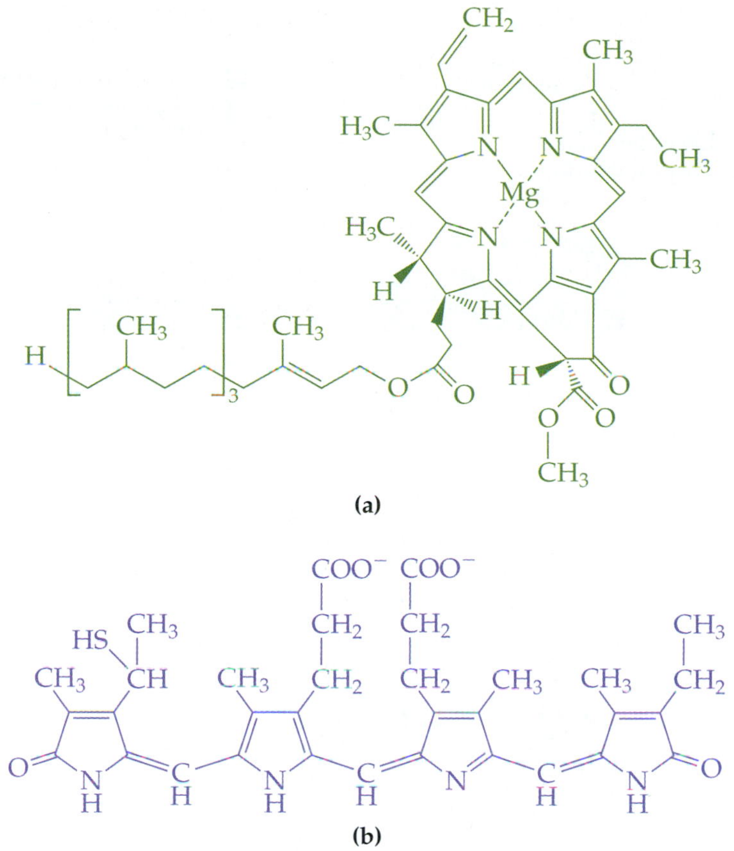

(a)

(b)

◀ **FIGURE 5.35** **Two common leaf pigments.** **(a)** The green pigment chlorophyll a. **(b)** The blue pigment phycocyanin.

How Are Colors Incorporated into Color Film?

STUDENTS OFTEN ASK

As you can imagine, the development of a color photograph is many times more complex than the relatively simple silver bromide chemistry we discussed for black-and-white photography. A piece of color film has a base that is similar to that of black-and-white film. Layered onto this base are three color-sensitive coatings, each containing a different conjugated organic molecule sensitive to a different range of colors (Figure 5.36).

Consider the color layer closest to the base. If you could see this layer, it would appear blue-green. (This color is called *cyan* in photography.) Now take a look again at the visible portion of the spectrum shown in Figure 5.21. We know that the colors reflected back to our eyes by this layer are at the higher-energy end of the visible spectrum, where the blues, violets, and greens are located. Now, if these colors are reflected back toward your eye, this means that *all the other colors* of visible light must be *absorbed* by the cyan layer. Thus, this layer of the film *looks* blue-green (cyan) to the eye because it is *absorbing* the *complementary colors* yellow, red, and orange.

The layer above the blue-green layer appears magenta to the human eye. The magenta reflected to your eyes is made up of reds, yellows, blues, and violets. This leaves green, the color the layer absorbs. Finally, the top color layer appears yellow to the eye. It is absorbing violets, blues, reds, and greens and reflecting oranges and yellows. When light from the original image hits the film, the various colors in the light are absorbed by the pigments in the various layers of the film. The result is a color negative in which all colors are complementary to those in the original image. These layers then undergo a complex series of chemical reactions that imprint the original colors onto the photographic paper when the film is developed.

QUESTION 5.20 Which color (or colors) do you see when you look at (*a*) a molecule that absorbs green, yellow, orange, and red light, (*b*) a molecule that absorbs violet, red, and orange light?

ANSWER 5.20 (*a*) Blue or violet, (*b*) yellow, blue, and green, which the eye sees as green.

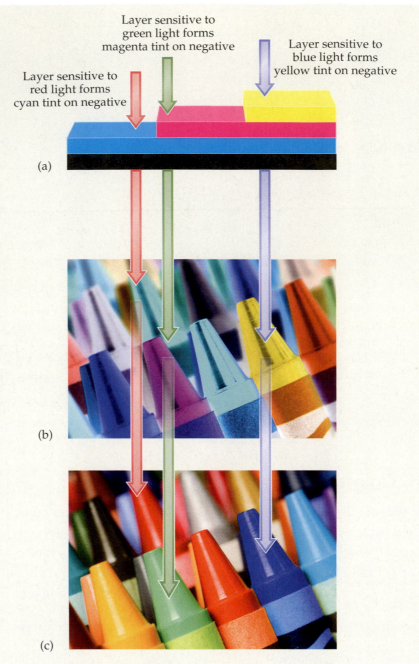

Layer sensitive to
green light forms
magenta tint on negative

Layer sensitive to
blue light forms
yellow tint on negative

Layer sensitive to
red light forms
cyan tint on negative

(a)

(b)

(c)

▶ FIGURE 5.36   **The structure of color film.**   **(a)** Color film is made by placing color-absorbing layers on top of one another. **(b)** Negative image. **(c)** Positive image.

Over the last few decades, photographic films have benefited from our most advanced scientific knowledge. As a result, the technology has matured from black-and-white film to exceptionally high-resolution color film. As this technology gets more and more sophisticated, we can be assured of one thing. Nicophore Niépce would be proud.

SUMMARY

This chapter is all about electron movement. We began with a discussion of the ground-state atom and the electrons in it. The hydrogen atom has discrete energy levels numbered 1, 2, 3, and so on. In multi-electron atoms, these energy levels are split into energy sublevels that are designated by the energy level number plus the letter *s*, *p*, *d*, or *f*. The periodic table serves as a map representing the order of increasingly energetic sublevels in multi-electron atoms (Figure 5.5). With an input of energy in the form of light or heat, a ground-state electron can move to an excited state. With additional energy, it can move entirely out of the atom, leaving behind a positively charged ion.

Redox reactions are the basis of the chemistry of black-and-white photography, in which silver ions are reduced to metallic silver atoms. In exposed film, tiny grains containing reduced silver create the photographic image. This image is enhanced during the development process, when more of the silver ions in certain grains are reduced.

The electrochemical cell is an apparatus that moves electrons as a result of differences in reduction potential between the two electrode half reactions. One of the two electrodes in an electrochemical cell has a greater affinity for electrons than the other, and this difference dictates the direction of electron flow. In the side of the cell having the smaller reduction potential, oxidation occurs and electrons move through the wire to the other side of the cell. These electrons reduce the metal at the other electrode, which has the larger reduction potential.

Batteries are based on the chemistry of ionization and utilize several redox reactions to create a current that can be used, for example, to light a light bulb. We looked at the design of dry-cell batteries and lithium batteries. Different batteries often use different redox reactions to produce a current, and the choice of reaction often is decided by the required voltage. The voltage is determined by the difference in the reduction potentials of the oxidation and reduction half reactions to be used: the greater the difference, the greater the voltage.

Organic molecules are bound together by covalent bonds, which can break homolytically to form free radicals. This type of reaction often occurs in the presence of oxygen, a strong oxidizer. In the body, free radicals can cause damage to both cell membranes and important biological molecules. Natural and artificial antioxidants can lessen the effects of unwanted free radicals by absorbing stray electrons.

The electromagnetic spectrum includes all types of electromagnetic energy, including infrared light, visible light, and ultraviolet light. How energetic a given type of radiation is determines the radiation's ability to, for example, penetrate human skin or promote chemical reactions. Once light causes an electron to move from its ground state to an excited state, the excited electron may then return to the ground state and give off energy in the form of light. If the electron is part of an elemental substance, discrete bands of light that make up that element's line spectrum are produced. If the electron is part of an organic molecule, the light absorbed when the electron is promoted to a higher-energy level tells us something about the molecule's structure. In general, molecules with extensive conjugation absorb longer wavelength light. Thus, many conjugated organic molecules absorb visible light, and this makes many of them brightly colored.

KEY TERMS

alcohol	dry cell	electron	ground state	reduction potential
anode	electrochemical cell	configuration	heterolytic cleavage	reversible chemical
antioxidant	electrode	energy level	homolytic cleavage	reaction
battery	electromagnetic	energy sublevel	infrared light	ultraviolet light
cathode	radiation	excited state	line spectrum	visible light
conjugated	electromagnetic	free radical	phenol	
molecule	spectrum			

QUESTIONS

The Painless Questions

1. Identify each half reaction as either reduction or oxidation:
   (a) $Fe^{2+} + 1 \text{ electron} \rightarrow Fe^{+}$
   (b) $Ag \rightarrow 1 \text{ electron} + Ag^{+}$
   (c) $Mo^{6+} + 3 \text{ electrons} \rightarrow Mo^{3+}$
   (d) $Cl^{-} \rightarrow 1 \text{ electron} + Cl$

2. Describe how silver bromide crystals are held in place on a piece of modern photographic film.

3. A household flashlight uses four 1.5-V batteries arranged in series. What is the total voltage available to this flashlight?

4. Describe one advance that has improved the performance of modern film. How does that advance allow film to absorb light more efficiently?

5. Which one of these chemical equations is written incorrectly? Rewrite it correctly.
   (a) $Cu^{2+} + 1 \text{ electron} \rightarrow Cu^{+}$
   (b) $I^{-} \rightarrow 1 \text{ electron} + I$
   (c) $Br \rightarrow 1 \text{ electron} + Br^{-}$
   (d) $Mg \rightarrow 2 \text{ electrons} + Mg^{2+}$

6. Thermal photographs allow the viewer to see areas of heat and cold in the photograph's subject. What type of electromagnetic radiation is detected by the film that produces such photographs?

7. Write an electron configuration for (a) F, (b) Si, (c) Be, (d) P.

8. Describe the difference between silver salt and metallic silver. Why is this difference crucial for the reactions of photography?

9. Describe the advances in photography made by Niépce and Daguerre and how these discoveries have led to the development of modern photographic film.

10. Identify each atom from its electron configuration:
    (a) $1s^2 2s^2 2p^4$, (b) $1s^2 2s^2 2p^6 3s^2 3p^2$,
    (c) $1s^2 2s^2 2p^6 3s^2 3p^6 4s^2 3d^1$.

11. What is wrong with this statement: *The only move electrons in an atom can make is from a lower energy level to the valence energy level, and only if the valence energy level has vacancies.*

12. For an atom of each of these elements, indicate the number of core electrons and the number of valence electrons: (a) aluminum, (b) lithium, (c) carbon, (d) neon.

13. For each object, indicate which color(s) of visible light the object absorbs: (a) a red book, (b) a green leaf, (c) a white dress, (d) a black rug, (e) a blueberry.

14. Ethane, H_3CCH_3, is a gaseous hydrocarbon used as a refrigerant. (a) Sketch the full structure of this molecule. (b) Draw the two fragments that result when the carbon–carbon bond in this molecule breaks heterolytically, showing the electrons around each carbon atom. (c) Repeat part (b) for a homolytic cleavage of the carbon–carbon bond.

15. A ground-state atom has three valence electrons and no electrons in the $4s$ and $4p$ energy sublevels. Which two elements fit this description?

More Challenging Questions

16. If a molecule absorbs the following colors of light, what color will it appear to the human eye: (a) red, orange, yellow; (b) green, orange, yellow, violet?

17. Consider the half reactions

    $$Mg \longrightarrow Mg^{2+} + 2 \text{ electrons}$$

    $$O + 2 \text{ electrons} \longrightarrow O^{2-}$$

    (a) Label each as reduction or oxidation. (b) Write a balanced redox reaction for this pair of half reactions. (c) What is the name of the salt formed between ions of magnesium and oxygen? (d) Write the chemical formula for this salt.

18. If photogray sunglasses use chemistry similar to that used in photography, how is it possible for the lenses to turn back to clear when the wearer goes indoors?

19. Some sliced fruit has a tendency to turn brown upon sitting in the open air, but other fruit does not. What is the difference between these two types of fruit? What ingredient is present in the fruit that does not turn brown? What is the full chemical structure of that ingredient?

20. A very long South American eel is swimming in a pool into which you are about to dive. What the eel does not know is that the water in the pool is absolutely pure and contains no dissolved ions or impurities. Why are you, being an ace chemistry student, not the least bit worried about any threat posed by the eel?

21. Describe the history of and challenges involved in the miniaturization of battery-operated devices. Provide one example not mentioned in the chapter of a device that has gotten smaller as a result of new battery technology.

22. The ions Fe^{3+} and V^+ both contain unpaired electrons, but are never referred to as free radicals. Why not? Where is the term *free radical* more aptly applied?

23. Write the electron configuration for (a) O^{2-}, (b) S^{2-}, (c) Na^+, (d) Be^{2+}.

24. Write the electron configuration for (a) H^+, (b) F^-, (c) V^{3+}, (d) K^+.

25. Name the parts of an electrochemical cell. If you were to put together a homemade cell, how could you tell if your cell is carrying a current?

26. Comment on this statement: *We must be able to see infrared light with our eyes because we can see that it exists using infrared photography.*

27. An electrochemical cell is assembled with a copper electrode immersed in a solution of Zn^{2+} ions and a zinc electrode immersed in a solution of Cu^{2+} ions. Why do no electrons move through the cell? How could the cell be made operational?

28. All of the hydrocarbons shown here absorb ultraviolet light. Rank them according to the wavelength (from shortest to longest) of ultraviolet light they absorb.

(a)

(b)

(c)

29. Free radicals are formed by metabolic processes that take place in our body cells. What natural defense mechanisms do our bodies have to counteract the damaging effects of free radicals? Name two naturally formed free radicals.

30. Consider the half reactions

$$Li(s) \longrightarrow Li^+(aq) + 1 \text{ electron}$$
$$Cl(g) + 1 \text{ electron} \longrightarrow Cl^-(g)$$

(a) Label each as reduction or oxidation. (b) Write a balanced redox reaction for this pair of half reactions. (c) What is the name of the salt formed between ions of lithium and chlorine? (d) Write the chemical formula for this salt.

The Toughest Questions

31. Write the half reaction for (a) the oxidation of a bromide ion, Br^-, to a bromine atom, (b) the oxidation of a magnesium atom to a magnesium ion that has a noble gas configuration of electrons, (c) the oxidation of a potassium atom in a half reaction that involves the transfer of one electron.

32. An electrochemical cell is designed with two electrodes made of copper, each sitting in a solution of Fe^{3+} ions. What criticism would you offer to the designer of this cell?

33. Given that the reduction potential for Cu^{2+} is greater than that for Fe^{2+}, propose a design for an electrochemical cell using copper and iron as electrodes. Sketch this cell and indicate which way the electrons move through it. Assume that each half reaction involves the gain or loss of two electrons.

34. Describe the various parts of a dry-cell battery. Identify each part, and describe the direction of electron flow through the battery.

35. Describe the various parts of a lithium battery. Identify each part, and describe the direction of electron flow through the battery. How does the design of this type of battery make it useful in miniaturized devices?

36. You would like to design a battery having the greatest voltage possible. If the reduction potentials for four imaginary electrodes decrease as $A > B > C > D$, which pair of electrodes would you want to use for your battery? Why?

37. Which phrase describes an attribute that is desirable in an antioxidant molecule:

    (a) it contains several electronegative atoms

    (b) it contains a hydrocarbon chain with no double bonds

    (c) it has a conjugated system of double and single bonds

    (d) all of the above

    (e) none of the above

38. Lipids are an important component of cell membranes in the human body. Describe how the integrity of a cell membrane could be compromised by the action of free radicals. What small molecule can contribute to the formation of free radicals and consequently to the destruction of cell membranes?

39. What features of the structure of tocopherols make them suitable antioxidants? Why might you find tocopherols on the list of ingredients in your breakfast cereal?

40. Your roommate is making a dish that requires peeled and sliced bananas. She has homemade tomato juice and apple juice on hand. Which of these two juices would you recommend she use to treat the banana slices to prevent browning? If she could make a trip to the grocery store to buy a more suitable alternative, what would you suggest she buy?

41. Given the information on silver halides in this chapter, would you conclude that silver iodide behaves more like silver chloride or more like silver bromide? How do you know?

42. A molecule absorbs visible light, but appears black to the eye. How is this possible?

43. When you heat an element and then allow it to cool, only one specific color of visible light is given off. Explain. How can this phenomenon be used to distinguish one element from another?

44. Describe two fates that can be met by an electron that has sufficient energy to leave its ground state. Describe each in a paragraph containing between four and six sentences.

45. Of the organic molecules shown here, one absorbs violet light from the visible portion of the electromagnetic spectrum, one absorbs UVB light, and one absorbs UVA light. Match each molecule with one type of light, and explain your choices.

(a)

(b)

(c)

46. Today's sunscreens tout UVA and UVB protection as well as waterproof formulations. Describe the ingredients that could give a sunscreen these attributes.

47. What features do organic pigments in leaves, sunscreen ingredients, and antioxidants have in common? In each case, how do these features help the molecules perform their job?

48. An herbal supplement currently on the market claims to reduce the damaging effects of sunlight on the skin. The formulation includes beta-carotene, vitamin E, and vitamin C. Is it possible for this formulation to live up to its claims? Why or why not?

49. Refer back to the structure of beta-carotene shown in Figure 5.29b. From what you know about this molecule, which colors of visible light must it absorb in order to give the color it presents to the human eye?

E-Questions

Go to **www.prenhall.com/waldron** to find these questions in electronic form, complete with hyperlinks directly to the various websites cited in the questions.

50. Go to **Cheresources** to answer the following questions: (a) What are three factors that determine the amount of silver deposited into a film emulsion during development? (b) What type of organic molecule described in this chapter is commonly used in the development of photographic film? (c) The gelatin emulsion used in making photographic film is described only briefly in this chapter. Describe in greater detail the type of gelatin used and how it is prepared. (d) Why would a surfactant be added to a film emulsion? (e) What is the difference between color sensitizers used in the production of film and the organic dyes used to color clothing?

51. Go to **Kodak** and read about Two-Electron Sensitization. Answer the following questions: (a) How is two-electron sensitization different from regular photographic chemistry? (b) What is added to the traditional photographic dispersion with the two-electron sensitization method?

52. The article at **Ray** is about electric fish. After reading it, answer the following questions: (a) What is the difference between weakly electric fish and strongly electric fish? (b) How do mature electrocytes build up charge? (c) What is the peak voltage attainable by electric light, and how does it compare to the peak voltages attained by the eels described in this article?

53. Read the article "MIT Team Creates New Battery Material" at **ScienceDaily**, and answer the following questions about this new technology: (a) What cathode material is used in the new battery? (b) What are the drawbacks of a traditional lithium battery? (c) How does the new battery designed by MIT engineers improve on these drawbacks?

54. Read more about the video pill at **CNNHealth**, and answer the following questions: (a) What four components make up a video pill? (b) Who designed the first video pill? (c) What is the primary risk associated with using a video pill?

WORLD WIDE WEB RESOURCES

As with the E-Questions, go to **www.prenhall.com/ waldron** to find these questions in electronic form, complete with hyperlinks directly to the various websites cited.

Some of the links that follow contain research articles related to the topics in this chapter and could be used as the basis for a writing assignment in your course. Other sites are of general interest.

- Read the article "Baghdad Battery" at **Battery**. This artifact, a vessel with a copper lining and possibly an iron electrode, may be the earliest known electrochemical cell.

- Read more about the history of gadgets used in Bond films since the mid-1960s at **JamesBond**.

- Harold E. Edgerton was a physicist who perfected the art of high-speed photography. Read about his work in the field of microsecond imagery and look at a selection of his photographs at **Eyestorm**.

- A detailed, illustrated account of the history of photography from 1816 to 1829 can be found at **Niépce**. Read about Nicophore Niépce, the photographers who further developed his inventions, and how the development of basic photographic principles during these years contributed to the film technology we know today.

- Read about the current debate surrounding sunscreen formulations at **FDA**. This easy-to-read article goes into depth about the importance of UVA versus UVB protection, the risks associated with sun exposure, and the ineffectiveness of SPF ratings.

CHEMISTRY TOPICS IN CHAPTER 6

- Chemical versus nuclear reactions, nuclear fusion [6.1]

- Elemental transmutations, formation of the elements, alpha particles [6.2]

- Beta particles, gamma radiation, radioactive decay, the belt of stability [6.3]

- Half-life of radioactive isotopes [6.4]

- Nuclear fission, chain reactions and critical mass [6.5]

- Ionizing radiation, radiation units [6.6]

"Who ever thought a nuclear reactor could be so complicated?"

HOMER SIMPSON

Sunshine

A STUDY OF NUCLEAR EVENTS AND THE INHERENT INSTABILITY OF (SOME) ATOMS

A curious researcher looking over patent applications for the past couple of centuries would find hundreds of inventions for gadgets that claim to produce energy out of nothing. These so-called *perpetual-motion machines* supposedly produce new energy continuously, which means the amount of energy put into the system is less than the amount coming out (Figure 6.1). Anyone trying to design a perpetual-motion machine is attempting to bypass an unwavering law of nature— the *first law of thermodynamics,* which says that energy is always *conserved*. That is, during any process, energy can be neither created nor destroyed. Time and again, machines purportedly used to create new energy have been debunked, and in each and every case, the energy input has been determined to be greater than the energy produced. Nevertheless, claims of energy-producing devices have continued right into the twenty-first century.

Modern scientists accept the facts that we must work with

the energy available to us and that creating new energy is a violation of everything we know about nature. Car engines tap the energy contained in the bonds of hydrocarbon fuel, for instance, and steam generators use the energy of water vapor to turn turbines that create electricity. Energy must be added to liquid water in order to turn it to the steam used to run a turbine, and that energy must come from somewhere. Often

Photo- is Greek for "light." We know that photosynthesis is the process plants use to get energy directly from sunlight. However, the word photosynthesis implies more. The light from sunlight (*photo-*) is used by plants to synthesize (*-synthesis*) new chemical compounds that they can use as fuels, in the same way that humans use food.

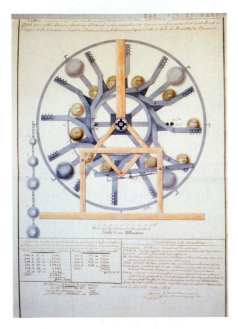

▲ FIGURE 6.1   **Getting something for nothing.**   This inventor's sketch shows a proposed perpetual-motion machine.

it comes from burning fossil fuels—like coal or petroleum—but fossil-fuel supplies, as you well know, are limited.

There is, however, one source of energy that provides a virtually unlimited supply for us: the Sun. Ultimately, it is the Sun that gives us the energy we need to sustain our bodies: we consume animals as one source of energy, the animals consumed plants before ending up on our tables, and before that the plants consumed sunlight. Through the extraordinary process of *photosynthesis*, plants are able to convert sunlight energy to chemical-bond energy. We do not worry about the limitations of this energy source. Our supply of fossil fuels will be depleted in perhaps 200 years (estimates vary from 75 to 500 years, depending on who is providing the numbers), but the Sun will continue to shine for the next few billion years.

The Sun is fueled by the extraordinarily powerful nuclear process called *fusion*, and today scientists are working to design equipment that will allow us to run controlled fusion reactions. In this chapter, we take a look at the reactions that occur when the energy in an atomic nucleus is tapped and the nuclear particles are manipulated. The enormous amounts of energy stored in the nucleus can be used peacefully, as in the production of power in nuclear power plants, or violently, as with nuclear-based weapons of mass destruction.

6.1 | Weight Gain for Atoms: Fusion

We cannot always depend on our friends or our cars or our appliances to do what they should. One thing we *can* always depend on, though, is the Sun. It comes up faithfully every morning and then, right on schedule, sets in the evening. It gives us a constant source of energy that we count on for our very existence on Earth, and sunrise is one of the most reliable events in our lives. The process by which the Sun provides us with a constant source of energy begins with the fusing together of hydrogen nuclei, and so let us see how this happens.

The temperature at the center of the Sun is estimated to be about 15,600,000 K. At this unimaginable extreme, the chemistry we know here on Earth no longer applies. On Earth, the existence of compounds and molecules tells us that Earth's ambient temperature is cool enough to allow atoms to interact with one another and form chemical bonds. It takes a finite amount of energy to break chemical bonds, and the temperature in the Sun is hot enough to disrupt even the most robust ones.

Solar temperatures are high enough to do more than just break chemical bonds between atoms. In fact, the temperatures are high enough to break up the atoms themselves. As you learned in Chapter 5, it is possible for electrons to be ejected from an atom whenever some threshold amount of energy is added to that atom. The temperatures that exist in the Sun are more than sufficient to ionize atoms because electrons cannot hang on to the nucleus in this intense environment. Thus, many elements found in the Sun, especially hydrogen and helium, exist in ionized form. There is even evidence that the Fe^{26+} ion exists in the hot solar environment—meaning that an iron atom is missing all 26 of its electrons!

The chemical reactions that take place here on Earth typically deal with the movement of electrons and not with changes in the number of protons and neutrons in the reacting atoms. In contrast to a chemical reaction, a **nuclear reaction** is a reaction in which the number of protons and/or neutrons changes in some way. There are two major classes of nuclear reaction: fusion and radioactive decay. We shall begin with the former.

The Sun is composed of about 75 percent hydrogen and 25 percent helium, the two lightest elements on the periodic table. A very, very small amount of the material in the Sun, less than 0.1 percent, is made up of heavier elements. Much of the energy produced by the Sun arises when two hydrogen nuclei come together to form a helium nucleus. This process, whereby nuclei combine to form a larger nucleus, is known as **nuclear fusion**. Two solar reactions convert hydrogen to helium via nuclear fusion:

Mass numbers

$$^{1}_{1}H + {}^{2}_{1}H \longrightarrow {}^{3}_{2}He + energy$$

$$^{3}_{2}He + {}^{3}_{2}He \longrightarrow {}^{4}_{2}He + 2\,{}^{1}_{1}H + energy$$

Atomic numbers

These equations may look strange to you because we do not ordinarily include mass numbers (sum of number of protons and neutrons in nucleus, shown as a *superscript* before the element symbol) in chemical equations. Atomic numbers (number of protons in nucleus, shown as a *subscript* before the element symbol) are also not customarily included in chemical equations. We include them here because we are discussing nuclear reactions, not chemical reactions. Remember, one element cannot be changed to another in a *chemical* reaction. Rather, a typical chemical reaction involves only the rearrangement of the *electrons* of the atoms taking part in the reaction and *never* the rearrangement of neutrons or protons. In nuclear reactions, however, atoms can change from one element to another because the number of protons in the nucleus can change, as shown in Figure 6.2.

One thing that equations representing chemical reactions have in common with equations representing nuclear reactions is that both types must be balanced. For chemical equations, this means equal numbers of each kind of atom on the two sides of the equation. For nuclear equations, this means that the mass numbers (superscripts) on the two sides of the equation must balance and that the atomic numbers (subscripts) must also balance. We shall use the aforementioned solar fusion reactions to illustrate this point. On the left side of the first equation, there are two isotopes of hydrogen shown as reactants, hydrogen-1 and hydrogen-2. (We call the first isotope simply *hydrogen*. The isotope hydrogen-2, which contains one proton and one neutron, has the special name *deuterium*.) The mass numbers on each side of the first equation sum to 3, and the atomic numbers sum to 2. Thus, this nuclear equation is balanced.

The second solar fusion reaction occurs when two identical helium-3 isotopes react to form a different helium isotope—helium-4—and two hydrogen-1 isotopes. The mass numbers and atomic numbers on each side of the second equation sum to 6 and 4, respectively. Thus, this nuclear equation is also balanced.

RECURRING THEME IN CHEMISTRY Chemical reactions are *nothing more than exchanges or rearrangements of electrons*, which is why we pay particular attention to electrons throughout this book.

If you read about nuclear chemistry in other books, you may notice that chemists often omit atomic numbers (subscripts) in equations for nuclear reactions. This is because atomic numbers are available on the periodic table. Mass numbers (superscripts) *are* written, though, because they identify specific isotopes, information that cannot be found on the periodic table. In this chapter, we shall include both superscripts and subscripts in our equations.

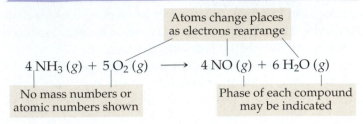

Typical CHEMICAL equation
All atoms appearing on left side must also appear on right side

Atoms change places
as electrons rearrange

$$4\,NH_3\,(g) + 5\,O_2\,(g) \longrightarrow 4\,NO\,(g) + 6\,H_2O\,(g)$$

No mass numbers or
atomic numbers shown

Phase of each compound
may be indicated

Typical NUCLEAR equation
Atoms appearing on left side do not have to appear on right side

$$^{241}_{95}Am + ^{4}_{2}He \longrightarrow ^{243}_{97}Bk + 2\,^{1}_{0}n$$

Mass numbers and
atomic numbers
usually shown

No phases
indicated

Particles such as
neutrons and electrons
are often reactants
or products

LIVE ART
Typical Chemical and Nuclear Equations

▶ **FIGURE 6.2 Chemical versus nuclear.** In a chemical equation, every atom shown on the left must also appear on the right. In a nuclear equation, the sum of the mass numbers (superscripts) on the left must equal that on the right, and the same holds true for the sums of the atomic numbers (subscripts). In the equation shown here, we have $241 + 4 = 243 + 2(1)$ for the mass numbers and $95 + 2 = 97 + 0$ for the atomic numbers.

Let's look at another example of a balanced nuclear equation:

$$^{209}_{83}Bi + ^{58}_{26}Fe \longrightarrow ^{266}_{109}Mt + ^{1}_{0}n$$

We know that it is balanced because the sum of the mass numbers on the left (267) equals the sum of the mass numbers on the right (267) and the sum of the atomic numbers on the left (109) equals the sum of the atomic numbers on the right (109). Note that one product of the reaction is a neutron, represented by the letter n with superscript 1 and subscript 0. As we move through this chapter, we shall see that neutrons, protons, and electrons all can act as reactants or products in nuclear reactions, as noted in Figure 6.2. In fact, as we shall see, they are included in *most* nuclear equations. These particles are treated like atoms in nuclear equations, and each has an associated superscript and subscript. Thus, when we check the mass-number sums and the atomic-number sums on the two sides of a nuclear equation, we always account for each neutron, electron, and proton in the reactant(s) and ensure that these particles are all represented in the product(s).

QUESTION 6.1 Indicate whether each nuclear equation is balanced or not:

(a) $^{1}_{1}H + ^{2}_{1}H \longrightarrow ^{3}_{2}He$

(b) $^{246}_{96}Cm + ^{12}_{6}C \longrightarrow ^{254}_{102}No + 4\,^{1}_{0}n$

(c) $^{14}_{7}N + ^{3}_{2}He \longrightarrow ^{17}_{8}O + ^{1}_{1}H$

ANSWER 6.1 Only the first two are balanced. The third is not balanced because the mass numbers are not balanced, with a sum of 17 on the left and 18 on the right.

QUESTION 6.2 Which equations represent nuclear reactions, and which represent chemical reactions?

(a) $^{218}_{84}Po \longrightarrow ^{4}_{2}He + ^{214}_{82}Pb$

(b) $^{12}_{6}C^{1}_{1}H_4 + 2\,^{16}_{8}O_2 \longrightarrow 2\,^{1}_{1}H_2{}^{16}_{8}O + ^{12}_{6}C^{16}_{8}O_2$

(c) $2\,^{3}_{2}He \longrightarrow ^{4}_{2}He + 2\,^{1}_{1}H$

ANSWER 6.2 Equations (*a*) and (*c*) represent nuclear reactions because new elements are formed. Equation (*b*) represents a chemical reaction because there is no change in the *identity* of the atoms during the reaction. Rather, the *arrangement* of atoms changes. You can see why atomic numbers and mass numbers are usually not written in chemical reactions. These numbers do not change, and they would add a great deal of clutter to the equation!

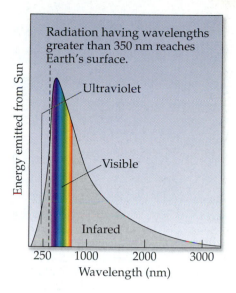

▲ **FIGURE 6.3 Sunlight reaching Earth.** All wavelengths longer than about 350 nm reach Earth's surface. Shorter wavelengths are filtered out in the upper atmosphere, mainly in the region called the *ozone layer*.

The energy produced in the solar fusion reactions shown on page 211 is released, in part, in the form of gamma rays, depicted by the symbol γ —the Greek letter gamma. Recall from Section 5.7 that gamma rays are extremely high in energy and have a very short wavelength. Gamma rays often are not included in nuclear equations because they are a type of electromagnetic radiation and not a type of particle, like a proton or a neutron. Most of the gamma rays produced by solar fusion are absorbed by Earth's atmosphere. Lucky for us, the solar radiation that makes its way to Earth's inhabitants is primarily in the ultraviolet and visible regions of the electromagnetic spectrum. Figure 6.3 shows the breadth of the solar spectrum as it arrives at Earth's surface.

To get a sense of how the energy associated with nuclear reactions compares with that of a typical chemical reaction, let us consider a familiar process, the combustion of methane:

$$CH_4 + 2\,O_2 \longrightarrow CO_2 + 2\,H_2O + energy$$

If you allow 1 mol (that is, 6.02×10^{23} molecules, or about 16 g) of methane to react with oxygen, and if the only chemical substances produced are carbon dioxide and water (as shown), this reaction gives off about 5.5×10^2 kJ of energy, equivalent to about 1.3×10^2 Calories, the number of Calories contained in your average banana. Solar fusion reactions produce about 6.6×10^8 kJ of energy per mole of helium produced (equivalent to about 1,200,000 bananas!). Thus, the nuclear reaction, which taps the enormous amount of energy contained in the atomic nucleus, produces about 1,200,000 times as much energy as a typical combustion reaction, which taps the energy of chemical bonds. In general,

You will learn in Chapter 12 why we put an uppercase *C* on Calories here.

> the amount of energy released in nuclear reactions is *much* larger than the amount released in chemical reactions because in nuclear reactions some of the energy holding together the particles in the nucleus is released.

We shall discuss the nature of this energy shortly.

QUESTION 6.3 If the combustion of a certain number of moles of methane produces 300 kJ of energy, how much energy would be produced for an equivalent amount of reactant undergoing a nuclear reaction in the Sun?

ANSWER 6.3 The solar reaction produces 1.2×10^6 times more energy. Thus, $1.2 \times 10^6 \times 300$ kJ $= 3.6 \times 10^8$ kJ of energy.

▲ FIGURE 6.4 **Controlling fusion.** These British scientists are building what they hope will be the first working fusion reactor.

When you compare the energy produced by methane combustion with that produced by fusion, you can see that the nuclear reaction would be a much better choice for producing power here on Earth. This being the case, why is it that we do not use nuclear fusion reactions to generate power? The answer is that fusion reactions run only at extremely high temperatures. This extreme heat energy is needed to increase the speed of reacting nuclei, which must be slammed together with tremendous force in order for a reaction to occur. Indeed, the fusion reaction shown on page 211 will take place in a fusion reactor only at a temperature of 1,000,000 K or more. This presents an enormous engineering problem because we do not have any materials that could withstand such heat. Thus, any kind of power plant built to house a fusion reactor would be destroyed by the heat.

Despite this problem, scientists are trying to find ways to use fusion reactions in a controlled manner. One solution to the problem of building a container for fusion reactions is to keep the heated hydrogen gas away from the reactor walls. Because extremely hot hydrogen gas is primarily in ionized form (H^+), a magnetic field can be used to hold the gas in a particular area of space not touching the reactor walls. Figure 6.4 shows the exterior of a magnetic-containment design worked out by British engineers. Inside this facility, hot hydrogen gas is heated to temperatures exceeding 1×10^6 K, enough to initiate a fusion reaction to create helium.

If the heat required for fusion and the energy released in the reaction do not have to be harnessed (the way they must be in a fusion reactor used by electrical utility companies), a fusion reaction can be used to make a powerful bomb. A weapon based on fusion technology is referred to as either a *thermonuclear bomb* or a *hydrogen bomb*. These bombs can produce blasts with 200 times the power of the nuclear bombs dropped on Hiroshima and Nagasaki during World War II. We shall discuss the anatomy of these types of bombs later in this chapter.

6.2 | Alchemy

▼ FIGURE 6.5 *The Alchemist*, by David Teniers, the Younger (1610–1690).

As citizens of the twenty-first century, we may look back on the efforts of early scientists and wonder what drove them to endless hours of seemingly futile laboratory work. Many experiments performed by early chemists, especially those of the medieval period, may seem outlandish to us now, but the focus of research efforts during any given period of history must be viewed in the context of the society at the time. For example, during medieval times, the intelligentsia widely believed in the Aristotelian notion of the "perfect substance." Gold was the most precious metal, and it was considered to be the most perfect element. Thus, many early chemists devoted themselves to achieving **transmutation**, the conversion of one element to another, with the goal of creating gold from other, less praiseworthy elements.

Today, we remember these early scientists not as chemists, but as *alchemists*—researchers devoted to finding a recipe for creating gold. (Figure 6.5 shows a typical alchemical laboratory.) When you look back through the early history of chemistry, though, you find that alchemists contributed much of the basic knowledge on which modern chemistry is based. Alchemists sought not only a recipe for creating gold, but also elixirs that would prolong human life. Although they did not reach their ultimate goal, along the way to that goal they developed processes for making useful materials, medicines, and dyes.

When the nineteenth century arrived, the scientists of the time were finally getting used to the idea that gold was not the perfect element and that trying to make one element from another was futile, as evidenced by centuries of unsuccessful attempts. Moreover, alchemists mixed scientific knowledge with spirituality and magic, and as researchers who focused solely on the science became more prominent, the alchemists began losing credibility. Throughout history, however, there have now and again been paradigm shifts that totally change our way of thinking. The twentieth century brought just such a paradigm—one that, in a sense, vindicated the work of the early prevailing theory about the formation of the heavier elements. It takes us back about 14 billion years to a huge explosion that is thought to have been the birth of our universe. According to this theory, called the *Big Bang*, the explosion of an enormous amount of energy that had been condensed into a very, very small volume of space created neutrons, protons, and electrons. These particles then came together to form hydrogen atoms, some of which, in turn, fused into helium atoms.

Learn more about the Big Bang in *A Brief History of Time: From the Big Bang to Black Holes* by Stephen W. Hawking.

All of this took place, it is thought, within the first 10 minutes of the life of the universe. During this time, newly formed hydrogen and helium atoms began to coalesce into very dense, very hot stars. The extreme temperatures present in these stars prompted the fusion of helium nuclei (also called **alpha particles** and symbolized with the Greek letter α), which are helium-4 atoms stripped of their electrons. The product of this fusion reaction is beryllium (four protons). Beryllium then fused with another alpha particle to create carbon (six protons), which could add another to create oxygen (eight protons), and so on, up to the element iron (26 protons). The nuclear equation representing the fusion of an alpha particle with silicon, for example, is written either

$$^{32}_{16}\text{Si} + {}^{4}_{2}\text{He}^{2+} \longrightarrow {}^{36}_{18}\text{Ar}$$

or

$$^{32}_{16}\text{Si} + {}^{4}_{2}\alpha \longrightarrow {}^{36}_{18}\text{Ar}$$

In addition to fusion with alpha particles, elements were also formed by fusion of newly formed isotopes of moderate size. For example, two carbon-12 atoms formed from successive alpha-particle fusion could fuse to create a neon atom plus an alpha particle:

$$^{12}_{6}\text{C} + {}^{12}_{6}\text{C} \longrightarrow {}^{20}_{10}\text{Ne} + {}^{4}_{2}\alpha$$

Odd-numbered elements were formed by other kinds of nuclear reactions that add or subtract one proton from the nucleus of a reactant.

QUESTION 6.4 Supply the missing isotope in the fusion equation

$$^{8}_{4}\text{Be} + {}^{4}_{2}\alpha \longrightarrow \text{?}$$

ANSWER 6.4 The symbol ${}^{4}_{2}\alpha$ stands for a helium-4 nucleus. Adding mass numbers for the reactants thus gives a total of 12, and adding atomic numbers gives 6. The atomic number defines the element, and atomic number 6 means carbon. The product is carbon-12, also written ${}^{12}_{6}\text{C}$.

The elements we know on Earth today originated before our own Sun formed. Elements heavier than hydrogen and helium formed as these lighter elements fused at the high temperatures in a dying star. These reactions were limited, however, by the temperature of that star, which was hot enough to produce only the elements from atomic number 2 (helium) to atomic number 26 (iron). Current thinking says that most of the elements heavier than iron were formed when that old star exploded and produced the heat necessary to fuse protons and neutrons from existing elements to form elements heavier than iron. On Earth, the abundances of elements heavier than iron gradually decrease with increasing atomic number, as shown in Figure 6.6. Thus, it is thought that all the elements we know today condensed from the debris created when that exploding star formed our own solar nebula, a flattened spinning disk from which our Sun and the rest of our solar system formed. We now know that there is some limit to the size of an element and that, consequently, the periodic table is finite. We shall see why this is so in the next section.

The events just described resulted in the formation of most of the atoms on the periodic table up to atomic number 92. Element 92, uranium, is the most massive naturally occurring element, which means that all the

▼ **FIGURE 6.6 Earth's elements.**   The abundances of the elements heavier than iron gradually decrease because these elements formed from one another in a stepwise fashion.

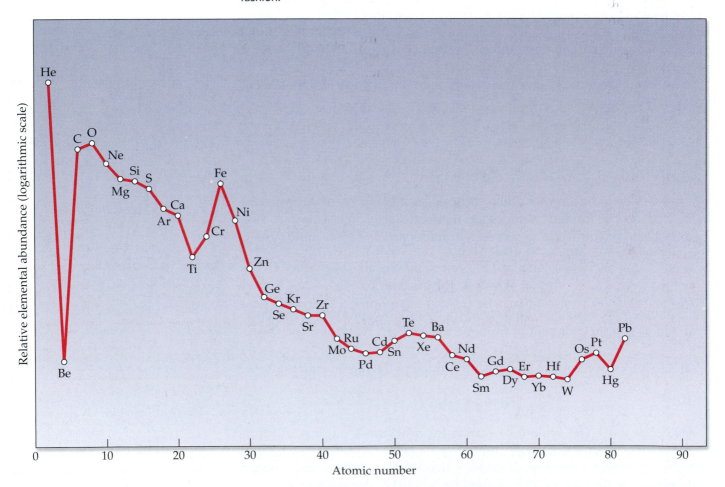

elements beyond atomic number 92 have been made artificially by the fusion of two smaller isotopes. The first artificial fusion reaction was performed by Ernest Rutherford in 1916. Rutherford bombarded nitrogen atoms with alpha particles to produce an isotope of oxygen:

$$^{14}_{7}N + ^{4}_{2}\alpha \longrightarrow ^{17}_{8}O + ^{1}_{1}H$$

Fusion reactions are difficult to perform because there is an inherent repulsion between atoms as their positively charged nuclei approach one another. That repulsion must be overcome before the atoms can fuse. The energy required to overcome the repulsion between atoms can be created if one of the atoms is accelerated to very high speed and then smashed into the other atom with enormous force. Such acceleration can be achieved in a *particle accelerator*, an apparatus in which particles travel faster and faster in a spiraling motion until they reach the desired speed. Because particles must travel long distances to reach the desired speed, particle accelerators are contained within enormous facilities, usually covering many acres of real estate, as shown in Figure 6.7. The accelerated particles are routed to a place in which they can crash into "regular" atoms and fuse with them. In this way, heavier and heavier elements have been synthesized.

At the time of this writing, there are 114 known elements: elements 1 through 112 plus elements 114 and 116. Elements 113 and 115 have yet to be created. We know that elements usually have properties similar to their vertical neighbors in the same column of the periodic table. Does that mean that element 108, which shares a column with iron, might be used in the production of a material like steel? The answer is no. All the artificially created elements are observed for only fractions of a second because they quickly fall apart and become other elements. In fact, element 114, which was created in 1999 in Russia, lasted for a full 30 s, which is a very long time for an element in the bottom row of the periodic table to exist (Figure 6.8). Because these elements exist for only seconds or fractions of a second, they have no practical uses. Nevertheless, the creation of each new element is a fresh technological challenge and the next frontier in nuclear science.

▲ FIGURE 6.7 **Aerial view of the Fermi National Accelerator Laboratory in Batavia, Illinois.** The underground ring is 4 miles in circumference, and particles can be accelerated in both the clockwise and counterclockwise directions within it. The laboratory pays about $18 million dollars per year to the local electric company.

▼ FIGURE 6.8 **A newspaper article describing the first observation of element 114.** Note the final sentence mentioning survival time.

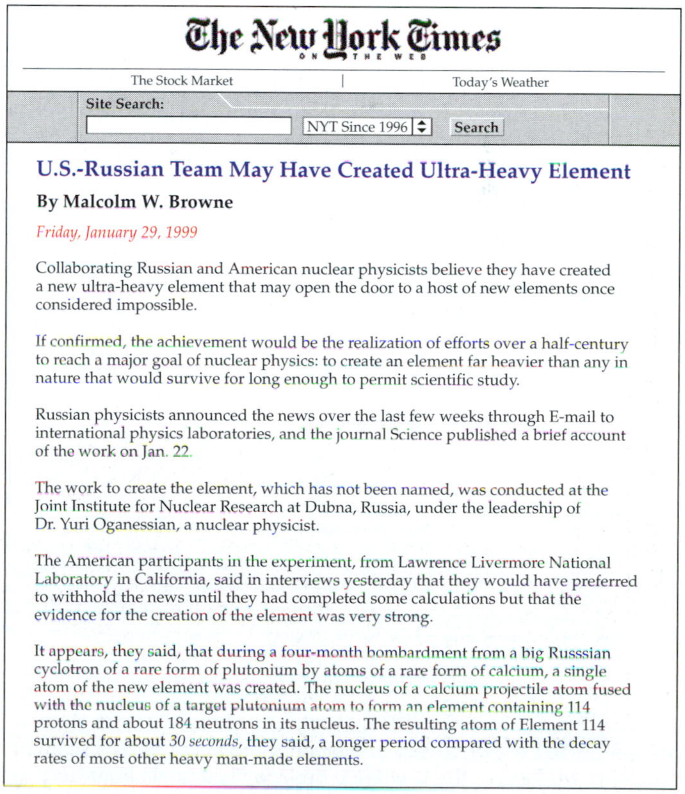

QUESTION 6.5 Imagine that element 117 has just been created and that it contains 130 neutrons. Its name will be noddackium, Nc. (*a*) Write the full symbol for this isotope, showing both superscript and subscript. (*b*) Under which element will this new element appear on the periodic table?

ANSWER 6.5 (*a*) With 117 as the atomic number and 117 + 130 = 247 as the mass number, the symbol is

$$^{247}_{117}Nc$$

(*b*) Under astatine, At, element 85 on the periodic table.

The New York Times

U.S.-Russian Team May Have Created Ultra-Heavy Element

By Malcolm W. Browne

Friday, January 29, 1999

Collaborating Russian and American nuclear physicists believe they have created a new ultra-heavy element that may open the door to a host of new elements once considered impossible.

If confirmed, the achievement would be the realization of efforts over a half-century to reach a major goal of nuclear physics: to create an element far heavier than any in nature that would survive for long enough to permit scientific study.

Russian physicists announced the news over the last few weeks through E-mail to international physics laboratories, and the journal Science published a brief account of the work on Jan. 22.

The work to create the element, which has not been named, was conducted at the Joint Institute for Nuclear Research at Dubna, Russia, under the leadership of Dr. Yuri Oganessian, a nuclear physicist.

The American participants in the experiment, from Lawrence Livermore National Laboratory in California, said in interviews yesterday that they would have preferred to withhold the news until they had completed some calculations but that the evidence for the creation of the element was very strong.

It appears, they said, that during a four-month bombardment from a big Russian cyclotron of a rare form of plutonium by atoms of a rare form of calcium, a single atom of the new element was created. The nucleus of a calcium projectile atom fused with the nucleus of a target plutonium atom to form an element containing 114 protons and about 184 neutrons in its nucleus. The resulting atom of Element 114 survived for about *30 seconds*, they said, a longer period compared with the decay rates of most other heavy man-made elements.

STUDENTS OFTEN ASK

How Are New Elements Confirmed and Who Gets to Name Them?

Because any new element created exists for only a very short time, you can imagine how hard it is for the members of a research team to prove that they have created something. Recall from Section 1.6 that important research results are always submitted to peer review for confirmation. With only a handful of

particle accelerators around the world, however, it is often not easy for peers to confirm the synthesis of a new element. Add to that the facts that only a handful of atoms of any new element are ever produced and that those atoms exist for only fractions of a second, and you have a field ripe for controversy and sometimes even fraud. In 1999, to take one example, an announcement came from Lawrence Berkeley Laboratory in California heralding the creation of elements 116 and 118 (Figure 6.9). After their work could not be reproduced by other scientists, the Lawrence scientists acknowledged that they had falsified information and had not created anything new.

This is not the first time this type of fraud has happened. There has been controversy surrounding the validity of new elements for decades. Thus, new elements are not printed on the periodic table until their

existence—however short lived—has been verified by independent members of the scientific community.

Artificial elements are usually named by the scientists who produce them. However, these names can be—and often are—vetoed by an international committee of scientists. Controversies surrounding the naming of new elements are commonplace, in many cases because it may not be clear who created the element first. In the case of element 106, the name seaborgium (Sg) was rejected at first because one of the men who played a key role in its creation, Nobel prize winner Glenn Seaborg (Figure 6.10), was still living. (Usually, the names are granted only posthumously.) After years of debate, the name seaborgium was finally made official. Referring to his new-found fame, Seaborg commented, "It's a greater honor than the Nobel Prize. A thousand years from now, these names will still be in use, while few people will recognize twentieth-century Nobel Prize winners."

During the writing of this book, element 110 was given the official name *darmstadtium* (Ds) after the town of Darmstadt, Germany, where the element was first created.

The New York Times

The Stock Market		Today's Weather

Site Search: [   ] [NYT Since 1996 ▼] [Search]

Team Adds 2 Elusive Elements to the Periodic Table

By Malcolm W. Browne

Wednesday, June 29, 1999

In a discovery that came as a complete surprise to most nuclear chemists, an international team of scientists at Lawrence Berkeley National Laboratory in California has added two new elements to the periodic table.

The laboratory announced on Monday that the two elements, with atomic numbers 118 and 116 on the periodic table, were created at the end of April using a cyclotron accelerator to bombard a lead target with projectiles of krypton atoms. In three of the collisions observed by the Berkeley team, target and projectile atoms fused to form gigantic atoms with nuclei containing 118 protons and 175 neutrons.

Each of the atoms of Element 118 survived for less than one ten-thousandth of a second before decaying radioactively into an atom of a second new element, Element 116, with 116 protons and 173 neutrons in its nucleus.

At Lawrence Berkeley, Physicists Say a Colleague Took Them for a Ride

By George Johnson (NYT) 2731 words

Late Edition - Final, Section F, Page 1, Column 3

Tuesday, October 15, 2002

ABSTRACT - Article on 1999 discovery, now discredited, of element 118, said to be the heaviest atom ever, by physicists at Lawrence Berkeley National Laboratory; physicists now say project leader Dr Victor Ninov, who had come there from Germany's Laboratory for Heavy Ion Research (GSI), misled them; photo; diagram (M) It's often said that the greatest thrill in science is to be first to observe a new phenomenon of nature. For nuclear physicists that means being present at the creation of an element, glimpsing for an instant a new kind of matter.

But science's most painful experience is having to withdraw a claim of discovery -- because of an honest mistake or, far worse, deliberate fakery.

▲ **FIGURE 6.9   Not to be believed.**   On June 19, 1999, the press reported the discovery of elements 116 and 118 by a team of physicists working at the Lawrence Berkeley National Laboratory. A little more than 2 years later, however, a different story was being reported.

▲ **FIGURE 6.10   Glenn Seaborg.**   The pioneering nuclear chemist Glenn Seaborg was a Nobel laureate before the age of 40 and an advisor to nine presidents. Add to that the fact that he was given the unique honor of having an element named after him while he was still alive (element 106, seaborgium, Sg). This was the first time an element was named after a living person.

6.3 | Weight Loss for Atoms: Radioactive Decay

The creation of heavy atoms from lighter ones through fusion is but one type of nuclear reaction. In many cases, nuclear reactions produce not only a new element, but also additional products in the form of either radiation or sub-atomic-sized particles. We refer to these products as **radiation**, and we define the emission of radiation as **radioactivity**. Take this nuclear equation as an example:

Formation of seaborgium, Sg, element 106

$$^{273}_{108}\text{Hs} \longrightarrow ^{269}_{106}\text{Sg} + ^{4}_{2}\alpha$$

This equation is similar to the nuclear equations we worked with earlier in that both its subscripts and its superscripts are balanced. However, this seaborgium equation differs from the earlier equations because it does not show the fusion of two nuclei to make a larger one. Instead, it represents the splitting of an unstable nucleus to form one or more smaller nuclei. This equation is an example of the process called **radioactive decay**, whereby an atom gives off some energetic product, either particle or radiation. In the seaborgium reaction, the energetic product is an alpha particle. Another way to say this is that hassium, element 108, is *radioactive* because it is undergoing **alpha decay**.

QUESTION 6.6 How many protons, neutrons, and electrons does an alpha particle contain?

ANSWER 6.6 An alpha particle is by definition a helium nucleus. Therefore, it contains two protons, two neutrons, and no electrons.

In the reaction

Formation of plutonium, Pu, element 94

$$^{239}_{93}\text{Np} \longrightarrow ^{239}_{94}\text{Pu} + ^{0}_{-1}\text{e}$$

an atom of neptunium decays to produce an atom of plutonium and an electron. The symbol $^{0}_{-1}\text{e}$ is often used by nuclear chemists to represent an electron. The negative subscript may seem odd to you. It represents the charge on the electron. By convention, whenever the symbol for a proton, electron, or neutron appears in a nuclear equation, the charge on the particle is written as a subscript, as shown in Table 6.1. Showing the subscripts allows us to use our usual method for balancing nuclear reactions. When a

TABLE 6.1 Symbols Used in Nuclear Equations	
	Symbol
Electron	$^{0}_{-1}\text{e}$ or $^{0}_{-1}\beta$
Neutron	$^{1}_{0}\text{n}$
Proton	$^{1}_{1}\text{p}$
Alpha	$^{4}_{2}\text{He}$ or α
Gamma	γ

value of -1 from an electron is part of a subscript sum, you simply add it in as a negative number, as this balanced equation shows:

$$^{40}_{19}K + ^{0}_{-1}e \longrightarrow ^{40}_{18}Ar$$

In this example, the subscripts on the left combine to give $19 - 1 = 18$, the atomic number for argon.

The emission of electrons is a type of radioactivity known as **beta decay**. Thus, we can say that neptunium in the reaction we looked at previously, namely,

$$^{239}_{93}Np \longrightarrow ^{239}_{94}Pu + ^{0}_{-1}e$$

is radioactive and that it is a *beta emitter*. For beta emitters, the electron in equations representing nuclear reactions is sometimes shown not as $^{0}_{-1}e$, but as either β or $^{0}_{-1}\beta$, called a **beta particle**. We shall always show beta particles as $^{0}_{-1}\beta$ so that you can use the superscript and subscript to balance nuclear equations.

On the basis of what you know about chemistry so far, you might think that the loss of an electron should signal the formation of a positive ion. However, many strange things happen in the world of nuclear chemistry. One of those strange things is that the electrons emitted in beta decay *do not come from the group of electrons that surround the nucleus.* In other words, electron loss in nuclear reactions is not the same as when a sodium atom, say, ionizes and loses its outermost electron to become a Na^+ ion. Instead, beta emission results from a special reaction in which the decay of a neutron in the nucleus produces one proton (which stays in the nucleus) and one electron (which is emitted):

$$^{1}_{0}n \longrightarrow ^{1}_{1}p + ^{0}_{-1}e$$

(Remember, in nuclear chemistry the symbol $^{0}_{-1}e$ is nothing more than an alternate symbol for $^{0}_{-1}\beta$.) Thus, although electrons do not exist in the nucleus per se, it is possible for an electron to be produced from a neutron in the nucleus. That electron is then ejected from the atom during beta decay. The net result is the loss of one neutron from the nucleus and the gain of one proton. Because the number of protons has changed, the identity of the atom has changed, as, for instance, in the beta emission we saw earlier for neptunium being converted to plutonium:

$$\overbrace{93\,p + 146\,n}^{} \qquad \overbrace{94\,p + 145\,n}^{}$$

$$^{239}_{93}Np \longrightarrow ^{239}_{94}Pu + ^{0}_{-1}\beta$$

If the neptunium atom loses a neutron, why does the product atom have the same mass number (superscript) as the reactant atom? The answer lies in the fact that the atom also gains one proton, $^{1}_{1}p$. Because mass number equals protons plus neutrons, the loss of the neutron is balanced by the gain of the proton; as a result, the mass number does not change. This idea is summarized in Figure 6.11. However, the atomic number (subscript),

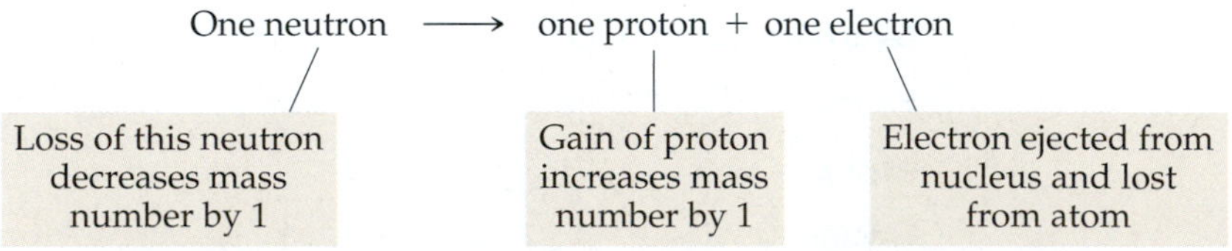

▶ **FIGURE 6.11 Beta emission.**
In beta decay, a neutron in the nucleus of the decaying atom splits into a proton and an electron. The proton stays in the nucleus, upping the atomic number by 1. The electron created by the splitting of the neutron is ejected from the atom.

One neutron $\longrightarrow$ one proton + one electron

| Loss of this neutron decreases mass number by 1 | Gain of proton increases mass number by 1 | Electron ejected from nucleus and lost from atom |

which counts only protons, increases by one unit as a result of beta emission because one proton is produced. Thus, the neptunium atom is converted to a plutonium atom of the same total mass.

The third and final kind of radioactivity we shall discuss is **gamma radiation**. As we saw in Section 6.1, some nuclear reactions produce very high energy radiation in the form of gamma rays—which are usually given off along with alpha particles or beta particles—as shown in the equation for the reaction used to produce seaborgium:

$$^{273}_{108}\text{Hs} \longrightarrow {}^{269}_{106}\text{Sg} + {}^{4}_{2}\alpha + \gamma$$

Although this equation is strictly correct, nuclear chemists often do not include the gamma symbol because it does not affect the balance of the nuclear equation. Thus, it is understood that gamma radiation may be present.

QUESTION 6.7 How many protons, neutrons, and electrons are contained in (*a*) a beta particle, (*b*) gamma radiation?

ANSWER 6.7 (*a*) No protons, no neutrons, one electron; (*b*) no protons, no neutrons, no electrons.

Many of the elements on the periodic table have at least one isotope that is radioactive and undergoes alpha, beta, or gamma emission. The isotope carbon-14, for instance, is radioactive. (Carbon-12 and carbon-13 are not.) Indeed, radioactive isotopes become more common as we move from the lighter elements to the more massive ones, and every isotope of every element beyond element 83 is radioactive. The fact that most radioactive elements have a high atomic number suggests that radioactivity is somehow related to atomic mass. If you were to compile a list of the most abundant isotopes of each element and show the number of protons and neutrons in each isotope, you would discover that the lighter elements have roughly equal numbers of protons and neutrons. As the elements get more massive, however, the ratio of neutrons to protons increases from about 1 to 1 to about 1.5 to 1.

Why is this the case? Think for a moment about the nucleus and the protons in it. Because all protons are positively charged, there is a *repulsive force* between them, meaning that they are eager to push away from one another. However, there is a counterbalancing force in the nucleus that permits protons to exist together in the same small space. This very strong *attractive force*, known as the **strong nuclear force**, is the force responsible for the attraction between protons and protons, between protons and neutrons, and between neutrons and neutrons.

In order for a nucleus to take advantage of the strong nuclear force, it must increase the number of neutron–neutron and neutron–proton interactions. In large nuclei, therefore, more neutrons are required to make the strong nuclear force more powerful than the repulsive force that pushes protons apart. The result is that the average number of neutrons per proton increases with increasing atomic size, and those neutrons help stabilize large nuclei.

When the ratio of neutrons to protons in the nucleus falls outside a certain narrow range of values, the atom is said to be *unstable* and decays to an atom of some other element. To see how this works, consider the nuclear reaction in which the radioactive isotope bismuth-210 decays to polonium-210:

$$^{210}_{83}\text{Bi} \longrightarrow {}^{210}_{84}\text{Po} + ?$$

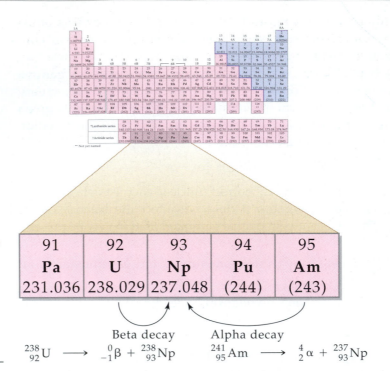

► **FIGURE 6.12 Transmutation in two directions.** The net result of beta emission is an increase in atomic number by one unit. The product element is one cell to the right of the reactant element. The net result of alpha emission is a decrease in atomic number by two units. The product element is two cells to the left of the reactant element.

91	92	93	94	95
Pa	**U**	**Np**	**Pu**	**Am**
231.036	238.029	237.048	(244)	(243)

Beta decay   Alpha decay

$$^{238}_{92}\text{U} \longrightarrow {}^{0}_{-1}\beta + {}^{238}_{93}\text{Np} \qquad {}^{241}_{95}\text{Am} \longrightarrow {}^{4}_{2}\alpha + {}^{237}_{93}\text{Np}$$

The type of radiation given off is not indicated, but you should be able to figure it out by balancing the equation. The decay must be beta emission because an electron (beta particle) placed on the right side balances the equation:

$$^{210}_{83}\text{Bi} \longrightarrow {}^{210}_{84}\text{Po} + {}^{0}_{-1}\beta$$

If you have been using the periodic table to follow the nuclear reactions we have looked at so far, you may have noticed that alpha emission changes an element to the element two places *to the left* (in other words, to the element having an atomic number smaller by 2), as indicated in Figure 6.12. Also, beta emission changes the reactant element to the element one place *to the right* (atomic number greater by 1).

The radioactive decay of bismuth-210 does not stop at polonium-210 because this polonium isotope is also unstable and decays to lead-206:

$$^{210}_{84}\text{Po} \longrightarrow {}^{206}_{82}\text{Pb} + ?$$

Can you identify the type of radiation given off by this nuclear reaction? The superscript must be 4, and the subscript must be 2, indicating that the decay is alpha emission and that the product is an alpha particle, ${}^{4}_{2}\alpha$. The isotope lead-206 is stable and does not undergo radioactive decay. Thus, this series of decay reactions stops when this isotope is formed.

QUESTION 6.8 In each equation, identify the species represented by the question mark:

$$(a) \ {}^{131}_{53}\text{I} \longrightarrow {}^{0}_{-1}\beta + ?$$
$$(b) \ ? \longrightarrow {}^{140}_{56}\text{Ba} + {}^{0}_{-1}\beta$$

ANSWER 6.8 (*a*) The missing product must have a mass number of 131 and an atomic number of 54. The missing product is $^{131}_{54}Xe$, xenon-131. (*b*) The missing reactant must have a mass number of 140 and an atomic number of 55. Thus, it is $^{140}_{55}Cs$, cesium-140.

QUESTION 6.9 In each equation, identify the species represented by the question mark:

$$(a)\ ^{222}_{86}Rn \longrightarrow ^{218}_{84}Po + ?$$
$$(b)\ ^{14}_{7}N + ? \longrightarrow ^{17}_{8}O + ^{1}_{1}H$$

ANSWER 6.9 (*a*) The missing product must have a mass number of 4 and an atomic number of 2. Hence, it is an alpha particle, $^{4}_{2}\alpha$. (*b*) The missing reactant must have a mass number of 4 and an atomic number of 2. Thus, it is also an alpha particle.

Strings of radioactive reactions are common because a radioactive isotope often forms another radioactive isotope, and the process continues until a stable isotope is reached. These sequential reactions are known collectively as a **radioactive decay series**. The radioactive decay series for the bismuth and polonium reactions we looked at earlier, for instance, would be written

$$^{210}_{83}Bi \longrightarrow ^{210}_{84}Po + ^{0}_{-1}\beta$$
$$^{210}_{84}Po \longrightarrow ^{206}_{82}Pb + ^{4}_{2}\alpha$$

We cannot write a radioactive decay series unless we know two things: (1) what type of radioactivity each radioactive atom emits and (2) where the series stops. How do we know when we have reached a stable isotope? This question is answered by Figure 6.13, in which the vertical axis represents the number of neutrons in an isotope and the horizontal axis represents the number of protons in the same isotope. The colored band is called the **belt of stability**. Any isotope that contains a neutron-to-proton ratio that falls anywhere in the colored band is stable. A radioactive decay series always begins with an isotope that lies outside the belt of stability and is therefore radioactive. If an isotope that forms during the decay series falls within the belt of stability, the radioactive decay series stops.

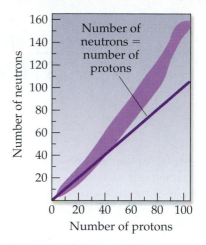

▲ FIGURE 6.13 **Belt of stability.** When the number of neutrons (vertical axis) in a nucleus is plotted against number of protons in the same nucleus, all the known stable isotopes form a belt of stability in the shape of a gradual upward curve, represented here by the shaded area.

QUESTION 6.10 What does the straight line in Figure 6.13 indicate? Why does the belt of stability curve upward away from it?

ANSWER 6.10 The straight line represents a ratio of one neutron to one proton. The portion of the band of stability representing the smaller elements (up to about atomic number 40) falls either on or very close to this line. This is because, in atoms of these elements, the numbers of protons and neutrons are roughly equal. As elements get larger, they tend to have more neutrons than protons; thus, the neutron-to-proton ratio increases. This makes the belt of stability gradually curve upward away from the straight line.

ISSUES IN CHEMISTRY

Three Women

In the introduction to Chapter 1, we outlined the events of the late nineteenth and early twentieth centuries that led to the discovery and characterization of radioactive materials. By the 1930s, the existence of the neutron had been documented, and dozens of chemists and physicists had begun to use beams of neutrons to perform experiments on the nature of radioactive decay. One of these

scientists, an Italian physicist named Enrico Fermi, decided to bombard each and every element, one at a time, with a beam of neutrons to see how each one responded. The most interesting result he obtained was with uranium, which, when bombarded, produced a confusing mixture of unidentifiable products. Fermi suspected that this mixture included element 93, now called neptunium:

$$^{238}_{92}U + ^1_0n \longrightarrow ^?_{93}Np + \text{other products}$$

However, no one who repeated this experiment, including Fermi himself, was able to identify neptunium or any other heavier-than-uranium elements as any of the products.

One researcher working in the field at the time, German chemist Ida Noddack (Figure 6.14), suggested in 1934 that rather than transmuting to a heavier element, perhaps the uranium was falling apart and making two lighter elements of similar mass. As Noddack expressed it, "When heavy nuclei are bombarded by neutrons, it would seem reasonable to conceive that they break down into numerous large fragments which are isotopes of known elements but are not neighbors of the bombarded elements." This idea was not accepted by the physics community, and Noddack did not pursue experiments that could validate her proposal.

QUESTION 6.11 If you assume that the "other products" in Fermi's equation are three neutrons, what must the mass number of neptunium be?

ANSWER 6.11 Because the combined mass number on the left is 239 and the mass number of each neutron is 1, neptunium must have a mass number of $239 - 3 = 236$.

In 1937, Marie Curie's daughter Irene Curie (Figure 6.15), a scientist like her mother, repeated Fermi's uranium experiments to try to figure out what the products might be. In conjunction with colleague Pavel Savitch, she found that one of the products was similar to lanthanum, element 57, and found no evidence for any elements heavier than uranium. Because all the nuclear reactions known at the time involved small increases in atomic number, it was reasonable to assume that uranium could increase (or even decrease) by one or two atomic numbers. However, a leap from atomic number 92 (uranium) to atomic number 57 (lanthanum) did not

▲ **FIGURE 6.14** **Ida Noddack.** In 1925, German chemist Ida Noddack was a co-discoverer of the element rhenium, Re, atomic number 75. The other member of the co-discovery pair was her husband, Walter Noddack.

▲ **FIGURE 6.15** **Irene Curie.** Becoming a physicist like her mother did, Irene Curie is shown here sitting with Albert Einstein. She and her husband were awarded the Nobel Prize in Chemistry in 1935 for their work on the synthesis of new radioactive isotopes.

▲ **FIGURE 6.16 Lise Meitner.** Lise Meitner (left) was the first woman professor of physics in Germany and the namesake of element 109, meitnerium.

make sense. Curie was baffled by her findings, but she maintained that her results were valid.

Another group working on this problem included Germans Otto Hahn, Fritz Strassmann, and Lise Meitner (Figure 6.16), the first woman professor of physics in Germany. According to biographical sources (such as *Lise Meitner: A Life in Physics* by Ruth Sime), Meitner was the theoretician of the group, although she spent many hours in the laboratory, while Hahn and Strassmann were experimentalists who had worked extensively in the area of nuclear chemistry. These three scientists began to work on unraveling the uranium data first collected by Fermi. They did not yet have an answer when Meitner, who was of Jewish ancestry, was forced to sneak out of Germany because of a new decree removing all Jewish workers from government-related jobs in Germany. She found a position in Sweden, but continued her contributions to the group by corresponding continuously with Hahn and Strassmann.

Upon hearing of the results obtained by Curie and Savitch in France, Meitner turned to her nephew, physicist Otto Frisch, as a sounding board for her thoughts about the puzzling results. Working together in 1938, Meitner and Frisch imagined the uranium nucleus to be like a drop of water. When hit with a beam of neutrons, the water drop might be blasted apart into two pieces of similar size, as suggested by Noddack 4 years before. Meitner believed that the two products were krypton and barium, and she communicated her idea to Hahn and Strassmann in Germany. They performed experiments verifying that uranium (atomic number 92) does indeed split to produce isotopes of krypton (atomic number 36) and barium (atomic number 56). At the same time, Frisch

also confirmed that isotopes of barium and krypton form in the bombardment of uranium with neutrons. The results were published separately by Hahn and Strassmann and by Meitner and Frisch. Nuclear *fission* had been discovered.

After the announcement of the discovery, Meitner became an instant celebrity, at least in the physics community in the United States. In 1944, however, Otto Hahn was awarded the Nobel Prize for "his discovery of the fission of heavy nuclei" (Figure 6.17). No other names appeared on the award. The Nobel committee claimed that Meitner was not included in the award because she was not present when the verifying experiments were performed. In retrospect, it seems clear that Meitner's Jewish heritage played an important role in the awarding of the prize, and many people believe that Hahn was under political pressure from the Nazi regime to exclude her when discussing fission research.

In the years since World War II, the controversy surrounding the 1944 Nobel Prize in Chemistry has been researched, and Meitner's critical role in the discovery of nuclear fission has been acknowledged. As a tribute to Meitner, now dead, element 109 was dubbed *meitnerium*.

▲ **FIGURE 6.17 Otto Hahn.** The 1944 Nobel Prize in Chemistry went to German physicist Otto Hahn for his "discovery of the fission of heavy nuclei."

6.4 | It's a Wonderful Half-Life

A workday in December 1985 began as usual for Stanley Watras of Colebrookdale Township, Pennsylvania. He drove to his job at the Limerick nuclear power plant at the usual hour and put in a full day's work. This day was different, though, because the radiation detectors at the plant exit screamed as Stanley passed through them on his way out. It might not seem unusual for a power plant fueled by nuclear reactions to have an occasional radiation alarm at the front door. At first, it was assumed that Stanley had somehow been exposed to radiation during the workday, and the plant was scrutinized for leaks and contamination. After none were found, the radiation levels in Stanley's house were checked and found to be higher than any reading ever recorded in a private residence. Indeed, the levels were so high that Stanley's clothing tripped the radiation alarms at his job. It was fortunate that Stanley happened to work at a job in which radiation levels are routinely checked. Otherwise, the super-high levels in his house would probably have gone undetected.

The high levels of radiation in the Watras home were monitored by detectors that pick up gamma radiation. Further testing in the house showed that the gamma radiation was due to the presence of the noble gas radon, Rn, atomic number 86. Radon does not interact chemically either with itself or with other substances, and because its atoms have no affinity for one another, it exists in the gas phase. Radon gas is highly radioactive, however, and decays to produce an alpha particle and gamma radiation:

$$^{222}_{86}Rn \longrightarrow {}^{218}_{84}Po + {}^{4}_{2}\alpha + \gamma$$

When we think of exposure to radiation, we normally worry about accidents at nuclear power plants or exposure from nuclear weaponry or X-ray machines, which are all sources of artificially produced radiation. Radiation is also a natural phenomenon, however, and radioactive materials are found in Earth's crust. For example, there is radioactive uranium in the ground in many regions across the United States. Areas with high uranium levels correlate with levels of radon found in homes: areas that have numerous deposits of uranium also have higher radon levels. A map of radon levels across the United States is shown in Figure 6.18.

What does uranium have to do with radon? The answer is that radon is one of the elements that forms as uranium-238 goes through its radioactive decay series:

$$^{238}_{92}U \rightarrow \rightarrow \rightarrow \rightarrow \rightarrow {}^{222}_{86}Rn + 4\,{}^{4}_{2}\alpha + 2\,{}^{0}_{-1}\beta$$

Once formed from uranium-238, radon-222 continues to emit radiation because it is itself radioactive, as we saw in the radon–polonium reaction above. This is not the end of the road, though, because the isotope polonium-218 is also radioactive and decays through seven more nuclear reactions until, at last, a stable isotope, lead-206, is formed (Figure 6.19).

QUESTION 6.12 In the equation shown above for the decay of uranium-238 to radon-222, are the mass numbers balanced? Are the atomic numbers balanced?

ANSWER 6.12 On the left, the mass number is 238. On the right, the mass numbers sum to $222 + 4(4) + 0 = 238$. The atomic number on the left is 92. On the right, the atomic numbers sum to $86 + 4(2) - 2 = 92$. The series is balanced as written.

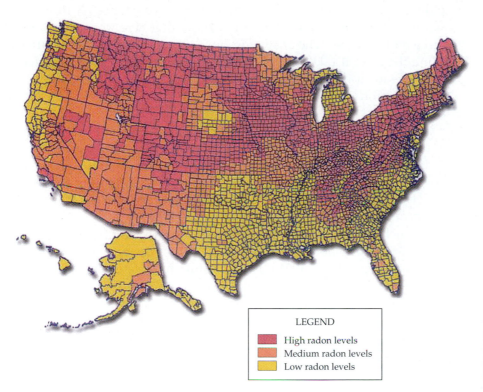

LEGEND

High radon levels
Medium radon levels
Low radon levels

◀ **FIGURE 6.18** **Levels of radon in the United States.** Red areas have the highest levels of radon, orange areas have moderate levels, and yellow areas have the lowest levels.

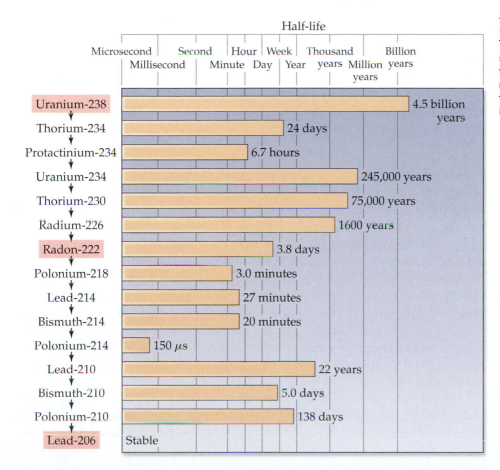

◀ **FIGURE 6.19** **Uranium decay.** The radioactive series for uranium-238, showing the steps that lead to the formation of the radioactive isotope radon-222. This isotope then decays, with the series finally ending at the stable isotope lead-206.

Loose Parts by Dave Blazek

© 2002 Dave Blazek • looseparts@comcast.net • Dist. by Tribune Media Services, Inc. • www.Comicspage.com

"GOOD MORNING! ARE YOU THE RADIOACTIVE ISOTOPE OF THE HOUSE?"

Ned Bollock: Half Life Insurance Salesman

How do we keep track of how radioactive a substance is? Practically speaking, is there a way to determine how long a radioactive element will give off radiation? The answer is yes. Scientists use the term **half-life** to define the time it takes for one-half of a radioactive sample to decay. The half-life of uranium-238 is 4.5×10^9 years, for instance, whereas the half-life of radon-222 is only 3.8 *days*, as Figure 6.19 shows. Thus, if you have 1 g each of these two isotopes at a certain moment, then 3.8 days later you will have only 0.5 g of radon left, but pretty much the whole 1 g of uranium-238. After 4.5×10^9 years, you will have 0.5 g of uranium. After another 4.5×10^9 years, 50 percent of that 0.5 g will have decayed, leaving 25 percent (0.25 g) of the original amount remaining, and so on. *The longer the half-life for a given radioactive element, the more slowly the element decays.*

With such a very long half-life for uranium-238, you may be wondering why we worry about radon forming from it. The answer is that the uranium deposits on Earth are plentiful, and there is enough in the ground to release substantial amounts of radon despite the fact that the uranium isotope has a half-life of 4.5×10^9 years. This uranium has been decaying for millions of years, producing thorium-234 at a slow, but conspicuous, rate. This isotope in turn has been decaying to protactimium-234, the next isotope shown in Figure 6.19, and on and on down to radon-222. So much uranium has decayed that there is a constant tide of radon-222 emitted from the ground in areas in which uranium deposits are found. Therefore, even though the half-lives of some of the isotopes in the uranium radioactive series are very, very long, these isotopes are still radioactive, and they decay at a measurable rate. The last isotope in the series, lead-206, does not emit radiation of any kind. Thus, the series ends when lead-206 is formed.

QUESTION 6.13   The half-life of iron-59 is 45 days. (*a*) How long will it take for one-half of a sample of this material to decay? (*b*) How much of this iron isotope will be left after 90 days?

ANSWER 6.13   (*a*) 45 days. (*b*) After 90 days, half of the half remaining after 45 days, or one-quarter, of the iron is left.

Of all of the isotopes in the uranium radioactive series, why do we worry most about radon-222? The answer is that only radon is a *gas*, and it seeps from uranium deposits in the ground and enters buildings through cracks and fissures in their foundations. All the reactions from radon-222 to lead-210 in Figure 6.19 are very fast. Thus, the radioactivity formed during these reactions, which includes three alpha particles and three beta particles, is released quickly as radon decays. If you inhale radon into your lungs and it decays before you can exhale it, the decay products can damage your lung tissue. The series slows down once lead-210 has formed because this isotope has a relatively long half-life, about 22 years. Consequently, exposure to high levels of radon gas will result in the deposition of this lead isotope in your lungs, which is both radioactive and highly toxic.

Should we be concerned about radon exposure? The answer is an unqualified *yes*. Figure 6.20 shows a graph of radiation exposure for various sources of natural and artificial radiation. According to these data, more

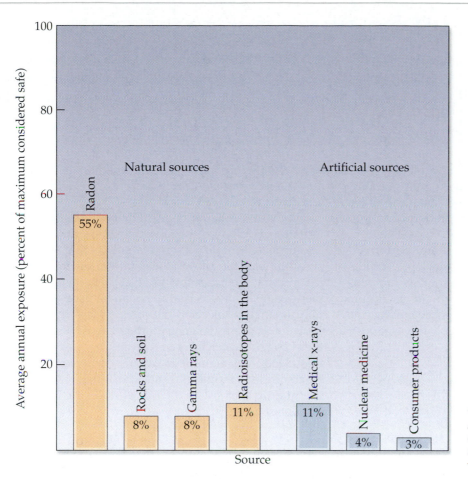

◄FIGURE 6.20 **Radiation exposure.**
A graph showing the average annual
exposure to radiation from four natural
and three artificial sources.

than half of the radiation to which an average person is exposed in a typical year can come from radon. This number is greater than the exposure from all other types of radiation combined. Further, it is estimated than nearly 10 percent of deaths due to lung cancer in the United States can be attributed to radon exposure.

Is it safe to assume that, if you live in an area shaded in yellow in Figure 6.18, you are safe from radon exposure? Experts agree that although people living in areas of known uranium deposits are probably more likely to have high levels of radon in their houses, the values from house to house can vary. In areas in which uranium is prevalent, it is not uncommon to find one house with a high level next door to a house with a very low level. This is because the radon seepage rate into a building depends on the construction materials used, the age of the building, its ventilation, and several other variables. Even for regions in which uranium deposits are few and far between, it is possible to find buildings with very high radon levels.

The only way to be sure about the radon level in your own house is to have it tested. Stanley Watras and his family, whose house is located on a uranium plume called the Reading Prong (Figure 6.21), lived with radon levels 675 times the allowable limit set by the U.S. government. The knowledge that radon levels were extraordinarily high probably saved their lives. The family moved out of the house immediately, and

▼ **FIGURE 6.21   Take along your
Geiger counter.** The shaded area is the
Reading Prong, a region of high radon
levels spanning parts of New Jersey, New
York, and Pennsylvania.

BEYOND THE ORDINARY:

A New Old Way of "Seeing" Alpha Particles

Ever wondered how radioactive substances are identified? After all, radiation is odorless and colorless. There are many ways that radiation is routinely detected. The most commonly used instrument is the *Geiger counter*. Geiger counters, which detect alpha, beta, and gamma radiation, have evolved from cumbersome instruments to miniature devices. Today's models can even be worn on the wrist (Figure 6.22).

The Geiger counter is not the only known portable, radiation-detecting device, however. Recent years have seen the development of a new radiation-detecting contraption that is worn as a goggle (Figure 6.23). The technology is based on a decades-old experiment performed by Ernest Rutherford, who observed that zinc sulfide, ZnS, glows when alpha radiation hits it. This happens because the electrically charged, high-energy alpha radiation dislodges electrons when it smashes into zinc sulfide. This high-energy process emits a tiny spark of light. Today, we refer to the emission as *scintillation* and use the term *phosphor* to describe any substance that scintillates.

Radiation-sensing goggles detect the wavelength of radiation given off by scintillating zinc sulfide. Spraying a solution of zinc sulfide onto a radioactive object that is an alpha emitter allows the scintillation to be seen by the person wearing the goggles. The big advantage these goggles have is that they can be used to see radiation in small nooks and crannies, places in which it is difficult to use a Geiger counter. The downside is that these goggles work only in the dark because their sensitive electronics cannot distinguish between light from zinc sulfide scintillation and sunshine. Thus, they are useful only at night in the out of doors or in very dark interior spaces.

▲ **FIGURE 6.22 Watching radioactivity.** This watch is a miniature Geiger counter.

▲ **FIGURE 6.23 Scintillating eyewear.** These goggles allow the wearer to see a radioactive substance that has been sprayed with zinc sulfide.

the Philadelphia Electric Company funded a remediation project in their house that included adding ventilation to the basement and sealing cracks in the basement floors and walls. These measures brought the levels of radon in the house down to acceptable levels, and they moved back in.

How Do Smoke Detectors Work?

You may not be aware that most smoke detectors contain the radioactive isotope americium-241, first produced by Glenn Seaborg in 1944. Figure 6.24 shows a diagram of a typical americium smoke detector. The device consists of two metal plates, a battery, and a source of the isotope. Even though the plates do not touch each other, there is an electric current because alpha radiation from the americium knocks electrons out of the atoms and molecules in the air between the plates. Movement of these electrons forms a current from the bottom plate to the top one. This is the default condition, and no alarm sounds as long as the electric circuit is complete. When smoke is present, the smoke particles prevent the electrons from flowing through the circuit. Therefore, the circuit is broken, and an alarm is triggered.

Any smoke detector that contains a radioactive material should not be thrown in the trash, and special collection points are often available in communities for disposing of this type of detector safely.

QUESTION 6.14   The isotope americium-241 has a half-life of 433 years. How long will it take for 75 percent of a sample of it to decay?

ANSWER 6.14   After 433 years, one-half of the sample will be gone. After another 433 years ($433 + 433 = 866$ years), one-half of the remaining half will be gone, and $0.5 \times 50\% = 25\%$. Thus, after 866 years, 75 percent of the americium will have decayed.

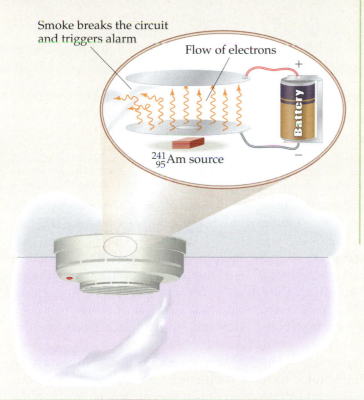

Smoke breaks the circuit and triggers alarm

Flow of electrons

Battery

$^{241}_{95}$Am source

◄ **FIGURE 6.24   A working smoke detector.**   A typical smoke detector is constructed with two metal plates and a small piece of radioactive americium that creates an electric current between the plates. In the presence of smoke, the current ceases, and an alarm is triggered.

6.5 | Power from the Nucleus: Fission

Earlier we compared the amount of energy produced in a combustion reaction with that produced in a typical nuclear reaction. We concluded that, as a power source, the nuclear reaction provides many orders of magnitudes more energy than do traditional fossil fuels—on a molar basis, 5.5×10^2 kJ versus 6.6×10^8 kJ. Like chemical reactions, nuclear reactions are driven by a move toward greater stability. When fusion reactions occur between small isotopes that combine to make one larger isotope, there is an increase in stability. For large nuclei, stability is a balance between the proton–proton repulsion and the strong nuclear force. As we move through the periodic table and increase the number of protons one by one with each successive element, we increase the contribution of the proton–proton repulsion. Very large nuclei tend to be radioactive because proton–proton repulsion starts to win out over the strong nuclear force. These unstable nuclei

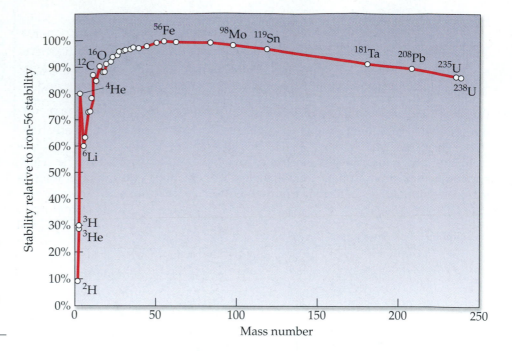

▶ **FIGURE 6.25** **Relative isotope stabilities.** To generate this graph, the isotope iron-56 was assigned a stability of 100 percent. Then the stability of the other isotopes shown was measured relative to iron-56. The isotope lithium-6, for example, is about half as stable as iron-56.

undergo radioactive decay, a process that relieves the stress in an overwhelmed nucleus.

Figure 6.25 demonstrates this push toward more stable isotopes. The vertical axis represents the stability of each element relative to the stability of iron-56, which is assigned a stability of 100 percent. The horizontal axis is mass number. According to these data, the greatest stability is found in the vicinity of iron on the periodic table. Thus, smaller atoms may undergo fusion to increase stability. (They fuse to become larger and thus closer to the size of iron-56.) In contrast, isotopes larger than iron may undergo fission in search of greater stability. (They split in two, each piece closer to the size of iron-56.)

Thus far, we have seen examples of radioactive decay in which an unstable nucleus transmutes to form a different element with an atomic number that is one or two positions away from the original element on the periodic table. However, as you may have read in the boxed section *Three Women* (page 224), it is also possible for an unstable nucleus to split into roughly equal parts to give two nuclei of similar size. This process, known as **nuclear fission**, is a typical nuclear reaction because it produces enormous amounts of energy. This is the type of reaction used to produce power in nuclear power plants today.

In both fusion and fission, the amount of energy released is enormous, and both reactions offer possibilities as energy sources. Earlier in this chapter, we looked at the use of fusion reactions to produce power, a technology that has yet to be realized. Fission, on the other hand, is easier to control and easier to initiate. The extraordinarily high temperatures required for fusion are not required for fission, and thus fission technology has advanced much more quickly.

In the Middle East, the 1970s brought political unrest that resulted in oil shortages all over the planet. France was particularly hard hit because it does not have natural fossil-fuel resources and must rely on other countries for much of her energy supply. To solve this problem, the French made a decision that would revolutionize the use of nuclear power: they vowed to reduce their dependence on foreign oil by investing significantly in nuclear energy. By 1999, about 76 percent of France's electricity was being generated by nuclear reactors. Today, nuclear plants dot the French landscape and seem to be as much a part of French culture as baguettes and champagne (Figure 6.26*a*). How can the French people live harmoniously in the shadow of a nuclear reactor? In fact, most French citizens surveyed find no reason to

◄ **FIGURE 6.26 Nuclear power around the world.** In France, nuclear plants have become part of the landscape. The graph shows the percentage of energy from nuclear power in various countries.

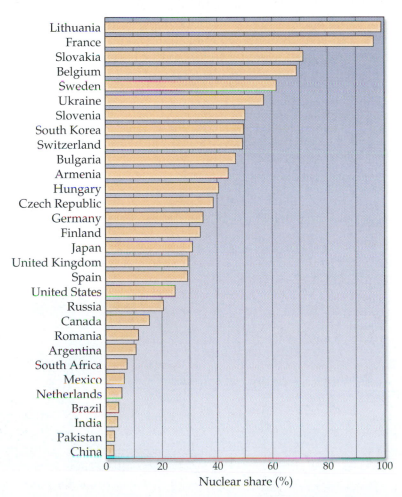

Nuclear share (%)

fear them. A standard retort from a French person asked about the potential menace posed by nuclear reactors is, "No oil, no gas, no coal, no choice."

How have the French been able to come to terms with their new dependence on nuclear power? For one thing, they decided to recycle their nuclear waste and have been doing so successfully for years. For another thing, they decided early on that they would use one construction plan for all their nuclear plants and would operate the same type of reactor (an American model) in each one. This uniformity allowed them to learn from their mistakes and to avoid repeating them at other installations. Furthermore, the use of identical plants made construction costs lower. In the United States, there is no one plan for nuclear reactors. Each is designed individually, and, as a result, each plant has its own unique problems.

Many countries, the United States among them, still rely on fossil fuels for most of their energy. As Figure 6.26 shows, however, other countries, such as Lithuania, have invested in nuclear power instead.

QUESTION 6.15   Rank these countries according to their use of nuclear energy, highest first: Switzerland, Canada, Sweden, United States, Japan, Hungary, Argentina, Belgium.

ANSWER 6.15   Belgium > Sweden > Switzerland > Hungary > Japan > United States > Canada > Argentina

A typical nuclear reactor is shown in Figure 6.27. Inside an outer confinement shell, an inner containment chamber holds the *reactor core*, which is essentially nuclear material in long tubes called *fuel rods*. A fission reaction produces enormous amounts of heat energy in the fuel rods, and this energy then heats either water or liquid sodium passing over the fuel rods. *Control rods* are lowered into the spaces between the fuel rods to dampen the intensity of the fission reaction and to prevent the overheating of the

▼ FIGURE 6.27   **Electrical power from fission.**   (a) Heat produced by fission reactions taking place in the reactor core is used to heat water. The heated water then generates steam that is converted to electrical energy with a steam turbine and electric generator. (b) A typical nuclear reactor.

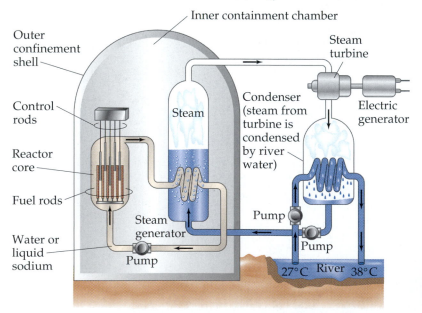

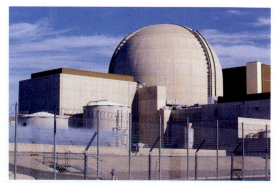

(a)

(b)

reactor core. The pipe routing the water (or liquid sodium) heated by the fission reaction runs through a steam generator holding a large volume of water. This water, heated by the hot liquid inside the pipe, is converted to steam. Reactors are designed with two separate systems of circulating water so that any steam released cannot be contaminated with radiation. The pressure of this steam then drives a turbine connected to an electric generator, which converts the mechanical energy of the rotating turbine to electrical energy.

One fuel commonly used in fission reactors is uranium-235, which undergoes several reactions that produce a variety of products. Among the possible fission routes for this isotope is

$$^{235}_{92}\text{U} + ^{1}_{0}\text{n} \longrightarrow ^{91}_{36}\text{Kr} + ^{142}_{56}\text{Ba} + 3\,^{1}_{0}\text{n} + \text{heat energy}$$

This reaction is initiated by a collision between one neutron and one atom of uranium-235 (Figure 6.28). Note from the equation that this initial collision produces *three* neutrons. When these three neutrons are released, they impinge on three more uranium-235 atoms, each one reacting to produce three more neutrons. Because this reaction amplifies itself by initiating more reactions, we refer to this series of fissions as a **chain reaction**, illustrated in Figure 6.29.

It is easy to see how such a chain reaction could rapidly go out of control. The number of fission reactions quickly escalates, and the reactor can easily overheat if the reaction is not kept in check. One way these reactions are controlled is through the control rods mentioned earlier, which work by absorbing some of the neutrons created during fission. The farther down into the fuel rods the control rods go, the more neutrons they absorb and the more the chain reaction is slowed. Another way the reaction is regulated is by the use of limited quantities of uranium-235. The acceleration of a chain reaction depends on the mass of radioactive material you begin with. If you start with a very small quantity of material, the chain reaction accelerates slowly because neutrons produced in a small mass tend to escape from the surface of the mass and dissipate. If you start with a large mass of uranium-235, however, the chain reaction accelerates rapidly. If the acceleration is sufficiently fast, the result is an explosion, an event that would not be desirable in a nuclear reactor facility.

The minimum amount of radioactive material required to sustain a chain reaction is referred to as the **critical mass**. Thus, the amount of uranium used in nuclear reactors is crucial and carefully regulated. Enough is needed to produce energy in the form of heat, but too much can cause big trouble.

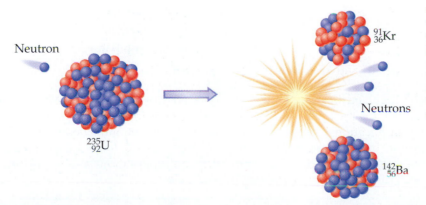

Neutron

$^{235}_{92}\text{U}$

$^{91}_{36}\text{Kr}$

Neutrons

$^{142}_{56}\text{Ba}$

◀ FIGURE 6.28 **Nuclear fission.** The fission decay of uranium-235 is initiated by a collision between the uranium and a neutron. The reaction produces an isotope of krypton and an isotope of barium, in addition to three neutrons. These three neutrons can then initiate three more reactions of this type.

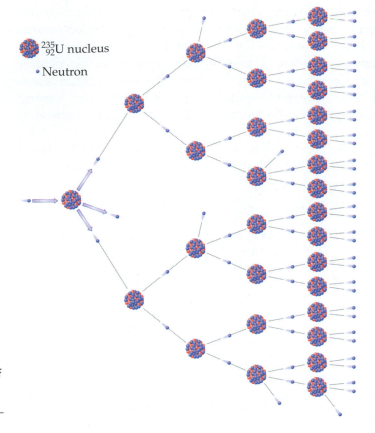

▶ **FIGURE 6.29** **A fission chain reaction.** When three neutrons are produced from one fission event, the subsequent reactions are multiplied and the reaction accelerates. (The reactions branching out from the middle neutron of the triplet on the left have been omitted for clarity.)

LIVE ART
U-235 Chain Reaction

QUESTION 6.16 Comment on this statement: *Critical mass is the minimum mass of a fissionable material that must exist in order for a nuclear reaction to take place.*

ANSWER 6.16 Not true. Nuclear reactions can occur in a sample of fissionable material for which the mass is less than critical. The critical mass refers to the mass required to *sustain* a chain reaction.

If the goal in using fissionable material is to create an explosion, as in an *atomic bomb*, then a *supercritical mass* of fissionable material is required in order to create a chain reaction that is out of control. The problem with

STUDENTS OFTEN ASK

What Is Weapons-Grade Uranium?

We know from our study of radon that uranium exists naturally on Earth. This natural uranium exists as three isotopes. Most of it—99.283 percent, to be exact—is uranium-238. The remainder is split between uranium-235 (0.711 percent) and uranium-234 (0.005 percent).

The isotope uranium-235 is the one involved in the chain reaction shown in Figure 6.29, and so it is the only isotope of the three used for military applications. In order for a chain reaction to occur, therefore, this isotope must be separated from the other two isotopes found in natural samples, as well as from any dirt and other minerals collected during mining. When uranium-235 exists in purified (*enriched*) form, it is referred to as *weapons-grade uranium*. You must begin with at least 142 g of the three-isotope mixture to get 1.0 g of weapons-grade uranium after you separate out any impurities (dirt and unwanted minerals).

creating a bomb with a supercritical mass is premature detonation. To circumvent this problem, fission bombs hold two separate portions of fissionable material, each portion having a subcritical mass, as shown in Figure 6.30. The two subcritical portions are placed one at each end of the bomb vessel, and a chemical explosive, such as TNT, is placed behind one of the portions. Until the two subcritical masses are combined, no explosion occurs (at least in theory), and the bomb can be safely transported. When the bomb is discharged, the chemical explosive is detonated, forcing the two subcritical masses to smash violently into each other. The total mass of fissionable material is now supercritical, and an explosive chain reaction is initiated.

The first fission bomb was dropped by the United States on Hiroshima, Japan, on August 6, 1945. The bomb's intensity was equal to that of 20,000 tons of TNT, and the devastation of Hiroshima was total. Figure 6.31 shows the shadow of a valve handle that was imprinted on the building behind it as a result of the powerful flash of light from the bomb. Every year on August 6, the citizens of Hiroshima gather to remember the victims of the bomb and to remind the world of the importance of peace (Figure 6.32).

QUESTION 6.17 About 4 kg of fissionable material constitutes a critical mass of uranium. If a bomb is designed to contain two portions of fissionable material, each with a mass of 6 kg, what engineering flaw may cause this bomb to function incorrectly? What undesirable event may occur?

ANSWER 6.17 The two fissionable portions used in the bomb must each be subcritical. In this case, both portions are greater than critical, and therefore a nuclear explosion may occur before the two portions are smashed together.

6.6 | Living Organisms and Radiation

The three types of radiation we have discussed have obvious differences. Gamma radiation is electromagnetic radiation of very high energy, whereas alpha and beta radiation are particles with positive and negative charge, respectively. Because they are dissimilar, it should come as no surprise that these three types of radiation all interact differently with human tissue. Figure 6.35 (p. 239) illustrates how each type of radiation has a different capacity for penetrating matter. Alpha particles are the largest and therefore the slowest. Because they are so slow, they do not penetrate matter, such as human skin, easily. Beta radiation, composed of electrons only, is faster and can penetrate up to a few centimeters into human flesh. Gamma radiation is the most dangerous because it can penetrate most kinds of matter, including human tissue. In fact, several inches of lead are needed to protect humans from gamma radiation and from X-ray radiation, which is almost as penetrating as gamma radiation.

The fact that gamma radiation has strong penetrating power is used in medicine for the treatment of tumors, which result when cells grow out of control (Figure 6.36, p. 240). One of the most commonly used isotopes in nuclear medicine is cobalt-60, which is produced in particle accelerators and emits both beta particles and gamma radiation. The cobalt isotope is

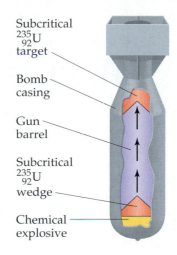

Subcritical $^{235}_{92}$U target

Bomb casing

Gun barrel

Subcritical $^{235}_{92}$U wedge

Chemical explosive

▲ **FIGURE 6.30** **The basis of an atomic bomb.** Two subcritical masses of fissionable material are initially kept separated from each other. When the chemical explosive is detonated, the two subcritical masses slam together and form a supercritical mass.

▲ **FIGURE 6.31** **Atomic bomb destruction.** The blast from the Hiroshima bomb created enough light on the wall of the building seen in the background to make an imprint of the image of a valve handle. In short, the building wall acted as a sheet of photographic film!

▼ **FIGURE 6.32** **Remembering Hiroshima.** A gathering held in Japan on August 6, 2000, to remember the victims of the Hiroshima bomb detonated in 1945.

ISSUES IN CHEMISTRY

Who's Going to Take the Trash Out?

The U.S. government is in much the same position as the obstinate teenager whose job it is to take the trash out: no one wants to do it, but somehow it has got to be done. The government is obliged to remove radioactive waste from nuclear reactors in the United States and dispose of it somehow. However, no one has yet figured out where to put it or how it can safely be stored.

In the 1980s, one site was chosen in the United States for all nuclear waste: Yucca Mountain, Nevada (Figure 6.33). About 80 miles northwest of Las Vegas, the site is desolate, and if an accident happened, projections are that only uninhabited desert areas would be contaminated. This storage plan has many Americans, especially Nevada residents, up in arms. Environmental groups have been vocal not only about the threat to the natural beauty of Nevada, but also about the hazards of transporting radioactive waste through 43 of the 48 contiguous states (Figure 6.34).

Yucca Mountain sits in an area prone to earthquakes, and the depth at which the radioactive waste is to

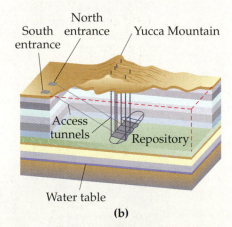

(a)

(b)

▲ **FIGURE 6.33 Taking out America's "trash." (a)** Yucca Mountain in Nevada, site of a proposed nuclear waste dump. **(b)** Relationship between the repository level, in which the waste containers will be stored, and the level of the water table.

be buried is close to the water table. These potentially serious threats to the environment prompted President Bill Clinton to veto the bill for the Yucca Mountain project in April 2000. In July 2002, however, President George W. Bush reversed this decision. Despite potential problems associated with environmental contamination, the U.S. government seems intent on developing the Yucca Mountain dump, which will not be prepared to receive waste until the year 2010 or later. Until then, 40,000 tons of nuclear waste are stored temporarily at nuclear reactor sites around the country until someone can take the trash out.

The French get more than three-quarters of their electricity from nuclear power, but their nuclear waste problem is minimal relative to ours. This is because the French recycle radioactive waste. This process produces some waste, but the amount is only a fraction of the 40,000 tons waiting for disposal in the United States. One French family using nuclear power over a 20-year period will produce only about 25 mL of nonrecyclable nuclear waste, a volume slightly smaller than an ice cube. What do the French do with any nuclear waste that cannot be recycled? They put it into temporary underground storage repositories so that when scientists of the future figure out a way to recycle it, they can take it out and reuse it.

In the meantime, in the United States, lawsuits are piling up over nuclear waste buildup around the country. Why is it that the United States cannot follow the French example? One reason is the government's concern that the plutonium produced in recycling could be stolen and used to make nuclear weapons. Additionally, some public figures believe that recycling nuclear material is too dangerous. It is probably safe to assume that the people who naysay the idea of nuclear waste recycling do not live in the vicinity of Yucca Mountain.

QUESTION 6.18   The isotope of plutonium produced in nuclear reactors is a result of the nuclear reaction

$$^{238}_{92}U + {}^{1}_{0}n \longrightarrow {}^{239}_{94}Pu + ?$$

What is the product represented by the question mark?

ANSWER 6.18   This product must have a superscript of 0 and a subscript of −2. Therefore, it must be two beta particles.

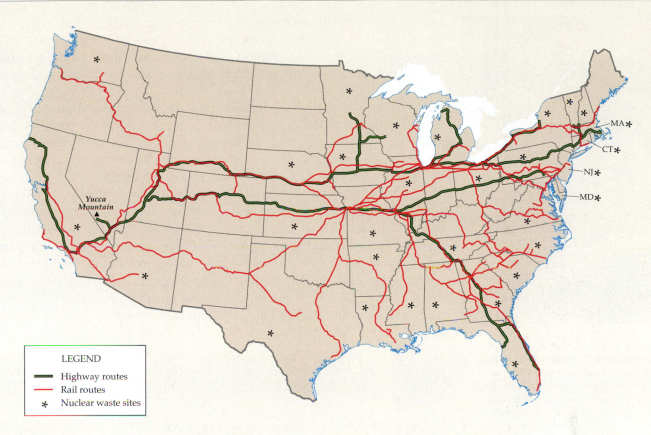

▲ **FIGURE 6.34   All roads lead to Yucca.**   By highway or by rail, the proposed transport route for nuclear waste will pass through every state except Rhode Island, Delaware, North Dakota, South Dakota, and Montana. States in which there is at least one operating nuclear reactor are indicated with an asterisk.

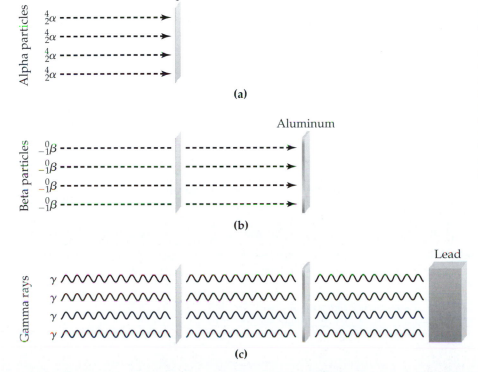

◄**FIGURE 6.35   Relative penetrating powers of radiation.**   (a) Alpha particles cannot penetrate paper, which means they also do not penetrate human skin. (b) Beta particles are more energetic and can penetrate paper. They are stopped by aluminum, however. Being this energetic, beta particles can penetrate skin to a depth of a few centimeters. (c) Gamma rays are able to penetrate paper and aluminum, but are stopped by a thick piece of lead. Being the most energetic form of radiation, they readily penetrate skin and are the most dangerous form of radiation.

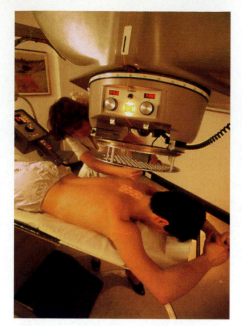

▲ FIGURE 6.36 Gamma radiation.
Radioactive cobalt, a gamma emitter, is
the most common source of focused
gamma rays. Here gamma radiation is
aimed at a specific area of the body,
possibly for the treatment of cancer.

put into a "gun" that directs radiation to the site of a growing tumor.
Because the beta particles penetrate only a very short distance into the skin,
it is mainly the gamma radiation that reaches the tumor. If the procedure
works well, the size of the tumor decreases as cells are destroyed by radiation exposure.

The irony of nuclear medicine is that, along with destroying cancerous
tissues, the process kills healthy cells as well, making many patients undergoing radiation therapy very ill. The symptoms exhibited by these patients
are the same ones exhibited by people accidentally exposed to radiation—
as a result of nuclear fallout, for example, or of a nuclear reactor accident.

Although alpha and beta particles cannot easily penetrate skin, it is possible for alpha-emitting and beta-emitting isotopes to be ingested or
inhaled into the body. Once inside the body they can damage healthy cells
and tissues by emitting radiation. We have already seen that radon, if
inhaled, can undergo a nuclear reaction that produces not only toxic lead
isotopes, but also alpha radiation. Inside the body, alpha radiation is much
more harmful than beta radiation, because the alpha particles interact with
tissues in one very localized area and can cause irreparable damage to that
tissue.

QUESTION 6.19 Radon gas is dangerous only when inhaled. We do
not worry about its effects on the outside of the human body. Why is this
true?

ANSWER 6.19 Radon gas emits alpha particles, which do not penetrate
skin.

Because our bodies are composed mainly of water, the interaction of
water with radiation accounts for most of the damage from radioactive isotopes. When water comes into contact with high-energy radiation—for
example, gamma radiation—the reaction is

$$\text{Energy} + H_2O \longrightarrow H_2O^+ + 1\,\text{electron}$$

Any radiation with an energy greater than 1200 kJ/mol causes water molecules (and most other molecules, for that matter) to ionize and is called
ionizing radiation. The cutoff wavelength for this energy level is about 100

▶ FIGURE 6.37 Creating free radicals.
Ionizing radiation has a wavelength of
100 nm or less. Longer wavelengths of
light do not promote the ionization of
atoms.

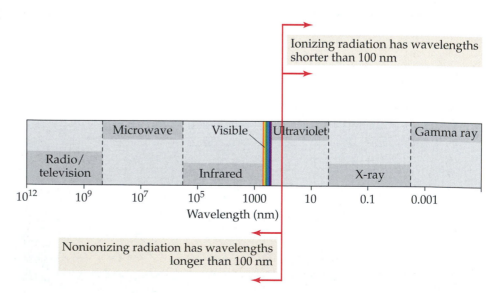

nm (Figure 6.37). Any radiation having a wavelength shorter than 100 nm qualifies as ionizing radiation. Alpha particles, beta particles, and gamma radiation all have energy above this value and so are all considered ionizing radiation.

The ionization of water in and of itself is not a problem. However, the reactive $H_2O^{+\cdot}$ ion undergoes a reaction that involves another molecule of water:

$$H_2O^{+\cdot} + H_2O \longrightarrow H_3O^+ + OH\cdot$$

The first product, H_3O^+, called the *hydronium ion*, is innocuous. It is a combination of a water molecule, H_2O, and a hydrogen ion, H^+. The second product, a *hydroxyl radical*, is a damaging free radical sometimes found in the body (as we saw in Chapter 5). In the presence of high levels of ionizing radiation, hydroxyl radicals are quickly produced in large quantities and initiate reactions that degrade cells and tissues.

QUESTION 6.20 Which of the following are considered ionizing radiation: (*a*) blue light, (*b*) X rays, (*c*) gamma radiation?

ANSWER 6.20 (*b*) and (*c*). Blue light is in the visible region of the electromagnetic spectrum and does not have enough energy to be ionizing, as seen in Figure 6.37.

Although radiation can be inadvertently taken into the body, there are some medical procedures in which the patient must intentionally ingest a radioactive isotope called a *radioactive tracer*, which works inside the body in a controlled manner. Many tracers produce gamma radiation and beta particles because potential damage from alpha particles is too difficult to control. For example, the tracer technetium-99 emits only gamma radiation. Its half-life is only 6 h, which means a patient who ingests it will be radiation free in a few days. During a *computerized tomography* (CT) examination, a patient who has been given a technetium-99 cocktail is scanned by a camera that detects gamma radiation (Figure 6.38). The camera moves around the body collecting images in three dimensions, images that are then converted by computer into slices from which the radiologist can diagnose abnormalities in the body. Technetium-99 moves quickly to the most metabolically active tissues in the body, and so it is especially useful for monitoring the brain and the heart, as well as regions of arthritis and bone cancer.

Radioactive isotopes other than technetium-99 are also used in tomographic examinations. For example, the radioactive isotope iodine-131 is administered to patients in the form of *radio-labeled* sodium iodide, $Na^{131}I$. Because iodine is transported to and used by the thyroid, this isotope is useful for monitoring a patient with suspected thyroid disease. Figure 6.39 shows a CT scan of a healthy thyroid and an enlarged one.

The SI unit for the amount of radiation a human being is exposed to is the **sievert** (Sv). The sievert is called an *effective unit* because it takes into account the differences in penetrating power of the three types of radiation. A person who ingests the same amount of radiation from an alpha source and from a beta source will experience more damage and more severe radiation sickness as a result of the alpha radiation for the reasons we already

MOLECULE
Hydronium Ion

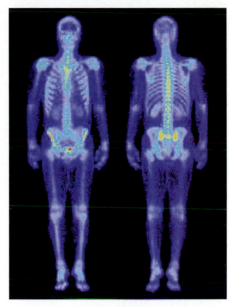

▲ FIGURE 6.38 **Radioactivity in medicine I.** The isotope technetium-99 is used in the diagnosis of disease. This isotope, which emits gamma radiation, is ingested by the patient, and then computerized tomography is used to detect gamma radiation and produce a scan of the body's tissues.

▼ FIGURE 6.39 **Radioactivity in medicine II.** A healthy thyroid (a) and an enlarged thyroid (b) are visualized using computerized tomography (CT).

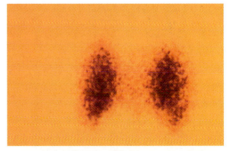
(a)

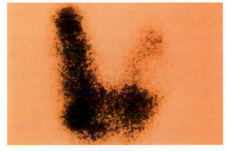
(b)

TABLE 6.2	Health Effects of Radiation on Human Beings
Dose (rem)	**Biological Effects**
0–25	No detectable effects
25–100	Temporary decrease in white blood cell count
100–200	Nausea, vomiting, longer-term decrease in white blood cells
200–300	Vomiting, diarrhea, loss of appetite, listlessness
300–600	Vomiting, diarrhea, hemorrhaging, eventual death in some cases
Above 600	Eventual death in nearly all cases

discussed. For example, a person may ingest a dose of radioactivity from a beta-emitting material and receive 0.6 Sv of radiation. If that source is an alpha emitter instead, the same dose will be about 60 Sv, an amount that is fatal for the average human being.

In nuclear medicine, the commonly used unit for effective radiation dose is the **rem**, which stands for _roentgen equivalent_ for _man_. By definition, 1 Sv = 100 rem. Table 6.2 lists the symptoms of radiation sickness resulting from exposure to different levels of radiation. In Figure 6.40, Figure 6.20 is reproduced with exposures expressed in _millirems_ (mrem), where 1 mrem = 0.001 rem. Figure 6.40 shows that the total radiation exposure of the average person during 1 year is 359 mrem, or 0.359 rem. According to Table 6.2, symptoms of radiation sickness begin at 25 rem. Thus, an individual would have to be exposed to 70 times the normal annual dose of radiation to exhibit signs of mild radiation sickness.

Radiation affects the fastest-growing tissues in the body because each new cell is exposed as it is created. Thus, the lymph nodes and the bone marrow are two of the regions of the body hit hardest by high

▶ FIGURE 6.40 **Radiation exposure.**
Figure 6.20 recast in units of millirems.

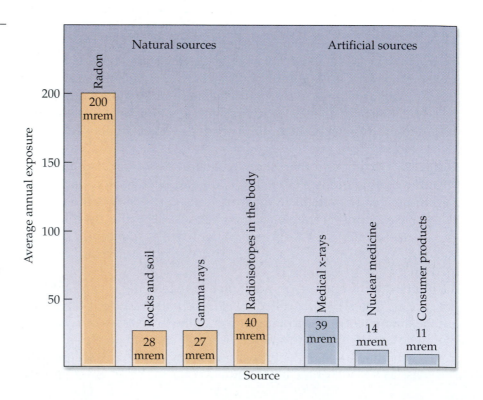

The Tragedy at Chernobyl

On April 26, 1986, the Chernobyl nuclear power plant in what is now Ukraine went out of control. In just over a week, the radiation given off as a result of a fiercely overheated reactor moved north and west, across most of continental Europe and over the North Pole to Asia and northern Canada (Figure 6.41).

The death count depends on whom you ask. Most estimates report the total number dead at the time of the accident at about 2500, in addition to tens of thousands who suffered (and continue to suffer) from various degrees of radiation exposure.

The most common health problems reported after the accident were a marked increase in babies born with deformities and very high incidences of thyroid cancer. Some studies estimate that as many as 12,000 people have died from radiation-related diseases in the years since the accident. These fatality figures remain a hot topic of debate.

Hard facts about the disaster are not easy to come by. Statistics on the meltdown and the number of citizens affected range widely, depending on the source. Although most people agree that the disaster brought devastating consequences for the environment and for people living under the cloud of radioactive contamination that spread over half the world, there are still places in which people espouse other points of view. What is the truth about the Chernobyl disaster? How much has the local population been affected? The answer is that we may never know the truth about the greatest nuclear disaster in history.

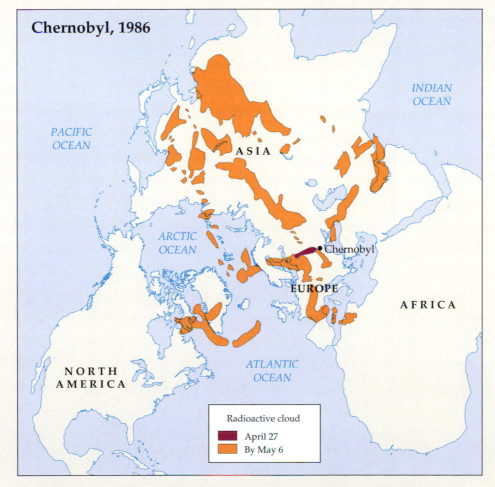

Chernobyl, 1986

INDIAN OCEAN

PACIFIC OCEAN

ASIA

ARCTIC OCEAN

•Chernobyl

EUROPE

AFRICA

NORTH AMERICA

ATLANTIC OCEAN

Radioactive cloud

April 27
By May 6

▲ FIGURE 6.41 **Fallout at Chernobyl.** The Chernobyl disaster occurred in what is now Ukraine on April 27, 1986. By May 6 of that year, the radioactive cloud had spread to an area spanning parts of Europe, Asia, North America, and Africa.

doses of radiation. Exposure of these tissues to radiation can lead to abnormalities in white blood cell production and leukemia. In fact, leukemia is the most common type of cancer associated with radiation exposure.

It is important to recognize that we humans are constantly exposed to radiation from both natural and artificial sources. Most of us will never accumulate enough radiation to cause radiation sickness, but it is advisable to limit your exposure to low-level sources of radiation whenever possible.

QUESTION 6.21 Homer Simpson is exposed to 0.50 Sv of radiation after an accident at the Springfield nuclear reactor. Will he exhibit symptoms of radiation sickness? If so, describe the symptoms.

ANSWER 6.21 Homer has been exposed to 0.50 Sv × 100 rem/Sv = 50 rem of radiation. Thus, a blood test will show that he has a temporary decrease in white blood cell count, but he is likely to have no other symptoms.

SUMMARY

We started our survey of nuclear reactions with the energy that comes from the Sun. All the energy we receive from the Sun is created by solar fusion reactions. In general, nuclear reactions release much greater amounts of energy than do chemical reactions. Chemical reactions differ from nuclear reactions in that the former always have equal numbers of atoms of each element before and after the reaction takes place. Conversely, nuclear reactions usually involve a transmutation in which atoms of one element are changed to atoms of some other element. When equations representing nuclear reactions are being balanced, the mass numbers (shown as superscripts) and the atomic numbers (shown as subscripts) must be balanced.

Most of the elements from hydrogen to uranium on the periodic table likely evolved through fusion reactions after the Big Bang, the name given to the event that marked the beginning of the universe. Elements coming after uranium on the periodic table can be synthesized by smashing atoms together in a particle accelerator.

In general, nuclear reactions give off large amounts of energy compared to chemical reactions because the strong nuclear force holding together the particles in the nucleus is released. Nuclear fission is the process whereby radioactivity is given off as an unstable isotope splits into two smaller nuclei. Radioactivity exists in three major forms. Alpha particles are helium nuclei carrying a charge of +2. Because they are more massive than other types of radiation, alpha particles move more slowly and do not penetrate matter readily. When an alpha particle is emitted during a nuclear reaction, the mass number of the decaying isotope decreases by 4 and the atomic number decreases by 2. Radon is an example of an alpha emitter that is highly dangerous because it is a gas and can be inhaled into the lungs, causing damage to lung tissue.

Beta particles are electrons ejected from an atom's nucleus; these nuclear electrons are formed as a neutron breaks apart into one proton and one electron. When a radioactive atom emits a beta particle, its mass number does not change, but its atomic number increases by 1. Beta particles move more quickly than alpha particles and so penetrate matter more easily.

Gamma radiation is not associated with any type of particle. Rather, it is very high energy electromagnetic radiation and is highly penetrating. Gamma radiation is a primary contributor to the deleterious effects when the surface of the human body is exposed to radiation. Because it penetrates flesh readily and destroys human cells, it is used in the treatment of tumors.

The half-life of a radioactive isotope is the amount of time it takes for a sample of the isotope to decay to half of its original amount. The belt of stability is a predictor of isotopic stability that plots the number of neutrons versus the number of protons for all the elements up to atomic number 83, bismuth.

Nuclear reactions have been harnessed on Earth for peacetime uses (such as nuclear power) and in war. Both endeavors result in the production of nuclear waste, which poses a threat to the environment because it contains radioactive isotopes that have long half-lives. The burial of nuclear waste presents an environmental hazard because there is risk of seepage into groundwater. Excessive exposure to radioactive isotopes adversely affects human health, and the effects of radiation sickness range from decreased white blood cell count at very low levels to death at high levels. Humans are exposed to natural as well as artificial sources of radiation, and excessive exposure to either source can result in severe radiation sickness.

KEY TERMS

alpha decay	chain reaction	nuclear fission	radioactive decay	strong nuclear force
alpha particle	critical mass	nuclear fusion	series	transmutation
belt of stability	gamma radiation	nuclear reaction	radioactivity	
beta decay	half-life	radiation	rem	
beta particle	ionizing radiation	radioactive decay	sievert	

QUESTIONS

The Painless Questions

1. In an equation representing a nuclear reaction, what symbol is used to represent (*a*) a helium nucleus, (*b*) a beta particle, (*c*) a neutron?

2. For each answer in Question 1, explain why each superscript and subscript you wrote is the correct one.

3. Identify each reaction as chemical or nuclear:
   (a) $^{174}_{77}\text{Ir} \rightarrow {}^{170}_{75}\text{Re} + {}^{4}_{2}\text{He}$
   (b) $\text{H}_2\text{O} \rightarrow \text{H}^+ + \text{OH}^-$
   (c) $\text{N}_2 + 3\,\text{H}_2 \rightarrow 2\,\text{NH}_3$

4. Indicate the total number of neutrons, protons, and electrons in each isotope:

(a) $^{10}_{5}B$

(b) $^{90}_{38}Sr^{2+}$

(c) $^{170}_{78}Pt^{2+}$

5. (a) According to Figure 6.20, what is the percentage of radiation exposure from all the natural sources named? (b) Of that total, what percentage can be attributed to gamma rays from the Sun? (c) What percentage of radiation exposure from artificial sources is attributable to medical X rays?

6. (a) Which type of radiation changes the atomic number of the reactant, but not its mass number? (b) Which type changes both the mass number and the atomic number of the reactant?

7. Comment on this statement: *Alchemists of medieval times were people who mixed magic with science. Their only goal was the creation of gold from other elements.*

8. Write the complete symbol (subscript and superscript) for the isotope that (a) contains 52 neutrons and has an atomic number of 48, (b) has a mass number of 249 and an atomic number of 98, (c) is the isotope of ytterbium containing 84 neutrons.

9. Beta particles penetrate human flesh more than alpha particles do. Based on your knowledge of the structure of these two particles, explain why this is so.

10. If you look at a typical nuclear equation, is it possible to tell whether or not the reaction produces gamma radiation? Why or why not?

11. If the combustion of a certain number of moles of methane produces 125,300 kJ of energy, how much energy would be produced for an equivalent amount of reactant undergoing a nuclear reaction in the Sun?

12. Lise Meitner (Figure 6.16) contributed much to our modern-day understanding of the atom. Outline her accomplishments in the field of nuclear chemistry.

13. According to Figure 6.34, which states will not have nuclear waste transported through them if the Yucca Mountain waste facility is used for storage: (a) Montana; (b) Tennessee; (c) Texas; (d) South Dakota; (e) Utah?

14. According to Figure 6.3, only a portion of the radiation leaving the Sun passes all the way through Earth's atmosphere and reaches the ground. At what wavelength is solar radiation cut off as it enters our atmosphere?

15. What is the percentage of electricity produced by nuclear reactors in (a) South Korea (b) Lithuania, (c) United States, (d) Ukraine?

More Challenging Questions

16. The Sun is composed primarily of the two lightest elements, hydrogen and helium. The particles that exist in the Sun are, for the most part, in ionized form. (a) Are the ions in the Sun cations or anions? (b) Why do the lightest elements exist in highest abundance in the Sun?

17. Identify what the question mark stands for in each reaction:

(a) $^{246}_{96}Cm + {}^{12}_{6}C \rightarrow 4\,{}^{1}_{0}n + ?$

(b) $^{82}_{34}Se \rightarrow 2\,{}^{0}_{-1}\beta + ?$

18. Which reaction would you expect to produce more energy? Why?

(a) $C_3H_8 + 5\,O_2 \rightarrow 3\,CO_2 + 4\,H_2O + energy$

(b) $^{1}_{0}n + {}^{235}_{92}U \rightarrow {}^{137}_{52}Te + {}^{97}_{40}Zr + 2\,{}^{1}_{0}n + energy$

19. If the radioactive series begins with uranium-238 runs all the way to formation of lead-206, (a) how many alpha emissions are there? (b) how many beta emissions?

20. Based on Figure 6.13 and what you know about the stability of radioactive isotopes, would you say that the isotope uranium-184 is stable or unstable? Give a reason for your answer. Where does this isotope fall on the graph shown in Figure 6.13?

21. Identify what the question mark stands for in each reaction:

(a) $^{64}_{29}Cu \rightarrow {}^{64}_{30}Zn + ?$

(b) $^{64}_{29}Cu + ? \rightarrow {}^{64}_{28}Ni$

22. Write the complete symbol (subscript and superscript) for the isotope that has (a) no charge, 56 electrons, and 82 neutrons, (b) a charge of 2−, 37 electrons, and a mass number of 76.

23. (a) What type of reaction do we use to harness the power of fossil fuels for energy? (b) Why do equal masses of fossil fuels and nuclear fuels provide different amounts of energy? (c) Why might an environmentalist discourage the use of nuclear fuel as a source of energy in the United States?

24. Comment on this statement: *Smoke detectors work by prompting the production of ions between the metal plates when smoke is present. This completes the detector's circuit and triggers the smoke alarm.*

25. Describe the Big Bang theory and the events that took place immediately after the initial explosion. Which type of nuclear reaction, fusion or fission, was responsible for the heavy elements (atomic numbers greater than 92) we know today?

26. The text talks about the minimum energy required for electromagnetic radiation to be considered ionizing.

If a certain kind of light possesses 500,000,000,000 μJ of energy, would it be able to ionize matter? Use dimensional analysis to arrive at your answer.

27. Explain why the field of chemistry that deals with the creation of elements with atomic numbers above 92 is often fraught with controversy.

28. We know that radon levels in buildings correlate somewhat with the presence of uranium in the ground under the buildings. (a) How is uranium related to radon? (b) How many nuclear reactions are required to convert uranium to radon? (c) In these steps, how many alpha emissions occur? (d) how many beta emissions?

29. These images are from a bone scan done on a patient who had drunk a solution containing a radioactive isotope. Two isotopes commonly used in nuclear medicine are cobalt-60 and technetium-99, which have half-lives of 5.26 years and 6 hours, respectively. One of the advantages of the technetium isotope is that it decays quickly and therefore does not remain in the body for long. If this is an advantage, why is cobalt also commonly used in nuclear medicine?

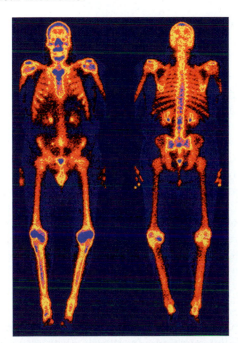

30. The emission of a beta particle involves the conversion of a neutron to a proton and an electron. If one beta emission occurs, and this emission eliminates one neutron from the nucleus of the decaying isotope, why does the mass number of the product species not decrease by 1?

The Toughest Questions

31. (a) Describe how a power plant using coal and a power plant using nuclear fuel work in similar ways.

(b) How is electricity produced in each? (c) How do the two types of plants differ from each other?

32. To date, a nuclear reaction for forming gold from some other element has not been found. One could imagine such reactions, however, even though they cannot be run in a laboratory because the appropriate radioactive isotopes do not exist. (a) Starting with your choice of platinum isotope, write an imaginary nuclear reaction that would produce gold-197 in addition to some type of radioactivity. (b) Do the same thing starting with your choice of thallium isotope.

33. (a) Describe what is meant by the term *strong nuclear force*. (b) If this force is felt between protons and protons, between neutrons and neutrons, and between protons and neutrons, why do nuclei become more unstable as they become larger?

34. After 936 years, the amount of isotope X remaining after fission is 25 percent of the original mass. What is the half-life of this isotope?

35. The reaction in a typical nuclear reactor produces the isotopes krypton-91 and barium-142 plus some neutrons. Imagine that instead of breaking up into these two isotopes, 1 mol of uranium in the reactor breaks up into 2 mol of a single isotope plus at least one neutron and no other type of radiation. (a) Using the minimum number of neutrons required, write a balanced nuclear equation showing this process. (b) Would this reaction promote a chain reaction? Why or why not?

36. Imagine that element 115, containing 145 neutrons, has just been created. Its name will be simpsonium, symbol Sp. (a) Write the complete symbol (subscript and superscript) for this isotope. (b) Under which element on the periodic table will this new element be placed?

37. The primary challenge in designing a fusion reactor is the high temperature required in order for fusion to occur. (a) Why must exceedingly high temperatures be used? (b) Why is it difficult to build a reactor that can house this type of reaction?

38. The half-life of polonium-214 is 1.6×10^{-4} s. How many microseconds will pass before 75 percent of a sample of this isotope has decayed?

39. TNT undergoes the combustion reaction

$$C_7H_5N_3O_6 + O_2 \longrightarrow CO_2 + H_2O + N_2$$

(*a*) Is this chemical equation balanced? If not, balance it. (*b*) This combustion produces 15 kJ of energy per gram of TNT. Given this information and the conversion factor 907 kg = 1 ton, calculate the kilojoules of energy produced by the Hiroshima atomic bomb.

40. The fusion reaction between an alpha particle and element X to make an isotope having an atomic number greater than that of element X must be performed in a particle accelerator. However, the bombardment of element X with *neutrons* to produce a product having an atomic number greater than that of element X does not require acceleration of the neutrons. Use your knowledge of atomic structure to speculate on why this is true.

41. A person working as a nuclear-weapons technician is exposed to 3.8 Sv of radiation over a 5-year period. Assuming the exposure was relatively constant over the 5 years, what symptoms of radiation sickness, if any, would this person exhibit? How many rems of radiation is this person exposed to annually?

42. Consider the belt of stability shown in Figure 6.13. In a radioactive decay series, why does only one isotope in the series fall into this belt?

43. A look at Figure 6.19 shows that many isotopes in the uranium radioactive decay series are quite stable relative to other isotopes in the series. If this is the case, why must we worry about radon exposure? Why is the notion of *relative stability* of radioactive isotopes important?

44. The isotope potassium-40 has a half-life of 1,280,000,000 years. (*a*) Express this half-life using scientific notation. (*b*) How many minutes will have passed when only 12.5 percent of a sample of this isotope remains? (*c*) Based on the position of potassium on the periodic table and the number of neutrons it contains, speculate on why this isotope has such as long half-life.

45. The dinnerware shown at the top of the page is made with a surface glaze that contains uranium. (*a*) If the isotope in the glaze is uranium-238, write a nuclear equation showing how this isotope decays via emission of one alpha particle. (*b*) Is it safe to have dinnerware like this in your home if it is never used to serve food? (*c*) If this dinnerware is used to serve food, how might it be a danger to the health of those who eat from it?

46. New element 114, created in Russia in 1999, is just below lead on the periodic table. Suppose the owner of some manufacturing company has decided that this new element should be made by the truckload to use in place of lead in things like batteries. As science consultant to the owner, what would you say to him about this idea?

E-Questions

Go to **www.prenhall.com/waldron** to find these questions in electronic form, complete with hyperlinks directly to the various websites cited in the questions.

47. Read the article "How Much Radon Is Too Much?" by Charles Atwood in **JChemEd**, and then answer the following questions with short essays: (*a*) What is the primary reason radon gas infiltrates houses? (*b*) How was the home radon survey performed by B.L. Cohen designed? Briefly describe the results obtained in this survey. (*c*) How does the risk posed by radon exposure compare with non-nuclear home risks, including risks from household chemicals? (*d*) At the time this article was written, what was the EPA recommendation on home radon levels? Use the Web to research this year's recommendations. How do the two compare?

48. Do a Web search for Geiger counter. Read information from reliable websites on the design and construction of a Geiger counter, and then answer the following questions: (*a*) Which types of radiation are detected by a Geiger counter? (*b*) Briefly describe the nuclear chemistry that takes place inside a Geiger counter. (*c*) What is the definition of a counting gas? (*d*) What are the advantages and disadvantages of using a Geiger counter versus the detector goggles described in this chapter?

49. Read about Arctic nuclear testing at **Acronym** and answer the following questions: (*a*) According to the article, what is the reason for U.S. accusations against

Russia regarding nuclear testing in the Arctic? (*b*) Do you perceive a bias in the way this article is written, or does it take a balanced approach to the issue? (*c*) Do a web search for a map showing the location of Novaya Zemlya. What country in the vicinity of Russia is closest to this testing site?

50. Read "The Science of the Silver Bullet" at **SciAm**. This story describes how radioactive waste in the form of bullets was used by allied forces in the dispute in Kosovo. (*a*) What makes these bullets different from the conventional variety? (*b*) What consequences have resulted from their use in Kosovo? (*c*) What types of isotopes and radiation are associated with them? (*d*) How many rounds were fired during the dispute in Kosovo?

51. Search the Web to learn more about the chemistry that takes place in the Sun, and then answer the following questions: (*a*) How is the solar wind created, and how is it related to the aurora borealis? (*b*) What is a plasma, and how is one created in the Sun? (*c*). The percentages of hydrogen and helium in the Sun based on mass are different from the percentages based on numbers of atoms. Why?

WORLD WIDE WEB RESOURCES

As with the E-Questions, go to **www.prenhall.com/waldron** to find these questions in electronic form, complete with hyperlinks directly to the various websites cited.

Some of the links that follow contain research articles related to the topics in this chapter and could be used as the basis for a writing assignment in your course. Other sites are of general interest.

- Learn more about radioactive waste in the oceans at **SciAm**. Read the article about nuclear submarines that litter the seabed and their potential hazards. Do you think it is safe to leave abandoned nuclear submarines on the sea floor?
- Find out how close you may be to a nuclear waste transport route at **NucWaste.** Read a thorough discussion of exactly how nuclear waste is transported at **UIC**, Australia's Uranium Information Centre.
- Go to **NuclearTourist** and click on *US locations*. This will bring up a map with states using nuclear power in blue and states with no nuclear power in white. If your state is blue, click on it, and then answer the following questions: (*a*) What percentage of the electricity generated in your state comes from nuclear power? (*b*) How many reactors currently operate in your state, and where are they located? (*c*) How is the waste produced at those plants stored? If your state is colored white, go to *state electricity profiles*, and click on the survey for your state. Look at Table 4 there to determine how electricity is produced in your state.
- Go to **WebElements**, click on your favorite element that has an atomic number greater than 83, and then answer the following questions: (*a*) What are the known isotopes for this element? Which of these isotopes is/are radioactive? What are their relative abundances? (*b*) Where has this element been observed? If it can be synthesized, what nuclear reaction is used?

ABC NEWS VIDEO
Cover Up at Ground Zero

ABC NEWS VIDEO
The Disposal of Nuclear Waste

CHEMISTRY TOPICS IN CHAPTER 7

- Isotopes of hydrogen [7.1]
- Dipole–dipole interactions, hydrogen bonding, surface tension [7.2]
- Density, freezing and melting points, nucleation centers [7.3]
- Specific heat [7.4]
- Boiling point [7.5]
- Evaporation, condensation, vapor pressure [7.6]
- Hydrophilic, hydrophobic, and amphipathic molecules, immiscible and miscible liquids, alcohols [7.7]

"In the bleak midwinter / Frosty wind made moan, / Earth stood hard as iron, / Water like a stone."

CHRISTINA ROSSETTI

Water

A LOOK AT INTERMOLECULAR INTERACTIONS, PARTICULARLY THOSE INVOLVING THE UNIQUE PROPERTIES OF WATER

Virtually every culture and religion has some form of ritual involving water. For example, the Japanese Shinto religion, which is steeped in an appreciation of nature, has a water ritual for almost every aspect of human life—staving off illness, improving eyesight, finding a new job, living a long life, having a healthy baby . . . . In all these rituals, water is a provider and a purifier. It provides health and good fortune, and it purifies by washing away bad luck or sin (Figure 7.1). The special significance of water carries over from culture and religion to the realm of chemistry. Chemists refer to it as the *universal solvent* because it is able to dissolve so many things. From a molecular point of view, water is unique, unlike any other substance on Earth. In this chapter, we shall figure out what makes water so extraordinary and why we cannot live without it.

You already know quite a bit about water. You know from Chapter 3 that water has a tetrahedral electron-pair geometry that gives it a bent molecular geometry (Figure 7.2). As you will see in this chapter, this bent shape gives water some very unusual characteristics and is responsible for some of its physical properties. You will learn that water molecules can have a unique kind of interaction with one another, an

▲ FIGURE 7.1 Water as purifier.

interaction that is strong and makes water suited to its role as the universal solvent. We shall explore what water does as it moves from one **phase** of matter to another—solid phase to liquid phase to gas phase and back again. As the temperature of water changes, these phase transformations are the result of the shifting interactions between water molecules. To understand these transformations, you must understand how water molecules shake and quiver and how these motions bring about melting, evaporation, condensation, and freezing. Finally, we shall discuss molecules that avoid water and the structures they form in order to exclude it. First, though, let us look once again at a single water molecule and explore the hydrogen and oxygen atoms it contains.

7.1 | Songbirds and Hydrogen Atoms

Consider a simple 10-oz glass of water (about 280 mL, Figure 7.3). If it comes from the tap in your kitchen, it contains dissolved ions and gases (even if you have an expensive water purifier). If you could somehow remove every impurity from your glass of water (which would be impossible to do), you would have a glass of pure H_2O. You might then assume every water molecule in the glass would be identical to every other water molecule. Or would it be?

Recall from Chapter 2 that oxygen has three isotopes (Figure 2.11). We saw how this fact was used to study the origin of the marble used in Greek statues. About 99.76 percent of oxygen atoms are oxygen-16, but the remainder is a mixture of the isotopes oxygen-17 and oxygen-18. Therefore, in your glass of water, there are going to be some water molecules that contain oxygen-17 and some that contain oxygen-18. The same is true for hydrogen. Hydrogen atoms come in three varieties: hydrogen-1, hydrogen-2, and hydrogen-3. By far the most abundant in nature is hydrogen-1, at 99.98 percent of all hydrogen atoms. This is the isotope that we commonly think of as plain old "hydrogen." Hydrogen-2 (known as *deuterium*) and hydrogen-3 (known as *tritium*) make up the remaining 0.02 percent of the isotopic abundance of hydrogen. In fact, we normally do not give much thought to deuterium and tritium because they are so rare. However, in the same way that the rare isotope carbon-14 can be used to date ancient objects (Chapter 2) and oxygen isotopes can be used to determine the origin of Greek marble, hydrogen isotopes can also be put to use.

Table 7.1 shows the isotopic distribution of hydrogen and oxygen in nature. In your 280-mL glass of water, there are about 9.4×10^{24} water

▼ FIGURE 7.2 **The shape of water.** The tetrahedral electron-pair geometry of the water molecule (top) produces a bent molecular geometry (bottom) because the two nonbonding pairs of electrons cause the hydrogen atoms to be at two corners of a tetrahedron instead of in a straight line with the oxygen atom.

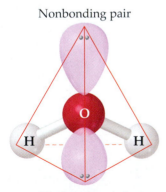

Nonbonding pair

Nonbonding pair

Tetrahedral electron-pair geometry

Bent molecular geometry

TABLE 7.1	Percentage of Each Isotope of Hydrogen and Oxygen Found in Nature	
Isotope		**Abundance in nature (%)**
^1H (hydrogen)		99.98
^2H (deuterium)		0.02
^3H (tritium)		1×10^{-16}
^{16}O		99.76
^{17}O		0.04
^{18}O		0.20

molecules, and so 0.04 percent of them (about 3.8×10^{21}) will contain oxygen-17 and 0.02 percent (about 1.9×10^{21}) will contain deuterium. Even though the percentages of these isotopes are very, very small, the enormous number of water molecules in the glass makes the total number of water molecules containing these isotopes quite large.

QUESTION 7.1 In your glass of water, how many tritium atoms would you find?

ANSWER 7.1 The glass contains 9.4×10^{24} water molecules, each containing two hydrogen atoms. Therefore, there are

$$\text{Number of H atoms} = 9.4 \times 10^{24} \text{ H}_2\text{O molecules} \times \frac{2 \text{ H atoms}}{1 \text{ H}_2\text{O molecule}}$$

$$= 1.9 \times 10^{25} \text{ H atoms}$$

From Table 7.1 you know that the abundance of tritium atoms is 1×10^{-16} percent in a given sample. Thus, you must multiply the number of H atoms by 1×10^{-18} (see Hint on Problem Solving for why the exponent changes):

$$\text{Number of }^{3}\text{H atoms} = (1.9 \times 10^{25} \text{ total H atoms}) \times (1 \times 10^{-18})$$

$$= 1.9 \times 10^{7} \text{ atoms of }^{3}\text{H}$$

Of the total number of hydrogen atoms, 1.9×10^{7} are tritium.

▲ FIGURE 7.3   **Not all just plain H's and O's.** Depending on where you live, water taken from a tap in your home will contain various amounts of the rarer isotopes deuterium, tritium, oxygen-17, and oxygen-18.

HINT ON PROBLEM SOLVING: In Question 7.1, we used a percentage, 1×10^{-16} percent, in calculating the number of tritium atoms in a sample of water. To use a percentage, in a calculation, you must divide the given percentage by 100 (because percentage means *parts per 100 parts*). Thus, when you want to take 30 percent of something, you multiply it by 30/100, or 0.30. Thus, 30 percent of 8.8 is $0.30 \times 8.8 = 2.6$. Students sometimes forget to divide by 100, and instead multiply by the percentage. In Answer 7.1, we multiply by $1 \times 10^{-16}/100 = 1 \times 10^{-18}$ to calculate the percentage.

The natural distribution of isotopes in water molecules was used by ornithologist Richard Holmes of Dartmouth University and his student Dustin Rubenstein to study bird migration patterns. This work is important because scientists who study breeding birds often lose track of them during the winter migration, and a better understanding of migratory and breeding habits can help to preserve a species. The traditional way to monitor migration is to tag the birds and try to recapture them as they move from place to place. This method can have success rates as low as 2 percent, however. Because the majority of birds are never caught, the number tagged must be very high. Another drawback to this method is that the process of tagging a bird is traumatic for the bird and can disrupt its natural pattern of migration.

▶ **FIGURE 7.4   For the birds.**
Their knowledge of isotope chemistry led Dustin Rubenstein (shown here) and advisor Richard Holmes of Dartmouth University to a study of the migration patterns of the black-throated warbler.

The Dartmouth biologists reasoned that it may be possible to use the distribution of isotopes in water to follow the migration patterns of the black-throated warbler, an insect-eating songbird that travels between the Caribbean and the eastern United States (Figure 7.4). The legwork had already been done: careful measurements of the ratios of hydrogen isotopes in water samples revealed that this ratio differs from one place to another. What does isotope distribution have to do with birds, though? Simple—birds eat insects, insects eat plants, and plants take up water from soil that has a distinctive isotope ratio. Thus, it should be possible to analyze the water in these birds and, from that analysis, determine where the birds have eaten and therefore their migration path. In a flock of birds migrating together, you will find traces of water from different places, and a map of these locations should provide a record of the migration path for the flock.

The Dartmouth biologists used measurements of deuterium to track the migration patterns of the black-throated warbler across the United States and the Caribbean, analyzing hydrogen-to-deuterium ratios in the water contained in the birds' feathers. Thankfully, this project did not involve a group of scientists chasing birds and pulling out their feathers! After mating, the birds molt, and the discarded feathers were collected and analyzed.

What Holmes and Rubenstein found was big news in the ornithology community. Birds that migrate to the western Caribbean come from the northern United States and southern Canada, whereas those that migrate to the eastern Caribbean come from farther south in the United States (Figure 7.5). In this study, the migration path of the black-throated warbler was established using measurements of hydrogen isotopes. This information will allow ornithologists to track these birds in the future and to assess the risks, such as those from environmental pollutants, that may threaten the black warbler population. Thus, we have now seen a third example of how the measurement of isotopes can be very valuable. (The other two, in case you have forgotten, were the Elgin marble study and carbon-dating of the prehistoric "iceman," both in Chapter 2.)

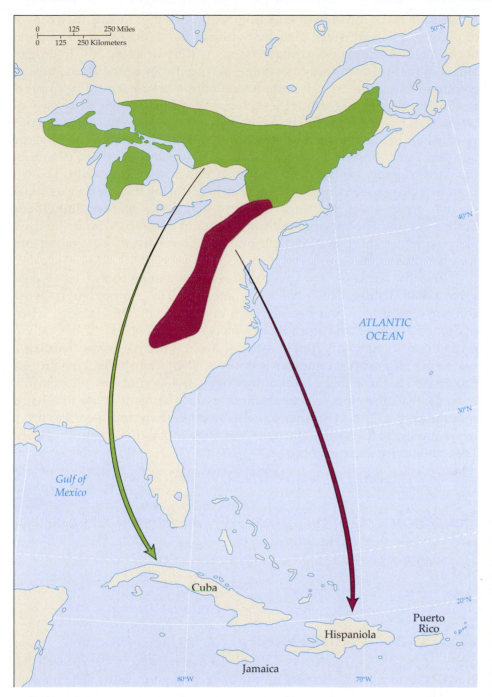

◄ **FIGURE 7.5** **South for the winter.** Isotope analysis of the water contained in the feathers of molting birds let Dartmouth University researchers identify the birds' migration destinations. Black-throated warblers who summer in the northern United States and southern Canada (green) spend the winter on islands in the western part of the Caribbean Sea. Those who summer in the southern United States (maroon) spend the cold months on islands in the eastern Caribbean.

The use of hydrogen isotope ratios goes beyond the study of flight patterns in black-throated warblers. For example, individuals who suffer from maladies related to water retention and loss, such as kidney failure, often need accurate measurements of their total water content. A new technique requires that they drink a known volume of water that has been artificially *enriched* with deuterium so that the percentage of deuterium isotopes is greater than normal. The water enters the muscle tissues after a given time and then is expired through the mouth with regular breathing. Water in the breath is the gaseous form of water, which we usually call **water vapor**, and this exhaled water vapor has a certain percentage of deuterium in it. Because the average ratio of ^{1}H to ^{2}H is relatively constant, that percentage reveals the total amount of water in the person's body.

Tritium is the rarest hydrogen isotope. It is used in the manufacture of luminous signs and watch dials that glow in the dark. The concentration of tritium in a sample of hydrogen is so low that for every 10^{18} hydrogen atoms in nature, only one, on average, is tritium. Thus, the abundance of tritium in water is extremely minute—so minute that it would be impractical to try to measure tritium levels in, say, a bird feather or a glass of water. This isotope is useful, however, when there is an enormous quantity of water being analyzed—the ocean, for example. Before the Nuclear Test Ban Treaty of 1962, large-scale testing of nuclear devices on islands in the Pacific Ocean produced, among other things, a lot of tritium. The tritium isotope released into Earth's atmosphere during these tests was taken up into ocean water and thus can serve today as a marker throughout Earth's oceans. The varying levels of tritium in oceanic currents are used to measure current speeds, and this information is used for navigation and weather prediction.

Now that we know how hydrogen isotopes can be useful, let us consider what distinguishes one from another. The isotope hydrogen-1 contains one proton and one electron, and that's it. This is why, when hydrogen-1 ionizes and loses its one electron to form $^1H^+$, it has only a proton remaining and is simply called a *proton*. Protons, as you recall, are the source of acidity in aqueous solutions. Deuterium must contain one proton also because it is a form of hydrogen. It also contains one neutron, however, and therefore its mass number is 2. (The name *deu*terium has a root indicating "two.") Tritium contains two neutrons and therefore has a mass number of 3. Its name (*tri*tium) contains a root indicating this number. The symbols for these isotopes are

$$^1_1H \text{ or } H \qquad \text{and} \qquad ^2_1H \text{ or } D \qquad \text{and} \qquad ^3_1H \text{ or } T$$

QUESTION 7.2 (*a*) How would you write the chemical formula for *deuterated* water? (*b*) for *tritiated* water?

ANSWER 7.2 (*a*) D_2O or ^2H_2O; (*b*) T_2O or ^3H_2O.

7.2 | Water with Water

Now let us focus our attention on the oxygen atom in water. The uniqueness principle (Chapter 3) tells us that, because of its small size, oxygen should have properties that are not representative of its column of elements on the periodic table. In Figure 7.6, we see the boiling points of four molecules having the formula H_2X, where X represents oxygen and the three elements under it in the periodic table—sulfur, selenium, and tellurium. We should expect to see that the more massive X gets, the more difficult it is to get H_2X to boil. In other words, the larger a molecule is, the more reluctant it is to leave the liquid phase and become a gas. (We shall look more closely at the phenomenon of boiling shortly.) Figure 7.6 shows exactly this trend for H_2S, H_2Se, and H_2Te: the boiling points get steadily higher as you make X larger. Now look at water. The trend line based on the elements S through Te predicts that the boiling point of water will be about $-120\,°C$! Lucky for us, it is $100\,°C$, a much higher value. Clearly, the molecule H_2X is very different when X is oxygen.

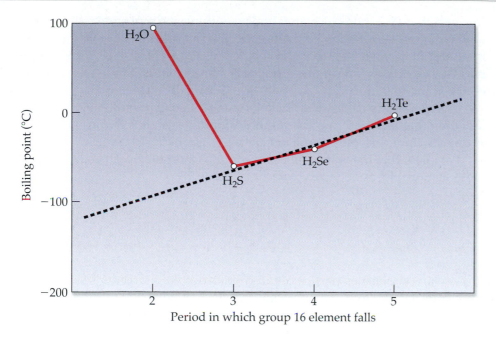

◀ **FIGURE 7.6   The uniqueness principle revisited.**   This graph plots the boiling points of four substances containing atoms from group 16 of the periodic table: oxygen, sulfur, selenium, and tellurium. The black trend line shows that if it were not for the uniqueness principle, water would boil at about $-120\ °C$.

How can water have a boiling point 220 °C higher than what it should be based on the periodic trend? To understand the answer to this question, you have to remember that when we are talking about boiling points, we are concerned with the interactions *between* molecules, as opposed to the covalent interactions *within* a water molecule that hold the one oxygen and two hydrogen atoms together. This is an important distinction because the covalent bonds within the molecule are much stronger than the noncovalent interactions (Section 4.3) that weakly bind water molecules to one another. The energy required to break the O–H covalent bond in a water molecule is 467 kJ/mol, whereas the energy needed to break a noncovalent interaction—such as one between two water molecules—ranges from about 3 to 86 kJ/mol, depending on the type of noncovalent interaction (and there are many types).

We have already seen examples of noncovalent interactions. For example, in Chapter 2, we examined ionic bonds, which exist because of the attraction between two oppositely charged ions. Because two full charges make up an ionic bond, this is the strongest type of noncovalent interaction. In Chapter 4, we looked at the interactions of ions with water. These ion–dipole interactions, which involve an attraction between the full charge on an ion and the dipole of water, are not as strong as ionic bonds because they include one full charge and one dipole, which is only a *partial* charge.

Now let us look at another type of noncovalent interaction. When two polar molecules interact and there is no full charge involved, the interaction is called a **dipole–dipole interaction**. In the molecule HCl, for instance, the two atoms are held together by a strong covalent bond. There is also an interaction *between* HCl molecules, though, because the chlorine atom is more electronegative than the hydrogen atom. Thus, we draw a partial negative charge on Cl and a partial positive charge on H:

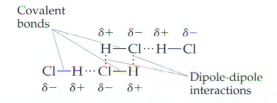

LIVE ART
Dipole-Dipole Interactions

In a group of HCl molecules, the dipoles within each molecule will align in a way that maximizes the interaction between opposite charges. Thus, the bond *within* each HCl molecule is covalent, but the interaction *between* any two HCl molecules is a dipole–dipole interaction, which is one type of noncovalent interaction.

In general, the energy required to break a dipole–dipole interaction is less than that required to break an ionic bond, which has two full charges. A typical dipole–dipole interaction energy is about 9.3 kJ/mol.

We know that the water molecule contains a dipole because the oxygen atom is more electronegative than the H atom:

$$\overset{\delta-}{:}\ddot{O} \overset{H^{\delta+}}{\diagup} \underset{H^{\delta+}}{\diagdown}$$

Thus, water molecules participate in dipole–dipole interactions with other water molecules. However, these interactions between water molecules are comparatively strong, on the order of 25 kJ/mol, much higher than the dipole–dipole average value given earlier (9.3 kJ/mol). Let us see why this is the case. Oxygen's position near the top of the periodic table means an oxygen atom has only a few electrons (eight, to be exact). Its position on the right side of the table means that those electrons are closely held to the nucleus. This accounts for the oxygen atom's small size and its high electronegativity, a concept discussed in Chapter 2.

Recall that a highly electronegative atom, such as oxygen, making a covalent bond to a less electronegative atom pulls electrons in the bond strongly toward itself. Thus, the four electrons involved in bonds in a water molecule are situated closer to the oxygen than to the hydrogen. In other words, each hydrogen atom's sole electron is pulled away from the H part of the water molecule. When hydrogen's only electron is pulled toward the oxygen atom, that leaves the hydrogen with only a proton (and possibly a neutron or two, depending on the isotope). Thus, the hydrogen has a partial positive charge, and this charge is left sticking out, exposed. The partial positive charge on the hydrogen atoms of a water molecule is significant because the low electron density around each H does not shield it from the environment:

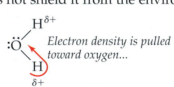

Electron density is pulled toward oxygen...

...leaving the opposite side of the hydrogen atom with an exposed positive charge.

In a glass of pure water, each water molecule is surrounded by other water molecules. Two neighboring water molecules interact via an attraction between the exposed partial positive charge on one of the hydrogen atoms in one water molecule and one of the nonbonding electron pairs on the oxygen of the other water molecule:

Hydrogen bond

This is a special type of dipole–dipole interaction called a **hydrogen bond**. It is stronger than a traditional dipole–dipole interaction because of the substantial partial positive charge on the hydrogen atom and the strong pull of the oxygen atom with which it interacts. Now we can understand the trend in Figure 7.6, which demonstrates the power of hydrogen bonding. Of the molecules in this graph, only water is capable of hydrogen bonding because it is the only one of the four molecules with a highly electronegative atom attached to a hydrogen atom.

This special kind of noncovalent interaction makes the attraction between water molecules *quite strong*. Hydrogen bonds are also unique because the small hydrogen and oxygen atoms can approach each other closely and can therefore interact more effectively than is typical for dipole–dipole interactions. Consequently, it takes a higher temperature to break those interactions, and therefore water's boiling point does not follow the periodic trend for H_2X molecules. As we shall see, to boil water requires a great deal of energy because hydrogen bonds must be broken before boiling can occur.

An extended network of hydrogen bonds forms in liquid water, as shown in Figure 7.7. Do hydrogen bonds form exclusively between water molecules, though? Can they also exist between a water molecule and some other type of molecule, or even between two molecules, neither of which is a water molecule? The answer is that hydrogen bonds can exist between any molecules that contain hydrogen bound to either oxygen, nitrogen, or fluorine, which are among the most electronegative atoms in the periodic table. For example, ethanol has the structure

Ethanol

Because ethanol has a very electronegative atom (oxygen) bound to a hydrogen atom, the hydrogen atom has nearly all of its electron density removed from it, and its exposed partial positive charge interacts strongly with a nonbonding pair of electrons on an electronegative atom in another molecule. Thus, ethanol molecules can interact with one another in much the same way water molecules do:

Ethanol molecules

Ethanol molecules can also interact with water molecules:

Ethanol and water

Thus, we can write a generic hydrogen bond as

> **RECURRING THEME IN CHEMISTRY** The individual bonds between atoms dictate the properties and characteristics of the substances that contain them.

LIVE ART
Hydrogen Bonding

▼ **FIGURE 7.7   Hydrogen bonding in water.**   The polar water molecules align to form hydrogen bonds in a sample of water. Note that hydrogen bonds (between water molecules) are indicated with dashed lines, whereas covalent bonds (within a given molecule) are indicated with solid lines. A ball-and-stick rendering of these hydrogen bonding interactions is shown at the bottom.

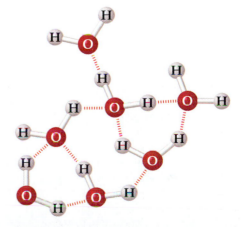

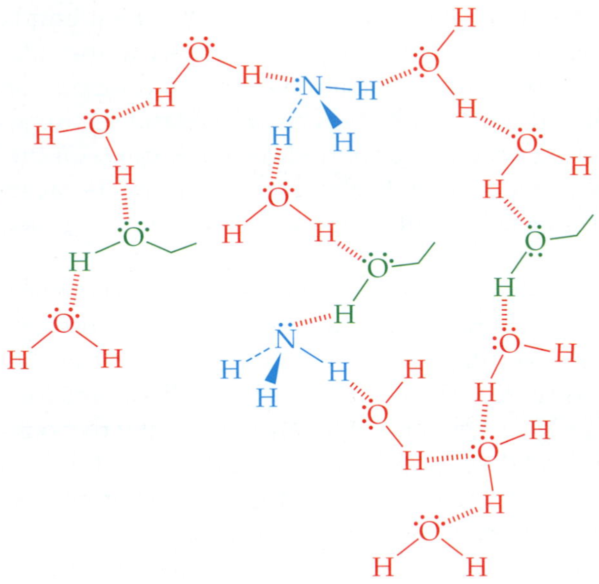

► FIGURE 7.8 **Hydrogen bonding in a mixture of polar liquids.** A network of hydrogen bonds forms in a solution containing water (red), ammonia (blue), and ethyl alcohol (green) molecules. If you need help with the dashed lines and wedges in the ammonia molecules, go back to Chapter 3 and review Question 3.2.

In general, the criterion for the formation of a hydrogen bond is the presence of an —X–H bond in one molecule and the presence of another molecule containing an X atom, where X is usually N, O, or F.

Figure 7.8 shows the hydrogen bonds between molecules in a mixture of water, ammonia, and ethanol. Notice that hydrogen bonds may form between molecules containing different X atoms.

The various noncovalent interactions discussed in this section are summarized in Figure 7.9.

QUESTION 7.3 Sketch several dimethylamine molecules, $(CH_3)_2NH$, in water and show the hydrogen bonds between a dimethylamine molecule and a water molecule, two dimethylamine molecules, and two water molecules.

ANSWER 7.3 Hydrogen bonds are indicated by the short red vertical dashes:

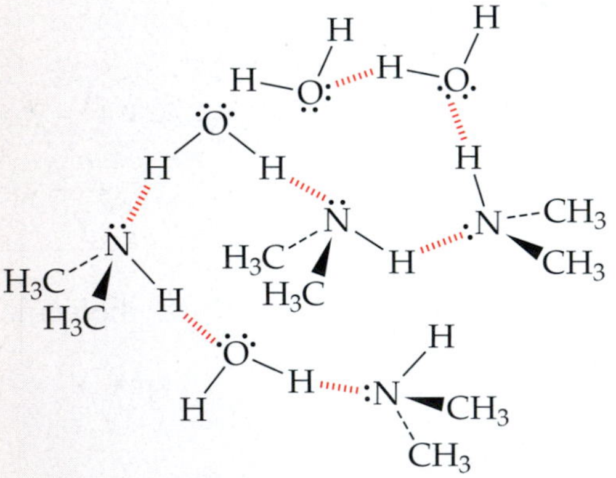

QUESTION 7.4 Do you think it is possible for a hydrogen bond to form within a molecule, such as this one?

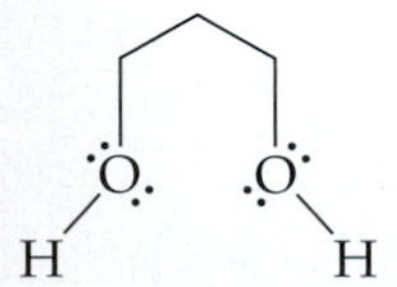

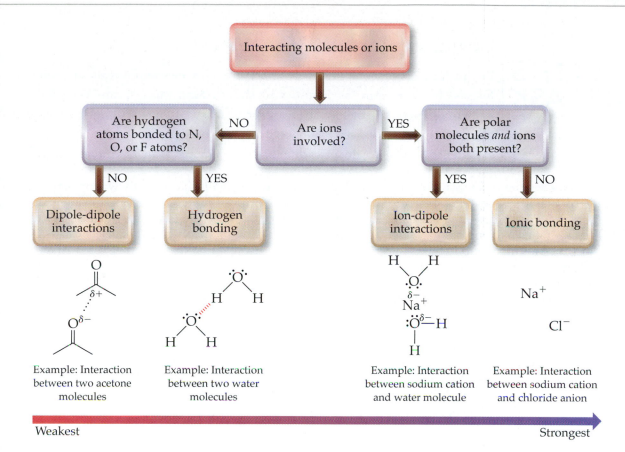

▲ FIGURE 7.9 Noncovalent interactions. This flowchart presents a summary of the four kinds of noncovalent interactions discussed in the text: dipole–dipole interactions, hydrogen bonding, ion–dipole interactions, and ionic bonding. Dipole–dipole interactions are the weakest of the four types, and ionic bonding is the strongest.

ANSWER 7.4 Yes. This molecule happens to contain both groups required for hydrogen bonding, H–X and X, where in this case X stands for an oxygen atom:

$$H\!-\!\underset{\cdot\cdot}{\overset{\cdot\cdot}{O}}\!:\text{''''''}H\!-\!\underset{\cdot\cdot}{\overset{\cdot\cdot}{O}}\!:$$

QUESTION 7.5 Which of the following organic molecules are capable of forming a hydrogen bond with a second molecule of the same compound?

(a)

(b)

(c)

(d)

(e)

(f)

ANSWER 7.5 Molecules *c*, *d*, and *f* because they contain either an O–H or an N–H bond.

STUDENTS OFTEN ASK

How Can Insects Walk on Water?

Except for a comic book hero here and there, humans cannot walk on water. Insects, however, can, as Figure 7.10 shows. Many insects are very light and have a high ratio of surface area to body mass. Moreover, although everything on Earth is subject to the pull of gravity, organisms with less mass experience a smaller downward pull from gravity than do more massive organisms (like people). For a human placing a foot on a water surface, the pull of gravity is so large that the strength of noncovalent interactions between water molecules on the water's surface pales in comparison. Thus, the foot breaks the surface and becomes submerged. For an insect, however, the gravitational force pulling it down is closer in magnitude to the forces between water molecules on the surface, which we refer to as **surface tension**.

We know how a water molecule interacts with its neighbors in three dimensions, but what happens to water molecules on the liquid surface? In a sample of water, every molecule (except those in the surface layer) is pulled in all directions as a result of its interactions with surrounding molecules, as the lower set of arrows in Figure 7.11 illustrates. For water, the most important type of interaction is hydrogen bonding, and hydrogen bonds are pretty strong. At the surface, though, the molecules are not pulled upward by any overlying water molecules, as the top set of arrows in Figure 7.11 shows. Instead these surface molecules are pulled only sideways and downward. As a result, the forces that would pull water in all directions are now distributed in fewer directions, making these interactions that much stronger. The result is that the surface of water is like a tightly woven blanket, and if an insect has a low mass, the surface tension holding the surface of the water together is stronger than the force of gravity acting on the insect. This insect can walk on water.

QUESTION 7.6   No other liquid has high surface tension the way water does. Acetone, for instance,

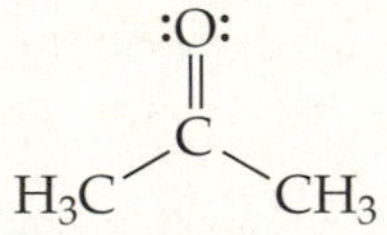

$$:\!O\!:$$
$$\|$$
$$C$$
$$H_3C \qquad CH_3$$
Acetone

has a low surface tension. (*a*) Is hydrogen bonding possible between two acetone molecules? (*b*) Would you expect an insect to be able to walk on the surface of acetone?

ANSWER 7.6   (*a*) No, because the acetone molecule contains no O–H or N–H bonds. (*b*) It is difficult to say, because we are not told anything about the possible interactions between acetone molecules. But because acetone does not hydrogen bond, we can guess that an insect would be more likely to be able to walk on water than on acetone.

▲ **FIGURE 7.10**   **Surface tension.**   Insects can sit on the surface of a liquid that has a high surface tension.

▲ **FIGURE 7.11**   **The source of surface tension.** When the molecules on the surface of a liquid are pulled sideways and downward, but not upward, the surface is like a tightly woven blanket.

7.3 | Solid Water

Now that we know something about how water binds to itself in the liquid phase, let us consider how water changes to the solid phase as the temperature is lowered. In Figure 7.12, notice how some of the water molecules take part in several hydrogen bonds. In liquid water, this network of hydrogen bonds is arranged willy-nilly. Discernable patterns within liquid water, if they exist at all, are short-lived. In the structure of solid water, or *ice*, however, the hydrogen bonds are arranged in a very specific hexagonal pattern:

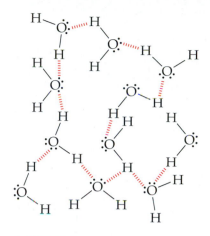

▲ FIGURE 7.12 **Hydrogen bonds in liquid water.** The arrangement of hydrogen bonds in a sample of liquid water is random and short-lived. These interactions are continuously forming and breaking, giving water its fluid nature.

If you extend this structure into three dimensions, you get the arrangement shown in Figure 7.13. Ice has a three-dimensional crystal lattice based on this repeating hexagonal motif, and the structure of ice has a lot of free space, making its *density* less than that of liquid water.

The mass of ice or water, or any other substance, squeezed into a given volume is known as the **density** of the substance. Density units contain a mass unit divided by a volume unit, and common examples are milligrams per milliliter (mg/mL), kilograms per liter (kg/L), and grams per milliliter (g/mL). Because of the spaces that exist in ice, liquid water has a greater density (1.00 g/mL) than does ice (0.92 g/mL). In other words, if you measure the mass of 100.0 mL of water and then measure the mass of the same volume of ice, you will find that the mass of the ice is slightly less (Figure 7.14). The variation in mass—0.08 g per milliliter of water— is significant, and the world would be a much different place were it not for this difference. In winter, for example, the water in a pond or river freezes at the surface first, and because ice is less dense than water, that surface ice does not sink to the bottom. The layer of ice acts as an insulator for the water below, allowing aquatic organisms to survive through the winter.

It is reasonable to assume that water is one of the best understood molecules around, for scientists have been studying it for centuries. However, the processes of freezing water to make ice and melting ice to make liquid water are complicated, and the computing capacity required to simulate the movements of a group of water molecules is daunting. Therefore, it was not until very recently that freezing was studied using computational methods that track every interaction in a group of virtual water molecules. In the mid-1990s, physicists Iwao Ohmine, Masakazu Matsumoto, and Shinji Saito at Nagoya University in Japan began a simulated freezing experiment

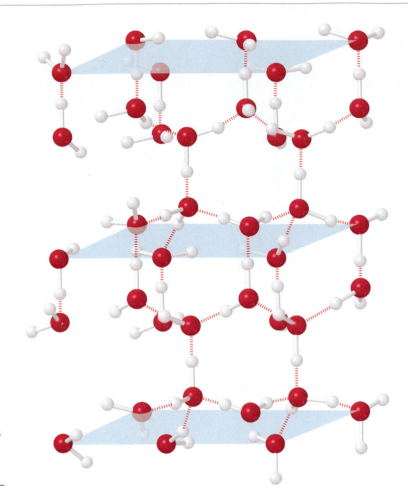

► FIGURE 7.13 **Hydrogen bonds in ice.** The H_2O molecules in ice form a hexagonal lattice held together by hydrogen bonds. Formation of a huge number of hexagons in the lattice creates a rigid structure. The formation of the lattice creates empty spaces within the ice crystal.

▼ FIGURE 7.14 **The densities of ice and water.** The beaker on the left contains exactly 100 mL of liquid water. The beaker on the right contains exactly 100 mL of ice. The fact that the liquid water presses down more on the scale tells you that the 100-mL volume of liquid water has a greater mass than the 100-mL volume of ice.

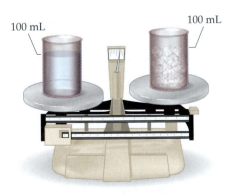

using a supercomputer and 512 virtual water molecules. Then they waited (Figure 7.15a). Six years later, they had the first data on how water freezes, and what they found was very interesting.

In liquid water, hydrogen bonds between molecules are being made and broken all the time, as noted earlier. The Japanese computational experiment was designed to study the behavior of water molecules when they reach their **freezing point**, the temperature at which any liquid becomes a solid. When we think about freezing point, we are imagining a liquid turning into a solid. You can also think of it from the opposite perspective, though. **Melting point** is the temperature at which a solid becomes a liquid. These two terms describe the same temperature, but that temperature is approached from opposite directions.

When the virtual molecules were cooled to just below water's freezing point, 0 °C, they remained randomized, the way they are in liquid water, for about 250 ns. During this brief time, however, a few molecules in this nearly frozen water formed long-lasting hydrogen bonds. The presence of these first few joined-together molecules acted as a **nucleation center** for the addition of more water molecules via hydrogen bonds, and the hexagonal lattice began to form over the next 40 ns (Figure 7.15b). Then, all at once, after 320 ns, freezing happened as the water molecules in the liquid locked into the hexagonal structure, with all the molecules then forming hydrogen bonds to one another.

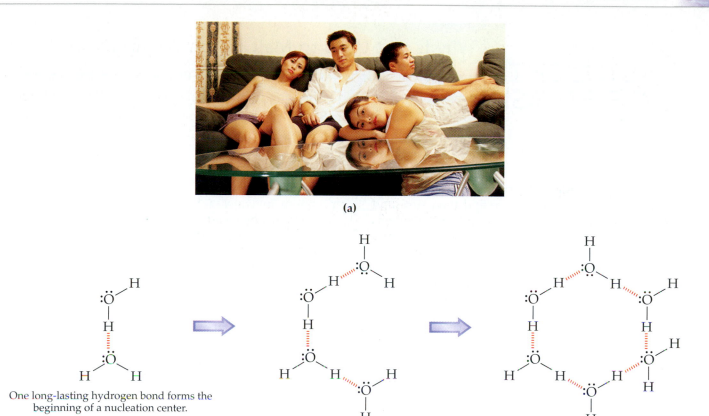

One long-lasting hydrogen bond forms the beginning of a nucleation center.

More water molecules join in the center.

(b)

The hexagonal lattice of ice forms quickly.

▲ **FIGURE 7.15   Hurry up and wait.   (a)** Elegant scientific experiments may require unlimited patience. **(b)** Nucleation centers form as ice freezes when long-lasting hydrogen bonds form between adjacent water molecules.

That this virtual experiment took six years to complete using a supercomputer seems unimaginable. This shows that studying a seemingly small collection of 512 molecules interacting with one another can present an exceedingly complex three-dimensional problem. These findings were big news in the scientific community, where results from very long computations such as this one are reported only occasionally.

A similar process occurs in rain clouds, in which there is a high concentration of water vapor just waiting to become rain. Water molecules in the water vapor form a nucleation center, the first step in the creation of tiny ice crystals. (Yes, it is strange, but true: both rain and snow begin as ice crystals.) As more and more water molecules adhere to this nucleation center, the collection of molecules eventually becomes heavy enough to fall toward Earth. If the temperature of the air near the ground is above freezing, the ice melts and turns to rain on its trip down. But if the air temperature is below freezing, the result will be sleet or snow.

Sometimes rain clouds form but produce no rain because the frequency of nucleation of ice crystals is very low. In these cases, farmers anxious for moisture wait in vain as dark storm clouds float by and no rain comes. When airplane flight became possible, a solution to this problem became clear: speed up nucleation by dropping something onto the clouds. In the process of *cloud seeding,* an airplane drops fine crystals of silver iodide into

STUDENTS OFTEN ASK

Why Is Salt Used to Melt Ice on Wintry Roads?

A salt spread on an icy road surface can cause the ice to melt even when the temperature is below the freezing point of water, but why? We know from Chapter 4 that water molecules surround cations and anions as any salt dissolves in liquid water, and pretty much the same thing can happen to water in its solid phase. When a salt such as sodium chloride is spread on an icy road, the water molecules that make up the hexagonal lattice in ice are pulled away from the lattice and then surround the cations and anions as they dissolve. This happens because the ion–dipole interaction between the salt and water is stronger than the hydrogen bonds holding the water molecules in the ice together.

Thus, even when the temperature is below freezing, water molecules from the ice on the road enter the liquid phase to take advantage of strong interactions with the ions of the salt. This disrupts the crystalline structure of the ice and replaces that structure with a disorganized liquid solution of the salt. The result is a briny, slushy mixture of salt water on the road surface, which provides better traction for cars. The salting of roads is not effective at extremely cold temperatures because eventually even salty water will freeze. Thus, when the mercury drops below about 15 °F, salting icy roads will not improve driving conditions.

QUESTION 7.7 Calcium chloride is often used rather than NaCl to melt icy roads. (*a*) Write a chemical equation to show how solid calcium chloride dissolves in water. (*b*) From what you know about interactions between ions and dipoles, why might calcium chloride be a more effective de-icer than sodium chloride?

ANSWER 7.7 (*a*)

$$CaCl_2(s) \xrightarrow{\ H_2O\ } Ca^{2+}(aq) + 2\,Cl^-(aq)$$

(*b*) Two reasons. First, calcium ions have a 2+ charge rather than the 1+ charge on sodium ions. The greater the charge, the more avidly the cation holds onto water molecules surrounding it. Second, when calcium chloride dissolves in water, it releases three ions, one calcium cation and two chloride anions. Sodium chloride releases only two. The more ions released, the more water molecules can be pulled away from the ice lattice and into solution, thereby lowering the freezing point even more.

▼ **FIGURE 7.16 Rainmakers.** These cloud-seeding planes drop silver iodide crystals into clouds to promote precipitation.

storm clouds (Figure 7.16). The crystals provide a surface onto which water molecules bind. As the molecules bind, a crystalline lattice of ice forms and an artificial ice nucleation center is created. As artificially nucleated ice crystals grow larger, they fall to Earth as rain or snow, depending on the temperature. In this way, farmers can tap the water in storm clouds that might otherwise have passed them by.

QUESTION 7.8 Chemists use an old trick to entice crystals to form in a saturated solution: they scrape the sides of the beaker vigorously with a glass rod, knocking tiny shards of glass into the solution. With any luck, crystals then form quickly. Speculate on how this technique can cause crystals to form.

ANSWER 7.8 The glass shards act the way silver iodide crystals do in clouds—they provide a surface to which molecules can bind and crystals can nucleate.

In a hurricane that is forming, ice-crystal nucleation occurs at the top of the brewing storm, where the temperature is lowest. This nucleation helps escalate the hurricane's intensity. Hurricanes are much more than just heavy rain, however. They produce strong winds that begin to swirl, producing the familiar hurricane eddy seen in satellite photographs. How does the formation of ice crystals contribute to the strong winds that characterize a hurricane? Consider for a moment the freezing of water and the melting of ice. One requires an input of energy, and the other produces heat, but which is which? If you want to melt a block of ice, you must add energy in the form of heat:

$$H_2O(s) + heat \longrightarrow H_2O(\ell)$$

Thus, if you want to freeze water, this process must go in reverse:

$$H_2O(\ell) \longrightarrow H_2O(s) + heat$$

In the top of a mounting hurricane, the second reaction takes place as ice crystals form nucleation centers and produce enormous amounts of heat. This heat causes the air to expand, and the expansion of the air in turn causes winds to rise. Thus, ice-crystal formation is responsible, in large part, for the devastating winds that accompany hurricanes.

In Figure 7.17, the number of billion-dollar natural disasters in the United States during the last two decades of the twentieth century is broken down by state. Hurricanes, which are most likely to occur in the southeastern part of the country, are the most expensive natural disasters. Figure 7.18a shows a 1992 satellite photograph of Hurricane Andrew, one of the most expensive natural disasters in the history of the United States. The hurricane began as a tropical storm off the western coast of Africa and built up energy as it moved over the Atlantic Ocean. By the time it reached the Florida coast, Hurricane Andrew was packing tides 17 ft high and winds in excess of 120 miles per hour! It caused 26 billion dollars in damage, took 26 lives, and destroyed more than 25,000 homes (Figure 7.18b).

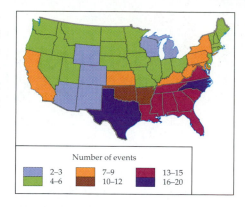

▲ FIGURE 7.17 Destructive nature. A map showing the number of billion-dollar disasters by state between the years 1980 and 2001.

▼ FIGURE 7.18 Hurricane Andrew. (a) In this satellite image taken on August 23, 1992, the eye of the hurricane is clearly visible. **(b)** Hurricane Andrew was one of the most costly natural disasters in United States history.

(a)

(b)

Can It Really Be Too Cold to Snow?

When it is *really* cold outside, you may hear people saying that it is too cold to snow. This saying is true. At very low temperatures, water "prefers" to stay as a solid, and very little of it enters the air from rivers, streams, and lakes. Less water vapor in the atmosphere means fewer clouds, fewer clouds mean the number of ice crystals forming in clouds is much lower, and those crystals are required in order for snow to form. When the temperature is warmer, but still below freezing, there is more water vapor present in the air, ice crystals are able to form from the water vapor, and these crystals adhere together and fall out of the sky as snowflakes.

7.4 | Shifting Phases

Let us do a virtual experiment of our own. Imagine you are an amateur chemist living in a bitterly cold and inhospitable climate. You have a glass of water at room temperature (about 25 °C) and a fairly accurate thermometer. You record the water temperature in your 25 °C kitchen and then take glass and thermometer outside, where the temperature is −25 °C. You carefully monitor the water's temperature as it drops to the freezing point, 0 °C. Then, a peculiar thing happens. The temperature stops dropping at the freezing point and stays exactly at 0 °C as ice crystals begin to form. Once the water has completely solidified into ice and the thermometer is held rigidly in place, the temperature once again begins to drop. What is going on here?

The phenomenon you just imagined happens with all liquids, not just water. It happens in the reverse direction, too. If you take your glass of ice back inside, the temperature begins to rise, but it ceases rising when the melting point (0 °C) is reached. Until all the ice has become water, the temperature stays constant at 0 °C. Then, once the ice is completely melted, the temperature begins to rise again until it reaches room temperature. What you are observing is related to the way molecules absorb and release energy.

The water molecules in ice are held in their hexagonal lattice by rigid hydrogen bonds. As the temperature increases toward the melting point, 0 °C, the added heat energy makes the molecules vibrate more and more vigorously. At the melting point, the available heat energy is used not to make the molecules vibrate more vigorously, but rather to break the hydrogen bonds between molecules. In fact, all the energy put into the water at the melting point goes to breaking hydrogen bonds. Now, the water molecules begin to exchange hydrogen bonding partners as they do in liquid water. In other words, all the energy absorbed is used to change the ice to liquid water, and there is no heat energy available to raise the temperature of your sample.

A diagram of this process is shown in Figure 7.19. At the lower left corner, where the temperature scale begins at −25 °C, the temperature rises as heat is added to the ice. When the melting point is reached, the line becomes flat. At this point, heat continues to be added, but the temperature does not increase. This is the phenomenon that you observed when you

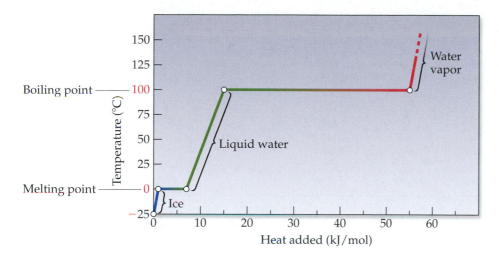

◀ FIGURE 7.19 **The three phases of water.** This diagram shows the phase changes water goes through as heat is added—from ice to liquid water to water vapor.

took your glass of water back inside. When melting is complete, the temperature begins to rise again because the heat absorbed goes toward increasing the motion and vibration of the molecules of water. When the boiling point is reached, the same phenomenon occurs again as the water starts to change from liquid water to water vapor: the temperature stays constant until every liquid water molecule has become water vapor. Once all molecules are in the gas phase, the temperature once again starts to rise, and we reach the upper right corner of the diagram.

There are three other features of Figure 7.19 you should notice. First, each horizontal line can be considered from two perspectives. The horizontal line at 0 °C represents melting if heat is being added (which means moving from left to right along the horizontal axis) and freezing if heat is being removed (moving right to left along the horizontal axis). Likewise, the horizontal line at 100 °C represents **evaporation** (phase change from liquid to gas) if heat is being added (moving left to right) and **condensation** (phase change from gas to liquid) if heat is being removed (moving right to left).

The second feature to notice in Figure 7.19 is that the two horizontal lines are of different lengths. Why is this so, you ask? Because these two lines represent the amount of heat required to change the phase of water either from solid to liquid or from liquid to gas. The longer length of the evaporation–condensation line tells us that, for a given mass of water, the amount of heat energy involved in evaporation/condensation is greater than that involved in melting/freezing. The molecules in ice hold onto one another relatively tightly, whereas the molecules in liquid water hold one another for short times and exchange hydrogen bonding partners. Thus, changing from solid to liquid requires a loosening of the bonds in the solid. The change from liquid to gas is much different because water must change from the liquid phase, where hydrogen bonding is extensive, to the vapor phase, where each molecule is completely separated from every other water vapor molecule. Therefore, evaporation/condensation requires much more heat because the molecules must completely relinquish all of their noncovalent interactions.

The third thing to notice in Figure 7.19 is that the blue line representing increasing temperature in the ice is steeper than the green line representing increasing temperature in the liquid water. The steepness of these lines tells us how much the temperature increases for a given input of heat. The steeper the line, the larger the temperature change. Thus, you can imagine

TABLE 7.2   Specific Heats for Several Substances

Substance	Specific heat (J/g·°C)
Air (dry)	1.01
Aluminum	0.902
Copper	0.385
Gold	0.129
Iron	0.450
Mercury	0.140
NaCl	0.864
Water(s)*	2.03
Water(ℓ)	4.179

*At =11 °C

measuring the temperature of a glass of ice and the temperature of the same volume of liquid water. If you add the same amount of heat energy to each, the temperature of the ice will increase more.

Looking at it from the opposite perspective, the steepness, or slope, of the blue and green lines tells us how much heat can be absorbed by ice or water for a given increase in temperature. Chemists define this quantity as **specific heat**, the amount of heat absorbed by 1 g of a substance as its temperature increases by 1 °C. The specific heats of several substances are shown in Table 7.2. Notice that the value for ice is much less than that for water, which reconciles our observations about line steepness in Figure 7.19.

The fact that liquid water has a relatively high specific heat has important consequences for our daily lives. If you live near an ocean or a large lake, you may have noticed that the temperatures in your area tend to fluctuate much less than in areas without a large body of water nearby. Take a look at Figure 7.20, which shows a map of the United States and southern Canada, along with data on rainfall and average temperature fluctuation (average high temperature minus average low temperature). Examine the temperature fluctuations first. Do you observe any correlation between fluctuation and location? Your answer should be yes. The temperature fluctuates little in cities located on an ocean or on one of the Great Lakes and varies more widely in cities that are landlocked. This difference can be attributed to the very high specific heat of liquid water. Cities close to the water enjoy more moderate weather and maintain a more constant temperature because the water absorbs and holds heat easily, and this keeps temperatures closer to one average value. (In other words, there is less fluctuation in temperature.)

The temperature fluctuations for landlocked cities vary from 21 °C to 34 °C. Why do some landlocked cities have much larger fluctuations than others? The answer can be found by looking at the average rainfall data. For cities with rainfall data in the pink range, the average June rainfall is quite high. This means that in those cities in June the air contains a great deal of humidity. The water held in the air maintains the temperatures in humid areas in the same way a nearby body of water maintains temperature. The cities coded green in the average-rainfall column have very low rainfall, and so these cities experience the most drastic temperature fluctuations. Thus, you can expect cities in dry areas, such as the desert southwest, to have wider extremes of temperature than landlocked cities located in humid areas.

The extraordinarily high specific heat of water also plays a role in keeping humans comfortable. Our bodies are mostly water, and so we are able

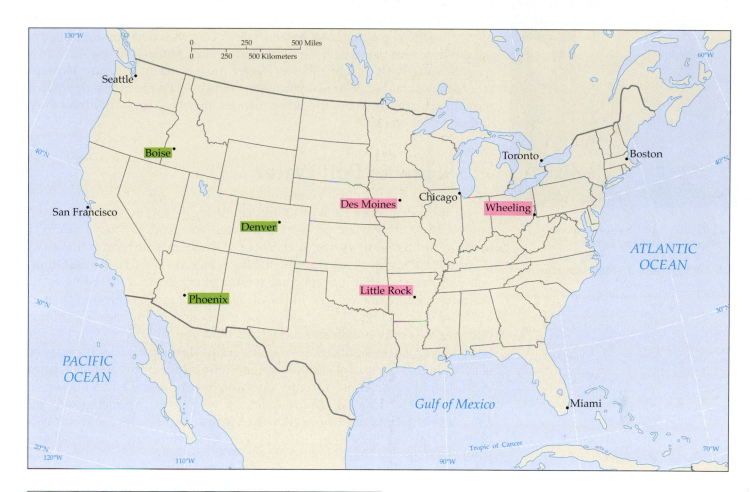

City	Average temperature fluctuation in June (°C)	Average June rainfall (in)
Landlocked		
Boise, Idaho	26	0.81
Denver, Colorado	34	1.79
Des Moines, Iowa	21	4.46
Little Rock, Arkansas	21	7.85
Phoenix, Arizona	27	0.13
Wheeling, West Virginia	21	3.59
On lake or ocean		
Boston, Massachusetts	18	
Chicago, Illinois	19	
Miami, Florida	15	
San Francisco, California	17	
Toronto, Canada	16	

▲ FIGURE 7.20   Average temperature fluctuations and June rainfall in landlocked and waterfront cities.

▲ FIGURE 7.21 **Keeping warm compliments of the Sun.** In this passive solar home, tubes of water absorb heat from the Sun during the day and release it at night.

to maintain a stable body temperature because the water in our bodies can absorb heat and buffer us against changes in temperature. The same principle is at work in passive solar homes, which can provide a comfortable living environment using only sunlight as an energy source (Figure 7.21). In a passive solar home, large tubes of water are placed in a sunny location where they can absorb large amounts of energy when the Sun is up. When the Sun sets, the water in the tubes releases that heat back into the house and provides warmth.

7.5 | How to Boil Water

If you have ever been on a camping trip at high altitude, you may have noticed the effect of air pressure on water. Imagine you have stopped to camp at 18,000 ft during a climb up Denali in south-central Alaska (Figure 7.22). If you are attempting to heat water at 18,000 ft, the first thing you will notice is that it takes considerable effort to get a fire going at an altitude at which the oxygen in the air is much less concentrated than at sea level. As you unpack your stove and try to light it, you may also notice that you are out of breath and more tired than you should be. This is your body telling you it wants more oxygen and to slow down. Once the stove is lit, though, your pot of water is boiling in minutes, and you feel a sense of pride and satisfaction. You mix up some hot chocolate and sit down to enjoy the view. To your disappointment, what you expected to be *hot* chocolate is instead only *warm* chocolate. You are quite sure the water was boiling, and surely boiling water is the same temperature everywhere, right? The answer is no.

To understand how water can boil at different temperatures, we must first think about how boiling happens in the first place. A pot of water that is becoming hotter and hotter contains water molecules that are vibrating and wiggling. The hotter the water gets, the more energetic the molecules become (Figure 7.23). As the temperature goes up, more and more of the

▶ FIGURE 7.22 **Cold hot chocolate.** If you are ever fortunate enough to climb Denali in Alaska and reach the 18,000-ft point, boiling water will not be as hot as what you are used to at much lower elevation.

Denali summit (20,320 ft)

You are here (18,000 ft)

water molecules get so animated that they have enough energy to escape from the liquid and become water vapor.

You know from everyday life that this happens to molecules at the surface of a liquid as the liquid evaporates. The interesting point for our discussion is that the same thing happens in the water below the surface—some interior molecules become energetic enough to become a gas. When these few molecules become water vapor for an instant, a tiny bubble forms. Then the pressure of the water above squashes the bubble, and it returns to the liquid phase. As the temperature continues to increase, more and more interior water molecules simultaneously leave the liquid phase and enter the gas phase, and a larger bubble forms. The pressure exerted by the gas molecules in these bubbles gets higher and higher as more water molecules take part in their formation.

Bubble formation is not the whole story, though. To understand boiling completely, we must also take the air outside the pot into account because it exerts a constant pressure on the water surface. At high altitudes, this pressure exerted by the air, called *atmospheric pressure*, is lower than the atmospheric pressure at sea level. In our pot of water, eventually a temperature is reached at which the pressure inside a submerged bubble equals the atmospheric pressure pushing down from above. At this temperature, called the **boiling point**, the pressure inside the bubble is equal to the atmospheric pressure, and the bubble stays intact and floats to the surface. This is why water that is about to boil contains small bubbles that appear and disappear. When full boiling begins, the bubbles rise and break the surface.

QUESTION 7.9   Would you expect the boiling point of liquids other than water to be lower at high altitude than at sea level?

ANSWER 7.9   Yes. All liquids become gases when heated to the boiling point. Furthermore, all liquids experience lower atmospheric pressures at high altitudes. Therefore, we should expect the boiling point of any liquid to be lower at high altitude than at sea level.

QUESTION 7.10   Cooks often save time boiling a large quantity of water by dividing the water into two pots and then recombining the water once boiling commences. Why do you suppose this method is faster than boiling the water in one pot?

ANSWER 7.10   It is faster because the heat from the burner flame contacts the entire surface area of the pot's bottom. If you divide the water into two pots, you have more of the water exposed to the heat of the flame.

It is important to recognize that water molecules are still water molecules when they move from one phase to another. They do not change composition: each one is still made up of two hydrogen atoms and one oxygen atom covalently bound together in the now-familiar bent shape. The only thing that changes as we move from gas phase to liquid phase to solid phase is the amount of interaction *among* the molecules. Thus, a group of H_2O molecules in the gas phase do not interact much with one another because they are very far apart. (We shall look closely at the behavior of gases in Chapter 8, Air.) These gas molecules may enter the liquid phase,

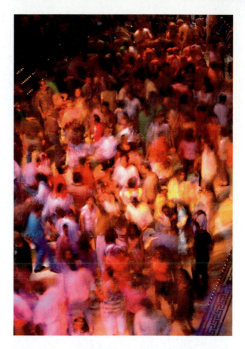

▲ FIGURE 7.23   **Reaching the boiling point.**   The crowd at a rock concert or dance club can get more and more energetic as the evening wears on. We can use this increase in activity as an analogy for the "scene" inside a pot of water being heated on the stove. As more and more heat energy goes into the water, the molecules get more and more energetic.

where they become close neighbors with other water molecules and interact with them via a disorderly network of hydrogen bonds. When the liquid water forms ice, the hydrogen bonds between molecules become more organized, but the molecules are still H_2O molecules.

It is also important to recognize that atoms that make up molecules can be involved in two types of interactions. As we just saw, noncovalent interactions, such as hydrogen bonding and dipole–dipole interactions, are those that occur either between two molecules or between two parts of a large molecule. These types of interactions are called *inter*molecular interactions. (*inter-* means "between.") The bonds *within* a molecule are covalent. They are the result of sharing of electrons between adjacent atoms in a molecule. This type of interaction is called an *intra*molecular interaction (*intra-* means "within.")

Now, we know that the boiling point of a liquid depends on the strength of the intermolecular interactions between the molecules in the liquid, but we do not know why the boiling point changes with altitude. The reason is that, as altitude *increases*, the atmospheric pressure pushing down on the surface of the liquid *decreases* because there are fewer molecules per given volume of air than at sea level. Another way to say this is that the air is less dense at high altitude than at low altitude. In fact, for every 540 ft of altitude *gain*, there is a 1 °C *decrease* in the boiling point of water. At high altitude, the pressure an underwater bubble must reach before it remains unbroken is lower than at sea level. As a result, it takes less time and a lower temperature to boil.

The opposite situation exists at depths far below sea level. At the bottom of the sea, pressures are very high. Underwater thermal vents created by volcanic activity on the ocean floor supply heat to the water at the ocean floor, but that water does not boil until it reaches, in some cases, 400 °C! Scientists who study the ocean floor have found that the plumes of exceedingly hot water that spew from these vents carry with them massive loads of dissolved minerals. The minerals are dark in color, and so these plumes of water are called *black smokers* (Figure 7.24). When the dissolved minerals hit the colder water outside the vents, they quickly cool down and return to solid mineral form, creating tall, hollow, mineral-laden columns called *hydrothermal vents*.

In 2001, an expedition to the middle of the Atlantic Ocean turned up a vast underwater landscape consisting of 30 hydrothermal vents. Now called Lost City, this collection of vents, which has been likened to Bryce Canyon in Utah (but under water), are the tallest yet on record, measuring up to 60 m tall (Figure 7.25). One scientist who was part of the discovery commented that if Lost City were terrestrial, it would be a national monument.

A pressure cooker works by simulating the pressure that exists under water, and this piece of cookware can help us understand the conditions that exist at hydrothermal vents. A typical pressure cooker is very sturdy and has a lid that seals the top tightly (Figure 7.26). When water is heated in the cooker, the pressure inside builds to a level greater than that outside the pot. This increase in pressure increases the pressure of water vapor above the water in the cooker and makes it more difficult for a water molecule to escape from the liquid phase into the gas phase. This, in turn, raises the boiling point of the water in the same way water near the ocean floor boils at high temperature. The elevated temperature cooks any food in the pressure cooker more quickly than it would cook in a regular pot. The

RECURRING THEME IN CHEMISTRY *Intra*molecular *bonds*, those *within* a molecule, are covalent. *Inter*molecular *interactions*, those either *between* two molecules or *between* two parts of a large molecule, are noncovalent *interactions*.

▲ **FIGURE 7.24 Black smokers.** These underwater hydrothermal vents spew out high-temperature water containing high concentrations of dissolved minerals.

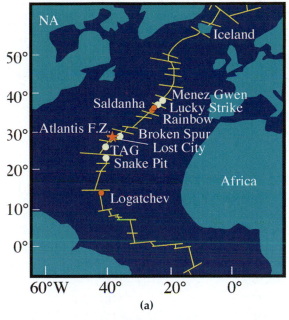

(a)

(b)

(c)

◀ FIGURE 7.25   **Atlantis lost and found?**   **(a)** The hydrothermal vent site called Lost City sits on the floor of the Atlantic Ocean, about halfway between northern Africa and the southern United States. **(b)** Researchers have described Lost City as looking like the landscape we can see in Bryce Canyon in Utah. **(c)** One of the vents from Lost City. These stone formations can reach heights of 60 m, about two-thirds the length of a football field.

increased pressure inside the cooker also helps infuse the food with water, keeping it moist and tender.

A summary of water's phase changes is given in Figure 7.27.

▼ FIGURE 7.26   **Cooking under pressure.**   Pressure cookers are sturdy containers built to withstand high internal pressures.

QUESTION 7.11   Knowing that a pressure cooker raises the temperature at which water inside the sealed vessel boils, can you think of a way to *lower* the boiling point of water in that same vessel?

ANSWER 7.11   To lower the boiling point, you must draw a vacuum to remove molecules from the air above the liquid. When you draw a vacuum, the pressure above the solution is lowered, and boiling occurs at a lower temperature, just as it does at high altitude. Chemists often take advantage of this technique to boil a liquid, such as water, that has a high boiling point.

Does Adding Salt to Water Make Pasta Cook More Quickly?

If you are a veteran Italian cook, the practice of throwing a handful of salt into the water being boiled for pasta is second nature. As we know, when salt encounters water, the water makes a shell of hydration around each cation and anion. A pot of boiling water will have a slightly elevated boiling temperature because the water molecules are involved in strong ion–dipole interactions with the ions. As a result, the water is more reluctant to leave the liquid phase and enter the gas phase. Therefore, the pot of water must be heated to a slightly higher temperature to make it boil, and the pasta in the pot will cook faster once boiling commences.

The addition of salt has another consequence, however: the time it takes salty water to boil is longer than the time needed for unsalted water to boil. Thus, the time saved in cooking the pasta is lost in the time spent heating the water. In the end, salting the water is not a time-saver. Culinary experts contend that the addition of salt is not about the time it takes to cook pasta, but rather the flavor the salt adds.

▶ **FIGURE 7.27   The phase changes in water.**   As energy is added to water in its solid form, the ice undergoes a phase change (melting) to become liquid water. Additional energy eventually brings about boiling, which changes the liquid water into the gas we usually call water vapor. This process can be reversed if energy is removed from the system, and the gas condenses to a liquid, which can then freeze to form a solid.

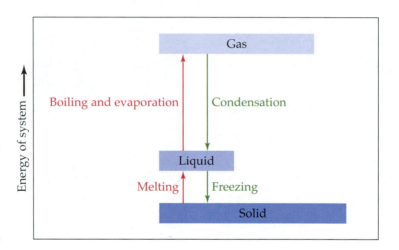

▼ **FIGURE 7.28   Evaporation and condensation at equilibrium.** Water added to a sealed flask evaporates as water molecules move from the liquid to the gas phase. As more and more molecules enter the gas phase, some of them begin condensing back into the liquid phase. Equilibrium is reached when the rate of evaporation equals the rate of condensation.

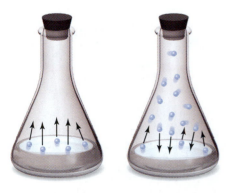

7.6 | From Aerosol Cans to Scuba Diving

We all know from experience that a container of water left open to the air will eventually become dry. This evaporation occurs when water molecules on the surface escape into the air, which contains few water molecules. If you take a flask partially filled with water and seal the top well, as in Figure 7.28, the air above the water will begin to become filled with gaseous water molecules that have escaped from the liquid surface. This process continues until a substantial number of water molecules are in the air above the liquid. As the air above the liquid becomes saturated with water molecules, some of the water molecules condense back into the liquid.

As more molecules evaporate to the gas phase, more condense to the liquid phase, and eventually the rates of these two opposing processes are equal:

$$H_2O(\ell) \underset{\text{condensation}}{\overset{\text{evaporation}}{\rightleftharpoons}} H_2O(g)$$

This double arrow should look familiar to you. In Chapter 4, we introduced the idea of equilibrium, the back-and-forth exchange of one thing for another at equal rates in a chemical reaction. When equilibrium is attained in your sealed flask of water, the air above the water is said to be *saturated* with water vapor. No more water molecules can enter the air unless the temperature of the liquid changes.

How much water is present in saturated air above the liquid surface is determined by the water's **vapor pressure**, defined as the pressure exerted by water vapor in the air above liquid water when the two phases are in equilibrium. Vapor pressure is always determined at a specific temperature. It can be measured for any liquid (not just water).

If you now remove the seal from your flask of water, what happens? Does the equilibrium between evaporation and condensation continue? The answer is no. When there is nothing to contain the water vapor molecules, they simply float away once they escape from the liquid surface. The odds that a water molecule will condense back into the liquid phase are small. When the flask is open to the environment, the evaporation rate is higher than the condensation rate, and eventually all the water evaporates. Condensation back into the flask and a return to equilibrium will occur at an appreciable rate only if the flask is again covered and the pressure of the water vapor above the liquid surface again equals the vapor pressure of the water.

QUESTION 7.12   If you put an uncovered glass of water on a rock in the desert where the temperature is 40 °C and an identical one on a stump in a rain forest where the temperature is also 40 °C, which one will evaporate more quickly? Why?

ANSWER 7.12   The one in the desert because in the desert there is very little moisture in the air. Therefore, water readily leaves the glass and enters the air. Under these hot, dry conditions, the rate of condensation is very small. In the very humid climate of the rain forest, however, there are already many water molecules in the air, some of which condense and enter the liquid in the glass. Because now the rate of condensation is much closer to the rate of evaporation, the water in the glass takes a very long time to evaporate.

The vapor pressure of water is low relative to the vapor pressure of other liquids. This is because the intermolecular forces in the water, which are the result of hydrogen bonding, are very strong. As a result, water is reluctant to evaporate and become a gas. Thus, hydrogen bonding gives water a high boiling point and a relatively low vapor pressure. Many other liquid substances have very high vapor pressures, meaning that these substances evaporate easily. Once a high-vapor-pressure liquid is placed in a sealed container, the molecules of the substance in the gas phase above the liquid surface exert a high pressure on the liquid surface.

As mentioned in our discussion about new-car smells in Chapter 3, we refer to molecules that evaporate easily as being *volatile*. An example of a volatile molecule is butane:

MOLECULE
Butane

Butane

RECURRING THEME IN CHEMISTRY The structure of carbon-containing substances is predictable because each carbon atom almost always makes four bonds to other atoms.

Butane has no capacity for hydrogen bonding because it does not contain any O–H, F–H, or N–H bonds. It is also nonpolar, meaning that its electrons are distributed fairly evenly across its structure. Because butane has no

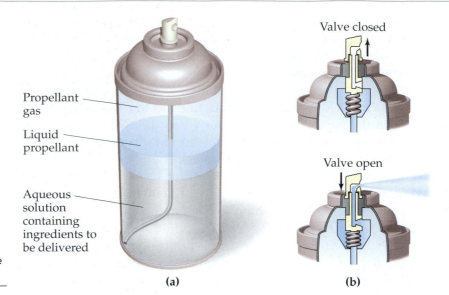

► FIGURE 7.29   Aerosols.
(a) An aerosol can contains a volatile liquid propellant above an aqueous solution. (b) The gas above the liquid propellant pushes the aqueous solution into the dispensing tube when the nozzle is pressed down.

electronegative atoms to upset the even distribution of electrons, it is not able to take part in dipole–dipole interactions with neighboring butane molecules. Thus, the intermolecular interactions between butane molecules are very weak. Because butane molecules interact only weakly with one another, a butane molecule at the top surface of a sample of liquid butane has a high propensity for leaving the liquid surface and entering the gas phase. Because the gaseous butane is readily formed as the liquid evaporates, this liquid is said to have a very high vapor pressure and to be highly volatile.

The fact that butane and other hydrocarbons, too, are very volatile, but water is relatively nonvolatile has been used in the design of aerosol cans (Figure 7.29a). Assume the product you want to get from the can is shaving cream, which is in the form of an aqueous solution. The other substance in the can is the *propellant*. The propellant is a volatile liquid, often a hydrocarbon, that will propel the product from the can. When the can is assembled, the aqueous solution of shaving cream is added through the valve at the top, followed by the propellant. The propellant, because it is volatile, has a very high vapor pressure. Thus, of all of the gas molecules in the space above the liquid propellant, most are propellant and only a few are water.

When you depress the nozzle of an aerosol can, an opening between the outside and inside of the can is made, and the contents can come out (Figure 7.29b). Because the propellant is so volatile, the pressure of the gas presses down on the liquid and pushes everything up the sipper tube. What comes out is a mixture of product and propellant. Because it moves from inside the can (where the pressure is high) to outside the can (where the pressure is lower), the expelled propellant in the mixture readily enters the gas phase, and this phase change causes the aqueous solution of product propelled with it to become foamy or misty. The nozzle, which has a fine hole, also helps create a foam or mist as the mixture is forced through it. As foam leaves the nozzle, empty space is created inside the can. This space immediately fills with gaseous propellant molecules, and pressure is maintained on the liquid inside until it is all expelled.

QUESTION 7.13   (a) Would $CH_3CH_2CH_3$ make a suitable propellant for an aerosol can? (b) Would ethane?

ANSWER 7.13   Because the choices are small hydrocarbons, both would make suitable propellants.

QUESTION 7.14   Why are we advised not to throw aerosol cans into incinerators?

ANSWER 7.14   There is always a small amount of propellant remaining in a spent aerosol can. Because the propellant is highly volatile, it will exist in the gas phase in the nearly empty can.

Recall from Chapter 6 that hydrocarbons react with oxygen in combustion reactions to produce carbon dioxide and water:

$$2\ C_4H_{10} + 13\ O_2 \longrightarrow 8\ CO_2 + 10\ H_2O$$
Butane

Thus, butane and other hydrocarbons, which are often used as propellants, are combustible and may explode if sealed in a bottle and thrown into a source of heat.

Soda cans, like aerosol cans, also use pressurized gas. At the factory, the aqueous mixture containing the soda syrup is placed in the can, and then carbon dioxide gas is forced in. Once the empty space above the liquid has filled with carbon dioxide gas, any more gas added is forced into the aqueous solution. The gas remains in the soda as long as the pressure in the can is maintained. When you open the can, the carbon dioxide gas escapes. Because the gas pressure inside the can is greater than the pressure of the outside air, the gas comes out of the can in an audible burst. Once the high gas pressure above the liquid in the can has dropped, the carbon dioxide gas dissolved in the soda leaves the liquid in the form of bubbles. When the can sits open long enough, the carbon dioxide floats away, and the soda becomes flat.

An understanding of the principles behind dissolving gases in liquids is critical for anyone who does any scuba diving. As a diver's depth increases, so does the pressure on the diver's body. The result is the same as with the soda can: gas under pressure is forced to dissolve in some liquid. In the case of a scuba diver, the gas is nitrogen (the primary component of air) and the liquid is blood, which is mostly water. Thus, a diver's blood is infused with more and more nitrogen as the diver goes deeper and the pressure increases. When the diver decides to resurface, she must do so very slowly because the rapid release of dissolved nitrogen from the blood causes *decompression sickness*. Symptoms of decompression sickness include severe joint pain (referred to as *the bends*), itchiness, brain damage, and heart attack. In fact, decompression sickness can be fatal if a diver resurfaces quickly from an extreme depth. It is treated by putting the victim in a chamber pressurized to the pressure at which the diver was swimming. The pressure is then released from the chamber slowly, and, in some cases, a serious case of decompression sickness can be averted.

7.7 | Like Dissolves Like

The term *universal solvent* makes it seem like water dissolves practically anything. We know, for example, that many salts dissolve well in water by virtue of the ion–dipole interactions between the water molecules and the salt anions and cations. In addition, small, electrically neutral polar organic molecules like ethanol also dissolve in water because the dipole of the organic molecule interacts with the dipole of the water molecule. Furthermore, because ethanol forms hydrogen bonds, it mixes with water in all proportions. We describe molecules like these that dissolve readily in water as **hydrophilic**, which means "water loving."

There are, however, thousands of substances that do not dissolve readily in water. If a molecule does not possess one of the attributes we have been discussing—a charge, a dipole, or the capacity for hydrogen bonding—then it probably does not dissolve in water. Let us take another look at our aerosol can in Figure 7.29a. Notice how the liquid portion of the propellant is not mixed with the aqueous solution of shaving cream. Rather, the liquid propellant sits on top of the aqueous solution and is separated from the water, just as oil (which is hydrocarbon based) separates from vinegar (which is water based) in salad dressing. Because butane, the propellant shown in Figure 7.29, has a lower density than water, the water is on the bottom and the butane is on top.

QUESTION 7.15 Why would it be inadvisable to use an aerosol propellant that has a density higher than that of water?

ANSWER 7.15 Because the propellant should be next to the air space at the top of the can. Therefore, it is preferable to have a less dense propellant so that it sits on top of the water.

Let us look again at butane and two other hydrocarbons:

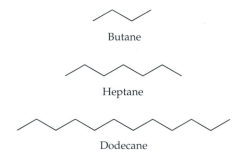

Butane

Heptane

Dodecane

These molecules do not have any full or partial electrical charges, nor do they have the capacity for hydrogen bonding. Clearly, none has a dipole because none contains an electronegative atom, such as O, N, or F. Thus, hydrocarbons do not interact with water in solution, and the result is that they form a separate layer. Two liquids that do not mix are referred to as **immiscible**. As you add carbon atoms to a hydrocarbon chain, the molecule's ability to interact with water decreases. Thus, although some very short hydrocarbons—ethane, C_2H_6, and propane, C_3H_8, are two examples—can mix to a small extent in water, longer hydrocarbons, beginning with hexane, C_6H_{14}, are immiscible.

QUESTION 7.16 Predict which substances dissolve readily in water: (a) $MgCl_2$; (b) ethanol, CH_3CH_2OH; (c) heptane, $CH_3(CH_2)_5CH_3$; (d) hexacosane, $CH_3(CH_2)_{24}CH_3$.

ANSWER 7.16 (a) dissolves in water because it is a salt. (b) dissolves in water because it is small and polar and can make hydrogen bonds. (c) and (d) are hydrocarbons and do not dissolve in water.

The phrase *like dissolves like* tells us that substances taking part in similar kinds of intermolecular interactions tend to dissolve in one another. For example, heptane, the seven-carbon saturated hydrocarbon, does not dissolve in water, but does dissolve in other nonpolar substances, such as melted beeswax. The structure of a wax molecule is shown in Figure 7.30. Notice that it contains two oxygen atoms. Because these two atoms are electronegative, they draw electron density toward themselves. However, the hydrocarbon chains of the molecule are so long that the polarity of the oxygen atoms is lost. Taken as a whole, the wax molecule is nonpolar because the hydrocarbon chains make up its bulk and the oxygen atoms are insignificant in comparison. Thus, beeswax is highly nonpolar and readily dissolves other nonpolar substances, such as heptane and dodecane. We refer to highly nonpolar substances as **hydrophobic** ("water fearing") because they avoid water at all costs.

Let us consider more examples of molecules that might meet the criteria for being either hydrophobic or hydrophilic. Ethanol, CH_3CH_2OH, is an alcohol. As we know from Chapter 5, alcohols are defined by the presence of an —OH group, and so they form hydrogen bonds with water.

Ethanol and water

Because ethanol and water mix together readily, we say that they are *miscible* in one another.

Now let us look at some bigger alcohols. Table 7.3 shows alcohols containing between four and ten carbon atoms, their chemical formulas, and the volume of the alcohol that mixes with 100 mL of water. (All of these alcohols are liquids at room temperature.) According to the table, you can mix up to 9.1 mL of butanol with 100 mL of water. If you add 9.2 mL, then the excess 0.1 mL will not mix. Pentanol is less miscible; you can add only 3.3 mL to 100 mL of water. Hexanol and heptanol are only very slightly miscible with water. The next alcohol, octanol, is completely immiscible with water, as are all the alcohols containing more than eight carbon atoms.

What is the basis of this trend in miscibility? Clearly, all these molecules, because they contain an —OH group, can form hydrogen bonds

RECURRING THEME IN CHEMISTRY Like dissolves like.

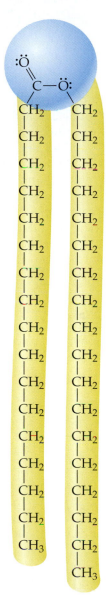

▲ FIGURE 7.30   **A typical wax molecule.**   Both hydrocarbon chains in the molecule are hydrophobic. The OCO portion of the molecule is hydrophilic because of the two oxygen atoms, but any hydrophilic character is overwhelmed by the hydrophobic character. For this reason, waxes are insoluble in water.

Miscible is the term used when discussing how two liquids mix. When we are talking about mixing a solid into a liquid, we use the term *soluble*.

TABLE 7.3	Selected Alcohols and Their Miscibilities in Water		
Name of alcohol	Number of carbons	Chemical formula	Miscibility with water
Butanol	Four	$CH_3(CH_2)_3$—O—H	9.1 mL butanol per 100 mL water
Pentanol	Five	$CH_3(CH_2)_4$—O—H	3.3 mL pentanol per 100 mL water
Hexanol	Six	$CH_3(CH_2)_5$—O—H	0.72 mL hexanol per 100 mL water
Heptanol	Seven	$CH_3(CH_2)_6$—O—H	0.11 mL heptanol per 100 mL water
Octanol	Eight	$CH_3(CH_2)_7$—O—H	Immiscible in water
Nonanol	Nine	$CH_3(CH_2)_8$—O—H	Immiscible in water
Decanol	Ten	$CH_3(CH_2)_9$—O—H	Immiscible in water

Chemistry

CHEMISTRY AT THE CRIME SCENE

Death by Drowning

The traditional method for investigating drowning involves analyzing the water in the lungs to look for *diatoms*, small marine organisms that live in all natural waters. If a person drowns, the water that is inhaled contains diatoms that can be identified by autopsy. However, if the victim was dead before going into the water, then water was not inhaled and fewer diatoms are found in the

lungs (Figure 7.31). Further, the particular species of diatom in a drowning victim can help place the site of drowning because diatom species vary from location to location. This technique has served the law enforcement community well for decades, but there is one significant problem with the test: there are cases in which a dead body thrown into natural water will have diatoms in the lungs because the organisms wander into the body as it sits under water. For this reason, forensic scientists have been searching for other ways to learn more about the deaths of submerged bodies.

One new technique involves analyzing surfactants in the lungs of the deceased. **Surfactants** are amphipathic molecules, and their name derives from the phrase surface active agents. The surfactant molecule shown in Figure 7.32*a* has a very polar end (shown in blue) and two long hydrophobic chains. Chemists often illustrate surfactants and other amphipathic molecules by drawing the polar part, or *head group*, as a circle and the hydrocarbon chains as long tails extending away from it, as shown in Figure 7.32*b*.

Surfactants keep their hydrophobic tails away from water. The inner lining of the lungs is coated with surfactants, with the hydrophobic chains sticking out into space, perpendicular to the lung lining, as shown in Figure 7.33. These surfactants crowd together inside the lung lining, and the hydrocarbon chains pack in firmly next to one another. As we breathe in and out, the lungs expand and contract, and the tight packing of these surfactant molecules is responsible for keeping the lungs from collapsing during exhalation.

Everyone has lung surfactants, and scientists have found that the surfactants in the lungs of a submerged

victim will vary depending on the circumstances of death.

QUESTION 7.17 The shape of surfactant molecules in which the hydrocarbon chains contain cis double bonds is different from the shape of surfactant molecules in which the chains are saturated:

Surfactant molecule with saturated hydrocarbon chains

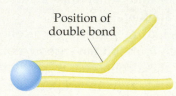

Position of double bond

Surfactant molecule with one cis unsaturated (top) and one saturated (bottom) hydrocarbon chain

Would a surfactant molecule containing cis double bonds in one of its chains make a suitable surfactant for the lining of the lungs?

ANSWER 7.17   No. The surfactants in the lungs work because they are able to pack together very tightly and prevent the lungs from collapsing. If the chains are not straight, they will not pack together closely.

with water. This enhances their solubility. Because alcohols contain both a hydrocarbon chain (hydrophobic) and an alcohol group (hydrophilic), they are referred to as **amphipathic,** a term used to describe molecules for which solubility is dictated by both hydrophobic and hydrophilic portions of the molecule. As you may already suspect, there is an offsetting factor: the length of the hydrocarbon chain. As the hydrocarbon chain gets longer, the —OH group becomes less and less important. The hydrocarbon chain dominates at a length of eight carbon atoms, making octanol (as well as larger alcohols) more hydrophobic than hydrophilic.

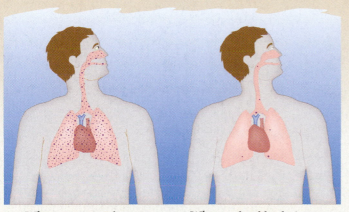

When a person drowns, water is inhaled and many diatoms are found in the body.

When a dead body is put into the water, few diatoms enter the body.

▲ **FIGURE 7.31** **Drowned or not?** The presence or absence of diatoms in the lungs of a body found in the water can provide clues about the cause of death.

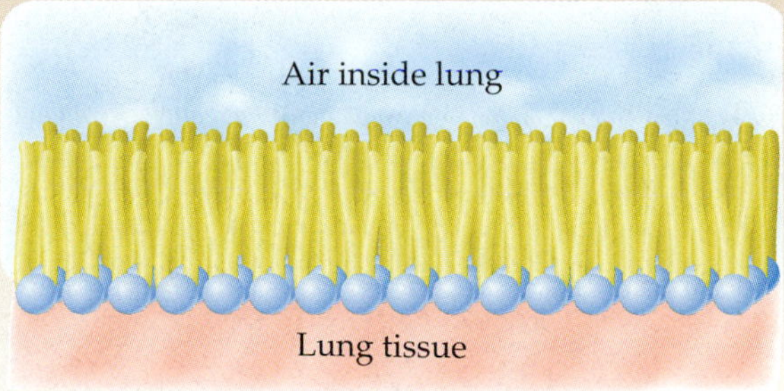

Air inside lung

Lung tissue

▲ **FIGURE 7.33** **Surfactants in the body.** At the surface of the lungs, hydrocarbon chains extend into the lung space and allow the tissue to expand and contract, but not to collapse.

▶ **FIGURE 7.32** **Surfactants.** **(a)** A typical surfactant molecule with the complete structure shown. **(b)** A surfactant molecule is often depicted in cartoon form, with a circle representing the polar head group and wavy tails representing the hydrocarbon chains.

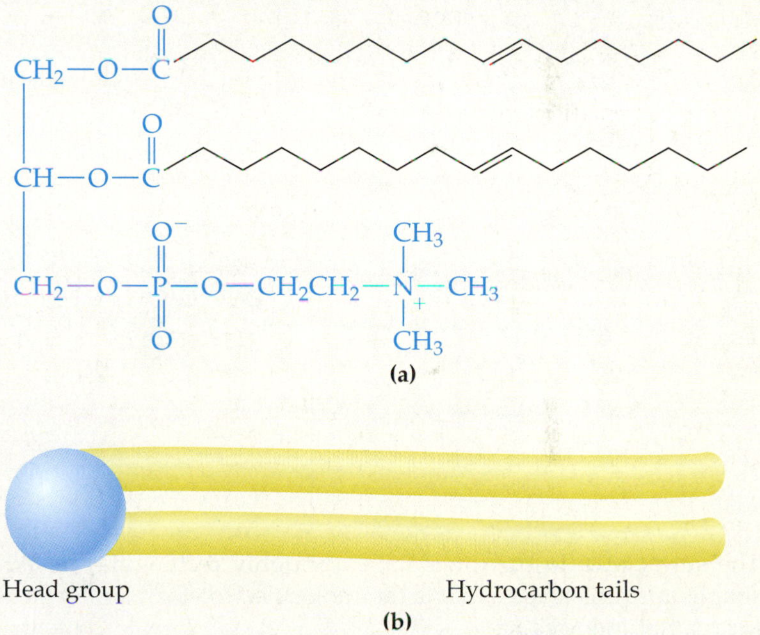

(a)

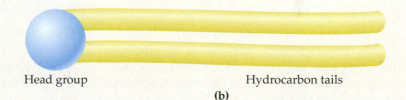

Head group

Hydrocarbon tails

(b)

For amphipathic substances, it is not always easy to predict whether or not a substance will mix with water. The dominant quality of the molecules making up the substance—whether they are predominately hydrophilic or hydrophobic—dictates the substance's miscibility in water.

QUESTION 7.18 Cholesterol is another example of a molecule that has some hydrophilic character and some hydrophobic character. Indicate the hydrophilic and hydrophobic portions of the molecule, and then predict whether cholesterol is miscible or immiscible in water.

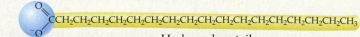

Cholesterol

ANSWER 7.18 Cholesterol is highly nonpolar and immiscible in water, despite its —OH group. The polarity of the —OH group is overwhelmed by the hydrophobic character of the hydrocarbon portion of the molecule.

Hydrophobic portion

Hydrophilic portion

Students
STUDENTS OFTEN ASK

How Do Detergents Get Greasy Dirt Out of Clothing?

The surfactants in the lungs have a roughly rectangular shape, and this shape allows the molecules to stack and form a flat membrane:

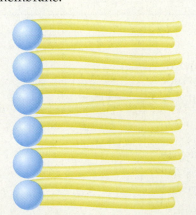

What happens, though, when you have a surfactant with a polar head group and only one hydrocarbon tail rather than two?

$$CCH_2CH_2CH_2CH_2CH_2CH_2CH_2CH_2CH_2CH_2CH_2CH_2CH_2CH_2CH_2CH_3$$

Polar head group   Hydrocarbon tail

Does this molecule have a rectangular shape? The answer is no because the molecule is wedge shaped. A collection of molecules having this shape assembles into three-dimensional spherical structures rather than flat membranes:

▲ **FIGURE 7.34** **Dirty work.** Indiana Jones could benefit from the action of detergent micelles.

These spherical structures, called **micelles**, can be made in the laboratory by mixing a certain minimum quantity of surfactant in water and setting the solution in a gently vibrating bath. The surfactants self-assemble into micelles that have the polar head groups exposed to the aqueous solvent and the hydrocarbon tails sequestered inside away from water. Thus, even though the interior of a micelle is highly nonpolar, these molecular aggregates dissolve in water. This type of structure is the basis for the action of **detergents**, which are amphipathic and contain a polar head group and a hydrocarbon tail.

What is the difference between a detergent and a surfactant? The difference between them lies in how and where they are used. Both are amphipathic molecules that can form organized structures in water, such as membranes and micelles. On one hand, surfactants often form structures on surfaces, such as the interface between lung tissue and air in the lungs. On the other hand, detergents are often associated with cleaning, and so we frequently think of them as micelles that dissolve oily substances.

If you have a pair of grease-stained jeans (Figure 7.34), the grease, which is made of oily, hydrocarbon-containing substances, will not dissolve in pure water because water has no means by which to interact with hydrocarbons. If you add detergent to the water, however, you now have

a place for the grease to go. The hydrocarbons in the grease encounter a detergent micelle and are pulled past the polar micelle surface into the hydrocarbon-containing center of the micelle, where they dissolve. Because the micelle is water soluble, it takes the grease away from the jeans and into the wash water.

QUESTION 7.19 Which molecules dissolve readily in the interior of a micelle: (a) ethanol; (b) ammonia, NH_3; (c) glucose, $C_6H_{12}O_6$; (d) hexane, C_6H_{14}?

ANSWER 7.19 Only (d) because it does not contain electronegative atoms and has a predominantly nonpolar character.

QUESTION 7.20 The structure of sodium dodecylbenzenesulfonate is

$$Na^+ \ \ ^-O-\overset{\displaystyle O}{\underset{\displaystyle O}{\overset{\|}{\underset{\|}{S}}}}-\langle\bigcirc\rangle-CH_2CH_2CH_2CH_2CH_2CH_2CH_2CH_2CH_2CH_2CH_2CH_3$$

Sodium dodecylbenzenesulfonate

(a) If this molecule were added to water, would it form micelles? (b) Would this molecule make an effective detergent? Why or why not?

ANSWER 7.20 (a) Because it has only one hydrocarbon tail, this molecule is wedge shaped. Thus, we expect it to form micelles in water. (b) Yes; molecules that form micelles make effective detergents because they can sequester greasy dirt away from water.

SUMMARY

This chapter looks at all aspects of the chemistry of water, including the phases it can adopt: solid ice, liquid water, and gaseous water vapor. We began our discussion with a look at the atoms a water molecule contains, hydrogen and oxygen. Hydrogen's three isotopes—hydrogen, deuterium, and tritium—exist in nature in specific ratios that can vary in natural waters from one location to another. This fact was used by scientists to study the migration patterns of birds. The rarest isotope of hydrogen, tritium, can be used when very large volumes of water are studied, as in the study of ocean currents.

Water molecules take part in dipole–dipole interactions both with other water molecules and with polar molecules other than water, which typically also contain an electronegative atom. In addition, the high electronegativity of the oxygen atom in H_2O causes the sole electron on hydrogen to be pulled toward the oxygen, leaving a bare partial positive charge on the hydrogen atom. This positive charge is attracted to the nonbonding pairs of electrons on other electronegative atoms, such as O, N, and F, in an interaction called a hydrogen bond, which is one type of noncovalent interaction. Hydrogen bonding allows molecules of water to bind very tightly together, and this makes the boiling point of water atypically high (based on the periodic trend). Hydrogen bonding also accounts for the high surface tension of water, which allows organisms with a high ratio of surface area to mass to walk on a water surface.

Hydrogen bonding allows water to form a hexagonal structure in the solid phase, which we know as ice. The spaces inside of ice make its density less than that of liquid water and make it possible for ice to float. When ice is warmed, it eventually reaches a temperature at which it begins to form liquid water. This point can be thought of as the melting point of ice or, alternately, as the freezing point of liquid water. In rain clouds, ice crystals begin to form around a nucleation center, and as they grow, they take on enough mass to fall to Earth in the form of snow, sleet, or, if the temperature is higher than the melting point of water, rain.

When ice being heated approaches its melting point, the temperature increases until liquid water begins to form and then stays constant until melting is complete. The processes of melting/freezing and boiling/condensation both occur at a constant temperature because the energy supplied to water during these processes is used to loosen (in the case of melting) or break (in the case of evaporation) the noncovalent interactions holding water molecules together. Water is able to absorb large amounts of heat with a correspondingly small increase in temperature, and thus we say that the specific heat of water is high.

Water heated to a temperature of 100 °C has reached its boiling point because, at that temperature, the bubbles transiently formed below the water surface have attained a pressure equal to the pressure pushing down on the surface. Boiling point varies with altitude because at high altitudes air is less dense and exerts a smaller force on the water surface than air does at sea level. Thus, water set to boil at high altitude will reach its boiling point quickly, and that boiling point will be lower than it would be at sea level.

When water molecules in a sealed container evaporate, they are trapped above the liquid. An equilibrium is set up between evaporating and condensing water molecules, and this process continues back and forth unless the seal on the container is broken. Once that happens, evaporating molecules leave the container, and the water eventually evaporates completely. The vapor pressure of a gas above a liquid is characteristic of that liquid, and water's vapor pressure is quite low. Liquids with very high vapor pressures are volatile and therefore readily leave the liquid phase to become a gas.

Because water molecules are very polar, water does not dissolve everything despite its reputation as the universal solvent. Very nonpolar molecules, such as waxes and long-chain hydrocarbons, avoid water by sequestering themselves away from it. Thus, when two immiscible liquids, such as oil and water, are brought together, they tend to form separate layers rather than mix. Amphipathic molecules have both hydrophilic character and hydrophobic character, and they tend to behave like the dominant character. Some amphipathic molecules, however, expose their hydrophilic portions to an aqueous environment and shield the hydrophobic portions from this environment. The amphipathic character of surfactants and detergents allow them to form organized structures such as micelles, which are the basis of detergent action.

KEY TERMS

amphipathic	dipole–dipole	hydrophilic	nucleation center	vapor pressure
boiling point	interaction	hydrophobic	phase	water vapor
condensation	evaporation	immiscible	specific heat	
density	freezing point	melting point	surface tension	
detergent	hydrogen bond	micelle	surfactant	

QUESTIONS

The Painless Questions

1. Why do people sometimes describe water as being the universal solvent?

2. Detergents are divided into three groups: anionic, cationic, and nonionic. Why is the detergent molecule shown in Question 7.20 (p. 285) classified as anionic?

3. Describe the way in which songbird migration routes can be tracked by taking advantage of the percentage of deuterium in water. How are the routes determined? How do the songbirds acquire deuterated water?

4. Comment on this statement: *Scientists have been studying the phase changes that water undergoes for centuries. Thus, we know all there is to know about how water freezes, evaporates, and melts.*

5. Show how hydrogen bonding can take place between a molecule of phenol (p. 189) and a molecule of the alcohol known as 2-propanol.

Phenol   2–Propanol

6. List four types of noncovalent interactions mentioned in this chapter. Briefly describe each one, and then rank them from strongest to weakest.

7. Which type of noncovalent interaction, if any, takes place between (*a*) a chloride ion and a water molecule, (*b*) a water molecule and an ammonia molecule, (*c*) an ammonia molecule and a butane molecule?

8. Which type of noncovalent interaction, if any, takes place between (*a*) two water molecules, (*b*) an H_2Se molecule and an acetone molecule (see Question 11 for the structure of acetone), (*c*) a magnesium ion and a water molecule, (*d*) a heptane molecule and a water molecule?

9. The molecule HF contains the highly electronegative fluorine atom. Draw the electron dot structure of HF, and indicate the direction of its dipole.

10. Describe the way strong winds rise during a hurricane.

11. While painting your nails, you have uncapped a bottle containing about 250 mL of nail polish remover, which is made from acetone. You also have a bottle

Acetone

containing about the same volume of water. You run to answer the phone and forget about both of them.

The next day, one of the containers is empty, but the other is only half empty. Which is which? Why did one evaporate completely and the other not?

12. Which molecules might make a suitable propellant for an aerosol can? (*a*) Ethane, CH_3CH_3; (*b*) ethanol, CH_3CH_2OH; (*c*) ammonia, NH_3. Briefly explain your choice(s).

13. Sketch ten water molecules in the liquid phase interacting with one another via hydrogen bonds.

More Challenging Questions

14. Rank these forms of water in order of their abundance in a typical sample of water, from most abundant to least abundant: D_2O, HTO, HDO, H_2O, T_2O.

15. A 0.35-mL sample of water is taken from a songbird wing. This volume of water contains 0.020 mol of water molecules. How many tritium atoms does it contain?

16. The alcohol hexanol is only slightly soluble in water. What experiment could you perform to determine exactly how much hexanol dissolves in water?

17. While you are ice fishing with a friend, she comments that it must be too cold to snow. You check the thermometer and find that it is –30.5 °C. Is it too cold to snow?

18. The vapor pressure of water and other substances decreases as temperature decreases. Based on your knowledge of gases and vapor pressure, explain why this must be true.

19. The directions for cooking oatmeal at high altitude suggest doubling the regular cooking time. If you live at high altitude, what would happen to your oatmeal if you ignore these instructions?

20. You are planning a trip to a landlocked city in central Africa, and you are planning to take clothes for various temperatures because areas far from water tend to experience large swings in temperature. What else should you know about the place you are visiting before you start to pack?

21. You and a friend who is a chef are planning a climb up Annapurna, one of the tallest peaks in the Himalayas. Your friend is planning an elaborate menu for the climb, including many foods that require heating. What advice will you give him regarding cuisine appropriate for such a trip?

22. Explain these situations: (*a*) A big cockroach jumps lightly onto the surface of a lake and sinks to the bottom. (*b*) A housefly lands on the surface of a solution of diethyl ether, $CH_3CH_2-O-CH_2CH_3$, and sinks to the bottom. (*c*) A housefly lands on the surface of a lake and skates across the surface.

23. The process $H_2O(\ell) \rightarrow H_2O(s)$ + heat occurs in your freezer. How does your freezer facilitate this reaction?

24. List the number of protons, neutrons, and electrons in $^1H^+$, $^2H^+$, ^3H, ^1H, $^3H^+$. Three of these symbols can be written using an alternative notation. Which symbols are they? Rewrite the symbols using this alternative notation.

25. Your friend is a graduate student in chemistry. She plans to study the levels of tritium in bumblebee wings to see if they vary from bee to bee. Why would you caution her to work on something else?

26. The woman shown here could use some cleaning up. Explain how detergents can remove the dirt and grease on her clothing.

27. The two atoms in C–H bonds do not take part in hydrogen bonding. Why do oxygen, nitrogen, and fluorine participate in hydrogen bonding, but carbon does not?

28. For each pair, state which noncovalent interaction is stronger: (*a*) calcium ion and bromide ion versus calcium ion and water molecule; (*b*) two water molecules versus water molecule and butane molecule, $CH_3CH_2CH_2CH_3$; (*c*) ammonia molecule and water molecule versus butane molecule and ammonia molecule.

The Toughest Questions

29. Comment on this statement: *Fevers are common for human beings, and the temperature of the human body can fluctuate widely around a healthy temperature of 98.6 °F (37.0 °C).*

30. The compound that has the condensed structure $CH_3CH_2CH_2CH_2$—O—H has the same chemical formula as the compound that has the condensed structure CH_3CH_2—O—CH_2CH_3. (*a*) What is that chemical formula? (*b*) What is the molar mass of both compounds? (*c*) Which compound would you expect to have the higher boiling point? Explain your choice.

31. Propanol, $CH_3CH_2CH_2OH$, methylamine, H_3CNH_2, and water all participate in hydrogen bonding, and all are miscible in one another. Sketch the hydrogen bonding in a mixture of these three substances, showing an example of a hydrogen bond between each type of molecule in the solution.

32. Some antifreeze formulations contain molecules capable of hydrogen bonding. Explain how the addition of this type of molecule can keep the water in your car radiator from freezing in cold weather.

33. Refer back to the wax molecule depicted in Figure 7.30. (*a*) Redraw this molecule using a line structure. (*b*) Calculate the molar mass of the molecule. (*c*) How many molecules of this wax are contained in a sample that has a mass of 2.85 g? (*d*) Use your knowledge of boiling points and intermolecular interactions to speculate on this observation: waxes, which have more than 20 carbon atoms per molecule, are solids at room temperature; smaller hydrocarbons, such as heptane, are liquids at room temperature; and the smallest hydrocarbons, such as propane, are gases at room temperature.

34. A diagram like the one shown for water in Figure 7.19 is available for substance Z. The line in the diagram for substance Z in the liquid phase is one-half as steep as the same line in the water diagram. After seeing both diagrams, you decide to empty the water from the tubes in your passive solar home and fill them with substance Z. Based on what you know about substance Z, is this a good idea? Why?

35. You may have noticed that a car tire gets warmer as you add air to it. Likewise, you may have felt an aerosol can cool down as you release its contents. Based on what you know about pressure and temperature, speculate on why these things happen.

36. Why is air in the mountains cooler than air at lower altitudes? As you move from lower altitudes, you get closer to the Sun. This being so, why is mountain air not *warmer* than air at lower altitudes?

37. The water in hydrothermal vents must reach nearly 400 °C before it begins to boil. Why are such high temperatures required for boiling at the extreme depths at which these vents are found?

38. (*a*) If you place 0.450 mol of butanol in 1 L of water, will it be completely soluble? (See Table 7.3 The density of butanol is 0.8 g/mL.) (*b*) If you place 3.00×10^{22} molecules of butanol in 400 mL of water, will it be completely soluble?

39. Acetone, which contains six hydrogen atoms and is shown in Question 11 (p. 287), is not capable of participating in hydrogen bonding. How is it possible for a molecule like acetone to contain hydrogen atoms, but not be capable of forming a hydrogen bond?

40. When ice is heated, its temperature increases until the melting point (0 °C) is reached. At that point, the temperature remains at 0 °C until all of the ice is melted and is in the liquid phase. Additional heat then raises the temperature of the water. Explain why this phenomenon occurs.

41. The energy associated with the interactions among all the water molecules that surround a lithium ion has been measured at about 400 kJ/mol. How would you expect this value to compare with the energy of interaction between two water molecules? How would you expect this value to compare with the energy of interaction between a lithium ion and a chloride ion?

42. We know of at least two salts that can be used to melt icy roads: calcium chloride and sodium chloride. Do you think any common salt could be used on winter roads? Think of one factor that would be used to determine which salt is used. (Hint: Think about the huge quantities required.)

E-Questions

Go to **www.prenhall.com/waldron** to find these questions in electronic form, complete with hyperlinks directly to the various websites cited in the questions.

43. How do you know when your Thanksgiving turkey is done? A new pop-up thermometer has replaced the older cooking thermometer, at least for turkeys. Read about how these devices work at **HowStuffWorks.** Use the chemical principles you have learned thus far, including the meaning of pressure and the composition of metallic solids, to write a short essay on how a turkey thermometer knows exactly when the meat is done.

44. Go to **TexasWater** and read the article titled "The Science of Cloud Seeding." After reading this brief discussion of cloud seeding, answer the following questions: (*a*) What percentage of a typical cloud is air? (*b*) Why does cloud seeding increase the size of clouds? (*c*) How much time must pass before a seeded cloud produces rain? (*d*) Why would a large hail-forming thunderstorm cloud be seeded?

45. Rain leaves a distinctive smell in the air. Go to **HowStuffWorks** to find out why. Look for at least three things that contribute to the sweet smell that accompanies a hard rain.

46. Go to the website **UniSci** to learn about gas hydrates. Answer the following questions after reading this two-page article: (*a*) What is a gas hydrate, and where are they found in nature? (*b*) Why would gas hydrates be a promising new source of fuel, and why have they not been tapped sooner? (*c*) What method is used to locate gas hydrate deposits?

47. At **SciAm** read the article titled "Water Molecules March Through Hydrophobic Nanotubes." After reading this brief news report, answer the following questions: (*a*) Describe the nanotube that was designed in this virtual experiment. (*b*) Why were the results obtained surprising? (*c*) How many water molecules were able to fit in one nanotube? (*d*) What is one practical application of this research?

WORLD WIDE WEB RESOURCES

As with the E-Questions, go to **www.prenhall.com/waldron** to find these questions in electronic form, complete with hyperlinks directly to the various websites cited.

Some of the links that follow contain research articles related to the topics in this chapter and could be used as the basis for a writing assignment in your course. Other sites are of general interest.

- The discovery of the Lost City hydrothermal vents is discussed at **ScienceNews.** Read about why this particular formation of hydrothermal vents is remarkable. What organisms populate Lost City? What do they live on, and how is this environment different from other marine environments?

- Find out more about rituals involving water in different cultures. Go to **HolyMountain** to learn more about Japanese rituals and the history of water in Japan.

- The songbird research discussed in Section 7.1 was published in the journal *Science* in 2002. A layman's summary of this research report can be found at **Songbird.**

- Watch an ABC News video called "Don't Drink the Water: Chromium-Tainted Water" at **ABCNews.** This story, reported by Diane Sawyer, recounts the court case that became the basis for the film *Erin Brockovich.*

ABC NEWS VIDEO
Chromium-Tainted Water

CHEMISTRY TOPICS
IN CHAPTER 8

- Real and ideal gases [8.1]

- Diffusion [8.1]

- Neurons and neurotransmitters [8.2]

- Open and closed systems [8.3]

- Pressure of a gas [8.3]

- Avogadro's law [8.3]

- Mean free path of gas molecules [8.4]

- Atmospheric pressure [8.4]

- Boyle's law [8.4]

- Molar volume of a gas [8.5]

- Standard temperature and pressure [8.5]

- Charles's law [8.5]

- Amontons' law [8.5]

- General gas law [8.6]

- Chlorofluorocarbons and the ozone layer [8.7]

- Global warming [8.7]

- The greenhouse effect [8.7]

And 'tis my faith that every flower
Enjoys the air it breathes.

"LINES WRITTEN IN EARLY SPRING," WILLIAM WORDSWORTH

Air

A STUDY OF THE GASEOUS ATMOSPHERE IN WHICH WE LIVE AND THE LAWS THAT DICTATE THE BEHAVIOR OF GASES

In his 1997 book *We Never Went to the Moon*, author Bill Kaysing tries to convince his readers that the July 1969 *Apollo 11* Moon walk by astronauts Armstrong and Aldrin never happened. According to "evidence" presented by Kaysing, the entire scene supposed to have taken place on the Moon was actually filmed in the Nevada desert, a ploy by the U.S. government to demonstrate its technological superiority over the then Soviet Union. One piece of "proof" cited by Kaysing involves the American flag planted into the lunar surface (Figure 8.1). According to Kaysing, the flag, which appears to be waving in the video footage of the Moon landing, should not be waving because there is no air on the Moon. The official response to Kaysing's flag theory is that the flag was not, in fact, "waving." Rather, it was held stiffly in place by horizontal wires and was wrinkled, which gave the impression of waving. Not many folks bought Kaysing's argument, although many

bought his book, which sold tens of thousands of copies.

We do not give much thought to waving flags here on Earth. In fact, we take for granted that air is all around us all the time, although it is mostly invisible. Try observing an outdoor scene for a few minutes, though, watching the people and the landscape. What evidence of air do you see? Chests lift and lower as people inhale

▲ **FIGURE 8.1   Proudly she waves?**   Some people believe that no one has ever walked on the Moon. These skeptics claim that this flag must have been photographed someplace on Earth because flags do not "wave" on the Moon.

▶ **FIGURE 8.2   Some unwelcome reminders of a constant presence.** **(a)** Air you can feel usually portends stormy weather. **(b)** Air you can see raises the alarm of pollution. **(c)** Air you can smell can bring pleasure—as when the fragrance from a pizza oven wafts your way—or pain—as when this cute, but very smelly, fellow crosses your path.

and exhale. Trees, flags, and grass shift slightly. An aircraft passing overhead and a hovering Frisbee float on an invisible cushion of air. The smell of baking bread or of exhaust fumes testifies to air's ability to broadcast molecules from one place to another. In fact, once you start paying attention, it is difficult *not* to notice air. Perhaps we are comforted by its constant presence. A continuous and unwavering supply of clean air is something we can almost always rely on.

Air we can feel or see or smell may be cause for alarm. Air we can feel may portend violent weather (Figure 8.2a). Air we can see may be laden with undesirable specks of airborne trash (Figure 8.2b). Air we can smell contains molecules that trigger the olfactory nerve, a signal of possible contamination (Figure 8.2c). Clean, dry air contains about 21 percent oxygen, 78 percent nitrogen, and small amounts of other benign gaseous atoms or molecules (Table 8.1). And in the same way that we would not want to wear dirty clothing or eat spoiled food or drink contaminated water, so we like our air to be as pure as possible. We don't want to see it before we breathe it.

In this chapter, we explore the unseen gases that constantly surround us. First we shall look at clean air, and then we shall see what can make it unclean and why. In order to understand these things about air,

(a)

(b)

(c)

TABLE 8.1 Composition of Clean Air at Sea Level	
Atoms or molecules	**Percentage**
Nitrogen molecules, N_2	78.1
Oxygen molecules, O_2	20.9
Argon atoms, Ar	0.934
Carbon dioxide molecules, CO_2	0.036
Trace gases	<0.03

though, we must first understand how air behaves. We do this by learning some of the rules we expect air to follow and then seeing what happens when it does not follow these rules.

8.1 | Trouble in Tokyo

March 20, 1995, promised to be a typical spring day in Tokyo as people left their homes to go to work. What they did not know was that members of the Aum Shinrikyo terrorist cult were en route to various points along the Tokyo subway system. At the height of the morning rush hour, members of this group punctured containers of *sarin*, a deadly nerve agent, which then spread through the air in several crowded subway cars.

When sarin changes from a liquid to a gas, which it does very readily, the gas can get into a victim's bloodstream by being absorbed through either the skin or the lungs. Sarin quickly disrupts the normal functioning of the nervous system, causing paralysis, convulsions, loss of vision, and violent headaches. In doses greater than 0.5 mg for a 70-kg adult, exposure is usually fatal. During the Tokyo subway attack, many people exposed to near-fatal levels of sarin escaped death, but have permanent neurological damage. (Those exposed to smaller quantities of sarin recovered completely.) All told, the attack resulted in the death of 12 people and injury of more than 5000 others. The fate of any one person riding the Tokyo subway on that horrifying day depended on the behavior of the gases around that person.

The components of clean air listed in Table 8.1 are completely mixed together because it is the nature of all gases to blend until the mixture they form is homogeneous. In this respect, gases are very different from solids and liquids. As we learned in Chapter 1, solids and liquids can be very well mixed together, but still be heterogeneous mixtures, as in the case of pulpy orange juice or solid concrete, which contains chunks of rock.

Why are gas mixtures usually homogeneous whereas solid and liquid mixtures oftentimes are not? The answer lies in the proximity of molecules and atoms to one another in each phase. Recall from Chapter 4 that solids pack into a dense mass as molecules, ions, or atoms stack up on one another. With ionic solids, the ions make a repeating, regular lattice that is held together by the mutual attraction of positive and negative ions. The situation is similar in liquids in that the molecules, ions, or atoms in the liquid come into close contact and commingle with one another through noncovalent interactions that can range from very weak to quite strong.

In order to understand why a solid or liquid exists and how it is held together, you must know what kind of substances it contains and how they interact with one another. This information can tell you, for example, how

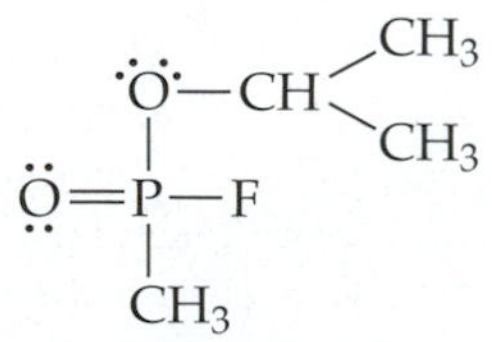

▲ **FIGURE 8.3** **The structure of the sarin molecule.**

▲ **FIGURE 8.4** **The three phases of matter.** **(a)** In water vapor, the molecules are far apart and fill all the available space in the container holding them. **(b)** In water, the molecules are close together but move relative to one another. The liquid conforms to the shape of the container, but does not fill the whole container. **(c)** In ice, the molecules are packed in a regular array and move very little relative to each other. The solid retains its own shape regardless of the shape of the container holding it.

easily the solid will melt or how volatile the liquid will be. Thus, intermolecular interactions are all-important for liquids and solids. With gases, though, individual units are so far apart that they come into contact with one another much less frequently. For this reason, it does not matter whether they do or do not have some kind of mutual attraction. In fact, the matter present in any volume of gas in a container takes up less than 0.1 percent of the total space. In a way, this makes it relatively easy to understand gas behavior.

In the real world, it turns out that there *are* interactions between the individual atoms or molecules in a gas. These interactions are usually so weak, however, that we are able to ignore them. When the interactions are not negligible, we refer to the gas as a **real gas**. For our purposes, though, we shall assume that all gases behave *ideally*. We define an **ideal gas** as one that is infinitely dilute and in which there are no chemical interactions between individual units. Because we will focus in this chapter on ideal gases only, *we do not have to concern ourselves with the intermolecular interactions between molecules of a gas.*

QUESTION 8.1 In our discussion of water in Chapter 7, we learned that hydrogen bonds between water molecules are responsible for many of the properties of liquid water. Would you expect hydrogen bonding to affect the behavior of water molecules in the gas phase?

ANSWER 8.1 No. We assume that each molecule in the gas phase is discrete and does not interact with other molecules because they are so far apart. For an ideal gas, no hydrogen bonding is possible.

We know from our discussion in Section 7.7 that liquids obey the rule *like dissolves like*. We do not need any such rule for gases, however. Instead, the rule for gases is simply that all gases dissolve in all other gases. To see an example, let us look at the structure of sarin shown in Figure 8.3. You should recognize sarin as an organic molecule because it contains a carbon-based framework held together by covalent bonds. If sarin can enter the gas phase, which it does readily, it will mix with other gases, regardless of their properties. So the fact that it is organic as well as polar is unimportant. It still dissolves easily in air, which is mostly composed of small, nonpolar, inorganic molecules.

QUESTION 8.2 Ethyl alcohol and liquid water mix in all proportions, but molecules less polar than ethyl alcohol are reluctant to mix with water. An example is the hydrocarbon butane, which is only slightly soluble in water. Given these facts, how will the behavior of a mixture of gaseous butane and water vapor differ from the behavior of a mixture of the same molecules when they are both liquids?

ANSWER 8.2 Butane and water do not mix readily in the liquid phase because they have very different properties—the former is nonpolar, and the latter is polar. In the gas phase, however, the rule *like dissolves like* does not apply because the polarity or nonpolarity of a gaseous molecule is irrelevant. The molecules in the gas phase are far apart and do not interact with one another. Therefore, butane and water mix readily in the gas phase.

Figure 8.4 highlights the differences among the three phases of matter. It shows that another difference among liquids, solids, and gases is in how they must be stored. You can leave most solids and liquids sitting out in the open indefinitely and, except for evaporation, not worry about losing any of your sample. With gases, though, the intermolecular forces of attraction are not strong enough to keep the individual atoms and molecules together. Thus, you must be careful to store a gas sample correctly! Most gases are stored in airtight containers that keep the gases from leaking out. Think about gas-filled things you might see or use everyday: a tire, a pipe carrying natural gas into a building, a balloon, a neon sign. In all these cases, the container must be leakproof. If a leak develops, the product or device will cease to do its job. Sometimes this is no big deal, as with a burst balloon or burst soda can, but when the gas in the container has the potential for harm—as in the case of highly flammable natural gas—a leak can lead to the asphyxiation of people near the leak and maybe even an explosion.

QUESTION 8.3 Imagine you have a glass containing ice cubes and some water. If the glass is tightly covered with a lid, how many phases of water exist in the glass?

ANSWER 8.3 All three phases. Liquid water is mixed with ice (solid water), and water vapor (gaseous water) exists in the space above the liquid surface.

All gas molecules move very, very fast. A typical nitrogen molecule at room temperature, for instance, moves at the supersonic speed of about 1850 km/h, nearly as fast as the late, lamented Concorde aircraft. If you could point a radar gun at the molecules in a sample of pure nitrogen, however, you would find that they are not all moving at the same speed. If you could measure the speed of each individual molecule and keep a tally of the number of molecules moving at each speed, you would end up with a curve with the average speed somewhere in the middle, as shown in Figure 8.5. Most of the molecules in the sample are moving at or near the average speed, but there are outliers that are moving either much faster or much more slowly.

Although the molecules in a sample of gas are very far apart and take up only a very small percentage of the total space of the container holding them, it is inevitable that they will come into contact with one another on occasion. The collisions that occur in an ideal gas are assumed to be completely *elastic*. That is, we assume that when two molecules in an ideal gas collide, they do not interact chemically with each other in any way. Instead, they simply bounce off each other. Thus, it is convenient to think of the molecules in a gas as tiny, high-speed billiard balls that are in continuous, random motion through space. This random motion leads to some interesting effects. For instance, although you easily see your uncle enter the house after cleaning fish in the backyard, it will be a while before you smell him because those smelly molecules he picked up from the fish do not travel directly from his body to your nose. Rather, they take a zigzag path across the room, all the while colliding with molecules and objects along the way (Figure 8.6).

RECURRING THEME IN CHEMISTRY *Intra*molecular *bonds*, those *within* a molecule, are covalent. *Inter*molecular *interactions*, those either *between* two molecules or *between* two parts of a large molecule, are noncovalent.

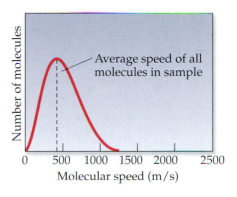

▲ **FIGURE 8.5** **The molecules in a gas sample move at various speeds.** This graph shows the speed of all the molecules in a sample of nitrogen gas at 300 K. The speeds are distributed in a curve because some molecules move at speeds lower than the average speed of all the molecules in the sample, while other molecules move at speeds higher than the average speed.

▲ **FIGURE 8.6** **A real stinker.** The fish odor that your uncle brought into the house will eventually make its way to your nose.

Gas molecules move into every space available and completely fill any container they occupy. Imagine you introduce a new molecule to a population of gas molecules speeding around in a container. The new molecule will swiftly become integrated into the existing mixture of fast-moving molecules, and it will be difficult to distinguish it from all the others. This process, known as **diffusion**, is analogous to a busy highway on which cars entering the highway must quickly attain the same speed as the other moving cars. Once a new car enters the highway, it becomes part of the traffic flow, in which all the cars are moving at slightly different individual speeds, but at a similar *average* speed.

The assimilation of one gas into another has important practical consequences because the air we breathe naturally mixes with other gases quickly and completely. The canisters of sarin used in the Tokyo subway released poisonous gas into the air. Just like cars merging into fast-moving traffic, the deadly molecules became mixed in with the rest of the molecules in the air breathed by thousands of commuters. Evidence gathered after the attack showed that people closest to the canisters had the most severe symptoms of nerve gas poisoning. This finding is reasonable because, as the poison gas molecules moved farther from a canister, they grew farther apart from one another as they diffused into the air and became diluted. Thus, a passenger farther from a canister would inhale air containing fewer sarin molecules than a passenger closer to the canister. The closer passenger would have to take fewer breaths to reach the fatal limit of sarin, which is about 0.5 mg for a typical adult.

8.2 | "Better Killing Through Chemistry"

The use of biological agents in warfare is not unique to today's high-tech civilization. In the fourteenth century, Tatars were catapulting corpses infected with bubonic plague into enemy settlements. In the eighteenth century, British troops in colonial America fighting Native Americans (who were allied with the French) handed out smallpox-infected blankets to tribes, killing thousands during what became known as the French and Indian War. The development of chemicals as military weapons began in the twentieth century with the extensive use of poisonous gases in World War I. The "improvement" of chemical weapons continued through the century, despite the 1928 Geneva Protocol prohibiting their use. The first *nerve agents*—chemicals that interfere with the release or uptake of neurotransmitter molecules at nerve synapses—were developed in Germany in the 1930s. Since that time, new airborne nerve agents that kill more efficiently have been developed as organic-synthesis techniques have become more and more sophisticated.

A nerve agent does its damage by interfering with the process illustrated in Figure 8.7. Nerve impulses travel from one nerve cell (neuron) to another by way of molecules called *neurotransmitters*. These molecules diffuse across the space between two neurons, the synaptic cleft, when an impulse is being sent and then are soon cleared out of the synaptic cleft before the next impulse arrives. The result is a controlled sequence of nerve signals. In the presence of a nerve agent, the neurotransmitter molecules are not cleared from the interneuron space once an impulse has been transmitted. The result is that the same impulse gets sent over and over again uncontrollably.

Sarin is one of the most potent nerve agents known. A liquid, sarin tends to enter the gas phase and dissipate rapidly in air. The structure of the

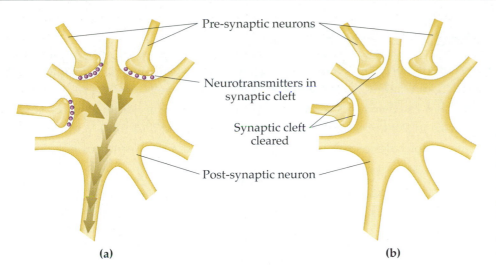

Pre-synaptic neurons

Neurotransmitters in synaptic cleft

Synaptic cleft cleared

Post-synaptic neuron

(a) (b)

◀ **FIGURE 8.7 Transmitting nerve impulses in the toxin-free brain. (a)** Nerve impulses are conducted from one neuron to another when neurotransmitter molecules traverse the space between neurons. **(b)** Once the impulse has been transmitted, the neurotransmitter molecules are cleared from that space.

sarin molecule is shown in Figure 8.3. An unusual feature in this molecule is the presence of a phosphorus atom. This unusual structural feature allows the molecule to interfere with normal neural function, and makes the sarin molecule a very efficient killer.

QUESTION 8.4 Write the chemical formula for sarin and calculate the molar mass to two decimal places.

ANSWER 8.4 Sarin: $C_4FH_{10}O_2P$; 140.09 g/mol.

You may be wondering how easy it would be for an average person with little training in chemistry to develop and use chemical weapons. Is it possible for a resourceful person to find a recipe for making sarin? The answer is yes. Much of the research on nerve agents has been done in university laboratories around the world. In the academic scientific community, chemists are expected to publish their results in peer-reviewed journals. Thus, anyone having access to a university library might be able to locate a report of the synthesis of a molecule like sarin. These journal reports are highly detailed because they are designed to provide exact instructions for anyone else in the scientific community who wishes to repeat the experiment. Moreover, many journals are now available on-line, and so a trip to the university library may not be necessary.

If someone seeking to synthesize a nerve agent has downloaded a protocol from the Internet, the only thing left is to get the ingredients. How could a person obtain the chemicals required to make a nerve agent? Figure 8.8 shows James Tour, a chemist at Rice University who, wondering the same thing, placed a telephone order to a major U.S. chemical house for the starting materials needed to synthesize sarin. The chemicals arrived by overnight shipping and cost a little more than $130. In an article titled "Better Killing Through Chemistry," editor George Musser of *Scientific American* described his own attempts to order the

▲ **FIGURE 8.8 Chemist was able to order dangerous chemicals by telephone.** James Tour, a chemist at Rice University, was able to send away for all the chemicals needed to synthesize the deadly nerve agent sarin.

▲ FIGURE 8.9 **Flying dust mops.** Jerry Bromenshenk, a bee expert at the University of Montana, has trained bees to look for specific molecules present in the air through which the bees fly.

BEYOND THE ORDINARY

Mind Your Own Bees-ness

Bees are devoted busybodies. Because they are small, they can fit into tiny nooks and crannies, all the while interacting with molecules in the air, molecules they can then take back to the hive along with the nectar they collect. Dust particles in the air naturally stick to bees as they fly, and those particles can contain molecules of any substances found in the air. In fact, bees have been described as "flying dust mops." When they return to the hive, they beat their wings furiously to cool off the temperature inside. This beating motion releases molecules that were stuck to the bees' bodies, molecules that then become concentrated in the hive space. The end result is that if a bee takes nectar from a plant containing a foreign molecule of some sort, or if the bee flies through air contaminated with some foreign molecule, those molecules will eventually end up in the hive, sometimes in the honey, and other times in the bee itself.

Scientists like bee expert Jerry Bromenshenk of the University of Montana are trying to exploit the natural talent bees have for taking home samples of the air through which they fly (Figure 8.9). Bromenshenk and his research group have taught bees to seek out specific non-nectar molecules, which he says is an easier task than training a bloodhound to follow a scent.

So how can this possibly be of use to humans? One way is in the detection of land mines, which typically contain the explosive trinitrotoluene (TNT). After being trained to sniff out TNT, bees are amazingly adept at locating buried land mines. In countries in which land mines are prevalent, such as Cambodia, Somalia, and Angola, mine removal is an urgent priority. With luck, the coming years will see mine-tracking bees working in these locations.

QUESTION 8.5 Honey and beeswax, both produced by honey bees, have very different chemical properties. One is hydrophobic, and the other is hydrophilic. Which is which, and how do you know?

ANSWER 8.5 Honey is a commonly used sweetener. When it is added to hot tea, it readily dissolves. This demonstrates that honey is hydrophilic. Beeswax, and all other waxes, will not dissolve in water. (Try putting a candle in a bowl of water and see what happens.) Thus, beeswax must be hydrophobic.

materials required to synthesize sarin. Musser reasoned that Tour, a reputable university chemist, might not have trouble ordering questionable chemicals, but could an average person (like Musser) place an order for the same chemicals without raising eyebrows? To find out, Musser placed an order. A package containing the ingredients for the synthesis of sarin was delivered within a few days.

8.3 | Keep a Lid on It

One of the pleasures of Italian cooking is the unmistakable smell of garlic that diffuses from the pot to your nose. The molecule in garlic that gives it its distinctive odor is called *allyl mercaptan* (Figure 8.10). The presence of sulfur in a molecule is often a sign that the molecule will have a conspicuous smell, usually a malodorous one. We can smell garlic all over the house when we cook with it because allyl mercaptan has an extremely low *odor threshold*, defined as the lowest concentration at which the human nose can detect a given molecule. In fact, the average person can detect the smell of garlic when there is only 100 ng of allyl mercaptan per 1 L of air. Because the gases in the pan containing sautéed garlic diffuse away, we say that the pan is an **open system**, which is a defined space that is always exchanging freely with the environment surrounding it. An open system can exchange gases with the surrounding air and can be heated or cooled by the surrounding environment.

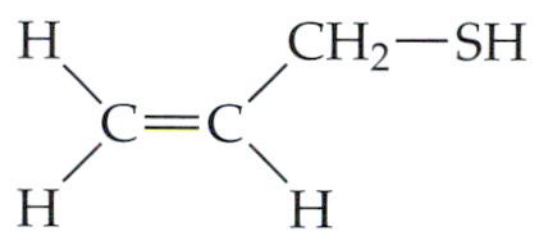

For scientists, open systems are a headache because the flow of materials into and out of an open system must be considered in any experiment performed on it. Therefore, scientists usually prefer to work with **closed systems**, for which the focus is on everything that is inside the system. Everything contained in a closed system is accounted for; and nothing is exchanged with the surroundings. Figure 8.11 shows examples of open and closed systems in our everyday world.

▼ FIGURE 8.10 **Allyl mercaptan.**
This is the molecule that gives garlic its characteristic odor and taste.

QUESTION 8.6 Natural gas, which is odorless, contains the odorant molecule *tert*-butyl mercaptan:

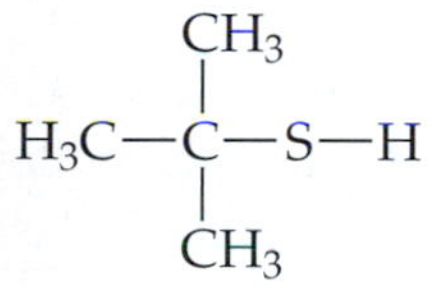

tert-Butyl mercaptan

This odorant molecule allows people to smell a natural gas leak before the leak can lead to an explosion. (*a*) What feature does the *tert*-butyl mercaptan molecule share with another very smelly molecule, allyl mercaptan? (*b*) The odor threshold for *tert*-butyl mercaptan is 80 ng/L. Which odor molecule is easier for the human nose to detect, *tert*-butyl mercaptan or allyl mercaptan? (*c*) A gas leak in a room that has a volume of 68,000 L releases 3.0 mg of *tert*-butyl mercaptan. Would the typical person be able to detect this gas leak?

ANSWER 8.6 (*a*) Both molecules contain sulfur. (*b*) The *tert*-butyl mercaptan is easier to detect because its odor threshold is lower. (*c*) We

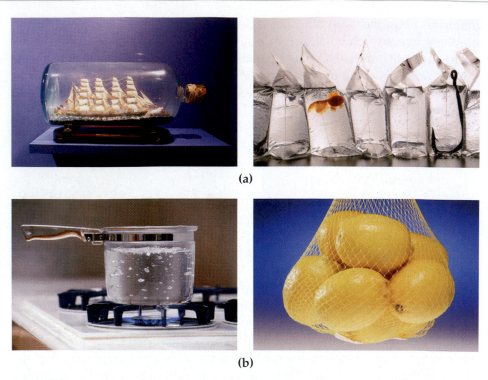

▲ FIGURE 8.11 **Closed or open?** **(a)** Closed systems are sealed, and no matter is exchanged with the surroundings. **(b)** Open systems freely exchange materials with the surroundings.

must calculate the number of nanograms per liter of air and compare this value with the odor threshold for *tert*-butyl mercaptan:

$$\frac{3.0 \text{ mg}}{68,000 \text{ L}} \times \frac{1 \text{ g}}{1000 \text{ mg}} \times \frac{10^9 \text{ ng}}{1 \text{ g}} = 44 \text{ ng/L}$$

This value is less than the odor threshold (80 ng/L). Therefore, the smell would not be detectable by the typical human nose.

QUESTION 8.7 Classify each of the following as an open system or a closed system: (*a*) a closed zipper-sealed plastic bag with a sandwich inside, (*b*) a styrofoam cup of hot coffee with a plastic drinking lid, (*c*) a teapot filled with steeping tea, (*d*) an unopened bottle of mineral water.

ANSWER 8.7   (*a*) and (*d*) are closed; (*b*) and (*c*) are open. The lid on the coffee cup has a hole through which to sip the coffee, and the system is therefore open. The teapot also has a hole (in the spout) and so is an open system.

Note that *closed* and *open* can be defined differently depending on the application. For example, a person performing a laboratory experiment that requires an absolutely leak-proof container might define the term *closed system* more rigorously than a person using Tupperware® to keep food fresh.

Living in an open system can be a godsend. When we are outdoors, for example, molecules in the air around us are free to diffuse to an infinite distance away from us. So when your friend who bathes herself in lilac perfume invites you to lunch, choosing an outdoor table will allow those perfume molecules to drift out and away. And when you are forced to ride in your chain-smoking uncle's car, you might wish he understood more about the behavior of gas molecules in closed systems. Because closed systems are understood and controlled, we shall often use them as examples in our discussion of gases. Also, even though your uncle's car is an interesting closed system, we shall look instead at a closed system that has more flexibility—a bicycle tire.

In many parts of the world, including many areas of the United States, bicycling enthusiasts cannot ride for many months of the year. During those months, bicycles are relegated to the garage, waiting until good weather returns and it is cycling season again. For most of us who are aching for a bicycle ride on the first hospitable day of spring, there is the inevitable trip to the air pump to replace any air that may have leaked out of the tires very, very slowly over the winter.

Let us consider for a moment the process of adding air to a tire. The loud sound the air pump makes as we use it tells us that energy is needed to squeeze the air down the hose and into the tire. This energy is in the form of electrical energy supplied to the compressor. How does the air in the jet going into the tire differ from the air in the gas station parking lot? The difference is in the number of molecules in a given volume. The air coming from the hose is *pressurized*, meaning that the molecules in it are compressed into a smaller-than-normal volume. There are more molecules in that volume than there are in the same volume of parking-lot air. The molecules in the pump move from the hose into the tire, and the tire fills up quickly because the pump is propelling a concentrated blast of air molecules (mainly N_2 and O_2) into it.

That the movement of air out of a tire proceeds naturally in only one direction should not surprise you. In previous chapters, we saw that natural processes tend toward lower energy. To go in the opposite direction—in our example, air entering the tire—energy must be added to the system. To understand why, we must consider that the air molecules in the tire are moving at high speeds. These molecules occasionally collide with one another and, more often, with the inner walls of our closed system—the tire. Each time a molecule hits the tire wall, the wall gets pushed outward ever so slightly. As the tire is filled, more and more molecules are added to the tire, and those molecules also collide with the tire wall. As the force against the wall becomes stronger, the wall is forced outward more and more (Figure 8.12). Because it is made of rubber—a flexible material—the tire begins to expand.

The phenomenon of molecules exerting a force against a surface is what we know as **pressure**. More collisions equal higher pressure; fewer collisions equal lower pressure.

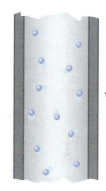

(a)

Wall of tire

(b)

▲ **FIGURE 8.12   Fix-a-flat.   (a)** A flat tire contains fewer gas molecules than a full tire. As a result, there is less pressure pushing against the inside tire walls. **(b)** A fully inflated tire contains enough gas molecules to press on the inside tire walls and keep the tire inflated.

STUDENTS OFTEN ASK

Why Are Tires Filled with Air Instead of Being Solid?

The primary job of a tire is to provide a cushion between the chassis of a car or bicycle and the road. A good tire will fill in the bumps and holes in the road temporarily as you move over them by changing shape, perhaps by compressing slightly. After a bump or hole has been passed, the tire must decompress and return to its original shape until the next obstacle comes along. A tire filled

with air is compressible. The air molecules in the tire are constantly in motion, and, if the tire changes shape suddenly, these molecules will adopt the new shape immediately, as gases always do. When a tire compresses as it hits a bump, the shape of the tire changes momentarily. Gases are good at this—they fill any shape that contains them, whether that space is a fully inflated tire in its natural shape or that same tire squashed down to traverse a bump. Once the bump has been crossed, the tire returns to its original shape.

Now consider a solid tire. Unlike gases, solids are not readily compressed because they are made up of densely packed atoms held rigidly together by chemical bonds. As the solid tire is forced to change shape as you ride over potholes and bumps, those chemical bonds will distort, and many of them will stay distorted. Thus, when the tire is supposed to switch back to its original shape, the bonds in the solid will not snap back into their original positions.

We are now ready to consider the first of our gas laws, which are nothing more than formal statements of commonsense observations about the way gases behave. You will read about several gas laws throughout the rest of this chapter, but there is no need to memorize them. Your intuition about the fundamental behavior of gases is enough to remind you about each of them, as you will see.

The first gas law is named for Avogadro, the same man whose number tells us how many units there are in 1 mol of a substance. Recall that 1 mol is equivalent to 6.02×10^{23} units of something. It is a number usually used to count extraordinarily tiny things like molecules and atoms. **Avogadro's law** uses the mole concept to describe how different numbers of atoms and molecules contribute to the properties of gases. To understand this law, we can imagine that we have a cylinder capped with an airtight piston and containing a small amount of gas (Figure 8.13). The cylinder also has an inlet valve through which gas can be added in much the same way that air is added to a tire. The piston is held in a position some distance up from the bottom of the cylinder because the gas molecules are colliding with the piston. These collisions keep the piston from falling to the bottom under the influence of gravity.

If we add more gas to the cylinder via the inlet, there are more gas molecules hitting the interior surface of the cylinder, and the piston moves upward to offset this. As the piston ascends, the interior surface area increases, giving the molecules more area with which to collide. As more and more gas molecules are added to the cylinder, the number of collisions of molecules with the interior surface increases, but so does the interior surface area. Thus, at every position of the piston, the number of collisions with the interior surface stays constant so the pressure stays constant too. Thus, we can say that the volume increases in the cylinder as the number of

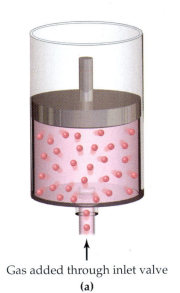

Gas added through inlet valve

(a)

(b)

◀ **FIGURE 8.13 Gas volume and amount of gas present. (a)** In a cylinder fitted with a piston, the pressure of the gas molecules inside determines the height of the piston and therefore the volume of the cylinder. **(b)** When more gas molecules are added to the cylinder, the piston rises and the volume increases.

molecules or number of moles of gas increases. Avogadro might have said it this way:

As the number of moles of gas increases, so does the volume.

We can write this relationship using symbols and a proportionality symbol, $\propto$:

$$V \propto n$$

where n represents the number of moles of gas and V represents the volume of the gas. The proportionality sign tells us simply this: as the number of moles goes up, so does the volume. Likewise, as the number of moles goes down, so does the volume.

QUESTION 8.8 The pressure exerted by a gas inside a container depends on the number of collisions the gas molecules make with the interior surface of the container. In the experiment we just described, the pressure inside the cylinder does not change. If more gas molecules are being added, how can this be true?

ANSWER 8.8 If the cylinder volume were unchangeable, the pressure in the cylinder would increase as molecules are added because there would be more collisions on the same amount of interior surface area. However, in this example, the cylinder volume changes because the piston is movable. As the piston moves upward, the interior area increases (and the volume also increases), and so the rate at which the gas molecules collide with the interior surface does not change.

As we explore the gas laws, you will see that they all have something in common. In each case, only two physical properties are permitted to change at one time. In the experiment represented in Figure 8.13, for example, the only two things that change as gas is added are the number of moles of gas and the volume. We refer to things that can change in an experiment as **variables**. As you will see, the four variables commonly manipulated in gas experiments are temperature T, pressure P, number of moles of gas n, and volume V, all depicted in Figure 8.14. In Figure 8.13, the number of moles of gas is increased by forcing gas molecules into the cylinder. Volume is the only other thing permitted to change, and the experiment is designed to tell us what happens to the volume (second variable) as the value of n (first variable) increases. The other two variables depicted in Figure 8.14, pressure and temperature, were not permitted to change, and we say that they were *held constant*.

QUESTION 8.9 A bicycle tire runs over a tack and deflates. How do the four variables that we usually keep track of in gas experiments change when this occurs?

ANSWER 8.9 As the tire deflates, it releases moles of pressurized gas. Thus, n and P decrease. Clearly, V decreases because the tire gets smaller, but T will not change substantially as the tire deflates.

We know from experience that Avogadro was right in saying that the volume of a gas increases when more gas is added to a container while the pressure is held constant. For example, we know that when we blow

The quantity n is the total number of **moles of gas** found in the **volume**, V, of the container.

Pressure, P, depends on the frequency of collisions with the walls of the container.

Temperature, T, depends on the average speed of the gas molecules.

▶ **FIGURE 8.14 The four variables of the gas laws.** Its pressure, temperature, volume, and number of moles all play a part in how a gas behaves.

up a balloon, the balloon gets bigger (Figure 8.15a). We know that when we push air from our lungs into our cheeks while keeping our mouth shut, our cheeks puff out (Figure 8.15b). We know that when we inhale air into our lungs, the size of our chest cavity increases (Figure 8.15c). It is second nature to us: as the number of moles of gas increases, volume also increases.

▶ **FIGURE 8.15 Avogadro's law in action.** In any flexible container, gas volume is proportional to the number of moles of gas present. **(a)** When you blow up a balloon, you are increasing the number of moles of gas (air) it contains. The volume of the (flexible) balloon increases. **(b)** Your cheek volume increases when the number of moles of gas (air) in your (flexible) mouth increases. **(c)** When you inhale a larger-than-usual amount of air, the volume of your (flexible) lungs increases noticeably.

(a)

(b)

(c)

QUESTION 8.10 To test Avogadro's law for yourself, you decide to use a bicycle pump to force air into a metal box that has a gas inlet valve. What is wrong with your plan? What variable have you mistakenly introduced into your experiment?

ANSWER 8.10 By using a metal box, you have not built flexibility into your container. Avogadro's law involves the variation of V with n, but there is no way for the volume of the box to change. As a result, the increase in the amount of air in the box will increase the pressure while the volume and temperature stay constant.

8.4 | Under Pressure

Imagine you live at sea level, say in San Francisco. Deciding to take a skiing vacation in the Colorado Rockies, you pack your bags and hop on a plane. When you arrive in Colorado and unpack your bags, you notice that your plastic shampoo bottle is *bigger* than it was when you packed it back at sea level (Figure 8.16). You carefully open it, and a gentle whish of air comes out. On the return trip, you arrive in San Francisco to find that your shampoo bottle is *smaller* than it was when you packed in Colorado. What's going on?

Although we cannot see them, atoms and molecules in the air are bouncing around and ricocheting off us and everything around us. Our bodies provide a surface for bouncing gas molecules that is no different from the interior surface of a cylinder with a movable piston or the interior surface of a tire. Every surface exposed to air is constantly bombarded by the gas molecules making up the air. However, surfaces exposed to air at different altitudes are bombarded with different numbers of molecules. To understand why, we must remember two things: (1) gravity pulls everything downward toward Earth's surface and (2) the pull of gravity gets stronger as you move closer to Earth. Atoms and molecules are not exempt from the pull of gravity. They are pulled downward just like everything else and they feel that pull more at sea level than at high altitude. Thus, we can imagine that, at sea level, the molecules in air are relatively close together because there are more of them in a given volume than in the same volume in, say, the mountains of Colorado. Another way to say this is air is *denser* at sea level than at high altitude, and its density (Section 7.3) changes through all points in between (Figure 8.17).

The mystery of the distorted shampoo bottle is solved by considering the density of air at the two altitudes. The denser air in San Francisco means that, relative to Colorado, there are more collisions between air molecules and the outside of the bottle. You could also say that the pressure on the outside of the bottle is greater in San Francisco than in Colorado. When you pack your shampoo in San Francisco, the density of air is the same on the inside and outside of the bottle, and so the bottle maintains its normal shape. When you carry that sealed bottle to the mountains, though, the pressure on the inside of the bottle (the "sea-level" pressure) is now greater than the pressure on the outside (the "high-altitude" pressure). Molecules in the air inside the bottle push outward, and the bottle expands.

▲ FIGURE 8.16 **An imbalance in air pressure.** When the cap was put on this plastic shampoo bottle at sea level, the air pressure inside the bottle was in equilibrium with the air pressure outside. Later, at an altitude at which the air pressure was lower than at sea level, the air trapped inside the bottle exerted enough pressure on the inside walls to distort the shape of the (flexible) container.

▼ FIGURE 8.17 **Relationship between altitude and air density.** An imaginary column of air starting at Earth's surface (sea level) and extending into the atmosphere will have fewer and fewer air molecules and atoms per unit volume as the distance from sea level increases. In other words, the air density decreases with increasing altitude.

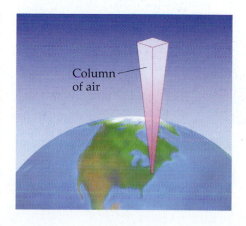

Column of air

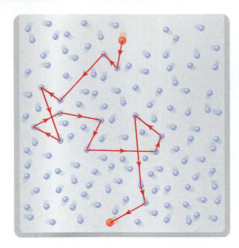

▲ FIGURE 8.18 **Gas molecules are always in motion.** They move at high speeds and change direction whenever they collide with other gas molecules or with the walls of their container. Here, a representative gas molecule (red) is shown on a typical path from one part of its container to another.

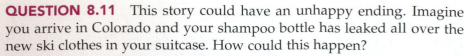

QUESTION 8.11 This story could have an unhappy ending. Imagine you arrive in Colorado and your shampoo bottle has leaked all over the new ski clothes in your suitcase. How could this happen?

ANSWER 8.11 When you get to Colorado, the pressure inside the bottle is higher than the pressure outside. The inside pressure pushes against the shampoo, forcing it up against the covered hole in the bottlecap. The pressure pushing on the cover from the outside is lower, and as a result of this difference in pressure the cover is pushed outward and pops off.

One way to envision gas density is to think about the motion of individual molecules in the gas phase. We know that they travel at very high speeds and randomly careen from one place to another, colliding occasionally with other gas molecules and with the walls of their container (Figure 8.18). It follows that those collisions will be more frequent in a high-density gas than in a low-density gas. This can be thought of quantitatively in terms of the **mean free path** of a molecule, defined as the average distance the molecule travels between collisions. For example, on average, the mean free path of nitrogen molecules at sea level is about 60 nm. At high altitude, though, the mean free path should be greater, because at high altitude there are fewer molecules in a given volume, and this is exactly what we find: the mean free path of nitrogen molecules on the summit of Mount Everest is about 180 nm, three times longer than the mean free path at sea level. At the top of Everest, molecules in the air are much farther apart than at sea level. This is why most climbers who scale Everest rely on supplemental oxygen to get them to the top (Figure 8.19).

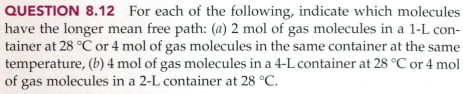

QUESTION 8.12 For each of the following, indicate which molecules have the longer mean free path: (*a*) 2 mol of gas molecules in a 1-L container at 28 °C or 4 mol of gas molecules in the same container at the same temperature, (*b*) 4 mol of gas molecules in a 4-L container at 28 °C or 4 mol of gas molecules in a 2-L container at 28 °C.

ANSWER 8.12 (*a*) The more moles of gas in the container, the more frequent are the collisions. Thus, the less-dense gas (2 mol in the container) will have the longer mean free path. (*b*) When the number of molecules is the same, the molecules encounter one another less frequently in a larger container. Consequently, the molecules in the 4-L container have the longer mean free path.

▲ FIGURE 8.19 **Climbing high mountains requires supplemental oxygen.** A 1953 photograph showing a climber preparing to climb Mount Everest with oxygen-containing tanks on his back.

Although we cannot see it and we really do not feel it, we have evidence that air is always colliding with our bodies. As we have just seen, however, the pressure pushing on us varies from location to location. It also changes with the weather. A weather system may bring high pressure (when the pressure of air molecules on us is greater than normal) or low pressure (when the pressure of air molecules is lower than normal). The pressure exerted on us by the air in the environment, called **atmospheric pressure**, fluctuates constantly, and part of the art of weather prediction involves keeping track of pressure fluctuations in the atmosphere.

Figure 8.20 shows a rudimentary, yet accurate, version of a *barometer*, a device for measuring atmospheric pressure. The pool of mercury in which a glass column sits is open to the air and therefore subject to the downward force of air molecules being pulled toward Earth by gravity. A force is most simply thought of as a push or pull (Figure 8.21a), and pressure is defined as the force exerted over a given surface area (Figure 8.21b). The higher the pressure on the surface of the mercury in the open dish in Figure 8.20, the farther up the tube the mercury moves. The height of the mercury column in the tube is therefore directly proportional to the atmospheric pressure. One of the most common units for expressing pressure is **millimeters of mercury (mmHg)**. This is nothing more than the distance, in millimeters, that the mercury in a barometer tube travels as a result of atmospheric pressure.

The **pascal** (Pa) is the SI unit for pressure, and 1 Pa is the pressure required to move mercury about 8 μm up a barometer tube, a value that is too small to measure. At sea level, atmospheric pressure is typically close to 760 mmHg, which is equivalent to 100,000 Pa. You probably will not encounter the pascal as often as you encounter units based on the distance mercury travels up a tube. The latter system of units is easy to use and is the one most often used by meteorologists. However, it is important to recognize the SI unit pascal because it is used in scientific work to measure gas pressures. Figure 8.22 shows two daily weather forecasts, one expressing pressure in inches of mercury and the other in kilopascals (kPa).

QUESTION 8.13 In the scientific literature, the *kilopascal* is sometimes used to report pressure. What is atmospheric pressure at sea level in kilopascals?

ANSWER 8.13 The prefix *kilo-* represents a factor of 1000. Therefore, there are 1000 Pa in 1 kPa. Using dimensional analysis and the sea-level value of 100,000 Pa from the text, we get

$$100{,}000 \text{ Pa} \times \frac{1 \text{ kPa}}{1000 \text{ Pa}} = 100 \text{ kPa}$$

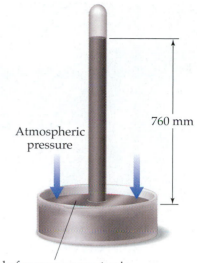

Atmospheric pressure

760 mm

Pool of mercury open to air

▲ **FIGURE 8.20** **A crude barometer.** Atmospheric pressure pushes on the surface of the mercury liquid in the bowl, forcing the mercury up into the evacuated tube. The height of the mercury column is one way to express the magnitude of the atmospheric pressure.

65 °F

Partly cloudy

UV index:	7 High
Dew point:	50 °F
Humidity:	58%
Visibility:	Unlimited
Pressure unit:	30.20 inches of mercury
Wind:	Variable at 6 mph

(a)

27 °C (80 °F)

A few clouds. Fog.

Wind chill:	42 °C (107 °F)
Wind:	7 km/h E
Relative humidity:	100%
Dewpoint:	27 °C (80 °F)
Pressure unit:	100.5 kilopascals
Visibility:	4.0 km
Ceiling:	unlimited

(b)

▲ **FIGURE 8.22** **And now for today's weather.** **(a)** In a forecast for Seattle, the unit used to report the atmospheric pressure is inches of mercury. **(b)** In a forecast for Hanoi, Vietnam, the unit used to report the atmospheric pressure is kilopascals.

(a)

8 in.

8 in.

(b)

▲ **FIGURE 8.21** **The difference between force and pressure.** **(a)** You exert a force on a door when you push it open. **(b)** Pressure is defined as force per unit of surface area. If you exert a force of 50 lb over a surface area of 64 in.^{2}, you are exerting a pressure of 0.78 pound per square inch (psi).

▲ **FIGURE 8.23 The mountain-biking capital of the world.**   In our thought experiment, we consider a tire moving over unforgiving terrain, like the rocks of Moab, Utah.

V decreases, T and n held constant

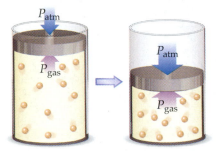

V larger, P lower   V smaller, P higher

▲ **FIGURE 8.24 Boyle's law.** With temperature *T* and number of moles *n* held constant, reducing the volume of a gas-filled container increases the pressure of the gas. Pressure is inversely proportional to volume.

Now that we have a clear understanding of pressure and how it can be measured, we can make pressure one of the two variables we change in our experiments on gas behavior. Let us return to our bicycle tire and imagine that it is on a mountain bike and you are riding over unforgiving terrain in the mountain-biking capital of the world, Moab, Utah (Figure 8.23). We already know what happens when a tire hits a bump and about the compressibility of a gas-filled tire: when the tire hits the bump and compresses, the tire volume gets a little bit smaller. In this thought experiment, the two variables we allow to change are volume and pressure, and we assume that the tire temperature and the number of moles of gas molecules in the tire remain constant.

In our thought experiment, we would like to know how the pressure in the tire changes as the volume changes. Before the tire hits the bump, its volume contains a fixed number of air molecules, and those molecules are hitting one another and the inside walls of the tire. The force they exert over the inside surface is the pressure of the tire. When the volume decreases slightly as the tire hits the bump, there is a decrease in the surface area with which the air molecules in the tire can collide. The number of molecules has not changed, though, and so the number of collisions they make with the inside surface of the tire must increase. In other words, the pressure in the tire must increase. We can express this observation, as we did before, in terms of a proportion:

$$P \propto \frac{1}{V}$$

This expression tells us that volume and pressure are *inversely proportional* to each other—as volume decreases, pressure increases; as volume increases, pressure decreases (Figure 8.24). This principle of the behavior of gases is known as **Boyle's law** after Robert Boyle, a seventeenth-century British scientist.

It is not necessary to memorize Boyle's law because the relationship between volume and pressure is already second nature to you. You already know, for example, that an inflated balloon compresses when you sit on it. Compression means the balloon volume decreases, and you may even worry that the balloon will burst because you sense that its pressure has increased as a result of the decrease in volume. Gas laws are really nothing more than common sense, and although you may not realize it, you already understand them.

8.5 | Turn Up the Thermostat

Back in Chapter 2, we learned about moles and how chemists use them as a counting tool for very large numbers of things, especially atoms and molecules. Up to this point, we have considered moles only in terms of liquids and solids, but the mole is also useful for keeping track of numbers of atoms and molecules in the gas phase. With gases, however, we always have four variables to keep track of. So if you would like to know what 1 mol of molecules looks like in the gas phase, you must define the conditions of temperature, volume, and pressure under which the gas will exist.

Let us consider conditions in Juneau, Alaska, as an example. This city is at sea level, meaning that we can take the normal atmospheric pressure there to be 760 mmHg (Section 8.4). In November of a typical year, the average temperature in Juneau is 0 °C. On a November day in Juneau, we are

How Does a Thermos Keep Cold Things Cold and Hot Things Hot?

We are all familiar with kitchen appliances that keep things cold—refrigerators—and others that keep things warm—ovens. Thermos bottles do both, but how? The answer is that thermos bottles cannot heat up or cool down liquids. If you put room-temperature coffee in your thermos, the coffee will not become

hot, because a thermos works only to *maintain* a temperature. One way a thermos does this is with its mirrored inside surface. Part of the energy given off by a hot substance is in the form of infrared radiation. This energy is bounced back into the interior of the thermos by reflection off the mirrored surface. As we are about to see, though, there are ways for heat to be transferred through the walls of a thermos bottle.

Let us revisit the fast-moving molecules of the gas phase to understand another feature of the thermos. As the temperature of a gas is increased, the molecules move faster. If the gas is in a closed container, the collisions the molecules make with the container walls impart energy to the atoms in the walls. If the walls are made of solid aluminum, for example, hot molecules striking the inner aluminum surface cause the atoms in the solid aluminum to vibrate a bit faster, although they stay in their solid lattices. Pretty soon, the aluminum atoms are all vibrating faster as the energy of the heat is passed though the

crystalline lattice from atom to atom. This process, whereby energy is transferred from one atom to its neighbor, is called *conduction*.

The increased vibrational energy in the aluminum walls is then imparted to the air molecules outside the container, so that the air gets a bit warmer. The transfer of energy to the air makes the air molecules near the aluminum surface move a bit faster. These warm, faster-moving molecules gradually move away from the surface, and this process, whereby heat moves as a result of the flow of molecules, is known as *convection*. Conduction and convection, both of which transfer energy from one place to another, happen naturally all the time. Hot things naturally cool down, and cold things get warmer. Eventually, warm things and cold things equilibrate with their surroundings and take on the same temperature as a result of convection and conduction (Figure 8.25).

The job of a thermos bottle is to keep hot coffee or cold lemonade inside from equilibrating with its

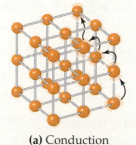

Heat energy moves through a crystalline lattice

(a) Conduction

◀ **FIGURE 8.25**   **Conduction versus convection.**
(a) In conduction, heat energy moves through a crystalline lattice from one atom in the lattice to another. **(b)** In convection, heat energy is transferred as heated gas molecules move away from a heat source. Here, the candle flame imparts some heat energy to the air molecules close to the flame. The warmed air molecules then move away from the flame, carrying heat energy with them.

Heat energy is transferred as warmer, faster moving gas molecules move from one location to another

(b) Convection

surroundings. A thermos is constructed from two containers, one inside the other (Figure 8.26). The space between the containers is sealed, and most of the air is removed from the space, leaving a **vacuum**, or absence of gas. As a result, there are only a few atoms and molecules in the space between the two containers. Thermal energy from hot coffee in the inner container can be *conducted* from the coffee to the inner wall of the thermos. However, that inner wall cannot effectively *convect* its heat to the evacuated space because the number of molecules in the space is almost zero. Convection through the evacuated space does not happen, and the coffee stays hot.

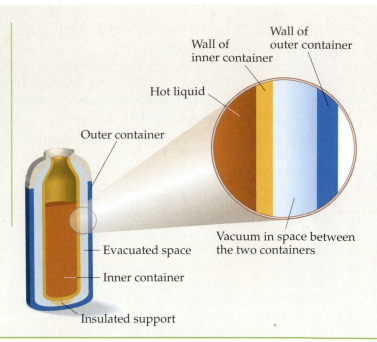

▶ **FIGURE 8.26 How a thermos bottle works.** A thermos keeps its contents warm because of an evacuated space between an inside container and an outside container. In the blowup shown here, hot areas of the thermos are brown and cool areas are blue.

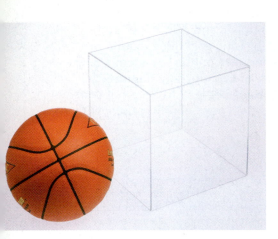

▲ **FIGURE 8.27 Molar volume.** A basketball will just fit into the box, which has a volume of about 22.4 L.

going to measure the volume of 1 mol of a gas that has a pressure of 760 mmHg and a temperature of 0 °C. Under these very specific conditions, 1 mol of gas occupies a volume of exactly 22.414 L. This volume is called the **molar volume**, and to give you an idea of its size, Figure 8.27 shows a transparent box with a volume of about 22.4 L. Scientists use the combination of 0 °C and 760 mmHg—known as **standard temperature and pressure**, or **STP**—often because it provides a convenient set of conditions for gas-phase experiments anywhere in the world.

The marvelous thing about molar volume is that 22.414 L is the space occupied by 1 mol of an ideal gas at STP everywhere, always. If you measure the volume occupied by 1 mol of a gas at STP in Thailand, that volume will be 22.414 L. If you perform the measurement in Green Bay or Mozambique or Rio de Janeiro or Vancouver or Honolulu, it will be exactly the same. If you perform the measurement on helium gas, nitrogen gas, or any other gas, it will be exactly the same.

QUESTION 8.14 If you have a box that has a volume of 11,200 mL at STP and that box contains nothing but air, how many moles of gas molecules will it contain?

ANSWER 8.14 With half the molar volume, there would be 0.500 mol of gas molecules in the box:

$$11{,}200 \ \text{mL} \times \frac{1 \ \text{L}}{1000 \ \text{mL}} \times \frac{1 \ \text{mol}}{22.414 \ \text{L}} = 0.500 \ \text{mol}$$

What happens to the volume of 1 mol of gas when its temperature is not 0 °C? We know from Boyle's law that if we change the pressure, the volume must change in the opposite direction (and vice versa), but what about temperature? How does changing the temperature affect the volume of a gas? If we think of 1 mol of gas at 0 °C as being enclosed in an expandable cube that is 28.2 cm on each side, then the total volume occupied by the gas is the molar volume:

$$28.2 \text{ cm} \times 28.2 \text{ cm} \times 28.2 \text{ cm} = 22{,}400 \text{ cm}^3$$

$$22{,}400 \ \cancel{\text{cm}^3} \times \frac{1 \ \cancel{\text{mL}}}{1 \ \cancel{\text{cm}^3}} \times \frac{1 \text{ L}}{1000 \ \cancel{\text{mL}}} = 22.4 \text{ L}$$

If we increase the temperature, what happens? Imagine that at 0 °C the gas molecules inside the expandable cube are moving at some average speed. As we increase the temperature, the average speed of the gas molecules will increase. This means that each molecule will collide with the inside surface of the cube more often than it did at 0 °C. The cube will expand as the number of collisions increases. Thus, we can say that as the temperature increases, the volume also increases. The opposite must also be true: if we lower the temperature, the volume of the cube will decrease. We can write this as a direct proportion:

$$V \propto T$$

What happens if the volume is held constant? That is, if our cube is not expandable, what happens as the temperature is increased? In this case, the molecules again speed up and collide with the walls more frequently. Because this time the volume is not able to change, however, the result is an increase in pressure inside the cube. Thus, at fixed volume,

$$P \propto T$$

QUESTION 8.15 What happens to the mean free path of the molecules in a sample of gas (*a*) as the temperature is increased and the volume is held constant, (*b*) as the volume is decreased and the temperature is held constant?

ANSWER 8.15 (*a*) An increase in temperature means molecules are moving faster and there is less time elapsed between collisions. Thus, the mean free path gets shorter. (*b*) When the volume decreases as the temperature stays unchanged, a molecule will collide with the walls of the container and with other molecules more frequently. Again, the mean free path gets shorter.

In each case just described, we were allowed to change only two variables at a time. In the first case, we allowed volume and temperature to change and found that volume is directly proportional to temperature. In the second case, we allowed temperature and pressure to change and found that pressure, too, is directly proportional to temperature. The first of these experiments was performed by a man named Jacques Charles. Charles, an eighteenth-century inventor, was best known by his contemporaries for his experiments on hot-air balloons. His most successful balloon flight took place in 1783, lasted 45 min, and landed him 15 miles from his starting point. Apparently, his arrival so startled the townspeople that they destroyed his balloon. He is famous for a remark he made when asked about the usefulness of his work on how gases are affected by changes in temperature. To this he replied, "And of what use is a newborn baby?"

Based on his experiments with volume and temperature, Charles formulated what we now call **Charles's law**:

When the temperature of a gas is increased, the volume also increases.

Charles's law is illustrated in Figure 8.28*a*.

LIVE ART
Amonton's Law and Charles's Law

▼ **FIGURE 8.28** **Charles's law and Amontons' law.** (a) Charles's law tells us that with pressure *P* and number of moles *n* held constant, increasing the temperature of a gas-filled container increases the volume of the gas. Volume is directly proportional to temperature. (b) Amontons' law tells us that with volume *V* and number of moles *n* held constant, increasing the temperature of a gas-filled container increases the pressure of the gas. Pressure is directly proportional to temperature.

(a) *T* increases, *P* and *n* held constant

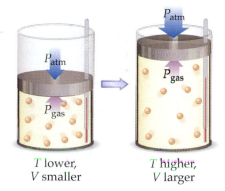

T lower, *V* smaller   *T* higher, *V* larger

(b) *T* increases, *V* and *n* held constant

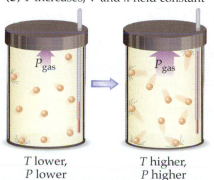

T lower, *P* lower   *T* higher, *P* higher

Why Is It More Difficult for Aircraft to Take Off at High Altitude or on a Hot Day?

Aircraft do not use thrust alone to push up and off the runway. The lift that occurs as a plane nears the end of the runway results from the movement of air molecules over each wing. Those air molecules are thrust downward as they flow over the rear edge of the wing. It is a law of physics that every force requires an equal and opposite force, and so the downward thrust of the air as it leaves the wing causes an equal upward thrust, and the plane lifts off the ground (Figure 8.29). How much power is created by the downward-thrusting air depends on the density of the air being pushed. The more molecules the air contains, the more effectively it can lift the aircraft.

Warm air is less dense than cool air. When an aircraft is taking off on a hot day, it must travel farther down the runway before it is able to lift away because the density of the air is lower. The same effect exists at high altitudes, at which the air is less dense than that at sea level. Telluride Airport in Colorado is the highest commercial airport in the continental United States. There are many summer days in Telluride when pilots elect to stay on the ground because the combination of high altitude and warm temperatures makes takeoffs extremely dangerous.

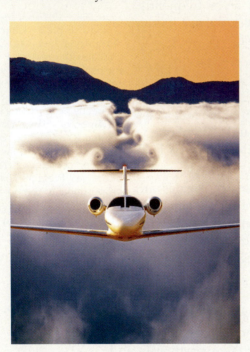

◄ **FIGURE 8.29** "Seeing" thrust. The downward thrust of air off the wings of the plane is evident in the cut this plane makes through a cloud.

The early work corresponding to the second experiment we discussed before, in which temperature was varied to see how that change affected pressure, was done by Guillaume Amontons, a little-known seventeenth-century scientist. His results are represented in Figure 8.28*b*, and we can summarize **Amontons' law** as follows:

If the volume of a gas is fixed, an increase in gas temperature results in an increase in gas pressure.

Up to this point, all our examples have involved gases mixed with one another thoroughly. Further, we saw that molecules introduced into an existing gas eventually blend with it because the gas molecules are moving at high speeds and do not interact with one another except to collide for an instant and then career away from one another. However, there are

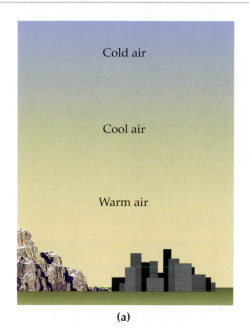

Cold air

Cool air

Warm air

(a)

Cold, clean air

Warm, still air

Cold, dirty air
(smog)

(b)

◀ **FIGURE 8.30 Thermal inversion.
(a)** Under normal environmental conditions, air is warmest near Earth's surface and becomes cooler as distance up from the surface increases. **(b)** In a smoggy area in which a thermal inversion occurs, the smog is trapped near the surface, and a cloud of polluting molecules hangs over the area.

times when gases do not mix together thoroughly. Instead, they can form layers, but this happens only under special temporary circumstances such as unusual weather conditions. Usually, the temperature of atmospheric gases decreases as their distance from Earth's surface increases, as shown in Figure 8.30a, because it is the ground that heats the air. Under normal conditions, the warm air near the ground moves steadily upward. This normal behavior can be reversed, however, in some instances. For example, in the wee hours of the morning, the air space located next to a slope will receive a downdraft of cool air from above. This incoming cool air will displace the warm air at the bottom of the slope, pushing that warm air upward. We refer to this juxtaposition of gas layers—warm above cool—as a *thermal inversion*.

Thermal inversions are a natural phenomenon. They can be caused by different sets of conditions, including the presence of a slope, as just described. These inversions cause problems, though, when they occur in areas with high levels of air pollution. In most polluted areas, the polluting molecules are lifted up with the normal upward flow of warm air and eventually diffuse into the upper atmosphere. When a thermal inversion exists in a polluted area, however, the toxic pollutant molecules sit in the cool layer and the warm layer stays positioned above it, preventing gases from escaping into the upper atmosphere. The result is a thick layer of smog stuck at Earth's surface, a layer unable to diffuse upward (Figure 8.30b).

One location prone to thermal inversions combined with smog is Salt Lake City. Alerts went out before the 2002 Olympic Games urging citizens and Olympic organizers to curtail automobile usage so that athletes arriving for the games would not be greeted with a thick layer of unhealthy smog. Despite these efforts, the city was mired in smog during the games. Figure 8.31a shows an aerial photograph of Salt Lake City three days before the games began. The presence of brown smog in the cool, lower layer of the thermal inversion makes it possible to see the boundary between the cool air layer below and the warm air layer above it. When the games ended, the city began a campaign to improve air quality (Figure 8.31b).

(a)

(b)

▲ FIGURE 8.31 **Salt Lake City air quality before and after 2002 Olympics.** **(a)** The brown cloud over Salt Lake City is due to a thermal inversion combined with pollution. **(b)** After the 2002 Olympic games in Salt Lake City, the city began an advertising campaign aimed at reducing pollution.

There are instances when the layering of gases has caused catastrophic problems. At Lake Nyos, Cameroon (Figure 8.32), on the evening of August 26, 1986, people who lived near the lake heard a deafening blast caused by an enormous release of carbon dioxide gas from the lake. Experts still do not agree on the source of the dissolved carbon dioxide that permeated the lake, although it was clearly related to local volcanic activity. They do agree that on that night in 1986, the gas suddenly reached its bursting point, and carbon dioxide was released in a gigantic whoosh, as though the surface of the lake were a champagne bottle that had just been uncorked. The carbon dioxide released as the lake burst open settled in a thick layer that covered the lake and houses in a valley below it. The carbon dioxide layer displaced the normal air and made breathing impossible. More than 1700 people were asphyxiated.

One question that may come to mind about Lake Nyos is this: if our gas laws tell us that all molecules are the same in a gas and that the identity of the molecules is unimportant, then why does carbon dioxide gas form a layer beneath normal air? What keeps the two gases from mixing? The answer lies in the molar masses and quantities of the gases under consideration. Think about clean air for a moment. According to Table 8.1, clean air is made up mostly of nitrogen molecules, molar mass 28 g/mol, and oxygen molecules, molar mass 32 g/mol. Because air is about 78 percent nitrogen and 21 percent oxygen, the average molar mass of clean air is about 28.5 g/mol. Because their molar masses are similar, nitrogen and oxygen molecules mix together in all proportions. If you add a small amount of a different gas to air, that gas will mix with the existing air easily. We have seen examples of this throughout this chapter: sarin gas in a Tokyo subway, a fishy odor in a closed room, the smell of cooking garlic, perfume, cigarette smoke, and so on. In these cases, a small quantity of a "foreign" gas diffuses easily into clean air.

◄ FIGURE 8.32 **Tragedy at Lake Nyos.** The location of Cameroon on the African continent. The inset shows the precise location of Lake Nyos in the western part of the country.

The situation changes, however, when the quantity of the foreign gas is very large. At Lake Nyos, the properties of the carbon dioxide molecule came into play because there were enough molecules to displace clean air and form a blanket beneath it. The molar mass of carbon dioxide is 44 g/mol, much greater than that of either nitrogen or oxygen molecules. Because they are heavier, carbon dioxide molecules move more slowly than the lighter oxygen and nitrogen molecules. Because they move more slowly, they collide with one another and with objects less frequently. This means that, in a sample of pure carbon dioxide gas, the molecules are not bouncing off one another as energetically as the molecules found in clean air. Because they are not as energetic and not bouncing off one another as vigorously, the molecules of carbon dioxide are closer together, and their density is greater. Add to this the fact that the heavier carbon dioxide molecule is pulled down by gravity more than the molecules of clean air, and it is reasonable that a layer of carbon dioxide can form beneath clean air.

Whenever there are large quantities of a gas that have a molar mass very different from that of clean air, you can expect that gas to form a layer either above or below the air. For example, ammonia gas, NH_3, floats on air, and air floats on butane gas, $CH_3(CH_2)_2CH_3$. In other words, gases that are ordinarily homogenous, like clean air, can form layers under special conditions, such as when there is a thermal inversion. Thus, we say that gases *usually* blend together to make homogenous mixtures, *but not always*.

QUESTION 8.16 If you are taking a laboratory chemistry course, you may have been reminded (repeatedly) to turn off the natural gas taps on the laboratory benches. Natural gas leaks are especially dangerous because the gas forms a highly flammable layer in the room. Would natural gas, which is mainly methane, CH_4, form a layer of gas at the floor or at the ceiling?

ANSWER 8.16 The molar mass of methane is about 16 g/mol, less than that of air (about 28.5 g/mol, as noted before). Therefore, methane floats on air and forms a layer at the ceiling.

8.6 | Designer Gas Laws

By now we have been introduced to four gas laws—Avogadro's, Boyle's, Charles's, and Amontons'—and have given some thought as to why each makes sense to us. At this point, you may not remember which law is which, but that is okay because you already have a fundamental understanding of how gases behave. In this chapter, we have simply asked you to acknowledge this and to practice thinking about gases from different perspectives.

Table 8.2 summarizes the gas laws introduced in this chapter. You can see that they are concerned with four variables: temperature T (in Kelvins), volume V, pressure P, and number of moles of gas n. For each gas law, two of these variables are permitted to change, and the other two are held constant. Rather than having to deal with four separate laws, though, is there any way to combine all these variables into one relationship? The answer is yes. The **general gas law** includes all four variables:

$$\frac{PV}{nT} = \text{constant}$$

What does it mean to say that this group of variables is equal to a constant? It means that, for any given set of conditions for which you are measuring the properties of an ideal gas, the value of PV divided by the value nT will always give you the same number. If you consider your bicycle tire on a cold day in January when the tire pressure is low and the tire looks flat (volume small), you can measure P and V and calculate PV and divide it by nT. This will give you some number. If you measure these variables for the tire on a blistering hot day in July when the tire is full and the pressure is maximal, dividing your new value for PV by your new value for nT will give you the same number you got in January. In fact, any measurement you do anywhere at any time will give you the same number for an ideal gas. Nature guarantees it.

This is a powerful concept that will allow us to manipulate our four gas variables in any way we please. Let us go back to our tire in January and July, and let us call the January conditions "before" (that is, before a change has taken place) and the July conditions "after" (after a change has taken place). We then represent the January tire pressure by P_b, the July tire volume by V_a, and so on, with the subscripts b and a standing for *before* and *after*, respectively. The general gas law tells us that the fractions for the before and after conditions are equal to the same constant value:

TABLE 8.2 Summary of Gas Laws			
Relationship	What question requires that you understand this relationship?	What's the answer?	Whose law is that?
$V \propto n$	If you're adding air to a tire, what happens to the volume of the tire?	Adding air increases n, the number of moles of gas. Increasing n increases the volume of the tire. Volume is directly proportional to amount of gas present.	Avogadro's
$P \propto \dfrac{1}{V}$	You sit on a balloon, and it does not pop. As you change the balloon's volume by sitting on it, what happens to the pressure of the gas in the balloon?	The pressure of gas in the balloon increases because molecules have a smaller area of the inner surface with which to collide. Pressure is inversely proportional to volume.	Boyle's
$V \propto T$	On the first really cold day of autumn, you notice that your car's tires all look flat and need air. What effect does temperature have on the size of your tires?	A decrease in temperature means that the molecules in your tire are less energetic and so collide less forcibly with the inner walls of the tire. Because the tire is made of flexible material, its volume decreases. Volume is directly proportional to temperature.	Charles's
$P \propto T$	On the first really cold day of autumn, you notice that your car's tires all look flat and need air. What effect does temperature have on the pressure of air in your tires?	Another way of looking at the answer for Charles's law is in terms of the pressure change. As the temperature goes down, collisions against the inner walls of the tire decrease, and thus pressure decreases. Pressure is directly proportional to temperature.	Amontons'

$$\frac{P_b V_b}{n_b T_b} = \text{constant} \qquad \frac{P_a V_a}{n_a T_a} = \text{constant}$$

Conditions that exist in your tire in January ("before")   Conditions that exist in your tire in July ("after")

This in turn means that the fractions are equal to each other:

$$\frac{P_b V_b}{n_b T_b} = \frac{P_a V_a}{n_a T_a}$$

Think for a moment about sitting on a balloon, the situation we used earlier to describe Boyle's law. In this situation, pressure and volume change, but the number of moles of gas and the temperature do not. If the value of nT is the same before and after you sit on the balloon, then the denominator $n_b T_b$ is the same as the denominator $n_a T_a$. For the sake of argument, let us say that $n_b T_b$ and $n_a T_a$ are both equal to 2. Because you are dividing both sides by the same value, the nT terms cancel and can be removed from the equation:

$$\frac{P_b V_b}{\cancel{n_b T_b}} = \frac{P_a V_a}{\cancel{n_a T_a}} \qquad \frac{P_b V_b}{2} = \frac{P_a V_a}{2} \qquad P_b V_b = P_a V_a$$

This tells us that, whenever the pressure and volume are changing but everything else is held constant, the product PV must also stay constant. In order for this to be true, P and V must adjust relative to each other. If PV must stay constant, volume must go *down* as pressure goes up and *up* as pressure goes down. Another way of saying this is to say that pressure and volume are inversely proportional to each other:

$$P \propto \frac{1}{V}$$

You might recognize this as Boyle's law.

Consider an example using real numbers. Suppose an inflated exercise ball you are about to sit on is filled with air to a pressure of 1000 mmHg and has a volume of 60 L. Now you sit on the ball and squash it down to two-thirds of its original size (meaning that its volume decreases to 40 L). What is the pressure in the ball now? To figure this out, we assign the values we know to the variables in our familiar equation $P_bV_b = P_aV_a$ and solve for P_a. From the values shown in Table 8.3, we have

$$P_bV_b = P_aV_a$$

$$P_a = \frac{P_bV_b}{V_a} = \frac{(1000 \text{ mmHg})(60 \text{ L})}{40 \text{ L}} = 1500 \text{ mmHg}$$

The volume units cancel, and the result is in pressure units, as we expect. The pressure after you sit on the ball is 1500 mmHg, a value greater than before the balloon was subjected to the mass of your bottom. We knew intuitively this had to be the case. Pressure and volume are inversely proportional. Boyle's law is common sense.

We have just seen that the general gas law is very convenient *because we can derive any of the other gas laws from it*. Every gas law is no more than a before-and-after situation in which only two variables are permitted to change. If your algebra is rusty, Appendix B offers a quick review of how to move things around in equations.

QUESTION 8.17 A cylinder with a movable piston can be characterized with Boyle's law, assuming that the variables are pressure and volume and that temperature and number of moles of gas are held constant. If a cylinder at STP has a volume of 78 L and the piston is compressed until the pressure equals 5200 mmHg, what is the new volume of the cylinder?

ANSWER 8.17 At STP, $P = 760$ mmHg. Solving our standard expression $P_bV_b = P_aV_a$ for V_a and plugging in the values provided, we get

$$V_a = \frac{P_bV_b}{P_a} = \frac{(760 \text{ mmHg})(78 \text{ L})}{5200 \text{ mmHg}} = 11 \text{ L}$$

TABLE 8.3 Values for Variables Used to Calculate P_a	
What we know	**What we do not know**
$P_b = 1000$ mmHg	$P_a = ?$
$V_b = 60$ L	
$P_bV_b = 60{,}000$ mmHg $\cdot$ L	
$V_a = \dfrac{2}{3} \times 60 \text{ L} = 40 \text{ L}$	

You may be wondering why we cannot simply consult Table 8.2 to find out how variables are related to one another. Why do we need the general gas law to figure things out? To demonstrate why this law is useful, let us try an example that is not in Table 8.2. Imagine you have taken a job as a clown to pay your rent for the summer (Figure 8.33). Working a child's birthday party, you begin the day with a full tank containing 270 mol of helium gas at a pressure of 110,000 mmHg. Dozens of balloons and 5 hours later, the tank gauge reads 33,000 mmHg. How many moles of helium gas remain in the tank at the end of the party?

This question asks us about P and n, but a look at Table 8.2 shows that there is no equation in which P and n are the variables that are allowed to change. Fortunately, we do not need any formula relating n and P because we can begin with the gas laws we know and figure things out. Table 8.4 lists the before and after variables for this situation. Two things are changing—P and n—and two things are not changing—V and T. Because V and T are constant, we can eliminate them from the equation. We begin, as before, with all of the before and after variables:

$$\frac{P_b V_b}{n_b T_b} = \frac{P_a V_a}{n_a T_a}$$

Next, we eliminate things that do not change:

$$\frac{P_b \cancel{V_b}}{n_b \cancel{T_b}} = \frac{P_a \cancel{V_a}}{n_a \cancel{T_a}}$$

$$\frac{P_b}{n_b} = \frac{P_a}{n_a}$$

This last equation tells us that the value of P/n at the beginning of the party is equal to the value of P/n at the end of the party. Because P/n does not change, n must increase as P increases. The reverse is also true: if we decrease P, which is what happens at the birthday party, then n must also decrease. Another way to say this is pressure and number of moles of gas are directly proportional to each other:

$$P \propto n$$

To see how much the value of n decreases, we plug in our numbers and solve for n_a:

$$n_a = \frac{n_b P_a}{P_b} = \frac{(270\text{ mol})(33,000\ \cancel{\text{mmHg}})}{110,000\ \cancel{\text{mmHg}}} = 81\text{ mol}$$

▲ **FIGURE 8.33** **What you did on your summer vacation.** A clown working a children's birthday party will gradually use the helium from a pressurized tank to fill balloons.

TABLE 8.4 Values for Variables Used to Calculate n_a	
What we know	**What we do not know**
$n_b = 270$ mol	$n_a = ?$
$P_b = 110,000$ mmHg	
$P_a = 33,000$ mmHg	
$V_b = V_a$	
$T_b = T_a$	

QUESTION 8.18   If one balloon holds 1.89 mol of helium gas when fully inflated, how many balloons can be filled using a tank holding 105.8 mol of gas?

ANSWER 8.18

$$105.8 \; \cancel{\text{mol He}} \times \frac{1 \; \text{balloon}}{1.89 \; \cancel{\text{mol He}}} = 56 \; \text{balloons}$$

Our clown example illustrates two things. First, it shows that you can use the general gas law to figure out any relationship between any number of gas variables. In this case, we changed P and n, and we were able to figure out how they are related mathematically. We then used this equation to arrive at an answer to the question posed. To do this, we did not need to refer back to some individual gas law we had memorized. There is no name that goes along with the equation relating pressure and moles of gas. In fact, if you were able to figure this out on your own as you were reading, why not name this relationship after yourself?

_____'s Law

The second thing our clown example illustrates is that, although there is no formal name for a law that describes how n and P are related, you can deduce their relationship from common sense. If you have a gas cylinder full of pressurized gas (P and n large) and you use the gas to fill up balloons (n gets smaller), then there are fewer gas atoms colliding with the walls of the tank after the balloons have been filled. The value of P has dropped, meaning P and n are directly proportional to each other.

QUESTION 8.19   The average adult male has a total lung capacity of 5.8 L. When he exhales, some air remains in the lungs. The volume occupied by this unreleased air is referred to as *residual volume* and has a typical value of about 1.2 L. If a man retains 0.049 mol of air in his residual volume, calculate the number of moles of air in his fully inflated lungs. Assume that the pressure and temperature of the air in the lungs remain constant.

ANSWER 8.19   Let us call the fully inflated lungs our before condition, so that V_b is 5.8 L and n_b is our unknown. After the man exhales is our after condition, with $V_a = 1.2$ L and $n_a = 0.049$ mol. Using the general gas law, we can solve for the unknown n_b by removing P and T from the equation:

$$\frac{\cancel{P_b} V_b}{n_b \cancel{T_b}} = \frac{\cancel{P_a} V_a}{n_a \cancel{T_a}}$$

$$\frac{V_b}{n_b} = \frac{V_a}{n_a}$$

$$n_b = \frac{V_b n_a}{V_a} = \frac{(5.8 \; \cancel{\text{L}})(0.049 \; \text{mol})}{1.2 \; \cancel{\text{L}}} = 0.24 \; \text{mol}$$

The behavior of gases and the laws governing that behavior are summarized in Figure 8.34.

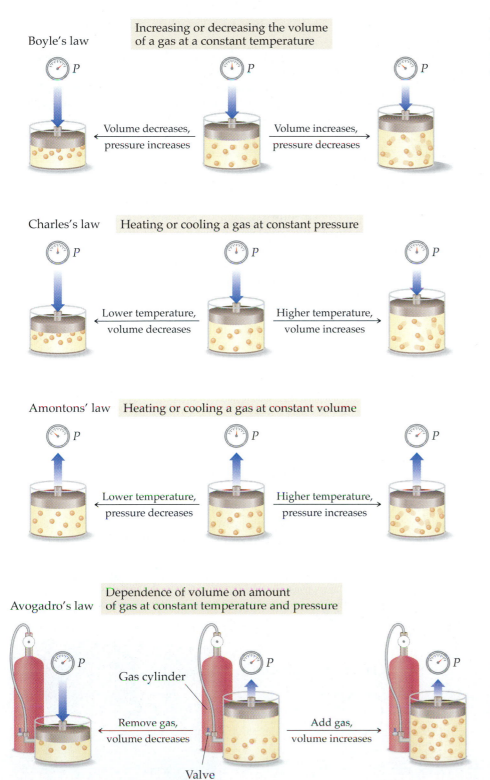

Boyle's law

Increasing or decreasing the volume of a gas at a constant temperature

Volume decreases, pressure increases

Volume increases, pressure decreases

Charles's law   Heating or cooling a gas at constant pressure

Lower temperature, volume decreases

Higher temperature, volume increases

Amontons' law   Heating or cooling a gas at constant volume

Lower temperature, pressure decreases

Higher temperature, pressure increases

Avogadro's law   Dependence of volume on amount of gas at constant temperature and pressure

Gas cylinder

Remove gas, volume decreases

Add gas, volume increases

Valve

◀ **FIGURE 8.34   Gas law summary.**
The four gas laws learned in this chapter: Boyle's, Charles's, Amontons', and Avogadro's.

ISSUES IN CHEMISTRY

Do Human Pheromones Really Exist?

That gases diffuse through the air might be more immediately relevant to your everyday life than you think. Although you probably will not notice when it happens, gases might even affect your love life. They certainly had a profound affect on the love life of a certain great peacock moth who had the misfortune of being hatched on the laboratory bench of French scientist Jean-Henri Fabre

in 1870. Fabre put the moth, a female, in a cage and thought no more about her until, a short time later, dozens of male giant peacock moths flew in the window. Fabre was convinced that some kind of chemical message was being dispatched by the female, although he was never able to identify any such agent.

Research on attractant molecules, such as the one given off by Fabre's moth, has continued, and, with modern technology, advances have been made in our understanding of them. In the 1950s, the term *pheromone* was coined. The roots of this word, *phero-* and *-mone*, derive from Latin words meaning *carry* and *excite*. A pheromone, therefore, is a molecule that is given off by an organism and moves though the air to be detected by other individuals of the same species. The response of those picking up the scent is often sexual.

Pheromones have been identified and characterized for dozens of species, from yeasts to mammals. Figure 8.35 shows a Colorado potato beetle, an insect that feeds on potatoes and destroys crops. The pheromone of this beetle has been identified and isolated, and is being used to lure beetles away from the crops.

There has been speculation about the existence of a human pheromone since before airborne chemical messengers were even known to exist. We have all experienced feelings of attraction or repulsion for other members of our species, and perhaps those feelings are the result of chemical messengers that diffuse through the air from person to person. There have been reports in the scientific literature claiming to have identified a human pheromone, and the latest research suggests that humans may manufacture steroid-like molecules and release them through their sweat glands. One such molecule is shown in Figure 8.36a. If you go to the World Wide Web and search for the name of this molecule, androstenone, you will not find many articles on its chemical structure and properties. Rather, you will find dozens of advertisements for products that contain it and other pheromone molecules (Figure 8.36b). These advertisements guarantee that you will be more attractive to men or to women, depending on which product you buy.

An informal ABC News investigation used a commercially available pheromone product on one woman, but not on her identical twin. (The story aired in March 1988 and was reported by Bill Ritter, an ABC News correspondent.) The women were sent to a bar and instructed to behave similarly and not approach men. The results surprised the investigators, who, it seems, were expecting to debunk the claims made by the product's manufacturer. The twin who did not wear the product was approached by 11 men, while the twin wearing the product was approached by more than 30 men!

This is only one unofficial experiment, true, but there have been other, more scientific studies that have shown that humans do produce some sort of messenger that is sensed by other humans. In fact, there has been speculation about the deleterious effects of constant bathing on the social interactions of overwashed Americans. Maybe our penchant for cleanliness is thwarting our bodies' natural plan for finding the right mate. So if you're trying to attract a mate, you may want to consider skipping your shower before your next date.

▼ FIGURE 8.35 **Beetle-mania.** The pheromone of this Colorado potato beetle has been identified and mass produced as a way to lure the beetles away from the crops they destroy.

▼ FIGURE 8.36 **Human pheromones.** **(a)** The structure of the androstenone molecule, a purported human pheromone. **(b)** Pheromones that will attract either women or men are available on line.

Androstenone

(a)

(b)

8.7 | Stay Cool: Ozone and Global Warming

There is nothing quite like spending a hot summer day at the pool or at the beach. You get really hot, you take a dip, you get out of the water, and then the cycle begins again. Sometimes it actually seems chilly when you leave the water, even on the hottest days. This phenomenon results from the evaporation of the water from your skin, a process that requires heat, as indicated in Figure 8.37. The heat used to evaporate the water comes from your skin, and when that heat is taken away, you feel cool.

This process, called *evaporative cooling*, can make a hot summer day bearable. It is also the basis on which refrigerators, air conditioners, and other kinds of chillers operate. In a refrigerator, for example, a liquid moves through a system of coils. This liquid, the *refrigerant*, is compressed into a small space, which increases its pressure (Boyle's law) and keeps it in the liquid phase. When the liquid is allowed to expand into a larger space, it becomes a gas, a process that requires heat. The heat for this process is taken from the inside of your refrigerator, and as heat is siphoned out of the refrigerator, the temperature inside drops.

You may be wondering what liquids can be used in refrigerators and air conditioners. As it happens, water is not an ideal refrigerant because its boiling point is relatively high. A more ideal refrigerant would have a lower boiling point and would lower the temperature inside the refrigerator more efficiently than water would. To see why this is true, think about what happens when you put a drop of ethanol and a drop of water on your skin. As it evaporates quickly because of its low boiling point, the ethanol makes your skin feel cold. The water, though, sticks around longer and evaporates more slowly. Therefore, even though evaporative cooling is happening, you do not feel the obvious sensation of cold that you do with ethanol. The liquids used in refrigerators and air conditioners have boiling points even lower than that of ethanol. Thus these refrigerants are able to cool down a space very efficiently by absorbing heat as they evaporate from liquid to gas.

In 1926, a class of molecules known as **chlorofluorocarbons** (CFCs) was discovered by a man named Thomas Midgely. (Midgely, incidentally, also developed leaded gasoline, prompting one historian to remark that Midgley "had more impact on the atmosphere than any other single organism in earth history.") CFCs are small, simple organic molecules that contain carbon, chlorine, fluorine, and sometimes hydrogen. Some examples are shown in Figure 8.38. These new wonder molecules, which were marketed under the name *Freon*, turned out to be ideal refrigerants. They replaced earlier refrigerant molecules, such as diethyl ether and ammonia, that were either toxic or highly flammable (and sometimes both). By the mid-1930s, super-cooling CFCs could be found in most household and industrial cooling appliances.

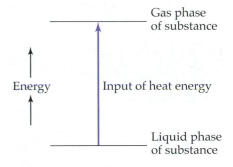

▲ **FIGURE 8.37** **Evaporative cooling.** The intermolecular interactions in a substance must be overcome in order for the substance to evaporate from the liquid state to the gas state. This process, called evaporative cooling, requires heat, and that heat is taken up from the surroundings. As a result, the temperature of the surroundings drops.

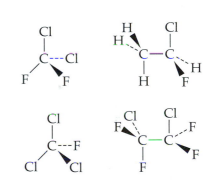

▲ **FIGURE 8.38** **Some representative CFC molecules.**

QUESTION 8.20 (a) Draw two examples of CFCs that contain one carbon atom and are not shown in Figure 8.38. (b) Describe the geometry around the carbon atom in each CFC molecule you have drawn.

ANSWER 8.20 (a) There are several that you may have drawn. For example,

▲ FIGURE 8.39 Ozone experts.
For their work on the depletion of the ozone layer, F. Sherwood Rowland, a scientist at the University of California, Irvine, and his post-doctoral fellow, Mario Molina, were two of the three recipients of the 1995 Nobel Prize in Chemistry.

▲ FIGURE 8.40 Structure of ozone.
The ozone molecule exists in two resonance forms.

Recall from Figure 5.21 (page 192) that ultraviolet light (200 to 400 nm) is more energetic than visible light (400 to 700 nm). The *longer* the wavelength of light, the *lower* its energy.

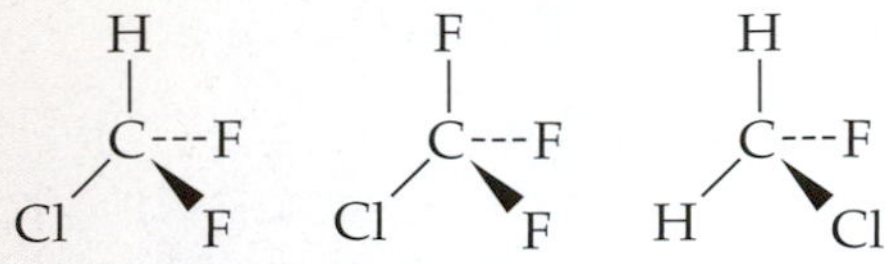

(b) Each molecule has a tetrahedral geometry around each central carbon.

Fast forward to 1974: Richard Nixon resigns from the Presidency, George Lucas writes the screenplay for *Star Wars*, disco rules the radio dial, and scientists are noticing atmospheric changes around the globe. Specifically, they observe a decrease in the thickness of the **ozone layer**, the thin, fragile shield that protects Earth from harmful ultraviolet rays. In an effort to explain this alarming phenomenon, two chemists at the University of California, Irvine—F. Sherwood Rowland and Mario Molina (Figure 8.39)—put forth a revolutionary postulate: maybe CFCs are causing the ozone layer to degrade.

In order to understand how CFCs could wreak havoc on the ozone layer, we must first consider the ozone molecule, which is like the oxygen molecule, but with another oxygen atom added. Thus the ozone molecule, O_3, is a string of three oxygen atoms connected together with covalent bonds (Figure 8.40). Ozone naturally forms in the *stratosphere*, which is the region of the atmosphere lying roughly 10 to 50 km above Earth's surface. Oxygen molecules in the stratosphere react with ultraviolet light that is emitted by the Sun and has wavelengths below about 242 nm:

$$O_2 \xrightarrow{\text{UV } \leq 242 \text{ nm}} 2\,O$$

The oxygen atoms that form from this reaction then react with oxygen molecules to make ozone:

$$O_2 + O \rightarrow O_3$$

The ozone formed in this way can then break down to an oxygen molecule and an oxygen atom, the reverse of the last reaction, in the presence of ultraviolet light that has with wavelengths of 320 nm or less:

$$O_3 \xrightarrow{\text{UV } \leq 320 \text{ nm}} O_2 + O$$

This is how the ozone layer intercepts ultraviolet rays on their way to Earth, shields plants from light damage, and protects us Earthlings from basal cell and squamous cell cancers.

The series of reactions we just looked at constitutes a cycle that naturally maintains the level of ozone in the stratosphere. Thus it made sense that a disruption of this cycle as a result of human activities could decrease the thickness of the ozone layer. Rowland and Molina reasoned that even though CFCs are extremely robust molecules, they can be broken down with the very-high-energy ultraviolet light found in the stratosphere. These researchers also reasoned that the documented leakage of CFCs from refrigerators and air conditioners, as well as aerosol cans, occurred on a scale broad enough to cause atmospheric damage.

Released CFCs find their way into the stratosphere and can break apart as they interact with light that has a wavelength of 220 nm or shorter. For example, the CFC known as Freon-11, CCl_3F, breaks down to produce a highly reactive chlorine atom, as shown in Figure 8.41. The chlorine atom is

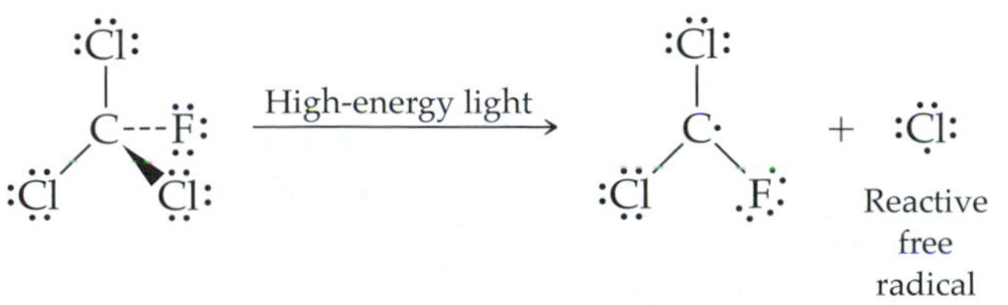

:Cl:
|
C---F: →(High-energy light)   :Cl:
|
:Cl   Cl: C·   +   :Cl:
                                  :Cl   ·F: Reactive
                                                  free
                                                  radical

◄ **FIGURE 8.41   Reaction of a CFC molecule.** CFCs break down in the stratosphere, where they are exposed to high-energy light from the Sun. The product of the breakdown reaction is the highly reactive chlorine atom, a free radical.

a very reactive free radical. In its presence, ozone can, in a complex series of reactions, be converted to molecular oxygen:

Refresh your understanding of free radicals by reviewing Section 5.6.

$$2\,O_3 \xrightarrow{\text{Presence of Cl atoms}} 3\,O_2$$

As a result of this reaction, the number of ozone molecules in the ozone layer decreases, along with the thickness of the ozone layer.

QUESTION 8.21   Which of these molecules could potentially cause ozone destruction via the reaction $2\,O_3 \rightarrow 3\,O_2$?

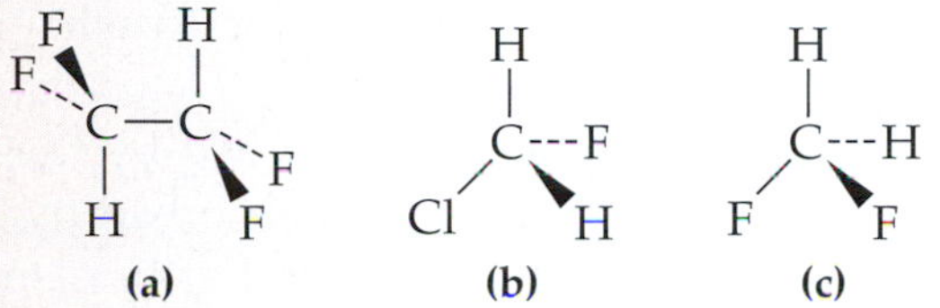

(a)   (b)   (c)

ANSWER 8.21   Compound (b), because it is the only one of the three that contains Cl atoms and so can produce reactive chlorine atoms.

When they published their ozone-destruction hypothesis, Rowland and Molina did not have the data to back it up. Over the next decade, though, scientists began to monitor the ozone situation more closely. In 1985, a team of British geophysicists led by Jonathan Shanklin measured the ozone layer above Antarctica and discovered an enormous hole in the layer, a hole whose area was greater than that of North America. At first, the scientists thought their instruments were faulty and sent off for new ones. Subsequent measurements confirmed their original data, however, and alarm began to rise over the quickly disappearing ozone layer and the subsequent potential for harm to living organisms and their environment.

In 1987, the Montreal Protocol, mandating that industrialized nations cease CFC use after 1997, was signed by 57 countries. Data show that, although the size of the hole in the ozone layer fluctuates throughout a given year, the average size of the hole has gradually increased. The years following 1998 showed a shrinking trend, but the measurements taken since 2002 show more expansion of the hole (Figure 8.42). In 1995, Rowland and Molina—along with Paul Crutzen of the Netherlands—were awarded the Nobel Prize in Chemistry for their groundbreaking work in atmospheric chemistry.

The search continues for new refrigerants that can replace Freon and other CFCs, and today some research is centered on *hydrofluorocarbons*, HFCs. These molecules contain only hydrogen, fluorine, and carbon and therefore cannot produce reactive chlorine atoms the way CFCs do. The HFC shown in Figure 8.43, pentafluoroethane, like all other HFCs, does not contribute to the destruction of the ozone layer at all. So why is it that pentafluoroethane is not on the list of refrigerant molecules for the future? The answer is that this molecule, and many others, contribute to the prob-

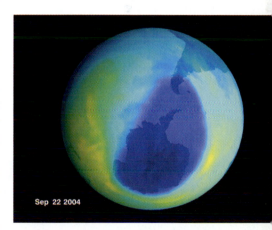

Sep 22 2004

▲ **FIGURE 8.42   The ozone hole.** This image from NASA shows the ozone hole (dark blue area) located over Antarctica (darker blue area in the background). In extreme cases like this one, the size of the hole extends all the way north to Patagonia at the southern tip of South America.

▼ **FIGURE 8.43   An improvement over CFCs.** The structure of the hydrofluorocarbon molecule known as pentafluoroethane.

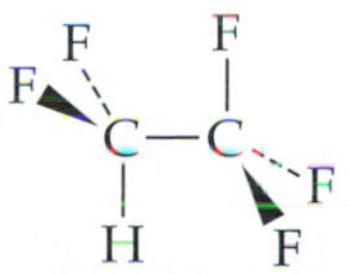

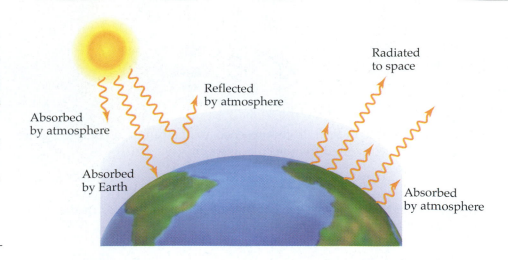

▶ **FIGURE 8.44** **The greenhouse effect.** Some of the Sun's energy makes its way to Earth's surface. The Earth radiates some of the absorbed heat outward, but much of the radiated heat is trapped by heat-absorbing molecules in the atmosphere. These heat-absorbing molecules—the greenhouse gases—enter the atmosphere as a result of human activities, mainly the burning of fossil fuels. This phenomenon is known as the greenhouse effect.

lem of **global warming**, the unnatural increase in the temperature of Earth's atmosphere as a result of human activities.

Figure 8.44 shows how this temperature increase happens. Some of the Sun's heat making its way to Earth is either reflected back toward the Sun by molecules in Earth's atmosphere or absorbed by these molecules. The rest passes through the atmosphere and is absorbed by Earth. To balance this heat gain, heat is radiated outward from Earth. A small percentage of this radiated heat leaves Earth's atmosphere, and the rest is absorbed by molecules in the atmosphere. This cycle, when left in a natural state, maintains the atmosphere at a cozy temperature that is agreeable to humans, plants, and wildlife.

Problems arise, however, when the atmosphere contains high quantities of molecules that absorb radiated heat. These gases increase the amount of radiated heat retained by the atmosphere and reduce the amount that escapes to outer space. Many of these gases come from natural sources, but since the Industrial Revolution the burning of fossil fuels and the release of pollutants into the environment have dramatically increased the levels of heat-absorbing gases in the atmosphere.

A florist's greenhouse works because radiation from the Sun loses energy once it is inside the structure. Incoming radiation has enough energy to pass through the glass of the greenhouse. This radiation is absorbed by objects inside, and when these warmed objects later release radiation, it has a longer wavelength and not enough energy to pass through the glass. The heat is therefore trapped inside the greenhouse, a boon to florists in the winter months. (The same thing happens to a closed-up car sitting in the Sun.) The presence of heat-absorbing gases in Earth's atmosphere make the same thing happen there: the gases remove energy from radiation that has been reflected off Earth's surface and is on its way back to outer space. The lower-energy radiation cannot pass through the atmosphere, remains where it is, and causes things to heat up. This phenomenon is called the **greenhouse effect**, and it is the primary cause of global warming. Fittingly, the heat-absorbing gases that contribute to global warming are referred to as **greenhouse gases**.

The scientific community has provided convincing evidence of global warming and the effects it will have on Earth's fragile ecosystem. The Kyoto Protocol, drafted in 1997 in response to such data, calls for a reduction in the release of greenhouse gases—carbon dioxide, HFCs (like pentafluoroethane), sulfur hexafluoride, nitrous oxide, perfluorocarbons, and

The Tuvalu Blues

A look at Figure 8.45 shows you how the average temperature of Earth has changed since 1880. Each year, temperatures fluctuate naturally, but the long-term trend is obvious: the planet is getting warmer. In fact, data suggest that Earth's temperature has increased by at least 0.6 C° since 1880. This may not seem like much, given that we are accustomed to much more dramatic *seasonal*

ISSUES IN CHEMISTRY

temperature fluctuations. Consider this, though: the rising temperature coincided with an increase of sea level by 10 to 20 cm during the twentieth century. Scientists estimate that sea level could rise by another 100 cm during the twenty-first century if steps are not taken to correct the warming trend.

Most of us live in areas where small increases in sea level will not affect our quality of life. Some people who live along an ocean coastline or on an island, though, are already experiencing the effects of global warming.

Consider the tiny British colony of Tuvalu, an atoll north of Fiji with a population of 11,000. The average elevation of Tuvalu is about 2 m, and the capital city, Fanafuti, sits on a piece of land that measures 400 m across at its widest point. Over the past decades, the country has gradually begun to disappear as the rising ocean engulfs it. Many residents of Tuvalu have been forced to relocate, and eventually the land that makes up this island nation will be completely uninhabitable.

◀ **FIGURE 8.45   Global temperature change.** Since 1880, the average temperature of Earth's atmosphere has gradually increased.

methane. According to this Protocol, participating countries must, by 2012, reduce their greenhouse gas emissions to 8 percent below their 1990 levels. The treaty, which was ratified by most countries of the developed world, has not been ratified by the United States. Those who question the existence of global warming complain that there are insufficient and imperfect data to support it. So where do we stand? Meteorological scientist Richard Sommerville puts it this way, "When you go to your doctor, and she says you're due for a heart attack, you don't turn around and say medicine is imperfect [because] she can't predict the date of your heart attack.... You take it seriously. I think climate science is in that position now."

The presence of HFCs on the list of greenhouse gases enumerated by the Kyoto Protocol requires that we continue our search for an ideal refrigerant. We do not yet have a candidate that meets all our needs: low toxicity, low flammability, and zero contribution to ozone depletion and global warming. The refrigerants in use today, which include HFCs, go much further toward meeting these requirements than, for example, CFCs. That we know that CFC and HFC molecules are potentially damaging to the environment is a step in the right direction, and the very existence of global treaties aimed at solving some of Earth's most urgent health problems should give us hope for her future.

SUMMARY

We began our study of gases with the idea of diffusion. Studying how gases diffuse can help us understand the constant motion of the atoms and molecules all around us. Ambient clean air is made up almost entirely of nitrogen and oxygen molecules moving at supersonic speeds. Small amounts of foreign gases introduced into this mixture are quickly assimilated and blended in. The atoms and molecules in a gas are very far apart, and that distance between particles for a gas under specific conditions is given by the mean free path. Because the atoms and molecules are so far apart, gaseous components are often compared to tiny billiard balls that undergo collisions with other billiard balls and with the walls of the container. For an ideal gas, the atoms and molecules do not interact with one another to a significant extent, and we often do not consider their chemical makeup. Instead, we can think of every particle in an ideal gas as being identical to every other particle in the gas.

When atoms and molecules in a gas collide with the walls of the container holding them, the container will expand if it is flexible and increase in volume. The greater the number of atoms or molecules, the greater the volume (Avogadro's law). The four variables that we commonly use to describe the conditions of a sample of gas are temperature T, pressure P, number of moles of gas n, and volume V. The gas laws in this chapter relate one of these four variables to one other variable while the other two are held constant.

Atoms and molecules in the gas phase are susceptible to gravity in the same way that liquids and solids are. Thus, air is denser at sea level and becomes less dense as you increase in altitude. Atmospheric pressure is the force that atoms and molecules in the air exert on a surface, and it changes with changing weather and with changing altitude. The atmospheric pressure can be mea-sured using a barometer in which mercury is pushed up a tube as atmospheric pressure presses down on a pool of mercury. Because barometers and other similar instruments are widely used, a common unit for pressure is millimeters of mercury. The SI unit is the pascal. Standard pressure is equal to 760 mmHg or 10^5 Pa, and standard temperature is 0 °C. At STP, the volume of 1 mol of a gas is exactly 22.414 L, the molar volume. Boyle's law tells us that the pressure exerted by a gas in a container is inversely proportional to the volume of the container. When volume increases, pressure decreases.

As the temperature of a gas is increased, the atoms and molecules begin to move faster. This increase in speed makes collisions more frequent, and this results in an increase in volume or pressure (or both). These relationships, which are known as Charles's law and Amontons' law, respectively, help us to understand how heat energy can move through a gas by convection and how diffusion occurs as a function of temperature.

The general gas law unites the four variables associated with gases into one relationship:

$$\frac{PV}{nT} = \text{constant}$$

$$\frac{P_b V_b}{n_b T_b} = \frac{P_a V_a}{n_a T_a}$$

This second equation is useful when you have a sample of gas in one situation, the "before" situation, and it changes in some way to give another situation, the "after" situation. To use this equation, first eliminate variables that do not change in your particular circumstance. Then solve for the variable that is unknown. Finally, plug in values for the remaining variables and calculate the answer.

The ozone molecule, O_3, exists in a thin layer around Earth, a layer that protects Earth from harmful ultraviolet rays; exposure to these rays can lead to skin cancer. Chemical reactions that break down ozone occur naturally, but chlorofluorocarbon pollutants produced from human activities have caused gradual degradation of this layer. Other pollutants are greenhouse gases, which are heat-absorbing molecules that trap heat in Earth's atmosphere and increase temperatures on the planet. These greenhouse gases are the primary contributors to global warming.

KEY TERMS

Amontons' law
atmospheric pressure
Avogadro's law
Boyle's law
Charles's law

chlorofluorocarbons (CFCs)
closed system
diffusion
general gas law
global warming

greenhouse effect
greenhouse gas
ideal gas
mean free path
millimeters of mercury (mmHg)

molar volume
open system
ozone layer
pascal
pressure
real gas

standard temperature and pressure (STP)
vacuum
variable

QUESTIONS

The Painless Questions

1. The molecules parathion and malathion are potent nerve toxins. What structural feature do they share with sarin?

Parathion

Malathion

2. You have a flask containing two gases. One of the gases is purple, the other is yellow, and they are separated into layers. The purple gas has a mean free path of 400 nm, and the yellow gas has a mean free path of 30 nm. Which gas is on top?

3. Why do gas-filled tires work more efficiently than solid tires or liquid-filled tires?

4. Your roommate is becoming more and more malodorous every day. You decide to adjust your thermostat to try to slow down the movement of her "fragrance" through your apartment. Will you turn on the heat or the air conditioning? Explain your answer.

5. Using your knowledge of gas behavior, discuss the factors involved in surviving a chemical weapon attack.

6. Figure 8.11 shows objects representing open and closed systems. If you move the lemons shown in the illustration from the mesh bag to a sealed plastic bag, you will have changed the system from open to closed. For each of the other objects in Figure 8.11, indicate how the system could be changed from open to closed or from closed to open.

7. Why is the phrase *like dissolves like* usually not used when discussing gases?

8. In the imaginary lakeside town of Goolikobruquik, the price of potatoes always goes up when the temperature of the lake water goes down. When the water temperature goes up, the price of cherries inevitably goes up. Pair each of these variables, and for each pair indicate whether the variables are directly proportional or inversely proportional to each other.

9. Imagine you have a 2-L sealed flask containing 1 mol of xenon atoms. You measure the average speed of these atoms to be 700 km/h. Is every atom of xenon in the sample moving at this speed? Explain.

10. In an attempt to keep your shampoo bottle the same size during your trip from San Francisco to the Rocky Mountains, you decide to remove air from the bottle as you pack it in San Francisco so that it collapses inward. In the mountains, you blow into the bottle so that it puffs outward and cap it quickly. Is this a reasonable plan? Why or why not?

11. Describe the properties of an ideal refrigerant molecule. Why are CFCs and HFCs not included on the list of ideal refrigerants?

12. For each of the following, indicate whether the substance described is a solid, liquid, or gas at room temperature: (*a*) a green substance with a melting point of 234 °C, (*b*) a clear substance that adopts the shape of the container you put it in but does not fill it completely, (*c*) a substance that maintains its shape when you remove it from the box in which it arrived, (*d*) a substance that disappears immediately when you uncork it.

13. Badwater, in Death Valley, California, is 282 ft below sea level. If you pack a plastic tube of sunscreen at the beach in San Diego and drive to Death Valley, will the tube be larger or smaller than it was in San Diego?

14. Comment on this statement:

All gases mix together completely regardless of the types of atoms or molecules they contain. This is true because the atoms or molecules in a gas are all very far apart from one another and interact only rarely during collisions.

15. Individual samples of molecules X, Y, and Z have mean free paths of 50 nm, 30 nm, and 250 nm, respectively. Which sample of gas is most dense? Which is least dense?

16. Describe the difference between force and pressure. What does the word *pressure* mean when it refers to a gas?

17. Figure 8.40 shows two versions of the ozone molecule. (*a*) Why is the molecule drawn in two ways? (*b*) Is one of the two oxygen-oxygen bonds in the ozone molecule stronger than the other?

More Challenging Questions

18. A sample of pure helium gas at STP occupies a volume of 22,400,000 μL. How many atoms of helium does it contain?

19. How are bees able to concentrate air pollutants in their hive? What practical uses have been found for bees as a result of this ability?

20. The *bar* is another unit of pressure used by meteorologists, with 1 bar = 10^5 Pa. Which of the following represents the highest pressure: (*a*) 1,000,000 Pa, (*b*) 100 kPa, (*c*) 1 bar, (*d*) 100,000 mbar?

21. A container with a piston begins with a volume of 2.85 L and a pressure of 1.25 kPa. If the piston is compressed until the volume is 1.92 L, what is the new pressure in the container?

22. An odor brought into a room by a malodorous person will take a while to travel from that person to your nose. Imagine this scenario on a hot day and a cold day. Will the travel time differ on those two days? On which day will the scent travel faster? Why?

23. Table 8.1 lists carbon dioxide as one of the molecules found in clean air. Given that the molar mass of carbon dioxide is much larger than the molar masses of the other components of air, why does the carbon dioxide mix completely with the other components?

24. Gas M has a mean free path of 50 μm, and gas P has a mean free path of 200 nm. (*a*) Which mean free path is longer? (*b*) How many times longer than the shorter path is the longer path?

25. Gas M has a mean free path of 50 μm, and gas P has a mean free path of 200 nm. Which of the statements that follow must be true about these gases? Choose all that apply.

(a) Gas P is denser than gas M.

(b) The containers that hold each gas may be exactly the same size.

(c) If their containers are equal in size, gas P must contain more atoms or molecules than gas M.

26. Describe the difference between conduction and convection. Give an example of each type of energy transfer.

27. Comment on this statement:

When you are hiking at 20,000 feet, breathing is very difficult because the ratio of oxygen to nitrogen gas decreases as you leave sea level and increase your altitude.

28. What is the volume of a balloon at STP that contains 1.5×10^{23} atoms of helium gas?

29. According to Table 8.1, how many molecules of nitrogen are there in a sample of air that contains (*a*) a combined total of 100,000 atoms and molecules? (*b*) 1 mol of gas?

30. For each pair of gas variables, describe an experiment that would allow only these two variables to change: (*a*) *P* and *V*, (*b*) *n* and *T*, (*c*) *n* and *P*, (*d*) *V* and *T*.

31. Gas A has a molar mass of 32 g/mol, gas B has a molar mass of 29 g/mol, and gas C has a molar mass of 56 g/mol. (*a*) Would you expect a small amount of gas A to mix with a large amount of gas B? (*b*) Would you expect a small amount of gas C to mix with a large amount of gas A? (*c*) Would you expect equal amounts of gases A and B to mix with each other? (*d*) Describe a scenario in which a layer would form between any two of the gases described.

32. Double-paned windows insulate houses from cold more efficiently than single-paned windows. Based on what you know about how a thermos bottle works, speculate on how double-paned windows are able to reduce heat loss.

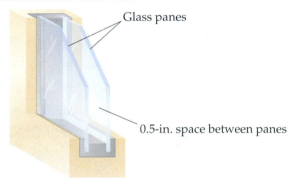

Glass panes

0.5-in. space between panes

33. Describe the greenhouse effect. In your explanation, relate the effect to a car that has been sitting in the sun on a hot day.

The Toughest Questions

34. A helium-filled balloon is released from Earth's surface. (*a*) What happens to its volume as it moves skyward? (*b*) How would your answer to part (*a*) change if the balloon were released in an area in which there was a thermal inversion? (*c*) Why would the balloon eventually pop?

35. We looked at several combinations of our four gas variables—*T*, *P*, *n*, and *V*. The only combination that we did not consider is how temperature changes as the number of moles of gas changes. (*a*) What does the general gas law tell you about the relationship between *T* and *n*? Are these two variables directly or inversely proportional to each other? (*b*) Based on your knowledge of gas behavior, can you explain this relationship?

36. Return to Question 1 and determine the molecular formulas and molar masses of parathion and malathion.

37. Figure 8.22b shows a weather forecast for Hanoi, Vietnam. The pressure reading on that day was 100.5 kPa. Convert this value to millimeters of mercury and to inches of mercury.

38. A container fitted with a movable piston is used to demonstrate Avogadro's law. As gas is introduced into the cylinder, the piston moves upward and the volume increases. Why does the pressure of the gas remain constant during this experiment?

39. Pilots know that hot days at high altitude make for a risky takeoff. They also keep an eye on the humidity in the air. Speculate on how humidity might affect the lift of an airplane. Would it be easier to take off on a hot day in the jungle or a hot day in the desert?

40. A three-year-old child receives for her birthday a balloon that has a volume of 12.2 L. Her birthday falls on a beautiful spring day when the temperature is 31 °C. Unfortunately, the helium-filled balloon is accidentally freed by the child and floats up and away. (*a*) When the balloon reaches an altitude at which the air temperature is 26 °C, what is the volume of the balloon? (*b*) Which variables are we assuming to be constant in this situation? (*c*) Which variables are we assuming to be changing? (*d*) In real life, would the two variables you named in part (*b*) stay constant? Why or why not?

41. The glass tube of a neon sign at a seaside resort contains neon gas at a pressure of 3000 mmHg. (*a*) What information would you require in order to calculate the number of neon atoms present in the tube? (*b*) Suppose the tube is punctured and all the neon gas escapes. What is the pressure in the tube now?

42. This chart shows the average molecular speeds of five molecules at 25 °C.

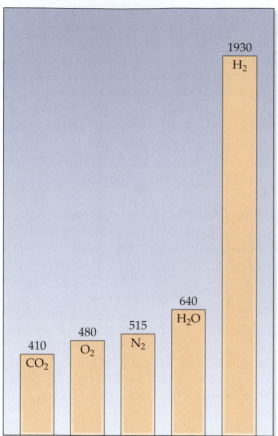

Average molecular speed (m/s)

(*a*) Calculate the molar mass of each gas listed in the graph. What correlation do you observe between molecular speed and molar mass? (*b*) What is the relationship between the density of a gas and its molar mass? (*c*) Estimate the average speed of a molecule of carbon monoxide, CO, at 25 °C.

43. Imagine you have a 22.414-L sample of each gas shown in Question 42. At 25 °C and sea level, which gas will have (*a*) the longest mean free path and (*b*) the lowest density?

44. A typical commercial jet flies at a cruising altitude at which the atmospheric pressure is 25 kPa. If the plastic bottle that you packed at sea level began with a volume of 800 mL, what is its new volume at cruising altitude? Assume several things: that the bottle is empty except for the air that filled it before it was sealed, that the luggage compartment is at a pressure of 25 kPa, that the temperature is held constant, and that the bottle can freely expand to any size.

45. In this chapter, we discussed the reasons thermos bottles are able to keep hot things hot and cold things cold. Eventually, however, even the best thermos will allow its contents to equilibrate with the outside temperature. Based on what you know about the behavior of gases, speculate on why a thermos would not be able to work indefinitely.

46. The unit pounds per square inch (psi) is often used to express tire pressure, with 1 psi = 52 mmHg. If a mountain bike tire holds 85 psi of air, what is this pressure expressed (*a*) in millimeters of mercury. (*b*) in kilopascals?

47. When a 350-lb man gets on the bicycle described in Question 46, the volume of each tire changes from 2.9 L to 2.7 L. What is the new pressure in each tire, in kilopascals?

48. If a flexible container at STP holds 4.0×10^{22} molecules of nitrogen, what is its volume?

49. Draw Lewis structures for the oxygen molecule, O_2, and the ozone molecule, O_3. Now reconsider these two reactions that occur in the natural ozone cycle:

$$O_2 \xrightarrow{\text{UV light } \leq 242 \text{ nm}} 2O$$

$$O_3 \xrightarrow{\text{UV light } \leq 320 \text{ nm}} O_2 + O$$

In each case, a bond between two oxygen atoms is broken. Which of these two reactions requires more energy? Using the Lewis structures you drew, explain why these two reactions are prompted by different wavelengths of light.

E-Questions

Go to **www.prenhall.com/waldron** to find these questions in electronic form, complete with hyperlinks directly to the various websites cited in the questions.

50. Go to the Weather Network website at **Weather Network**. Go to Weather Maps and then select "System" from the top of the U.S. map. This should show you a map of the United States with current high and low pressure areas and fronts indicated. Select two cities, one in a high-pressure area and one in a low-pressure area. Type their names in the search box at the top of the screen to view the local forecast. What units are used to express pressure? What is the atmospheric pressure in each location, in millimeters of mercury? How do they compare? What is the atmospheric pressure in your home town?

51. Go to the website **HowStuffWorks**. Read the four-page report "How Gas Masks Work," and then answer the following questions: (*a*) How is air filtered through a gas mask to protect against biological versus chemical agents? (*b*) Which component of a gas mask is designed to absorb organic molecules such as sarin? (*c*) What is the difference between the terms *adsorb* and *absorb*? Which is taking place in the gas mask, adsorption or absorption?

52. Read more about the use of bees to detect trace amounts of airborne heavy metals at **InformNauka**. Read this two-page article from Moscow State University, and then answer the following questions:

(*a*) What did the researchers learn about how the heavy metal concentrations in the center of Moscow compared with concentrations in the countryside? (*b*) Where do bees accumulate heavy metals in highest quantity? (*c*) What unexpected result did the researchers obtain when they analyzed the honey?

53. The University of Washington and Washington University in St. Louis have developed a tutorial on the chemistry of airbags at **Airbags**. This 14-page tutorial offers a more advanced approach to the gas laws you learned in this chapter, including a movie about airbag deployment. Read the article, and then answer the following questions: (*a*) What makes airbags unsafe? (*b*) What chemical reactions take place in an airbag? (*c*) What is the name given to the distribution of molecular speeds in a gas?

54. Read about a recent discovery in the area of human pheromone research at **JunkScience** and **CNN**. After reading these short news reports, answer the following questions: (*a*) How did Martha McClintock first become interested in the study of human pheromones? (*b*) What experiment did she perform to study human pheromones? (*c*) When these articles were written, what was known about the chemical structure of human pheromones?

WORLD WIDE WEB RESOURCES

As with the E-Questions, go to **www.prenhall.com/waldron** to find these questions in electronic form, complete with hyperlinks directly to the various websites cited.

Some of following links contain research articles related to the topics in this chapter and could be used as the basis for a writing assignment in your course. Other sites are of general interest:

- The links **ESS** and **Gulflink** provide additional information on the history of biological and chemical weapons. The first site includes a link to a copy of the Geneva Protocol, signed in 1928. The second link discusses the uses of biological and chemical weapons during the first Gulf War.

- Go to the Australian EPA's website at **EPA**. This link takes you to a page on inversion layers and how they

form. The page includes a short experiment that you can perform in the laboratory or even in your kitchen. Perform this experiment, and report on the location of the food coloring in the layers that exist. How does this experiment mirror the formation of inversion layers at Earth's surface?

- Read about another carbon dioxide-linked catastrophe at **PBS**. The article titled "Permian Triassic Extinction" explains the role of carbon dioxide in the mass extinction that occurred during this era.

- If you are interested in reading more about how air density affects flight, try one of the following websites: the first gives a detailed explanation of how humidity, temperature, and altitude affect takeoff; the second is a general discussion of how lift is generated: **USAToday**; **Washington**.

ABC NEWS VIDEO
Carbon Monoxide Poisoning

CHEMISTRY TOPICS IN CHAPTER 9

- Combustion reactions and explosions [9.1]

- Endothermic and exothermic reactions [9.1]

- Alkanes, alkenes, and alkynes [9.2]

- Hydrocarbon nomenclature [9.2]

- Cis–trans isomerism [9.2]

- Ethers [9.3]

- Bond energies [9.4]

- High explosives and low explosives [9.5]

- Entropy and the second law of thermodynamics [9.6]

- Gas chromatography [9.7]

"For my part, I wish all guns with their belongings and everything could be sent to hell, which is the proper place for their exhibition and use."

ALFRED NOBEL, INVENTOR OF DYNAMITE

Explosives

A STUDY OF ORGANIC MOLECULES WITH SIMPLE FUNCTIONAL GROUPS AND THE FORCES WITHIN THEM

The November 5 celebration in England, Canada, and New Zealand known as either Guy Fawkes Day or Bonfire Night commemorates the evening in 1605 when the Roman Catholic rebel Guy Fawkes was caught red-handed in an attempt to blow up the House of Lords in London while Parliament was in session with King James I in attendance. On that fateful night, Fawkes was found in a room below the House of Lords preparing 36 barrels of gunpowder that would have incinerated the building and killed the king, who was loyal to the Church of England. Today, the holiday is still celebrated with enormous bonfires in which figures of Guy Fawkes (Figure 9.1) are burned in effigy, although the modern festival is more of an excuse for merriment than a rebellion against the Catholic Church.

Explosives have been known since pre-Biblical times, and their use has punctuated history with acts of violence and rebellion. The technology of explosives has come a long way since those early times, when explosions were driven by relatively simple mixtures of pure elements and simple salts. The nineteenth century saw the discovery of more sophisticated explosive mixtures, a development that coincided with the advancement of organic chemistry and the synthesis of carbon-based molecules. As we already know from Chapter 6, the twentieth century brought with it the

▲ FIGURE 9.1   **Effigy of Guy Fawkes.** Children posing with Guy, who later will have an encounter with a bonfire.

capability for manipulating the atom's nucleus and an entirely new class of explosive device, the nuclear bomb. In this chapter, we shall look at the chemistry of non-nuclear explosive materials and, in particular, what makes a given substance unstable and therefore capable of undergoing a rapid, and sometimes unpredictable, reaction. We shall look at why the end result of an explosive reaction is chaos, and why increased disorder is inevitable when a detonation occurs.

The chemistry of explosive and combustible materials takes us back to the realm of organic chemistry, and we will learn which features of organic molecules make them prone to instability. We begin our look at the evolution of explosives with Guy Fawkes and his plot to undermine the authority of the British throne.

9.1 | The Smoking Gun

A seventeenth-century anarchist like Guy Fawkes had limited explosive resources. At the time, the most common explosives were made by mixing finely ground mixtures of pure elements and salts into a powder that was highly combustible. A modern-day mixture of gunpowder (or *black powder* as it is also called) consists of about 15 percent wood charcoal, essentially pure carbon; 10 percent sulfur; and 75 percent potassium nitrate, KNO_3, which is also known as *saltpeter*. Both wood charcoal—obtained by charring wood in the absence of oxygen—and sulfur serve as the fuel component of the gunpowder. Potassium nitrate provides a source of oxygen. The combustion reaction is

$$10\,KNO_3(s) + 11\,C(s) + 4\,S(s) \longrightarrow 5\,CO(g) + 5\,CO_2(g) + 5\,N_2(g)$$
$$+ K_2CO_3(s) + 3\,K_2SO_4(s) + K_2S(s)$$

This chemical equation may seem intimidating, but we can derive some useful information about the nature of explosions from it.

The first point to consider is that in any sample of our two fuels, solid carbon or solid sulfur, only one type of atom is present. This means that every carbon atom is bound to identical carbon atoms and every sulfur atom is bound to identical sulfur atoms. This is not true on the product side of the equation, however. There, the carbon and sulfur atoms are combined with oxygen in several of the products. For example, carbon monoxide, carbon dioxide, and potassium carbonate contain a carbon atom bound to one, two, and three atoms of oxygen, respectively. Potassium sulfate contains a sulfur atom with four oxygen atoms bound to it. Thus, we recognize the fuels in this reaction—carbon and sulfur—as the substances that interact with oxygen to make new compounds in which the oxygen is bound directly to the fuel atoms.

QUESTION 9.1   If you run the gunpowder reaction with 40 molecules of saltpeter (KNO_3), 44 atoms of carbon, and 16 atoms of sulfur, how many molecules will you produce of (*a*) nitrogen, (*b*) carbon monoxide, (*c*) carbon dioxide?

ANSWER 9.1   (*a*) 20, (*b*) 20, (*c*) 20.

The second thing to consider about this reaction is that some of the products are gases and others are solids. The first three products are the gases carbon monoxide, carbon dioxide, and nitrogen, which exists in nature in the diatomic form N_2. It is the production of gases that causes the blast we associate with gunpowder and with explosive materials in general. The other three products are the solids potassium carbonate, potassium sulfate, and potassium sulfide. Together, these three solid products produce a spray of particulates that we know as smoke.

Early gunpowders produced a great deal of smoke. Although the smoke was not a problem if the gunpowder was being used for blasting rock or razing buildings, it was a drawback for shooters using the gunpowder to expel a bullet from the barrel of a gun. If you wanted to fire your gun while concealing yourself, the plume of smoke produced by the solids in the reaction would make your location obvious (Figure 9.2). Instead of the saltpeter–carbon–sulfur reaction, modern gunpowders use reactions that do not produce telltale smoke.

A *combustion reaction*, like the one shown for gunpowder, is a heat-producing chemical reaction that involves the interaction between a fuel source and oxygen. With gunpowder, the fuels are sulfur and carbon, and the oxygen comes from potassium nitrate. This combustion reaction is also an **explosion**, defined as a violent expansion of gases and release of energy produced by a fast chemical reaction. Not every combustion reaction is an explosion. For example, our mixture of saltpeter, charcoal, and sulfur leads to an explosion only if the correct conditions exist. If the reactants are confined to a small space (such as the barrel of a gun) and pressure is exerted on them, the gases produced during the reaction will build up until the gun barrel can no longer hold them. Then the buildup of pressure causes the bullet to be expelled from the barrel with great force, resulting in an explosion. If, on the other hand, the reactants are not confined to a small space, the reaction proceeds more slowly and the gaseous products build up gradually. In this situation, the buildup of pressure required for an explosion may not exist. Thus, the term *combustion* refers to the type of chemical reaction taking place, and the term *explosion* refers to the end result.

By the seventeenth century, it was obvious that gunpowder could be used not only for weapons, but also to blast away rock. In addition, scientists at the time reasoned that the power of gunpowder might also be harnessed to make some sort of engine that could be used to convert chemical energy to movement. The earliest attempts at a gunpowder engine proved futile, and continued work on this invention showed that the solids produced during combustion fouled the engine mechanisms quickly, dooming the gunpowder engine to failure. The basic idea of a combustion engine was sound, but a fuel that did not produce a solid product was needed. Modern-day automobile engines use gasoline as fuel, and we will briefly discuss how an automobile engine works later in this section as well as in the next section. We will also talk about where gasoline comes from and how it is obtained from crude oil. Before we learn about how gasoline is manufactured, though, we must discuss some of its properties—in particular, what makes it a fuel and what characteristics a substance must have in order to be used as a fuel.

Gasoline is a mixture of hydrocarbons, which, as we learned in Chapter 3, are the simplest class of organic compound because they contain only two elements—carbon and hydrogen. In previous chapters, we learned about the covalent bonds that hold molecules together in hydrocarbons and other organic molecules. In organic molecules, every carbon atom makes exactly four bonds, and every hydrogen atom makes only one. To take just

▲ **FIGURE 9.2 A smoking gun.** In the old days, guns produced smoke when fired, giving away the location of the shooter.

RECURRING THEME IN CHEMISTRY The structure of carbon-containing substances is predictable because each carbon atom almost always makes four bonds to other atoms.

one example, the skeletal structure for the hydrocarbon octane, C_8H_{18}, is

$$
\begin{array}{ccccccccc}
& H & H & H & H & H & H & H & H & \\
& | & | & | & | & | & | & | & | & \\
H- & C- & C- & C- & C- & C- & C- & C- & C & -H \\
& | & | & | & | & | & | & | & | & \\
& H & H & H & H & H & H & H & H &
\end{array}
$$

Octane (as well as all other hydrocarbons) undergoes a combustion reaction readily in the presence of oxygen to make carbon dioxide and water:

$$2\,C_8H_{18}(\ell) + 25\,O_2(g) \longrightarrow 16\,CO_2(g) + 18\,H_2O(g)$$

This reaction has some things in common with the gunpowder combustion we saw earlier. First, the fuel, in this case octane, reacts with a source of oxygen, called the *oxidizer*, in this case diatomic oxygen, O_2. Second, this reaction produces gases that cause a buildup of pressure that can be explosive if the reaction occurs quickly enough and produces large volumes of gas. However, this reaction is also different from the combustion of gunpowder in several respects. For one thing, the combustion of gunpowder produces solids in addition to gases, making gunpowder a poor engine fuel because it leaves behind solid residues after combustion. As the previous equation shows, however, the combustion of octane produces only gases, which means that an engine could be designed around the combustion of a hydrocarbon because there is no solid residue. Furthermore, because all the products of hydrocarbon combustion are gaseous, the pressure created when this reaction occurs is much larger than it would be if some of the products were in the solid state. Thus, using a hydrocarbon as a fuel is more efficient than using gunpowder because every one of the products in the hydrocarbon reaction contributes to the buildup of pressure. This pressure is desirable in a combustion reaction when the purpose of the reaction is to do some kind of mechanical work, such as moving a car.

QUESTION 9.2   Which of these hydrocarbon combustion equations are balanced?

(a) $C_3H_8 + 5\,O_2 \rightarrow 3\,CO_2 + 4\,H_2O$

(b) $C_{12}H_{24} + 15\,O_2 \rightarrow 12\,CO_2 + 12\,H_2O$

(c) $2\,C_6H_{12} + 16\,O_2 \rightarrow 12\,CO_2 + 12\,H_2O$

(d) $C_6H_6 + 7\,O_2 \rightarrow 6\,CO_2 + 3\,H_2O$

ANSWER 9.2   (*a*) only.

QUESTION 9.3   Write a balanced equation for the combustion of acetylene, C_2H_2, a molecule commonly used as a fuel for welding.

ANSWER 9.3 $2\,C_2H_2 + 5\,O_2 \rightarrow 4\,CO_2 + 2\,H_2O$

It is important to recognize that chemical reactions do not always occur as written in balanced equations. Chemical equations are, after all, expressions of ideal situations, and the way they are expressed often applies only under theoretical conditions. One example of this phenomenon is the combustion of hydrocarbons, which, ideally, produces only carbon dioxide and water. In the real world, however, this is not the way it happens. In your car engine, for instance, there are other combustion products—most notably carbon monoxide. One way carbon monoxide can be produced in addition to carbon dioxide and water is via the reaction

$$2\,C_8H_{18}(\ell) + 21\,O_2(g) \longrightarrow 8\,CO_2(g) + 8\,CO(g) + 18\,H_2O(g)$$

This reaction tells us that 50 percent of the carbon atoms that begin as octane end up in carbon dioxide and 50 percent end up in carbon monoxide. Compare this reaction with the idealized *complete combustion* of octane shown earlier, in which carbon dioxide is the only carbon-containing product. You see that when carbon monoxide is produced, less oxygen takes part in the reaction because the carbon dioxide on the product side has been partially replaced with carbon monoxide, which contains only one oxygen atom instead of two. When this occurs, we refer to the reaction as *incomplete combustion* because not all of the carbon ends up in the form of carbon dioxide. Incomplete combustion occurs in your car engine either because there is insufficient oxygen to make combustion complete or because there is not ample time for the reactants to react completely.

The incomplete combustion we just saw exaggerates the amount of carbon monoxide formed in a typical car engine. In reality, less than 5 percent of the carbon that reacts ends up as carbon monoxide. This percentage varies depending on the age of your car and on several variables related to your car's performance. Although both CO and CO_2 are known environmental pollutants, CO has deleterious effects on human health and is therefore controlled by emissions standards. Emissions standards vary by state, but most states require that levels of carbon monoxide leaving your tailpipe be less than about 2 percent of the total carbon burned. To help reach increasingly strict emissions standards, automakers have added *catalytic converters* to car exhaust systems. One reaction that takes place in a catalytic converter gives the carbon monoxide another opportunity to react with oxygen to form carbon dioxide:

$$O_2(g) + 2\,CO(g) \longrightarrow 2\,CO_2(g)$$

Thus, carbon monoxide formed in your engine may leave your tailpipe as carbon dioxide.

We now know that hydrocarbons make good fuels, but how can we know which other types of molecules are suitable fuels and which are not? Could cars be designed using other organic molecules, such as rubbing alcohol or glucose or acetone? One part of this answer lies in the bonds that connect the carbon atoms in an organic molecule. When bonds form to make a fuel molecule, energy is released. Or, looking at it from the opposite perspective, energy is required to break these same bonds. Let us consider a very simple combustion reaction, that of the hydrocarbon ethane:

$$2\;
\begin{array}{c}
\text{H} \;\; \text{H} \\
| \;\;\; | \\
\text{H}-\text{C}-\text{C}-\text{H} \\
| \;\;\; | \\
\text{H} \;\; \text{H}
\end{array}
\; + \; 7\,O_2 \longrightarrow 4\;
\begin{array}{c}
\text{O} \\
\| \\
\text{C} \\
\| \\
\text{O}
\end{array}
\; + \; 6\,H_2O + \text{heat}$$

Ethane

MOLECULE
Carbon Monoxide

MOLECULE
Carbon Dioxide

MOLECULE
Octane

Chemists refer to any reaction that gives off heat, such as this one, as an **exothermic reaction**. Any reaction that requires heat is referred to as an **endothermic reaction**. As you can imagine, most of the reactions discussed in a chapter about explosives will be of the exothermic variety.

These words are derived from the Greek *endo-*, "taking in"; *exo-*, "producing"; and *thermo-*, "heat."

Combustion of ethane (C—C bond strength 348 kJ/mol)

$$2 \; H-\underset{\underset{H}{|}}{\overset{\overset{H}{|}}{C}}-\underset{\underset{H}{|}}{\overset{\overset{H}{|}}{C}}-H + 7 \, O_2 \longrightarrow 4 \, CO_2 + 6 \, H_2O$$

Also produces 2854 kJ of energy

MOLECULE
Ethene

▶ **FIGURE 9.3** **Three combustion reactions.** In each of these reactions, a two-carbon hydrocarbon undergoes a combustion reaction in which the sole products are carbon dioxide and water. The bond strength of each fuel is shown, along with the energy produced in the reaction.

Combustion of ethene (C=C bond strength 614 kJ/mol)

$$2 \quad \overset{H}{\underset{H}{}}C=C\overset{H}{\underset{H}{}} + 6 \, O_2 \longrightarrow 4 \, CO_2 + 4 \, H_2O$$

Also produces 2644 kJ of energy

Combustion of ethyne (C≡C bond strength 839 kJ/mol)

$$2 \; H-C{\equiv}C-H + 5 \, O_2 \longrightarrow 4 \, CO_2 + 2 \, H_2O$$

Also produces 2510 kJ of energy

QUESTION 9.4 Which of these reactions are endothermic, and which are exothermic:

(a) $TiCl_2(g) + heat \rightarrow TiCl(g) + Cl(g)$

(b) $2 \, C_5H_{10}(g) + 15 \, O_2(g) \rightarrow 10 \, CO_2(g) + 10 \, H_2O(g) + heat$

(c) $CO(g) + Cl_2(g) \rightarrow COCl_2(g) + heat$

ANSWER 9.4 Reaction (a) is endothermic; (b) and (c) are exothermic. Note that (b) is a combustion reaction: the hydrocarbon fuel C_5H_{10} reacts with oxygen. All combustion reactions are exothermic.

In order for the ethane reaction to take place, each carbon atom must break all its bonds to other atoms so that it can bind to oxygen and form carbon dioxide. The stronger the bonds in the fuel molecule, the more difficult they are to break. To see this, consider a group of three related reactions, including the combustion of ethane (Figure 9.3). Notice the similarities in these three reactions: all are hydrocarbon combustion reactions (and therefore exothermic), and each reactant contains two carbon atoms. The only difference from one reaction to the next is the type of bond connecting the two carbon atoms in the hydrocarbon. In ethane, ethene, and ethyne, these are a single, a double, and a triple bond, respectively (Section 3.2). Figure 9.3 gives the strength of each carbon–carbon bond, and these data show that, as the number of bonds between carbon atoms increases, the strength of the bond increases. Notice, too, that as the strength of the bond increases, the amount of energy given off by the combustion reaction decreases. This makes sense because more energy is required to break a stronger bond, and this takes away from the overall energy produced by the reaction.

Figure 9.3 also shows you that the combustion reactions for ethane, ethene, and ethyne are all very exothermic, meaning that any of these molecules would make a suitable fuel. In fact, though, ethane and ethene are not typically used as fuels, but ethyne, also known as *acetylene*, is used as a fuel in blowtorches. Acetylene torches reach temperatures of more than 3300 °C and can be used to solder and cut metals in various environments, including under water (Figure 9.4).

▼ **FIGURE 9.4** **Underwater torch.** A diver repairs an underwater structure using an acetylene torch.

Disappearing Ships

We have all heard tales about disappearing ships, the Bermuda Triangle, and other strange goings-on on the high seas. At least some of these "tall tales" have a basis in fact, however, as there are many well-documented incidences of vessels missing at sea when the usual causes, such as bad weather, were not to blame. Investigations of the sea floor have turned up sunken vessels that look

completely normal—they have no holes in their hulls and they are sitting upright on the sea floor (Figure 9.5). Recent evidence has revealed that these unexplained sinkings usually occur in areas of high methane activity. In these areas, methane gas, which is produced by the decay of organic matter in the seabed, builds up into pockets of high-pressure gas. When the pressure reaches a certain critical point, the gas is released in an enormous underwater burst that suddenly infuses the water with methane molecules. The methane-laden water becomes less dense than normal water, and the unlucky ship sitting on the water's surface simply drops out of sight.

Alan Judd, a marine geologist at the University of Sutherland in England, describes it like this: "Any ship caught above would sink as if it were in a[n elevator] shaft." If you are on board ship and can somehow get quickly into the water, you are still out of luck. Everything that could once float on water loses it buoyancy and goes straight down. Fortunately, it is possible to identify areas of high methane activity on the sea floor by a characteristic pock-marked appearance. At some point in the future, maps showing these areas may be available so that marine vessels can avoid them.

▲ FIGURE 9.5   **Methane pockets in the ocean may cause ships to sink.**   Stories of disappearing ships in the area of the Bermuda Triangle may now have a basis in fact. Some scientists believe that the release of pockets of methane gas may create a "hole in the water" that causes ships to drop to the ocean floor.

9.2 | Keeping Track of Hydrocarbons

By now, it should be obvious to you that hydrocarbons exist in a variety of lengths and types. To keep track of these, chemists use a standard naming system—formally called **nomenclature**—that is agreed upon worldwide. The name of a hydrocarbon is based on a Greek or Latin word root indicating the number of carbons it contains. For example, from Table 9.1, we see that any hydrocarbon containing seven carbon atoms has the root *hept-* somewhere in the name. Likewise, any hydrocarbon containing ten carbon atoms contains the root *dec-*.

Table 9.2 lists the names of some hydrocarbons made up solely of single bonds. In this list of similar organic molecules, the only difference from one member to the next is that each molecule contains one carbon and two hydrogens more than the molecule immediately before it on the list. This type of list is known as a **homologous series**. We shall see that a homologous series can be written for every class of organic molecules and also that the Greek roots in Table 9.1 are used throughout organic chemistry to name molecules. The **alkanes** are a class of organic molecules that have only single bonds, and the ending *-ane* indicates that only single bonds exist between the carbon atoms. When a hydrocarbon consists of a straight chain

MOLECULE
Heptane

TABLE 9.1 Root Names Used in Organic Nomenclature

Number of carbon atoms	Root name	Name of *n*-alkane	Name of substituent group
1	*Meth-*	Methane	Methyl
2	*Eth-*	Ethane	Ethyl
3	*Prop-*	Propane	Propyl
4	*But-*	Butane	Butyl
5	*Pent-*	Pentane	Pentyl
6	*Hex-*	Hexane	Hexyl
7	*Hept-*	Heptane	Heptyl
8	*Oct-*	Octane	Octyl
9	*Non-*	Nonane	Nonyl
10	*Dec-*	Decane	Decyl

of carbon atoms, the name of the hydrocarbon is often preceded by the letter *n-*, as you can see in Table 9.2.

QUESTION 9.5 Draw a condensed structure and a line structure for (*a*) decane, (*b*) octane, (*c*) butane, (*d*) ethane. (Return to Chapter 3 if you need to review the various types of chemical structures we use to represent molecules.)

ANSWER 9.5 $CH_3CH_2CH_2CH_2CH_2CH_2CH_2CH_2CH_2CH_3$

(a)

$CH_3CH_2CH_2CH_2CH_2CH_2CH_2CH_3$

(b)

$CH_3CH_2CH_2CH_3$

(c)

CH_3CH_3

—

(d)

TABLE 9.2 The Homologous Series of *n*-Alkanes

Name	Chemical formula	Condensed structure
n-Methane	CH_4	CH_4
n-Ethane	C_2H_6	CH_3CH_3
n-Propane	C_3H_8	$CH_3CH_2CH_3$
n-Butane	C_4H_{10}	$CH_3CH_2CH_2CH_3$
n-Pentane	C_5H_{12}	$CH_3CH_2CH_2CH_2CH_3$
n-Hexane	C_6H_{14}	$CH_3CH_2CH_2CH_2CH_2CH_3$
n-Heptane	C_7H_{16}	$CH_3CH_2CH_2CH_2CH_2CH_2CH_3$
n-Octane	C_8H_{18}	$CH_3CH_2CH_2CH_2CH_2CH_2CH_2CH_3$
n-Nonane	C_9H_{20}	$CH_3CH_2CH_2CH_2CH_2CH_2CH_2CH_2CH_3$
n-Decane	$C_{10}H_{22}$	$CH_3CH_2CH_2CH_2CH_2CH_2CH_2CH_2CH_2CH_3$

All hydrocarbons are not based on a simple straight chain of carbon atoms. Take this molecule, for example:

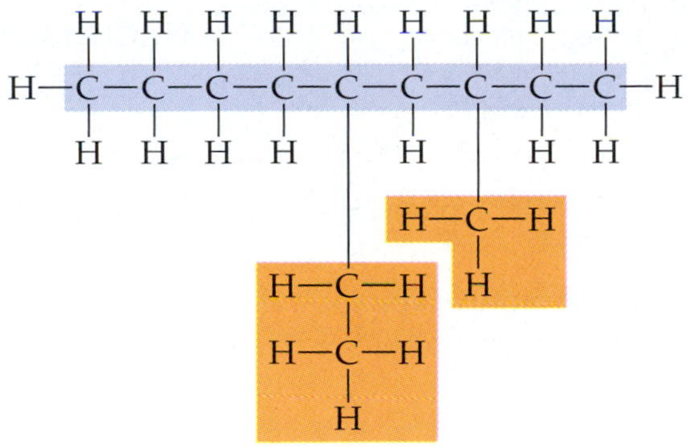

The orange highlights the two areas where the *main chain*, which is the longest carbon chain in the molecule, has *branches* consisting of shorter carbon chains. This type of hydrocarbon is called a **branched hydrocarbon**, and when naming one, the first thing you must do is identify the longest chain in the molecule. The root name of the molecule is determined by the length of this chain. The longest chain in this molecule, highlighted in blue, has nine carbon atoms, and thus we call it *nonane*. The remaining carbon atoms branch off of the main chain, and we call these appended groups **substituents**.

We name a substituent according to the number of carbon atoms it contains. The root names are given in the far right column of Table 9.1. They are formed by adding the ending *-yl* to the root name given in the second column. Thus, we call the one-carbon substituent group in our branched molecule a *methyl group* and the two-carbon substituent group an *ethyl group*. Substituents are indicated in the hydrocarbon name by numbering the main chain from one end to the other, with the beginning of the chain being, by convention, the end closest to a substituent group, so that the substituents have the lowest numbers possible. The numbering for our branched alkane is therefore

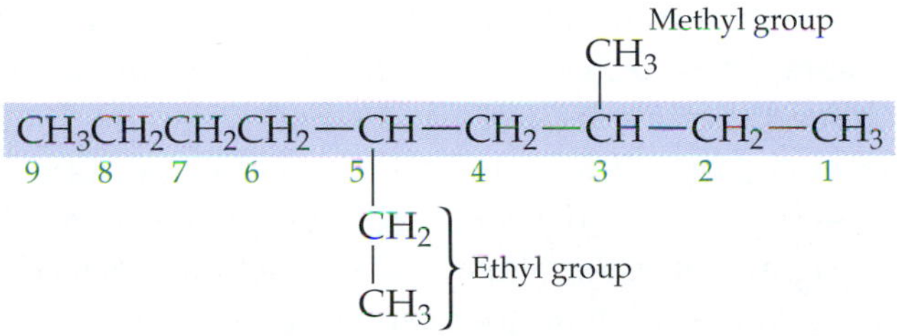

The two substituent groups are at positions 3 and 5 of the main hydrocarbon chain, and we use these numbers to identify those positions. The complete name for this molecule is then *5-ethyl-3-methylnonane*. Notice a few things about this name. First, alphabetization of the substituent names takes precedence over the numbers that indicate their positions. Thus, the ethyl group is listed first in the name because alphabetically *E* (for ethyl) precedes *M* (for methyl). Notice, too, that the name is one word with no spaces. Hyphens are used only to separate letters from numbers.

QUESTION 9.6 Name these branched alkanes:

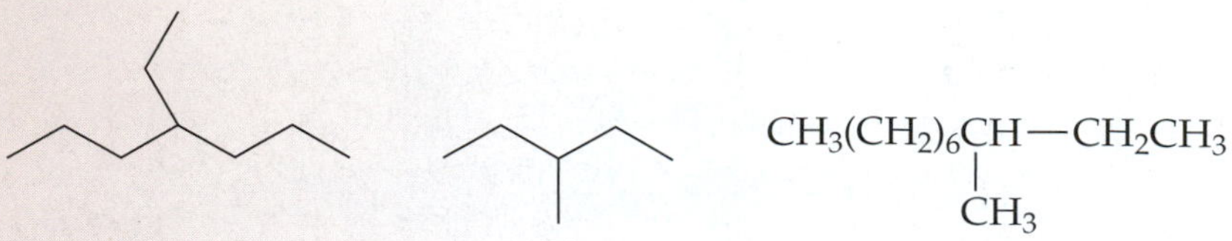

(a) (b)

$$CH_3(CH_2)_6CH—CH_2CH_3$$
$$|$$
$$CH_3$$

(c)

ANSWER 9.6 (*a*) 4-ethylheptane, (*b*) 3-methylpentane, (*c*) 3-methyldecane.

Let us look at a few more examples before moving on. Before looking at the answers in the next paragraph, see if you can name these three hydrocarbons correctly:

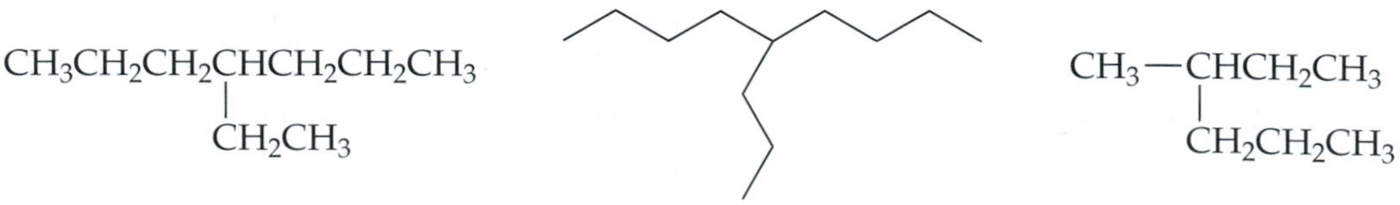

$$CH_3CH_2CH_2CHCH_2CH_2CH_3$$
$$|$$
$$CH_2CH_3$$

$$CH_3—CHCH_2CH_3$$
$$|$$
$$CH_2CH_2CH_3$$

Beware of a common mistake: organic molecules can be drawn so that the main chain is not written on one line; instead, it may be drawn bent, as in the third example:

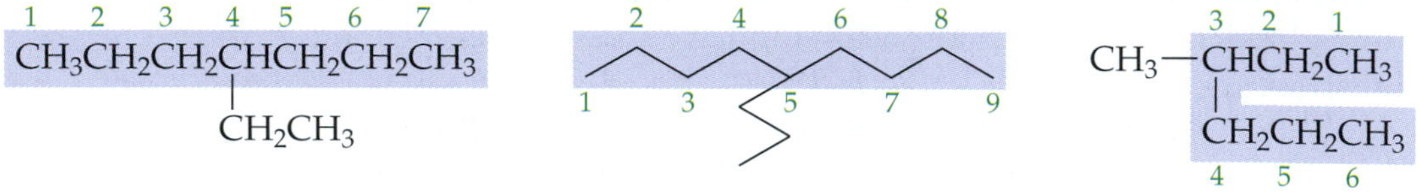

1 2 3 4 5 6 7
$$CH_3CH_2CH_2CHCH_2CH_2CH_3$$
$$|$$
$$CH_2CH_3$$
5

3 2 1
$$CH_3—CHCH_2CH_3$$
$$|$$
$$CH_2CH_2CH_3$$
4 5 6

Locating the main chain in the first two examples is relatively easy. The numbering is also easy for these two because the substituent is at the same numbered position whether you start numbering at the left end of the main chain or at the right end. Thus, we call the first molecule *4-ethylheptane* and the second molecule *5-propylnonane*. The third molecule is a bit trickier. First of all, the main chain is not in a straight line, but it is clearly made up of the six consecutive carbons indicated. The methyl group is placed at position 3 if you begin numbering as shown. However, if you begin numbering at the other end, then the methyl group is at position 4. As we have noted, we name organic molecules in a way that makes the substituent numbers as small as possible. Thus, we call this molecule *3-methylhexane*.

As we already know, hydrocarbons may contain double and triple bonds between carbons. If a hydrocarbon contains a double bond, it falls into a class of organic molecule called **alkenes**. Just as with alkanes, we can write a homologous series for alkenes, with each member of the series having one more carbon atom than the member preceding it. The presence of a double bond is indicated in the name with the suffix -*ene*, and the position of the double bond is indicated by the main-chain numbering scheme we already know. Hydrocarbon molecules containing triple bonds are referred to as **alkynes**, and these molecules are named with an -*yne* suffix.

We can figure out some of the rules for naming alkenes and alkynes by looking at these molecules:

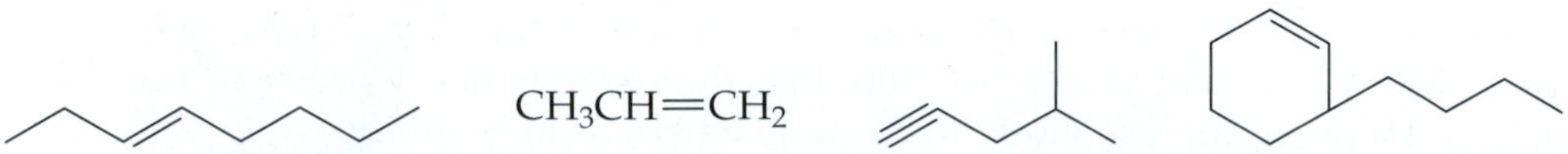

$$CH_3CH=CH_2$$

The first molecule is named *3-octene*. The 3 tells us the position of the double bond. Using the numbering scheme that we already are familiar with, the double bond is placed between carbons 3 and 4, and the number in the name indicates the position *at the beginning of the double bond*.

The second molecule contains three carbon atoms (prop-) and a double bond (-ene), and so its root name is *propene*. In fact, this is the complete name, with no number necessary to indicate the position of the double bond. Because there are only three carbons, the double bond must be at the first position. Check this for yourself. The name you give to an organic molecule should always pass this test: if you begin with the name of the molecule, you should be able to draw only one molecule. That is, *the name must describe only one specific structure*.

The third molecule has a main chain with five carbon atoms and a triple bond. Thus, its root name is *pentyne*. There is also a methyl substituent, and so we have to decide from which end we number the main chain. With the leftmost carbon numbered 1, the name is 4-methyl-1-pentyne. With the rightmost carbon numbered 1, the name is 2-methyl-4-pentyne. Which is correct? Fortunately, we have a rule to tell us the answer: double and triple bonds are always given the *lowest number possible* in the numbering system, even if doing so means giving the substituent groups a higher number. Therefore, the correct name is *4-methyl-1-pentyne*.

The fourth molecule may look odd to you because the hydrocarbon chain makes a ring. Rings are common structures in organic chemistry, and naming this molecule simply requires that we use the prefix *cyclo-* before the root name. The numbering of the main chain, which is the string of carbons in the ring, begins with the two carbons of the double bond because starting here places the double-bond carbons at the first two positions. We consider the ring to be a separate entity from the butyl group, which is named as a substituent on the ring. The numbering proceeds clockwise because going in this direction gives the butyl group a lower number than it would have if we numbered counterclockwise:

Correct   Incorrect

We call this molecule *3-butylcyclohexene*. Note that, when we are naming rings that contain one double or one triple bond, it is not necessary to identify the position of the bond with a number, because it is always assumed to be at the first position in the ring, as the next exercise demonstrates.

QUESTION 9.7 Identify each hydrocarbon as an alkane, alkene, or alkyne, and write the name of each:

(a)   (b)   (c)   (d)

ANSWER 9.7 (a) alkene, 1-methylcyclohexene; (b) alkyne, 3-heptyne; (c) alkene, 4-propylcyclohexene; (d) alkane, 5-ethyl-3-methylnonane.

Let us look at one example in which we must draw a molecule from a name. The compound *2,2,4-trimethylpentane* is a component of gasoline that burns smoothly in your car's engine. What is its structure? We begin with the root name pentane and draw a hydrocarbon with all single bonds and five carbons in a row:

$$C–C–C–C–C$$

The numbers 2, 2, and 4 tell us the locations of the substituents. The Greek prefix tri-, meaning three, tells us there are three identical substituents, in this case methyl groups. In organic nomenclature, we use a Greek prefix whenever the same substituent appears more than once in a molecule. A list of these prefixes is provided in Table 9.3. So far, our structure looks like this:

Now we can fill in the hydrogen atoms around each carbon to show each carbon forming four bonds:

2,2,4–Trimethylpentane

TABLE 9.3	Greek Numerical Prefixes Used to Name Multiple Substituent Groups in Organic Molecules
Number	**Prefix**
2	*di-*
3	*tri-*
4	*tetra-*
5	*penta-*
6	*hexa-*
7	*hepta-*
8	*octa-*
9	*nona-*
10	*deca-*

The nomenclature system for hydrocarbon molecules we have outlined here will be useful as we study other classes of organic molecules. The basic method is the same for all organic molecules, and because all are based on a carbon framework, the nomenclature of hydrocarbons is the basis for all the other nomenclature of organic chemistry.

QUESTION 9.8 Draw a line structure and a shortened condensed structure for (*a*) 3,3-dimethylnonane, (*b*) 5-butyldecane, (*c*) 3,3-diethyloctane.

ANSWER 9.8

$CH_3CH_2C(CH_3)_2(CH_2)_5CH_3$

(a)

$CH_3(CH_2)_3CH(CH_2CH_2CH_2CH_3)(CH_2)_4CH_3$

(b)

$CH_3CH_2C(CH_2CH_3)_2(CH_2)_4CH_3$

(c)

If you count all the hydrogen and carbon atoms in 2,2,4-trimethylpentane, you should arrive at a molecular formula C_8H_{18}. However, that molecular formula is not unique to this compound, as there are many, many other structures, including *n*-octane, you could draw that would have this same molecular formula. Some examples are shown in Figure 9.6. Because all these molecules have the same molecular formula, but clearly have different structures, we say they are **structural isomers** of one another. All the molecules have the same number of carbon atoms and the same number of hydrogen atoms, but those atoms are arranged differently in each molecule. Consider the two molecules

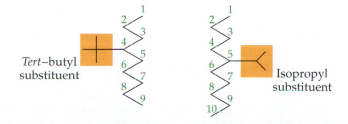

These alkanes both have the molecular formula $C_{13}H_{28}$ and are structural isomers of each other. Naming them is not a simple matter because each has a branched hydrocarbon substituent. Fortunately, these substituents are very common and have been given special names to simplify their nomenclature. The cross-shaped substituent at the fourth position in the molecule on the left is referred to as a *tert-butyl group*, and so we call the molecule 4-*tert*-butylnonane. (Note that we write the *tert* in italic type. This is merely

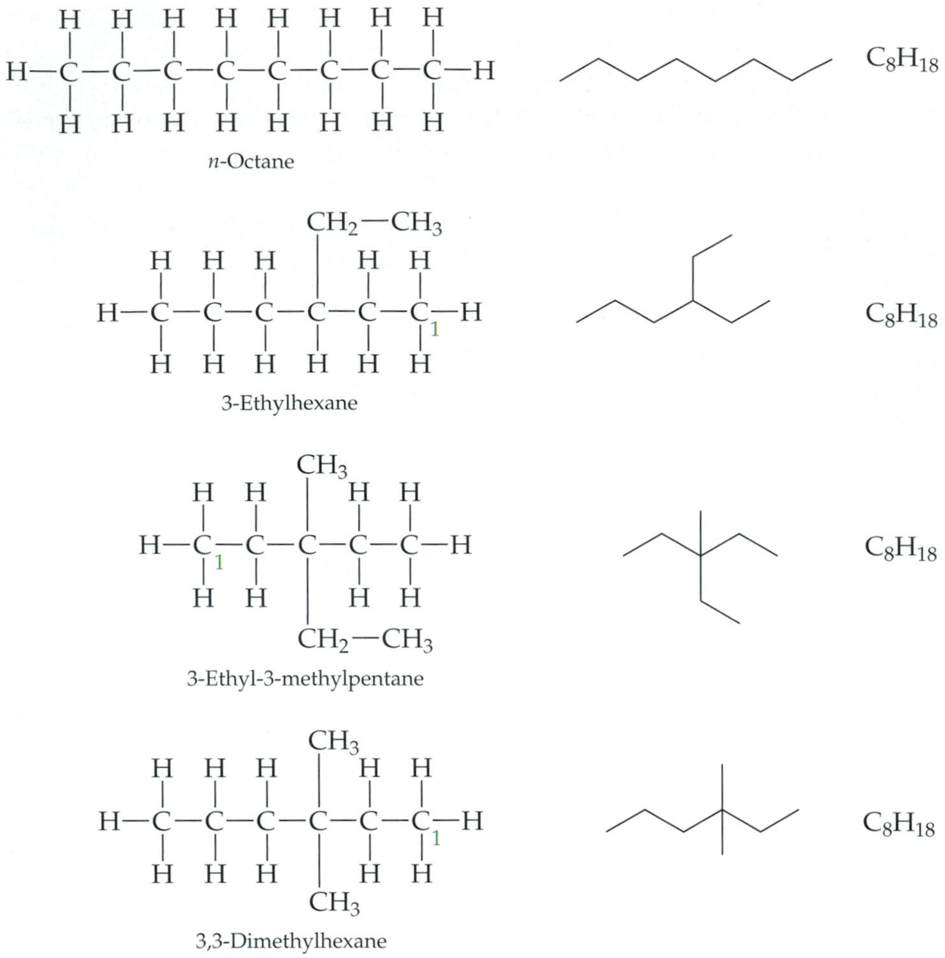

n-Octane C_8H_{18}

3-Ethylhexane C_8H_{18}

3-Ethyl-3-methylpentane C_8H_{18}

▶ **FIGURE 9.6 Four structural isomers.** Each of these four alkanes has the molecular formula C_8H_{18}. Each one, though, has its carbon atoms connected into a different framework, as you can see from either the full structures or the line structures.

3,3-Dimethylhexane C_8H_{18}

a style convention.) The substituent at the fifth position of the molecule on the right is referred to as an *isopropyl group* and so we call the molecule 5-isopropyldecane.

One type of structural isomer is unique to alkenes because there can be more than one way to draw the groups surrounding a double bond. Let us take the molecule 2-hexene as an example. Based on this name, it is possible to draw the two structures shown in Figure 9.7. Alkenes are a special case in organic nomenclature because the presence of a double bond between carbon atoms fixes the positions of those atoms with respect to each other. Single C–C bonds are free to rotate about the axis that connects the two atoms. No rotation is possible with double bonds, however. This restriction means that we cannot twist one of our 2-hexene molecules to make the other one. Instead, the two carbon atoms of a double bond are held in fixed position relative the horizontal axis through the bond, shown

▼ **FIGURE 9.7 Cis or trans?** **(a)** The structure of *cis*-2-hexene. The two hydrogen atoms attached to the double-bond carbons are on the same side of the horizontal axis running through these two carbons. **(b)** The structure of *trans*-2-hexene. Here, one hydrogen atom is on one side of the horizontal axis, and the other hydrogen atom is on the other side of the axis.

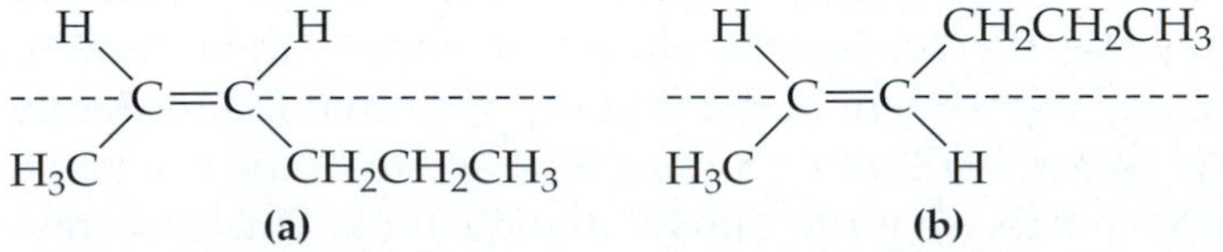

(a)   **(b)**

as a dashed line in Figure 9.7. The groups attached to the two carbons are locked into a specific orientation with respect to the double bond.

In Figure 9.7a, the hydrogen atoms attached to the double-bond carbons are on the same side of the horizontal axis, and the methyl and propyl groups are on the other side of the axis. In Figure 9.7b, the hydrogen atoms are on opposite sides of the horizontal axis, and the same is true for the two hydrocarbon groups. This type of isomerism, which is unique to alkenes, is known as **cis–trans isomerism**. The prefix *cis-* is used for the isomer shown in Figure 9.7a, in which the hydrogen atoms are located on the same side of the horizontal axis. The term *trans* refers to the structure shown in Figure 9.7b, in which the hydrogen atoms are located on opposite sides of the axis. Thus, the correct names for these molecules are *cis*-2-hexene (Figure 9.7a) and *trans*-2-hexene (Figure 9.7b). These two molecules have different physical properties, such as boiling point and melting point, which tells us that they are two different, distinct compounds.

The process of ranking of groups attached to a double bond can be very complex. In all examples of cis–trans isomerism in this text, there are two –H groups and two hydrocarbon groups attached to the double bond. While the method of nomenclature presented here will not allow cis–trans assignment in complicated cases in which three or four hydrocarbon groups are attached to a double bond, it works well for the examples given in this text.

QUESTION 9.9   Which of these names is sufficient to describe the structure of an organic molecule: (*a*) 1-octene, (*b*) 3-octene, (*c*) *trans*-2-octene, (*d*) cyclopentene? For each complete name, draw a line structure.

ANSWER 9.9   (*a*) 1-octene is a complete name because there can be no cis or trans isomers when the double bond is at the end of the molecule. The line structure is

(*b*) The name 3-octene requires a cis or trans prefix to be complete; (*c*) and (*d*) are complete names. The line structures are

No cis or trans prefix is required for cyclopentene because the ring fixes the double bond into the cis orientation.

In the name cyclopentene, no number is needed to specify the position of the double bond because it is, by default, at position 1 on the ring.

What Is the Meaning of the Octane Number Reported for Gasoline?

Although most hydrocarbons undergo combustion reactions in the presence of a source of oxygen, not all of them are suitable for use in your car's engine. In the engine, air, which contains about 21% oxygen, is mixed with the fuel inside a combustion chamber called a *cylinder*. The cylinder has a piston that moves

STUDENTS OFTEN ASK

up and down and thereby expands the cylinder volume when the explosion of the fuel and oxygen produces gases that increase the cylinder pressure (Figure 9.8). The car's transmission then *transmits* this motion to the axles, and the wheels begin to turn.

In order for this process to work efficiently, the explosions must occur at the appropriate time as the fuel and air are mixed. If the explosion occurs prematurely, the engine makes a sound that we know as *knocking*. Fuels are designed to minimize knocking, and various hydrocarbon mixtures have been tested to see which ones create explosions that are timed so that the engine runs smoothly. As it turns out, one of the best fuels for a smooth-running engine with minimal knocking is the hydrocarbon we met earlier in this section, 2,2,4-trimethylpentane, also known as *isooctane* (because it is a

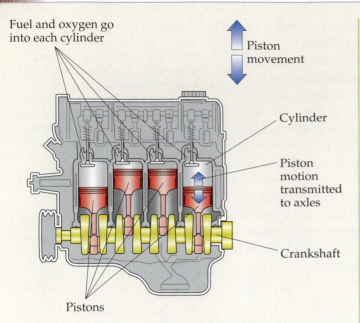

Fuel and oxygen go into each cylinder

Piston movement

Cylinder

Piston motion transmitted to axles

Crankshaft

Pistons

▲ **FIGURE 9.8** **Harnessing combustion.** In your car's engine, combustion reactions cause the cylinders to move, and this movement is translated to the axles, which then move the wheels.

structural *iso*mer of *n*-octane). This molecule has been assigned an arbitrary number, called the **octane number**, of 100 to indicate that it does not cause any knocking. Other molecules that do cause knocking are given lower numbers, depending on the amount of knocking they cause. For example, *n*-heptane causes constant knocking in a car engine and so has an octane number of zero.

Gasoline is a mixture of hydrocarbons and octane-boosting additives, each with its own particular octane number, and the overall octane rating given to a sample of gasoline tells you that its complex hydrocarbon mixture has the same antiknock properties as a sample containing that percentage of isooctane. For example, say

you have a sample of gasoline with an octane rating of 92. This rating tells you that this gasoline behaves like a mixture of 92 percent isooctane and 8 percent *n*-heptane. Why not simply use pure isooctane as fuel for cars? The answer is that isooctane makes up only a small fraction of crude oil, the source of gasoline, and therefore pure isooctane would be a very expensive fuel.

Most of today's gasoline mixtures have octane ratings of at least 85, meaning that they have a high percentage of low-knock hydrocarbons. Most gasoline stations offer you a choice of gasolines with different octane numbers (Figure 9.9). Why would a person choose any gasoline other than the one with the highest octane number? For one thing, high-octane gasolines are generally more expensive than lower-octane products. Second, the higher the octane number, the hotter the fuel burns. Cars designed to run cooler require lower-octane fuel to prevent the engine from overheating.

▼ **FIGURE 9.9** **Octane ratings in gasoline.** Most gasoline stations in the United States give you a choice of octane ratings when you buy gasoline.

9.3 | Cocktails and Anesthetics

The year 1939 brought trouble to the placid country of Finland. It was in August of that year that the Soviet Union and the Nazi regime of Germany signed an agreement called the Molotov–Ribbentrop pact, in which they agreed on which countries each would invade. According to the agreement, the Soviets would be free to invade Lithuania, Estonia, Latvia, and Finland, and the Germans would not interfere. Lithuania, Estonia, and Latvia fell quickly to Soviet forces, but these troops had a surprise waiting for them in Finland. The Finnish army, well versed in the art of snow skiing, captured Soviet tanks by arming Finnish soldiers on skis with homemade bombs that were nothing more than fuel inside a bottle with a rag hanging out. The Finns proudly named this potent little firebomb the *Molotov cocktail* after the Soviet diplomat responsible for the treaty between the Soviet Union and Germany. Liquid hydrocarbons work well as the fuel in Molotov cocktails because they undergo combustion reactions readily in the presence of air, which is about 21 percent oxygen. As we shall see next, however, not all hydrocarbons are liquids at room temperature.

In order to understand why different hydrocarbons exist in different phases at a given temperature, we must consider their structures. Recall from Section 7.7 that because hydrocarbons contain no electronegative atoms, they do not interact strongly with polar molecules (like water). However, hydrocarbon molecules do have weak attractions for one another, and the strength of these weak interactions increases with the number of carbon atoms in the molecule, as Figure 9.10 shows. Because their interactions are very weak, hydrocarbons boil at low temperatures, and this fact allows hydrocarbons with different chain lengths to be separated from one another, as we shall see shortly. The smallest hydrocarbon—methane, CH_4—is a gas at room temperature and atmospheric pressure because it has a small mass and because the methane molecules do not interact with one another strongly enough to allow it to be a liquid under these conditions. As hydrocarbons get larger and more massive, their ability to interact with one another increases, and the intermolecular attractions between molecules are strong enough to allow the hydrocarbons to exist in the liquid state at room temperature and atmospheric pressure. Continuing with this logic, you can see why very massive hydrocarbons with still stronger intermolecular attractions, like waxes, are solids at room temperature.

Table 9.4 lists the first 18 alkanes and some of their physical properties, including boiling point and freezing point. The trends are clear: as the mass of the alkane increases, the boiling point and freezing point increase.

(a)

(b)

▲ **FIGURE 9.10   Forces of attraction.** For hydrocarbon molecules, the attraction between individual chains is very weak, but it does exist. **(a)** n-Decane molecules, each ten carbon atoms in length, interact with one another to some limited extent. This interaction is represented by the way these line structures align with one another. **(b)** The smaller the number of carbon atoms in the hydrocarbon chain, the weaker the intermolecular attraction is. Thus, the interaction between these n-hexane molecules is weaker than the interactions between the n-decane molecules.

TABLE 9.4   Some Properties of the First 18 Alkanes

Hydrocarbon name	Number of carbon atoms	Molar mass (g/mol)	Boiling point (°C)	Freezing point (°C)	Phases at room temperature (25 °C)
Methane	1	16.05	−164	−183	Gas
Ethane	2	30.08	−89	−183	Gas
Propane	3	44.11	−42	−189	Gas
n-Butane	4	58.14	0	−138	Gas
n-Pentane	5	72.17	36	−130	Liquid
n-Hexane	6	86.20	69	−95	Liquid
n-Heptane	7	100.23	98	−91	Liquid
n-Octane	8	114.26	126	−57	Liquid
n-Nonane	9	128.29	151	−51	Liquid
n-Decane	10	142.32	174	−30	Liquid
n-Undecane	11	156.35	196	−26	Liquid
n-Dodecane	12	170.38	216	−10	Liquid
n-Tridecane	13	184.41	235	−5	Liquid
n-Tetradecane	14	198.44	254	6	Liquid
n-Pentadecane	15	212.47	271	10	Liquid
n-Hexadecane	16	226.50	287	18	Liquid
n-Heptadecane	17	240.53	303	23	Liquid
n-Octadecane	18	254.56	317	28	Solid

RECURRING THEME IN CHEMISTRY *Intra*molecular *bonds*, those *within* a molecule, are covalent. *Inter*molecular *interactions*, those either *between* two molecules or *between* two parts of a large molecule, are noncovalent.

Hydrocarbons with four or fewer carbon atoms are gases at room temperature, and those with 18 or more carbon atoms are solids. Everything in between is a liquid. Because there are only weak intermolecular attractions holding the molecules together, liquid hydrocarbons are highly volatile, with volatility increasing as molecule size decreases. Thus, hydrocarbons that are liquids at room temperature will tend to have many of their molecules in the space above the liquid. Or if a liquid hydrocarbon is in an open container, it will tend to evaporate readily.

QUESTION 9.10   At 0 °C and 100 °C, what is the phase of (*a*) pentane, (*b*) dodecane, (*c*) hexadecane? Refer to Table 9.4.

ANSWER 9.10   (*a*) liquid at 0 °C, gas at 100 °C; (*b*) liquid at both temperatures; (*c*) solid at 0 °C, liquid at 100 °C.

Figure 9.11 shows some hydrocarbon-based products with which you may be familiar. These products differ widely in appearance and consistency, from butane found in disposable lighters to motor oil to petroleum jelly. All of these products come from *petroleum*, otherwise known as *crude oil*. Crude

▼ **FIGURE 9.11   A hydrocarbon collage.**   Everyday items made from fossil fuels are shown from top left (clockwise): petroleum jelly; motor oil; stove fuel; dog apparel made from fossil fuel-based materials; a butane lighter; and natural gas (methane).

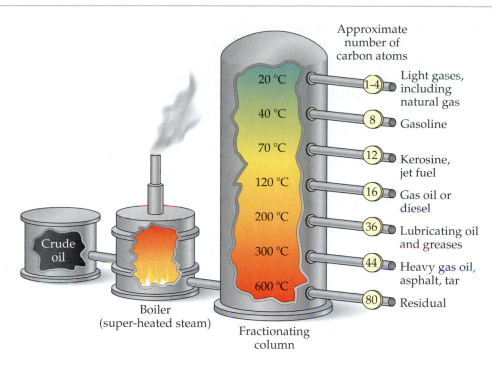

Approximate number of carbon atoms

20 °C	1-4 Light gases, including natural gas
40 °C	8 Gasoline
70 °C	12 Kerosine, jet fuel
120 °C	16 Gas oil or diesel
200 °C	36 Lubricating oil and greases
300 °C	44 Heavy gas oil, asphalt, tar
600 °C	80 Residual

Crude oil

Boiler (super-heated steam)

Fractionating column

◄ FIGURE 9.12 **Fractionation of crude oil.** After crude oil is heated, it moves up a fractionating column. The lightest hydrocarbons move all the way up to the top of the column, and the heaviest ones remain close to the bottom.

oil is a fossil fuel because it is derived from organic matter (fossils) that, over the eons, has decomposed to form a mixture of carbon-based molecules. The oil is pumped out of the ground as a dark, viscous liquid, and it must be *refined* to separate the various components on the basis of how many carbon atoms each component contains. This is done by a process called **distillation**, in which the mixture is heated to a high temperature and then allowed to travel up a tall column in the gas phase (Figure 9.12). This structure is called a **fractionating column** because it separates the hydrocarbon mixture into *fractions* containing hydrocarbons having similar masses and physical properties. The fractionating column gets cooler toward the top, and this drop in temperature causes the hydrocarbons that distill off the crude oil to condense back into the liquid phase. The heavier the hydrocarbon, the shorter its trip up the column and the sooner it is converted from gas to liquid. Thus, smaller hydrocarbons travel farther up the column before condensing. Methane, the smallest hydrocarbon, travels to the top of the column, where it is collected and sold as natural gas. So a fractionating column converts a complex mixture of hydrocarbons to a collection of products that are useful in our daily lives.

Organic molecules other than hydrocarbons also undergo combustion reactions and have the high volatility required for ignition as we shall see next.

▼ FIGURE 9.13 **A sampling of ethers.** Ethers are composed of two hydrocarbon chains connected to one oxygen atom. The top three ethers shown here are ethyl methyl ether, butyl propyl ether, and methyl pentyl ether. The bottom notation, with its R group and its R' group, is the generic formula used to represent all ethers.

QUESTION 9.11 Rank these alkanes according to the height reached in a fractionating column, highest first: butane, heptadecane, dodecane, ethane, decane.

ANSWER 9.11 ethane > butane > decane > dodecane > heptadecane.

Another class of organic molecules that undergoes combustion reactions is **ethers**, which are similar to hydrocarbons except that they have an oxygen inserted into the middle of the hydrocarbon chain, as Figure 9.13 shows. The

$$CH_3CH_2-O-CH_3$$

$$CH_3CH_2CH_2-O-CH_2CH_2CH_2CH_3$$

$$CH_3-O-CH_2CH_2CH_2CH_2CH_3$$

$$R-O-R'$$

last ether on the list represents the generic way to write any ether. Using this notation, the *R* represents any hydrocarbon chain and the *R'* represents any other hydrocarbon chain. Thus, you can think of an ether as an oxygen atom with two hydrocarbon chains attached. Throughout the coming chapters, we shall be talking about groups within molecules, like the oxygen in an ether, that are a modification of a simple hydrocarbon structure. We refer to such groups as **functional groups**.

Ethers are easy to name if you remember the roots used to name the substituent groups of Table 9.1. To name an ether, simply put the two names of the substituent groups in alphabetical order, and add the word *ether* to the end. For example, the ether

$$H_3CCH_2CH_2CH_2-O-CH_3$$

has a butyl group and a methyl group attached to the oxygen, and so it is called butyl methyl ether. The names of the ethers shown in Figure 9.13 are, from the top, ethyl methyl ether, butyl propyl ether, and methyl pentyl ether. (Notice that the substituent names and the word *ether* are all separated by a space.)

It is not necessary to have two different hydrocarbon groups in an ether. When an ether has two identical hydrocarbon groups attached to the oxygen atom, the Greek prefix *di-* is placed in front of the hydrocarbon group name to show that there are two of that group in the molecule. The ether

$$H_3C-O-CH_3$$

for instance, is called dimethyl ether.

Diethyl ether, $CH_3CH_2-O-CH_2CH_3$, was once used widely as a medical anesthetic. In 1846, it became the first anesthetic available for surgical work and was used for many years thereafter. In fact, in nineteenth-century America, diethyl ether was commonly used as a recreational drug. The development of more effective and less flammable anesthetics and recreational drugs soon made the use and abuse of diethyl ether obsolete.

Let's compare the molecular formula of an ether with the formula of the alcohol containing the same number of carbons. Notice that the two formulas are the same, telling you that this alcohol and this ether are structural isomers of each other. Sometimes, as in this case, two molecules with the same molecular formula have markedly different properties because they contain different organic functional groups. For example, let us compare diethyl ether and *n*-butanol, both of which have the molecular formula $C_4H_{10}O$:

Diethyl ether, $C_4H_{10}O$

n-Butanol, $C_4H_{10}O$

Although both compounds are liquids at room temperature, their boiling points are very different: 35 °C for diethyl ether and 118 °C for *n*-butanol. Why are the boiling points so different when the molecular formulas are exactly the same? The answer lies in the position of the oxygen atom. For

Ethers may be named using a more formal system of nomenclature. Because we are naming only simple ethers, we shall use only the method described in the text.

MOLECULE
Dimethyl Ether

Look at Section 7.7 if you need to review the structure of alcohols. The alcohol functional group is –OH.

ethers, the oxygen atom is inserted between two hydrocarbon chains. In this position, the oxygen atom, even though it is electronegative, gives the molecule a polarity that is much lower than the polarity of the butanol molecule. In fact, ethers are, in general, not much more polar than their hydrocarbon counterparts. The alcohol *n*-butanol, on the other hand, has its oxygen atom attached to the hydrocarbon chain, not inserted in it, and so the oxygen is always attached to a hydrogen atom. Recall from Section 7.2 that when an oxygen atom is attached to a hydrogen atom, the molecule can take part in hydrogen bonding. The hydrogen bonds that connect molecules of *n*-butanol make the liquid less volatile because the molecules are less likely to break their hydrogen bonds in order to enter the gas phase. This explains the much higher boiling point for *n*-butanol.

Every hydrogen bond involves two groups: (1) a hydrogen atom attached to an electronegative atom and (2) an electronegative atom like O, N, or F. When talking about hydrogen bonds, we sometimes refer to the electronegative atom connected to the H atom as being the *hydrogen bond donor*. Then the other electronegative atom is referred to as the *hydrogen bond acceptor*. This terminology allows us to see how hydrogen bonding accounts for *n*-butanol's relatively high solubility in water. The solubility is high because the –OH group forms hydrogen bonds with water molecules readily (Figure 9.14a). The oxygen atom's lone pair of electrons allows the –OH group to behave as a hydrogen bond acceptor. The hydrogen atom in the –OH allows the same group to act as a hydrogen bond donor. In diethyl ether, the oxygen atom can accept hydrogen bonds from water, but the ether molecule cannot donate hydrogen bonds because the oxygen atom is not bound to a hydrogen atom (Figure 9.14b). As a result, the solubility of the ether in water is far less than the solubility of the alcohol in water.

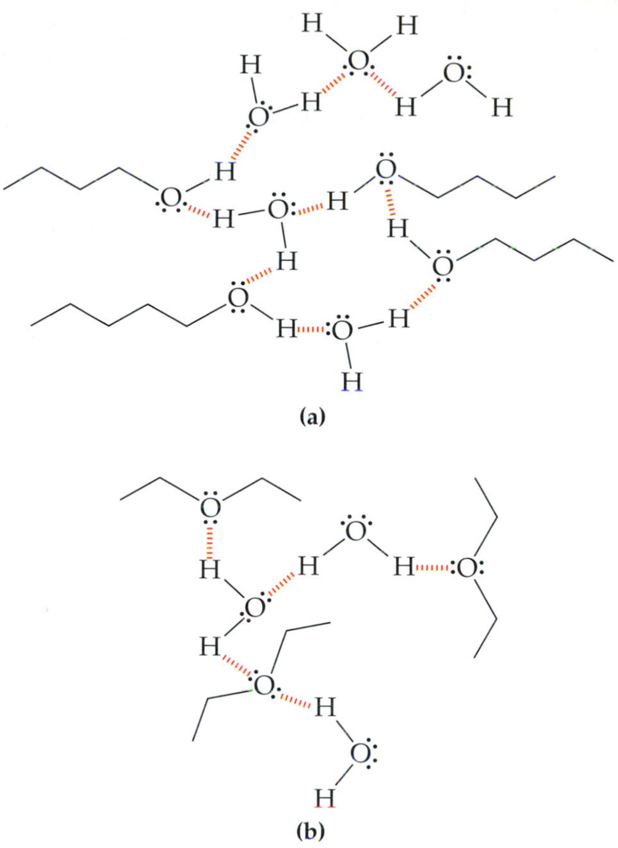

(a)

(b)

▲ **FIGURE 9.14 The universal solvent? (a)** In the hydrogen bonding between *n*-butanol and water, the alcohol acts both as a hydrogen bond donor through the H on the –OH group and as a hydrogen bond acceptor through the O. **(b)** Diethyl ether can accept a hydrogen bond from water's hydrogen atom, but it cannot donate a hydrogen bond because it has no H atoms attached to O atoms.

RECURRING THEME IN CHEMISTRY   Like dissolves like.

QUESTION 9.12   Identify each molecule as an ether or an alcohol and name each ether.

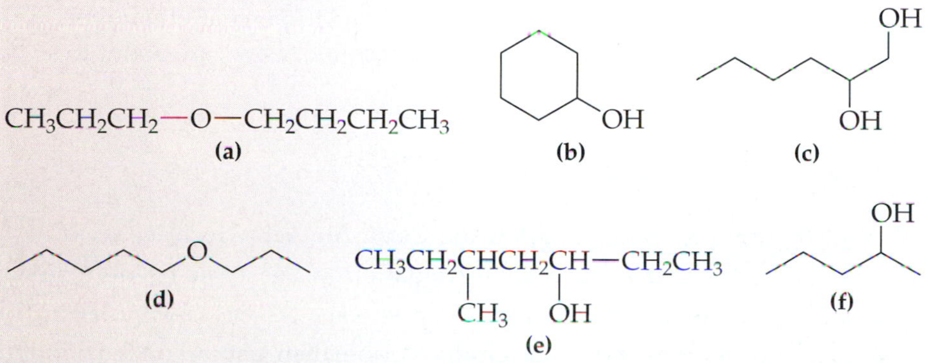

$CH_3CH_2CH_2$—O—$CH_2CH_2CH_2CH_3$
(a)

(b)

(c)

(d)

$CH_3CH_2CHCH_2CH$—CH_2CH_3
         | |
        CH_3 OH
(e)

(f)

ANSWER 9.12   (*a*) Butyl propyl ether, (*d*) pentyl propyl ether. The molecules in (*b*), (*c*), (*e*), and (*f*) are alcohols.

QUESTION 9.13 Sketch the hydrogen bonding interactions between water and cyclopentanol:

ANSWER 9.13

The very volatile diethyl ether molecule is suitable as an anesthetic because it readily enters the gas phase and can be administered to patients through a mask. As we have noted, however, it can undergo combustion reactions in the same way that hydrocarbons do, a fact that makes diethyl ether less than desirable as an anesthetic, as you can easily imagine. The combustion reaction is

$$2\,H_3CCH_2{-}O{-}CH_2CH_3 + 12\,O_2 \longrightarrow 8\,CO_2 + 10\,H_2O$$

Alcohols can also undergo combustion reactions, as in this reaction for *n*-butanol:

$$2\,H_3CCH_2CH_2CH_2{-}OH + 12\,O_2 \longrightarrow 8\,CO_2 + 10\,H_2O$$

However, the volatility of alcohols is lower than that of ethers or hydrocarbons because alcohols form hydrogen bonds in the liquid phase, and so alcohols are not often used in applications in which volatility is desirable, such as anesthesia.

The alcohol ethanol is often used as a gasoline additive because it lowers the amount of carbon monoxide produced during incomplete combustion of the hydrocarbon fuel. In some countries, such as Brazil, many cars run on fuels containing a very high percentage of ethanol. In the United States, our gasoline mixtures usually contain no more than 10 percent ethanol because ethanol dramatically increases the price of gasoline, as we are about to see.

ISSUES IN CHEMISTRY

MTBE in Gasoline

In 1991, the U.S. government passed the Clean Air Act, which included new, mandatory requirements for the formulation of gasoline—requirements designed to reduce hazardous emissions from cars. As we saw earlier in the chapter, the combustion of gasoline produces more than just carbon dioxide and water. Because there is limited oxygen available and because there is limited time for the reactants to mix in a car engine, combustion is incomplete, and CO is formed in addition to CO_2:

O₂ gas present in
limited amount
in car engine

Octane + oxygen ⟶

carbon monoxide + carbon dioxide + water

CO forms because of
limited availability
of O₂

The CO represents a direct and significant health risk to the breathing public, and catalytic converters have been added to car exhaust systems to reduce the amount of CO leaving the tailpipe.

The Clean Air Act sought to reduce emissions by improving gasoline formulations in a way that would reduce the amount of incomplete combustion, thereby addressing the problem at the engine rather than at the tailpipe. Because there is no practical way to infuse gasoline with oxygen molecules, the next best thing is to add a source of oxygen to the gasoline so that it will have an opportunity to undergo more complete combustion. The source of oxygen for this reaction, the **oxygenate**, should contain a high percentage of oxygen and should be inexpensive and easy to come by. The combustion reaction then becomes

Octane + oxygen + oxygenate ⟶

carbon monoxide + carbon dioxide + water

Amount of CO formed
should decrease in
presence of oxygenate

Amount of CO₂ formed
should increase in
presence of oxygenate

To fulfill the oxygenate requirement dictated in the Clean Air Act, the oil companies had two viable choices for oxygenate: ethanol, which would raise the price of gasoline because it is costly to produce, and methyl *tert*-butyl ether (MTBE), an additive that was already being used in some gasoline formulations:

Two depictions of methyl *tert*-butyl ether (MTBE)

MTBE was a good candidate for oxygenate because it has one oxygen atom and a small molar mass, making the percentage of the molar mass contributed by oxygen relatively high. It is also cheap and easy to come by because it is made from methanol, a by-product of crude oil refining.

QUESTION 9.14 Name these compounds:

(a)

(b)

(c)

ANSWER 9.14 (a) *n*-butyl *tert*-butyl ether, (b) *tert*-butylcyclohexane, (c) 4-*tert*-butylnonane.

MTBE is an ether and therefore highly volatile. It also mixes readily with gasoline and burns easily in a car engine. MTBE's one oxygen atom, however, makes the molecule slightly more polar than its hydrocarbon equivalent because of the oxygen's high electronegativity. Although MTBE molecules cannot hydrogen bond because they do not possess an –OH group, the presence of the electronegative oxygen atom makes it possible for the molecule to take part in weak dipole–dipole interactions with water. In contrast, consider the molecule 2,2-dimethylbutane, which has the same structure as MTBE except that the oxygen atom has been replaced with a carbon atom, as Figure 9.15a shows. Because it lacks a polar atom, 2,2-dimethylbutane cannot interact with water molecules. Figure 9.15b shows dipole–dipole interactions between MTBE and water. These dipole–dipole interactions are weak but, along with hydrogen bonding, these interactions give MTBE some solubility in water. Whereas 2,2-dimethylhexane has no solubility in water at 25 °C, it is possible to mix 4.8 g of MTBE with 100 g of water at that temperature, a considerable amount.

The solubility of MTBE in water has caused some unexpected problems for gasoline companies. The underground tanks in which gasoline is stored at gas stations are considered relatively leakproof because they do not release detectable amounts of gasoline into groundwater

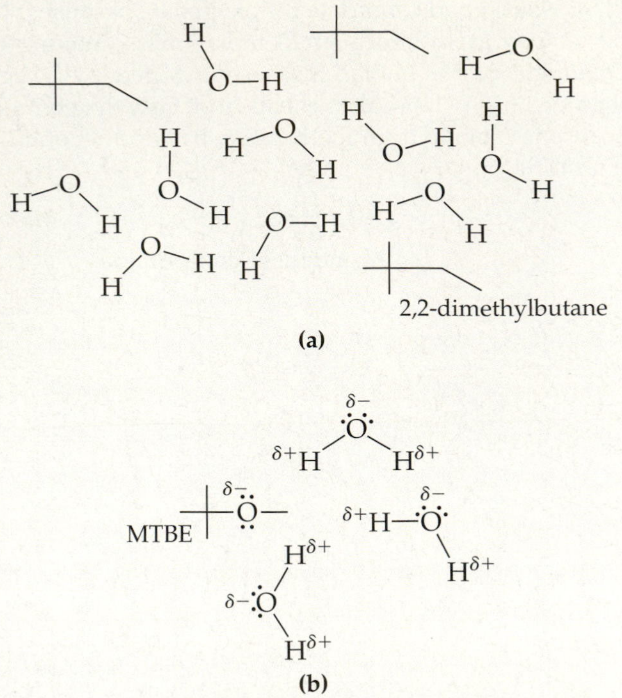

2,2-dimethylbutane

(a)

(b)

▲ FIGURE 9.15 **Like dissolves like.** The solubilities of two molecules in water are explained by considering their ability to form hydrogen bonds. **(a)** 2,2-Dimethylbutane is not able to hydrogen bond because it does not contain any polar atoms. For this reason, it is not soluble in water. **(b)** MTBE has a structure that is almost identical to that of 2,2-dimethlybutane. The only difference is the presence of the oxygen atom. This one difference allows MTBE to form hydrogen bonds with water and enhances its polarity. Thus, it is water soluble.

that comes into contact with the tanks. However, because the solubility of gasoline in water is almost zero, the mixing of gasoline with groundwater would be unlikely even if the tanks actively leaked gasoline into the surrounding soil. When MTBE was added to gasoline formulations, the gasoline now had a mildly water-soluble component, and this component—MTBE—leaked slowly into groundwater from underground gasoline storage tanks. Suddenly, the wisdom of storing gasoline in underground tanks, a nonissue in the past, became a source of controversy as MTBE began to show up in public drinking-water supplies and in rainwater.

If ethanol had been chosen as a gasoline additive rather than MTBE, leakage would not have caused significant problems for the environment because, although ethanol is water soluble, it has low toxicity. Comparatively little was known about the toxic effects of

MTBE when refineries began adding it to gasoline. Reports of deleterious health effects quickly started to roll in, however, as MTBE became infused into drinking water. In formal studies, it has since been established that MTBE is a lethal carcinogen in laboratory rats, causing a variety of cancers ranging from lymphomas to kidney tumors.

Another drawback to MTBE is that the molecule does not biodegrade once released into the environment. Thus, we cannot wait out the breakdown of the molecule and expect it to go away on its own. After more than a decade of MTBE use in gasoline formulations countrywide, the problem of groundwater contamination has reached mammoth proportions. In the Long Island region of New York, for instance, more than 3 million people live in areas with MTBE-contaminated drinking water. To make matters worse, the high volatility of MTBE allows it to enter the gas phase when household water is heated, making it possible for people to inhale MTBE as well as drink it. In South Lake Tahoe, California—one of the worst areas for MTBE contamination—tourists are staying away for fear of drinking water that tastes like paint thinner and taking a shower with water that fills the air with a toxic odor.

The gasoline companies claim to be between a rock and a hard place. On one hand, the Clean Air Act required them to put an oxygenate into their gasoline formulations, and in 1991 the deleterious health effects of MTBE were not widely known. When faced with the choice between ethanol and MTBE, the obvious choice was MTBE, as ethanol use would raise the price of gasoline dramatically. Any company that chose to use ethanol would have had lonely gas pumps because consumers would choose gas stations with cheaper MTBE-laden gasoline. Bob Campbell, CEO of Sunoco, says, "There's a law out there that requires reformulated gasoline to contain [an oxygenate]. The only practical alternative I have in order to comply with that law, is that we use MTBE."

Should the oil companies be held responsible for the decision to include MTBE in gasoline? Apparently, many people feel they should, and today lawsuits are being won by water utility companies who are suing oil companies over the contamination of their groundwater. Some people believe that the EPA is to blame because it approved the use of MTBE in gasoline formulations when the Clean Air Act was passed.

Ironically, environmental measurements have shown that the use of MTBE has not resulted in a reduction in carbon monoxide emissions and is thought to actually increase the emission of another toxin associated with car emissions—nitrous oxide.

(a)

(b)

9.4 | Bigger Bangs

Thus far, we have seen that a fuel mixed with some oxidizer, which may be the oxygen in air or in a substance such as KNO_3, undergoes combustion to produce a great deal of heat in addition to carbon dioxide, carbon monoxide, and water. This chemistry has been known about and used for centuries. In the mid-1800s, however, several discoveries came along that revolutionized the use of explosives for blasting and for warfare. In 1867, a Swedish inventor by the name of Alfred Nobel patented the use of the molecule nitroglycerin for this purpose (Figure 9.16*a*). Nobel found that nitroglycerin itself was impractical for use in industry or as a weapon because it would detonate unexpectedly with the slightest shock. His insight was to mix the nitroglycerin with an inert, solid substance such as charcoal or porous silica to make the much more stable explosive **dynamite**. In total, Nobel received more than 400 patents for various inventions related to the use of explosives, and those patents brought him considerable financial gain. In his last will and testament, he allocated a substantial portion of his wealth to initiate a series of Nobel prizes for achievements in the sciences and the humanities. Nobel's legacy carries on today in the annual Nobel prizes and in the use of dynamite more than 120 years after it was first patented.

The structure of nitroglycerin is shown in Figure 9.16*b*. Unlike the fuels we have considered thus far, nitroglycerin does not require a source of oxygen for detonation. In fact, the term *combustion* is really reserved only for molecules that interact with an oxygen source to form water and carbon dioxide or carbon monoxide. Nitroglycerin is a self-contained, all-in-one explosive, as this reaction shows:

$$4\,C_3H_5N_3O_9(\ell) \longrightarrow 6\,N_2(g) + 12\,CO_2(g) + 10\,H_2O(g) + O_2(g)$$
$$+ \text{ lots of heat energy}$$

Although it is not a combustion reaction, the detonation of nitroglycerin meets all of our requirements for an effective explosion. First, it produces huge volumes of gas. Four molecules of liquid nitroglycerin molecules produce an impressive 29 molecules of product in the gas phase. This enormous gas pressure is the source of the blasting power that we associate with dynamite. Second, the reaction is strongly exothermic. The heat given off makes the product gases expand that much faster and contributes to the immense gas pressures supplied by the reaction.

QUESTION 9.15 Comment on this statement:

The equation shown for the detonation of nitroglycerin is not balanced because there are unequal numbers of molecules on the two sides of the equation.

RECURRING THEME IN CHEMISTRY Chemical reactions do not create or destroy matter. Therefore the types and numbers of atoms must be the same on the two sides of a chemical equation.

ANSWER 9.15 Chemical reactions are rearrangements of atoms, and the atoms may regroup in any number of ways. Several molecules may become one molecule, or vice versa. Thus, in order for a reaction to be balanced, it is the number of *atoms* on the two sides of the equation that must be equal, not the number of *molecules*.

Clearly, the nitroglycerin reaction is highly exothermic. Why, though, are some reactions exothermic while others are endothermic? To answer this question, we must compare the bonds that hold together the molecules in an explosive reaction. The atoms in a nitroglycerin molecule are held together by several bonds, many of which are weak. Figure 9.17a shows the full structure of this molecule and the bond strength of each type of bond. You can see that the strongest bond is the $N{=}O$ covalent bond. The weakest bond, which is about one-third as strong as the $N{=}O$ bond, is the N–O covalent bond. In general,

a molecule that contains relatively weak bonds is more likely to undergo a reaction because those bonds are easily broken.

Compared with many other molecules, nitroglycerin has many weak bonds, especially the six single bonds between nitrogen and oxygen.

The strength of the bonds in the reactant is not the end of the story. The bonds in the products drive the reaction because

the stronger the bonds are in the products relative to the bonds in the reactants, the more energy the reaction will produce.

There are several products of the nitroglycerin reaction, and Figure 9.17b shows their bond strengths. As you can see, the weakest bonds in the products—the O–H bonds in water—are stronger than most of the bonds in the reactant nitroglycerin. The triple bond in diatomic nitrogen is most impressive with a whopping 940 kJ/mol.

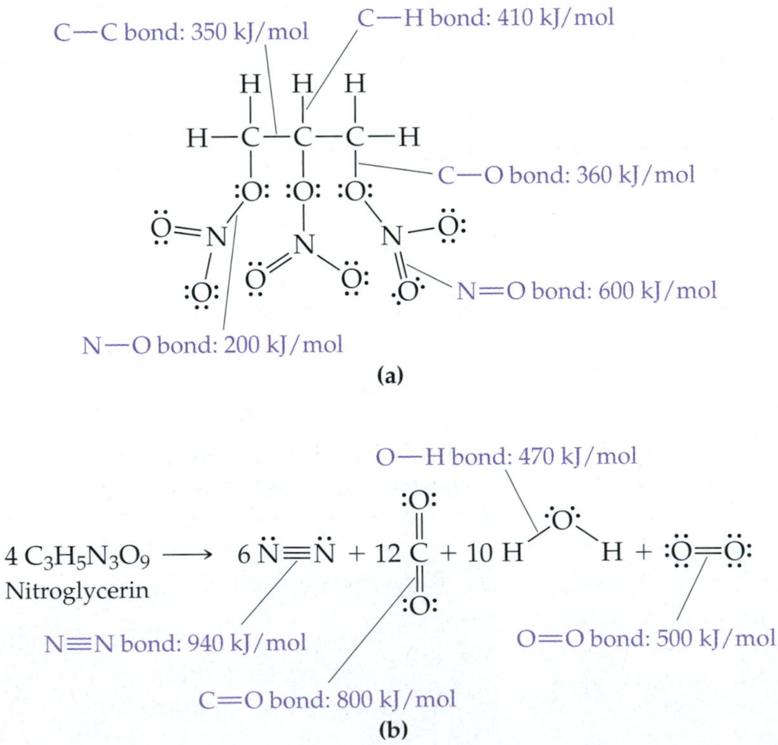

▶ **FIGURE 9.17** **Why are explosives unstable?** (a) Bond energies for all bonds in the nitroglycerin molecule. (b) This balanced equation for the detonation of nitroglycerin shows that a few molecules containing weak bonds will react violently to form a larger number of molecules containing stronger bonds.

C—C bond: 350 kJ/mol
C—H bond: 410 kJ/mol
C—O bond: 360 kJ/mol
N—O bond: 200 kJ/mol
N=O bond: 600 kJ/mol

(a)

$$4\ C_3H_5N_3O_9 \longrightarrow 6\ N{\equiv}N + 12\ C + 10\ H_2O + O_2$$
Nitroglycerin

O—H bond: 470 kJ/mol
N≡N bond: 940 kJ/mol
C=O bond: 800 kJ/mol
O=O bond: 500 kJ/mol

(b)

QUESTION 9.16 To what can the incredible strength of the bond in the N_2 molecule be attributed?

ANSWER 9.16 The bond between nitrogen atoms in a nitrogen molecule is triple. In most cases, triple bonds are stronger than double bonds, which are stronger than single bonds.

Let us do some bond bookkeeping for the nitroglycerin reaction, using the balanced chemical equation as a guide. According to this equation, four molecules of nitroglycerin react. When this happens, every bond in these molecules must break because none of the bonds present in nitroglycerin is found in any of the products. Thus, we can sum all the bond strengths for four molecules of nitroglycerin to determine how much energy is required to break every bond:

$$2 \text{ C–C bonds} \times \frac{350 \text{ kJ}}{\text{C–C bond}} + 5 \text{ C–H bonds} \times \frac{410 \text{ kJ}}{\text{C–H bond}} + 3 \text{ N=O bonds}$$

$$\times \frac{600 \text{ kJ}}{\text{N=O bond}} + 6 \text{ N–O bonds} \times \frac{200 \text{ kJ}}{\text{N–O bond}} + 3 \text{ C–O bonds}$$

$$\times \frac{360 \text{ kJ}}{\text{C–O bond}} = \frac{6830 \text{ kJ}}{\text{nitroglycerin molecule}} \times 4 \text{ molecules} = 27{,}320 \text{ kJ}$$

is the amount of heat energy contained in all bonds in the four reactant molecules. Consequently, it takes 27,320 kJ of energy to break all the bonds in nitroglycerin. Now let us look at the products and add up the energy gained by their formation:

$$6 \text{ N≡N bonds} \times \frac{940 \text{ kJ}}{\text{N≡N bond}} + 24 \text{ C=O bonds} \times \frac{800 \text{ kJ}}{\text{C=O bond}}$$

$$+ 20 \text{ O–H bonds} \times \frac{470 \text{ kJ}}{\text{O–H bond}} + 1 \text{ O=O bond} \times \frac{500 \text{ kJ}}{\text{O=O bond}} = 34{,}740 \text{ kJ}$$

is the amount of heat energy contained in all bonds in the 29 product molecules.

If you subtract the energy required to break the bonds in nitroglycerin from the energy produced when the products are made, you have a difference of $34{,}740 - 27{,}320 = 7420$ kJ. What this exercise shows is that if the products of a reaction have stronger bonds, on average, than the reactants, there is energy given off during the reaction, and the reaction is exothermic. In this case, the amount of energy produced is substantial, and it is this energy, in the form of heat, that makes this explosive reaction so remarkable.

In general, making bonds releases energy because bonds form when atoms are more stable together than they are apart. Breaking bonds requires an input of energy because the atoms of the bond benefited by being together, and breaking them apart is an energetically unfavorable process. A chemical reaction is endothermic when the energy that goes into breaking reactant bonds is greater than the energy produced by making product bonds. A chemical reaction is exothermic when the energy that goes into breaking reactant bonds is less than the energy produced by making product bonds. This idea is summarized in Figure 9.18.

RECURRING THEME IN CHEMISTRY Making bonds releases energy; breaking them requires an input of energy.

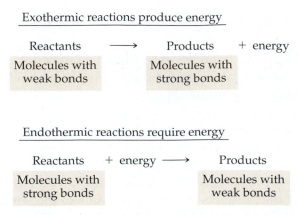

▲ **FIGURE 9.18 Energy as reactant and product.**   When the bonds in reactant molecules are weaker than those in product molecules, a reaction is exothermic because energy is given off during the reaction. When the bonds in reactant molecules are stronger than those in product molecules, a reaction is endothermic because we must add energy to the reactants to get things going.

We can now add another criterion for an explosive reaction to our list: the sum of the energies of all bonds in the product molecules should be larger than the sum of the energies of all bonds in the reactant molecules. As we look at other examples of unstable molecules, you will notice that many contain N–O single bonds, which are among the weakest bonds found in organic molecules. These bonds make a reactant unstable and more likely to undergo an explosive reaction. In addition, when N–O single bonds exist in a reactant, nitrogen gas, N_2, is often one of the products formed. The extraordinarily strong triple bond in this little molecule makes a large contribution to the overall energy given off by the formation of products, and this makes the reaction that much more exothermic.

STUDENTS OFTEN ASK

Why Do People with a Heart Condition Take Nitroglycerin?

It is often assumed that Alfred Nobel invented nitroglycerin, but in fact this molecule was first synthesized by Italian inventor Ascanio Sobrero in 1847. A physician as well as an inventor, Sobrero was interested in creating new drug molecules rather than molecules that could be used as explosives. Thus, his characterization of nitroglycerin involved taste tests, and his experiments showed how the molecule affected the human body. Through these tests, he found that nitroglycerin caused relief from the discomfort of angina, intermittent chest pain resulting from insufficient flow of blood through the heart. These findings were later confirmed in a study of workers at nitroglycerin factories who suffered from angina. The angina symptoms experienced by these workers decreased during the work week, when they were exposed to nitroglycerin, and increased during the weekends, when they were at home.

Today nitroglycerin is used to treat angina patients, and a patch has been developed that allows diffusion of the drug through the skin. There have been a few reports of exploding patches, but these events are rare. The health benefits of nitroglycerin therapy presumably outweigh the risk of patch explosion, which is usually harmless and produces nothing more than a mild pop.

9.5 | High Explosives, Low Explosives

At the end of the nineteenth century, another class of organic molecule emerged as the next new explosive. Unlike nitroglycerin, this new type of molecule contained a benzene ring with substituent groups attached to various positions around the ring. The first molecule of this type to be used was *picric acid*:

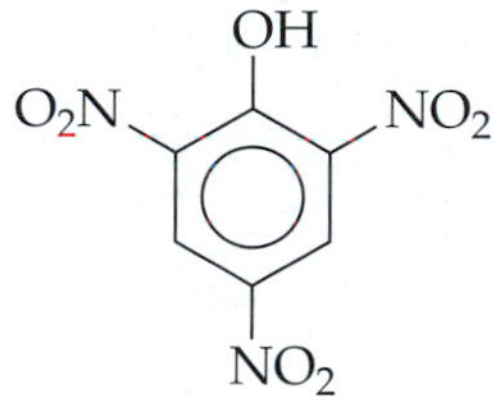

That this molecule is powerfully explosive should come as no surprise because you now know that the presence of N–O single covalent bonds is often a telltale sign of instability.

Picric acid is a phenol, which is an alcohol in which the –OH group is attached to a benzene ring rather than to a hydrocarbon chain. The name *picric acid* comes from the Greek word *pikros*, meaning "bitter." The fact that we know the taste of such a dangerous molecule may seem alarming, but early chemists often used flavor as a way of characterizing chemicals.

Over the decades that followed the discovery of picric acid, many similar molecules were made and their properties tested. All had two things in common: a benzene ring and at least one $-NO_2$ group. The $-NO_2$ group is a substituent in the same way a methyl group is, and the root name *nitro* is used to indicate the presence of $-NO_2$ substituents. Picric acid, for example, has the formal name *trinitrophenol*. The presence of three nitro groups is indicated by the Greek prefix *tri-*, and the framework of the molecule, along with the –OH group, is indicated with the name *phenol*.

If you subject the name trinitrophenol to our test for naming organic molecules, you will find that it fails because there are many different ways to place the three nitro groups and one –OH group around the ring, as shown in Figure 9.19a. Which one is picric acid? For molecules with substituents attached to a benzene ring, we number the ring either clockwise or counterclockwise and indicate the position of substituent groups using these numbers. In this case, we number the phenol ring beginning with the

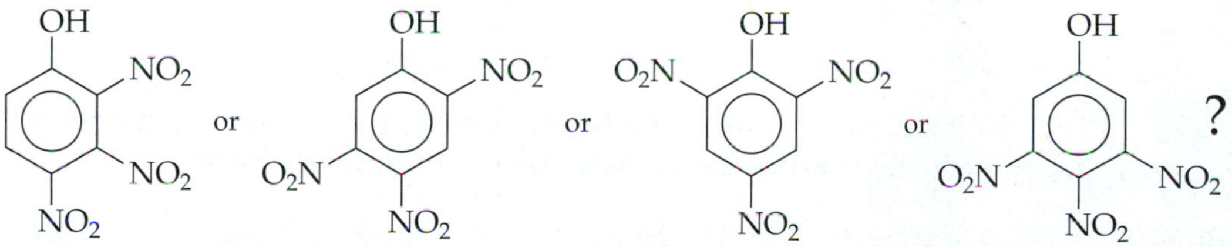

(a)

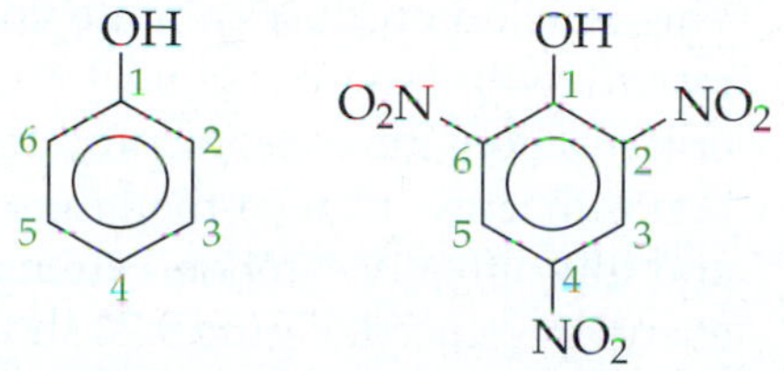

(b)　　(c)

◀ FIGURE 9.19 **Will the real picric acid please stand up?** (a) A molecule named trinitrophenol could be any of the molecules shown here. (b) Numbering of the benzene ring begins at the –OH group. (c) The three nitro groups in picric acid are correctly placed at positions 2, 4, and 6 on the ring. Thus, the "real" picric acid in part (a) is the third structure.

–OH because this group is part of the phenol's fundamental structure (Figure 9.19*b*). Using this numbering system, the three nitro groups in picric acid are attached at positions 2, 4, and 6 on the ring (Figure 9.19*c*). So we can refine our formal name for picric acid to *2,4,6-trinitrophenol*.

QUESTION 9.17 Draw the line structure for (*a*) 3-methylphenol, (*b*) 2,3-diethylphenol, (*c*) 2, 3, 5-trimethylphenol.

ANSWER 9.17

(a)   (b)   (c)

Picric acid, as it turns out, was an extremely unpredictable explosive for use in guns and bombs. It would explode unexpectedly, causing more damage to the army using it than to the enemy. To solve this problem, researchers had to make modifications to the picric acid structure. The goal was to find a molecule that had picric acid's power, but not its unpredictable behavior. The most promising candidate for the new explosive was the molecule we call TNT, more formally known as *2,4,6-trinitrotoluene*. The TNT molecule is a modification of a molecule called *toluene*, which is the common name for methylbenzene:

Methylbenzene (toluene)

Three nitro groups are attached at positions 2, 4, and 6 on the toluene ring to give TNT:

2,4,6-Trinitrotoluene (TNT)

The chemical reaction that occurs when TNT explodes is similar to that for nitroglycerin:

$$2\ C_7H_5N_3O_6(s) \longrightarrow 3\ N_2(g) + 12\ CO(g) + 5\ H_2(g)$$
$$+ 2\ C(s) + \text{lots of heat energy}$$

This reaction produces a large volume of gas that expands rapidly with the intense heat created. As in the nitroglycerin reaction, the gaseous products in a TNT reaction have bonds that are stronger than those in the reacting molecule, and this difference in bond strength again accounts for the energy produced. Figure 9.20 shows British women assembling TNT-based munitions for use in World War I.

▼ **FIGURE 9.20 Women doing their part for the war effort.** British women in a munitions factory assembling TNT-containing explosive devices during World War I.

Because it would not detonate as easily and could be transported and manipulated more safely, TNT proved to be a more practical explosive than picric acid. In fact, TNT is usually detonated using another, smaller charge made up of a less explosive material called a *primer*. Over the years, various substances have been used as primers, including *1,3-dinitrobenzene*:

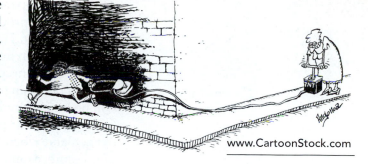

www.CartoonStock.com

1,3-Dinitrobenzene

This molecule is less explosive than either TNT or picric acid because it contains fewer nitro groups on the benzene ring. The small explosion that a small mass of this and other primers make is sufficient to detonate a larger mass of TNT.

In reading other chemistry books, you may encounter another system of nomenclature for molecules containing two ring substituents, such as 1,3-dinitrobenzene. According to this older, but still commonly used, system, the name contains the prefix *ortho-*, *meta-*, or *para-* depending on the orientation of the two substituents relative to each other. As Figure 9.21 shows, *ortho-* means the substituents are next to each other on the ring, *meta-* means the two substituents are separated by one position on the ring, and *para-* means they are across from each other on the ring. Often you will hear chemists say things like "The two nitro groups are meta to each other" or "The ring has para substituents." Although this is an old-fashioned way of expressing the nomenclature of benzene rings, it is extremely convenient and is still widely used today.

ortho-dinitrobenzene
(a)

meta-dinitrobenzene
(b)

para-dinitrobenzene
(c)

▲ **FIGURE 9.21 Ortho-, meta-, para-names for benzene compounds.** When there are two substituents on a benzene ring, they can be named using the terms *ortho-*, *meta-*, or *para-* instead of according to the numbers of the carbons in the ring.

QUESTION 9.18 Provide two names for each molecule. (Hint: The halogen substituent –Cl is named *chloro*.)

(a) **(b)** **(c)**

ANSWER 9.18 (*a*) 1,2-dimethylbenzene, *ortho*-dimethylbenzene; (*b*) 1,3-diethylbenzene, *meta*-diethylbenzene; (*c*) 1,4-dichlorobenzene, *para*-dichlorobenzene.

The explosives we have discussed thus far can be divided into two categories: low explosives or high explosives. Low explosives usually must be confined so that the high pressure created in a small space contributes to the blasting effect and accelerates the chemical reaction. Most hydrocarbon fuels, therefore, are classified as low explosives. High explosives typically consist of only one molecule. They do not require a separate source of oxygen, as do hydrocarbon low explosives, and there is no need to confine these substances to a small space in order to have an effective explosion. Thus, TNT and picric acid classify as high explosives.

If you compare the heat produced by a typical low explosive combustion with that produced by a high explosive detonation, you will find that they are similar. The difference is in the rate of the chemical reaction. A combustion reaction involving a low explosive—a hydrocarbon, say—may take place over a long period of time, and measures must be taken to make the reaction fast if an explosion is desired. In addition, the speed of the reaction will be limited by the availability of oxygen. Thus, hydrocarbons are made to explode when they are mixed thoroughly with air or another source of oxygen and when they are confined in a way that permits a rapid buildup of pressure, as in an internal combustion engine. High explosives typically produce energy immediately and forcefully, and the reaction is completed within the time it takes for the explosion to occur. With high explosives, there is no need to manipulate the reactant so that detonation will occur. When they go off, they detonate completely. In fact, the problem with high explosives is their tendency to explode prematurely. Thus, the conditions under which high explosives are detonated are often aimed not at making the reactant more explosive, but rather at making the reactant more stable until an explosion is desired.

STUDENTS OFTEN ASK

What Is the Most Explosive Compound Known?

In the early 1980s, University of Chicago chemist Phil Eaton and his colleagues began the task of making the molecule known as *octanitrocubane*, a modified version of *cubane*, a cube-shaped hydrocarbon first synthesized by Eaton in 1964. Nearly 20 years and two unexpected explosions later, the molecule was completely synthesized in the year 2000. Eaton says that his interest in this molecule, a molecule that has been described as beautiful, was mostly aesthetic even though it was always known from the structure of the molecule that it would be extremely explosive. It is, in fact, about 20 times more explosive than the most powerful non-nuclear explosive in use today. Eaton's research was funded, in large part, with funds contributed by the U.S. Army and Navy.

Octanitrocubane

What makes octanitrocubane so incredibly unstable? First of all, the presence of nitro groups in organic molecules is often a hallmark of instability, as we learned in the previous section. This molecule contains eight of them. Second, the cubane structure is inherently unstable in and of itself, which explains why it is so difficult to synthesize.

In a typical straight-chain hydrocarbon, the bend angle formed by any C–C–C sequence is about 110°. In cubane, in which carbon atoms make up the corners of a cube, the bend angle formed by any C–C–C sequence has to be 90°:

Propane   Cubane

Because the carbon atoms would prefer to have 110° angles and not be forced into 90° angles, the cubane molecule reacts quickly in an effort to alleviate the strain that these tight bond angles impose. Thus, this molecule has two good reasons to undergo a reaction as quickly as possible—many NO_2 groups and strained C–C bond angles—and the result is an extraordinarily potent explosive. The U.S. military is looking for ways to use octanitrocubane in ultralight missiles.

9.6 | Chaos

Thus far, we have established that reactions are likely to occur when the bonds in the reactants are weaker than those in the products. When this is the case, energy is given off, often in the form of heat in an exothermic reaction. This disparity in bond strength is part of what drives these reactions forward. However, there are many examples of reactions that naturally go forward even though they are endothermic. There are also examples of exothermic reactions that do not go forward. Clearly, there must be some other factor that dictates whether a reaction happens.

To better understand why certain chemical reactions happen and others do not, we must take into account a fundamental law of nature: the **second law of thermodynamics**. According to this law, the disorder in the universe is always increasing. Or, put another way, disorder—also known as **entropy**—must increase whenever any event takes place, even though the increase in entropy may not always be easy to recognize. To better understand entropy, we shall begin with an everyday example and then will look at how entropy can be used to characterize a group of atoms or molecules.

Although entropy is an esoteric topic, it has important consequences for many things we do. Consider a simple situation that illustrates what entropy means. Imagine that your house is a mess, with dirt scattered all over the floor. Fed up with your living conditions, you decide to vacuum up the dirt, which has a very high entropy because it is scattered randomly and is highly disordered. Your vacuum cleaner imposes order upon the dirt particles by sucking them into a bag (Figure 9.22). Once the dirt is in the bag, it has a lower entropy because it has been collected into one place. Thus, you have decreased the entropy in your house; but does this mean you have violated the second law of thermodynamics, which demands that entropy increase for any process? The answer is no, because somewhere, somehow, disorder must have been created as a result of your vacuuming. Moreover, the disorder created must be greater than the order you created by collecting the dirt particles into a bag. But how can this be?

To solve this conundrum, you have to consider every aspect of your vacuuming job. Think for a moment about your vacuum cleaner. Where does it get the energy required to suck up dust particles? The answer is that the electricity that powers your vacuum cleaner comes from a power plant of some sort. If that plant uses steam, for example, to generate power, then you must consider the entropy changes of that process in addition to the changes that took place with the dirt on your floor. Steam plants heat liquid

(a)   (b)

◄ **FIGURE 9.22 High and low entropy.** (a) This messy room is a high-entropy situation. (b) Once the room has been straightened up, it becomes a low-entropy situation. In order to decrease entropy here, entropy had to *increase* someplace else in the universe.

water to make steam, the pressure of the steam is used to turn a turbine, and that motion is converted to electricity that eventually makes its way to your electrical outlet. Without the production of steam in the power plant, your floor would still be dirty. Now consider for a moment the entropy associated with making steam from liquid water. The molecules in steam are highly disordered and fast moving. In comparison, the molecules in liquid water are relatively immobile and better organized. Thus, when liquid water is converted to steam, there is a substantial increase in the entropy of the water molecules. If you could somehow measure the entropy increase that takes place at the power plant, you would find that it is greater than the entropy decrease that occurs for the dirt on your floor.

Clever readers might try to devise a way around the second law. For example, what if you used a broom rather than an electric vacuum cleaner? This way, the dirt is collected, but there is no power plant involved. (Or is there?) The idea is a good one, but you must consider the energy required to push the broom. After all, brooms do not sweep your floor on their own. The power for the broom must come from your body, and this power ultimately comes from the food you eat, which is no more than a collection of molecules, ions, and atoms. Your body breaks down the bonds in food molecules, and the energy of those broken bonds provides power for your muscles (to move the broom) and your brain (to tell your muscles what to do), in addition to maintaining your body temperature above room temperature. As you sweep your floor, your body gets slightly warmer, and the molecules around your body get slightly more disordered as their entropy increases.

The primary fuel used by the human body, glucose, is broken down in a process that should look familiar to you by now:

$$C_6H_{12}O_6(s) + 6\,O_2(g) \longrightarrow 6\,CO_2(g) + 6\,H_2O(g) + \text{heat energy}$$

If you recognize this as a combustion reaction, you are correct. Although cells in the body use literally thousands of chemical reactions to metabolize food, the combustion reaction shown here is the central, overall reaction common to all animals that breathe air. The primary differences between this reaction and the one that occurs in your car are the fuel used (glucose rather than gasoline) and the efficiency with which the products are used. While car engines make use of only about 25 percent of the energy available in gasoline, the cells of the body perform combustion reactions slowly and efficiently and make good use of most of the energy that they produce (Figure 9.23).

QUESTION 9.19   Which member of each pair has higher entropy?

(a) 50,000 people marching in file or 50,000 people at a peace rally

(b) cars in a parking lot or cars in traffic on a freeway

(c) two butterflies mounted in a display case or two butterflies flying through a field of wildflowers

(d) three coins in a fountain or three coins stacked on top of one another

ANSWER 9.19   (a) people at a peace rally, (b) cars in traffic, (c) butterflies in a field of flowers, (d) three coins in a fountain.

The glucose reaction is driven in the forward direction by the production of heat energy in an exothermic reaction. However, the reaction is also

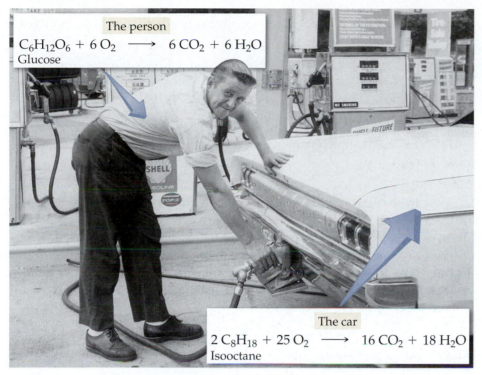

The person
$$C_6H_{12}O_6 + 6\,O_2 \longrightarrow 6\,CO_2 + 6\,H_2O$$
Glucose

The car
$$2\,C_8H_{18} + 25\,O_2 \longrightarrow 16\,CO_2 + 18\,H_2O$$
Isooctane

Similarities
• Both require fuel and oxygen to function.
• Both use combustion reactions to create energy.
• Both produce waste products.

Differences
• Car uses gasoline, person uses food.
• Car combustion happens quickly, food breaks down slowly.
• Car combustion very inefficient, most of food energy tapped.

◀ FIGURE 9.23 **Comparison of combustion in humans and machines.** Both the man and the car use a combustion reaction to provide energy.

driven by the second law of thermodynamics: entropy increases. To see why this is true, we must consider the relative entropy levels on the two sides of the equation. On the left side, we have one molecule of glucose in the solid phase, a very low entropy condition. Oxygen has a much higher entropy because it is in the gas phase, and there are six molecules of oxygen gas on the left. On the right side, there are 12 molecules of gas. Clearly, the entropy of the product molecules is greater than the entropy of the reactant molecules, and this difference in entropy drives the reaction in the forward direction.

The same phenomenon helps us understand the spontaneity of explosive reactions. Consider again the detonation of nitroglycerin:

$$4\,C_3H_5N_3O_9(\ell) \longrightarrow 6\,N_2(g) + 12\,CO_2(g) + 10\,H_2O(g)$$
$$+ O_2(g) + \text{lots of heat energy}$$

The reactant nitroglycerin is a liquid, and four molecules undergo the reaction. The products include 29 molecules, all in the gas phase. Thus, in this reaction, a few molecules in the liquid phase (low entropy) react to give many molecules in the gas phase (very high entropy). This reaction happens because it is exothermic and because disorder increases enormously when the reaction occurs.

QUESTION 9.20 For each reaction, indicate whether entropy is increasing or decreasing:

(a) $SO_2(g) + SrO(s) \rightarrow SrSO_3(s)$
(b) $Ba(OH)_2(s) \rightarrow BaO(s) + H_2O(g)$
(c) $C_3H_8(g) + 5\,O_2(g) \rightarrow 3\,CO_2(g) + 4\,H_2O(\ell)$
(d) $C_3H_8(g) + 5\,O_2(g) \rightarrow 3\,CO_2(g) + 4\,H_2O(g)$

ANSWER 9.20 Entropy increases in reactions (*b*) and (*d*) and decreases in (*a*) and (*c*).

We can see now that there are two factors that contribute to whether or not a chemical reaction will take place. If a reaction is exothermic and involves an increase in entropy, it will go forward. If a reaction is endothermic and involves a decrease in entropy, it will not go forward. It is possible for these two effects to counteract each other, and when this happens, it can be difficult to predict whether or not a reaction will occur. Consider, for instance, the following reactions:

$$CaCO_3(s) + heat \longrightarrow CaO(s) + CO_2(g) \qquad \text{endothermic, entropy increases}$$

$$Si(s) + 2\,Cl_2(g) \longrightarrow SiCl_4(\ell) + heat \qquad \text{exothermic, entropy decreases}$$

The first reaction is endothermic and produces an increase in entropy. The second reaction is exothermic and produces a decrease in entropy. In each case, these two effects compete against each other. Do these reactions go forward or not? The answer is maybe. In cases in which the energy change in the reaction counteracts the entropy change, the reaction will go forward under some conditions (that is, of temperature, pressure, and so forth), but not under other conditions. As conditions change, the dominance of these factors with respect to each other shifts. Thus, each case is unique, and very often the only way to know for certain whether a reaction will proceed is to put on a lab coat and try it in the laboratory.

9.7 | Arson and the Analysis of Explosives

▼ **FIGURE 9.24 Looking for clues at a possible arson site.** At the scene of an unexplained fire, porous materials such as upholstery and wood tend to absorb fuel molecules best. These kinds of items are often taken for analysis.

John Orr had a reputation for luck in the California law enforcement community. Orr and his group of arson investigators had a knack for knowing how and where fires were started. His unprecedented success rate in pinpointing the cause of arson cases began to raise some eyebrows in the higher ranks of the California criminal investigations community. Eventually evidence showed that, despite his position as the leader of the arson squad, Orr was a devoted arsonist. His success rate in determining the method used in so many arson crimes was simply explained—he was setting the fires himself. When the investigation into Orr's criminal activities was complete, he was held responsible for setting more than 2000 fires and for the deaths of four people. He is serving a life sentence.

Most arson cases involve a flammable liquid fuel. As you might suspect, the most common fuel used is gasoline because of its ready availability, but other liquids with high volatility that undergo combustion can be used. At an arson scene, investigators collect pieces of porous materials, such as upholstery, carpeting, and wood, because the fuel used to start the fire may have soaked into these soft substances (Figure 9.24). During the fire, any fuel that has soaked into a piece of wood, for example, will not have ready access to oxygen from the air and may remain unburned.

Evidence collected at the arson scene is taken to a laboratory for analysis. Because the fuel soaked into a piece of evidence is volatile, it will tend to escape from the article and into the air. Thus, pieces of evidence are placed in sealed containers, and the *headspace* above the article in the container becomes permeated with molecules of the fuel in the gas phase. In a method referred to as *headspace analysis*, samples of the headspace air are removed from the container and analyzed for the presence of fuel molecules.

QUESTION 9.21   It is more difficult to perform a headspace analysis on a water-based sample than on a sample of hydrocarbon fuel. Why do you think this is so?

ANSWER 9.21   Watery samples will have a very low volatility because water molecules hold one another tightly via hydrogen bonds. Thus, the number of water molecules in the headspace above a solution that is water based is much smaller than the number of fuel molecules in the headspace above a solution of fuel.

Most fuels used in arson crimes are highly complex mixtures of hydrocarbons, including alkanes, alkenes, alkynes, cyclic hydrocarbons, and hydrocarbons with benzene rings. Yet methods of fuel analysis have become so advanced that it is often possible to identify the specific gas station from which a sample of fuel originated. How is it possible to disentangle such a complex concoction? Many methods are available for the analysis of complex mixtures, but one of the most well-known and reliable methods is **gas chromatography**, which is based on chemical principles you already know.

Figure 9.25*a* shows a diagram of a typical gas chromatograph. To begin the analysis of headspace gases in a sealed container suspected to contain fuel molecules, a gastight syringe is inserted into the container and a small amount of gas is removed. This gas is then injected into the chromatograph injector port, which is heated to ensure that the molecules in the sample remain in the gas phase. The sample makes its way to the entrance of a very long, very thin tube called a *chromatographic column*, which is packed with a porous material, called the *solid support*, with which the molecules in the sample can interact.

The type of gas chromatography we are interested in here is based on the fact that the different kinds of organic molecules in the sample will interact with the solid support to differing extents (Figure 9.25*b*). For example, a long, straight hydrocarbon molecule such as $C_{16}H_{34}$ will tend to interact with the solid support more often and for longer periods of time than a shorter hydrocarbon like *n*-butane, C_4H_{10}. The more massive the molecule and the larger its carbon framework, the more opportunity it will have to stick to the solid support. Smaller molecules have fewer atoms with which to interact with the solid support.

Movement through this type of chromatographic column is a simple matter of traffic flow. You can think of a chromatographic column as a forest in which the trees are pieces of solid support. For a bus and a motorcycle both trying to traverse this expanse of forest, it will be much more difficult for the bus than for the motorcycle to drive through. You would expect the motorcycle to emerge on the other side of the forest sooner than the bus. It is the same way with molecules. Larger, bulkier molecules must carefully navigate through the maze of solid support while a smaller molecule can glide through more easily. The upshot is that the different varieties of organic

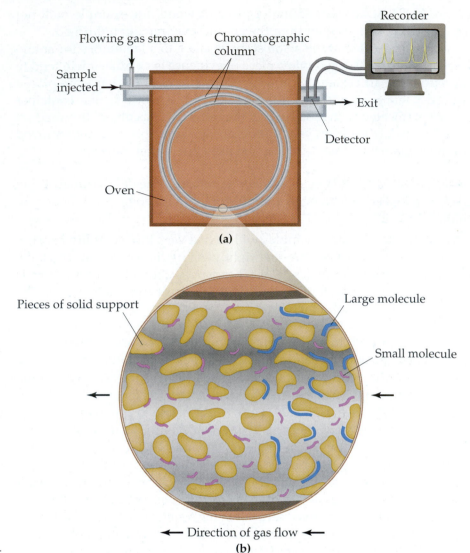

Recorder

Flowing gas stream

Chromatographic column

Sample injected →

Oven

Exit

Detector

(a)

Pieces of solid support

Large molecule

Small molecule

← Direction of gas flow ←

(b)

▶ **FIGURE 9.25** **Gas chromatography.** (a) In a gas chromatograph, the mixture to be analyzed is injected into one end of a very long column containing a solid support to which the molecules in the mixture bind to some extent. (b) The fact that different molecules bind to the solid support to different extents means that they will emerge from the column at different times. As molecules move through the solid support of the column, larger molecules are caught up and take longer to emerge from the other end. Thus, smaller molecules emerge first and larger molecules emerge last.

molecules in the sample mixture can be separated according to size and bulk in much the same way that they are separated in a fractionating column during petroleum refining (Figure 9.26). The result of a gas-chromatography analysis is a printout called a *gas chromatogram* with time plotted on the horizontal axis and quantity printed on the vertical axis.

Figure 9.27 shows a gas chromatogram for a sample of gasoline. If you begin reading the chromatogram at the left, you can see that there are no peaks for about the first 3 min. At about 3.5 min, the first molecules begin to emerge from the other end of the chromatographic column. As molecules continue to emerge, you recognize individual peaks, each representing a group of molecules having similar properties. It is possible that one peak represents a pure hydrocarbon and another represents a mixture of hydrocarbons that have very similar properties. For example, the leftmost peak shown in Figure 9.27 is likely to be a small molecule (or a group of similar small molecules) that moved quickly through the column. Peaks on the far right were produced when the heaviest, bulkiest molecules emerged from the column. The area under each peak is proportional to the number of molecules that come out of the end of the chromatographic column at that point in time. Thus, a taller, fatter peak means more molecules contributed to it.

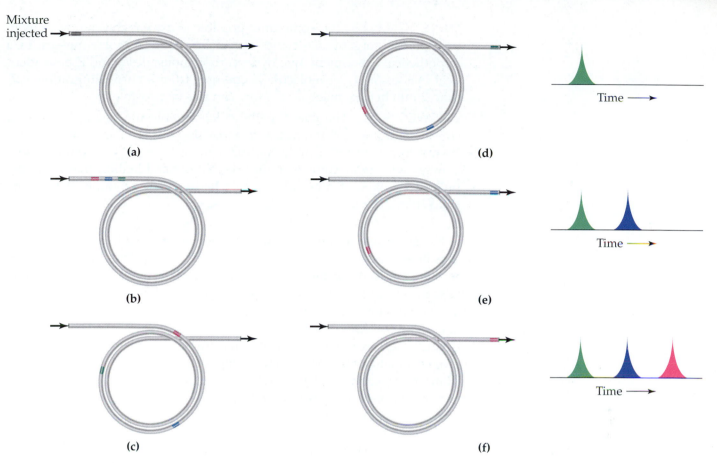

Mixture injected ➝

(a)

(b)

(c)

(d)

Time ➝

(e)

Time ➝

(f)

Time ➝

▲ FIGURE 9.26 **Three molecules, one column.** (a and b) A mixture of three molecules is injected into a gas-chromatography column and begins to move through the column. (c) The farther the molecules move down the column, the more separated they become. (d) The green molecule reaches the end of the column first, and a peak on the chromatogram marks its emergence from the column. (e and f) The blue molecule emerges next, followed by the red molecule.

 **LIVE ART**
Gas Chromatography

▼ FIGURE 9.27 **The complexity of gasoline.** Each peak in this chromatogram of gasoline represents a different molecule or group of molecules contained in the mixture.

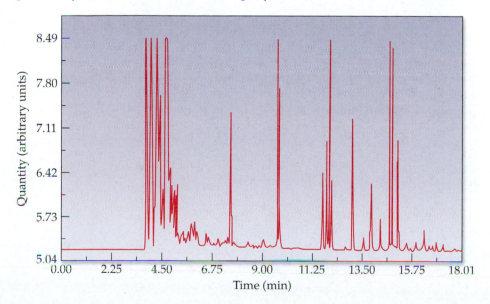

QUESTION 9.22   It is sometimes possible to use gas chromatography to separate two organic molecules that have identical molar masses. In a hypothetical experiment, two hydrocarbon molecules X and Z have identical masses, and yet molecule Z emerges from a chromatographic column 2 min before molecule X. How can this be possible?

ANSWER 9.22   The speed with which a molecule moves through a chromatographic column depends on its shape as well as its mass. Thus, a very long, extended molecule may move through the column more slowly than a more compact molecule that has an identical mass, but less bulk. The shape of molecule Z must be much more compact than the shape of molecule X.

A gas chromatogram of a complex mixture of organic molecules, such as gasoline, may have hundreds of peaks, as Figure 9.27 demonstrates, but it is the complexity of the gas chromatogram that makes it useful. If you analyze several samples from the same sealed container, you will find that the chromatograms are very similar to one another, with all the major peaks present in the same positions and at the same relative heights. Because gasoline is an extremely complex mixture, it provides a detailed chromatogram that can be considered a "fingerprint" for that very specific formulation of gasoline. Just as no two human fingerprints are alike, it would be highly unlikely that gasolines taken from two sources would have identical chromatograms. Thus, a sample of gasoline from gas station A will be distinguishable from a sample of gasoline from gas station B. It is even possible to distinguish between two deliveries of gasoline to the same gas station. Therefore, it is sometimes possible for arson investigators to determine not only the fuel used for starting a fire, but also the specific source of that fuel and when it was purchased.

Gas chromatography can be used to detect more than hydrocarbon mixtures. For example, recent increases in airport security across the United States have brought on a demand for a fast and simple way of detecting explosives. If you are asked to remove your shoes at the airport, it is likely

▶ FIGURE 9.28   **Gas chromatography as an airport security aid.**   In the twenty-first century United States, it is not uncommon to be asked for your shoes when you pass through airport security to test for explosives. Fast and efficient gas-chromatography technology allows for the routine detection of such explosives as dinitrotoluene and nitroglycerin, as this chromatogram shows.

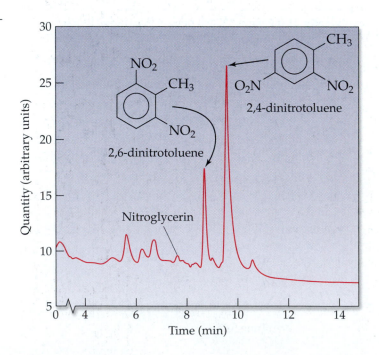

they are being swabbed to remove traces of explosives, such as TNT and nitroglycerin, that could be present. The paper swab used is inserted into a gas chromatograph, which quickly analyzes the gases drawn off of the paper. A computer checks the peaks on your shoe-swab chromatogram against the chromatogram of known explosive mixtures and gives the examiner a quick "yes" or "no" answer. Figure 9.28 shows a chromatogram that is positive for the presence of the explosive mixture that includes nitroglycerin, 2,6-dinitrotoluene and 2,4-dinitrotoluene.

"We'll need to declaw the cat."

CHEMISTRY AT THE CRIME SCENE

The Dog Nose Knows

The discussion on headspace analysis and gas chromatography might make it seem like detecting arson is a piece of cake, but these methods, although they work well, are often not sensitive enough to detect the presence of minute quantities of fuel. Difficulties arise in choosing pieces of evidence that contain the greatest number of fuel molecules. In many cases, this involves nothing more than an investigator's picking up pieces of material at a fire site and sniffing for the smell of fuel. The randomness of this approach, an unavoidable fact in many arson investigations, often means that important evidence is overlooked.

Another problem for arson investigators is that it can be difficult to establish whether a fire was set intentionally. When it is possible to establish that a fire started in two locations, however, it can usually be assumed that it was started intentionally. In order to demonstrate that a fire was set in more than one location, it is necessary to identify specific areas in which the fuel was deposited in high concentrations by an arsonist. Although a gas-chromatography analysis can tell you yes, this is a flammable liquid, or no, this is not, it is not an easy matter to determine from a headspace analysis which samples have higher concentrations of fuel than others because different pieces of fuel-soaked evidence might release vapor into the headspace at very different rates.

Luckily, there is a piece of equipment that can outwit even the most sophisticated scientific instrumentation nearly every time: the nose of a dog. After about 8 weeks of training, certain breeds of dog can enter a suspected arson scene to search for evidence. These dogs are specially trained to detect specific molecules, and many dogs are able to distinguish among as many as 20 very similar volatile molecules. Furthermore, arson dogs can be trained to locate areas of concentrated fuel or explosive, and this can often reveal whether the fire was an arson (multiple high-concentration areas) or an accident (single high-concentration area). In jurisdictions in which humans rather than dogs collect the evidence, only about 50 percent of the samples submitted from a suspected arson scene contain evidence of fuel. In jurisdictions in which arson dogs are used, nearly 100 percent of samples submitted for analysis come up positive for the presence of fuel.

Just as impressive is the fact that dogs can detect fuels up to 18 days after a fire. Charlotte, shown in Figure 9.29, is a 7-year-old black Labrador retriever and one of the most highly respected members of the Portland Fire Bureau in Portland, Oregon. She lives with Greg Keller, her handler, who admits to spoiling her rotten. According to Keller, "I'm with Charlotte more than I'm with my wife."

◀ FIGURE 9.29   A dog's nose can detect minute amounts of fuel and/or explosives.   This proud Labrador retriever is an esteemed member of the Portland Fire Bureau. Her sniffer is many, many times more effective than those of her human colleagues in detecting explosive residues.

Chapter 9

SUMMARY

Many varieties of chemical compounds are explosive, and in this chapter we have categorized them and looked at the factors that make a molecule unstable and therefore useful as an explosive. We began with a discussion of three classes of hydrocarbons: alkanes, alkenes, and alkynes. The nomenclature for these molecules is based on a system of Greek prefixes that denote the number of carbon atoms in each molecule, suffixes to specify the presence of multiple bonds, and *cis* or *trans* in the case of many alkenes. For each class of organic molecule, it is possible to write a homologous series in which successive carbon atoms are added to the structure to make the next molecule in the series. The homologous series for alkanes is provided in Table 9.2.

For a given molecular formula, it is possible for several organic molecules to be drawn with different configurations of the carbon chain. The carbon framework in organic molecules may be branched or straight, and we refer to two molecules having the same molecular formula, but different shapes, as structural isomers. When an organic molecule is branched, we refer to the branching groups as substituent groups, and we name each one according to its length and position on the carbon framework. Beginning with a hydrocarbon framework, it is possible to build organic molecules with atoms other than carbon and hydrogen, such as oxygen, nitrogen, and sulfur. For example, the ethers are a class of organic molecule with an oxygen atom connecting two hydrocarbon groups.

Hydrocarbon molecules undergo combustion reactions in the presence of oxygen. The products of the reaction are carbon dioxide and water when complete combustion occurs. However, in real reactions—in your car's engine, for example—incomplete combustion occurs, with the formation of some carbon monoxide. Gasoline is a complex mixture of hydrocarbons derived from the distillation of crude oil. In the refining process, crude oil is heated, causing organic molecules to enter the gas phase.

They move up a fractionating column that is cooler at the top than at the bottom. When a molecule reaches a point on the column at which the column temperature causes the molecule to condense into a liquid, it is removed from the column as a fraction, and each fraction contains molecules of similar size with similar physical properties.

Small quantities of hydrocarbon mixtures can be analyzed to determine their composition using headspace analysis, a procedure in which a sample of gas above a volatile liquid is removed and injected into a chromatographic column. The resulting gas chromatogram has peaks that represent different fractions within the hydrocarbon sample. This type of analysis is routinely used in forensic investigations of arson crimes.

Certain molecules—such as dynamite, TNT, and nitroglycerin—are able to undergo explosive reactions without oxygen. These molecules meet the requirements for an explosive reaction: production of large volumes of gas and fast reaction times. These reactions are also exothermic, which means that, on average, the bonds in the reacting molecule(s) are weaker than those in the product molecules. When the reverse is true, heat must be added, and the reaction is endothermic. Many highly explosive chemical compounds contain nitro groups, which contain several very weak bonds. These compounds often contain benzene rings. Trinitrotoluene (TNT) is one well-known example.

The release of energy in the form of heat is one of the driving forces for chemical reactions. The other is an increase in disorder, or entropy, as stated in the second law of thermodynamics. When a reaction is exothermic and entropy increases, we know that the reaction goes forward. When the reverse is true, the reaction does not go forward. Thus, it is sometimes possible to predict whether or not a chemical reaction will take place if you know whether it is endothermic or exothermic and whether there is an increase or decrease in entropy.

KEY TERMS

alkane	cis–trans isomerism	ether	gas chromatography	second law of
alkene	distillation	exothermic reaction	homologous series	thermodynamics
alkyne	dynamite	explosion	nomenclature	structural isomer
branched	endothermic reaction	fractionating column	octane number	substituent
hydrocarbon	entropy	functional group	oxygenate	

QUESTIONS

The Painless Questions

1. Identify each molecule as alkane, alkene, or alkyne:

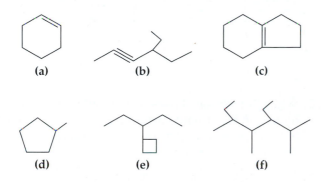

(a)   (b)   (c)

(d)   (e)   (f)

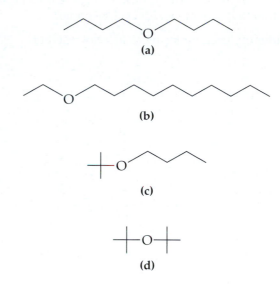

(a)

(b)

(c)

(d)

2. For parts (b) and (f) in Question 1, indicate the length of the longest chain. What is the root name for each of these two molecules?

3. Draw full structures and provide molecular formulas for (a) 4,4-diethyltetradecane, (b) methylcyclooctane, (c) 4-ethyl-5,5-dipropylhexadecane, (d) 3,3,7,7-tetramethylnonane.

4. Is the word *combustion* synonymous with the word *explosion*? If not, what is the difference between the two?

5. Methane is the primary component of natural gas, a fuel that many of us use in our homes. Write a chemical reaction for the complete combustion of methane.

6. Indicate whether each reaction is endothermic or exothermic:

   (a) $2 H_2O(g) + heat \rightarrow 2 H_2(g) + O_2(g)$

   (b) $CH_4(g) + 2 O_2(g) \rightarrow CO_2(g)$
   $+ 2 H_2O(g) + heat$

   (c) $N_2(g) + O_2(g) + heat \rightarrow 2 NO(g)$

7. Identify whether each event involves increasing entropy or decreasing entropy: (a) a building is demolished with a wrecking ball, (b) a marathon runner eats a Power Bar® for energy, (c) two molecules in the gas phase come together to make one molecule in the liquid phase, (d) one molecule in the solid phase breaks into three liquid-phase molecules.

8. Name each ether and write the molecular formula for each:

9. Comment on this statement: *Combustion reactions go forward because they are exothermic and always cause an increase in entropy.*

10. Explain why organic molecules that contain nitro groups have a tendency to be unstable and undergo explosive reactions.

11. Carbon dioxide is an environmental toxin and a greenhouse gas. Given this, why is it desirable to produce as much carbon dioxide as possible during the combustion of gasoline in your car engine?

12. The hydrocarbon *trans*-4-octene is an alkene that can be found in trace quantities in gasoline. Draw the line structure of this molecule and the line structure for its *cis* isomer.

13. Comment on this statement: *In organic chemistry, a homologous series is a group of organic molecules having the same number of carbon atoms and successively increasing numbers of bonds between two carbon atoms, as, for example, pentane; 2-pentene; 2-pentyne.*

14. Which of these alkenes does *not* require the prefix *cis* or *trans* in the name: (*a*) 1-octene, (*b*) 4-nonene, (*c*) 3,3-dimethyl-1-pentene, (*d*) cyclopentene?

15. Identify each molecule as alcohol or ether:

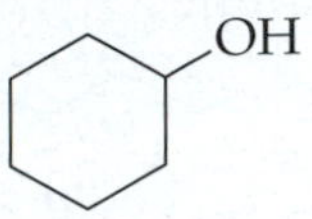

(a)

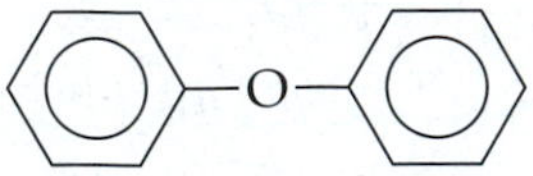

(b)

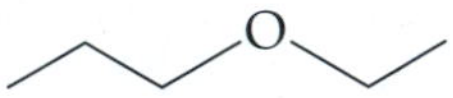

(c)

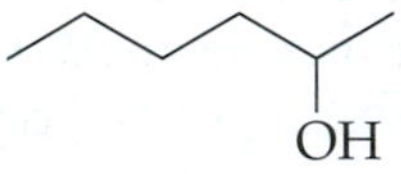

(d)

More Challenging Questions

16. What is the composition of a typical sample of gunpowder? Indicate what role is played by each ingredient you name.

17. Provide a name for each alkene:

(a)

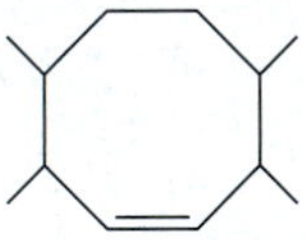

(b)

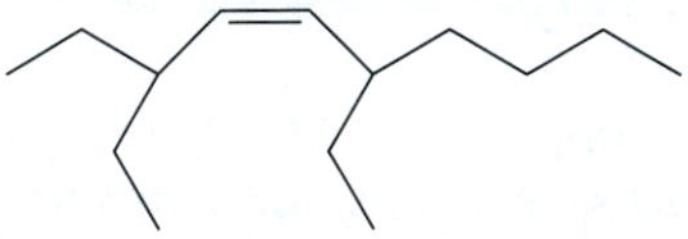

(c)

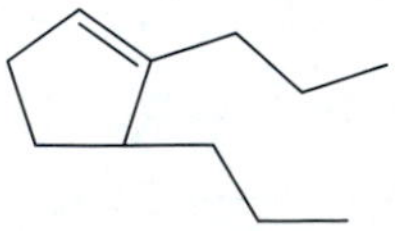

(d)

18. Draw a line structure and a full structure for (*a*) 3,3-dimethyl-1-butene, (*b*) 5,6-diethyl-3-methyl-4-propyl-nonane, (*c*) cyclohexene, (*d*) 1,4-dipentylcyclohexane.

19. Make a sketch showing the interactions between five molecules of water and five molecules of *tert*-butyl alcohol:

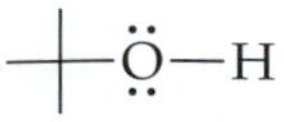

$$\text{} \bignleftbracket \ddot{\text{O}} - \text{H}$$

tert-Butyl alcohol

20. Which of these classes of organic molecule can hydrogen bond with each other: (*a*) ethers and alcohols, (*b*) phenols and alcohols, (*c*) alkanes and ethers, (*d*) alkenes and alkanes?

21. (*a*) Locate the main chain in this molecule:

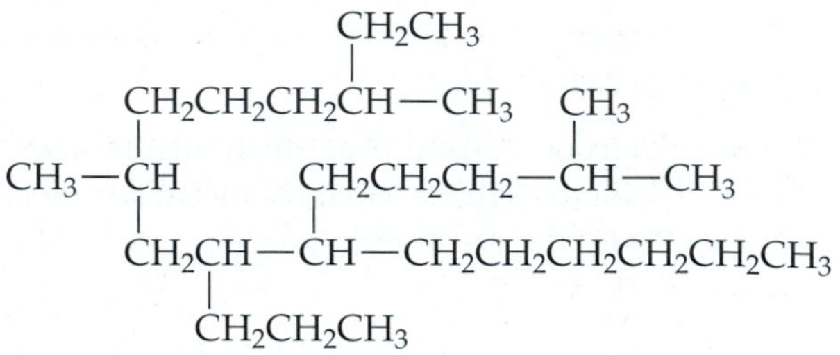

$$CH_2CH_3$$
$$CH_2CH_2CH_2CH-CH_3 \quad CH_3$$
$$CH_3-CH \qquad CH_2CH_2CH_2-CH-CH_3$$
$$CH_2CH-CH-CH_2CH_2CH_2CH_2CH_2CH_3$$
$$CH_2CH_2CH_3$$

(*b*) Is this molecule an alkane, alkene, or alkyne?
(*c*) What is its root name?

22. Which name is correct for the organic molecule shown in Question 21?

(**a**) 2-ethyl-6-methyl-9-(4-methylpentyl)-8-propyl-pentadecane

(**b**) 6-hexyl-2,9,13-trimethyl-7-propylpentadecane

(**c**) 3,7-dimethyl-10-(4-methylpentyl)-9-propylhexa-decane

(**d**) 9-propyl-3,7-dimethyl-10-(4-methylpentyl)hexa-decane

23. Draw a line structure for (*a*) 2,4-dimethyltoluene, (*b*) *ortho*-dinitrobenzene, (*c*) *meta*-diethylbenzene, (*d*) 2,3,4-trichlorophenol.

24. (*a*) Write the equation for the combustion of pentane. (*b*) Draw the full structure for each carbon-containing reactant and product in your equation. (*c*) Do the carbon compounds in this reaction follow our rule about carbon atoms in organic molecules making four bonds to other atoms? (*d*) Which bonds in the pentane molecule are left intact and which are broken when the molecule undergoes combustion? (*e*) Which do you think are stronger, the bonds in the carbon-containing reactant or those in the carbon-containing product?

25. There are many occasions in organic nomenclature in which rules for naming organic molecules contradict one another. For example, when a molecule contains a benzene ring, the ring is sometimes named as a substituent group attached to something else. When this is the case, the benzene substituent is called a *phenyl-* group. Name this molecule using the phenyl nomenclature:

26. (*a*) Draw all possible structural isomers having the molecular formula C_4H_{10}. (*b*) Name each one.

27. (*a*) Does the reaction of saltpeter, carbon, and sulfur combined together as gunpowder constitute an explosion? (*b*) Is it also a combustion reaction? (*c*) Why was gunpowder eventually replaced with other explosive materials?

28. Oil companies considered using ethanol and MTBE in gasoline formulations as a way of responding to the Clean Air Act. Why was MTBE the more practical choice for oil companies at the time? What are the pros and cons of using ethanol and MTBE in gasoline?

29. Name each molecule:

(a)

(b)

(c)

(d)

30. (*a*) What is meant by the terms *high explosive* and *low explosive*? (*b*) Why do we refer to TNT as being a self-contained explosive? (*c*) Why does TNT produce a powerful blast that has enormous destructive force?

The Toughest Questions

31. Which of the following types of reactions will always go forward? Which will not? Provide a brief explanation for each of your choices: (*a*) an exothermic reaction with an increase in entropy, (*b*) an endothermic reaction with an increase in entropy, (*c*) an exothermic reaction with a decrease in entropy, (*d*) an endothermic reaction with a decrease in entropy.

32. Each of these names is either incorrect or incomplete. Correct the mistake in each name in any way you wish, and draw a line structure for each molecule: (*a*) 2-pentene, (*b*) 1-propyne, (*c*) 1-methylnonane, (*d*) 3-methyl-4-propyl-5,6-diethyldodecane.

33. True or false: The second law of thermodynamics tells us no process can result in a decrease in entropy. Thus, it is impossible to clean up dirt scattered on a floor because doing so would constitute a decrease in entropy. Explain your answer.

34. If hydrocarbons X, Y, and Z have octane ratings of 0, 50, and 100, respectively, what is the overall octane rating of gasoline made up of 15 percent X, 20 percent Y, and 65 percent Z?

35. The isomers *cis*-2-pentene and *trans*-2-pentene are both flammable liquids. However, their melting points are −180 °C and −140 °C, respectively. (*a*) Draw a line structure for each molecule. (*b*) Why are the physical properties, such as melting point, different for these two molecules?

36. In a gas-chromatography analysis, molecules P and Q were separated and produced the two peaks shown in this chromatogram. The sample injected into the chromatograph contained equal amounts of P and Q, and yet the peak for Q is only half the height of the peak for P. How is this possible?

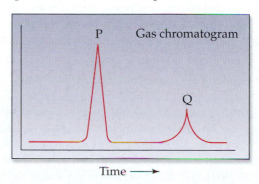

37. Consider again the detonation of nitroglycerin:

$$2\ C_7H_5N_3O_6(s) \longrightarrow 3\ N_2(g) + 12\ CO(g)$$
$$+ 5\ H_2(g) + 2\ C(s) + \text{lots of energy}$$

(a) If you want to detonate 1 mol of nitroglycerin, how many kilograms do you need? (b) If you begin with 1 mol of nitroglycerin, how many atoms of elemental carbon, C(s), are formed? (c) What is the mass, in kilograms, of the C(s) produced?

38. The synthesis of the molecule nitrocubane preceded the synthesis of octanitrocubane. (a) Draw the line structure of the moderately explosive nitrocubane molecule. (b) Why is nitrocubane less explosive than octanitrocubane? (c) Why is nitrocubane explosive at all? (d) Why is it unnecessary to indicate the position of the nitro group in nitrocubane?

39. (a) Write equations showing the combustion of n-butane, trans-2-butene, cis-2-butene, and 2-butyne. (b) Which of these reactions uses the most molecules of oxygen per molecule of fuel? (c) Does your equation for trans-2-butene differ from your equation for cis-2-butene? Why or why not?

40. Draw all possible structural isomers for organic molecules having the molecular formula C_6H_{14}. Name each one.

41. Two hydrocarbon molecules both have the molecular formula $C_{10}H_{22}$. When a mixture of these two molecules is passed through a chromatograph, one molecule emerges 8 minutes before the other. If this chromatographic column separates molecules on the basis of size, explain how two molecules with identical formulas can emerge from the column at strikingly different times.

42. (a) Use the formal name of picric acid, 2,4,6-trinitrophenol, to draw its structure. (b) When picric acid explodes, it produces nitrogen gas, carbon dioxide gas, solid carbon, and hydrogen gas. Given this, propose a balanced reaction for the explosion of picric acid. Use the explosion reactions for TNT and nitroglycerin in this chapter as a guide.

43. In a short essay, describe the process by which crude oil is separated into fractions. In your description, use the terms *fractionating column*, *fraction*, and *distillation*. Give two examples of fractions that are collected at points along a fractionating column, and indicate where the average person might encounter them.

44. For each of these situations, indicate whether entropy is increasing or decreasing. For each instance of decreasing entropy, explain why the second law of thermodynamics is not violated: (a) a volcanic eruption, (b) a dam is formed in a river to make a reservoir, (c) confetti is thrown from a Mardi Gras float, (d) the World Trade Center twin towers are rebuilt.

45. Each of these names is incorrect: (a) 1-methyl-2,3-dinitrotoluene, (b) propyl methyl ether, (c) 1-methyl-cyclobutane, (d) trans-cyclohexene. Correct the mistake in each name in any way you wish, and draw a line structure for each molecule.

46. It is usually safe to assume that combustion reactions will go in the forward direction. However, there is a special instance when they do not. For example, only one of these reactions happens under all conditions. The other reaction happens under some conditions, but not others. Which is which? Explain.

$$2 C_8H_{18}(\ell) + 25 O_2(g) \longrightarrow 16 CO_2(g) + 18 H_2O(g)$$
$$2 C_8H_{18}(\ell) + 25 O_2(g) \longrightarrow 16 CO_2(g) + 18 H_2O(\ell)$$

E-Questions

Go to **www.prenhall.com/waldron** to find these questions in electronic form, complete with hyperlinks directly to the various websites cited in the questions.

47. Read the article "Fill'er Up…With Veggie Oil" at **ScienceNews**. Select the SEARCH button, type the words *veggie oil* in the search box, and then answer the following questions: (a) What is the primary drawback to using vegetable oils as automotive fuels? (b) What types of plant oils are mentioned as being potentially useful as fuels? (c) How does a fatty acid molecule differ from a hydrocarbon molecule? (d) What are the disadvantages of saturated and unsaturated fats when used as fuels?

48. The article "Powerful Explosive Blasts Onto Scene" describes the explosive power of cubane-based molecules. Access this article at **ScienceNews**. Answer the following questions: (a) What is the density of the cubane molecule? Why does the high density of octanitrocubane give the molecule more explosive force? (b) What products are produced when octanitrocubane is detonated? (c) Why does the military find this reaction the ideal basis for a weapon?

49. Read two articles about detectors for locating underground land mines. The first article can be found at **UniSci**. The second site is **ScienceNews**. After reading these two short articles, answer the following questions: (a) How many land mines are thought to be buried worldwide, and in which countries can they be found most often? (b) Summarize the problems associated with locating land mines. (c) In the second article, what type of new device has been developed for the detection of land mines? How does it work? (d) How are living microorganisms being used in land-mine detection?

50. Read two perspectives on the MTBE issue. First look at the government's report on MTBE and underground storage tanks at **EPA**. Then read the article "Pollution by Gasoline-Containing Hazardous Methyl Tertiary Butyl Ether (MTBE)" in **Archives**.

After reading the two articles, answer the following questions: (*a*) Do you perceive a bias in either of the articles? (*b*) What purpose was served by each article? (*c*) Do both articles contain credible references? (*d*) Do the articles contain complementary or conflicting views on the issue?

51. Take a look at the short article "Researchers Unveil New Approach to Cleaning Fuel" at **Scientific American**. Answer the following questions: (*a*) What EPA goal are the scientists described in this article trying to address? (*b*) What has been the problem with removing sulfur from fuels in the past? (*c*) How is the SARS method developed by Penn State scientists an improvement over the old method? (*d*) What type of chemical compound is used to remove sulfur with the SARS method?

WORLD WIDE WEB RESOURCES

As with the E-Questions, go to **www.prenhall.com/ waldron** to find these questions in electronic form, complete with hyperlinks directly to the various websites cited.

Some of following links contain research articles related to the topics in this chapter and could be used as the basis for a writing assignment in your course. Other sites are of general interest:

- Read about an interesting application of explosives in the article "Ka-Boom! A Shockingly Unconventional Meat Tenderizer" at **ScienceNews**. This article introduces you to Morse B. Solomon of the USDA, who has developed a method of tenderizing meat by submerging it in water and blasting it with an explosive.

- In this chapter, we read about how methane deposits can spell disaster for ships at sea. However, methane trapped in the oceans can also be tapped as an energy source. Read a discussion about underwater methane deposits from the Department of Energy website at **DOE**.

- Read more about the work that arson dogs do at **WorkingDogs**, **ArsonDogs**, **FireOrg**, and **FireDogs**.

- Read a summary of the effects oil spills have on the environment at **Restoration**. You can also read about oil spills and their effects on specific ecosystems at **SierraMag** and **BioSci**.

ABC NEWS VIDEO
Cold Fusion Patterson Cell

ABC NEWS VIDEO
Hydrogen Fuel Cell Cars

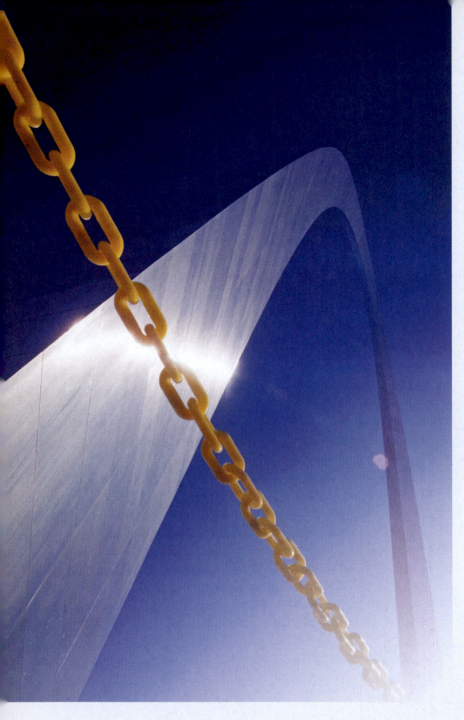

CHEMISTRY TOPICS IN CHAPTER 10

- Monomers and polymers [10.1, 10.2]
- Properties of polymers [10.2]
- Addition polymers [10.2]
- Nucleophiles and electrophiles [10.2]
- Esters [10.3]
- Copolymers [10.3]
- Condensation polymers [10.3]
- The carbonyl group [10.3]
- Ketones [10.4]
- Amides [10.4]
- Steric hindrance [10.5]
- Biodegradability [10.6]
- Hydrolysis [10.6]

"Thought dissolves the material universe by carrying the mind up into a sphere where all is plastic."

RALPH WALDO EMERSON

Chains 10

A STUDY OF SYNTHETIC POLYMERS AND THE WAYS THEY IMPROVE EVERYDAY LIFE

Take a moment to think about words beginning with *poly-*. The prefix *poly-*, as in *polyglot, polygamist,* and *polygraph,* multiplies the root word many times. A polyglot speaks many languages. A polygamist has many spouses. A polygraph measures many things—blood pressure, respiration, heart rate—all at once. In chemical terms, the prefix *poly-* has the same meaning: many. Thus, a polysaccharide contains many saccharides (*saccharide* is an upscale word for sugar), polystyrene contains many styrene molecules, and polyester contains many ester linkages.

As we shall see, there is an enormous list of chemicals that begin with the prefix *poly-*. These molecules are all made up of small organic molecules, called **monomers**, that are connected together into a long chain, or **polymer**. There are no hard-and-fast rules about how many monomers must be linked together to constitute a polymer, although most polymer chains contain at least 100 monomers and often many thousands. In addition, the variety of polymers is nearly as vast as the number of monomers they contain. Polymer chains can be long or short, branched or straight, hydrophobic or hydrophilic. Their mechanical properties are also widely varied, as we shall see. There are polymers designed for strength that are tougher than steel. There are polymers designed for stretchiness, water

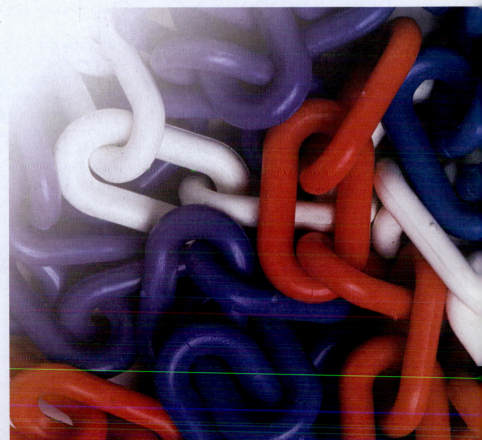

▶ **FIGURE 10.1   A plastic world.**
When you look around, you may be surprised at how much of our world is made from plastic.

resistance, biodegradability, and even antibacterial action. In this chapter, we survey the polymer realm to learn how these molecules are made and how they can be designed for specific functions. To do this, we must rely on some of the common-sense chemistry fundamentals we learned in the core chapters: like dissolves like, opposite charges attract, nature tends to lowest energy.

Examples of everyday items made from polymers are shown in Figure 10.1. Perhaps the most common term that describes them all is **plastic**. As used in the quotation by Emerson on p. 382, this word is an adjective meaning "mutable, or capable of change," and in chemistry the word means the same thing: items made from plastic are changeable and can be shaped in many ways. From a plastic material, you can fashion a food container, a compact disk, or a weather-resistant coat. Thus, plastics are indeed *plastic*—capable of being molded, stretched, or flattened into any conceivable shape. As we shall soon see, what puts the plasticity into plastics is the fact that they are composed of polymers, that diverse class of organic molecules that provides an almost endless variety of *macro*molecules.

10.1 | Dubble Bubble

The obituary for Walter Diemer, who died in 1998, described him as the inventor of bubble gum (Figure 10.2a). Diemer, an accountant for the Fleer Chewing Gum Company in Philadelphia, had a flair for mixing up new gum recipes in addition to keeping the books. For more than 20 years, the company had been working on a new gum called Blibber-Blubber that could be blown into bubbles, but the formula was always too sticky, meaning that if a bubble popped onto a face, it would be a challenge to remove. One day in 1928, Diemer came upon the perfect combination of ingredients, producing a bubble-making chew that could be peeled off the face and returned to the mouth for further chewing enjoyment. As it happened, the only food dye handy that day was pink, and so Diemer's invention, Dubble Bubble®, was and still is pink today. That day, Diemer took 5 lb of the stuff down to the local candy store and sold out the whole batch in hours. He continued working at the Fleer Chewing Gum Company for decades and never collected a penny for the invention. According to his wife, however, Diemer never missed the royalty money. He claimed that the hours he spent teaching kids to blow bubbles and holding bubble-blowing contests were payment enough (Figure 10.2b).

Chewing gum, in particular bubble gum, might seem like a brainchild of the twentieth century. After all, how often do you read about a character chewing gum in an historical novel or see someone chewing gum in a film set in the past? You may be surprised to learn, though, that the first documented instance of gum chewing takes us back to A.D. 50 in Greece, where the locals would chew resin collected from mastic trees. (This is the origin of the word *masticate*, which means "to chew.") Historians also know that less than a century later, Mayan gum chewers were chomping on *chicle*, the dried sap of the sapodilla tree. Much later, Native Americans would introduce the Pilgrim settlers in the northeastern United States to the pleasures of chewing resin from spruce trees.

In order to understand what makes gum gummy, we must consider what "gumminess" looks like at the molecular level. All modern gum contains a polymeric *gum base*, the material you chew on. The other ingredients in chewing gum provide pliability, flavor, and freshness. Figure 10.3 shows some of the materials that can be used as a starting material in chewing-gum recipes. Natural polymers used in gum base are those obtained from living sources, most often trees, and synthetic polymers are those made in bulk in a factory.

Notice a few things about the condensed structure of the polymers shown in Figure 10.3. First, it is usually impossible and unnecessary to show the entire structure of a polymer in one drawing. Because polymers are composed of a constantly repeating sequence of monomers, it is necessary to show only one (or just a few) monomer unit of a polymer. The brackets around the monomer indicate that it repeats indefinitely. Very often a subscript *n* is included to indicate that there is some large, but unknown, number of monomers in the polymer chain. Second, it is usually impossible to tell whether a polymer is natural or synthetic by looking at its chemical structure. The structure for the polymer rubber, for example, shown in Figure 10.3, looks exactly the same whether the polymer is derived from a rubber tree or made in a laboratory.

(a)

(b)

▲ **FIGURE 10.2 Walter Diemer, inventor of bubble gum.**
(a) This employee of a chewing gum company came up with the perfect formulation for gum that could be used to blow great bubbles but would not stick to the bubble-blower's face when the bubble inevitably burst. **(b)** Bubble blowing is often a required social skill for kids.

QUESTION 10.1   Which of the molecules in Figure 10.3 contain (*a*) an alkene group, (*b*) a methyl group, (*c*) a benzene ring?

ANSWER 10.1   (*a*) Rubber, (*b*) rubber and polyvinlyacetate, (*c*) styrene butadiene.

In the nineteenth century, chewing-gum base was always natural because synthetic polymers had not been invented. Thus, the first commercial

$$\left[CH_2-\underset{\underset{CH_3}{|}}{C}=CH-CH_2\right]_n$$

Rubber (natural or synthetic)

(a)

$$\left[CH_2-CH_2-CH_2-CH_2-CH_2-CH\right]_n$$

Styrene butadiene (synthetic)

(b)

$$\left[CH_2CH_2\right]_n$$

Polyethylene (synthetic)

(c)

$$\left[CH_2-\underset{\underset{O-\underset{\underset{O}{||}}{C}-CH_3}{|}}{CH}\right]_n$$

Polyvinylacetate (synthetic)

(d)

▶ FIGURE 10.3 **Gummy molecules.** All of the polymer molecules represented here are used as a foundation in gum formulations.

▼ FIGURE 10.4   **The manufacturing of chewing gum.**   **(a)** Gum factory workers handle freshly made chewing gum. **(b)** A man holding two slabs of newly manufactured gum base.

(a)

(b)

chewing gums, introduced at the end of the nineteenth century, were natural products derived from plants or insects. With the advent of synthetic polymers in the early twentieth century, chewing-gum formulations changed, and synthetic polymers gradually replaced natural ones. Why would a gum manufacturer prefer to use a synthetic polymer over a natural one? One reason is that synthetic polymers can be made in controlled surroundings, in a factory in which conditions are constant, the weather never changes, and each batch of polymer is nearly identical to every other batch. If a polymer comes from a natural source, such as a rubber tree, the quality of the gum will depend on the weather, on where the tree was grown, and on dozens of other impossible-to-control variables. In addition, a factory can produce virtually unlimited supplies of a polymer if the monomer is readily available.

Figure 10.4*a* shows bubble gum being produced in large quantities in a factory. Today's chewing gums are made of gum bases that are mostly synthetic, with some natural polymer mixed in. Unfortunately, we cannot tell you exactly what they contain because gum recipes are closely guarded secrets. The gum-making process begins at a gum-base company, where natural and synthetic gum bases are converted to a usable form. According to the Gum Base Company, a British firm, "Customers may choose from a range of more than 200 different gum-base formulations, traditionally supplied in five different shapes: slabs, sheets, drops, pearls, and pellets, . . . " (Figure 10.4*b*).

Throughout history, humans have found other uses for polymers besides mastication. For example, shell lacca, a resinous insect secretion and the principal component of shellac, was used as a varnish by early civilizations. Frankincense and myrrh are aromatic, resinous, tree-derived polymers that were used as incense and perfumes. Gutta-percha, another resin from trees, has also been known since ancient times and was used as a coating on the first underwater cable that crossed the English Channel.

As the craft of alchemy began to change into the science of chemistry in the eighteenth and nineteenth centuries, new polymers were designed from natural products. In many cases, these changes were the result of shortages of such natural materials as ivory, silk, and horn. The twentieth century saw an explosion of patents for synthetic polymers, most notably rubber, polystyrene, polyethylene, polyvinylchloride (PVC), and various polyesters. Figure 10.5 shows worldwide consumption rates for the most common plastics. Take a look around you. How many plastic items do you see? Peek at the tags on your clothing. Does the fabric contain polyester, polypropylene, or some other synthetic polymer, or is it made from natural polymers such as cotton, wool, or linen? Are you using a writing implement made from plastic? Do your shoes have plastic soles? It is hard to imagine a life devoid of plastics.

Despite our dependence on plastics and the obvious benefits they provide, they are not benign. Our environment is paying a high cost for the convenience of plastics, and there is a movement afoot to move back to natural materials. The "old-school" plastics developed early in the twentieth century were, for the most part, not easily decomposed and will exist in garbage dumps for centuries. Thus, while the twentieth century saw dramatic progress in the design and utilization of synthetic polymers, the new century will see a taming of the polymers that have been developed. The new trend is toward polymers that are easily degraded and recycled. We shall explore degradable polymers in the last part of this chapter, but before we can understand what might make a polymer decompose, we must understand the structure of polymers and how they are made.

▼ **FIGURE 10.5   World plastic consumption in the year 2000.**   Asia led the world in the consumption of plastics in the year 2000.

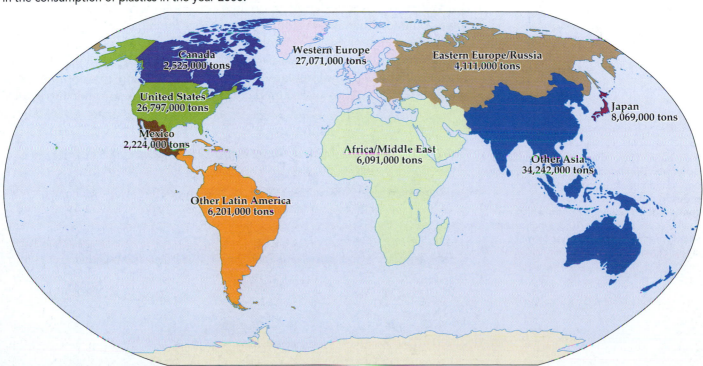

10.2 | Monomers to Polymers

In previous chapters, we saw a few examples of reactions between organic molecules. We shall use polymerization reactions to further demonstrate how organic molecules react with one another. Let's begin with the question of how the monomers in a very long polymer chain can be connected to one another in the laboratory or factory. Do they connect together one at a time at the end of an existing chain, as shown in Figure 10.6a, or do they assemble into small pieces and then into a long chain, as depicted in Figure 10.6b? As we shall see, both pathways are possible.

First let us consider a chain to which monomers are added one by one to form a macromolecule known as an **addition polymer** (Figure 10.7). A monomer that adds to the end of a growing polymer chain must be able to react with that end in some way. For organic molecules, this reaction is usually the result of the simplest type of interaction: positive attracted to negative. In most cases, these regions of depleted and excess electron density are located on different molecules. However, they can also be located on different parts of the same molecule. In either case, the reaction results in a connection between two oppositely charged regions and a product molecule that has a new size and shape. Thus, this type of addition polymerization is simply *the attraction of a positive region in one molecule to a negative region in another molecule.* In the example shown in Figure 10.7, a full positive charge on the polymer is attracted to a region of partial negative charge on the monomer being added.

The reaction shown in Figure 10.7 illustrates a general point about studying chemistry. Part of mastering organic chemistry is knowing where charge is scarce in a molecule and where excess charge accumulates. What makes organic chemistry straightforward is that many reactions work in exactly the way shown in Figure 10.7. If you surveyed reactions between organic molecules, you would find that most of them fall into this category. It boils down to this: two molecules are attracted to each other when positive and negative regions within them are drawn together.

When we are considering an organic reaction in which positive and negative come together, it is customary to say that the region of negative charge *attacks* the region of positive charge. We refer to the region of negative charge as a **nucleophile**, which means, literally, "nucleus loving." We use this term because the nuclei of atoms are positively charged, and nucleophiles seek out and attack regions of accumulated positive charge on other molecules.

LIVE ART
Polymer Formation

▶ FIGURE 10.6 **Two ways to make a polymer.** **(a)** In addition polymerization, one monomer adds to the end of a growing polymer chain. **(b)** In condensation polymerization, which we shall discuss in Section 10.3, small chains of polymer connect together to make larger chains.

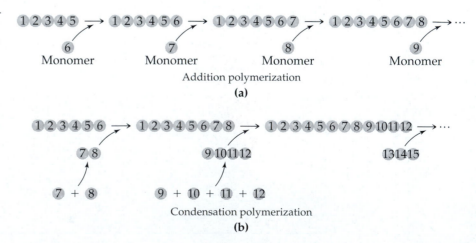

▲ FIGURE 10.7 How addition polymers form. Addition polymers form when a positive (or partial positive) charge is attracted to a negative (or partial negative) charge. In the reaction shown here, the monomer has a partial negative charge. It is attracted to the end of the growing polymer chain because that end has a positive charge. One monomer unit in the chain is shown in green to help you keep track of the length of the chain as monomer units are added.

We refer to the region of positive charge as an **electrophile**, literally meaning "electron loving," because it is attracted to regions of negative charge. Thus, we can think of the addition reaction shown in Figure 10.8a as a nucleophilic attack by the nucleophile (blue) on the electrophile (red). The monomer in this reaction is a molecule called vinyl acetate. The polymer it forms—polyvinylacetate (PVA)—is the primary ingredient in white glue (Figure 10.8b).

QUESTION 10.2 Predict which region of Molecule A is most likely to react with which region of Molecule B.

Molecule A   Molecule B

ANSWER 10.2 The negatively charged region in Molecule A (the nucleophile) will interact with the positively charged region of Molecule B (the electrophile).

The beauty of polymer chemistry is that you can tweak the structure of a monomer to get a polymer with completely new, hopefully interesting,

THE ELECTROPHILE
has either a full or a
partial positive charge

THE NUCLEOPHILE
has either a full or a
partial negative charge

(a)

◀ FIGURE 10.8 Nucleophiles and electrophiles. **(a)** Two monomers of vinyl acetate are bonded together when their regions of positive (red) and negative (blue) charge come together. We often refer to the positive group as the electrophile and the negative group as the nucleophile. The addition of a monomer leads to the formation of a new region of positive charge, which prepares the polymer for the addition of the next monomer. **(b)** White glue is made primarily from polyvinylacetate, the product of the reaction shown in part (a).

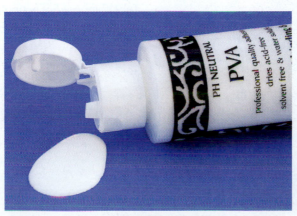

(b)

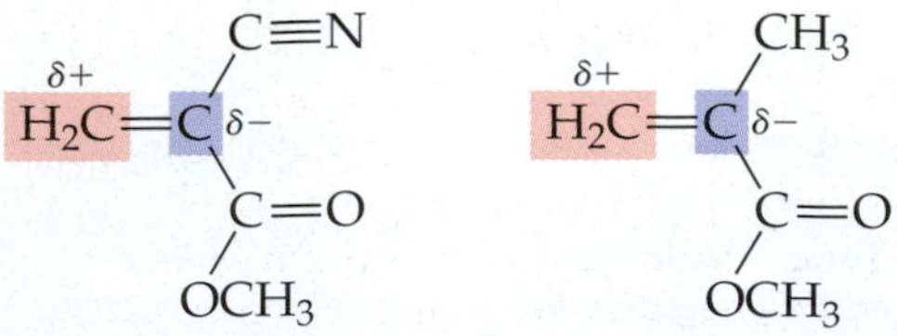

C≡N
$\delta+$
$H_2C = C\ \delta-$
C=O
OCH₃

Methylcyanoacrylate
monomer

(a)

CH₃
$\delta+$
$H_2C = C\ \delta-$
C=O
OCH₃

Methyl methacrylate
monomer

(b)

▲ FIGURE 10.9 **Structure dictates function.** Two similar monomers polymerize to make polymers that have very different properties. **(a)** The methylcyanoacrylate monomer polymerizes to make polymethylcyanoacrylate, the main ingredient in super glue. **(b)** The monomer methyl methacrylate differs from methylcyanoacrylate in only one detail: the C≡N group of cyanoacrylate has been replaced by a methyl group. This monomer polymerizes to form the polymer polymethyl methacrylate, used in the very strong plastic we know as Plexiglas®.

properties. Let us look at a new monomer to learn more about this tweaking. The monomer called methylcyanoacrylate (Figure 10.9a) polymerizes to make polymethylcyanoacrylate. This polymer is the primary component in super glue (Figure 10.10). When the glue comes into contact with water (or with your fingers), the polymerization reaction is initiated, and the glue forms a super-strong bond between any two items it happens to come into contact with (such as your thumb and forefinger).

If you replace the cyano group (—C≡N) of polymethylcyanoacrylate with a methyl group (—CH₃), you get methyl methacrylate (Figure 10.9b). This seemingly minor change in substituent group has an enormous effect on

► FIGURE 10.10 **A nonstick container for super glue.** **(a)** The polymer polymethylcyanoacrylate contains the highly electronegative atoms nitrogen and oxygen. These atoms make the molecule a very polar one. **(b)** Any material used to coat the inside of the tube holding the super glue should be nonpolar so that the glue product will not stick to the inside of the tube.

$$\left[CH_2 - \underset{\underset{OCH_3}{\overset{C=O}{|}}}{\overset{\overset{C\equiv N}{|}}{C}} - CH_2 - \underset{\underset{OCH_3}{\overset{C=O}{|}}}{\overset{\overset{C\equiv N}{|}}{C}} - CH_2 - \underset{\underset{OCH_3}{\overset{C=O}{|}}}{\overset{\overset{C\equiv N}{|}}{C}} - CH_2 - \underset{\underset{OCH_3}{\overset{C=O}{|}}}{\overset{\overset{C\equiv N}{|}}{C}} - CH_2 - \underset{\underset{OCH_3}{\overset{C=O}{|}}}{\overset{\overset{C\equiv N}{|}}{C}} \right]_n$$

(a)

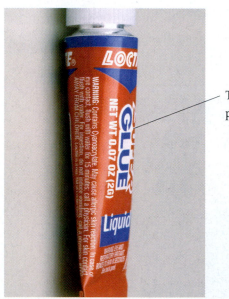

Tube composed of nonpolar polymer
polyethylene $\left[CH_2CH_2 \right]_n$

(b)

STUDENTS OFTEN ASK

If Super Glue Adheres Immediately to Everything It Touches, Why Doesn't It Adhere to the Inside Walls of the Super-glue Tube?

If you are the person charged with finding a container for methylcyanoacrylate, the monomer for super glue, you should worry about two things. First, you must keep the monomer from sticking to the walls of the container. Note in Figure 10.9a

that this monomer contains two electronegative atoms—nitrogen and oxygen. As you learned in Chapter 4, the presence of these atoms makes this monomer a highly polar molecule. It interacts easily with other polar molecules (we can invoke our now-familiar rule and say *like interacts with like*), but not with nonpolar molecules. Thus, you should choose a nonpolar material for the super-glue container if you want to keep the glue from adhering to it. The super-glue containers used commercially are made from polyethylene, shown in Figure 10.10. Polyethylene, the most widely produced polymer, is highly nonpolar because it is a hydrocarbon and contains no electronegative atoms.

Your second concern should be that the monomer does not polymerize while it is in the container. As we already

know, methylcyanoacrylate polymerization is initiated when the monomer comes into contact with water. Thus, you must be sure that the environment in which the monomer is prepared and packaged is perfectly water-free.

QUESTION 10.3 Pretend for a moment that super glue is a highly nonpolar molecule like polyethylene. Would a polyethylene container still be the best choice for this glue? If not, what type of material should be used for a container?

ANSWER 10.3 If the glue and the container are both nonpolar, they will stick to each other. A highly polar material should be used as a container for the nonpolar glue.

the polymer formed. The polymerization of methyl methacrylate makes Plexiglas®, quite a different material (Figure 10.11). Plexiglas, the brand name for a material made from polymethyl methacrylate, is used to make everything from bulletproof glass to pulpits to tissue boxes. Throughout this chapter, we shall see further examples of how polymer structure dictates function.

RECURRING THEME IN CHEMISTRY The individual bonds between atoms dictate the properties and characteristics of the substances that contain them.

▼ **FIGURE 10.11 Polymethyl methacrylate.** The polymer polymethyl methacrylate is used to make everything from protective shields to see-through pulpits to tissue boxes.

(a)

(b)

▲ **FIGURE 10.12 Polyester cloth is wrinkle-resistant.** **(a)** The popularity of polyester clothing derives, in part, from its wrinkle-free look. **(b)** If you like the wrinkled look, then cotton is a better choice for you.

10.3 | Groovy, Baby: The Science Behind Polyester

Not many decades are associated with a specific fabric, but the 1970s and polyester are as intertwined as ancient Rome and the toga or the Baroque era and the harpsichord. Polyester clothing from this era was blatantly artificial, clearly a close cousin to such other plastic items as camera film and shower curtains. In the years since the rise of polyester fabric, its popularity has waned, but polyesters are still used in dozens of fabrics today, and its plasticlike properties are usually disguised by blending the polymer with natural fabrics. The beneficial qualities of polyester—durability and crease resistance—are undeniable, as Figure 10.12 shows. Fabrics that contain even small amounts of this polymer are often classified as "wash and wear" or "permanent press" because they come out of the dryer wrinkle-free. In contrast, natural fibers—such as cotton and its upscale cousin linen—are well known for having that unironed look.

Polyester is, as the name implies, a polymer of *esters*, and thus it is now time for us to add the ester group to our list of organic functional groups. The functional groups we have studied so far are shown in Figure 10.13 as a quick review. As explained in Section 9.3, each functional group in an organic molecule contributes to the character of the molecule and helps determines its physical properties. As you can see by looking back at Figure 10.8 on page 389, an alkene functional group often is a site of excess electron density. These sites in a molecule are most often the sites that take part in chemical reactions.

QUESTION 10.4 Identify the functional groups in this molecule:

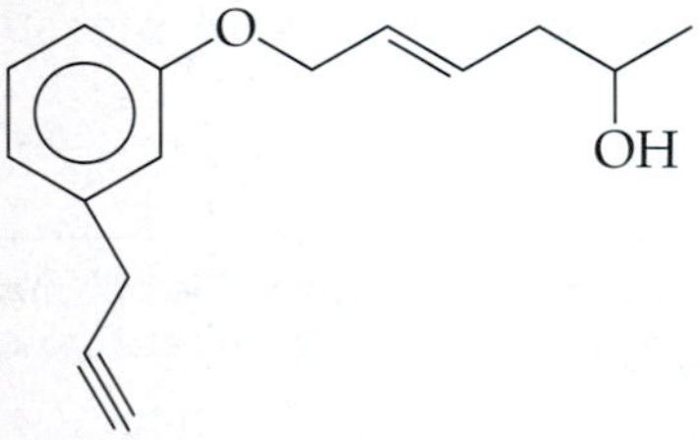

ANSWER 10.4

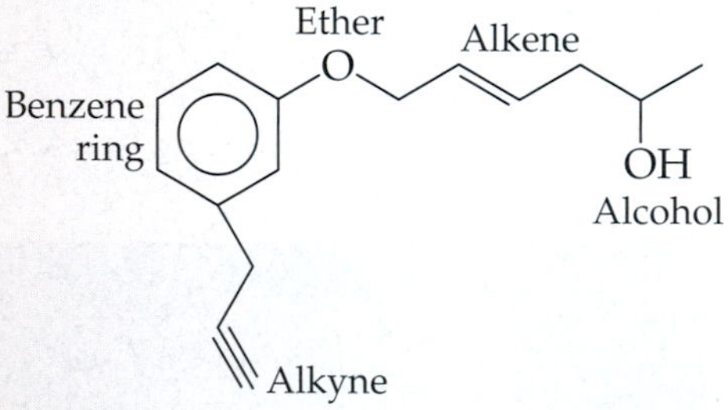

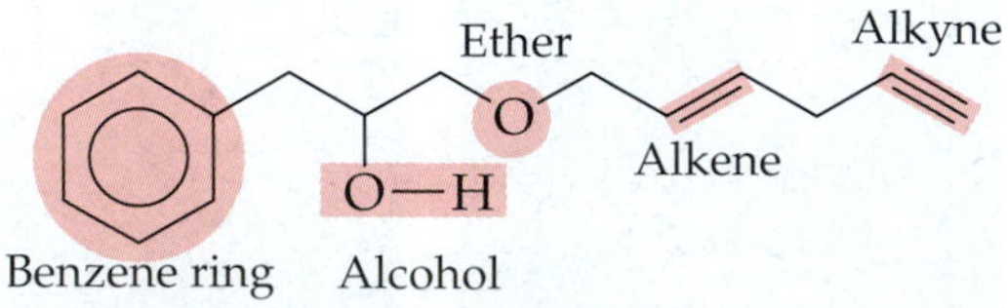

Ether

Alkene

Benzene ring

OH
Alcohol

Alkyne

▶ **FIGURE 10.13 Summary of functional groups we know.** Thus far, we have learned about alkenes, alkynes, and benzene rings, all of which are hydrocarbons. We were also introduced to ethers and alcohols, which are oxygen-containing functional groups.

Ether

Alkyne

Alkene

O—H

Benzene ring   Alcohol

Figure 10.14 shows our newest functional group, the *ester functional group*: a carbon atom plus two oxygen atoms. Molecules containing this functional group are called **esters**, and they consist of two hydrocarbon chains linked by an ester functional group. In Figure 10.14, the two hydrocarbon chains are represented by R and R'. The ester group is very much a part of our lives, as Figure 10.15 shows, because esters are flavoring molecules. In addition to their role as flavoring agents, esters are the main ingredient in many perfumes. A few of the most recognizable esters are shown here. The orange flavoring molecule, for example, contains a methyl group and an octyl group attached to an ester functional group. You should be able to name the two hydrocarbon chains in all the other esters shown in Figure 10.15.

Ester names have two parts (Figure 10.16). The hydrocarbon chain attached to the oxygen in the main chain of the molecule is named first. In Figure 10.16, this is an ethyl chain. To name the second part, the remaining carbon atoms (*including* the one in the ester functional group) are counted and used to determine the root name for the ester. The ending -*e* is removed from this root name, and the suffix -*oate* is added. In Figure 10.16, the hydrocarbon chain on the left is a propyl chain, but we have to include the carbon of the functional group, which means that this group has four carbons. Our root name is therefore butyl. Thus, the ester, which is the pineapple flavoring molecule from Figure 10.15, is ethyl butanoate.

Following this logic, we can name the orange flavoring molecule shown in Figure 10.15. Going through the procedure shown in Figure 10.16, you should come up with the name octyl ethanoate. The common name for ethanoate is *acetate*, and for that reason this ester is more frequently called octyl acetate.

Of course, we can work in the opposite direction, using an ester's name to figure out its structure. For instance, one of the esters shown in Figure 10.15 is named pentyl butanoate. Which one is it? To figure this out, think about the number associated with each part of the name. The *pentyl* part, which represents the hydrocarbon chain attached to the oxygen atom in the main chain, contains five carbon atoms. The *butanoate* portion, which is derived from the root name *butane* for four carbons, represents the remaining carbons in the molecule. Thus, this ester must be the apricot flavoring molecule.

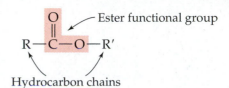

▲ FIGURE 10.14 **The ester functional group.** This functional group, highlighted in red, consists of a carbon atom bonded to an oxygen atom through a double bond and to a second oxygen atom through a single bond. The symbols R and R' represent hydrocarbon chains attached to the functional group.

MOLECULE
n-Butane

$$CH_3CH_2CH_2C-O-CH_2CH_3$$

Second part of name: First part of name:
    Butane   Ethyl

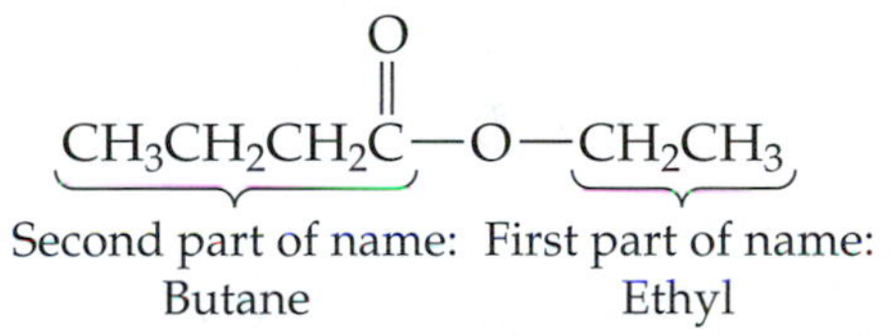

Butane − e = butan
Butan + -oate = butanoate
Whole name: ethyl butanoate

▲ FIGURE 10.16 **Naming esters.** The first part of an ester's name comes from the number of carbon atoms in the hydrocarbon chain attached to the oxygen atom that is part of the main chain. The second part of the name comes from the other hydrocarbon chain. The number of carbon atoms in this chain plus the one carbon of the ester group give you the hydrocarbon root name of this second part of the name. Remove the final -*e* from that name and then add the suffix -*oate*. The molecule shown here is named ethyl butanoate.

▼ FIGURE 10.15 **Esters found in six fruits.** Esters are notoriously fine-smelling molecules.

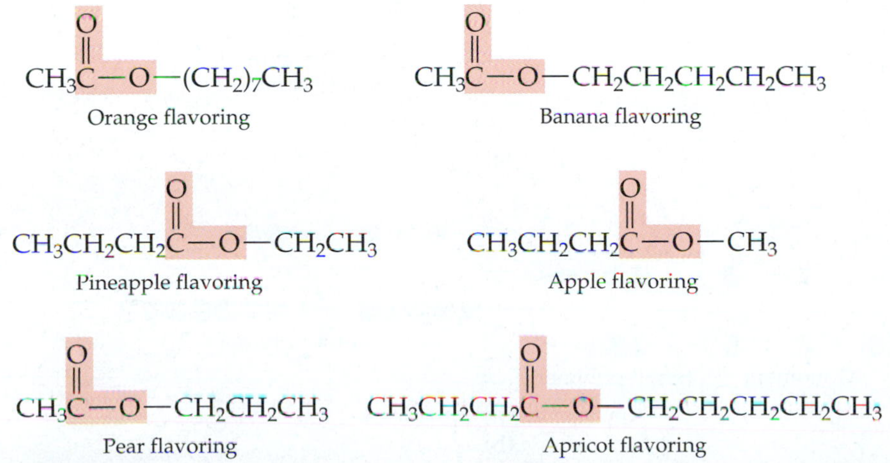

$$CH_3C-O-(CH_2)_7CH_3$$
Orange flavoring

$$CH_3C-O-CH_2CH_2CH_2CH_2CH_3$$
Banana flavoring

$$CH_3CH_2CH_2C-O-CH_2CH_3$$
Pineapple flavoring

$$CH_3CH_2CH_2C-O-CH_3$$
Apple flavoring

$$CH_3C-O-CH_2CH_2CH_3$$
Pear flavoring

$$CH_3CH_2CH_2C-O-CH_2CH_2CH_2CH_2CH_3$$
Apricot flavoring

QUESTION 10.5   Name the (*a*) banana, (*b*) pineapple, and (*c*) pear esters shown in Figure 10.15.

ANSWER 10.5   (*a*) either pentyl ethanoate or pentyl acetate, (*b*) ethyl butanoate, (*c*) either propyl ethanoate or propyl acetate.

QUESTION 10.6   In an ester molecule, it is possible for the group attached directly to the carbon atom in the ester functional group to be not a hydrocarbon chain, but rather only a single hydrogen atom, as in the structure

$$H-\overset{\overset{\displaystyle O}{\|}}{C}-O-CH_2-CH_3$$

When this is the case, this portion of the molecule is called *formate*. Given this information, what is the full name of this molecule, which is used as a flavoring for artificial rum?

ANSWER 10.6   Ethyl formate is the common name for this molecule. The name ethyl methanoate is strictly correct, but it is not used.

The sweet smell of esters does not carry over to their polymers, which means that polyesters are fragrance-free. In fact, they are touted for their ability to avoid odors of all kinds, including body odor, which is one reason polyesters are still used in modern fabrics. The best-selling polyester today is still the one that was used most widely for fabrics in the 1970s: Dacron® (Figure 10.17). This particular polyester is classified as a **copolymer**, which means a polymer in which the repeating unit is made up of two or more different monomers. In Figure 10.17, one monomer of the Dacron repeating unit is shown in blue and the other is shown in red.

The method of polymer formation shown in Figure 10.6*b* (page 388) is the way the copolymer Dacron is formed. Copolymers formed in this way are referred to as **condensation polymers**. The two mechanisms for polymer formation—addition polymerization and condensation polymerization—are compared in Figure 10.18. Both produce long polymer chains, but in

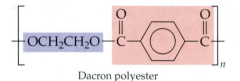

▲ FIGURE 10.17 Base unit of the copolymer we know as Dacron. This polyester is a copolymer made of two monomers (red and blue) that are repeated in a chain.

▶ FIGURE 10.18 Two ways to make polymers—revisited. The two mechanisms we have discussed for polymer formation are addition and condensation. These mechanisms differ in the way monomers become part of the chain. **(a)** In addition polymerization, monomers add one at a time to one end of the growing polymer chain. **(b)** In condensation polymerization, small polymers form from the condensation of monomers, and then these small polymers condense into larger ones. (For clarity, a small molecule formed during each step is not shown.)

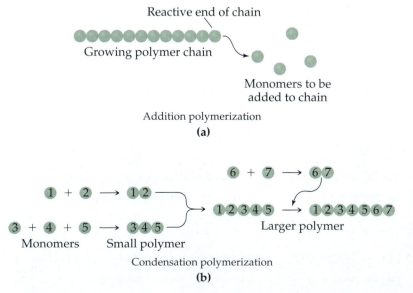

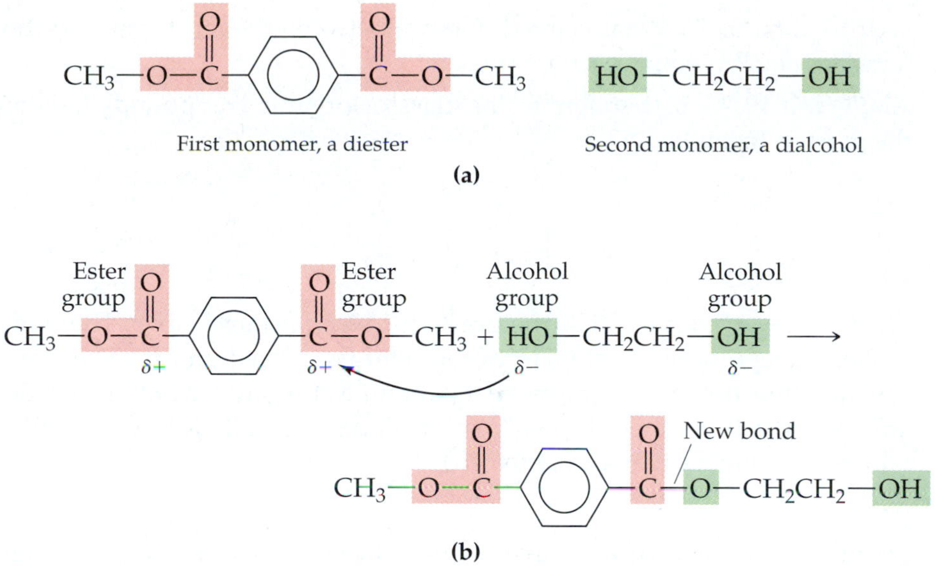

First monomer, a diester · Second monomer, a dialcohol

(a)

Ester group · Ester group · Alcohol group · Alcohol group

New bond

(b)

◀ FIGURE 10.19 The synthesis of Dacron polyester. (a) Two monomers go into creating the polymer. The first monomer is a diester, and the second is a dialcohol. (b) The condensation polymerization takes place when one alcohol group from the dialcohol attacks the carbon of one of the ester groups. The O from the ester group and the H from the alcohol group are split off, and a new ester bond is formed between the C and the O of the alcohol group.

addition polymerization, the only way the polymer chain can get longer is by the addition of a monomer to one end. In this mechanism, it is impossible for two medium-length chains to combine to form one long chain. In condensation polymerization, such joining of polymer chains is possible, as Figure 10.18 shows.

In the production of the copolymer Dacron, the two monomers shown in Figure 10.19a are in the initial mix, and the resulting polymer molecule has an alternating pattern of these two monomers. Both monomers have a feature found in many molecules that participate in condensation polymerization reactions: the same functional group on both ends of the monomer. In our example, the first monomer has an ester group on each end and the second monomer has an alcohol group on each end. It should not surprise you that the two-ester unit is called a *diester*, and the two-alcohol unit is called a *dialcohol*. As polymerization begins, one end of the diester monomer reacts with one end of the dialcohol monomer to form a two-monomer chain having an alcohol group on one end and an ester group on the other end (Figure 10.19b). Now, a new diester monomer can attach to the alcohol end of the two-monomer chain, and small polymers can link together to make a larger polymer.

What drives the reactions that form a Dacron chain? As we have noted, chemical reactions most often take place because there is an attraction between a region of positive charge and a region of negative charge. This reaction is no exception. The diester monomer contains a region of positive charge at each ester carbon, as shown in Figure 10.19b. This positive charge is more specifically described as belonging to the carbon atom of the $C\!=\!O$ part of the ester group. This $C\!=\!O$ combination is called a **carbonyl group** (pronounced kar-bo-NEEL), and it is found in several other functional groups, as we shall see in later chapters. The partial positive charge on the carbon atom of the carbonyl groups makes the ester group susceptible to attack by any group with a full or partial negative charge. The alcohol groups on the dialcohol monomer fit the bill. Alcohols act as nucleophiles because they have an accumulation of electron density on the oxygen atom. In the synthesis of Dacron, one of the two alcohols of the dialcohol monomer attacks one of the two ester groups on the diester monomer. The result of this nucleophilic attack is a linkage between two monomers that forms the beginning of a Dacron chain.

RECURRING THEME IN CHEMISTRY Many chemical reactions, especially organic reactions, are driven by the attraction between positive and negative parts of molecules.

QUESTION 10.7   What is the difference between the term *ester* and the term *carbonyl*? When is each term used?

ANSWER 10.7   *Ester* refers to the class of organic compounds having the generic formula

$$R-\overset{\overset{\displaystyle O}{\|}}{C}-O-R'$$

The $O=C-O$ part of this structure is the *ester functional group*. The term *carbonyl* refers to the $C=O$ part of the ester functional group. (There are several other functional groups that contain a carbonyl group, including amides, which we meet formally in Section 10.4, and carboxylic acids, which we will take up in Chapter 11.)

At this point, you should have a sense of the ways that monomers can come together to make long chains, as well as a sense of the differences that result in the physical properties of the finished polymer product. (Table 10.1 lists four commercially important polymers and the benefits and drawbacks of including them in fabrics.) We shall continue to explore unique polymers that can be made using the straightforward reactions that we have just considered. In Chapter 11, we shall see examples of how natural polymers—such as DNA, proteins, and cellulose—are formed by polymerization reactions.

TABLE 10.1   Properties of Polymers Used in the Manufacture of Synthetic Fabrics

Polymer	Monomer structure	Benefits	Drawbacks
Acetate fiber		Looks like silk; quick drying; doesn't shrink; doesn't wrinkle	Melted by several chemicals; melted by heat (cool iron)
Acrylic fiber		Bulky like wool; dries quickly; nonallergenic; washable	Very heat sensitive (very cool iron)
Nylon fiber	(nylon 6)	Strong and elastic; dries quickly; resists chemicals and perspiration	Somewhat heat sensitive; white nylon may turn grey or yellow
Polyester		Doesn't crease or wrinkle; strong and resilient; often blended with natural fibers	Somewhat heat sensitive

10.4 | Design-Your-Own Plastics

The Mayans of pre-Columbian Mesoamerica invented a game akin to modern-day soccer, except that the players could use only their heads and torsos to move the ball—no hands, no feet. The game involved two teams on a court with a goal at each end and something that Europeans did not yet have—a rubber ball. We take for granted that balls are bouncy, but imagine trying to make a bouncing ball out of leather or wood or cotton. In medieval Europe, balls used for sport were either pig bladders covered in leather or animal skulls. In Mesoamerica, though, there were trees that produced *latex*, an aqueous solution of rubber particles that could be processed to make a flexible, pliable, bouncy material. The Mesoamerican ball game, which was a popular sport from Paraguay in the south to what is now Arizona in the north, was played with a ball made of this "miracle" substance.

Ball courts found in Mexico have been carbon-dated to 1600 B.C. (Figure 10.20). The *Popol Vuh*, the primary religious text of the Mayans, is a story about the Mayan gods using a game played with a rubber ball to decide disputes. The humans who played this game were often warriors who would vie in a fight-to-the-death spectacle. The heads of the losers would then be covered in rubber and used to play subsequent games. By the time the Spanish conquistadors arrived on the scene in the sixteenth century, the game had evolved into a more innocuous sport in which the winners would chase spectators and take their clothing and jewelry as a reward. The Spaniards were so taken by this game that they took two teams of players and several of the bouncy balls back to Europe. The Europeans also took latex-producing *Hevea braziliensis* seedlings and planted them in warm climates such as present-day Sri Lanka and Malaysia, where rubber plantations expanded the availability of this new material to the rest of the world.

The white latex that drips from a rubber tree is mostly water (Figure 10.21). Once the amount of water is reduced, it takes on a creamy consistency and a new name: *rubber*. Rubber can be used as a coating, and the early Mesoamericans coated many things with rubber, including clothing and shoes. They recognized that rubber was the best protection against the elements because it could seal out wind and rain. What gives rubber these properties? How can it be waterproof and at the same time extraordinarily flexible? To figure this out, let us now turn, as we always do, to the chemical structure of the molecule to see why it behaves as it does. Figure 10.22a shows the line structure of *cis*-isoprene, a monomer, and the product of its polymerization, *cis*-polyisoprene. Take note of the fact that every double bond in this structure has a cis orientation, which is shown in Figure 10.22b just in case you need a reminder.

(a)

(b)

▲ FIGURE 10.20 **Rubber use in ancient Mesoamerica.** **(a)** Mesoamerica was the first civilization to use rubber. Shown here is a Mayan ball court used to play a game in which the ball was a human head covered with rubber. **(b)** This frieze shows two warriors playing the game.

▲ FIGURE 10.21 **The source of latex.** Latex is a white liquid that is tapped from the rubber tree.

QUESTION 10.8 Identify each molecule as being a cis isomer, a trans isomer, or neither. How can a molecule contain a carbon–carbon double bond but be neither cis nor trans?

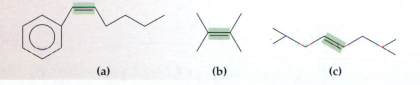

(a)   (b)   (c)

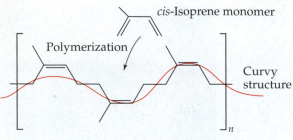

cis-Isoprene monomer

Polymerization

Curvy
structure

Structure of natural rubber (all cis double bonds)

(a)

H_3C H

(b)

▲ **FIGURE 10.22   The synthesis of rubber.**
(a) Rubber is made from the polymerization of the *cis*-isoprene monomer, a monomer that yields a very curvy polymer containing all cis double bonds. **(b)** This isoprene monomer is classified as a cis isomer because the two smallest groups on the double-bond carbons are on the same side of the horizontal axis through the double bond.

▼ **FIGURE 10.23   Gutta-percha, a streamlined polymer.**   **(a)** The structure of the gutta-percha polymer is less curvy than the structure of the rubber polymer and therefore more extended. The reason for this difference is that in gutta-percha all the groups on the double-bond carbons are in the trans orientation. **(b)** This isoprene monomer is classified as a trans isomer because the two smallest groups on the double-bond carbons are on opposite sides of the horizontal axis through the double bond.

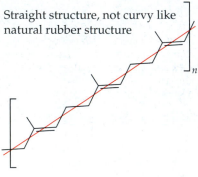

Straight structure, not curvy like natural rubber structure

Structure of gutta-percha
(all trans double bonds)

(a)

H_3C

H

(b)

ANSWER 10.8   Molecule (*a*) is a cis isomer because two H atoms are on the same side of the double bond. Molecule (*c*) is a trans isomer because the two H atoms are on opposite sides of the double bond. Molecule (*b*) is neither because it has four identical groups around the double bond. Whenever the two groups on either C in the double bond are identical, the terms *cis* and *trans* cannot be used.

Gutta-percha is a water-resistant, natural polymer that has a structure almost, but not quite, the same as that of rubber. The difference is that the double bonds in gutta-percha are in the trans configuration, as shown in Figure 10.23. This polymer, a *trans*-polyisoprene, is a much harder, much less flexible material than the cis form found in natural rubber. To see why, consider the shape of each chain. In the cis form, the chain is curvy and therefore less extended. When the long strands of *cis*-polyisoprene exist together in solid rubber, they are a jumbled mess of intertwined strands (Figure 10.24*a*). Because the individual strands are not fully extended, they can be pulled apart and stretched out. When released, the chains go back to their original position, and the material snaps back into place. The less curvy and therefore more extended shape of *trans*-polyisoprene results from the trans orientation around the double bond. This orientation makes the *trans*-polyisoprene chain more streamlined. Although it is still a tangled mess, *trans*-polyisoprene is more rigid, less tangled, and less stretchy than *cis*-polyisoprene (Figure 10.24*b*).

The rubber produced on tropical rubber plantations caused problems for rubber manufacturers because the *cis*-polyisoprene polymer is tacky and sticky when hot, and stiff and fragile when cold. Early experiments on the chemistry of rubber in the nineteenth century were directed toward firming it up and making it a stronger, more resilient, better-behaved material. This is when an inventor named Charles Goodyear took the stage. While experimenting with different rubbery concoctions in his kitchen one day in 1839, Goodyear purportedly spilled sulfur into some rubber on the stove and serendipitously came across a way to make rubber more manageable.

Figure 10.25 shows the result of adding sulfur to rubber, a process Goodyear called **vulcanization** after Vulcan, the Roman god of fire. The sulfur created what we now know as **cross-links**, which are connectors of one

▼ **FIGURE 10.24   Comparison of molecular arrangement in rubber and gutta-percha.**   **(a)** In rubber, long strands of *cis*-polyisoprene molecules form a jumbled muddle. This chaotic arrangement of the strands makes this polymer stretchy. **(b)** In gutta-percha, long strands of *trans*-polyisoprene molecules, because they are not as curvy as the molecules in rubber, have more extended strands that form not quite so jumbled a mix. This less chaotic arrangement of strands makes this polymer more rigid and less stretchy than natural rubber.

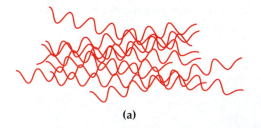

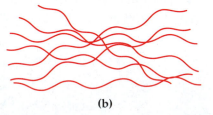

(a)   **(b)**

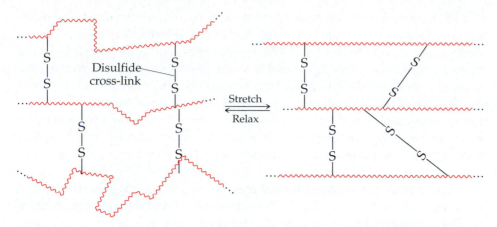

or more atoms that join two polymer chains together. Cross-links can be
made with atoms of various elements, but those in vulcanized rubber are
made up of two sulfur atoms. Therefore, we refer to them as *disulfide cross-
links*. The cross-links do two things. First, they add strength by creating a
new, tough bond between the individual strands in rubber. Second, they
maintain the flexibility of the rubber by connecting different chains at specific points. When the vulcanized polymer is stretched out, the disulfide
cross-links limit how far the molecules can stretch. When the material is
allowed to snap back, it goes back to the original position, which is also dictated by the cross-links. Thus, vulcanized rubber stays stretchier longer
than unvulcanized rubber.

The rubber we know today is made in the laboratory. Synthetic rubber
provides a more consistent raw material than natural rubber; as a result, the
raincoats and boots shown in Figure 10.26 will have little variation from one
to the next. Synthesizing rubber in the laboratory meant that rubber manufacturers no longer had to depend on getting their raw material from rubber
trees. This was of tremendous advantage because now the dozens of variables involved in the process could be controlled. Furthermore, the flexibility and rigidity of the rubber could be varied in a predictable way by changing the amount of sulfur added during vulcanization. For example, the soles
of crepe shoes, made of unvulcanized rubber, are extremely pliable and tacky
to the touch. Rubber bands are lightly vulcanized with about 2 parts sulfur to
100 parts rubber. Harder rubber, such as that used in car tires, is more extensively cross-linked, with up to about 10 parts sulfur to 100 parts rubber.

Regardless of the amount of sulfur added, there is one thing that all formulations of rubber have in common: they are all waterproof. To understand
why, look again at the structure of rubber shown in Figure 10.22*a*. Notice
that the polymer contains no electronegative atoms and therefore no regions
of excess or depleted charge. Thus, rubber is a typical hydrocarbon in that it
is extremely hydrophobic and does not form hydrogen bonds with water
molecules. If you are designing something to be water-repellent, such as a
tire, wetsuit, or shoe sole, rubber may be the material of choice.

**RECURRING THEME IN
CHEMISTRY**   The individual
bonds between atoms dictate the
properties and characteristics of the
substances that contain them.

▲ **FIGURE 10.26** **A consistent
product.**   The variation in the material
used to make raincoats and boots is low
because the rubber used to make them is
synthesized in a factory under controlled
conditions.

**RECURRING THEME IN
CHEMISTRY**   Like dissolves
like.

QUESTION 10.9 Do you think gutta-percha is hydrophobic or
hydrophilic? Why was this polymer useful as a coating for underwater
cables?

ANSWER 10.9 Gutta-percha has the same composition as rubber, but
different orientation about its double bonds. Like rubber, it is very
hydrophobic and repels water, making it an ideal polymer for water-
proofing underwater cables.

We have seen that it is possible to design a polymeric material for rigidity, for resistance to water, and for flexibility. What if we would like to design, say, a soft contact lens? What properties would be desirable in the material we choose? A soft contact lens should be flexible, like rubber, but it should also be hydrophilic because it must be accepted comfortably onto the surface of the eyeball, which is coated with an aqueous solution. Thus, rather than using a hydrocarbon, we would want to choose a polymer that is water-friendly, but does not dissolve in water. Ideally, this material would be able to make hydrogen bonds, a property that ensures that a substance will be eyeball-compatible.

A polymeric material commonly used in the manufacture of soft contact lenses is shown in Figure 10.27. This material is bendable because it has a flexible carbon backbone composed entirely of single bonds between carbon atoms. This structure allows the chain to twist and bend because free rotation about the single bonds is permitted. The material is not especially stretchy, however, because it cannot be pulled out and released back the way curly or cross-linked rubber can. Stretchiness would be an undesirable feature for a contact lens, which must maintain its shape so that it remains fitted to the surface of the eye. There are several electronegative oxygen atoms in this polymer, and these atoms make the molecule both polar and hydrophilic.

Because it is attached to two hydrocarbon groups, the carbonyl group in the molecule shown in Figure 10.27 is not part of an ester functional group. Rather, a carbonyl group attached to two hydrocarbons is referred to as a **ketone**, another organic functional group we can add to our growing list. Ketones have the generic formula

MOLECULE
Propanone

▶ FIGURE 10.27 **See-through polymer.** The polymer used to make contact lenses is very flexible because of its bendable carbon backbone. The alcohol groups of the polymer, highlighted in blue, allow the polymer to form hydrogen bonds with water.

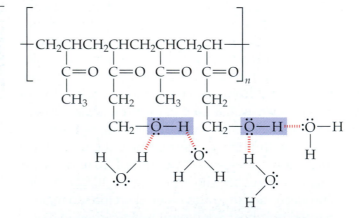

Because they contain a carbonyl group—which has a locus of positive charge at the carbon atom—ketones are susceptible to attack by nucleophiles, although they generally are not nearly as quick to react as esters are.

To see how to name ketones, we begin with a simple molecule that has no branched hydrocarbon chains, such as the one shown in Figure 10.28a. The first step is to count the number of carbons to get the root name of the ketone; four carbons in our example means that the root name is *butane*. Remove the *-e*, and add the suffix *-one* to get *butanone* (pronounced BEAUT-a-known). To indicate the location of the carbonyl group, number the carbons, starting at the end that gives the lower number to the carbonyl carbon. In our example, this means the left end. Place the number of the carbonyl carbon at the beginning of the name, with a hyphen: *2-butanone*. The name of the molecule shown in Figure 10.28b—cyclohexanone—demonstrates how no number is needed in the name of a cyclic ketone. It is understood that carbon 1 is the carbonyl carbon.

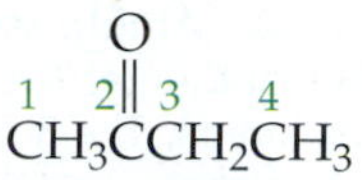

QUESTION 10.10 Indicate whether the name for each ketone is correct or incorrect. If incorrect, provide the correct name:

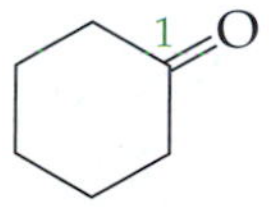

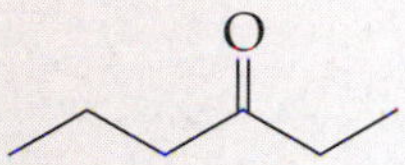

4-Hexanone
(a)

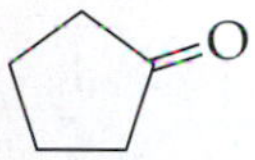

1-Cyclopentanone
(b)

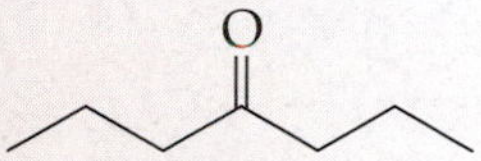

4-Heptanone
(c)

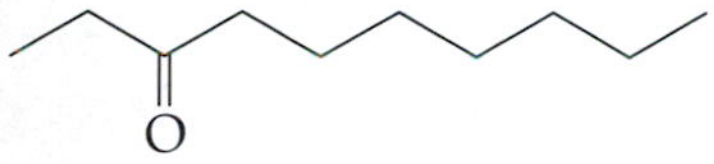

3-Decanone
(d)

ANSWER 10.10 Molecules (c) and (d) are named correctly. The correct name for molecule (a) is 3-hexanone. The correct name for molecule (b) is cyclopentanone. No number is needed in a cyclic ketone because the molecule looks the same wherever the carbonyl group is placed.

Ketones are used as solvents because they easily dissolve many other organic molecules, especially other polar molecules. As we have seen, ketones are also one of the functional groups found in synthetic polymers, such as the material used to make soft contact lenses. The ketone group gives the molecule shown in Figure 10.27 polarity, while the long hydrocarbon chain holds the polymer together and keeps it from breaking apart.

Let us work on the design of one more polymer, this time one that is incredibly strong and resilient. This was the challenge faced by chemist Stephanie Kwolek, who worked as a polymer expert at I.E. DuPont de Nemours for 45 years (Figure 10.29a). During this time, she experimented with new polymers that had exceptional strength. In particular, she was interested in polymers that were rigid enough to pack together as crystals, in much the same way as the ions in solid salts pack together into a crystalline lattice. She came up with the polymer shown in Figure 10.29b, which contains an alternating pattern of benzene rings (which are rigid) and

O
1 2‖ 3   4
CH₃CCH₂CH₃

2-Butanone
(a)

Cyclohexanone
(b)

▲ **FIGURE 10.28** **Naming straight-chain and cyclic ketones.** When numbering the carbons in a ketone molecule, begin at the end that gives the lower number to the carbonyl carbon. **(a)** If carbons were numbered from the right end of the chain, this ketone would be 3-butanone, an incorrect name. Starting from the left gives the correct name: 2-butanone. **(b)** In a cyclic ketone, there is no need to write a number in the name because it is understood that the carbonyl carbon is carbon 1.

▼ **FIGURE 10.29** **Indestructible.** **(a)** Stephanie Kwolek, a polymer chemist, led the development of the Kevlar polymer. **(b)** Kevlar is a polymer of amides, which are functional groups containing a carbonyl group attached to a nitrogen.

(a)

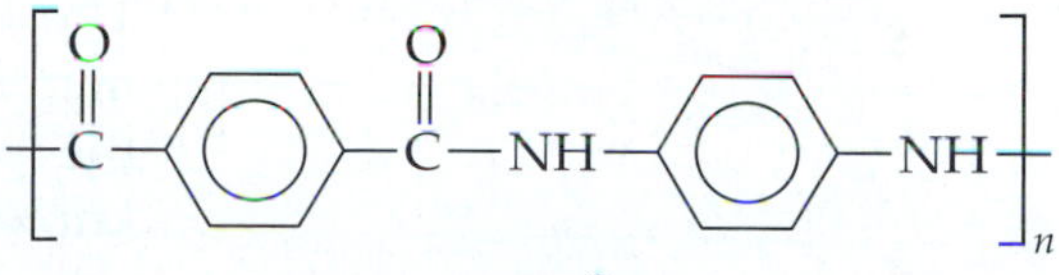

Kevlar®
(b)

► FIGURE 10.30 **Amide structure and the C–N bond.** **(a)** The generic structure of an amide. The R stands for one hydrocarbon chain, and the R′ stands for another hydrocarbon chain. **(b)** The C—N bond in an amide functional group cannot rotate freely about its own axis.

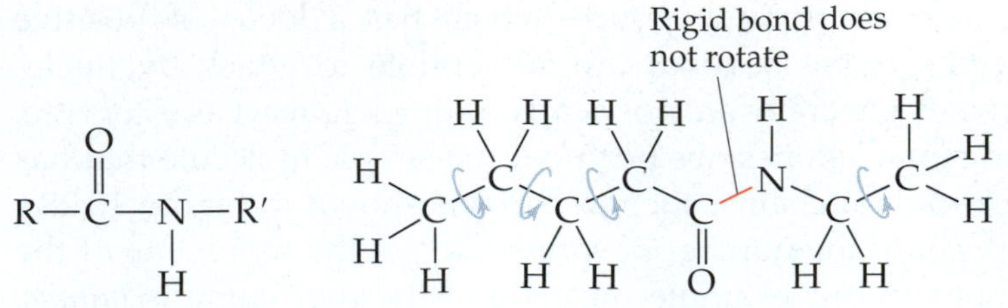

functional groups known as **amides**. From the generic structure shown in Figure 10.30*a*, you can see that, like esters and ketones, amides contain a carbonyl group. However, amides have a nitrogen atom attached to one side of the carbonyl group and a hydrocarbon attached to the other side. We shall revisit the amide functional group in Chapter 11 when we discuss proteins, which are natural polymers made up of amide linkages.

QUESTION 10.11 Identify all the organic functional groups in each molecule:

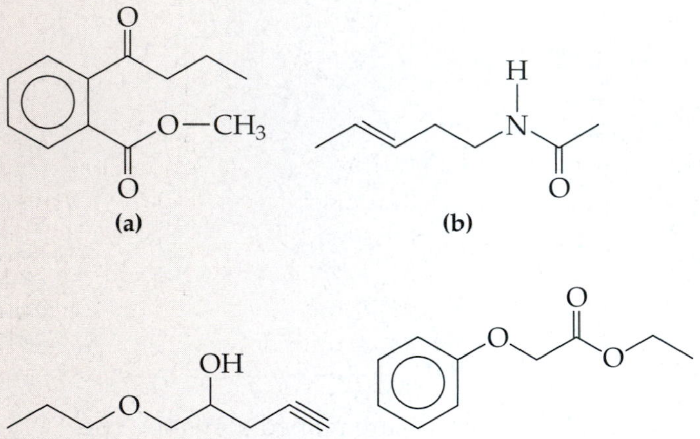

ANSWER 10.11 (*a*) benzene ring, ketone, ester; (*b*) alkene, amide; (*c*) ether, alcohol, alkyne; (*d*) benzene ring, ether, ester.

The bond that nitrogen makes to the carbon atom in the amide functional group is relatively inflexible because, while free rotation is possible in singly bonded carbon atoms in a hydrocarbon chain, it is not possible around the C–N bond (Figure 10.30*b*). For this reason, the amide group reduces the flexibility of Kwolek's polymer chain and, along with the inflexible benzene ring, contributes to its rigidity. Because the polymer flexes very little, each strand is relatively straight, meaning that strands can pack together into a regular, three-dimensional crystalline matrix. When they pack together closely, the individual strands can interact with one another via hydrogen bonds, as shown in Figure 10.31. These hydrogen bonds act as cross-links between strands and make the material even stronger. This three-dimensional crystalline structure gives the polymer extraordinary strength. In fact, this polymer, which we know as *Kevlar*®, is five times stronger than steel. It is used to make cables for suspension bridges, bulletproof vests, protective helmets, and sails, among other things.

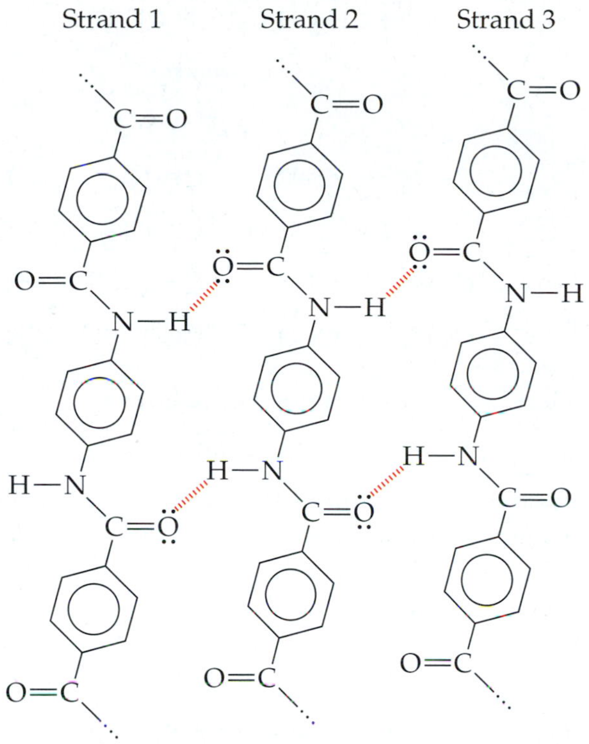

Strand 1   Strand 2   Strand 3

◀FIGURE 10.31 **FIGURE 10.31   Cross-linked and rigid.**   The rigidity of amide bonds and the hydrogen bond cross-links they make to one another across polymer strands make Kevlar an extraordinarily heavy-duty, hard-wearing polymer.

In this brief survey of some interesting polymers, we have seen that it is possible to incorporate specific characteristics into the design of new polymers. We can, for example, design a water-resistant polymer by using a purely hydrocarbon-based structure or a hydrophilic polymer by adding groups that can take part in hydrogen bonding. Polymers can be designed with flexibility or rigidity, and cross-links can be included to add stretchiness or strength. In the next section, we shall look at yet another variable that can be manipulated in polymer recipes: temperature.

10.5 | Get Out of My Way!

We have seen that the polymer Kevlar owes its extraordinary strength, in part, to the ability of the polymer strands to stack together in a crystalline pattern. This phenomenon occurs to some extent in many other polymers, and the extent to which a polymer is crystalline depends on its ability to align itself into a well-ordered matrix. Any highly ordered, crystalline region in a polymer is called a **crystallite**. The more crystallites a polymer contains, the harder it is and the more resistant to external influences, such as increases in temperature. For example, the number of crystallites in a given volume of the relatively rigid polymer polyethylene is much higher than the number in the same volume of the liquid polymer polymethylcyanoacrylate, which is found in super glue. A look at the structures of these two polymers (shown in Figure 10.32) tells us why. Notice that polyethylene has a streamlined, linear structure, but polymethylcyanoacrylate has branches leaving the main chain. These branches get in the way of one another and prevent the polymer chains from readily aligning into crystallites.

Organic chemists call this phenomenon, when the size and bulk of a molecule hinder its range of motion, **steric hindrance**. Imagine the two

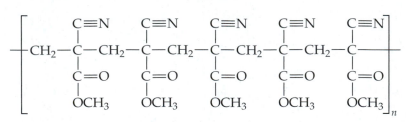

$$\left[\!\!\!\begin{array}{c}CH_2CH_2CH_2CH_2CH_2CH_2\end{array}\!\!\!\right]_n$$

Polyethylene

(a)

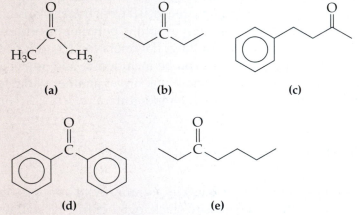

Polymethylcyanoacrylate

(b)

▶ **FIGURE 10.32 Crystallite formation. (a)** Polyethylene's streamlined structure permits the formation of crystallite structures in the polymer. **(b)** The branch groups in polymethylcyanoacrylate make it difficult for molecules of this polymer to pack into a regular, repeating crystallite form.

▼ **FIGURE 10.33 Steric hindrance.** The concept of steric hindrance in chemistry is like the situation in which two people—a small girl and a large man—try to make their way through a crowded subway car. Smaller objects have lower steric hindrance.

(a)

(b)

people shown in Figure 10.33 are trying to move through a crowded subway car. Steric hindrance will cause the man to collide with other passengers, and his bulk will retard his ability to move. The steric hindrance that the little girl experiences will be comparatively small, and as a result she can easily weave in and among the crowd. Molecules have size and bulk, too, and the larger and bulkier the molecule, the more steric hindrance it experiences.

QUESTION 10.12 Which of the following molecules will experience the highest degree of steric hindrance during a chemical reaction involving the carbonyl group?

(a)   **(b)**   **(c)**

(d)   **(e)**

ANSWER 10.12 (*d*), because of the bulky benzene rings.

When heated, some polymers gradually change composition as the energy supplied by the added heat breaks crystallites apart and changes them to **amorphous regions**, which are regions that lack specific shape and order. As the percentage of crystallites decreases, the polymer becomes more pliable and more malleable. This is a very useful characteristic because it means that the hot polymer can be molded into a comb, say, or a toothbrush. When cooled, these **thermoplastic polymers** re-form crystallites and therefore retain their shape. Most of the plastic stuff you encounter every day—things like plastic forks, takeout containers, pens, and cases for contact lenses—are made of thermoplastic polymers. If you heat one of these items, its crystallites will once more revert to amorphous regions and the object will melt into a fluid mass that will have a new shape when cooled. As we shall see in the next section, the ability to manipulate thermoplastic polymers makes them recyclable, meaning that we can use these polymers over and over again.

Synthetic Fibers

Practically all forensic science is rooted in one simple theory known as *Locard's principle*: any time one thing comes into contact with another, there is always an exchange of something, whether it is dirt, sweat, fibers, hair, paint, or skin. Whenever one car hits another, there is always something to be learned about the first car by looking at the second car. Any thief entering a room always

leaves something behind, even if it is only a single piece of hair or a single flake of skin. Forensic scientists count on these exchanges, and, with twenty-first-century technology, smaller and smaller pieces of evidence can be analyzed.

The criminal element is no doubt aware of Locard's principle (though probably not by this formal name) and knows it is possible to reduce the amount of exchange that occurs by wearing gloves (no oily fingerprints) and a cap (no stray hairs). One of the most difficult kinds of exchanges to prevent is the transfer of fibers to and from clothing, upholstery, and carpets. Figure 10.34 illustrates the ease with which fibers from a sweater can be transferred to another article of clothing after only incidental contact.

Fibers from various sources can be unique, and when they are, they can be used to link a crime scene with the perpetrator. For example, in a landmark case involving serial killings in Georgia during 1979, 1980, and 1981, fiber and hair evidence was the most important proof used to convict Wayne Williams. Carpet, bedspread, and blanket fibers from Williams's bedroom, hairs from his dog, and fibers from his car were found on at least six of the 12 victims of Williams's killing spree. This evidence was used to convict him of the killings.

Fibers collected from a crime scene are often viewed under a microscope before being subjected to any kind of chemical analysis, because the chemical testing might destroy the fiber. This visual inspection can usually reveal whether the fiber is natural or synthetic. Figure 10.35 shows fibers made from cotton, nylon, cellulose, and acrylic. Microscopy can be used to link the size, shape, and color of an unknown fiber to fibers collected from a known location, such as Williams's bedroom carpet. In cases where the identity of a fiber made from a synthetic polymer is in question, chemical tests may be done to determine the fiber's solubility in solvents of different polarities. For example, a synthetic fiber that dissolves in a highly nonpolar solvent is probably a nonpolar fiber, possibly a hydrocarbon. One that dissolves in a more polar solvent, as a polyester would, must contain more polar groups.

Another fiber characteristic used in identification is melting point. The melting point of a polymer depends on the strength of the bonds holding the atoms together, the strength of interaction between strands, and the degree of crystallinity. Polymers with a high percentage of crystallites and a low percentage of amorphous regions have high melting points. Consider nylon, for

▼ **FIGURE 10.34   Locard's principle in action.**   The sleeve of a freshly laundered blazer is shown on the left. In the center, the sleeve briefly comes into contact with a fuzzy white sweater. The transfer of lint to the blazer (right) is a demonstration of Locard's principle.

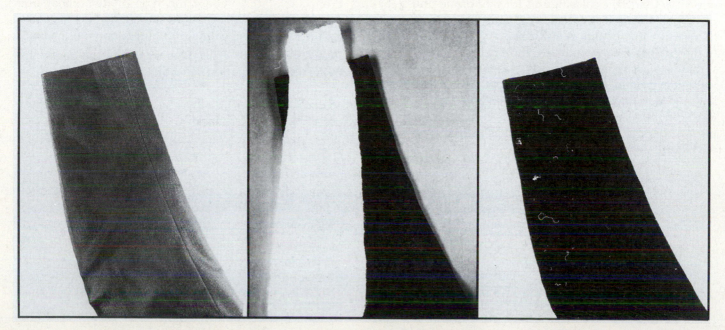

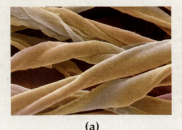
(a)

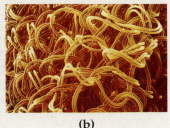
(b)

(c)

(d)

▲ **FIGURE 10.35** **A close look at some everyday fibers.** Microscopic views of **(a)** cotton, **(b)** nylon, **(c)** cellulose, and **(d)** acrylic. Of these, only cotton is a natural fiber.

instance, the generic name for a class of thermoplastic polymers containing amide linkages (Figure 10.36). The slight differences in structure in the three forms of nylon give each polymer a different resistance to heat. In addition, these variations in structure lead to differences in the way individual fibers pack together. This means that each type of nylon has a different percentage of crystallites and a different melting point.

Table 10.2 shows each of these polymers and their melting points. The melting points are determined by watching one fiber under a microscope as the temperature of the microscope stage is gradually increased. The melting point is the temperature at which the inside of the fiber begins to liquefy, and this usually occurs over a range of many degrees Celsius. Even though each polymer has a melting point *range* rather than a *single* melting point, this evidence is useful if these ranges do not overlap, as in the case of the three nylons listed in Table 10.2.

When you combine the fiber evidence from visual inspection, solubility tests, and melting-point determination, it is usually possible to identify the exact polymer type. Once this is known, investigators can begin the tedious process of tracking down the origin of the fiber. For example, in the Williams case, the investigators were able to show that the bedroom carpet was made for only one year. Using the sales figures for that brand and color of carpet, the investigators were able to calculate that there was a one in almost 8000 chance that another home in the Atlanta area would have the same carpet as that in Williams's bedroom. When this evidence was combined with fiber evidence from the car carpet, bedspread, blanket, and dog hair, it created an undeniable link between the 12 victims and Williams. He was convicted of multiple murders in February 1982 and is currently serving two life sentences.

$$\left[NH(CH_2)_{10}\overset{O}{\overset{\|}{C}}-NH(CH_2)_{10}\overset{O}{\overset{\|}{C}}-NH(CH_2)_{10}\overset{O}{\overset{\|}{C}} \right]_n$$

Nylon 11

$$\left[NH(CH_2)_{5}\overset{O}{\overset{\|}{C}}-NH(CH_2)_{5}\overset{O}{\overset{\|}{C}}-NH(CH_2)_{5}\overset{O}{\overset{\|}{C}} \right]_n$$

Nylon 6

$$\left[\overset{O}{\overset{\|}{C}}(CH_2)_4\overset{O}{\overset{\|}{C}}-NH(CH_2)_6NH-\overset{O}{\overset{\|}{C}}(CH_2)_4\overset{O}{\overset{\|}{C}}-NH(CH_2)_6NH \right]_n$$

Nylon 6,6

▲ **FIGURE 10.36** **The structures of three nylon polymers.** The slight variations in structures give these polymers different melting points.

QUESTION 10.13 The melting-point range of a nylon fiber found at a crime scene is 253 to 256 °C. From Table 10.2, choose the type of nylon that best matches the evidence. Write the condensed structure for the smallest repeating unit of this polymer chain.

ANSWER 10.13 This fiber matches the range for nylon 6,6. The smallest repeating unit is

$$\left[\overset{O}{\overset{\|}{C}}-(CH_2)_4-\overset{O}{\overset{\|}{C}}-NH(CH_2)_6NH \right]_n$$

TABLE 10.2	Properties of Three Nylons	
Name	**Melting-point range (°C)**	**Used to make**
Nylon 11	182–186	Tubing, coatings for dishwasher racks, and clothes hangers
Nylon 6	210–230	Tires
Nylon 6,6	250–264	Carpeting, clothing

What if, instead of combs, food containers, or some other molded plastic product, your business is manufacturing plastic parts for jet engines—parts that will have to be stable no matter how hot the engines run? Obviously thermoplastic polymers, even the toughest ones, would not be suitable because, by definition, a thermoplastic polymer must become fluid at high temperatures. For your jet engines, you need a polymer that will maintain its shape when heated. **Thermosetting polymers** are extremely tough polymers that do not change shape when heated. The polymer shown in Figure 10.37a, which is the thermosetting polymer known as *Bakelite*, has been produced longer than any other synthetic polymer in history. In the 1940s and 1950s, it was used to make everything from telephones to airplane parts to jewelry to radios (Figure 10.37b). Notice that Bakelite does not have the long individual chains present in all the other polymers we have studied so far in this chapter. Rather, bonds between benzene rings that exist throughout the structure create a blanket that is essentially one enormous molecule.

Today, Bakelite has been replaced by other thermosetting polymers, and old Bakelite items have become collectors' items. According to Gloria Lieberman, an expert on antique jewelry, "The true Bakelite collector doesn't want other plastics. This plastic could take extremely bright colors that don't fade over time. It's also hard and durable and can be carved. Most Bakelite sells in the $100 to $400 range."

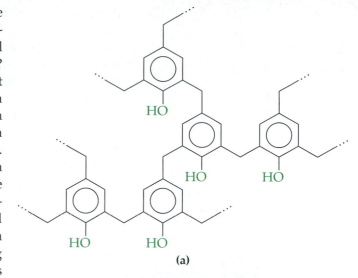

(a)

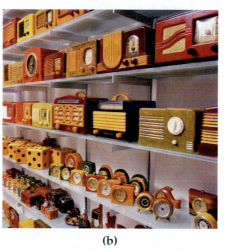

(b)

▲ **FIGURE 10.37   A permanent polymer.   (a)** Like Kevlar, Bakelite is a polymer that is virtually indestructible. Because Bakelite does not contain sites that are susceptible to nucleophilic attack, it is not biodegradable. **(b)** A collection of clocks and radios made of Bakelite.

10.6 | The Weakest Link

The qualities that made Bakelite a marvel during the 1940s and 1950s—strength, durability, heat resistance—also caused a serious problem: how do we get rid of the Bakelite telephones and radios we no longer need? Once a Bakelite object is formed, that is what it will continue to look like for posterity. If you cannot melt and remold Bakelite, what happened to all the Bakelite that was produced? Mostly, it sits in attics and in landfills and in antique stores, where it will continue to defy the aging process. That is the problem with thermosetting polymers in general: once they are set, they are not easily converted to some other shape or used for some other purpose. The challenge facing polymer chemists nowadays is to design tough, resilient, strong plastics that can resist the elements, chemicals, and heat, but are also susceptible to some form of breakdown, whether by sunlight, bacteria, or chemicals.

Unlike thermosetting polymers, thermoplastic polymers are more easily reshaped and reused. The word *thermoplastic* means "heat moldable." Therefore, thermoplastic polymers are, by definition, capable of being melted, but this does not mean that they all have the same melting point. Small differences in structure give each polymer a different melting point, and, as we have already seen, these differences are exploited by forensic scientists in analyzing crime-scene evidence. These differences in melting point can also be used to create plastics that serve different purposes. Six thermoplastic polymers that are ubiquitous in our lives are listed in Table 10.3. In the first column of the table is a number known as the *resin ID code*. These codes tell us how easily a thermoplastic polymer melts and therefore how recyclable the polymer is. The numbers range from 1 to 6, with 1 being the most readily recycled. There is also a miscellaneous seventh category that includes other thermoplastic polymers and plastics that are a blend of more than one polymer.

Take a few moments to look at some of the plastic items around you. If you are reading in the park, take a look at a plastic Frisbee® or a plastic soda bottle if you can find one. If you are in the library, look at a plastic wastebasket, stapler, or pencil holder. If you are at home, look at plastic containers of shampoo, cleanser, saline, margarine, milk, or detergent. Many of these items will have a resin ID code stamped on the bottom surrounded by a triangular arrow indicating that it is recyclable. Can you find examples of all seven resin ID codes?

Because most municipal recycling programs have us combine all our plastic items in one container, the first step at a recycling facility is usually separation into the seven categories listed in Table 10.3 (Figure 10.38). Once the items have been separated, they are sent to a chipper in which they are cut into small flakes that are then thoroughly washed to remove

▶ FIGURE 10.38 **Steps in the recycling process.** After an initial separation step, plastic items of the same resin ID code are chipped into small flakes, washed, and sterilized to produce recycled plastic pellets.

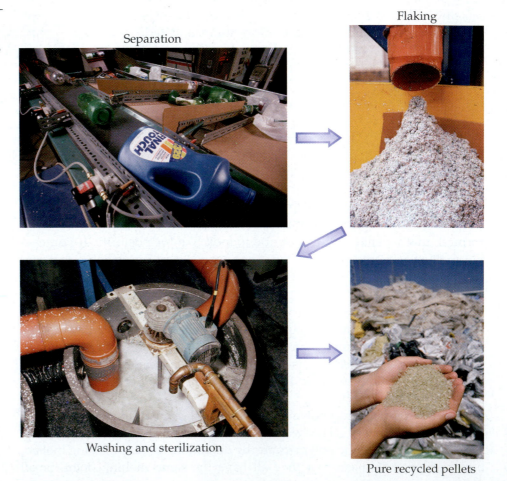

Separation

Flaking

Washing and sterilization

Pure recycled pellets

TABLE 10.3   Resin ID Codes for Recyclable Polymers

Resin ID Code	Polymer Name	Monomer Structure	Used for
1	PET (polyethylene terephthalate)	Ethylene glycol *and* Terephthalic acid	Soft drink, salad dressing, and peanut butter containers; fleece clothing; luggage
2	HDPE (high density polyethylene)	Ethylene	Milk, juice, and water containers; toys; liquid detergent containers
3	PVC (polyvinyl chloride)	Vinyl chloride	Raincoats; packaging film; shower curtains; credit cards; loose-leaf binders; mud flaps; traffic cones
4	LDPE (low density polyethylene)	Ethylene	Bags for bread; grocery bags; zipper seal bags; shipping envelopes; compost bins
5	Polypropylene	Propylene	Food packaging; injection molding; automobile interior parts; ice scrapers
6	Polystyrene	Styrene	Styrofoam; fast-food containers; CD cases; cafeteria trays; insulation
7	Mixture	Mixture	Plastic lumber; bottles; water carboys; citrus juice bottles; catsup bottles

contaminants and labels. After the flakes have been sterilized, the sterile, label-free plastic in each category is sold to a plastics manufacturer who then melts it down and re-forms it into some new shape.

> **QUESTION 10.14**   It is possible to buy very inexpensive recycled plastic that contains a mixture of thermoplastic polymers. Why do you suppose this mixed type costs a lot less than the same amount of a recycled plastic containing only one type of polymer?
>
> **ANSWER 10.14**   At the recycling plant, the initial separation step was not needed for the mixed polymer. Skipping this step lowers the price of the recycled product.

During the 1980s, only a small percentage of Americans recycled plastics. The process was time consuming because there was no central infrastructure in place to collect and handle plastics for recycling. Today, more than 80 percent of cities in the United States have recycling programs, and nonrecyclers have become the minority. Part of the success of the recycling effort can be attributed to legislation at the state level requiring that rigid plastic containers be manufactured with a certain minimum percentage of recycled and post-consumer plastic. The *Oregon Recycling Act,* passed in 1991, requires a minimum of 25 percent recycled plastic in new containers. On average, the plastic in such a container is used five times, meaning that a shampoo bottle may well have been a motor oil container or a soda bottle in a previous life cycle. Many other states are adopting similar legislation.

Because the raw material for creating thermoplastic polymers is petroleum, our demand for plastic products is part of our dependence on petroleum. Thus, our depletion of crude-oil resources is reduced with increased recycling. In addition, research on the use of recycled plastics for new purposes is yielding innovative products. For example, in 1993, an outdoors-clothing company introduced a fleece fabric made from recycled plastic soda bottles. According to the company's literature, the production of 150 pieces of clothing uses 3700 2-L plastic bottles.

> **QUESTION 10.15**   If the clothing-company figures just cited are correct, how many 2-L plastic bottles are used, on average, to make one piece of clothing?
>
> **ANSWER 10.15**
>
> $$1 \text{ piece of clothing} \times \frac{3700 \text{ bottles}}{150 \text{ pieces of clothing}} = 25 \text{ bottles}$$

Despite the success of recycling programs, there is still an enormous amount of plastic that ends up in landfills and as solid pollutant waste. In a 1995 article in the journal *New Scientist* called "Dead in the Water," scientists described data collected on the effects of plastic waste on marine animals. For every 30 whale, dolphin, and porpoise carcasses examined, one had choked to death on a piece of plastic. When plastic ends up as a solid pollutant in the ocean, the recyclability of the polymer used to make that plastic has little import for the animal that swallows it. Unfortunately, most of the plastics in Table 10.3 do not degrade once they are thrown out of a car window or into the ocean. Rather, much like thermosetting polymers, they can survive in the environment for decades. The energy from sunlight and the best efforts of bacteria are not able to break them down at a significant rate.

For this reason, today there is an implicit directive for the design of new polymers. Not only must they meet the criteria for the particular job they must do, they must also be **biodegradable**, which means that they must be able to be broken down into smaller pieces by something in the environment, such as microorganisms and sunlight. This is very different from the mechanical recycling we discussed earlier.

When a thermoplastic polymer is recycled, it is melted down and changes from solid to liquid. Melting, freezing, boiling, and condensation are all physical changes, changes of *phase*, which, as you learned in Chapter 7, do not change the identity of a material. When a plastic is melted, it is still composed of long polymer chains, and these chains are simply moving more fluidly in the liquid form at a higher temperature. When cooled, the polymer reverts to the solid phase and takes on any new shape it has been given. When a polymer is degraded chemically (as with heat or light energy) or biologically (as with microorganisms), however, the chains are broken, and smaller pieces result. If the process continues long enough, it is possible to convert the polymer back to its constituent monomers.

Now, let us consider how a synthetic polymer could be broken down chemically. Earlier in this chapter, we saw examples of chemical reactions used to make synthetic polymers. These reactions, like many other chemical reactions, can also work in the reverse direction, and we can take clues from the way polymers are made to see how they might be broken down. Figure 10.39 shows the polymers polypropylene and polylactic acid. The process of breaking down a polymer is similar to the synthesis of a polymer in that the reaction usually involves a site of positive electrical charge seeking out a site of negative charge, or vice versa. If you look closely at the structure of polypropylene, you will see there are no electronegative atoms. There are no sites of excess positive charge that could be targets for attack by another molecule acting as a nucleophile. Thus, in general, long chains of carbon atoms are usually more inert to chemical degradation than are polymer chains that contain electronegative atoms. This is true of polypropylene.

Polylactic acid is made up of ester linkages and is made via condensation polymerization. Recall from our earlier discussion that the carbon atom of the carbonyl group, $C{=}O$, carries a partial positive charge and is therefore susceptible to attack by nucleophiles. Ester groups, because the carbonyl carbon is attached to an oxygen atom, are especially susceptible to attack by nucleophiles. For this reason, polylactic acid is a highly biodegradable polymer.

In nature, the nucleophile attacking a polymer backbone is often the ubiquitous water molecule, which, as you know, has a dipole of its own. With its partial negative charge, the oxygen atom of a water molecule is able to approach the partial positive charge of a carbonyl group, as shown in Figure 10.40. The result is a break in the main chain of the polymer. Because the agent causing the polymer chain to split was a water molecule, we refer to this reaction as **hydrolysis**, literally meaning "splitting by water." In this case, we say that the polyester molecule is *undergoing ester hydrolysis*, which means that its ester linkages are being broken by a reaction with water.

▼ FIGURE 10.39 **Comparison of biodegradability in polypropylene and polylactic acid.** Can I recycle this? The polypropylene polymer is not easily broken down because it contains no electronegative atoms and no sites that might be susceptible to nucleophilic attack. The polylactic acid polymer contains carbonyl groups, which are susceptible to nucleophilic attack. As a result, this polymer is more easily broken down.

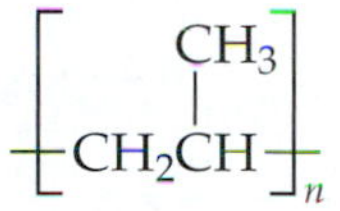

Polypropylene

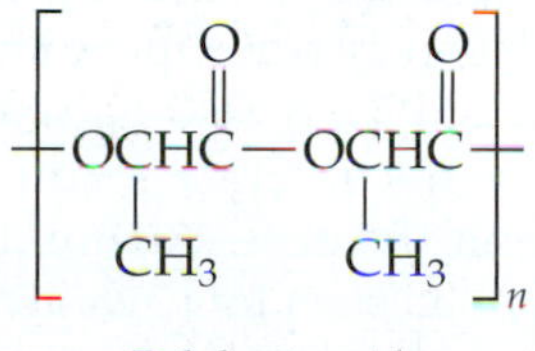

Polylactic acid

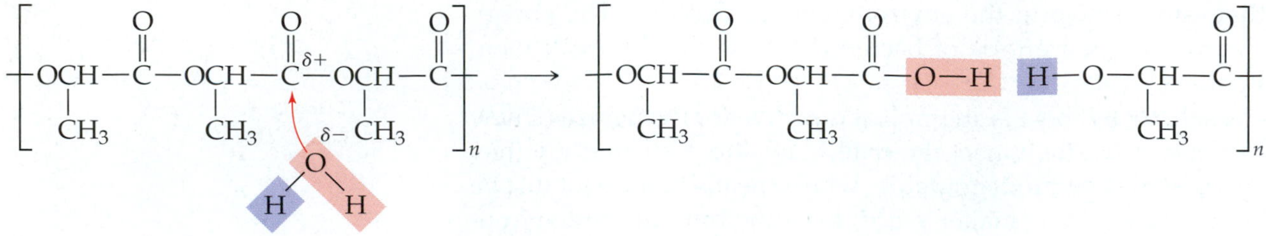

▲ FIGURE 10.40 **Hydrolysis of polylactic acid.** Water molecules act as nucleophiles that attack the carbonyl carbon atom of polylactic acid. This mechanism is referred to as hydrolysis because the polymer chain is split by the action of water.

QUESTION 10.16 Which of these ester hydrolysis reactions does not show the correct product molecules?

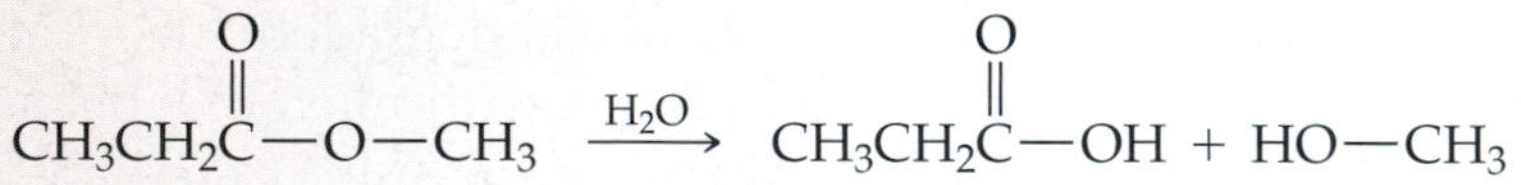

(a)

(b)

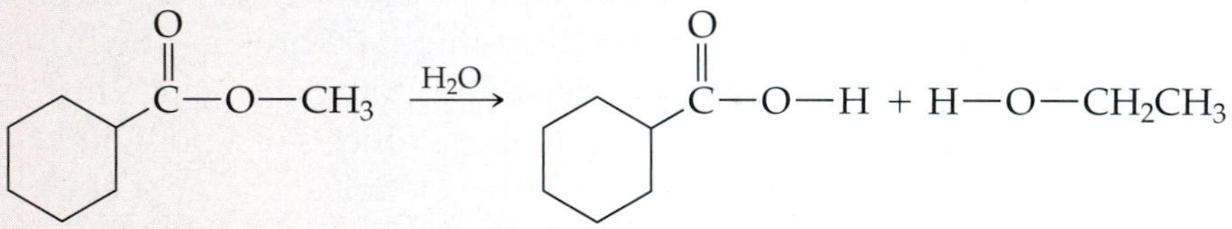

(c)

ANSWERS 10.16 Reaction (*c*), because it shows eight carbon atoms in the reactant but nine in the products.

STUDENTS OFTEN ASK

If Esters Are Hydrolyzed by Water, Why Can We Wash Polyester Clothes?

To answer this question, let us consider two polyesters: Dacron, a fabric made from polyethylene terephthalate (PET, shown in Table 10.3), and polylactic acid, a polyester that biodegrades easily (Figure 10.39). Although you cannot see this from the structures, PET polyester molecules have numerous crystalline regions, and these regions make it difficult for a nucleophile—a water molecule when we are talking about washing things—to penetrate the polymer and split it via ester hydrolysis. PET is also naturally hydrophobic and therefore does not interact readily with water. Thus, if you wore your favorite Austin Powers costume (made of Dacron polyester) to a sweaty disco party and now want to wash it, there is no need to worry about its dissolving in your washing machine.

Polylactic acid, on the other hand, has an amorphous structure and is more easily accessed by water molecules. Thus, this polymer is more biodegradable. In a landfill, polylactic acid breaks down in only about 12 months, on average.

University of Arizona Garbage Project

For many of us, garbage is something we reluctantly drag to the curb or to the dumpster on a regular basis. Most of us do not worry about what happens to our garbage once it leaves the house, but that is not true for some scientists at the University of Arizona (UA). The Garbage Project, begun at UA in 1973, was a data collection effort that spanned two decades, and some of its findings may surprise you.

When we think of landfills, many of us conjure up an image of discarded baby diapers and plastic wrappers, imagining that anything made from organic materials decomposes relatively quickly. The data show, however, that plastic items make up only about 16 percent of average landfill volume. What the UA researchers *did* find in great quantities was paper, accounting for about 40 percent of the volume. They found buried newspapers dating back dozens of years with the type still clearly legible. Apparently, paper does not biodegrade as quickly as we thought. In fact, the researchers were surprised to find that the landfills that had been degrading undisturbed for fifty years still contained undegraded organic materials, including paper. They found that the only things that did degrade reliably were food scraps and grass clippings.

The Garbage Project also revealed that the mass of plastics at landfills has not changed much over the 20 years of the study, which is surprising given the increased use of plastics during the same time period. The researchers say that this phenomenon is the result of *lightweighting*, the process of reducing the mass of plastic used to make a container to hold a given volume of something. For instance, a typical soda bottle had a mass of 67 g in 1974, but only 48 g today. Typical 1-gal milk jugs have gone from 120 g in 1974 to 65 g today, and the reason your grocery bags fall apart is that their thickness has gone from 33 μm in 1976 to 18 μm today (probably why you might hear a request for "double bags, please" at the checkout counter).

Workers on the UA Garbage Project concluded that our landfills are not doing their job on schedule, and they predict that landfills are going to start filling up. When this happens, we will have no choice but to wait out the decomposition. Recycling is common sense because one day there may not be a place to take the garbage you put out at the curb.

The twentieth century saw the development of every conceivable new kind of plastic. We went from wearing wrinkly cotton and linen to unwrinkled synthetic fabrics that offer stretchiness and breathability. Plastics became stronger, lighter, more flexible, more rigid, more colorful, and more durable. As millions of tons of plastics have been produced, though, we have been forced to think about their disposal. What can we do with all of this plastic, especially the super-strong types that cannot be remolded to make something else? Clearly, we must design polymers that will break down in a landfill in weeks or months, rather than years or decades. This is one goal of today's polymer chemists, but another interesting new avenue is the rediscovery of natural fabrics and fibers, which are also made of polymer chains. Such natural polymers are the focus of our next discussion of chains in Chapter 11.

SUMMARY

We started this chapter with the chemistry of gum, a hydrophobic polymer that does not easily degrade in the mouth (or the stomach). Polymers can be hydrophobic, like gum, or hydrophilic, like the material used to make contact lenses. There are different ways monomers can be linked to make a polymer; two such mechanisms— addition polymerization and condensation polymerization—are discussed in this chapter. Addition polymerization works by adding a monomer to the end of a growing polymer chain. This reaction occurs when the end of the chain carries either a full positive or negative charge. We refer to the negative group in this type of reaction as a nucleophile and the positive group as an electrophile. Condensation polymerization occurs when a monomer has functional groups that can react with other monomers. These monomers make a covalent bond and split off some small molecule.

In an ester, the carbon of the carbonyl group is attached to a hydrocarbon chain on one side and to an oxygen atom and hydrocarbon chain on the other side. In a ketone, both groups attached to the carbon of the carbonyl group are hydrocarbon chains. In an amide, one group attached to the carbonyl carbon is a hydrocarbon and the other is a nitrogen atom.

We looked at numerous synthetic polymers having a wide range of physical properties. Polymers can be designed to be hydrophilic or hydrophobic, biodegradable or not biodegradable, cross-linked or not cross-linked, flexible or rigid, crystalline or amorphous. In each case, the chemical structure of the polymer dictates the physical properties of the resulting material. Some polymers can be spun into fibers to make fabrics. Others are fashioned into plastic wrap, food containers, bridge cables, or any number of other shapes. For this reason, plastics are considered to be the most versatile materials known.

Polymers can be divided into two broad categories depending on their ability to be melted down. Thermoplastic polymers are easily melted and changed into new shapes. Therefore, products made of these polymers are easily recycled. Thermosetting polymers cannot easily be melted. These polymers are extremely tough and resilient and are used for high-temperature applications.

Despite the successes of recycling across the United States, millions of tons of thermoplastic polymers still end up in landfills, in which they take decades or even centuries to decompose. Research is focused on the development of polymers that degrade under ambient conditions, by the action of either sunlight or bacteria. The ester functional group is a site of potential bond cleavage through the process of ester hydrolysis. In this reaction, the carbonyl portion of an ester group is attacked by water to split the ester bond, breaking the polymer chain. This is just one of the ways that a biodegradable polymer can be developed.

KEY TERMS

addition polymer	condensation	electrophile	nucleophile	thermoplastic
amide	polymer	ester	plastic	polymer
amorphous region	copolymer	hydrolysis	polymer	thermosetting
biodegradable	cross-link	ketone	steric hindrance	polymer
carbonyl group	crystallite	monomer		vulcanization

QUESTIONS

The Painless Questions

1. Is the following statement true or false?

   Esters, ketones, and amides are all organic compounds that contain a carbonyl group. They differ in the identity of the two groups attached to the carbonyl group.

2. Describe three advantages to producing rubber in a factory rather than cultivating rubber from tree plantations.

3. This radio is made from Bakelite. Describe the fate of this radio based on what you know about the structure of thermosetting polymers.

4. Natural rubber and synthetic rubber are often used to make a base for chewing gum. Should you worry about the gum dissolving in your mouth? What feature of rubber's structure makes it an appropriate material for chewing gum?

5. Why might a fabric made of polylactic acid be difficult to wash? How does this polymer differ from the polyesters used to make Dacron?

6. Name three organic functional groups that contain a carbonyl. Write the formula for each.

7. Liquid latex straight from the tree is allowed to sit undisturbed for a period of time until it reaches a creamy consistency. What happens during this time to change the consistency of the latex?

8. How do natural rubber and gutta-percha differ from each other structurally? How do these differences affect their physical properties?

9. Recycling has the potential to reduce dramatically the amount of crude oil our country uses each year. (a) How is recycling linked to the use of crude oil? (b) What happens to recyclable plastics that are thrown away as litter? (c) If recycling is becoming more and more successful, why are efforts being made to make biodegradable plastics?

10. Identify each organic functional group in these molecules:

    (a)   (b)   (c)

    (d)   (e)

11. Which step in the recycling process have these milk jugs already been through? Which steps have they not yet undergone?

12. The milk jugs from Question 11 have a resin ID code of 2. (*a*) What type of polymer was used to make the jugs? (*b*) Based on the structure of this polymer, would you expect milk jugs to be hydrophilic or hydrophobic? Why?

13. Thermoplastic polymers and thermosetting polymers are structurally very different from each other. (*a*) Describe this difference. (*b*) Which is more easily recycled and why? (*c*) Name one example of a thermosetting polymer. (*d*) Does this polymer have a resin ID code?

14. One of the names that follow does not match its line structure. Which is it? What is the correct name?

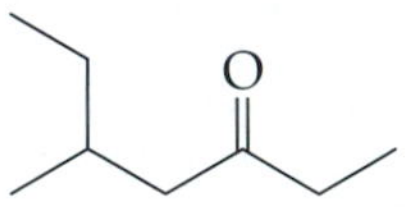

2-Ethyl-4-hexanone
(a)

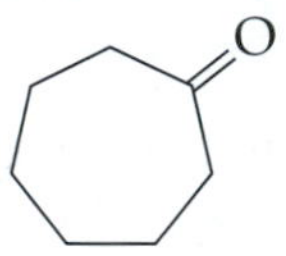

Cycloheptanone
(b)

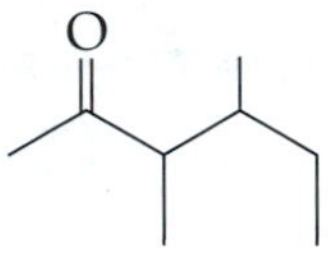

3,4-Dimethy-2-hexanone
(c)

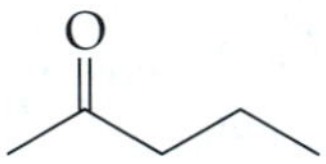

2-Pentanone
(d)

15. Two fibers have been collected from a crime scene. The investigator is planning to check each fiber's solubility in different solvents, determine melting ranges, and examine the fibers under a microscope. In what order should the investigator do these three experiments? Why does it matter which she does first?

More Challenging Questions

16. Comment on this statement:

The subscript n in the structural formula of a polymer indicates the number of monomers in the polymer chain for artificial polymers. In nature, the value of n is always a fixed number for a given polymer.

17. Draw the full structure and the line structure for (*a*) polyethylene and (*b*) nylon 6. Which chain has the greater flexibility?

18. Classify each polymer as polar or nonpolar:

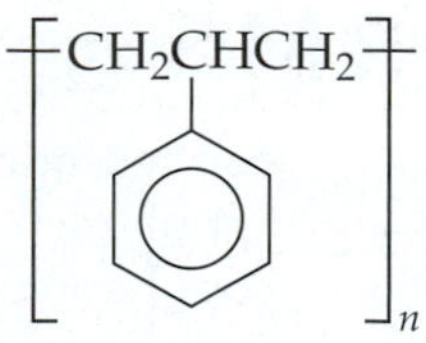

(a) $+CH_2CHCH_2+_n$

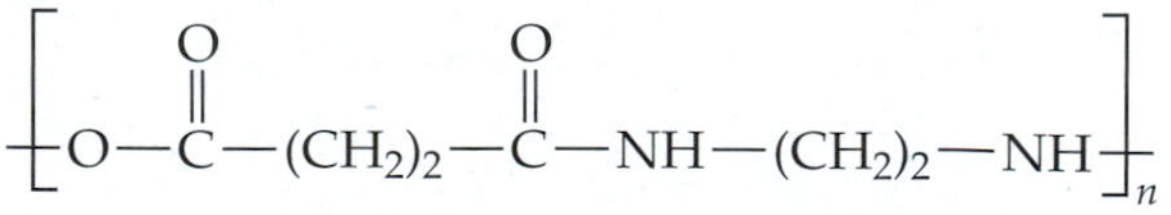

(b) $+O-C(=O)-(CH_2)_2-C(=O)-NH-(CH_2)_2-NH+_n$

(c)

19. Name these esters:

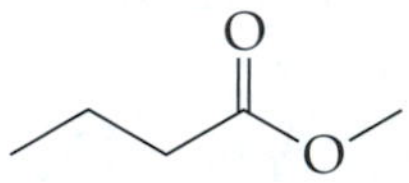

(a)

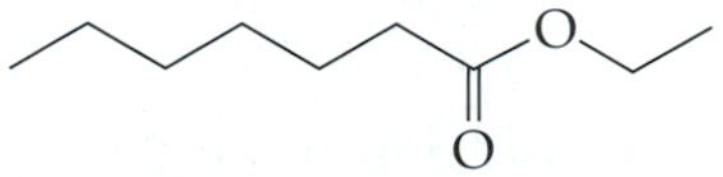

(b)

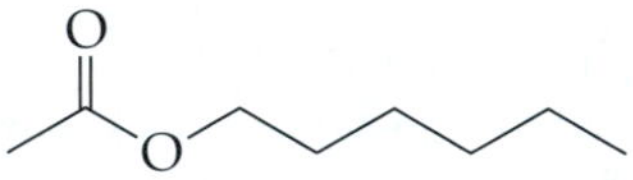

(c)

20. We saw several examples of polymers that have very similar chemical structures, but very different physical properties. Describe two such pairs of these polymers in detail.

21. What myths about the contents of landfills were dispelled by the Garbage Project? Did this project paint an optimistic or pessimistic view of our efforts to manage our garbage?

22. (*a*) Draw a polymer chain that has eight repetitions of this structure:

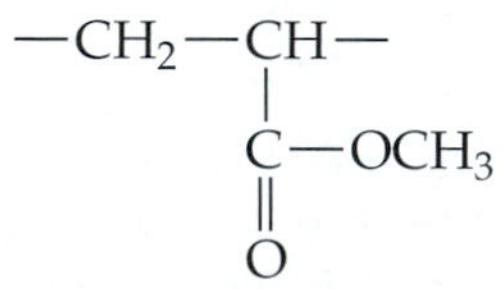

$$-CH_2-CH-$$
$$|$$
$$C-OCH_3$$
$$||$$
$$O$$

(*b*) What is the name of this polymer?

23. Describe the differences between a polymer that can be recycled and a polymer that is biodegradable.

24. What is an electrophile? To what are electrophiles attracted? What charge or partial charge do electrophiles carry?

25. What is a nucleophile? To what are nucleophiles attracted? What charge or partial charge do nucleophiles carry?

26. A polymerization reaction is taking place in a test tube in which a solution of a monomer has been placed. After 1 h, there are few monomers left and many short chains containing between 10 and 15 monomers. Which polymerization mechanism discussed in this chapter is at work in this test tube?

27. A polymerization reaction is taking place in a test tube in which a solution of a monomer has been placed. After 1 h, the solution is 30 percent monomers and 70 percent long polymeric chains. Which polymerization mechanism discussed in this chapter is at work in this test tube?

28. Identify each organic functional group in these molecules:

(a)　　　　　(b)

(c)　　　　　(d)

29. (a) Describe how vulcanization can add strength and stretchiness to a polymer. (b) The formation of thermosetting polymers also strengthens polymers. What does this process have in common with vulcanization?

30. Describe the polymers used to make these two kinds of glue. How do the structures of these two polymers differ from each other?

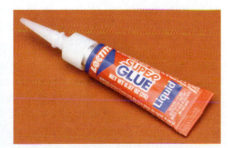

The Toughest Questions

31. Draw a line structure and a full structure for (a) 6-methyl-3-decanone, (b) 4-ethyl-2-octanone, (c) cyclooctanone.

32. Draw a line structure and a full structure for (a) octyl butanoate, (b) butyl formate, (c) heptyl nonoate.

33. When super glue leaves the tube, what event immediately takes place? In terms of chemical structure, what is the difference between the liquid in the tube and the glue after it leaves the tube?

34. (a) Draw the line structure of Kevlar, showing the smallest unit of the polymer repeated three times. (b) What is the molar mass of the smallest unit that makes up Kevlar? (c) Give two reasons Kevlar is so much stronger than most other polymers.

35. (a) What is the literal meaning of the word *hydrolysis*? (b) Which of these molecules will undergo ester hydrolysis in the presence of water?

Molecule A　　Molecule B　　Molecule C

(c) For each molecule you chose for part (b), draw the line structure for the products of the ester hydrolysis reaction.

36. Point out the error in each drawing:

(a)　　　　　(b)

37. (a) Identify each functional group in this fictional organic molecule:

(b) What is the molecular formula of this molecule? (c) What is the mass of 1 mol of this molecule?

38. (a) What are the resin ID code and three-letter abbreviation for the polymer polyethylene terephthlate? (b) What two monomers are used to make this copolymer? (c) What is the molar mass of the smallest repeating unit in this polymer?

39. (a) Predict which area of Molecule A will interact with which area of Molecule B:

Molecule A   Molecule B

(b) Are two molecules of A likely to react with each other? (c) Are two molecules of B likely to react with each other?

40. Based on your understanding of condensation polymerization, suggest the chemical structure of each of the two monomers used to make the copolymer

41. If an 897-g Bakelite telephone is composed of one molecule, what is the molar mass of that molecule?

42. Using your knowledge of condensation polymerization, predict the product of the chemical reaction

$$CH_3CH_2CH_2\overset{O}{\overset{\|}{C}}-OCH_3 + HO-CH_2CH_2CH_2CH_3 \longrightarrow ?$$

43. (a) Identify the functional groups in the two reactants in Question 42. (b) What small molecule is produced during this condensation reaction?

E-Questions

Go to **www.prenhall.com/waldron** to find these questions in electronic form, complete with hyperlinks directly to the various websites cited in the questions.

44. The website **PlasticResource** provides details about the Garbage Project. Under the search box, type in the words *garbage* and *myths* to find "Five Major Myths About Garbage and Why They're Wrong." Then answer the following questions: (a) When this project was conceived, what were the investigators attempting to do? (b) What is expanded polystyrene foam used for, and what myth is associated with it? (c) What are some of the reasons plastics are not as problematic in landfills as most people think? What are some of the real problems associated with landfills? (d) How many pounds of garbage does the average American throw away each year? Is this number much higher now than at any other time in

our history? (e) Why should we be skeptical of the word *recycled*? How does the law define this word? How can we know when a plastic product is truly recycled?

45. Read about biodegradable medical devices in the article "Synthetic Biodegradable Polymers as Medical Devices" at **Devicelink**. (a) How long does it take for a typical polymer to degrade in the human body? What lengths of time are required for the biodegradation of the new polymers discussed in this article? (b) List five types of biodegradable medical device that can be made from plastic. (c) Look at the chemical reactions used to produce these polymers. Are they made via addition polymerization or condensation polymerization? (d) Several copolymers are discussed in the article. Choose one, and draw the chemical structures of its two monomers. Identify all the functional groups contained in the monomers and in the final polymer.

46. "Shape-shifters" are polymers that can remember their original shape and return to it. Read about how this new technology is used to make sutures at **Azom** and answer the following questions: (a) What problems arise when traditional sutures are used? (b) What are SMAs? (c) What are the new "smart polymers" and why are they ideal for use as a suturing material? (d) How might shape-memory polymers be used in the automotive industry?

47. Did you ever wonder why you can wash some fabrics while others must be dry cleaned? Did you ever wonder why some fabrics can be ironed with high heat, and others require a cool iron? Read more about the physical properties of some fabric polymers and why they must be handled according to specific manufacturer's instructions at **Fabrics** and **DryCleaning**. After reading these two articles, answer the following questions: (a) What is involved in the dry-cleaning process, and why are some fabrics ruined by it? (b) When you do laundry in a washing machine, what types of molecules are being removed? How does washing clothes with water differ from dry cleaning? (c) What types of fabrics cannot be ironed at high temperature? Why?

48. Read a detailed account of the history of plastics from before the turn of the twentieth century up to 1960 at **AmericanPlastics**. After reading this five-page article, you will be able to answer questions such as these: (a) What was the first plastic wrap, and who invented it? (b) What are Saran™ and Teflon® and why do they have the properties they have? (c) What year was the "birth of Velcro®?" What material is used to make Velcro?

WORLD WIDE WEB RESOURCES

As with the E-Questions, go to **www.prenhall.com/ waldron** to find these questions in electronic form, complete with hyperlinks directly to the various websites cited.

Some of following links contain research articles related to the topics in this chapter and could be used as the basis for a writing assignment in your course. Other sites are of general interest:

- Believe it or not, your everyday plastic trash bag may be helping NASA build better spacecraft. At **NASA**, read about how polyethylene is being modified for use in plastic spaceships that may one day be able to carry humans to Mars.

- The article "World's fastest camera shows how things break apart" at **Break** talks about how one scientist has been able to watch as a material is hit with a projectile. Super-strong plastics, like Kevlar, are filmed to see how they can be strengthened to further resist impact.

- Read more about the history of fabrics at **Inventors**. Each fabric has a description that includes the type of polymer the fabric contains and why that fabric is unique.

- The following website describes the evolution of Astroturf® into the material it is today: **Blastroturf**. Write a short essay describing the type of polymer used in Astroturf. What properties were sought for this material? How does the material used today differ from the first artificial grass?

CHEMISTRY TOPICS IN CHAPTER 11

- Natural polymers from plants and animals, genetic engineering [11.1]

- Amines and carboxylic acids [11.2]

- Amino acids and proteins [11.2]

- Primary, secondary, and tertiary structure in proteins [11.3]

- Nucleic acids [11.4]

- DNA versus RNA [11.4]

- Complementary base pairs [11.4]

- The genetic code and protein synthesis [11.4]

- Recombinant DNA [11.5]

"Nature is an endless combination and repetition of a very few laws. She hums the old well-known air through innumerable variations."

RALPH WALDO EMERSON

Chains

11

A SURVEY OF NATURAL POLYMERS, INCLUDING PROTEINS AND NUCLEIC ACIDS

Picture in your mind's eye a trip to your favorite grocery store. Take a look at the containers that hold the food and other products you decide to buy. What are all these containers made of? For the most part, your virtual stroll along the store's aisles will reveal that the majority of foods, cleaning products, and other household items are packaged in plastic, which is made of *polymer* molecules hundreds of thousands of *monomer* units long. In some cases, you may find that the packaging is made from recycled plastic that has been scavenged and reshaped from discarded plastic products. You may also notice that packaging made from paper products is comparatively rare.

Once you arrive at the checkout counter, you may be asked which type of bag you prefer—paper or plastic. While debating which to choose, you may consider the fact that paper uses up a natural resource, trees. Then again, a paper bag might be easier than a plastic bag for a landfill to digest—or maybe not, because some new kinds of

paper degrade slowly. The plastic bag might be made from recycled plastic, which is a good thing. Virgin or recycled, however, the plastic probably was manufactured originally from crude oil, and that means an increased reliance on another dwindling natural resource. Confounded by your knowledge of recycling and the scarcity of natural resources, you stand there perplexed. Paper or plastic?

▲ FIGURE 11.1 **Drink your milk.** At your local grocer or health food store, you can find milk packaged in glass, cardboard, and plastic.

Consider another grocery-store dilemma. A trip to the dairy section reveals that milk can be packaged in a variety of ways (Figure 11.1). You may see plastic bottles made from polyethylene (HDPE, resin ID code 2), waxed cardboard containers, or even glass bottles. On the one hand, you may decide that the plastic bottle is the most environmentally responsible choice because it is recyclable. On the other hand, even if it is recycled dozens of times, it is still derived from crude oil. So you think about the glass bottle. Glass comes from sand, which seems to be an almost unlimited resource, but consider this: glass bottles are many times heavier and thicker than plastic and cardboard ones. That extra weight and volume mean that trucks delivering the milk have to work harder and use more fuel to bring it to market. Plus, glass bottles do not grow on glass bushes. They must be manufactured, and that process uses energy—lots of it. So you opt for the cardboard container. Then, though, you remember that cardboard comes from trees and that not only does the production of cardboard consume trees, but energy is required to cut the trees down and fashion them into paper stock.

Whew! Your innocent trip to the grocery store has become a stress-filled battle of conscience. You want to be an upstanding, environmentally concerned consumer, but all your options seem to compromise mother nature somehow.

The twentieth century saw an explosion of research on synthetic plastic products, but today synthetic plastic is slowly losing its reputation as a supermaterial. Landfills are filling up with unusable plastic waste, and efforts are now under way to do something about the problem. A cohort of researchers in diverse fields in the scientific community is working on ways around these resource and landfill problems. In the twenty-first century, the garbage we add to landfills will be more prone to breakdown by bacteria and sunlight—in other words, more *biodegradable*. One way to accomplish this goal of biodegradability is to take a fresh look at natural polymers, those made by living organisms. After all, if a living organism can make a polymer, then some other living organism, such as a microorganism in a landfill, can probably break it down. Natural polymers may take only weeks or months to degrade at the landfill, whereas some synthetic polymers (such as many of those we discussed in Chapter 10) can stick around for decades.

In this chapter, we consider polymer chains designed by mother nature. We will look at how these polymers are made and what they are made of, and we will see that much of what we know about synthetic polymers has come from studying the design of natural biomolecular chains. Let us begin by considering how new technology can be used to study natural polymers.

11.1 | Tapping Mother Nature

Consider the milk-container dilemma again. The root of the problem is not that we must use natural resources to make the containers, but rather that we must use natural resources *that are in finite supply*. For example, our supply of trees to make cardboard milk containers is limited by the time it takes to grow one tree—about 30 years. Likewise, when we consume crude oil to make plastic containers and wax coatings for cardboard containers, we are whittling away at our crude oil supply. Most estimates predict that U.S. oil reserves will run dry before the year 2100. Thus, long-term solutions to the problem of energy use should be based, as much as possible, on alternative resources.

If our aim is to avoid depleting natural resources, why would we want to return to natural polymers as a replacement for synthetic ones? Won't natural polymers require the use of limited natural resources? The answer is no. Rather than using natural resources that are in limited supply the way, say, trees are, these new polymers can be made from natural resources that regenerate frequently and regularly. In this section, we consider four approaches to using natural, renewable resources to fill the world's demand for plastics.

The *first approach* uses natural molecules that are already polymerized. Walter Schmidt, a research scientist interested in environmental problems and the use of chemistry to help solve them, wanted to figure out a use for the 10,000 pounds of chicken feathers produced at a typical chicken-processing plant every hour. He focused on the barb portion of the feathers, which contains a more useful protein-based material than the quill (Figure 11.2). After figuring out how to separate quill from barb, Schmidt shredded the barbs and found ways to process them into everything from plastic films to car dashboards. Because these materials are derived from living organisms, they are completely biodegradable.

Chickens are not the only organisms that produce what was once considered useless natural material. Ground-up clamshells have been developed into a fibrous solution that can be used as "scaffolding" for growing new skin on wounds. Although neither chickens nor clams are grown for the purpose of producing plastics, these examples are proof that materials which usually end up in the garbage can be fashioned into useful products.

In the *second approach*, a small molecule produced by an organism in large quantities is harvested and polymerized to make a biodegradable plastic. For example, the small organic molecule lactic acid, shown in Figure 11.3a, is derived from corn, one of the most abundant and inexpensive crops to grow. With a bit of manipulation, the lactic acid molecule can be used as a monomer to create the polymer *polylactic acid* (PLA, Figure 11.3b). We think of PLA as a natural polymer because its monomer, lactic

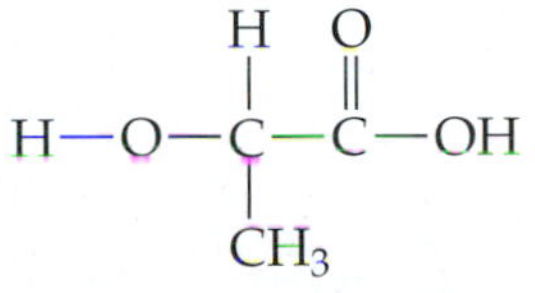

▲ FIGURE 11.2 **Naked chickens, new plastics.** Unused chicken feathers can be recycled to make plastic products.

◄ FIGURE 11.3 **A biodegradable polymer.** (a) The lactic acid monomer obtainable from corn. (b) Lactic acid monomers link together to form the polymer polylactic acid (PLA).

Lactic acid, a monomer
(a)

One repeating unit

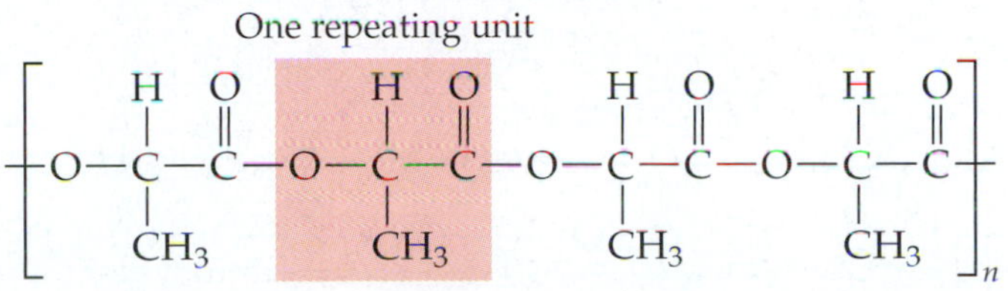

Polylactic acid, a polymer
(b)

acid, is derived from a natural source, corn. This polymer is biodegradable, taking an average of about two weeks to reach 80 percent degradation in a landfill. One commercially available PLA plastic has physical properties that resemble those of some popular plastics that we see all over the planet. In fact PLA-based plastic can be fashioned into everything from trash bags to wedding dresses. When used as fabric for clothing, PLA has the benefit of being wrinkle resistant, but its biodegradability means that the fabric is sensitive to heat and salt water. So buyer, beware! If you purchase a swimsuit made of PLA and wear it in the ocean, you will end up wearing your birthday suit instead.

QUESTION 11.1 (a) What is the molecular formula of the lactic acid monomer? (b) What is its molar mass? (c) Would you classify this monomer as polar or nonpolar? Explain your answer.

ANSWER 11.1 (a) $C_3H_6O_3$. (b) $3(12.011 \text{ g/mol}) + 6(1.01 \text{ g/mol}) + 3(16.00 \text{ g/mol}) = 90.09 \text{ g/mol}$. (c) Polar because of the three highly electronegative oxygen atoms.

The idea of harvesting monomers and polymers from food crops is ingenious, but our *third approach* makes use of a resource that can be cultivated even more quickly than corn: bacteria. A British chemical company has figured out a way to get one particular bacterium, *Alcaligenes eutrophus*, to mass-produce a novel natural plastic. Many bacteria, including *A. eutrophus*, store their energy not in the form of fatty tissue, as humans do, but in the form of hydroxyalkanoate monomers that can be polymerized to *polyhydroxyalkanoates* (PHAs, Figure 11.4a, b). PHAs are natural polyesters, with the identity of the R group dictated by the food used to feed the bacteria. By regulating what the bacteria eat, it is possible to design a variety of useful bacteria-derived polymers.

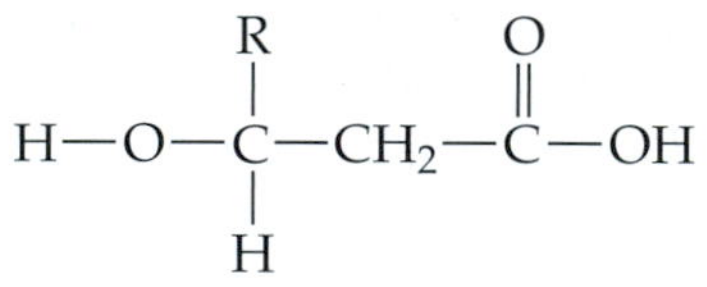

▶ FIGURE 11.4 **Another biodegradable polymer.** (a) Bacteria are a natural source of the monomer hydroxyalkanoate. The R represents a hydrocarbon chain that can be up to 13 carbons long. The identity of R is determined by the diet fed to the bacteria. (b) Polyhydroxyalkanoate polymer, the polymer created from the monomer shown in part (a). (c) A PHA sequence in which the R groups are a one-carbon chain and a two-carbon chain. These monomers contain a total of four and five carbons, respectively.

Hydroxyalkanoate, a monomer
(a)

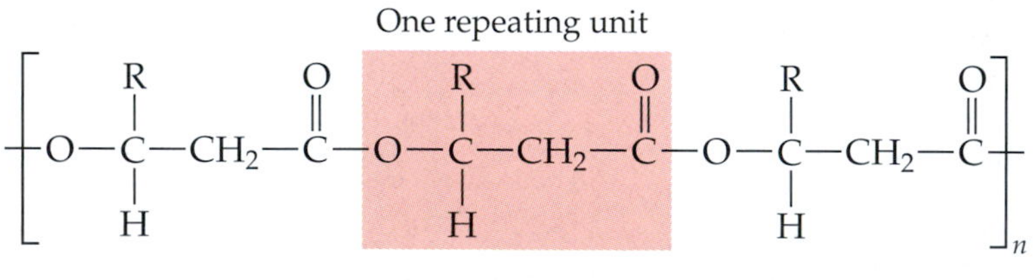

One repeating unit

Polyhydroxyalkanoate, a polymer
(b)

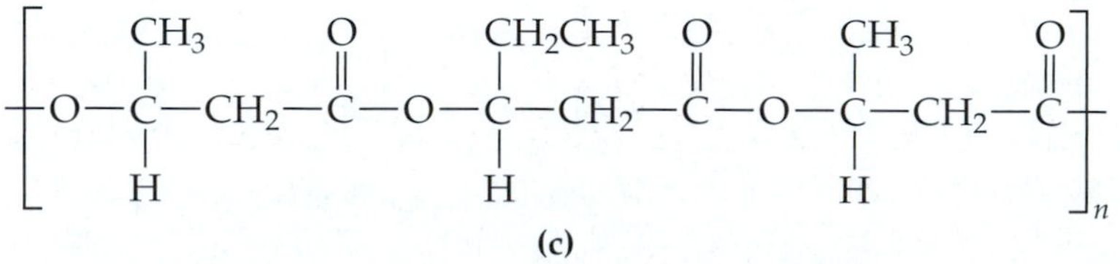

(c)

After serving bacteria a smorgasbord of monomeric fare, researchers at the British company came up with a polymer made of two different repeating units, one containing four carbons and the other containing five carbons (Figure 11.4c). Next, the chemists figured out a way to trick the bacteria into making more of the polymer than they ordinarily would. This trick is based on the fact that the bacteria make lots of polymer when they sense that they are going to be deprived of food (just as a bear getting ready to hibernate stores up fat for a long winter). The bacteria are first given unlimited amounts of food and then switched to a nutrient-deficient diet. Immediately, they begin to manufacture oodles of the desired polymer, which is then harvested and made into things like shampoo bottles. Unlike PLA (the polymer derived from corn), PHA polymers degrade only in the presence of specific bacteria. One study showed that products made from PHAs can be 80 percent degraded by landfill bacteria in only 19 weeks.

QUESTION 11.2 (a) How does the structure of the hydroxyalkanoate monomer differ from the structure of lactic acid? (b) Draw a three-monomer length of a PHA polymer in which R is a propyl group.

ANSWER 11.2 (a) The hydroxyalkanoate monomer has a $-CH_2-$ group not found in lactic acid, and the methyl group ($-CH_3$) of lactic acid is replaced by an R group in hydroxyalkanoate.

(b)

When bacteria are used to produce polymers, the economy of the process is diminished by the fact that a source of carbon must be fed to the bacteria. However, what if there were a way to make the polymer without having to provide a carbon source? Glad you asked, because the *fourth* (and most technologically sophisticated) *approach* provides just such a way: use plants instead of bacteria. Plants get all the carbon they need from the carbon dioxide in the air, meaning that there is no need to feed them. If *plants* make the polymer, they can obtain carbon from the air in the form of carbon dioxide. In this process, known as **fixation**, an organism takes some or all of the atoms in a gaseous molecule and *fixes* them by incorporating them into a larger, nongaseous molecule. Thus, because plants use carbon fixation to get their carbon from carbon dioxide gas, it is not necessary to provide a carbon source. (Of course, they do need sunlight and water.)

Researchers working in this area of polymer research decided to show corn plants how to polymerize lactic acid monomers to polylactic acid polymers. The polymerization takes place only in the leaves of the plants, which means that the corn kernels are unchanged and can still be used as food. How do they do this, though? How do you entice an organism into producing a molecule that it does not normally produce? How do you make a crop of corn into a biorefinery? This is when the techniques of molecular biology come into the picture. The advent of **genetic engineering**, technology that allows us to manipulate the hereditary material contained in the cells of living things, now gives us more options for working with organisms. We can take a portion of hereditary material from one organism

It is called *genetic* engineering because *genes* are being manipulated. As we shall learn in Section 11.4, a *gene* carries both the instructions for how offspring inherit traits from parents and instructions for creating biomolecules in an organism.

and put it into another, a technique that expands the palette of molecules the modified organism can make. Working with *A. eutrophus*, scientists were able to remove the genetic material that tells the bacterium how to make plastic. Then the scientists modified this material to work in plants and incorporated it into the cells of corn plants. The modified corn plants are referred to as **transgenic**, because they contain hereditary material from another organism.

If all this is possible, why are there no vast fields of transgenic corn growing plastic across the country? The answer to this question has to do with the practical difficulties associated with harvesting the plastic. Getting the plastic out of the plant has proved to be an energy-consuming and tedious process. At the time of this writing, the process uses more energy than it saves, and it is still cheaper to make plastic the old-fashioned way, using petroleum feedstocks. There is no doubt, though, that transgenic plants will one day become a staple of modern agriculture.

All of these routes to making plastics without using petroleum products are summarized in Figure 11.5. We have seen that monomers, like those grown naturally in corn, can be mass-produced in crops and polymerized to make biodegradable plastics. We can also look to bacteria as a source of natural polymers, which they produce as part of their life cycle. These are all examples of **sustainability**, a word that comes from *sustainable*, defined as "relating to a method of harvesting or using a resource so that the

▶ **FIGURE 11.5 Four approaches to "natural" plastics.**   This flowchart shows increasingly sophisticated methods for using organisms to produce environmentally friendly plastics that do not require petroleum products as a starting material.

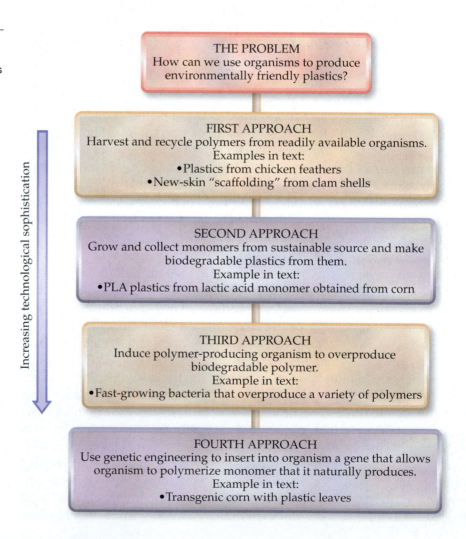

Increasing technological sophistication

THE PROBLEM
How can we use organisms to produce environmentally friendly plastics?

FIRST APPROACH
Harvest and recycle polymers from readily available organisms.
Examples in text:
• Plastics from chicken feathers
• New-skin "scaffolding" from clam shells

SECOND APPROACH
Grow and collect monomers from sustainable source and make biodegradable plastics from them.
Example in text:
• PLA plastics from lactic acid monomer obtained from corn

THIRD APPROACH
Induce polymer-producing organism to overproduce biodegradable polymer.
Example in text:
• Fast-growing bacteria that overproduce a variety of polymers

FOURTH APPROACH
Use genetic engineering to insert into organism a gene that allows organism to polymerize monomer that it naturally produces.
Example in text:
• Transgenic corn with plastic leaves

resource is not depleted or permanently damaged." By choosing methods that do not deplete a limited natural resource, such as crude oil, we are setting up a long-term solution to the problem of our ever-increasing demand for plastics. An added benefit is that plastics which come from living organisms are likely to be biodegradable. In the future, perhaps our landfills will be able to do what they were designed to do—break down our garbage—and perhaps even allow us to reuse it.

The examples we have just looked at all involve natural polymers. In this chapter, we shall explore two important natural polymers—DNA and protein—and see how the former provides an "instruction manual" for making the latter. In order to understand how this process takes place, we must, as we often do, first look at the structures of these molecules to see how they do what they do.

11.2 | Amino Acids: Nature's Building Blocks

The development in the 1930s of the synthetic polymer we know as *nylon* caused a revolution. As soon as it became available in the early 1940s, nylon could be found in everything from toothbrushes to fishing line and, of course, ladies' hosiery. (Sixty-four million pairs of nylon stockings were sold the first year they were on the market.) Then came World War II, and all the nylon the United States could produce was used for parachutes, tents, and other military applications. In the first months after the war ended, women waited in line for hours to get their hands on a single pair of stockings (Figure 11.6), and angry mobs of nylon-seeking women forced stores to close to avoid riots. Over the next several decades, nylon stockings evolved into the modern pantyhose we know today. This evolution of hosiery is attributed in part to Julie Newmar, who played Catwoman in the Batman television series and who holds a patent on supersheer pantyhose (Figure 11.7). There was even a moment in history—a very brief moment—when men were offered the convenience of this new form of hosiery. In a memorable advertisement in 1974, football star Joe Namath created quite a sensation when he posed in a pair of what were humorously called "mannyhose."

◄ **FIGURE 11.6 Women lining up to purchase nylon stockings in the years following World War II.** Lines formed to buy nylon stockings on April 8, 1946, in Washington, DC. The caption on this photograph read, "If you can't get stockings, nylon, or rayon, don't blame the Office of Price Administration. According to the OPA Textile Division, stockings are now turned out at a rate of 54,000,000 pairs per month—enough to give every female American over the age of 14 one pair per month. OPA blames the 'shortage' on the 'Hose Hogs' who grab more than their share in one line after another."

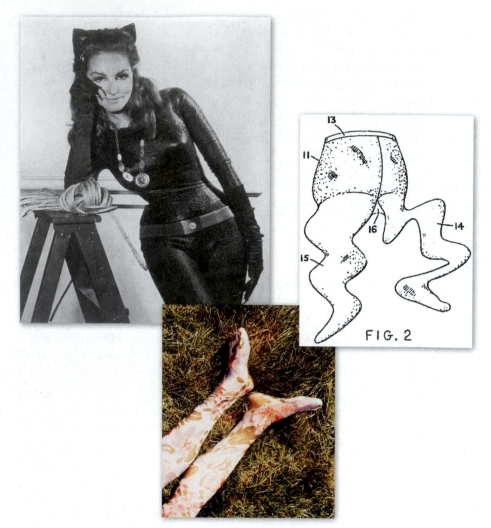

▶ **FIGURE 11.7 Julie Newmar, a developer of pantyhose.** Julie Newmar, one of the innovators in the development of pantyhose, as Catwoman, along with a detail from one of her patent applications. Julie Newmar U.S. Patent 4,003,094 Jan. 18, 1977 Fig. 2 drawing/ http:///www.uspto.gov/patft/index.html). Early pantyhose offerings ranged from the simple to the psychedelic.

Recall that an alkyl group can be any hydrocarbon group such as ethyl, pentyl, decyl, and others.

$$R-\overset{\overset{\displaystyle O}{\|}}{C}-O-H$$

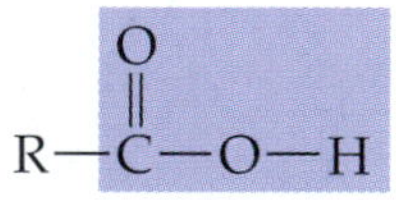

A generic carboxylic acid

(a)

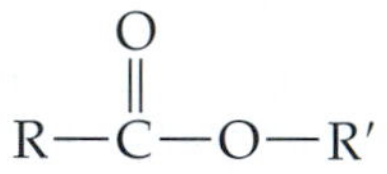

$$R-\overset{\overset{\displaystyle O}{\|}}{C}-O-R'$$

A generic ester

(b)

$$R-O-H$$

A generic alcohol

(c)

$$R-O-R'$$

A generic ether

(d)

Before we talk about how nylon polymers are formed, we need to learn about two more organic functional groups. The first, the *amine functional group*, $-NH_2$, will be covered in detail in Chapter 14. The other is the *carboxylic acid functional group*, highlighted in Figure 11.8a. This functional group is made up of a carbonyl group, $-C=O$, and a hydroxyl group, $-OH$. Organic molecules containing the $-COOH$ group are known as **carboxylic acids**. The R in Figure 11.8a stands, as usual, for any hydrocarbon group. Notice an important property of carboxylic acids: the $-COOH$ functional group is always at the end of the molecule. This must be so because if you replace the H of $-COOH$ with some alkyl group, represented by R' in Figure 11.8b, you no longer have a carboxylic acid; you have an ester instead. (Alcohols work the same way. As you learned in Chapter 9, if you replace the H of an alcohol's $-OH$ group with some alkyl group, you get an ether, as illustrated in parts c and d of Figure 11.8.)

◀ **FIGURE 11.8 Reminder of some organic functional groups.** (**a** and **b**) Replacing the H in a carboxylic acid functional group with a hydrocarbon group R' yields an ester. (**c** and **d**) Replacing the H in an alcohol functional group with a hydrocarbon group R' yields an ether.

QUESTION 11.3 Identify each of these molecules as a carboxylic acid, an ester, an alcohol, or an ether:

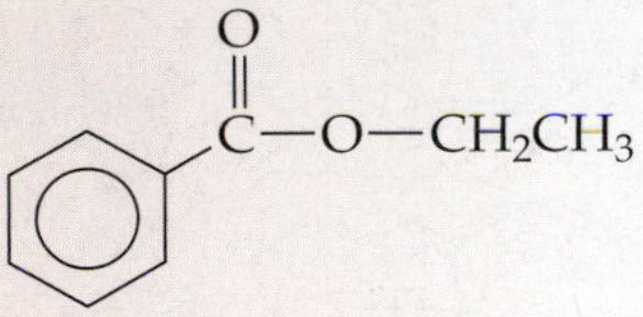

(a)

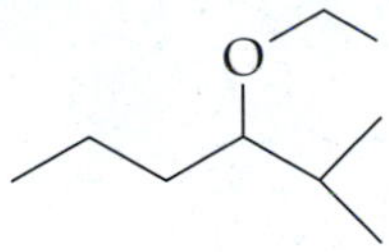

(b)

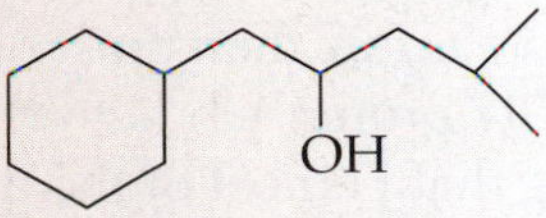

(c)

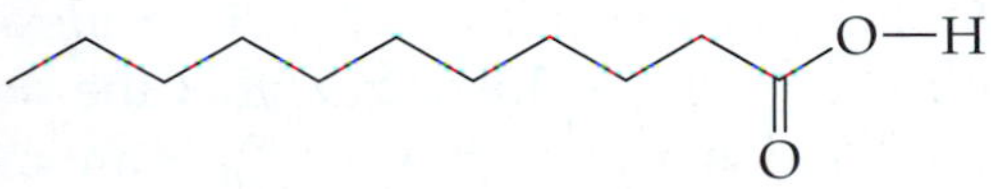

(c)

ANSWER 11.3 (*a*) Ester, (*b*) ether, (*c*) alcohol, (*d*) carboxylic acid.

To name a simple carboxylic acid, you replace the terminal "e" in the name of the hydrocarbon base with "-oic acid." Figure 11.9 provides two examples. The molecule shown in Figure 11.9a has seven carbon atoms in its chain. (Note that you must count the carbon in the —COOH group when determining chain length.) If this were a simple hydrocarbon with seven carbon atoms, it would be called *heptane*. Because it has a —COOH group at one end, we change the name to *heptanoic acid*. When numbering the carbons, you must begin at the —COOH end, and you must find the longest chain, as the molecule shown in Figure 11.9b illustrates. The longest chain in this molecule has ten carbon atoms, and there is an ethyl group attached to carbon 7. We therefore name this carboxylic acid *7-ethyldecanoic acid*.

QUESTION 11.4 Name these carboxylic acids:

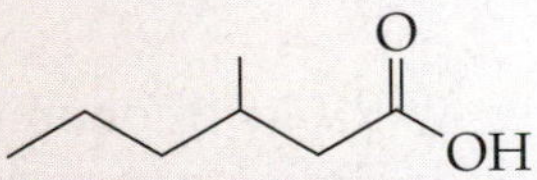

(a)

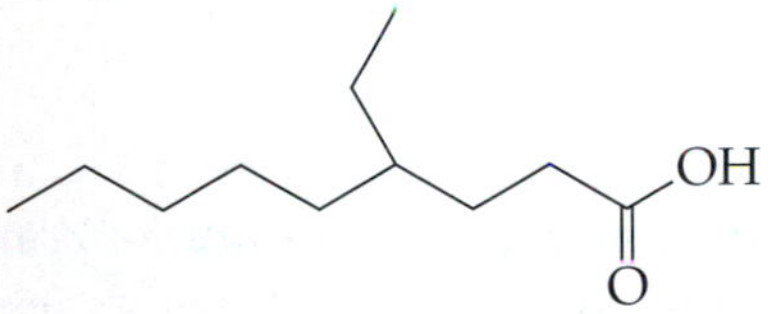

(b)

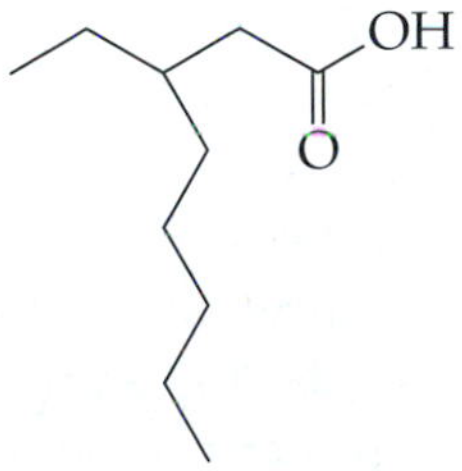

(c)

ANSWER 11.4 (*a*) 3-Methylhexanoic acid, (*b*) 4-ethylnonanoic acid, (*c*) 3-ethyloctanoic acid.

MOLECULE
Propanoic Acid

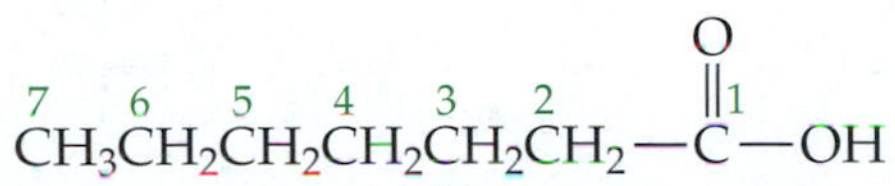

$$\underset{\text{(a)}}{\overset{7\quad6\quad5\quad4\quad3\quad2\quad\overset{\displaystyle O}{\overset{\|}{1}}}{CH_3CH_2CH_2CH_2CH_2CH_2-C-OH}}$$

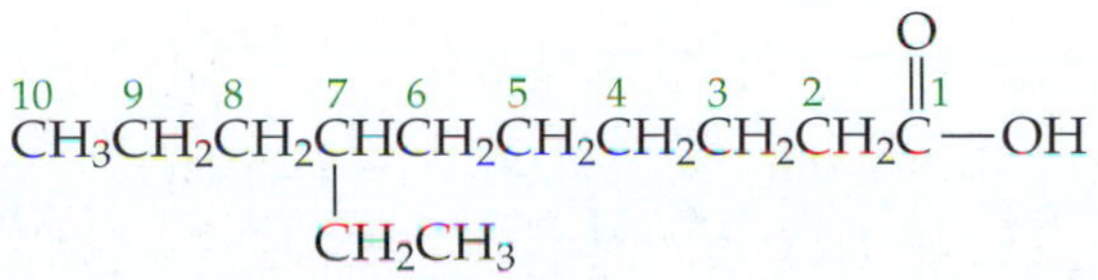

$$\underset{\underset{\underset{\text{(b)}}{CH_2CH_3}}{|}}{\overset{10\quad9\quad8\quad7\quad6\quad5\quad4\quad3\quad2\quad\overset{\displaystyle O}{\overset{\|}{1}}}{CH_3CH_2CH_2CHCH_2CH_2CH_2CH_2CH_2C-OH}}$$

◀ **FIGURE 11.9 Naming carboxylic acids.** Carboxylic acids are named by replacing the "e" at the end of the hydrocarbon root name with the ending *–oic acid*. **(a)** For this seven-carbon carboxylic acid, *hexane* becomes *hexanoic acid*. **(b)** The —COOH carbon is always carbon 1, and the longest chain provides the root name. Because the longest chain here is ten carbons long, the name of the molecule is 7-ethyldecanoic acid.

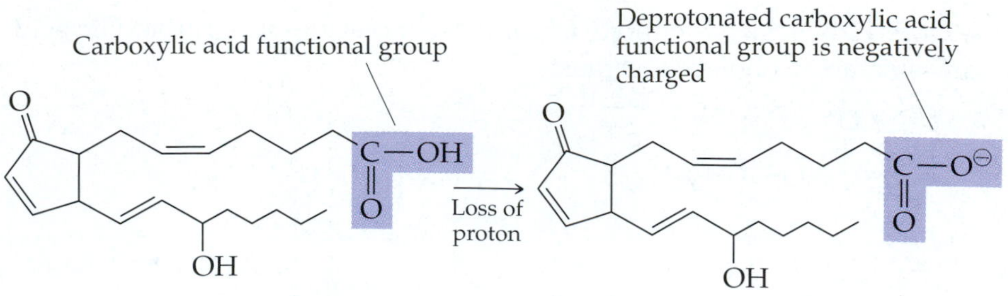

► FIGURE 11.10 **Proton loss in the carboxylic acid functional group.** The hormone prostaglandin A₂ illustrates that the carboxylic acid functional group can lose its proton under certain conditions.

► FIGURE 11.10 **Proton loss in the carboxylic acid functional group.** The hormone prostaglandin A₂ illustrates that the carboxylic acid functional group can lose its proton under certain conditions.

As their name implies, carboxylic acids are more acidic than most other organic compounds. This means that the —COOH proton tends to come off easily, and, in fact, carboxylic acids are missing that H most of the time. Figure 11.10 shows the loss of a proton by the carboxylic acid known as prostaglandin A₂, a hormone that may play a defensive role in certain species of coral. As you can see, a negative charge is left behind when a carboxylic acid loses its proton. Because it usually carries this full negative charge, the carboxylic acid portion of a molecule imparts a polar character to the molecule. Add to that property the ability of this anion to form hydrogen bonds with water (Figure 11.11), and you find that this polar region confers some water solubility to a molecule that is otherwise nonpolar due to its nonpolar hydrocarbon chain. Thus, the prostaglandin molecule has both polar and nonpolar character.

Now we are ready to talk about creating nylon polymers. All forms of nylon are *polyamides*, because they are made from amide monomers. Recall from Section 10.4 that an amide is a molecule containing an amide functional group, which consists of a nitrogen bonded to a carbonyl group:

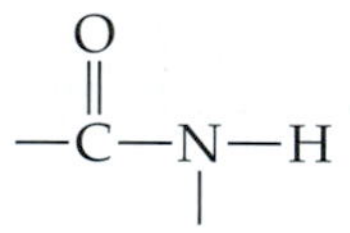

Figure 11.12 shows the polymer nylon 6, along with its monomer *6-aminohexanoic acid*. Note that this monomer contains an amine functional group, —NH₂, at one end and a carboxylic acid functional group at the other end. We call molecules that contain these two functional groups **amino acids**. As you can see from Figure 11.13, the nylon polymer forms when the amine group of one 6-aminohexanoic acid molecule bonds with the carboxylic acid group of a second 6-aminohexanoic acid molecule, splitting off a water molecule in the process.

► FIGURE 11.11 **Water-loving molecules.** Carboxylic acids, shown here in their anionic form, are able to form hydrogen bonds with water molecules.

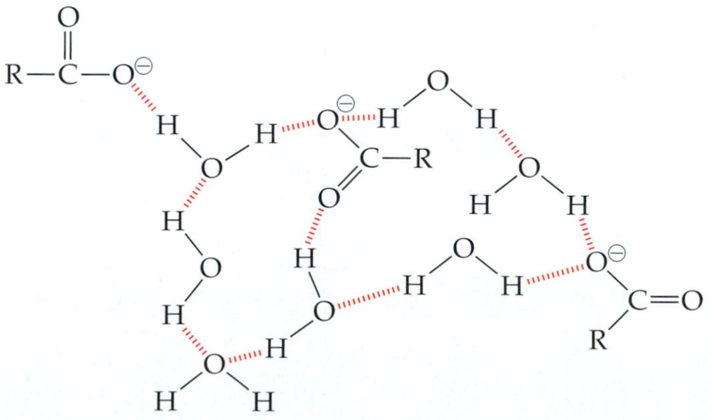

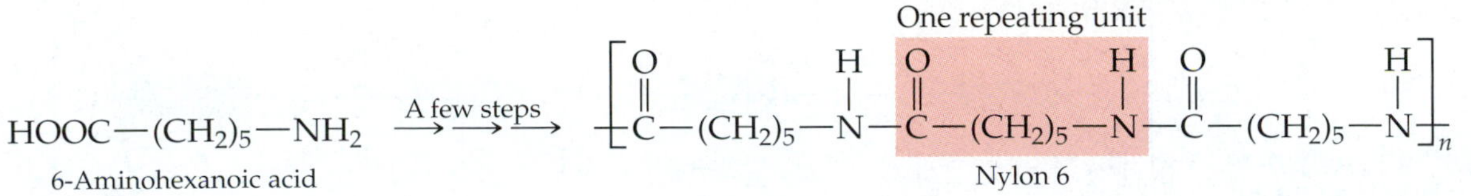

One repeating unit

$$\text{HOOC}-(\text{CH}_2)_5-\text{NH}_2 \xrightarrow[\text{A few steps}]{} \left[\underset{\text{C}}{\overset{\text{O}}{\parallel}}-(\text{CH}_2)_5-\underset{\text{H}}{\overset{|}{\text{N}}}-\underset{\text{C}}{\overset{\text{O}}{\parallel}}-(\text{CH}_2)_5-\underset{\text{H}}{\overset{|}{\text{N}}}-\underset{\text{C}}{\overset{\text{O}}{\parallel}}-(\text{CH}_2)_5-\underset{\text{H}}{\overset{|}{\text{N}}}\right]_n$$

6-Aminohexanoic acid   Nylon 6

▲ **FIGURE 11.12 Creating a nylon polymer.** The synthesis of nylon 6 starts with the monomer 6-aminohexanoic acid. The —NH₂ group of one monomer forms a bond with the —COOH group of a second monomer, and on and on.

The amine and carboxylic acid functional groups come together as a result of attraction between opposite charges. The amine group, with its nonbonding pair of electrons, possesses a partial negative charge, as indicated in Figure 11.13. In the carboxylic acid group, the electronegative oxygen of the carbonyl group pulls electron density away from the carbonyl carbon atom. The result is a partial positive charge on the electron-deficient carbon atom, which makes it susceptible to attack by the amine.

RECURRING THEME IN CHEMISTRY Many chemical reactions, especially organic ones, are driven by the attraction between positive and negative parts of molecules.

QUESTION 11.5 The name *nylon 6* comes from the fact that each monomer contains six carbon atoms. (*a*) Draw a monomer for nylon 4. (*b*) Draw the product of the reaction between two nylon 4 monomers.

ANSWER 11.5

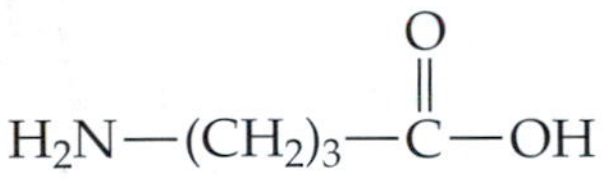

$$\text{H}_2\text{N}-(\text{CH}_2)_3-\overset{\text{O}}{\overset{\parallel}{\text{C}}}-\text{OH}$$

(a)

$$\text{H}_2\text{N}-(\text{CH}_2)_3-\overset{\text{O}}{\overset{\parallel}{\text{C}}}-\underset{\underset{\text{H}}{|}}{\text{N}}-(\text{CH}_2)_3-\overset{\text{O}}{\overset{\parallel}{\text{C}}}-\text{OH}$$

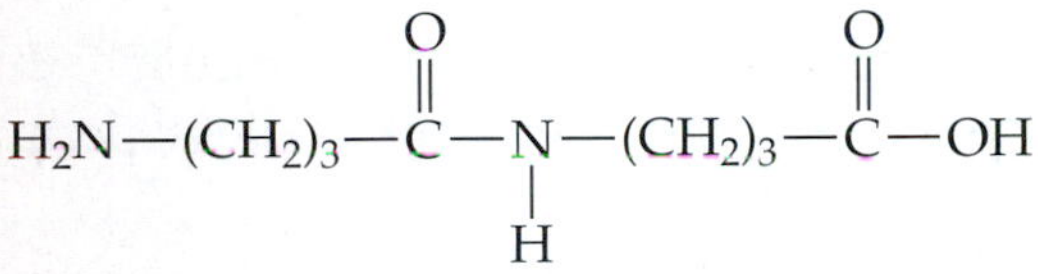

(b)

Although the synthesis of nylon was truly revolutionary, the idea of a polymer made up of amino acids strung together is nothing new. In fact, mother nature has been making such polyamides all along, and she can do it better and more creatively than humans will ever do in either a test tube or a factory. We refer to natural polyamides as **proteins**—polymer chains composed of amino acids linked together.

▼ **FIGURE 11.13 Opposites attract.** The partial positive charge on the carbon of a —COOH group is attracted to the partial negative charge on the nitrogen of an —NH₂ group. A molecule of water is split off as the two groups come together to form the beginning of a polyamide.

LIVE ART
Polyamide Synthesis

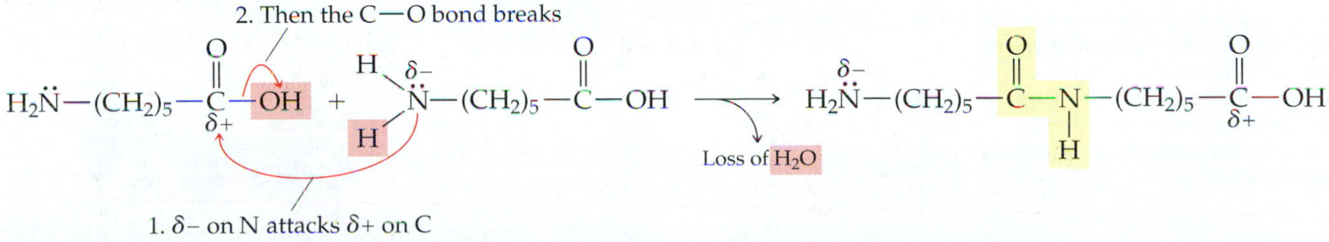

2. Then the C—O bond breaks

$$\text{H}_2\overset{..}{\text{N}}-(\text{CH}_2)_5-\underset{\underset{\delta+}{}}{\overset{\text{O}}{\overset{\parallel}{\text{C}}}}-\text{OH} + \underset{\underset{\text{H}}{|}}{\overset{\text{H}}{\overset{\delta-}{\text{N}}}}-(\text{CH}_2)_5-\overset{\text{O}}{\overset{\parallel}{\text{C}}}-\text{OH} \longrightarrow \underset{\text{Loss of H}_2\text{O}}{\longrightarrow} \text{H}_2\overset{\delta-}{\text{N}}-(\text{CH}_2)_5-\overset{\text{O}}{\overset{\parallel}{\text{C}}}-\underset{\underset{\text{H}}{|}}{\text{N}}-(\text{CH}_2)_5-\underset{\underset{\delta+}{}}{\overset{\text{O}}{\overset{\parallel}{\text{C}}}}-\text{OH}$$

1. δ− on N attacks δ+ on C

$$\left[-\overset{H}{\underset{\underset{}{|}}{N}} - (CH_2)_5 - \overset{\overset{O}{\|}}{C} - \overset{H}{\underset{\underset{}{|}}{N}} - (CH_2)_5 - \overset{\overset{O}{\|}}{C} - \overset{H}{\underset{\underset{}{|}}{N}} - (CH_2)_5 - \overset{\overset{O}{\|}}{C} - \overset{H}{\underset{\underset{}{|}}{N}} - (CH_2)_5 - \overset{\overset{O}{\|}}{C} - \right]_n$$

Nylon 6

Amino acid 1 · Amino acid 2 · Amino acid 3 · Amino acid 4

$$H_2N - \overset{R_1}{\underset{}{C}H} - \overset{\overset{O}{\|}}{C} - \overset{H}{\underset{}{N}} - \overset{R_2}{\underset{}{C}H} - \overset{\overset{O}{\|}}{C} - \overset{H}{\underset{}{N}} - \overset{R_3}{\underset{}{C}H} - \overset{\overset{O}{\|}}{C} - \overset{H}{\underset{}{N}} - \overset{R_4}{\underset{}{C}H} - \overset{\overset{O}{\|}}{C} - OH$$

A generic protein

▶ **FIGURE 11.14 Synthetic polymers mimic natural ones.** The structures of the synthetic polymer nylon 6 and the natural polymer we call a protein. Each yellow area is an amide functional group, and each green area is a hydrocarbon unit located between two amide functional groups. In the protein, each blue R area represents the side chain on one of the amino acid monomers making up the protein polymer.

Compare the structure of nylon 6 with that of a generic protein (Figure 11.14), and you will see that they have two things in common. First, both polymers contain amide functional groups (yellow). Second, a hydrocarbon unit (green) separates any two adjacent amide groups. In the nylon 6 polymer, the separating unit is a hydrocarbon chain made up of five $-CH_2-$ groups. In the protein polymer, the separating unit is a single $-CH-$ group (green) with an R group attached (blue).

The protein polymer in Figure 11.14 is made up of four amino acid monomer units. Note that the R group is different in each monomer (R_1, R_2, R_3, R_4). Each R group is called a **side chain** of its amino acid. There are 20 side chains in the amino acids found in nature. This means that there are 20 different amino acid monomers available for making protein molecules. This multiplicity is extremely important, and it is the reason proteins are extraordinarily diverse. In fact, they are arguably the most diverse class of biological molecules, and there are endless permutations of protein sequences, as we are about to see.

The 20 amino acids found in proteins are shown in Table 11.1. For each one, the side chain is highlighted in red. The part shown in black includes the amine functional group and the carboxylic acid functional group, both illustrated in their electrically neutral form here. When these groups come together to make a protein chain, the black portions come together to make the polyamide *protein backbone*, and the red portions hang off the backbone like charms on a bracelet (Figure 11.15). The specific reactions by which amino acids join together to form a protein are exceedingly complex. However, the end result is the same as the end result for the synthesis of nylon: the amine group of one monomer bonds with the carboxylic acid group of another monomer, and a molecule of water is split off in the process (Figure 11.16). In proteins, this amide bond between an amine group and a carboxylic acid group is given the specific name **peptide bond**.

Depending on pH, amino acid functional groups can carry electrical charge: COO^- and NH_3^+. In this text, we shall ignore these charges and concern ourselves only with the charges, if any, on the side chains.

▶ **FIGURE 11.15 Amino acid jewelry.** The side chains of the amino acids making up a protein are attached to the main chain—in other words, to the protein backbone—the way charms are attached on a charm bracelet.

TABLE 11.1 The 20 Amino Acids Used to Make Proteins

Name	Three-letter abbreviation	Essential?	Structure, with side chain shown in red
Glycine	Gly	No	$\begin{array}{c} \text{H} \\ | \\ \text{CH}-\text{COOH} \\ | \\ \text{NH}_2 \end{array}$
Alanine	Ala	No	$\text{CH}_3-\underset{\underset{\text{NH}_2}{|}}{\text{CH}}-\text{COOH}$
Phenylalanine	Phe	Yes	$\bigcirc-\text{CH}_2-\underset{\underset{\text{NH}_2}{|}}{\text{CH}}-\text{COOH}$
Valine	Val	Yes	$\text{CH}_3-\underset{\underset{\text{CH}_3}{|}}{\text{CH}}-\underset{\underset{\text{NH}_2}{|}}{\text{CH}}-\text{COOH}$
Leucine	Leu	Yes	$\text{CH}_3\text{CHCH}_2-\underset{\underset{\text{NH}_2}{|}}{\text{CH}}-\text{COOH}$ with CH_3
Isoleucine	Ile	Yes	$\text{CH}_3\text{CH}_2\text{CH}-\underset{\underset{\text{NH}_2}{|}}{\text{CH}}-\text{COOH}$ with CH_3
Proline	Pro	No	$\begin{array}{c} \text{CH}_2-\text{CH}_2 \\ \text{CH}_2 \qquad \text{HC}-\text{COOH} \\ \text{N} \\ \text{H} \end{array}$
Methionine	Met	Yes	$\text{CH}_3-\text{S}-\text{CH}_2\text{CH}_2-\underset{\underset{\text{NH}_2}{|}}{\text{CH}}-\text{COOH}$
Serine	Ser	No	$\text{HO}-\text{CH}_2-\underset{\underset{\text{NH}_2}{|}}{\text{CH}}-\text{COOH}$
Threonine	Thr	Yes	$\text{CH}_3\text{CH}-\underset{\underset{\text{NH}_2}{|}}{\text{CH}}-\text{COOH}$ with OH
Asparagine	Asn	No	$\text{H}_2\text{N}-\underset{\underset{}{\overset{\overset{\text{O}}{\|}}{\text{C}}}}{}-\text{CH}_2-\underset{\underset{\text{NH}_2}{|}}{\text{CH}}-\text{COOH}$
Glutamine	Gln	No	$\text{H}_2\text{N}-\underset{\underset{}{\overset{\overset{\text{O}}{\|}}{\text{C}}}}{}-\text{CH}_2\text{CH}_2-\underset{\underset{\text{NH}_2}{|}}{\text{CH}}-\text{COOH}$

(Continued)

TABLE 11.1 (*Continued*)

Name	Three-letter abbreviation	Essential?	Structure, with side chain shown in red
Cysteine	Cys	No	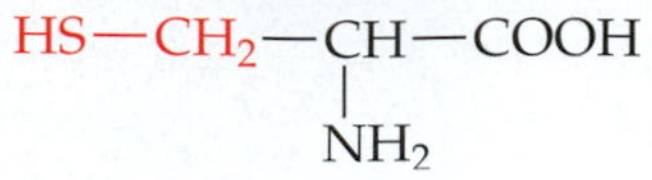 $HS-CH_2-CH-COOH$ with NH_2
Tyrosine	Tyr	No	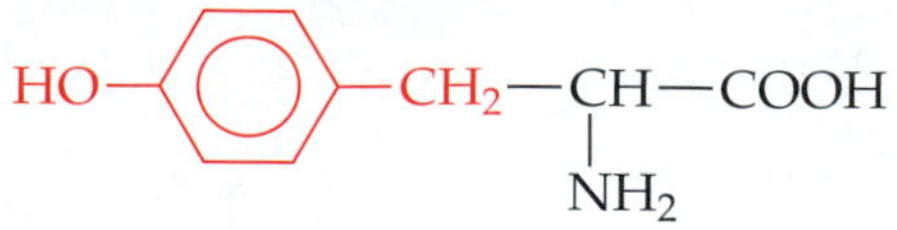 HO—(ring)—$CH_2-CH-COOH$ with NH_2
Tryptophan	Trp	Yes	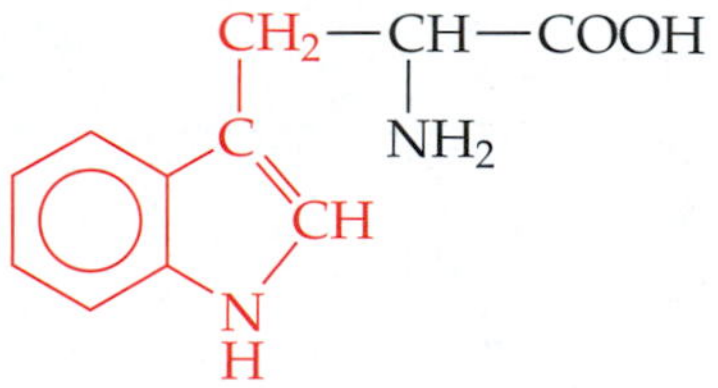 $CH_2-CH-COOH$ with NH_2 (indole ring)
Lysine	Lys	Yes	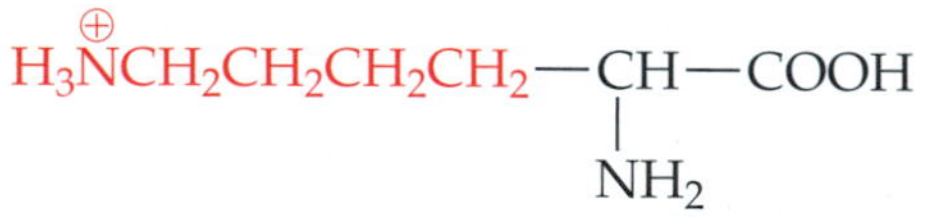 $\overset{\oplus}{H_3N}CH_2CH_2CH_2CH_2-CH-COOH$ with NH_2
Arginine	Arg	Yes	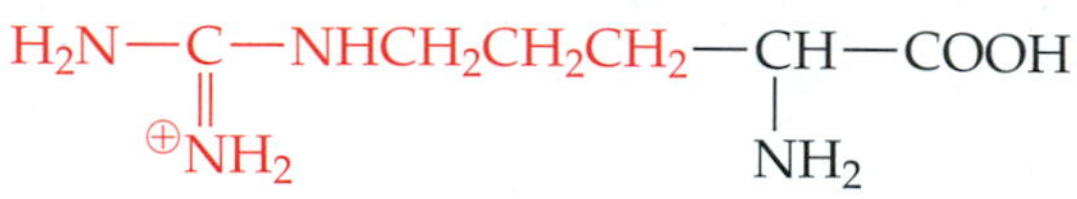 $H_2N-C-NHCH_2CH_2CH_2-CH-COOH$ with $\overset{\oplus}{NH_2}$ and NH_2
Histidine	His	Yes	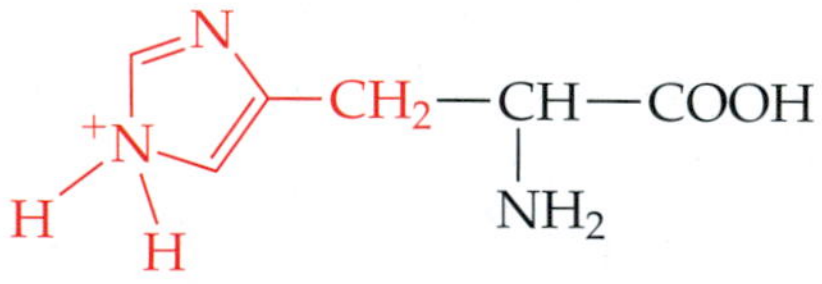 $CH_2-CH-COOH$ with NH_2 (imidazole ring)
Aspartic acid	Asp	No	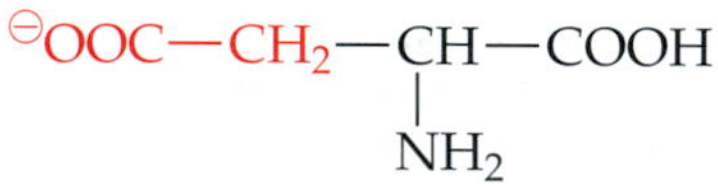 $\overset{\ominus}{O}OC-CH_2-CH-COOH$ with NH_2
Glutamic acid	Glu	No	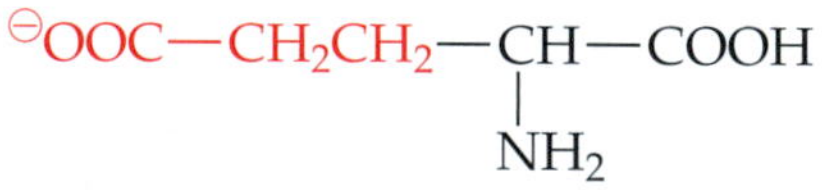 $\overset{\ominus}{O}OC-CH_2CH_2-CH-COOH$ with NH_2

MOLECULE
Cysteine

▼ **FIGURE 11.16 Monomer to polymer.** Two amino acids come together to form a peptide bond as the nitrogen on one amino acid attacks the carbon of the carboxylic acid on the other. As the process repeats over and over, the polymer protein is built.

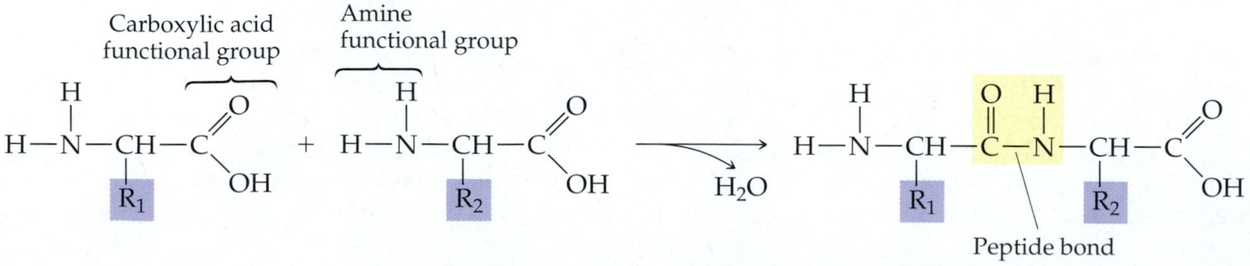

QUESTION 11.6 Use Table 11.1 to answer these questions. (*a*) Which amino acids have a side chain that is a hydrocarbon—that is, composed only of carbon and hydrogen atoms? (*b*) Which amino acids contain sulfur? (*c*) Which contain a benzene ring? (*d*) Which has a methyl group as its side chain?

ANSWERS 11.6 (*a*) Alanine, phenylalanine, valine, leucine, isoleucine, proline; (*b*) cysteine, methionine; (*c*) tryptophan, phenylalanine, tyrosine; (*d*) alanine.

QUESTION 11.7 Use Table 11.1 to answer these questions. (*a*) Which amino acid has an R group that is just a hydrogen atom? (*b*) Which amino acids contain a hydroxyl group? (*c*) Which have an amide in their side chain? (*d*) Which have a carboxylic acid in their side chain?

ANSWERS 11.7 (*a*) Glycine; (*b*) serine, tyrosine, threonine; (*c*) glutamine, asparagine; (*d*) aspartic acid, glutamic acid.

An example of a short protein (often called a **peptide** or **polypeptide**) is shown in Figure 11.17. In this case, we could call the molecule a *tripeptide* because it contains three amino acids. Note two things about this structure. First, we normally write the amine end of the peptide, called the *N-terminus*, on the left and the carboxylic acid end, called the *C-terminus*, on the right. Second, remember that even though we write the molecular structure in two dimensions, the peptide is not flat. In fact, peptide chains are reasonably flexible, and they often orient themselves in ways that minimize unfavorable crowding between side chains.

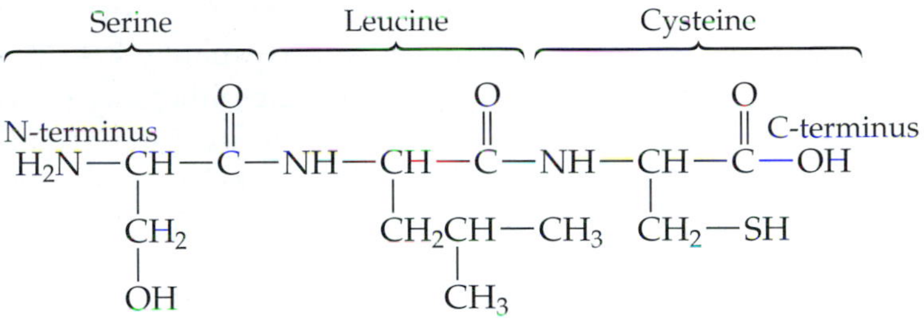

▲ **FIGURE 11.17 A tripeptide.** Peptide and protein chemical structures are usually written with the N-terminus on the left and the C-terminus on the right. This tripeptide contains the amino acids serine, leucine, and cysteine.

What Is the Difference Between a Peptide and a Protein?

A peptide is officially defined as a short protein chain, but exactly how long is short? The answer is not clear cut, but 100 amino acids is often used as a cutoff point between protein and peptide. For example, *insulin*, a small protein containing 51 amino acids, would be best described as a peptide. However, because peptides are a subset of proteins, it is not incorrect to call insulin a protein. Some people define proteins as polyamide chains that are long enough to fold up into a three-dimensional shape. From this definition, insulin would be considered a protein because it *does* form a three-dimensional shape.

Very short peptides are often identified with a Greek prefix to indicate the number of amino acids, as we saw for the *tripeptide* in Figure 11.17. Thus, you might see *hexapeptide* or *dodecapeptide* or even *oligopeptide*, all names which tell you that the polyamide chains contains some small (possibly unknown, in the case of *oligo-*) number of amino acids.

TABLE 11.2 Amino Acids Classified According to Side-Chain Polarity		
Degree of side-chain polarity	Amino acids	Amino acid class
Carries full positive or negative charge	Arg (+), Asp (−), Glu (−), His (+), Lys (+)	Charged
Partial separation of positive and negative charge	Asn, Gln, Cys, Ser, Thr, Trp, Tyr	Polar
No separation of charge	Ala, Gly, Pro, Ile, Leu, Met, Phe, Val	Nonpolar

There are almost as many ways to categorize amino acids as there are amino acids. For example, you can divide the 20 amino acids according to side-chain polarity, as in Table 11.2. Five of the amino acids carry full charges and are classified as *charged*. Seven are classified as *polar*, because they contain electronegative atoms and therefore carry partial charges (δ+ or δ−), and eight are *nonpolar*, because their side chains are mostly hydrocarbons.

The five charged amino acids are either positively or negatively charged at the pH of a typical cell. That pH is called *physiological pH* and is approximately 7.4. Most of the seven polar (but uncharged) amino acids contain either oxygen or nitrogen atoms in their side chains, both of which contribute to molecular polarity. The eight nonpolar amino acids are hydrophobic. In the next section, we will see how side-chain polarity dictates the three-dimensional shape of any protein molecule.

Issues

ISSUES IN CHEMISTRY

Transgenic Plants—Friend or Foe?

Transgenic plants make up more than 60 percent of the food crops grown in the United States today. For cotton and soybeans, Figure 11.18 shows even higher numbers: 80 percent for soybeans and just over 70 percent for cotton. Very few transgenic crops are devoted to unusual products, such as the corn with the plastic-producing leaves described in Section 11.1. Rather, almost all of these crops are genetically engineered to ward off pests or to make plants resistant to herbicides.

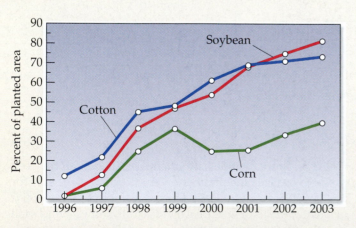

▲ FIGURE 11.18   **Rise of genetically modified crops.**   The rate at which transgenic cotton, soybean, and corn crops were planted in the United States has increased steadily since 1996.

For transgenic soybeans, hereditary material for herbicide resistance is incorporated into each cell. Soybean fields are often overcome by weeds, and the trangenic crops can be sprayed with an herbicide that kills all plants *except* those with resistance. One herbicide that plants can be designed to tolerate is Roundup™. The active ingredient in Roundup is *glyphosate*, a molecule similar to the amino acid glycine. This little molecule is a potent inhibitor in one of the steps that plants and bacteria use to make three amino acids essential to growth (Figure 11.19). (This inhibition of synthesis occurs only in plants and bacteria, not in animals.) Roundup applied to a soybean crop kills most weeds and leaves the herbicide-resistant transgenic soybeans untouched.

That the development of herbicide-tolerant and pest-resistant crops has markedly improved crop yields is undeniable. Furthermore, farmers growing pest-resistant transgenic crops are using a dramatically lower volume of conventional pesticides, which are usually toxic to animals

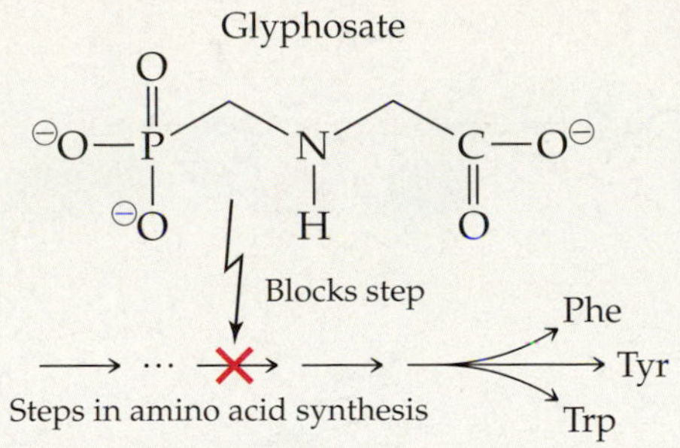

Glyphosate

▲ FIGURE 11.19 Killer weed killer. Glyphosate, the active ingredient in the herbicide Roundup, works by blocking one of the steps that lead to the synthesis of three amino acids in plants and bacteria.

and often linked to long-term health problems in humans. In fact, Roundup is touted as environmentally friendly. According to Monsanto, Roundup's manufacturer,

> Regulatory agencies around the world have concluded that glyphosate-based herbicides pose no unreasonable risks to human health and the environment when used according to label directions.

It's a rosy picture. Healthy crops mean more food for the hungry. Soon, new genetic engineering technology may bring us canola oil with a higher vitamin E content, bananas spiked with edible vaccines against tropical diseases, coffee without caffeine, and tobacco without nicotine. Who could be opposed to such advances?

The answer is that scores of people are opposed to the growth of transgenic crops. There are religious protesters who insist that modifying the hereditary material of organisms is morally wrong. Environmental groups fear that there are inadequate controls being used to monitor the design of transgenic plants. Health

advocates refer to transgenic foodstuffs as "frankenfood" and say such products are adulterated and unnatural. And there are those who worry about the development of new strains of pests that resist all our antibiotic weapons. What is worse, there have been reports concerning possible deleterious effects of Roundup on human health.

There were no protests when scientists used genetic engineering techniques to modify and improve the insulin given to diabetic patients, because this transgenic product provides an obvious medicinal benefit. For many people, though, the genetic modification of plants used to feed the general public is taking things too far. These people believe that transgenic plants will cause ecological disaster.

The battle over transgenic plants continues, and it is not difficult to find passionate pleas on both sides of the issue. According to writer and environmental activist Paul Goettlich, for instance,

> Unlike the *Exxon Valdez* oil spill, a spill [of transgenic material] is permanent. What that means is that once natural crops have been mixed with genetically engineered crops, it is permanent and cannot be undone.... This genetic pollution is already occurring.

So whom should you believe? Who knows what is best? The answer is that the voting public *should* know best. After reading this chapter, you will have a basic understanding of transgenic crops and how they are developed. You can locate reports from known experts in the field, and you can seek out opposing viewpoints and evaluate their relative merits. Then, when you go to cast a vote for or against the use of transgenic plants, you will be able to make an informed decision based on your own opinions. This is the best that any of us, as citizens, can do.

11.3 | Protein: Nature's Jack of All Trades

Because proteins are polymeric and have 20 building blocks to use as monomers, an enormous number of different proteins are found in nature. The variety of amino acids is not simply a way to produce random proteins, however. The placement and patterning of the amino acids in a given protein provide specific instructions that tell the protein what three-dimensional shape to adopt. Thus, mature proteins are not just long, floppy chains floating aimlessly in solution. Rather, regions of the protein backbone form special, folded structures as the amino acid side chains interact with one another. These interactions stabilize a protein molecule in the form of a folded, three-dimensional structure.

In Figure 11.20, the amino acid sequence of the small protein insulin is shown. We refer to the sequence of amino acids in a protein as the protein's **primary structure**. This is the backbone of the protein and contains all the

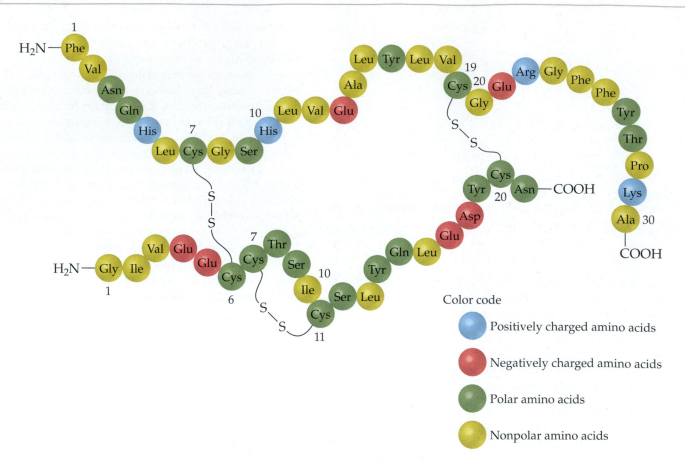

▲ **FIGURE 11.20** **The amino acid sequence of insulin.** The protein insulin is composed of two chains held together by disulfide bonds.

covalent bonds found in the molecule. Notice two things about the primary structure of insulin that are true of many other proteins. First, there is no regularly repeating pattern in the order in which the amino acids are linked in the primary structure; instead, the various amino acids are connected in a seemingly random order (*seemingly* only because in a moment we shall see how this order determines the three-dimensional shape of the molecule). Second, there is no regularity in the number of times each amino acid appears. For example, leucine appears six times in insulin, but tryptophan is not present at all.

Notice in Figure 11.20 that cysteine (Cys) appears six times and that pairs of cysteines are connected. (To spot these three Cys–Cys pairs, look for amino acids numbered 7 and 19 on the top chain and 7 on the bottom chain.) The link connecting cysteine amino acids in proteins is called a **disulfide bond.** We first saw disulfide bonds in Chapter 10 where we discussed their role in the stabilization of vulcanized rubber. Disulfide bonds in proteins serve the same purpose: they act as reinforcements that hold the overall structure together.

QUESTION 11.8 Redraw this peptide with a shape that allows a disulfide bond to form. Indicate the position of the disulfide bond.

ANSWER 11.8

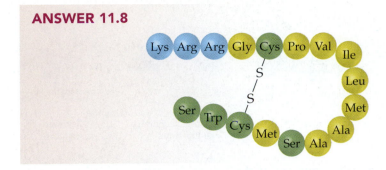

Proteins that curl up as we have just described are referred to as **globular proteins**; they make up the bulk of known proteins. Each kind of globular protein has a slightly different shape as a result of its three-dimensional structure, and each molecule of the same protein folds up into exactly the same three-dimensional shape every time.

How does a protein "know" how to fold up into the same three-dimensional shape time and time again? The answer is that every protein molecule first forms a **secondary structure** that then continues to coalesce into a compact three-dimensional shape. There are two common kinds of secondary structure. One, the **alpha helix**, is a right-handed coil with three to four amino acids making up each turn (Figure 11.21). Because most proteins are enormous, and thus drawing a separate sphere to represent each atom would be impractical, the alpha helix is usually depicted as either a coiled ribbon or a cylinder, as shown on the far right in the figure.

The other secondary structure found in proteins is the **beta pleated sheet**. In this type of structure, each amino acid chain is configured in a pleated shape that zigzags back and forth. When several of these zigzagging chains are placed side by side, they look like a sheet that has been folded into pleats, as shown in Figure 11.22. Note in the drawing how adjacent chains of the pleated sheet are held together by hydrogen bonds. For convenience, each chain of a beta pleated sheet is usually depicted as a

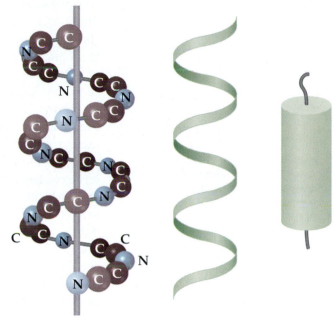

▲ **FIGURE 11.21 Secondary protein structure: alpha helix.** An alpha helix is a helical chain of connected atoms. Its symbol is either a curled ribbon or a solid cylinder.

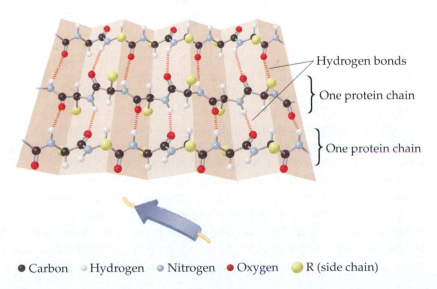

Hydrogen bonds

} One protein chain

} One protein chain

● Carbon   ○ Hydrogen   ○ Nitrogen   ● Oxygen   ○ R (side chain)

◀ **FIGURE 11.22 Secondary protein structure: pleated sheet.** A beta pleated sheet is a zigzag arrangement of amino acid chains. Its symbol is a fat, slightly curved arrow.

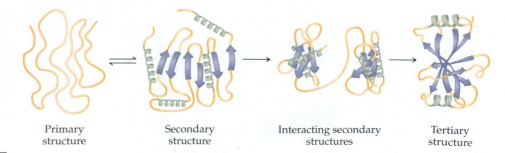

Primary structure   Secondary structure   Interacting secondary structures   Tertiary structure

fat arrow, as shown at the bottom of the figure. Protein molecules may be composed entirely of alpha helices, entirely of beta pleated sheets, or, more often, of a mixture of the two. In all globular proteins, short segments of the chain known as **random coil** are required to connect segments of alpha helix and beta pleated sheet.

Once the secondary structures have formed locally within a protein sequence, they begin to make brief contacts with one another, as indicated in Figure 11.23, and these contacts start to stabilize the emerging three-dimensional structure. Eventually, the regions of secondary structure settle into a fixed arrangement with respect to one another. The fully folded, compact three-dimensional shape of the globular protein is known as its **tertiary structure** (pronounced TER-she-air-ee). This entire process of making and breaking contacts and coalescing to create tertiary structure is known as **protein folding**. The tertiary structure of the protein is stabilized not only by disulfide bonds, but also by the intermolecular interactions we know from previous chapters: hydrophobic (nonpolar) amino acid side chains interact with one another, and an extensive network of hydrogen bonds forms throughout the protein structure. Hydrogen bonds form between turns in an alpha helix, between chains in a beta pleated sheet, and anywhere else where there happens to be an —NH or —OH group near a highly electronegative atom. As an example, Figure 11.24 shows the positions of hydrogen bonds in a bovine protein.

RECURRING THEME IN CHEMISTRY *Intra*molecular *bonds*, those *within* a molecule, are covalent. *Inter*molecular *interactions*, those either *between* two molecules or *between* two parts of a large molecule, are noncovalent.

▼ **FIGURE 11.24 Holding a protein in its three-dimensional shape.** For a fully folded protein, interactions between protein secondary structures join one part of the convoluted amino acid chain to another.

> **QUESTION 11.9** Indicate whether each description pertains to primary, secondary, or tertiary protein structure: (*a*) the folding of some short segment of a protein chain into a beta pleated sheet, (*b*) the coiling of a short segment into an alpha helix, (*c*) the coalescence of alpha helices into a compact globular shape, (*d*) the sequence of amino acids in the protein.
>
> **ANSWER 11.9** (*a*) and (*b*) secondary, (*c*) tertiary, (*d*) primary.

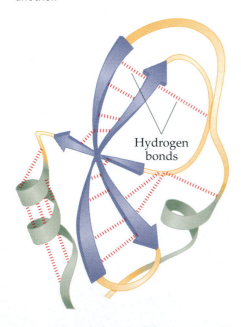

Hydrogen bonds

Now we are ready to look at the tertiary structure of a protein. Consider the tertiary structure of *bovine chymotrypsinogen*, a cow protein that breaks down other proteins, shown in Figure 11.25*a*. Take a look at another view of the same protein, shown in Figure 11.25*b*. This *space-filling model* shows you the surface of the protein rather than the interior. In this view, you can see that this enormous molecule is roughly spherical, a lump. When you look at a protein in this way, you cannot see the secondary structure, but you can determine which amino acids are on the protein's surface. In a living cell, the surface amino acids are exposed to salty water, the medium in which most proteins find themselves. Because salty water is extremely polar, the most polar amino acids in the protein tend to remain exposed to the solvent as the protein folds up into its tertiary structure. At the same time, the hydrophobic amino acids tend to avoid the salty water.

(a)

(b)

◀ FIGURE 11.25 **Proteins as art.**
(a) A symbolic rendering of the protein bovine chymotrypsinogen. Each cylinder represents an alpha helix, and each fat arrow represents a beta pleated sheet.
(b) A space-filling model of the same protein.

Therefore, they become buried in the interior, a water-free environment, as the protein folds up. *This tendency for amino acids to locate themselves either at the surface or in the interior is one of the most important factors driving the folding of the protein into a three-dimensional shape.*

RECURRING THEME IN CHEMISTRY Like dissolves like.

QUESTION 11.10 Choose the amino acid in each pair that is more likely to be located on the surface of a protein: (*a*) tryptophan or aspartic acid, (*b*) phenylalanine or lysine, (*c*) glutamic acid or valine.

ANSWER 11.10 The more polar amino acid of each pair is more likely to be at the surface: (*a*) aspartic acid, (*b*) lysine, (*c*) glutamic acid.

When we use the term *globular protein*, we mean that the three-dimensional shape is somewhat spherical (in other words, *globe*-shaped), but the shape can also be oblong, ellipsoidal, or flattened. Globular proteins exist in a variety of shapes because it is often the protein surface that determines what the protein does. For example, some globular proteins have surface indentations to which smaller molecules can bind. In some cases, the protein envelops the incoming molecule. In other cases, the protein binds to a cell's hereditary material to cut or repair it. In fact, in living systems, globular proteins do most of the "grunt work." When they are chemical factories, we call them *enzymes*. When they are chemical messenger molecules that travel through the blood, we call them *hormones*. When they are produced and sent out in response to an infection, we call them *antibodies*. The biomolecules that carry and store oxygen, *hemoglobin* and *myoglobin*, respectively, are proteins, too.

Proteins owe their diverse capabilities to the fact that the 20 amino acid building blocks are arranged in different sequences. It is this sequencing of the amino acids that determines the specific secondary structures a given protein attains along the length of its backbone. These secondary structures then coalesce into one single tertiary structure which is that protein chain's final folded structure.

Figure 11.26 highlights the other major group of proteins, **structural proteins**, which have mechanical roles. In living organisms, structural proteins have a protective role (in skin and fingernails) and are the basis of such hard-wearing materials as the horn of a rhinoceros and the ivory of an elephant's tusks. Structural proteins also can be used to do mechanical work, as in contractile systems, which are the basis of muscle movement in

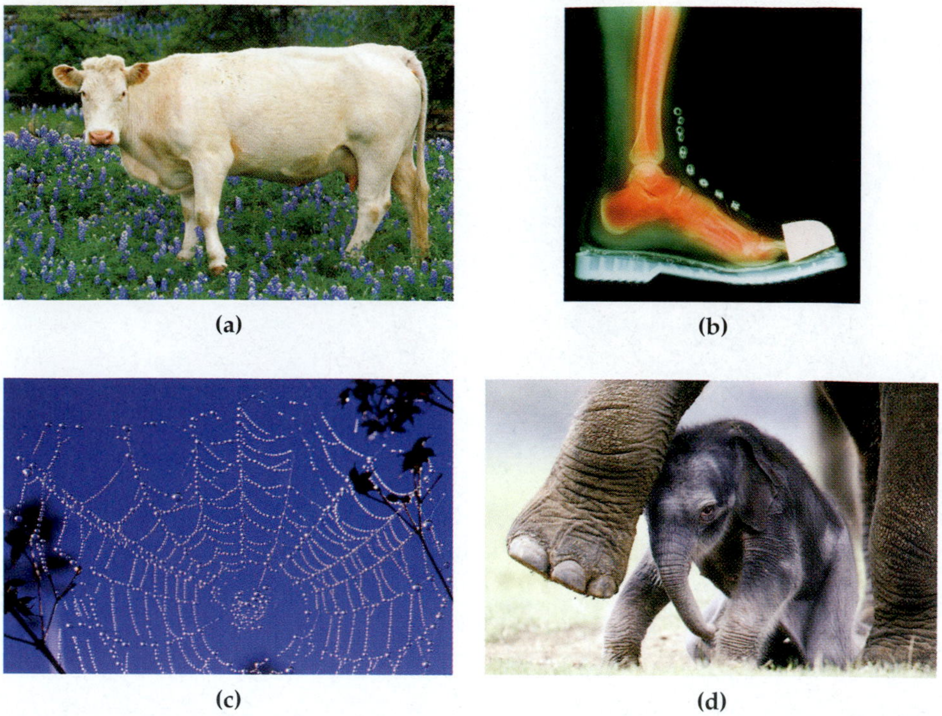

▶ **FIGURE 11.26** Nature's scaffolding: the many jobs of structural proteins. **(a)** The hide on an animal (as well as the skin covering your own body) is a protective barrier made of pliable structural protein. **(b)** Structural proteins give bones the strength needed to support your body. **(c)** A spider's web is possible because of the extraordinary strength of the structural proteins in spider's "silk." **(d)** The toughness of some structural proteins is evidenced in the nails, antlers, and horns of some animals, as well as in your own fingernails.

humans and other animals. Structural proteins can be stretchy and flexible, as in *resilin*, the biomolecule found in the hinges of insect wings. The diversity of structural proteins mirrors the diversity of synthetic polymers discussed in Chapter 10. For almost any application, nature has found a way to make a protein suited to it.

As you might expect, structural proteins do not have a globular shape. Rather, like many synthetic polymers, they are often fibrous. As an example, let us look at the structure of the silk protein. The primary structure is an alternating pattern of glycine and alanine amino acid units (Figure 11.27*a*), and the secondary structures are all beta pleated sheets. Recall from our

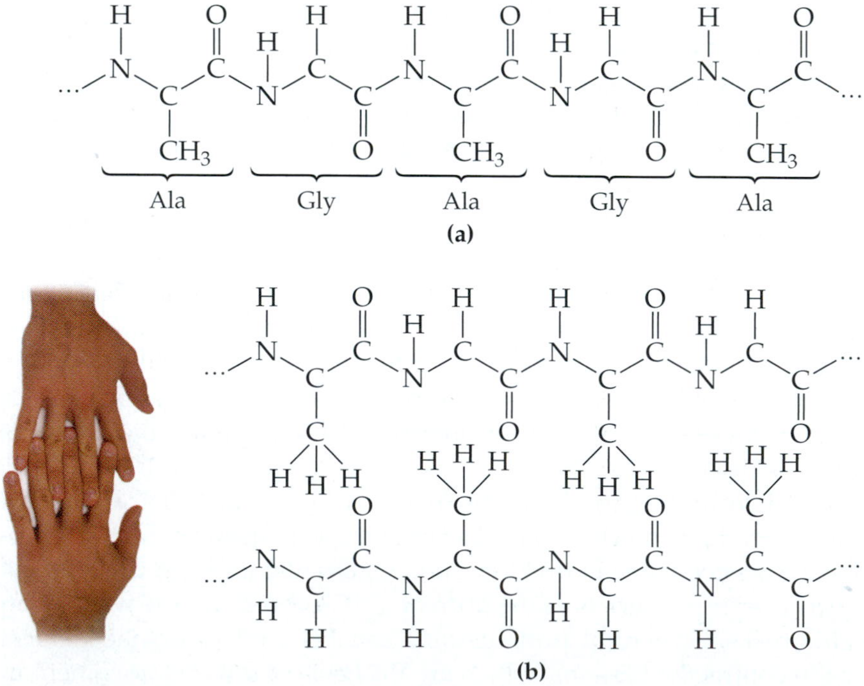

▶ **FIGURE 11.27** Interlocking beta pleated sheets. **(a)** The amino acid sequence in silk protein is an alternating pattern of alanine and glycine. **(b)** The secondary structure is beta pleated sheets. The zigs and zags of this secondary structure allow the alanine methyl groups of a chain in one pleated sheet to fit snugly into a chain in an adjacent pleated sheet. In this way, neighboring sheets are able to interlock with one another the way you can interlock the fingers of your two hands. Some hydrogen atoms have been omitted from this structure for clarity.

discussion of Figure 11.22 on page 439 that the individual chains in a beta pleated sheet are held together by hydrogen bonds. Notice in Figure 11.27a that the side chain on glycine is a hydrogen atom and that on alanine is a methyl group, —CH_3. Because both side chains are small, each protein chain in a pleated sheet can get quite close to its neighboring sheets above and below. The alanine methyl groups in one pleated sheet interdigitate with the zigs and zags of a protein chain in the sheet below or above, forming the interlocking pattern represented in Figure 11.27b. The result is that silk is a very strong fiber.

Silk is known both for its strength and for its tendency to tear. How can a material possess these two seemingly incompatible—indeed, contradictory—properties? We know that the individual pleated sheets in silk are strong because the polymer chains in each sheet are held together by a tight-knit system of hydrogen bonds (Figure 11.22). We also know that the interdigitation of alanine groups between sheets gives silk strength in the third dimension. However, the protein chains in beta pleated sheets, such as those in silk, are completely stretched out. Consequently, when you pull on a piece of silk from opposite directions, you do not feel any stretchiness because there is no way for its protein chains to be extended farther. If you pull hard, silk will always tear rather than stretch. The sound of silk tearing is the sound of covalent bonds in the protein chains being broken by a force that exceeds the strength of the bonds. (Later in this section, we shall see examples of materials that *do* stretch as a result of their protein secondary structure.)

QUESTION 11.11 The amino acid serine sometimes replaces some of the alanine in silk protein chains, and fabric made from this silk is stronger than fabric made from silk that does not contain serine. Why do you suppose serine would give silk added strength?

ANSWER 11.11 The hydroxyl group in the serine side chain is able to form hydrogen bonds that act as strong bridges between adjacent sheets in a given beta pleated sheet in the silk.

Recall that most globular proteins do not have any regularly repeating pattern in their amino acid sequence. This is because the sequence provides instructions for folding into a globular three-dimensional shape, and a great deal of sequence variety is needed in the primary structure in order to attain the desired secondary and tertiary structure. Unlike globular proteins, structural proteins are often fibrous, and fibrous proteins frequently do have repeating patterns of amino acids. Let us consider the most ubiquitous protein in vertebrates: the structural protein *collagen*. In the human body, collagen is like mortar. It is the stuff that holds things together and is a major component of bone and skin. For example, tendons, the tissue that connects our bones to one another, are nearly all collagen. Like other structural proteins, collagen has a repeating sequence of amino acids: glycine–X–proline, with X typically being proline or alanine. Take a look at the proline molecule in Table 11.1 on page 433. Because the side chain is part of a ring, this amino acid is unusually rigid and inflexible. As a result, proline causes kinks and turns in a protein chain. When two proline amino acids are found close together in a protein sequence, they create a rigid, curved shape. Thus, more than one-third of the amino acids in collagen are proline, and it forms a tough, inflexible left-handed helix.

MOLECULE
Collagen

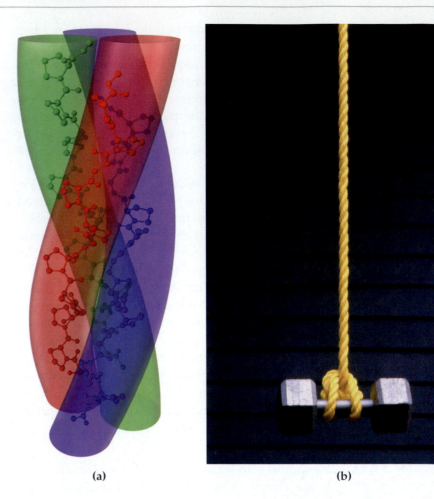

▶ **FIGURE 11.28** **Protein ropes.**
(a) A strong fiber can be made by wrapping several protein chains coiled in one direction into a larger protein chain coiled in the opposite direction. **(b)** This strategy is used to make strong ropes.

(a) (b)

In a collagen fiber, three left-handed collagen chains come together to make a triple helix that has a right-handed twist (Figure 11.28*a*). This motif—individual chains twisted in one direction and then twisted together in the opposite direction—is one way strong rope is made (Figure 11.28*b*). The opposite-twists pattern creates a very tight weave, and its use in rope is just one more way we humans make use one of nature's original designs.

Collagen derives its strength from several factors. First, it is very tightly wound, and every third amino acid is squeezed into the interior of the helix. For this reason, no amino acid with a side chain larger than an —H group can fit into that spot, explaining why tiny glycine is placed at every third position along the protein chain. The primary structure of collagen is thus a repeat of gly–X–Y, where X is often proline and Y is either proline or a modified version of proline. After collagen is made in the body, many of the proline side chains undergo a chemical reaction that adds an —OH group to the side chain (Figure 11.29). In order for this reaction to take place, ascorbic acid (vitamin C, Section 5.6) is required. The —OH group added to the proline side chain allows each collagen chain to make hydrogen bonds with other collagen chains, and the right-handed triple helix of collagen fibers is superstrong as a result. When vitamin C is absent from the diet, the —OH group is absent from the proline in collagen and the strength of the collagen fibers is significantly reduced.

▼ **FIGURE 11.29** **Role of vitamin C in strengthening collagen.** In collagen, one of the H atoms in the amino acid proline is replaced by an —OH group. This substitution requires the presence of vitamin C, a vitamin obtained from citrus fruits such as oranges, lemons, and limes.

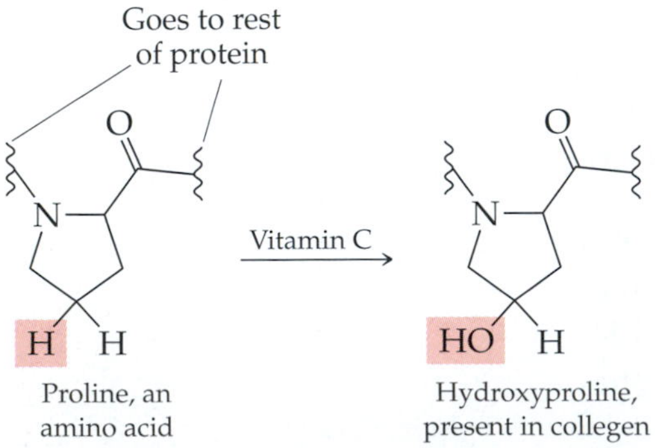

Goes to rest of protein

Vitamin C →

Proline, an amino acid

Hydroxyproline, present in collegen

The importance of the —OH group on proline is underscored in the following story: Sailors in the eighteenth century lived under rough conditions. For the many who survived the onslaught of violent weather and the unsanitary conditions aboard ship, there was a paltry selection of things to eat. Very few fruits or vegetables were available, because they could not be kept fresh on long voyages. After a few months at sea with no fruit or vegetables, the sailors became weak and listless. Lesions formed on their skin, and their teeth began to fall out. These are symptoms of a deficiency disease called *scurvy*. One surgeon working on a ship in the British navy observed that there were far fewer cases of scurvy among sailors who included lime juice in their diets. As a result of this finding, doctors began to recommend that sailors drink lime juice while at sea to prevent the onset of scurvy. (This is the origin of the British epithet "limey," used to refer to a sailor.)

It turns out that one substance found in lime juice, vitamin C, is crucial for the correct formation of collagen connective tissue in the body. Thus, the symptoms of scurvy, such as bleeding gums and skin lesions, are related to loss of tissue integrity. It took more than 150 years to determine that the cause of scurvy is a deficiency of vitamin C, which is found in high concentrations in many fruits.

The structural proteins we have looked at here—silk and collagen—are both fibrous, and these two examples demonstrate that the amino acid sequence can bestow rigidity and strength to the protein fibers. Not all structural proteins are fibrous, however. Some structural proteins need to have mechanical strength, but not rigidity. For instance, the structural proteins found in blood vessels must be able to stretch and flex. Likewise, lung tissue should be able to expand and contract. For these purposes, we have the structural protein *elastin*, which is unusual as proteins go because it does not have much secondary structure and is mostly random coil. Elastin is also unusual because the chains are able to form cross-links to one another. Thus, the structure of elastin looks something like a bowl of spaghetti, except that here and there the strands are linked together. The result is a very stretchy, yet strong, polymer in which the structure is reminiscent of that of vulcanized rubber, shown in Figure 10.25 on page 399. The rubber has the same unruly jumble of polymer strands with occasional cross-links to add strength. Elastin is sold as a nutritional supplement that claims to add flexibility to joints, even in dogs (Figure 11.30). This is yet another example of our duplication of nature's best ideas.

▲ **FIGURE 11.30   Protein bungee cords.**   Elastin can be found in products that claim to give added flexibility to tendons and joints.

11.4 | The Secret Language of Chains

Now that we know a bit about what proteins look like, let us consider how a protein is made by living organisms. The process is exceedingly complex, and scientists' current understanding of it is incomplete. One thing known for sure, though, is that the instructions for making every protein are stored in molecules of **deoxyribonucleic acid**, or **DNA**, the hereditary material present in the nucleus of every living cell. It is through the copying of DNA that traits are passed from parents to offspring.

DNA belongs to the class of biological molecules called **nucleic acids**, which are polymers that exist primarily in a cell's nucleus (hence the name). Even though both are polymers, DNA is very different from protein, and the main difference is this: instead of amino acid building blocks, the building blocks in DNA are molecules called **nucleotides**. As

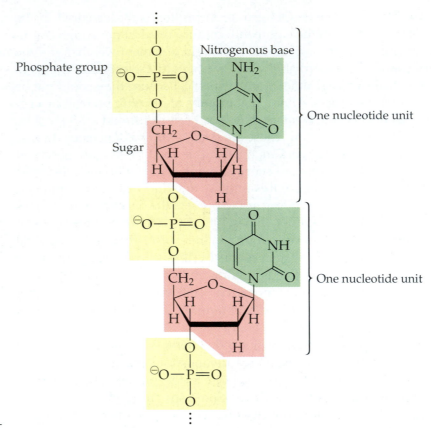

Phosphate group

Nitrogenous base

Sugar

One nucleotide unit

One nucleotide unit

▶ **FIGURE 11.31   Nucleotide structure.**
Nucleotides are the monomer units used
to make the polymers called nucleic acids,
mainly deoxyribonucleic acid (DNA) and
ribonucleic acid (RNA). Every nucleotide
contains a phosphate group, a sugar, and a
nitrogenous base.

▼ **FIGURE 11.32   What a difference
an oxygen atom makes, eight little
electrons.**   The nucleic acid
deoxyribonucleic acid (DNA) contains the
sugar deoxyribose. The nucleic acid
ribonucleic acid (RNA) contains the sugar
ribose. These two sugars differ from each
other by only one —OH group on one
of the ring carbons. This group is present
in ribose, but absent in deoxyribose
(*deoxy-*, "without oxygen").

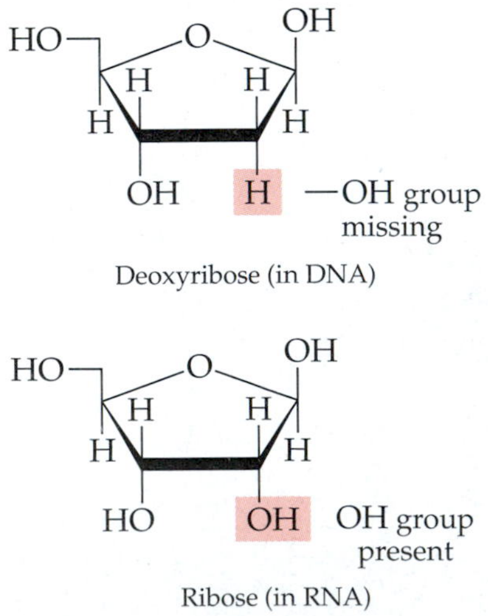

HO

OH

H   H

H   H

OH   H —OH group
              missing

Deoxyribose (in DNA)

HO

OH

H   H

H   H

HO   OH   OH group
                present

Ribose (in RNA)

Figure 11.31 shows, a nucleotide has three parts: a phosphate group (colored
gold in the drawing), a sugar part (red), and a nitrogen-containing part
called a **nitrogenous base** (green).

You can recognize the sugar part of a nucleotide by looking for an ether
functional group (R–O–R′) that is looped into a ring. Two sugars are commonly found in nucleic acids. The nucleic acid DNA contains the sugar
deoxyribose, and the nucleic acid **ribonucleic acid** (**RNA**) contains the sugar
ribose (Figure 11.32). (We shall look at RNA in more detail shortly.)

Notice in Figure 11.31 how the phosphate groups and sugars alternate
down the length of the nucleic acid chain, with the nitrogenous bases off to
the side. This alternating pattern of phosphate groups and sugars is called
the **sugar–phosphate backbone** of the polymer chain, and the bases are
attached to that backbone.

There are four possibilities for the nitrogenous bases in DNA: thymine,
adenine, cytosine, and guanine (Figure 11.33). This means that there are
only four nucleotide building blocks for DNA, rather than the 20 amino
acid building blocks for proteins. These four bases form a seemingly random pattern along a DNA sugar–phosphate backbone.

QUESTION 11.12   Which of the four nitrogenous bases has the greatest
molar mass?

ANSWER 11.12   Guanine.

QUESTION 11.13   Sketch a DNA chain segment that is two monomer
units long, with one of the units containing the nitrogenous base cytosine
and the other containing the nitrogenous base adenine.

Thymine   Adenine   Cytosine   Guanine

◀ FIGURE 11.33 **The four nitrogenous bases in DNA.** Each base attaches to the sugar of a nucleotide at the nitrogen highlighted in green.

▼ FIGURE 11.34 **A with T, C with G.** The nitrogenous bases in a DNA molecule are attached to a sugar–phosphate backbone. When complementary bases pair up with each other via hydrogen bonds, the result is a long chain consisting of two sugar–phosphate backbones enclosing the base pairs. The bases act as "rungs" in a DNA "ladder." Notice that two-ring adenine (A) always pairs with one-ring thymine (T) and two-ring guanine (G) always pairs with one-ring cytosine (C).

ANSWER 11.13

Cytosine

One nucleotide unit

Adenine

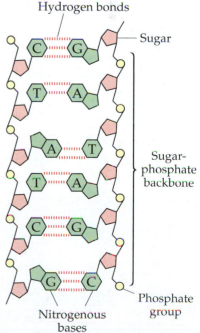

Notice in Figure 11.33 that adenine and guanine each have two rings fused together and cytosine and thymine have a single ring each. This is an important point because, in a DNA molecule, the bases come together into pairs and link via hydrogen bonds. Because all base pairs in the molecule should be approximately the same width across, adenine (two rings) always pairs with thymine (one ring) and guanine (two rings) always pairs with cytosine (one ring). We say that the two bases in a **base pair** are **complementary** to each other: adenine is complementary to thymine (and vice versa), and cytosine is complementary to guanine (and vice versa).

Each nitrogenous base in DNA is attached to a sugar in the sugar–phosphate backbone, which means that each base is part of a long polymeric chain. When speaking of nucleic acids, chemists usually call the chains *strands*, and so we shall use that term here in our discussion. A DNA molecule consists of two polymer strands, each resembling one vertical side of a ladder. When two DNA strands come together to form *double-stranded* DNA, the bases on the two strands form hydrogen bonds that create the rungs of the ladder (Figure 11.34).

QUESTION 11.14 A single strand of DNA has the sequence AGGTAC-CTGGTA. What is its complementary sequence?

ANSWER 11.14 TCCATGGACCAT.

QUESTION 11.15 How would the appearance of the ladder representing double-stranded DNA change if adenine paired with guanine and cytosine paired with thymine?

ANSWER 11.15 The two sides of the ladder would not be parallel to each other. Instead, there would be bulges and indentations formed by wide adenine–guanine base pairs and narrow cytosine–thymine base pairs:

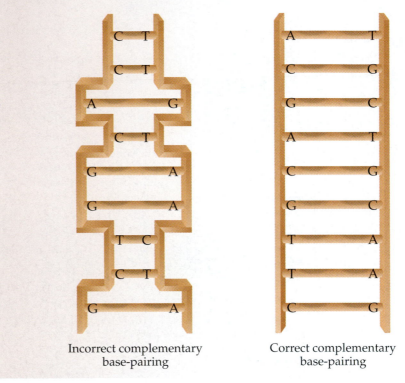

Incorrect complementary base-pairing

Correct complementary base-pairing

RECURRING THEME IN CHEMISTRY *Intra*molecular *bonds,* those *within* a molecule, are covalent. *Inter*molecular *interactions,* those either *between* two molecules or *between* two parts of a large molecule, are noncovalent.

The ladder in Figure 11.35 represents a segment of double-stranded DNA. Imagine that this ladder is flexible. You and a friend stand at opposite ends of the ladder and start twisting it until there are about ten rungs every time the ladder makes one full revolution. This is the shape of most DNA found in living organisms, a *double helix*. Each bond *within* a strand is covalent and strong, but the hydrogen bonds *between* the two strands are comparatively weak and easily broken. Because the intermolecular bonds between strands are weak, the two strands of the double helix can unzip, and it is possible for new strands of DNA to be made when this happens. One separated strand acts as a *template*, and complementary bases are added to it (A with T, G with C). In this way, DNA is able to make copies of itself, so that when a cell divides, identical DNA is present in both new cells (Figure 11.36).

We know the purpose of the amino acid sequence in a protein: to dictate the protein's three-dimensional structure. Thinking along parallel lines, we can ask, What is the significance of the nucleotide sequence in DNA? The answer is that any strand of DNA can be thought of as being divided into many small segments, each referred to as a **gene**. The significance of DNA's nucleotide sequence is that *the sequence in a given gene is what provides instructions for making a specific protein* needed by the cell. Every nucleus of every cell in a human being contains DNA, and that DNA includes enough genes to create nearly 100,000 proteins. Let us see how this is done.

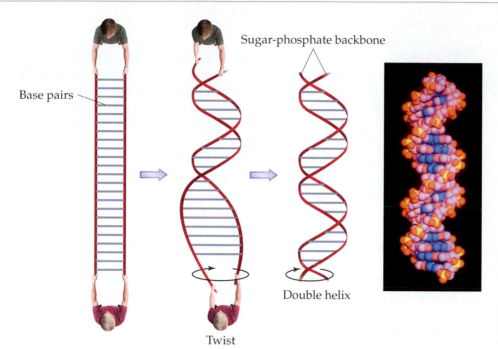

Sugar-phosphate backbone

Base pairs

Double helix

Twist

◀ **FIGURE 11.35 Ladder with a twist.** Two people holding the ends of a flexible ladder can create a structure like a DNA double helix by twisting the ladder to create a right-handed spiral. The structure at the right is a space-filling model of a DNA molecule.

Protein synthesis requires the presence of the other common nucleic acid found in the body—RNA—which is made from a DNA template in a process known as **transcription**. RNA differs from DNA in two ways: the sugar in RNA is ribose (Figure 11.32), and RNA uses the nitrogenous base uracil (U) in place of thymine. These two bases are the same, except that uracil is missing the methyl group present on the thymine ring (Figure 11.37). When a DNA strand is used to create RNA, every adenine on DNA pairs with a uracil on RNA. The other base-pairings remain the same: DNA guanine with RNA cytosine, DNA thymine with RNA adenine, and DNA cytosine with RNA guanine. Thus, the DNA sequence ATTGCA is transcribed into the RNA sequence UAACGU.

Say your body needs a specific protein molecule to fight off an infection. The strands of the double-stranded DNA segment responsible for making this protein separate, as they do when DNA makes copies of itself, but in protein synthesis the new strand produced is a strand of RNA rather than DNA. The newly made RNA, called **messenger RNA** (or **mRNA**), travels to a *ribosome*, a protein-making center located in the cell, but outside the nucleus.

QUESTION 11.16 Give the complementary RNA sequence for the DNA sequences (*a*) GCGCATTC and (*b*) TTCTAGGCTTACTG.

ANSWER 11.16 (*a*) CGCGUAAG, (*b*) AAGAUCCGAAUGAC.

The question of how you get from mRNA, a molecule built from various combinations of only four nitrogenous bases, to a protein, a molecule built from various combinations of 20 amino acids, stumped scientists for decades until, in the 1960s, the answer was found. In order to translate a

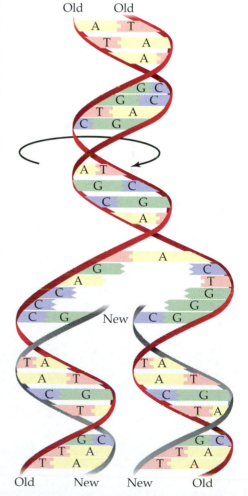

▶ **FIGURE 11.36 DNA copy machine.** Living organisms make new DNA by unraveling the two strands of a DNA molecule and then creating a complementary copy of each. After the copy is made, each new strand (shown dark gray in the drawing) and its old partner (maroon) twist back up into a double helix. What started out as a single double helix has become two double helices.

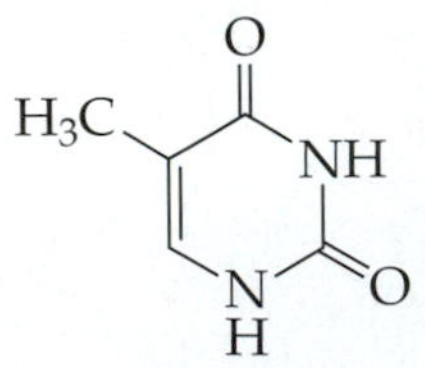

▲ FIGURE 11.37 Thymine versus uracil. The only difference between these two nitrogenous bases is that uracil does not have a methyl group attached to the ring. Thymine is found in DNA, uracil in RNA.

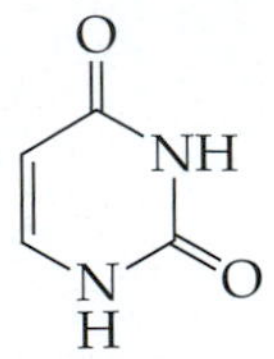

message that uses four "letters" to one that uses 20, the body uses what is called the **genetic code**.

The genetic code works by equating a group of three mRNA nitrogenous bases, called a **triplet**, with one specific amino acid, as shown in Figure 11.38. For example, both the triplet UUU and the triplet UUC, found at the top of the first column in the figure, code for the amino acid phenylalanine, and the triplets UAU and UAC (third column, top) both code for the amino acid tyrosine.

Let us translate the mRNA sequence GAUUGGUUCGGG into a protein amino acid sequence. To make our job easier, we can rewrite the sequence separated into triplets:

GAU UGG UUC GGG

Beginning with the first triplet, GAU, we look in the red-shaded column on the left in Figure 11.38 for the first letter of the triplet: G. We place a finger on this red-shaded G and then move the finger to the right until it rests under the letter in the yellow-shaded row at the top that matches the second letter of our triplet: A. We are now in the white portion of the chart, showing the four possibilities for triplets beginning with GA: GAU, GAC, GAA, and GAG. As our third nitrogenous base is U, we see that our triplet GAU codes for aspartic acid, the first amino acid in the protein we are building. The next amino acid in the sequence comes from UGG. Repeating our finger exercise for this triplet shows us that it codes for tryptophan. You should be able to do this twice more and end up with the sequence Asp-Trp-Phe-Gly. This amino acid sequence could be the beginning of a large protein being formed.

A summary of protein synthesis is shown in Figure 11.39. To keep things simple, the protein being made is a dipeptide, only the beginning of what will eventually be a long protein chain. As the chain of amino acids forms, the RNA strand is threaded across the ribosome, as shown at the bottom of the figure. For every three nucleotides that pass by, one amino acid is specified and added to the growing protein chain. This process of **translation** takes one sequence—that of nitrogenous bases in RNA—and *translates* it into another—that of amino acids in protein. In the example we just finished, the RNA sequence GAU UGG UUC GGG is translated to the amino acid sequence Asp-Trp-Phe-Gly.

► FIGURE 11.38 The genetic code. Each triplet of nitrogenous bases codes for a specific amino acid.

	U	C	A	G
U	UUU ⎫ Phe UUC ⎭ UUA ⎫ UUG ⎭	UCU ⎫ UCC ⎬ Ser UCA ⎪ UCG ⎭	UAU ⎫ Tyr UAC ⎭	UGU ⎫ Cys UGC ⎭ UGG Trp
C	CUU ⎫ CUC ⎬ Leu CUA ⎪ CUG ⎭	CCU ⎫ CCC ⎬ Pro CCA ⎪ CCG ⎭	CAU ⎫ His CAC ⎭ CAA ⎫ Gln CAG ⎭	CGU ⎫ CGC ⎬ Arg CGA ⎪ CGG ⎭
A	AUU ⎫ AUC ⎬ Ile AUA ⎭ AUG Met	ACU ⎫ ACC ⎬ Thr ACA ⎪ ACG ⎭	AAU ⎫ Asn AAC ⎭ AAA ⎫ Lys AAG ⎭	AGU ⎫ Ser AGC ⎭ AGA ⎫ Arg AGG ⎭
G	GUU ⎫ GUC ⎬ Val GUA ⎪ GUG ⎭	GCU ⎫ GCC ⎬ Ala GCA ⎪ GCG ⎭	GAU ⎫ Asp GAC ⎭ GAA ⎫ Glu GAG ⎭	GGU ⎫ GGC ⎬ Gly GGA ⎪ GGG ⎭

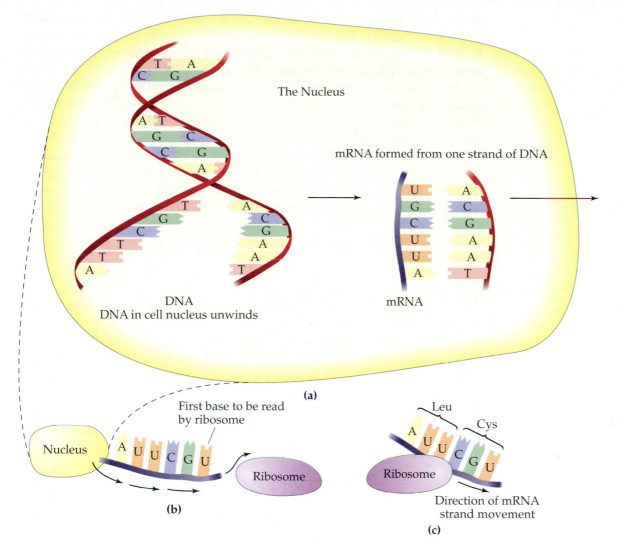

▲ FIGURE 11.39 DNA to mRNA to protein. **(a)** An unwound section of DNA is used to create mRNA. **(b)** mRNA leaves the nucleus and travels to the ribosome, where protein is made. **(c)** Each triplet of nitrogenous bases on the mRNA codes for one amino acid on the protein chain, according to the genetic code shown in Figure 11.38.

QUESTION 11.17 List all the triplets that code for (*a*) Trp, (*b*) Cys, (*c*) Leu, (*d*) Arg.

ANSWER 11.17 (*a*) UGG; (*b*) UGU, UGC; (*c*) UUA, UUG, CUU, CUC, CUA, CUG; (*d*) CGU, CGC, CGA, CGG, AGA, AGG. Notice that triplets for arginine are listed in two areas of Figure 11.38, as the third amino acid in the fourth column and again as the fifth amino acid.

QUESTION 11.18 What peptide sequence is coded for by the RNA sequence UUACGGCAUUCG?

ANSWER 11.18 Leu-Arg-His-Ser.

Of the 20 amino acids your body uses to make protein molecules, only ten of them can be made by the body's cells. These amino acids are called *nonessential amino acids*. The other ten are also needed, but your body's cells do not have the instructions to make them. Thus, these *essential amino acids* (so called because they are *essential* in the human diet) must be obtained from the food you eat. (Each amino acid is classified as essential or nonessential in Table 11.1 on pages 433–434.)

In the same way that garbage accumulates in a landfill, proteins and other biomolecules accumulate in the body and must be broken down. Thus, the peptide bonds holding them together are susceptible to degradation by design. DNA is another story. Because each cell in your body has only one copy of the instructions for making all of your proteins, you do not want that copy to degrade. (You can think of DNA as the information stored on your computer's hard disk and of RNA as information copied from it onto an unlimited number of compact discs.) Therefore, the sugar–phosphate bonds in DNA must be especially robust. These bonds are so robust, in fact, that intact DNA sometimes can be extracted from fossils and its sequence determined.

11.5 | Genetic Engineering: The DNA Shuffle

Now that we have learned how DNA is used to make proteins, let us go back to a topic mentioned at the beginning of the chapter: transgenic plants. We learned that the genetic material in corn can be manipulated so that it will grow plastic in its leaves. Let us consider how a transgenic plant can be created by genetic engineering. How can we make a plant produce a protein that it does not naturally produce?

The secret behind any kind of genetic engineering is the ability to cut and paste pieces of DNA (in other words, cut and paste genes) from one organism into another organism. The technology of genetic engineering—the capacity to cut a strand of DNA into fragments and then reattach the fragments in a different order—was developed in the 1970s and is a mainstay of scientific work today.

One of the key players in most genetic engineering experiments is a class of proteins called restriction enzymes. Enzymes, as you know from Section 11.3 are large biomolecules, usually made of protein, that carry out chemical reactions. A **restriction enzyme** is an enzyme that cuts DNA. Genetic engineering is possible because many restriction enzymes cut at very specific locations in a DNA sequence. Here is an example: The restriction enzyme *Eco*RI cuts a DNA strand *only* between guanine and adenine and *only* when the sequence is GAATTC. Given this, see if you can locate the position of the two cuts *Eco*RI makes in this piece of DNA:

AGAGGTCGAATTCATTGAATCCGTGTCGACTTGAATTCGTCGAC

Here they are:

Note that the GA pairs numbered 1 through 4 in the blue-shaded regions are not cut by *Eco*RI, because in these regions the sequence after the A is not ATTC. The two GA pairs in the red-shaded regions *are* cut by this restriction enzyme because the six-base sequence is GAATTC in each case.

Figure 11.40 shows two segments of DNA obtained from two different organisms coded blue and gold for the purposes of this discussion. Suppose

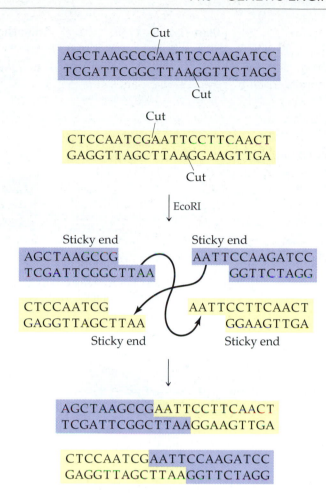

◀ **FIGURE 11.40 Making sticky ends meet.** The restriction enzyme *Eco*RI cuts two sequences of DNA between G and A in the GAATTC sequences. The resulting four fragments can now "trade partners," so that the two new DNA segments formed each contain part of the blue sequence and part of the gold sequence.

we want to use genetic engineering techniques and the *Eco*RI restriction enzyme to move a gene from the blue organism into the DNA of the gold organism. The position of each *Eco*RI cut is indicated on the four strands. Once the cuts are made, we get four double-stranded fragments, each with one strand longer than the other. The uneven end of each fragment is called a **sticky end** and can be used to stick the DNA molecules back together. The four fragments can be pasted together because the linear ends are complementary to each other. This can happen in two ways: either in their original sequence—blue with blue, gold with gold—or by trading partners to make two new sequences, each containing one blue fragment and one gold fragment, as shown at the bottom of the figure.

QUESTION 11.19 (*a*) Write the complementary DNA sequence to TTG-GAATCCAATCGG. (*b*) Will the restriction enzyme *Eco*RI cut a DNA strand containing this sequence? If so, indicate the position of cutting.

ANSWER 11.19 (*a*) AACCTTAGGTTAGCC. (*b*) No, because this restriction enzyme cuts only in the sequence GAATTC.

Genes can be engineered by taking advantage of sticky ends, as shown in Figure 11.41. In this example, the gene from human DNA that provides instructions for making a specific protein is cut out of a piece of human DNA. It is then spliced into a small, circular piece of bacterial material called a *plasmid*. Once the gene from the human DNA has been added to the bacterial plasmid, the plasmid DNA has been *genetically altered* and is often referred to as **recombinant DNA** (pronounced re COM bi nent), because the

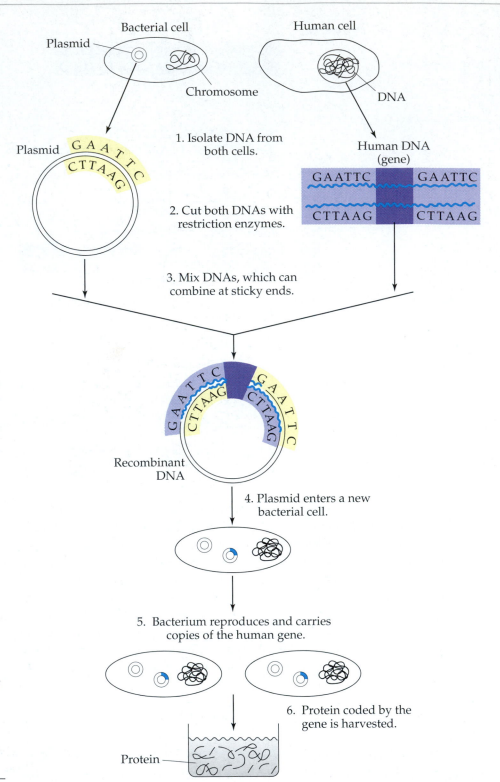

▶ FIGURE 11.41   **Recombining DNA.** Through genetic engineering, bacteria can be used to produce large amounts of a desired protein. A plasmid from a bacterial cell is cut, and the gene from human DNA that codes for the protein is inserted. The genetically altered plasmid is incorporated into a new bacterial cell, and the new protein is produced.

plasmid circle has broken and then *recombined*. Now the plasmid containing the human gene is introduced into a new bacterial cell. The bacterium can then be instructed to make copies of the plasmid, and it is therefore able to produce the protein coded for by the inserted human gene.

Now we are prepared to look at an example of how this process can produce genetically engineered plants. The bacterium *Agrobacterium tumefaciens* infects berries, cherries, and roses. When intentionally infected with this bacterium, the plants develop a disease called crown gall, a kind

DNA—True to Life

As you probably have read in the newspaper time and again, DNA collected from a crime scene can be powerful evidence in deciding guilt or innocence. Because the amount of DNA collected from a crime scene is, in many instances, very small, it is important to be able to make copies of the collected DNA so

CHEMISTRY AT THE CRIME SCENE

that there is enough material for the forensic chemist to work with and get accurate results.

If you have a sample of DNA from a crime scene, you can copy it millions of times in the same way a copy machine can make an unlimited number of copies of a master document. The method used to make DNA copies is the *polymerase chain reaction* (PCR). The first step in PCR is to decide which part of the human DNA sequence will be copied—in other words, which gene in the crime-scene DNA will be copied. Ideally, the gene chosen should have a known sequence that varies somewhat from one person to another. The entire gene can be duplicated using PCR, with each copying cycle taking around 90 seconds. Typically, about 20 to 30 rounds are done for a DNA sample, and this produces more than 1 million DNA copies from one original.

Now you have at least 1 million copies of your gene, enough to allow you to analyze its sequence. Imagine (a) that there are 20 common sequences for this gene in the human population and (b) that each of these 20 variants is found in about 2 to 8 percent of the population. If you can determine the sequence of your crime-scene gene, you can figure out which variant it is. If you collected the DNA at a crime scene where a suspect is known to have (unknowingly) left a DNA sample behind, you can test suspects to see if they have the same variant of the gene. If they have a variant different from the crime-scene sample, you can eliminate them from the list of suspects. If they have the same variant, they stay on the list. Clearly, because 2 to 8 percent of the population will carry that variant, this evidence would not be sufficient to convict the suspect of the crime, and further testing would be required to demonstrate guilt.

DNA analysis often can narrow down possible suspects much more severely. For example, when we hear about DNA cases in the courtroom, it is often something like this: "The DNA of Ms. X was found to match the DNA found at the scene. Because there is a one-in-50 million chance that another person would have this DNA match, the jury found her guilty of the crime." Seeing as there are about 6,000,000,000 people on planet Earth, in this case only $6{,}000{,}000{,}000/50{,}000{,}000 = 120$ people could have committed the crime, and it is highly unlikely that any other suspect will have exactly the same DNA as Ms. X (unless she has an evil identical twin). This kind of overwhelming, persuasive DNA evidence is often enough to convince a jury.

---

of plant cancer. The plants can live with the infection, which produces nodules (called *galls*) on the plant stem, but why would anyone want to give a plant cancer? The answer to this question is that *A. tumefaciens* has a useful genetic engineering purpose. The cancer this bacterium creates does not threaten the life or productivity of most plants. The important point is that *A. tumefaciens* works by transferring some of its own DNA into the DNA of the plants it infects. If the bacterial DNA is genetically altered to contain the gene for some trait we want to see in the plants, then the bacterium provides a convenient way to deliver that gene to the plant.

The *A. tumefaciens* DNA can be modified by cutting and pasting, just as outlined in Figure 11.41. For example, if you would like to grow cherry trees that glow in the dark, you can splice the gene for *green fluorescent protein*, a brightly glowing green protein, into the *A. tumefaciens* DNA. Then you can allow the bacterium to infect your cherry trees. Part of the DNA from the bacterium is incorporated into the DNA of the trees, and if all goes as planned, the gene for the green protein goes along with it.

As this example illustrates, genetic engineering technology is limited only by the imagination of the scientists who use it. As a member of the voting public, you may be asked to play a role in the responsible application of genetic engineering know-how.

SUMMARY

We started this chapter with a discussion of natural polymers and how new technology allows scientists to use them as replacements for synthetic polymers. When we take advantage of natural polymers from sustainable resources, the waste we create can be either easily degraded or recycled for further use. Some of the new products based on natural sources are genetically engineered, and there is currently a vigorous debate over the ethics of gene manipulation.

We discussed the structure of nylon and compared it with that of protein. Both contain amide functional groups and both exist in long chains. That is where the similarities end, however. Protein molecules are extraordinarily diverse because there are 20 amino acid building blocks that can be used to create a protein sequence. Each amino acid contains an amine group, a carboxylic acid group, and a side chain. When peptide bonds connect amino acids to one another, the resulting polymer has a polyamide backbone with side chains that dangle from the main chain like charms on a bracelet.

There are two major classes of protein. Globular proteins are mainly spherical and are formed by the systematic folding of the amino acid chains that make up protein molecules. This shaping begins as the amino acid sequence dictates the local folding of the chain into secondary structure. The two most common secondary structures are the alpha helix and the beta pleated sheet. When these secondary structures form, they are connected by regions of random coils. As the secondary structures interact with one another, often via hydrophobic interactions, protein tertiary structure begins to form. The final globular protein structure contains nonpolar amino acids in the interior and polar amino acids on the surface.

Structural proteins constitute the other major class of protein. Like globular proteins, structural proteins have primary, secondary, and tertiary structure. Unlike globular proteins, structural proteins are often fibrous, and this fibrous structure is due, in part, to repeating sequences of amino acids. In the human body, structural proteins make up everything from bone, to hair, to connective tissue, to muscle fiber.

We introduced two new organic functional groups in this chapter: the carboxylic acid functional group, —COOH, and the amine functional group, —NH$_2$. The carboxylic acid functional group is always located at the end of a chain. The nitrogen of an amine functional group can attack the carbon atom of a carboxylic acid functional group to produce an amide functional group,

$$\overset{\displaystyle O}{\overset{\displaystyle \|}{-C}}-NH- .$$ This is the general way polyamide polymers, such as nylon and protein, are made.

Protein synthesis in a living cell begins with DNA, a polymer made up of the nitrogenous bases adenine, cytosine, guanine, and thymine. The backbone of DNA is made up of alternating phosphate and sugars groups to which the nitrogenous bases are attached. Each nitrogenous base can make hydrogen bonds with a complementary base. Guanine forms hydrogen bonds to cytosine, and adenine forms hydrogen bonds to thymine. These hydrogen bonds connect two strands of DNA into a double helix. When the two strands separate, each strand can act as a template for the synthesis of either a complementary DNA strand or an RNA strand. When proteins are synthesized, RNA made in the cell nucleus from a DNA template leaves the nucleus and travels to a ribosome. At the ribosome, the genetic code is used by the cell to translate the information encoded in the RNA into an amino acid sequence that becomes the new protein molecule. The genetic code equates three-base sequences of RNA to one amino acid. Thus, every triplet of RNA bases codes for a specific amino acid.

We now have the ability to cut a segment of DNA from the gene of one organism and splice it into the DNA of another organism. This can be accomplished through the use of restriction enzymes, which cut DNA only at specific sequences. It is possible to make cuts in DNA so that sticky ends are left, and sticky ends can be used to connect one DNA piece to another. The recombinant DNA that is formed is now a mixture of DNAs from two different organisms. Recombinant DNA technology is used extensively in agricultural applications.

KEY TERMS

alpha helix	deoxyribonucleic	genetic engineering	nucleotide	protein folding
amino acid	acid (DNA)	globular protein	peptide	random coil
base pair	disulfide bond	messenger RNA	peptide bond	recombinant DNA
beta pleated sheet	fixation	(mRNA)	polypeptide	restriction enzyme
carboxylic acid	gene	nitrogenous base	primary structure	ribonucleic acid
complementary	genetic code	nucleic acid	protein	(RNA)

secondary structure
side chain
sticky end

structural protein
sugar–phosphate
backbone

sustainability
tertiary structure

transcription
translation

transgenic
triplet

QUESTIONS

The Painless Questions

1. (a) Draw the line structure for the amino acids leucine and isoleucine. (b) What is the molecular formula for each? (c) How does the structure of leucine differ from that of isoleucine? (d) Would these amino acids be found inside a globular protein molecule or on its surface?

2. Name these carboxylic acids:

$$CH_3-\overset{\overset{\displaystyle O}{\|}}{C}-O-H$$
(a)

$$CH_3-CH_2-CH_2-\overset{\overset{\displaystyle O}{\|}}{C}-O-H$$
(b)

$$CH_3CH_2CH_2CH-CH_2-\overset{\overset{\displaystyle O}{\|}}{C}-O-H$$
$$\underset{\displaystyle CH_3}{|}$$
(c)

3. (a) Identify each component of this DNA double helix:

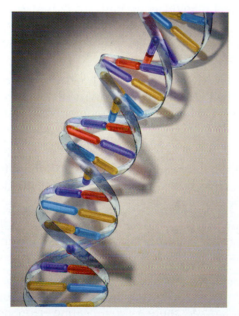

(b) Is the number of colors used for the nitrogenous bases appropriate? (c) In a real DNA molecule, what is the molecular structure represented by each blue strand in the illustration?

4. (a) Identify the sugar, the phosphate group, and the nitrogenous base in this nucleotide monomer:

(b) Which nitrogenous base is shown in this nucleotide? (c) In a DNA molecule, with which other nitrogenous base does this one pair?

5. Does the nucleotide shown in Question 4 belong in DNA or RNA? Explain your answer.

6. Name the two main forms of secondary structure in globular proteins. Briefly discuss how the two structures differ from each other.

7. The film *GATTACA*, starring Ethan Hawke and Uma Thurman, shows what might happen if genetic engineering techniques are applied to human beings. (a) In an RNA molecule, what is the sequence complementary to GATTACA? (b) Which dipeptide is produced by the first six letters of the film title?

8. The opening photograph for this chapter shows rows of silk jackets worn by horse jockeys. (a) What are some of the physical properties of silk? (b) Which detail of the molecular structure of silk is responsible for each physical property you named in (a)? (c) Which type of protein secondary structure is found in silk?

9. Tryptophan, the amino acid blamed for post-Thanksgiving-meal sleepiness, is found in high levels in turkey meat. (a) Draw the full structure of tryptophan. (b) What is its molecular formula? (c) What is its molar mass?

10. Which amino acid in each pair is more likely to be located in the interior of a protein: (a) leucine or arginine, (b) lysine or methionine, (c) proline or asparagine, (d) phenylalanine or glutamic acid?

11. Identify each of the following as carboxylic acid, ester, alcohol, or ether:

(a)

(b)

(c)

(d)

12. Which of the 20 amino acids is (a) the only one with two rings fused together, (b) the only one that is a

phenol, (c) the smaller of the two that contain an amide functional group in the side chain?

13. (a) What is the difference between the words *protein* and *peptide*? (b) Are the compounds denoted by either one of these terms a subset of the other? (c) If you are not sure which term to use, which choice is safer?

14. (a) Describe the job performed by restriction enzymes. (b) How have these compounds helped to revolutionize genetic engineering?

15. Write the RNA sequence complementary to (a) CGT-AATTCGG, (b) CGGTATTTACGG, (c) CCTAAT-CGTTAA.

More Challenging Questions

16. The amino acid sequence shown here contains several cysteines. Sketch as many structures as you can

that include each cysteine participating in one disulfide bond. (The chain shape shown here is not meant to suggest any kind of protein secondary structure.)

Thr Leu Cys Ile Lys Gln Ala Cys Gly Thr His Trp Ser Phe Cys Asn

Met

Cys

Ile Asp Gln Cys Ala

Cys

Leu

Val Ala His Glu Cys Gly Glu Phe Tyr Ser Gly Gly Pro

17. Look again at the sequence shown in Question 16. Where in this sequence would a tight turn be most likely to occur? Explain your reasoning.

18. Which are sustainable energy sources for our planet: (a) the Sun, (b) crude oil, (c) the wind, (d) coal?

19. What feature do the amino acids cysteine and methionine have in common?

20. (a) Describe the biomolecule that makes up fingernails and toenails. (b) What properties should a molecule in this role have? (c) How does the structure of that biomolecule give it these properties?

21. The temperature required to separate two strands of DNA containing only guanine and cytosine is much higher than the temperature required to separate two strands containing only adenine and thymine. Using Figure 11.34 on page 447 as a reference, explain why this is the case.

22. Figure 11.4b on page 424 shows the generic structure of the PHA molecule. (a) Write the molecular formula for the highlighted monomer repeating unit if R is $(-CH_2)_5CH_3$. (b) What is the molar mass of this monomer?

23. (a) Why are carboxylic acids considered to be acids? (b) Give two reasons carboxylic acids increase the polarity of an organic molecule.

24. How is it possible for crops to withstand the application of Roundup? Why does Roundup kill only weeds (and not crop plants) in some cases?

25. Decide whether each statement refers to a protein's primary, secondary, or tertiary structure: (a) In protein A, there is a repeating sequence of Gly-Pro-His-Gly. (b) The alpha helices in protein B come together in bundles of four to make a tightly packed group. (c) Protein C has regions of local structure typified by a zigzagging chain that has a pleated shape.

26. If you wanted to use a living organism to produce an organic molecule, why would it be preferable to use a plant rather than an animal microorganism?

27. In a two-paragraph essay, describe the role of hydroxyproline in the structure of healthy collagen. In your discussion, explain how hydroxyproline is created from proline amino acids naturally present in collagen, and discuss the effects that a lack of hydroxyproline have on the structure of collagen.

28. Name these carboxylic acids (Hint: In organic compounds, "Cl" is named "chloro-"):

(a)

(b)

(c)

29. True or false: Nonessential amino acids are so named because it is not necessary for the body to synthesize them. Essential amino acids are those the body must create. Explain your choice.

30. (a) What is the numeric value of physiological pH? (b) Fluctuations in the pH of the blood can cause conditions called alkalosis and acidosis. One involves a pH greater than physiological pH, and the other involves a pH less than physiological pH. Which is which?

The Toughest Questions

31. One strand of a double-stranded piece of DNA has the sequence CAAAAAAATCCTTATCGGAAC. (a) Write the complementary DNA sequence. (b) Write the complementary RNA sequence to the original DNA sequence. (c) Write the amino acids coded for by the RNA sequence of part (b). (d) Where would you expect to find this amino acid sequence, on the surface of a protein molecule or in its interior?

32. (a) What does the technique known as PCR do to a piece of DNA? (b) Why is it necessary to perform a PCR in some cases where DNA is to be used as forensic evidence? (c) If you begin with four copies of double-stranded DNA, how many double-stranded

DNA molecules will you have after eight rounds of PCRs? (d) After 12 rounds?

33. (a) Write the molecular formula for each molecule shown in Question 11. (b) Calculate the molar mass for each molecule. (c) Which of these molecules can contribute H atoms to a hydrogen bond? (d) Which can contribute the electronegative atom that pairs with H in a hydrogen bond?

34. In the body, proteins are bathed in water most of the time. However, there is one place in the body—the cell membrane—where proteins are embedded in a nonpolar medium. Such proteins are referred to as *membrane proteins*. How would you expect the amino acid composition of a membrane protein to differ from the amino acid composition of a protein that spends its time in water?

35. In a brief essay, use the examples of nylon and protein to compare and contrast the similarities and differences between natural polymers and synthetic ones. Include an examination of the structure of each, and comment on the ways in which each polymer can be modified.

36. People who suffer from sickle-cell anemia have red blood cells that are misshapen and tend to stick together. The cause of sickle-cell anemia is a single amino acid substitution—valine instead of glutamic acid—in the gene that codes for hemoglobin. Using the genetic code shown in Figure 11.38, determine the minimum number of nitrogenous bases that must be altered in an RNA sequence in order to change the code for glutamic acid to that for valine.

37. The structural protein keratin, which is the protein in human hair, has this type of coiled structure:

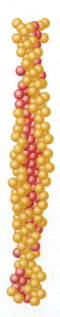

This same structure is found in other proteins. (a) Overall, the molecule twists to the left, but the individual coils twist to the right. Explain why this arrangement makes a strong fiber. (*b*) The red amino acids in the illustration are nonpolar, and the gold ones are polar. How does this pattern in the primary structure promote the formation of an opposite-twists pattern?

38. Describe how hydrogen bonding is critical for the stability of proteins and nucleic acids. Describe where hydrogen bonding contacts are made in each of these biomolecules.

39. (*a*) Identify as many organic functional groups as you can in the penicillin G molecule:

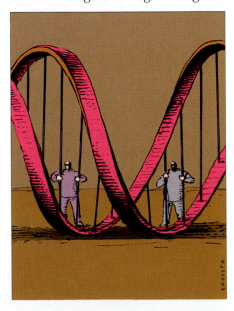

Penicillin

(*b*) Does this molecule contain a carboxylic acid functional group? (*c*) If the answer to part (*b*) is yes, write a chemical reaction that demonstrates the acidic nature of the penicillin molecule.

40. (*a*) We say that the genetic code is *redundant*, which means that, for most amino acids, there is more than one triplet code (Figure 11.38). Given this redundancy, write two RNA sequences that result in the peptide Gln-Trp-His-Gly-Pro-Gly-Pro-Arg-Ala-Thr. (*b*) Where in a protein that has tertiary structure would you be likely to find this sequence?

41. (*a*) What point is this cartoon making about our biochemical makeup? (*b*) Why are the men behind bars? (*c*) Would the cartoon have been appropriate before the advent of genetic engineering?

42. Examine the protein structure in Figure 11.25*a* (p. 441). Starting at the red cylinder at the bottom right of the illustration (the N-terminus), trace the chain from beginning to end. Along the way, identify each feature, including each region that connects secondary structures to each other. By trying to imagine this as a space-filling structure, speculate on which areas will be buried in the interior of the protein and which will be located on the surface.

43. In some instances, alpha helices have been known to have a repeating structure that can be represented as (nonpolar-x-polar)$_n$, where x is any amino acid. If such an alpha helix is located partially on the surface of a protein, what is the purpose of this alternating sequence?

44. In Section 11.1, we discuss several ways in which natural polymers can replace synthetic ones. (*a*) Summarize two of these examples and compare them. (*b*) Do the examples make use of plant crops or animal microorganisms? (*c*) Do these approaches require genetically engineered organisms? (*d*) Do the examples illustrate sustainable processes?

45. In a short essay, discuss one example of a transgenic organism described in this chapter. Include information on how a transgenic organism is produced from another organism's genetic material.

E-Questions

Go to **www.prenhall.com/waldron** to find these questions in electronic form, complete with hyperlinks directly to the various websites cited in the questions.

46. Use the website **colostate** to get to the article "Transgenic Crops Currently on the Market." After reading this article, answer the following questions: (*a*) In 2000, what country had the highest acreage of transgenic crops? (*b*) What are the top three transgenic crops planted in the United States? (*c*) What is Bt, and how is it used in the genetic engineering of crops? (*d*) Why have genetically altered papayas been developed?

47. The article "Bacterial Plastics" can be found at **bact.wisc**. Read this article, and then answer the following questions: (*a*) What three types of biodegradable plastics are discussed in the article? (*b*) Why are synthetic plastics referred to as xenobiotic? What does this term mean? (*c*) What drawbacks are associated with PHA-based plastics?

48. The article "Seafood into Super Glue" discusses how the glue on the "feet" of mollusks is one of the strongest adhesives known. Go to the **sciencedaily** website, and search for *seafood* and *glue*. Read the article, and then answer the following questions: (*a*) What is mollusk glue is made of? (*b*) How successful have researchers been at engineering the genes required to make this glue? (*c*) What special advantage does glue from mussels provide over traditional glues?

49. Read the article "Prion Diseases" at **Prion**, and then answer the following questions: (*a*) What is the formal definition of the term "prion"? (*b*) Name two examples of prion diseases that infect wildlife. (*c*) Name two examples of prion diseases that infect humans. (*d*) What are two ways that humans can become infected with a prion diesase?

50. After reading the short article "Wheat Wrappers for Fast Food," by H. Frazen, at the **Scientificamerican** website, answer the following questions: (*a*) What are two drawbacks to using polystyrene "clamshell" containers for fast food? (*b*) What are two drawbacks to using cardboard containers for fast food? (*c*) What alternative material does the article suggest? Has this idea been tried before? Explain.

WORLD WIDE WEB RESOURCES

As with the E-Questions, go to **www.prenhall.com/waldron** to find these questions in electronic form, complete with hyperlinks directly to the various websites cited.

Some of following links contain research articles related to the topics in this chapter and could be used as the basis for a writing assignment in your course. Other sites are of general interest.

- Read an excellent article on spider silk called "An Airbus Could Tiptoe on Spider Silk" at **Biocom**, and answer the following questions: (*a*) What is dragline silk made of and why has its popularity increased recently? (*b*) Describe the secondary structure of dragline silk. (*c*) What feature of dragline silk's structure makes it difficult to use for human activities?

- Some detractors of new plastics technology claim that "green" plastics are more costly to make than traditional ones. Read the article "How Green Are Green Plastics?" at **Mindfully** (August 20, 2000) to see the other side of this issue.

- The article, "Lawyer: New DNA tests point to killer in Sheppard case", talks about the case of Dr. Sam Sheppard who was accused of killing his wife in 1954. Find it at **USNews**. The case, upon which the movie and television series *The Fugitive* were based, is still making news. Read about the history of this case and the role DNA technology has played in it.

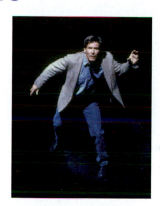

- Read two point-of-view articles on the transgenics debate. At the site **Monsanto**, the article "Backgrounder: History of Monsanto's Glyphosate Herbicides" is an overview of the Roundup product line. For an alternative point of view, go to **Mindfully**, a site devoted to promoting environmental awareness, specifically the benefits of sustainable resources and organic farming. Search for the article "The Green Revolution: A Critical Look," by P. Goettlich. After considering what each article has to say, think about the biases inherent in these two reports. Consider also the style of each report. (*a*) What is your best guess as to the audience reading each? (*b*) How are the tone and style of each article appropriate for the article's intended audience? (*c*) After reading these two articles, do you know where you stand on this ethical question?

CHEMISTRY TOPICS IN CHAPTER 12

- Lipids and triacylglycerides [12.1]
- Phospholipids [12.2]
- Catabolism and anabolism [12.3]
- Acid-catalyzed hydrolysis [12.3]
- Enzymes as catalysts [12.3]
- Reaction coordinates and transition states [12.3]
- Mono-, di-, and polysaccharides [12.4]
- The aldehyde functional group [12.4]
- Energy units for foods [12.5]
- Fuel values of foods [12.5]

"There is no sincerer love than the love of food."

GEORGE BERNARD SHAW

Groceries

THE CHEMISTRY OF THE FOODS WE EAT

For many of us, a trip to the grocery store can be a confusing and frustrating experience. Walking down the aisles, we are tempted to buy a food that is appealing, but then we remember something about that food which steers us away. We are fortunate that nowadays most items in the grocery store include nutritional information, but understanding that information can be a complicated task. We are bombarded with data from the health community telling us which foods are "bad" and which are "good." For instance, we know that foods high in cholesterol and high in fat are bad, but we are also told that there are good and bad kinds of cholesterol and good and bad kinds of fats. Fiber presents another problem. We know that a diet high in fiber is good, but fiber is a carbohydrate and we also hear that an excess of carbohydrates in the diet is bad. We sometimes hear about the benefits of a high-protein diet, but a look at the food pyramid shown in Figure 12.1 shows that protein is only one of the components of a healthy diet. Further, everyone knows that eggs, steaks, and hamburgers are "bad," and so how can a high-protein diet be healthy? On the vitamin–mineral front, everyone knows that we must be sure to take in enough calcium to keep bones strong and healthy, but should you buy the orange juice spiked with added calcium,

Fats, oils (lipids)
Bread, cereal, rice, pasta
(complex carbohydrates)
Meat, poultry, fish, dry beans,
eggs, nuts (primarily protein)
Fruits
(carbohydrates, vitamins)
Vegetables
(vitamins, fiber)
Milk, yogurt, cheese
(primarily protein)

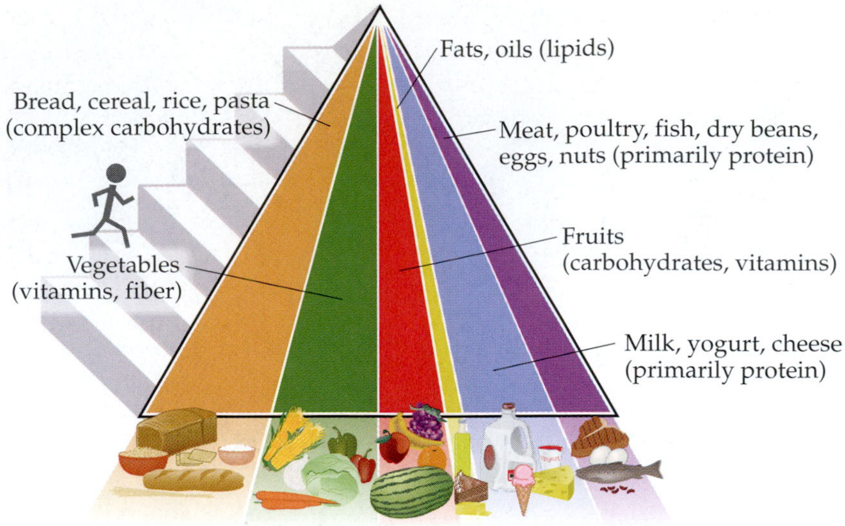

▶ FIGURE 12.1 The food pyramid. This schematic published by the U.S. Department of Agriculture is designed to teach us which foods should be eaten sparingly (thinner wedges) and which should be a substantial part of a healthy diet (thicker wedges).

or is it healthier to buy the juice the way nature made it, with no added calcium?

Faced with a plethora of information about nutrition, it is no wonder that many of us leave the grocery store feeling guilty, bewildered, or annoyed. The irony is this: despite all we know about good and bad food, we are still making the wrong choices, according to reputable studies on American health and nutrition. For example, one measure of ideal weight is a number called the *body mass index* (BMI). Anyone with a BMI above 25 is considered overweight, and anyone with a BMI greater than 30 is considered obese. Healthy individuals have BMI values between 18.5 and 25. According to the most recent BMI study from the National Institutes of Health (NIH), only 36 percent of Americans have a healthy weight, with 34 percent classified as overweight and another 30 percent as obese.

You can calculate your BMI from the formula

$$\text{BMI} = \frac{(\text{mass in pounds})(704.5)}{(\text{height in inches})^2}$$

Figure 12.2 shows weight data for Americans during five periods between 1960 and 2000. These data indicate that the number of overweight individuals is very high, but has stayed fairly constant over these 40 years. The more alarming data are in the number of obese persons, which has more than doubled over the same period. Whom can we blame for the fact that 130 million of us are overweight? Are we ignoring advice from the health community? Or are we simply befuddled by the deluge of information we are given? How can we know what information to use and what to ignore? In short, why can't we eat well?

Percent of U.S. population
0
10
20
30
40
50
38.8
42.5
40.1
40.4
39.3
1960–1962
1971–1974
1976–1980
1988–1994
1999–2000
24.5
24.3
24.9
24.7
26.0
10.7
12.2
12.3
20.6
27.7
15.7
15.8
17.1
25.0
34.3
Men
Women
Men
Women
BMI 25–29.9 (overweight)
BMI ≥ 30 (obese)

▶ FIGURE 12.2 Body mass index values for overweight and obese U.S. women and men, 1960–2000. Percentages of overweight and obese Americans, broken down by sex and divided chronologically into five groups that span the years 1960 through 2000.

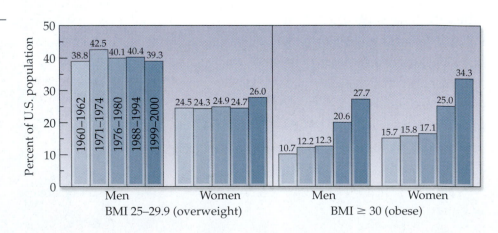

In this chapter, we shall demystify some of the claims made about food and about what constitutes a healthy diet. We shall look at the three major food groups: protein, carbohydrate, and fat. Back in Chapter 11, we considered the structure and function of proteins. In this chapter, we shall look closely at the structure of the other two—carbohydrates and fats—and at how they interact with proteins. Then we shall consider how foods are processed in our bodies and why some foods are categorized as "bad" and others as "good." We cannot possibly clarify every aspect of modern nutrition, but an understanding of the basic composition of food and of how food is used as fuel will provide you with the background necessary to understand nutritional information and to make well-informed decisions at the grocery store.

12.1 | The Fat Tax

When you consider the way American lifestyles have changed over the past few decades, it does not seem surprising that bathroom-scale readings are creeping higher and higher each year. Twenty-first-century Americans enjoy diversions that require little or no physical activity, and with the Internet, video games, and plain old television, entertainment that burns few calories is all too easy to come by. Couple this with the convenience of fast-food restaurants and the increasing size of their fat-laden meals, and it is no wonder the recent NIH study just described showed that about one-third of Americans are overweight (BMI > 25) and almost another one-third are obese (BMI ≥ 30).

The government invests substantial resources into collecting these data, but why should the government care about the mass of the American public? Why are our bodies, and the mass of matter they contain, anybody's business but our own? The answer is that obesity takes a significant toll on our health care system. Estimates suggest that as much as 10 percent of our annual health care budget is spent on treating problems related directly to obesity and that, for all causes, the risk of death for obese people is 50 to 100 percent higher than for the normal-weight public.

QUESTION 12.1 Use Figure 12.2 to answer these questions: (*a*) In 2000, was there a higher percentage of overweight American men or American women? (*b*) What percentage of men and women were obese in 1962? (*c*) In any given year, is there a larger percentage of American men or women with BMI between 25 and 29.9? (*d*) If there were 143 million American women in 2000, how many of them were of normal or below-normal weight (BMI < 25) that year?

ANSWER 12.1 (*a*) More men (39.3 percent) than women (26.0 percent). (*b*) 10.7 percent of men, 15.7 percent of women. (*c*) Men. (*d*) The total percentage of women with BMI > 25 was 26.0% + 34.3% = 60.3%. Thus, the percentage of women with BMI < 25 was 100.0% − 60.3% = 39.7%. This equals 0.397 × 143,000,000 = 56,800,000 women.

Can America afford to be fat? According to the Center for Advancement of Public Health:

No one is suggesting the creation of a refrigerator police, but so long as the government is spending $360 billion per year at the federal level on health through Medicare, Medicaid, and the Children's Health Insurance Program, the government's interest in trying to prevent needless illness and death from obesity is . . . simple [to understand].

There have been many suggestions for addressing the problem. One idea is to tax junk foods such as soft drinks and French fries. A more radical suggestion comes from investigative journalist Jonathan Rauch in his article "The Fat Tax" in the *Atlantic Monthly*. According to Rauch, Americans should be taxed by the pound, and cash bonuses should be given to Americans who maintain healthy, fit bodies.

The food pyramid is the U.S. government's way of guiding us toward healthier eating and leaner bodies. It is interesting to note that nutrition guidelines in other developed countries are much the same as the American guidelines, although they are often presented in a format other than the pyramid of Figure 12.1. A comparison of Figures 12.1 and 12.3 tells you that, whether you live in Portugal, China, or the United States, your daily intake of fat should be minimal.

If we are eating foods that contain all three major food biomolecules—fat, carbohydrate, and protein—then why is it only fat that is added to our bodies when we overeat? The answer is that fat is the way we humans store energy. Your body requires a certain amount of energy to fuel a certain level of activity. To maintain a steady weight, ideally you should consume exactly that amount of energy in the form of food. The problem arises when you consume excess calories, and those extra calories are most often in the form of fats and carbohydrates. When excess fat is consumed, the portion not used as fuel is stored in adipose (fatty) tissue. When excess carbohydrates are consumed, the leftovers are converted to fat. This fat also goes into long-term storage in adipose tissue.

Of the four major classes of biomolecules—DNA and RNA, proteins, carbohydrates, and fats—fats are the only ones that are not polymeric. That is, they are not made up of repeating units of a smaller molecule. The word *fat* is colloquial, and it is used to describe the subset of biological molecules called **lipids**, which are defined as biomolecules that dissolve readily in nonpolar solvents. In this chapter, we shall explore different classes of lipids that can be found in the human body or are part of the human diet.

▶ FIGURE 12.3 **Ideal-diet graphics for Portugal and China.** Other countries also use graphics to display the ideal diet for their citizens. In Portugal they use a Roda dos Alimentos ("Wheel of Food"), while in China they use a pyramid that is pagoda-shaped.

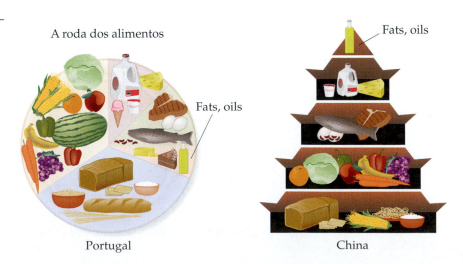

A roda dos alimentos

Fats, oils

Fats, oils

Portugal

China

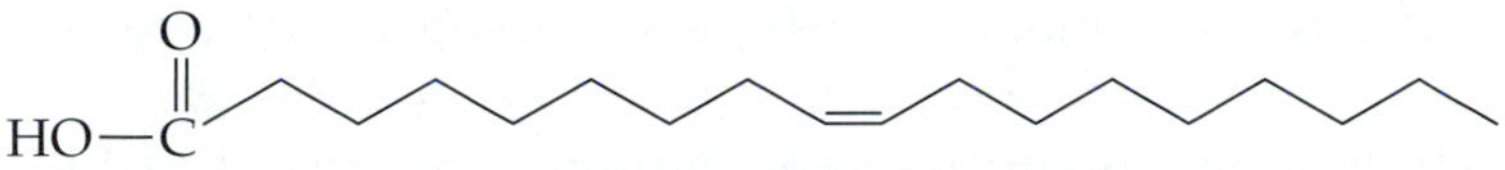

▲ **FIGURE 12.4** **The simplest lipids.** Every fatty acid molecule consists of a carboxylic acid functional group at one end of a long hydrocarbon chain. Because they are lipids, all fatty acids are hydrophobic.

The simplest lipids are the **fatty acids**. Every fatty acid molecule is composed of a long hydrocarbon chain with a carboxylic acid functional group (—COOH) attached at one end (Figure 12.4). The presence of a carboxylic acid group is usually a signal that a molecule is polar, because this functional group contains electronegative atoms that cause the electron density in the molecule to be lopsided. As we know from previous chapters, however, molecules that contain hydrocarbon groups are decidedly nonpolar. Why, then, are fatty acids classified as nonpolar lipids if they have both polar and nonpolar character? The answer is that, because the hydrocarbon chains in fatty acids are so long, the nonpolar character of the chain is dominant over the polar character of the carboxylic acid functional group. Because the nonpolar character is dominant, fatty acids dissolve in nonpolar solvents and are classified as lipids.

Table 12.1 lists some of the fatty acids found in the human diet. Notice that they are divided into two groups: saturated and unsaturated. These identifiers tell us whether there are any carbon–carbon double bonds in a fatty acid molecule. Fatty acids having only single bonds between carbon atoms are referred to as *saturated*, while those which contain carbon–carbon double bonds are *unsaturated*. Monounsaturated fatty acids contain one double bond, and polyunsaturated acids contain more than one double bond.

RECURRING THEME IN CHEMISTRY Like dissolves like.

TABLE 12.1 Some Biologically Important Fatty Acids

Common name	Shortened condensed structure	Number of carbon atoms in fatty acid	Melting point (°C)
Saturated fatty acids			
Capric	$CH_3(CH_2)_8COOH$	10	31.6
Lauric	$CH_3(CH_2)_{10}COOH$	12	44.3
Myristic	$CH_3(CH_2)_{12}COOH$	14	53.9
Palmitic	$CH_3(CH_2)_{14}COOH$	16	63.1
Stearic	$CH_3(CH_2)_{16}COOH$	18	69.6
Arachidic	$CH_3(CH_2)_{18}COOH$	20	76.5
Behenic	$CH_3(CH_2)_{20}COOH$	22	81.5
Lignoceric	$CH_3(CH_2)_{22}COOH$	24	86.0
Cerotinic	$CH_3(CH_2)_{24}COOH$	26	88.5
Unsaturated fatty acids			
Palmitoleic	$CH_3(CH_2)_5CH{=}CH(CH_2)_7COOH$	16	0
Linoleic	$CH_3(CH_2)_4CH{=}CHCH_2CH{=}CH(CH_2)_7COOH$	18	−5
Linolenic	$CH_3CH_2CH{=}CHCH_2CH{=}CHCH_2CH{=}CH(CH_2)_7COOH$	18	−11
Arachidonic	$CH_3(CH_2)_4CH{=}CHCH_2CH{=}CHCH_2CH{=}CHCH_2CH{=}CH(CH_2)_3COOH$	20	−50

Remember that, for any substance, melting point and freezing point are the same temperature. We use two terms because sometimes we are thinking in terms of going from liquid to solid (freezing) and other times we are thinking in terms of going from solid to liquid (melting).

▼ FIGURE 12.5 **Temperature makes the difference.** As is true of all other substances, fatty acids are solid at any temperature below their melting point and liquid at any temperature above their melting point. **(a)** The melting point of linoleic acid is −5 °C, meaning that it is a liquid at 20°C (room temperature) and can be poured over your salad. **(b)** At any temperature below −5 °C, linoleic acid is a solid.

Linoleic acid in liquid phase
(a)

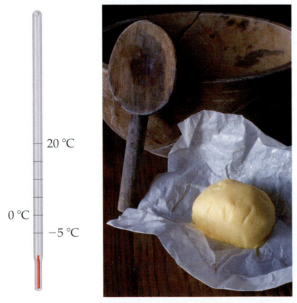

Linoleic acid in solid phase
(b)

The number of carbon–carbon double bonds a fatty acid contains has a dramatic effect on its physical properties. As an example, take a look at the melting points shown in the far right column in Table 12.1. Notice the correlation between number of double bonds and melting point. All the saturated fatty acids listed have a melting point above room temperature (about 25 °C). The four unsaturated examples in the table have one, two, three, and four double bonds, and the melting point gets lower as the number of double bonds increases. Additionally, the melting point of every unsaturated fatty acid in the table is below room temperature, which means that they all are liquids at room temperature. This relationship between temperature and physical phase is illustrated in Figure 12.5 for linoleic acid. When the temperature is above −5 °C, that fatty acid is a liquid; when the temperature is below −5 °C, it is a solid.

QUESTION 12.2 Have you ever noticed that some oily substances congeal when cooled in your refrigerator? When an oil that is a liquid at room temperature is cooled to below its melting point, it becomes a solid. (*a*) Say your refrigerator is set at 3 °C. Which of the unsaturated fatty acids in Table 12.1 will congeal if refrigerated overnight? (*b*) Which of the unsaturated fatty acids in Table 12.1 will congeal if stored overnight in a freezer at −22 °C?

ANSWER 12.2 (*a*) Any fatty acid that has a melting point below the temperature of the refrigerator will not congeal. Therefore, none of the four unsaturated fatty acids will congeal. (*b*) Any fatty acid that has a melting point above the temperature of the freezer will congeal. Therefore, the palmitoleic, linoleic, and linolenic acids will congeal, but the arachidonic acid will not.

It is clear that as the number of double bonds in a fatty acid increases, the melting/freezing point decreases, but why does such a correlation exist? How do double bonds affect the temperature at which a fatty acid changes phase? Figure 12.6 shows us the answer. As you can see, any double bond in a fatty acid chain causes a kink. When a fatty acid (or almost any other substance, for that matter) changes from the liquid phase to the solid phase, the molecules get closer together and pack into a firm solid. When there is a kink in the fatty acid molecule, this packing process is less efficient. Usually, the more kinks there are in a fatty acid, the more difficult it is for the molecules to pack in tightly against one another.

As an example, let us begin by comparing two of the four 18-carbon acids in Figure 12.6: stearic acid and oleic acid. When the saturated stearic acid solidifies, all the (unkinked) hydrocarbon chains can line up adjacent to one another and pack very tightly together. The closely packed chains make for a solid that has a relatively high density, and a high temperature is required to melt such high-density substances. In oleic acid, the kink in the hydrocarbon chain keeps the molecules from packing easily into

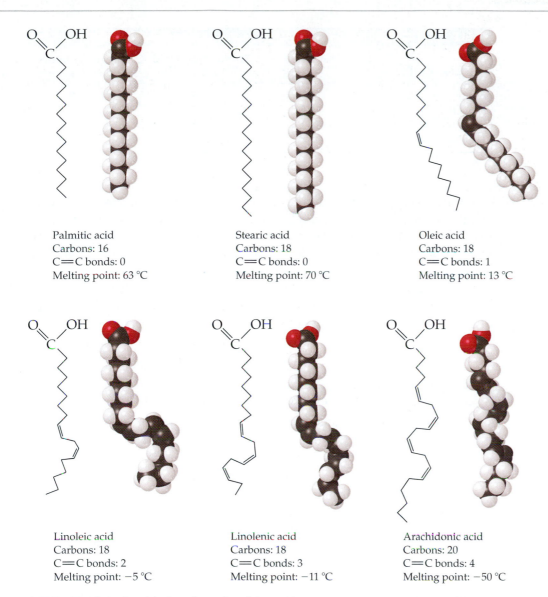

Palmitic acid
Carbons: 16
C=C bonds: 0
Melting point: 63 °C

Stearic acid
Carbons: 18
C=C bonds: 0
Melting point: 70 °C

Oleic acid
Carbons: 18
C=C bonds: 1
Melting point: 13 °C

Linoleic acid
Carbons: 18
C=C bonds: 2
Melting point: −5 °C

Linolenic acid
Carbons: 18
C=C bonds: 3
Melting point: −11 °C

Arachidonic acid
Carbons: 20
C=C bonds: 4
Melting point: −50 °C

▲ FIGURE 12.6   **Double bonds and melting point.**   Line structures and space-filling models for six fatty acids. Notice how the molecules are affected by the presence or absence of double bonds in the hydrocarbon chain. The more C=C double bonds in a fatty acid molecule, the lower the melting point of the acid.

a compact arrangement. Consequently, solid oleic acid has a density lower than that of solid stearic acid. Less heat is required to melt this less dense solid, and so the melting point of oleic acid is lower than that of stearic acid. You should be able to extend this argument to the other two 18-carbon fatty acids in the figure. The more double bonds in the hydrocarbon chain, the more kinks there are in the chain, the less compact the solid, and the lower the melting point.

   This phenomenon has another effect for fatty acids: the higher the degree of unsaturation, the softer the solid. Suppose you put our four 18-carbon fatty acids in your freezer tonight, and all are cooled to below their melting points. When you check in the morning, you will find that the solids have different degrees of firmness. The most unsaturated one (linolenic) will be easiest to move a knife through, and the saturated one (stearic) will be the firmest and therefore hardest to cut.

TABLE 12.2 Comparison of Fatty Acid Content of Selected Fats and Oils			
Product	Saturated (%)	Monounsaturated (%)	Polyunsaturated (%)
Butter	68.2	27.8	4.0
Hard (stick) margarine	18.2	50.5	31.3
Soft (tub) margarine	17.1	36.5	46.4
Vegetable oil	10.8	20.4	68.8

RECURRING THEME IN CHEMISTRY The individual bonds between atoms dictate the properties and characteristics of the substances that contain them.

Butter, margarine, salad oils, and any other lipids we commonly consider to be food all contain a mixture of fatty acids, and we refer to these mixtures in nonscientific jargon as *oils* if they are liquids at room temperature and as *fats* if they are solids at room temperature. Table 12.2 lists some common oils and fats and the percentages of saturated and unsaturated fatty acids they contain. These numbers show us that, the firmer the product, the higher its percentage of saturated fatty acids. Thus, the percentage of saturated fatty acids is lowest for the vegetable oils, which are completely liquefied at room temperature. This trend is the direct result of molecular structure.

QUESTION 12.3 Comment on this chain of reasoning:

Oils are composed mainly of monounsaturated and polyunsaturated fatty acids. Thus, they are always in the liquid phase. Fats, however, contain a higher proportion of saturated fatty acids. Therefore, they are always in the solid phase.

ANSWER 12.3 In order for the preceding to be sound, the second sentence and the final sentence should include the qualifying phrase *at room temperature*. It is always possible to change liquids into solids by cooling them. Likewise, it is possible to change solids into liquids by heating them. Oils are liquids and fats are solids *at room temperature*.

Many Americans recognize the terms *saturated* and *polyunsaturated* because we are constantly bombarded with information about how fats affect our health. What's more, the products we buy in the grocery store are advertised in a way that appeals to our health concerns, as Figure 12.7 shows. It is a well-established fact that saturated fats are correlated with increased cholesterol levels and have been linked to heart disease, hypertension, and diabetes, among other things. Given this fact, why do we eat saturated fats at all? Why not just boil our food rather than cook it in butter or oil? The answer lies in the difference between the boiling point of water (100 °C) and the boiling points of fats and oils (260 to 400 °C). When you cook something in boiling water, the maximum possible cooking temperature is only 100 °C. When you cook something in oil or fat, it is possible to cook at a much higher temperature. High temperatures allow foods to brown, and the flavor associated with browning is one of the tastes many of us love. Frying and sautéing add a delicious buttery or oily flavor to boot. (We shall look at the reactions involved in browning later in this chapter.)

If you have spent some time in the kitchen, you may have noticed that overheated oil or fat produces smoke that fills your kitchen with

▼ FIGURE 12.7 **Many food labels offer nutritional information.** At your local grocer, you will find dozens of lipid-based products. Advertisements, such as those which say "no genetically engineered ingredients," help us to decide which products fit our dietary requirements best.

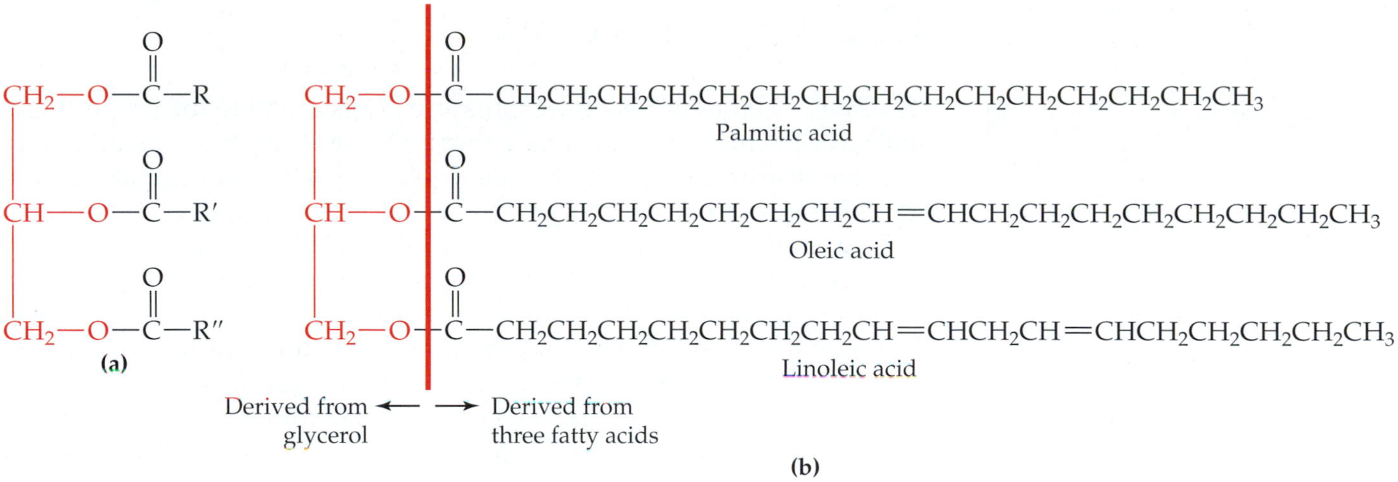

$$CH_2-O-\overset{\displaystyle O}{\overset{\|}{C}}-R$$
$$CH-O-\overset{\displaystyle O}{\overset{\|}{C}}-R'$$
$$CH_2-O-\overset{\displaystyle O}{\overset{\|}{C}}-R''$$
(a)

$$CH_2-O-\overset{\displaystyle O}{\overset{\|}{C}}-CH_2CH_2CH_2CH_2CH_2CH_2CH_2CH_2CH_2CH_2CH_2CH_2CH_2CH_2CH_3$$
Palmitic acid

$$CH-O-\overset{\displaystyle O}{\overset{\|}{C}}-CH_2CH_2CH_2CH_2CH_2CH_2CH_2CH=CHCH_2CH_2CH_2CH_2CH_2CH_2CH_2CH_3$$
Oleic acid

$$CH_2-O-\overset{\displaystyle O}{\overset{\|}{C}}-CH_2CH_2CH_2CH_2CH_2CH_2CH_2CH=CHCH_2CH=CHCH_2CH_2CH_2CH_2CH_3$$
Linoleic acid

Derived from ⟷ Derived from
glycerol three fatty acids

(b)

▲ **FIGURE 12.8 Triacylglycerides.** **(a)** This generic molecule shows how a triacylglyceride is a combination of one glycerol molecule (red) and three fatty acid molecules (black). **(b)** The triacylglyceride made up of palmitic acid, oleic acid, and linoleic acid.

acrid, unpleasant fumes. To understand what causes the smoking, you have to know that the fatty acids in oils and fats are connected together in groups of three in molecules called **triacylglycerides** (**TAGs**, also sometimes called *triglycerides*). The structure of a generic TAG is shown in Figure 12.8*a*, and Figure 12.8*b* shows an actual TAG molecule. Note several things about this structure. First, the three fatty acids contained in a TAG need not be all the same. In fact, they are usually different from one another and may be a mixture of saturated and unsaturated fats. Second, the portion of the structure that connects the fatty acids is the molecule *glycerol*, an alcohol containing three —OH groups (Figure 12.9). When three fatty acids combine with a glycerol molecule, the carboxylic acid groups on the fatty acids attach to the glycerol —OH groups, and three ester groups result.

This process of attachment is reversible, which means that when your body needs energy, it can metabolize TAGs to release their fatty acids. In a frying pan, the same process occurs as oils and fats are heated: the ester bonds break, and the fatty acids and glycerol separate from one another. The smoke observed when fats and oils are overheated results when the glycerol molecules undergo a reaction to form a noxious little molecule called *acrolein*, which is to blame for the pungent smell of smoking oil and fat (Figure 12.9).

▼ **FIGURE 12.9 Decomposing glycerol molecules cause overheated cooking oil to smoke.** When triacylglycerides in food are heated, they break down into glycerol plus three fatty acids. At high temperatures, the glycerol molecule continues to degrade, forming the smelly acrolein molecule and water.

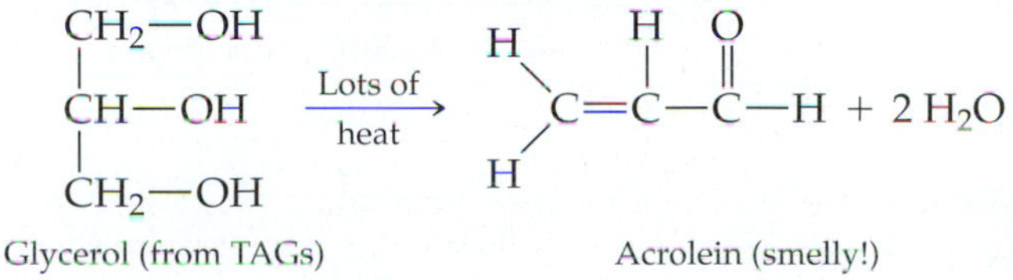

$$CH_2-OH$$
$$CH-OH \quad \xrightarrow[\text{heat}]{\text{Lots of}} \quad \overset{H}{\underset{H}{>}}C=\overset{\overset{\displaystyle H}{|}}{C}-\overset{\displaystyle O}{\overset{\|}{C}}-H + 2\,H_2O$$
$$CH_2-OH$$

Glycerol (from TAGs)   Acrolein (smelly!)

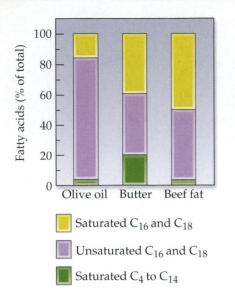

▲ FIGURE 12.10 Fatty acid comparison in three cooking lipids. These three common lipids vary from one another in their ratios of saturated to unsaturated fatty acids and in their ratios of long-chain to short-chain fatty acids. As usual, the degree of saturation determines the phase of each substance. At 25°C (approximately room temperature), olive oil is a liquid and butter and beef fat are solids.

12.2 | Organized Fats

The different types of fat we see in nature are essential to the proper functioning of the organism containing them. Figure 12.10 shows us how fatty acids are distributed in three familiar lipids: olive oil, butter, and beef fat. The differences in fatty acid composition in these substances, in terms of both saturation and chain length, are clear. One thing all three lipids have in common, though, is that the fatty acids in them are, for the most part, stored in the form of TAGs. TAGs are a way to combine fatty acids into groups of three, and different combinations of saturated and unsaturated fatty acids can be grouped into one TAG molecule for storage.

Fatty acids in the body are also incorporated into molecules known as **phospholipids**, which are composed of a glycerol backbone (like a TAG), but with only two fatty acid molecules attached. The third position on the glycerol backbone is occupied by a phosphate group connected to some small organic group (represented by X in Figure 12.11a). An example of a phospholipid molecule is shown in Figure 12.11b. If we think of TAGs as three-tailed lipids because they contain three fatty acid chains, we can think of phospholipids as two-tailed lipids. These two-tailed lipids are the primary component of the membrane that encloses every cell in your body. Lined up side by side and in two layers (or a *bilayer*), as illustrated in Figure 12.12, phospholipids form a water-resistant barrier that encloses the contents of the cell. The molecules in one layer of the bilayer have their polar phosphate ends facing the exterior of the cell, which is an aqueous environment and therefore very polar. The molecules in the other layer of the bilayer have their polar phosphate ends facing the interior of the cell, also a watery environment. With this arrangement, the nonpolar hydrocarbon chains of both sets of molecules in the bilayer are facing toward the membrane interior, which is hydrophobic.

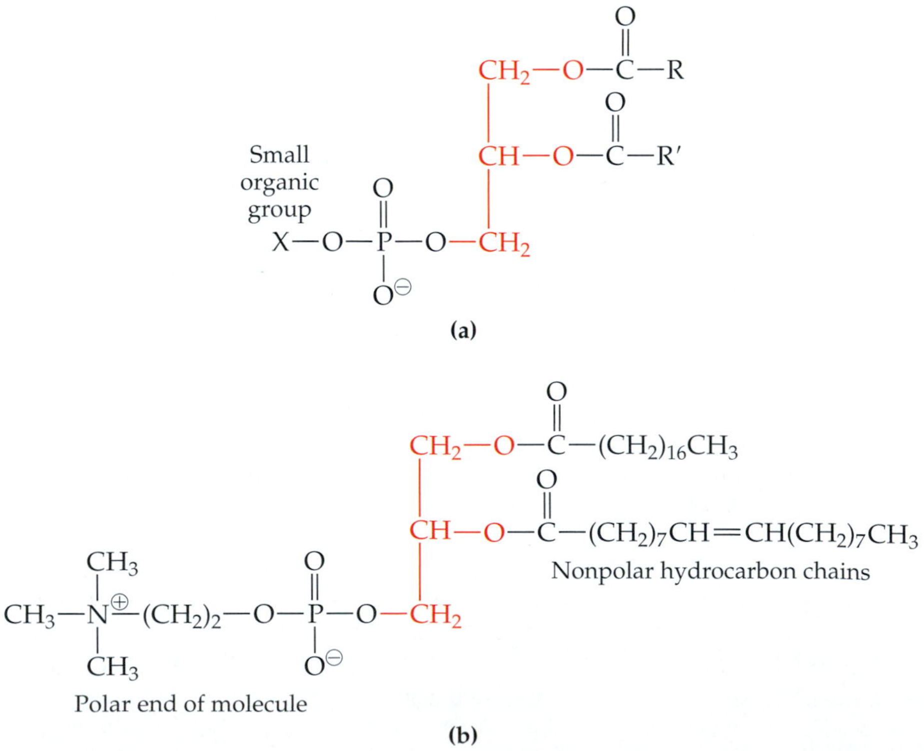

► FIGURE 12.11 Phospholipids. (a) Phospholipids are based on the glycerol molecule, to which two fatty acids are attached. At the third position on the glycerol molecule, a phosphate group is attached, and a small organic group is attached to that. **(b)** A specific phospholipid. The charged part of the molecule gives it a polar character, and the long hydrocarbon chains give it a nonpolar character.

The fluidity of the cell membrane is dictated, in part, by the types of fatty acids in its phospholipids. In the same way that saturated fatty acids form a firm solid in butter by packing tightly together, a high proportion of saturated fatty acids in a cell membrane's phospholipids increases the rigidity of the membrane. Conversely, a high percentage of unsaturated phospholipids make a membrane that is fluid and flexible. One reason for the diverse collection of different fatty acids in the body is to enable different phospholipids to be added to and removed from cell membranes as environmental conditions change. For example, studies on ground squirrels (that is, squirrels that live in the ground, not ground-up squirrels) show that, as the squirrels prepare to hibernate for the winter, the composition of their membrane lipids changes dramatically to allow the membranes to remain fluid at very low temperatures. The same phenomenon has been observed in reindeer: the percentage of unsaturated lipids increases as you move from the top of a reindeer's leg down to the hoof (Figure 12.13). This adaptation allows the cell membranes near the hoof, where it is colder, to have the same fluidity as the cells membranes in the rest of the leg, where it is warmer. Without the ability to change lipid composition, a cell membrane runs the risk of becoming rigid, in the same way that cooking oil may solidify when put into the refrigerator.

QUESTION 12.4 Use Table 12.1 on page 467 to answer this question: Which of the two in each pair of fatty acids is more likely to be at the top of a reindeer's leg, and which is more likely to be closer to the hoof: (*a*) myristic acid and arachidonic acid, (*b*) linolenic acid and lauric acid, (*c*) palmitic acid and palmitoleic acid?

ANSWER 12.4 (*a*) Myristic top, arachidonic bottom; (*b*) lauric top, linolenic bottom; (*c*) palmitic top, palmitoleic bottom.

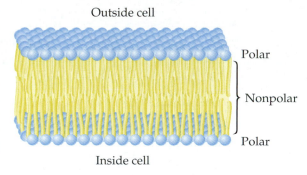

▲ **FIGURE 12.12 The phospholipid bilayer.** The membrane enclosing the cells in the human body comprises a double layer—a *bilayer*—of phospholipid molecules.

Outside cell

Polar

Nonpolar

Polar

Inside cell

▲ **FIGURE 12.13 Nature's clever ways.** The lipid composition in the leg of a reindeer varies from top to bottom. At the top, where the leg is warmer, there is a larger percentage of saturated lipids. Towards the hoof, where it is colder, there is a larger percentage of unsaturated lipids. These unsaturated molecules, because they have relatively low melting points, stay flexible and pliable even when the hoof is in contact with the very cold ground.

What Are Trans Fats, and Why Should I Worry About Them?

STUDENTS OFTEN ASK

Trans fats have been making news since the mid-1990s, when consumer groups began to petition the government to pay attention to the presence of these disease-promoting fats in commercial food products. In order to understand what makes trans fats troublesome, we must turn to their molecular structure,

as we often do when thinking about how a substance behaves. In Chapter 9, we learned about cis and trans double bonds in alkenes. Figure 12.14 shows both as they appear in a typical lipid molecule. A look back at Figure 12.6 on page 469 will show you that most naturally occurring fatty acids are in the cis orientation. In this orientation, the kink produced by the double bond distorts the hydrocarbon chain. As we know from the previous discussion, that distorted shape causes unsaturated lipids to

be relatively soft at room temperature because the hydrocarbon chains cannot pack together closely.

A visit to the dairy case of a grocery store tells you that margarine can be very soft or quite firm, and is sold in tubs or sticks. Modern margarines are typically derived from plants and contain a high percentage of monounsaturated and polyunsaturated lipids that are either oils or very soft solids at room temperature (Table 12.2). These products are appealing to consumers who want to buy

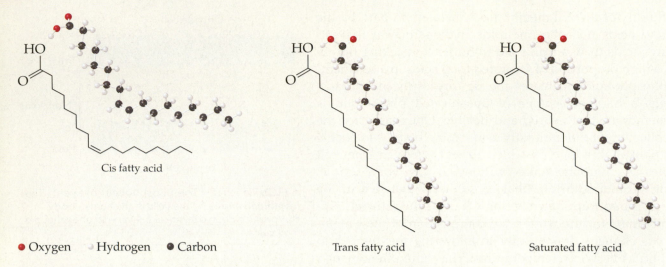

● Oxygen   ◦ Hydrogen   ● Carbon   Trans fatty acid   Saturated fatty acid

▲ **FIGURE 12.14   Cis and trans double bonds in lipids.**   A cis bond causes a kink that leads to a noticeable bend in the hydrocarbon chain of a lipid molecule. Because the kink produced by a trans double bond does not create a noticeable bend in the hydrocarbon chain, a trans fatty acid and a saturated fatty acid look similar.

heart-healthy products containing low percentages of saturated fats.

Some margarines contain a high percentage of unsaturated fats and are therefore very soft. Consumers love butter, though, and so margarine manufacturers have sought ways to make soft margarines more like butter. Nowadays, we have margarine that is more butterlike. These products are available as a result of some creative chemistry. When some of the cis double bonds in the unsaturated lipid molecules of margarine are converted to single bonds, a firmer, more easily molded product that has a more butterlike taste is formed. This process also extends the product's shelf life, because fewer double bonds mean a lower tendency towards rancidity.

The reaction used to create saturated lipids from cis unsaturated lipids is an *addition reaction* (Section 3.2). In this reaction, hydrogen molecules are added to the unsaturated lipid, and the two atoms of the hydrogen molecule add across the cis double bonds, changing them from double to single:

$$CH_3CH_2C\!\!=\!\!CCH_2CH_3 + H_2 \longrightarrow CH_3CH_2\underset{\underset{H}{|}}{\overset{\overset{H}{|}}{C}}-\underset{\underset{H}{|}}{\overset{\overset{H}{|}}{C}}CH_2CH_3$$

This type of addition reaction is referred to as **hydrogenation** (hy drodge eh NAY shun), because it is hydrogen atoms that are added across the double bond. There is one problem, though: When the reaction is carried out, some of the cis double bonds are converted not to single bonds, but rather to trans double bonds. Trans fats found in food products are usually produced through this side reaction.

The point of hydrogenation is to produce saturated hydrocarbons from unsaturated hydrocarbons in which

the double bonds are mainly cis double bonds. Given this, why are trans double bonds permitted to remain in the product? The answer is that the kinks in trans double bonds are not as pronounced as those in cis double bonds, and therefore trans chains are straighter and more like saturated chains (Figure 12.14). This means that lipids made from trans fats are packed more tightly against one another and have desirable higher melting points and a firmness like butter. Thus, it is all right that not all of the products of the hydrogenation reaction are saturated chains. The trans chains accomplish the same goal (a product that is solid at room temperature), and so there is no reason to spend more money to remove the trans fats from the final product.

This seems like an auspicious result for margarine manufacturers who have been using this addition reaction to create all manner of new margarine products. However, in recent years, alarming statistics about the effects of trans fats on the human body have arisen. Part of the problem is that trans fats are not broken down readily in the body (as cis fats are) and, as a result, their levels build up. Reports from the medical community consistently show that trans fats increase the risk of heart disease as much as, if not more than, the consumption of saturated fats. Thus, consumers who choose margarine containing trans fat over butter because of the risks posed by butter's overload of saturated fats are unwittingly choosing an even more deleterious alternative.

Because there is no law that requires the reporting of trans fat values on food packaging, how much of these substances are to be found in foods is information mostly hidden from consumers. The worst culprits, besides margarines, are processed baked goods, such as cookies, muffins, and pastries. The good news is that, in the summer of 2003, the U.S. Food and Drug Administration (FDA) passed legislation requiring that trans fat content

INGREDIENTS: ENRICHED FLOUR (WHEAT FLOUR, NIACIN, REDUCED IRON, THIAMINE MONONITRATE {VITAMIN B1}, RIBOFLAVIN {VITAMIN B2}, FOLIC ACID), SUGAR (BEET OR CANE), PARTIALLY HYDROGENATED SOYBEAN OIL, HIGH FRUCTOSE CORN SYRUP, ARROWROOT FLOUR, CORNSTARCH, SOY LECITHIN (EMULSIFIER), SALT, LEAVENING (BAKING SODA).

INGREDIENTS: ENRICHED FLOUR (WHEAT FLOUR, NIACIN, REDUCED IRON, THIAMINE MONONITRATE {VITAMIN B1}, RIBOFLAVIN {VITAMIN B2}, FOLIC ACID), SOYBEAN OIL, DEFATTED WHEAT GERM, SUGAR, CORNSTARCH, HIGH FRUCTOSE CORN SYRUP, CORN SYRUP, SALT, MALT SYRUP, LEAVENING (CALCIUM PHOSPHATE, BAKING SODA), MONOGLYCERIDES, VEGETABLE COLORS (ANNATTO EXTRACT, TURMERIC OLEORESIN).

◄ **FIGURE 12.15   Checking for the bad fats.**   In the list of ingredients on the left, partially hydrogenated soybean oil is the third ingredient, a clue that this product may be high in trans fats. In the list on the right, there is no mention of any hydrogenated products, telling you that this product is comparatively low in trans fats.

be added to nutrition labels beginning in 2006. Until then, there are a few ways to figure out whether a product might contain a trans fat. First, look for the words *hydrogenated* or *partially hydrogenated* on the list of ingredients (Figure 12.15). Because most trans fat is formed as a by-product of hydrogenation reactions, it is a good bet that products which have been hydrogenated contain trans fats. Second, look at the position of the hydrogenated ingredient on the list of ingredients. Because the ingredients are ranked in order of their percentage of the content of the product (from highest to lowest), ingredients at the beginning of the list are present in greatest quantity. When the word *hydrogenated* appears in one or more of the first four ingredients, you can bet that the product is loaded with trans fats.

With all the ruckus about trans fats in the news, many food manufacturers are making efforts to reduce or eliminate the percentage of trans fats in their products. Some margarine manufacturers have even figured out a way to produce margarine that is trans fat free (Figure 12.16).

The package in Figure 12.16 mentions omega-6 and omega-3 fatty acids, which are polyunsaturated fats believed to play a role in the prevention of heart disease, cancer, and arthritis. Omega-3 fatty acids first came to the attention of scientists in the 1970s, when a study was released on the nutritional habits of the Inuit Eskimos living in Greenland. This population had a very low incidence of heart disease, a finding that was attributed to a diet which includes lots of whale, salmon, and seal. We now know that cold-water fish, which also include

▲ **FIGURE 12.16   Margarine containing omega-3 fatty acids.**

mackerel and tuna, are especially high in the omega-3 fatty acid known as DHA (Figure 12.17*a*). Fatty acids are usually numbered starting at the carboxylic acid (the alpha end), but they can also be numbered from the opposite end (the omega end). Omega-3 fatty acids are those which have a double bond between the third and fourth carbons, counting from the omega end of the molecule. If you would like to start including more omega fatty acids in your diet, but are not crazy about cold-water fish, you can obtain the omega-3 fatty acid known as ALA (Figure 12.17*b*) from green leafy vegetables and flaxseed oil.

Omega-3 double bond

DHA (from fish oil)

(a)

Omega-3 double bond

ALA (from green leafy vegetables and flaxseed oil)

(b)

◄ **FIGURE 12.17   Omega-3 fatty acids.**   The end of a fatty acid opposite the carboxylic acid functional group is called the omega end of the molecule. Omega-3 fatty acids have a double bond between carbons 3 and 4, counting from this end. **(a)** The omega-3 fatty acid DHA (docosahexenoic acid). **(b)** The omega-3 fatty acid ALA (alpha-linolenic acid).

MOLECULE
Cholesterol

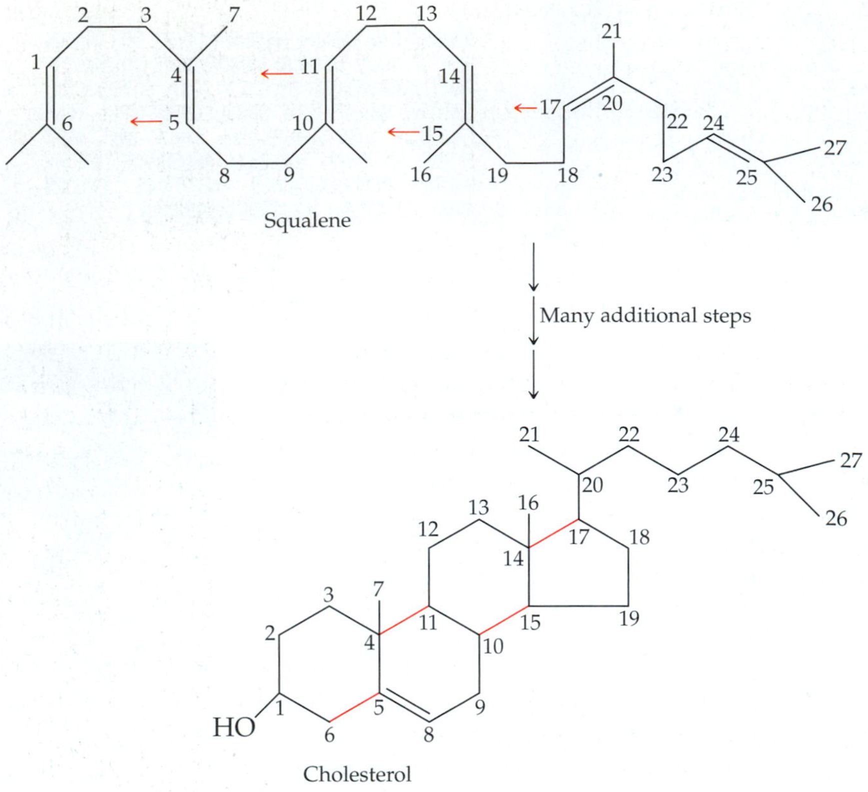

▶ **FIGURE 12.18 Conversion of squalene to cholesterol.** Cholesterol is formed from the polyunsaturated hydrocarbon squalene. The bending back and forth of the squalene hydrocarbon chain allows the carbon atoms involved in the double bonds to come together and form the rings of cholesterol. The cholesterol molecule is a combination of four rings attached to an eight-carbon chain. Note the presence of the one double bond and the alcohol functional group.

By now, we know that saturated fats are "bad" foods, as are trans fats, but what makes them unhealthy? Data have shown conclusively that the consumption of trans fats or saturated fats is correlated with high incidences of heart disease and is linked to elevated cholesterol levels. To better understand how cholesterol, the last lipid we shall study, can wreak havoc in our arteries, we turn to its structure, shown in Figure 12.18. Cholesterol does not look like the other lipids we have seen, which all contain long hydrocarbon chains. Cholesterol is indeed a lipid, however, because it is readily soluble in nonpolar solvents. Although its structure may seem complex, it is really just a long hydrocarbon molecule formed into four rings plus one double bond and one alcohol group (—OH). Cholesterol is made, ultimately, from two-carbon units that are added together to form a long hydrocarbon called *squalene.* If you draw the structure of squalene so that it winds back and forth, as shown at the top in Figure 12.18, it is easy to see how the cholesterol molecule forms through the creation of new bonds that result in a series of closed rings. Because the carbon atoms in cholesterol are part of closed rings, their movement is much more restricted than the movement possible in, say, the carbon atoms in a long saturated hydrocarbon chain. As a result, cholesterol is a rigid, elongated, relatively flat molecule.

Cholesterol is produced in the body and plays an essential role in the structure of cell membranes. A cholesterol molecule inserted into the lipid bilayer of the cell membrane (Figure 12.12, page 473) causes a pronounced change in the membrane's pliability. Because it has an inflexible structure, cholesterol inserted into a cell membrane can increase the rigidity of the membrane and is one way the membrane can decrease its fluidity, for example, in response to very warm temperatures. Given this seemingly important role it plays in our bodies, why is cholesterol a problem?

The answer lies in the flattened, rigid shape of the cholesterol molecule and in the fact that this molecule must be transported through the body. In general, lipids must be attached to carriers in order to move through the bloodstream, because their highly nonpolar nature makes them insoluble in blood. Cholesterol is carried through the bloodstream by protein molecules in enormous composites called **lipoproteins**. There are two important varieties of lipoprotein for cholesterol transport: *low-density lipoproteins* (LDLs), which have a low protein-to-lipid ratio, and *high-density lipoproteins* (HDLs), which have a high protein-to-lipid ratio. The job of LDLs is to take part in the transfer of lipids *from* the liver, where they are manufactured, to various locations in the body. Because they have a comparatively low density, LDLs have a tendency to form artery-clogging plaque in the blood vessels. HDLs are responsible for moving cholesterol *to* the liver, where it is broken down. Because HDLs have a higher density, they are like miniature snowplows that can plow though the circulatory system. There is evidence that this plowing action cleans out blood vessels and reduces the buildup of plaque.

When we eat foods that contain lipids, those foods influence the ratio of LDLs to HDLs present in the bloodstream. Thus, the goal when choosing lipid-containing foods should be to select those which promote the formation of HDLs (the "good" lipoproteins) and decrease the formation of LDLs (the "bad" ones). On one hand, well-established evidence from the medical community suggests that unsaturated fats contribute to increased levels of HDLs and decreased levels of LDLs, a beneficial situation all around. Saturated fats, on the other hand, are known to increase levels of HDLs, which is a good thing, but also increase levels of LDLs, which is a bad thing. The worst offenders, trans fats, are suspected of decreasing HDLs, which is bad, and increasing LDLs, which is worse. Of course, there are scores of spry octogenarians who have eaten unhealthy fats all their lives. Clearly, warnings about food are just pieces of cautionary advice for healthy living and not absolute, hard-and-fast rules that must be followed at all times. (In the case of cholesterol, it turns out that heredity is another major factor in determining whether a person will have problems.)

One way to remember which lipoproteins are good and which are bad is to change *bad* to *lousy*. You can remember that *lousy* cholesterol is the LDL form, because both begin with the letter L. Also, HDLs are *healthier*.

QUESTION 12.5 Pretend you are an organism that can dictate the composition of your cell membranes. Which of the following modifications to the membranes will benefit you most in very hot weather? (*a*) addition of unsaturated fatty acids, (*b*) addition of saturated fatty acids?

ANSWER 12.5 Remember from our discussion of ground squirrels and reindeer that it is important that cell membranes maintain a certain degree of fluidity at all times (fluid enough to allow molecules to pass into and out of the cells, but not so fluid as to melt and cease acting as membranes). Very high temperatures mean that low-melting-point lipids (the unsaturated ones) will be liquids, causing the membranes to be too fluid. Thus, you want your choice to be the one having the highest melting points, which is (*b*).

12.3 | Bite, Chew, and Swallow

Each of us can think of our body as containing a long tube that is an extension of our external selves, but on the inside. When we eat, food travels through this tube—more formally called the digestive tract—and along the way, the food is broken down. The body absorbs what it needs and sends the rest to the other end of the tube. Whenever you eat something, digestion begins immediately in your mouth. Thus, your food arrives in your stomach partly digested, and the breakdown into smaller molecules that can be absorbed into the bloodstream continues there and in the small intestine. The intestinal tract does not work optimally on its own. To aid digestion, it harbors millions of bacteria that help with digestion in a symbiotic exchange that is beneficial for them and beneficial for you.

We can divide the nutrients we eat and drink into three categories: lipids (fats and oils), carbohydrates, and proteins. All of these biomolecules are built in a modular design. The lipids are modules composed of either two (in phospholipids) or three (in TAGs) fatty acids attached to one molecule of glycerol. Proteins (Chapter 11) are modular polymers made up of amino acids. Carbohydrates also are polymeric, as we shall see in a moment. In order to be absorbed into the bloodstream, these biomolecules must be broken down into their component parts. Once absorbed into the cells that line the digestive tract, many of these small molecules are broken down further into two-carbon units.

All of the reactions that take place in a living organism are collectively referred to as **metabolic pathways**. Those metabolic pathways in which the large biomolecules in food are broken down into smaller molecules (and ultimately to carbon dioxide and water in human cells) are called **catabolic pathways**, and the general term for all the processes that take part in these pathways is **catabolism**. Those metabolic reactions in the body that build up new biomolecules from smaller ones are referred to as **anabolic pathways**, with the general name for the processes being **anabolism**. Because we are concerned with the breakdown of food molecules in this chapter, we shall focus primarily on catabolic pathways—those which dismantle what we eat and use the resulting bits and pieces as fuel.

When we eat something, we rely on our teeth to start breaking it down into smaller and smaller pieces. This mechanical action continues in our stomachs, where the churning process called *peristalsis* helps to homogenize our meal and make it easier to break down further. At this point, there are two chemical processes that occur in the stomach and the small intestine to break down what is left of the food into fragments that can be absorbed into the body. The first process occurs primarily in the stomach and involves the breakdown of biomolecules at extreme pH values. Recall from Chapter 4 that pH is an indicator of acidity or basicity, and the lower the pH, the more acid there is. When we think about pH, we are usually concerned with substances dissolved in water. Thus, the environment in the stomach, where the pH is about 2, is made up of water with acid dissolved in it. The breakdown of food in the stomach at low pH occurs when the water and acid work simultaneously to attack large molecules and promote cleavage of the bonds in them.

Imagine that a TAG from a grilled cheese sandwich has just arrived in your stomach. Recall from Section 12.1 that TAGs are held together by ester linkages created when the carboxylic acid groups (—COOH) of three fatty acids join with the three alcohol groups on glycerol (Figure 12.8, page 471). The breakdown of a TAG is the reverse of this process, and in the stomach, where the pH is very low, esters break apart easily. Let us focus on the TAG ester groups, highlighted in red in Figure 12.19. Under the low-pH conditions found in the stomach, both water and acid are present. In acidic solutions, ester functional groups, such as the three found in a TAG, are broken down to carboxylic acids and alcohols, as shown in Figure 12.19. This process is initiated in a hydrolysis reaction when a water molecule (carrying a partial negative charge on the O atom) attacks the carbon atom of the ester (which carries a partial positive charge). For a TAG, when the three fatty acids are released in this type of reaction, the triple alcohol glycerol forms as the alcohol-containing product. The carboxylic acids are the three

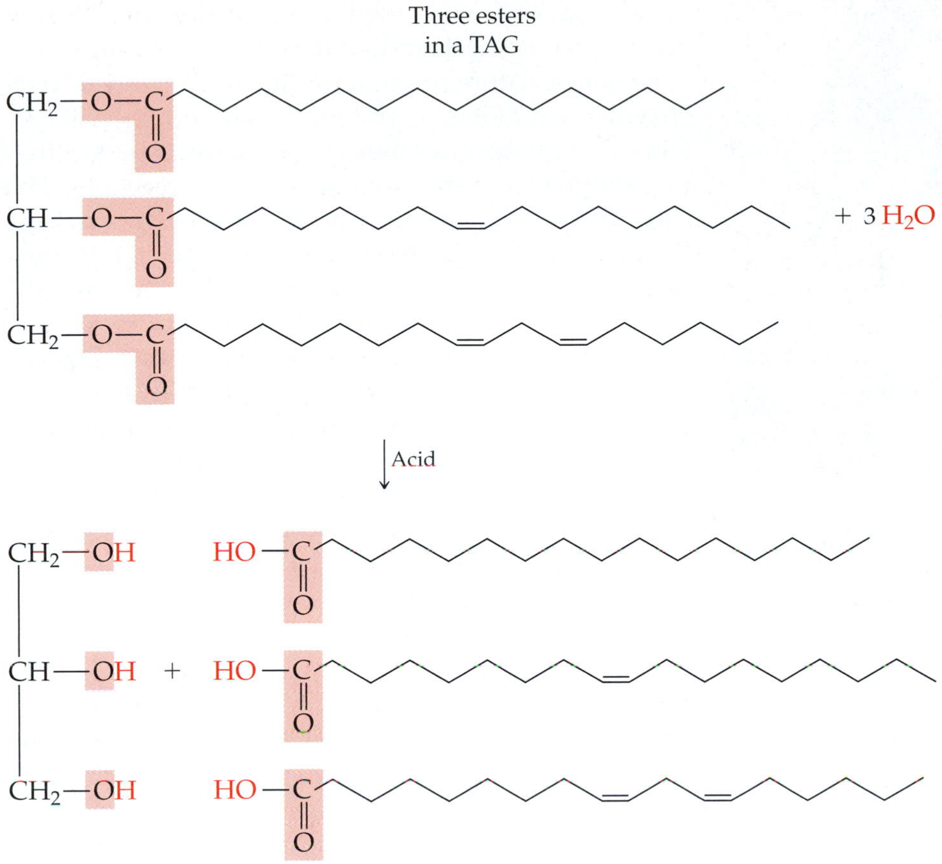

Three esters
in a TAG

+ 3 H_2O

Acid

LIVE ART
Hydrolysis of a Triacylglycerol

◀ **FIGURE 12.19 Hydrolysis of a triacylglyceride.** In the presence of water and acid, the three ester groups in the TAG molecule undergo hydrolysis to form a trialcohol (glycerol) and three fatty acids.

fatty acids that are also formed. This process of breaking a bond by using water is referred to as hydrolysis (*hydro* = water and *lysis* = break), as we learned in Section 10.6. Thus, the low pH of the stomach encourages digestion by promoting acid-catalyzed hydrolysis reactions that contribute to the complete breakdown of a food biomolecule.

QUESTION 12.6 Write the products of these acid-catalyzed ester hydrolysis reactions:

(a)  $H_3C-\overset{\overset{\displaystyle O}{\|}}{C}-O-CH_2CH_3 \xrightarrow{H^+} ?$

(b) 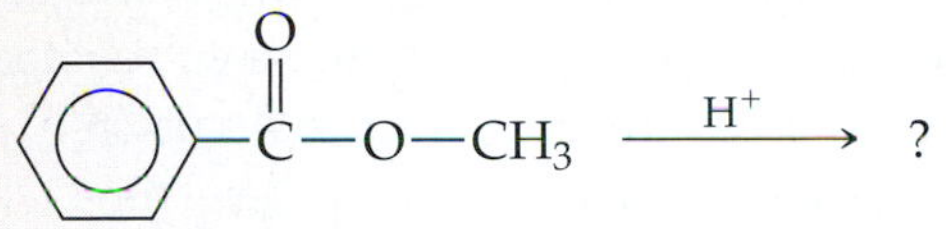 $\xrightarrow{H^+} ?$

ANSWER 12.6

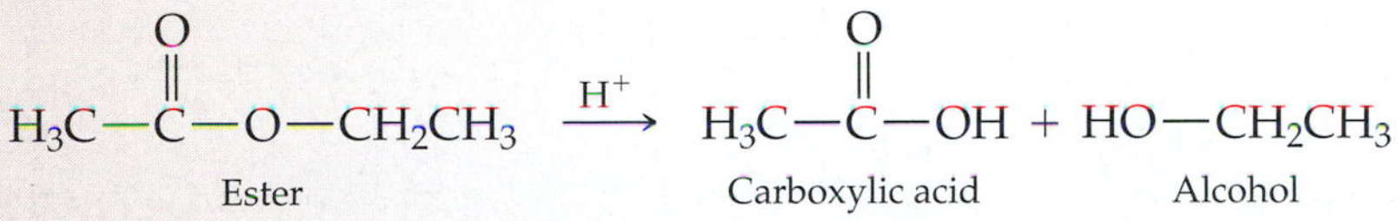

$H_3C-\overset{\overset{\displaystyle O}{\|}}{C}-O-CH_2CH_3 \xrightarrow{H^+} H_3C-\overset{\overset{\displaystyle O}{\|}}{C}-OH + HO-CH_2CH_3$

Ester   Carboxylic acid   Alcohol

(a)

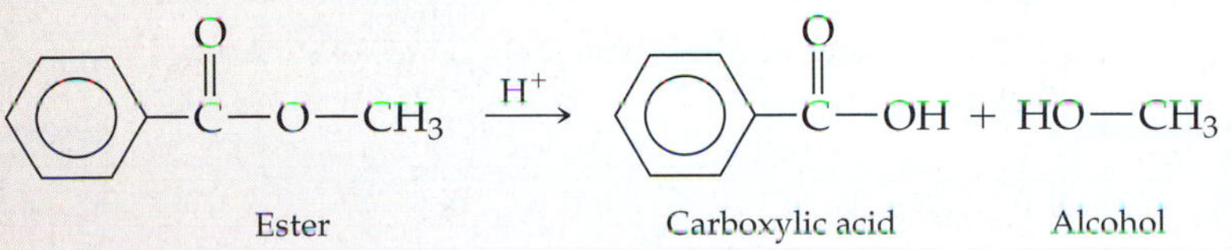

Ester   Carboxylic acid   Alcohol

(b)

▲ **FIGURE 12.20** **Bow-tying analogy for enzyme action.**
The action of an enzyme is akin to the finger used to help tie a
ribbon into a bow. The finger and the enzyme each hold
everything in place while the process is completed. In the case of
the ribbon, the result is a bow; in the case of an enzyme, the
result is a successful chemical reaction.

The second way food biomolecules are broken
down in the stomach and small intestine is by the action
of **enzymes**, nature's miniature chemical factories. Most
enzymes are globular proteins. They are special proteins, though, because they act as **catalysts** by facilitat-
ing, and therefore speeding up, chemical reactions. The
situation is analogous to tying a bow on a gift package
(Figure 12.20). If you try to do it yourself, the process
can be frustrating and time consuming because the
ribbon will not stay in place so that you can make a
tight knot. When you have the help of another person's
finger, however, the process is simple and quick. The
finger that comes to your rescue is like an enzyme
catalyst, because it works by holding the reactants
(the pieces of ribbon) securely in one place while something else happens (the knot is tied).

Consider the chemical reaction shown in Figure 12.21,
a reaction that is the first step in the pathway by which the
body breaks down glucose molecules to produce energy.
With enzymatic reactions, we refer to the reactants as
substrates. In this reaction, the substrates are a molecule
of glucose and some source of phosphate ion (PO_4^{3-}), and there is only one
product: glucose-6-phosphate. In the absence of the appropriate enzyme, this
reaction proceeds at a pace that is unacceptably slow. In the presence of the
enzyme *hexokinase*, however, the reaction hums along at a speed that keeps up
with the metabolic activity of the cell.

Figure 12.22 shows a space-filling model of hexokinase, and we can use
it to look more closely at the reaction of Figure 12.21 between glucose and
phosphate ion. In Figure 12.22*a*, the enzyme has an open cleft when glucose
is absent. A glucose molecule that happens by moves into the cleft and
docks at a specific site that accommodates its shape *perfectly*. We refer to the
site of substrate binding in an enzyme as the *active site*. You can see that the
glucose sits at the pivot point between the two green lobes of the enzyme.
When the glucose binds at the active site, the cleft closes and the white and
green regions clamp down on the glucose. The phosphate group that will
react with the captive glucose molecule is already situated at another position inside the active site, and the enzyme facilitates the interaction
between the two substrates by holding them in the correct orientation for
reaction. Once the glucose and phosphate have reacted to form glucose-6-
phosphate, the cleft opens and the product floats free.

When scientists first started to study enzymes, they quickly learned
that these catalysts are extraordinarily specific. That is, most enzymes will

MOLECULE
Glucose

▶ **FIGURE 12.21** **The hexokinase
reaction.** The enzyme hexokinase
catalyzes the reaction that adds a
phosphate group to glucose to make
glucose-6-phosphate.

CH_2OH

H, O, H
H
OH H
HO, OH

H, OH

Glucose

+ source of phosphate ion

→ Hexokinase →

$CH_2OPO_3^{2-}$

H, O, H
H
OH H
HO, OH

H, OH

Glucose-6-phosphate

The Most Amazing Enzymes Known

There are different ways enzymes can be impressive. If you are thinking about speed, the fastest enzyme known to date, *carbonic anhydrase*, catalyzes a reaction that allows carbon dioxide to be transported from the body's tissues to the lungs, where it is eliminated by exhalation. Every second, this enzyme carries out its reaction 600,000 times!

If your criterion for most amazing is how much a reaction rate is increased, the award goes to the enzyme *OMP decarboxylase*, which carries out the last step in the metabolic pathway that makes one of the building blocks of ribonucleic acid (RNA). Without OMP decarboxylase, this reaction occurs about once every 114,000,000 yr. *With* OMP decarboxylase catalyzing the reaction, it occurs 39 times per second, which is 140,000,000,000,000,000 times faster! It would be difficult to decide which of these two supercatalysts is more remarkable.

QUESTION 12.7 If the enzyme OMP decarboxylase turns over its substrate 39 times per second, how many substrate molecules would it turn over in one day? One year? Express your answer in reactions per day and reactions per year.

ANSWER 12.7

$$\frac{39 \text{ rxn}}{1 \text{ s}} \times \frac{3600 \text{ s}}{1 \text{ h}} \times \frac{24 \text{ h}}{1 \text{ day}} = 3,400,000 \text{ rxn/day}$$

$$\frac{3,400,000 \text{ rxn}}{1 \text{ day}} \times \frac{365 \text{ days}}{1 \text{ yr}} = 1,200,000,000 \text{ rxn/yr}$$

fit only one particular substrate, and enzymes that react with several different molecules are the exception. One notable example of a group of enzymes that are not very specific is the group involved in digestion in the stomach. Because digestive enzymes must break down a wide variety of foods, they tend to react with more than one specific molecule—often, dozens.

High specificity is usually the rule for enzymes, however. Until the late 1950s, the explanation for the high degree of specificity was the *lock-and-key model*, which said that an enzyme and its substrate are designed to fit only each other, like a lock for which there is only one key. As scientists have learned more and more about enzymes, though, it has become clear that the situation is more complicated than two static objects fitting each other exactly. The problem with the lock-and-key model is that a key is not changed in any way when it is used in a lock: when it comes out of the lock, the key has exactly the same shape it had before it was inserted into it.

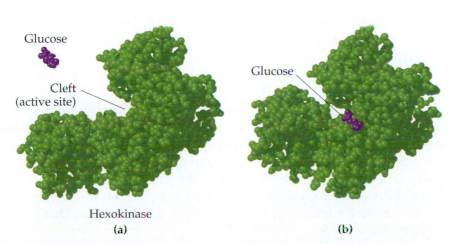

Glucose

Cleft
(active site)

Glucose

Hexokinase

(a)

(b)

◀ **FIGURE 12.22** The hexokinase reaction of Figure 12.21 represented by space-filling models. **(a)** The enzyme hexokinase has a cleft (the active site) flanked by two green lobes. The shape of the cleft matches the shape of a glucose molecule. **(b)** Once a glucose molecule (the substrate) enters the cleft, the two sides of the enzyme molecule clamp down on it.

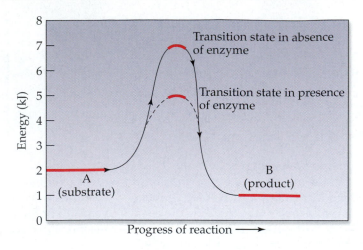

▶ **FIGURE 12.23 Climbing an energy hill.** The reaction coordinate for a chemical reaction shows energy on the vertical axis and the progress of the reaction on the horizontal axis. The transition state is attained by the substrate at the highest point on the energy hill. In the presence of the appropriate enzyme, less energy is required to get to the transition state, and this makes the reaction occur more frequently than it would in the absence of the enzyme.

Thus, the lock-and-key model does not account for the fact that the substrate and product are different molecules. Furthermore, any model that seeks to explain the action of enzymes should account for the astonishing rates at which enzymes can form products. The lock-and-key model does not do this.

To understand better how enzymes operate, we can envision chemical reactions in terms of a **reaction coordinate**—a diagram that plots energy (on the vertical axis) versus the progress of a reaction (on the horizontal axis). Figure 12.23 shows a reaction coordinate for a hypothetical chemical reaction A → B. The reaction proceeds from start to finish as you move from left to right across the horizontal axis. At the beginning of the reaction, the energy of the substrate A is 2 kJ. (The energies of biochemical reactions are often expressed in kilojoules; 1 kJ = 1000 J.) In order for the reaction to proceed and for the product B to form, the substrate must become distorted into a shape called a **transition state** that is partway between the shape of A and the shape of B. Energy must be added until A has enough energy to form this transition state. When a lot of energy is needed to get up the "energy hill" to the transition state, the process is slow and tedious, and the reaction occurs infrequently.

Now the enzyme enters the picture. The job of the enzyme is not only to hold the substrate in a specific position, but also to distort it so that it becomes more like the transition state. In other words, when an enzyme is present, it is easier for the substrate to get to the transition state and therefore easier for product to form. To put it another way, in the presence of the enzyme, the reaction coordinate has an energy hill that is not as tall. In our hypothetical reaction, the substrate energy must change from 2 kJ to 7 kJ when no enzyme is present, but only from 2 kJ to 5 kJ when the enzyme is present. Thus, the reaction occurs more frequently in the presence of enzyme because the energy hill is easier to scale.

Today we no longer talk about a lock-and-key model. Instead, we say that the substrate interacts with the enzyme by an **induced-fit model** whereby the substrate is distorted as it binds to the enzyme's active site in a way that forces the substrate's shape to be more like the transition state's shape. The induced-fit model explains the impressive rate enhancements that have been measured for most enzymes. In fact, most enzymes make a reaction run anywhere from 1000 to 10,000,000,000,000 times faster than the uncatalyzed reaction.

12.4 | Sugar, Sugar

The USDA food pyramid of Figure 12.1 (page 464) provides for several servings of complex carbohydrates each day. In fact, the central metabolic pathways in aerobic cells are designed to break down carbohydrate molecules and tap the energy stored in their bonds. Thus, it should come as no surprise that these "carbs" are a mainstay of a healthy diet.

All carbohydrates are polymers composed of monomers called **saccharides**. As with other polymers we have studied, prefixes are used to indicate the number of saccharide units present in a molecule. Carbohydrates made up of only one monomer are called **monosaccharides**, and those containing two monomer units are **disaccharides**. For molecules containing from a few to about 100 monomers, we use the term **oligosaccharide**, and for those containing hundreds or even thousands of monomer units, the term is **polysaccharide**. We use the name **carbohydrate** to refer to all saccharides, large and small, and we give a special name, *sugar* (or *simple sugar*), to one-unit and two-unit carbohydrates. (In other words, *sugar* used in this context is a synonym for both *monosaccharide* and *disaccharide*. As we shall see in a moment, when we say *sugar* in everyday life, we usually mean one specific disaccharide, the one called *sucrose*, usually referred to as *table sugar*.) Supersweet foods, such as candies, are composed mainly of simple sugars. When you hear the words *complex carbohydrate*, they usually refer to a polysaccharide containing hundreds or thousands of individual units. Examples of complex carbohydrates are foods that are starchy, such as potatoes and breads. Because this terminology can be confusing, Figure 12.24 provides a summary of carbohydrate jargon.

When you take a bite of carbohydrate and begin to chew, breakdown begins as soon as the food hits the saliva in your mouth. Saliva contains an enzyme called *amylase* that takes polysaccharides and starts to cut them up into small pieces. Try this experiment the next time you are eating a piece of bread: chew the bread much longer than you normally would, giving it ample opportunity to mix thoroughly with the saliva in your mouth.

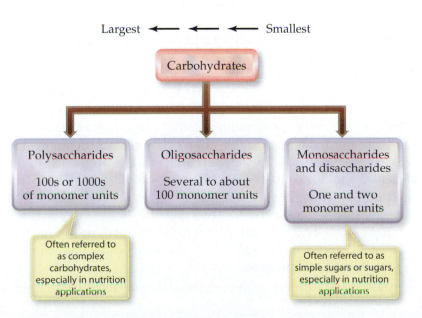

◄ **FIGURE 12.24 Summary of carbohydrate names.** The term *carbohydrate* applies to several kinds of saccharides, including polysaccharides, oligosaccharides, disaccharides, and monosaccharides. The determining factor is the number of monomer saccharide units in each polymer molecule.

Largest ← ← ← Smallest

Carbohydrates

Polysaccharides
100s or 1000s of monomer units

Oligosaccharides
Several to about 100 monomer units

Monosaccharides and disaccharides
One and two monomer units

Often referred to as complex carbohydrates, especially in nutrition applications

Often referred to as simple sugars or sugars, especially in nutrition applications

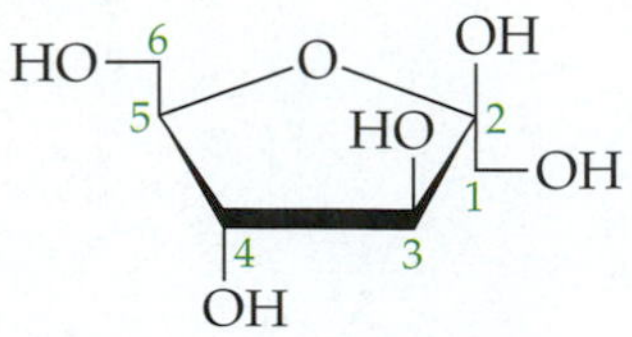

Fructose, $C_6H_{12}O_6$

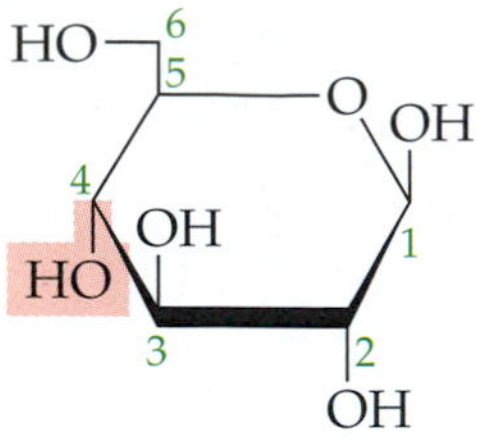

Glucose, $C_6H_{12}O_6$

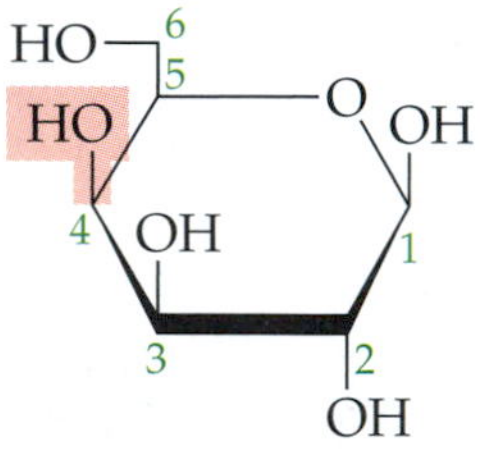

Galactose, $C_6H_{12}O_6$

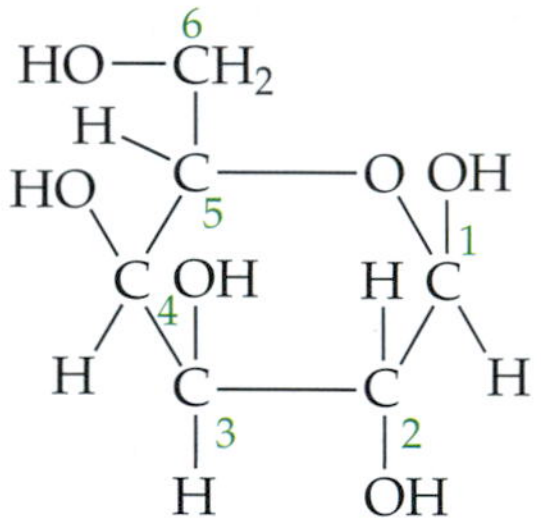

Galactose full structure

▲ **FIGURE 12.25 Three simple sugars.** Structures of the three most common monosaccharides in the human body: fructose, glucose, and galactose. Notice that glucose and galactose differ only by the orientation of one — OH group (highlighted red).

Do you notice any change in the flavor of the bread; specifically, does it seem to get sweeter? If you do notice such a change, it is due to the action of amylase, which chops the long-chain polysaccharides in the bread into small pieces. Some of the small pieces produced are disaccharides. The disaccharide you taste after chewing bread is called *maltose*. This common simple sugar tastes sweet, as do most other disaccharides as well as most monosaccharides.

The simple experiment you did chewing a piece of bread tells us two things about carbohydrates. First, carbohydrate molecules contain bonds that can be broken by enzymes. We shall look at examples of these enzymes shortly. Second, the properties of polysaccharides, such as those in a piece of bread, are different from the properties of smaller carbohydrates, such as maltose. The bread experiment tells us that the quality we perceive as sweetness is a result of the presence of simple sugars that form from the breakdown of a polysaccharide.

When the well-chewed piece of bread hits your stomach, the amylase from your saliva keeps breaking it down until eventually the amylase is inactivated by the extremely low pH in the stomach. When the remains of the bread make their way to your small intestine, amylase is once again added to the mix, this time from the pancreas. As the amylase continues to make the disaccharide maltose from polysaccharides in the bread, other enzymes act on the increasingly smaller and smaller remnants of the original polysaccharide. These enzymes are capable of breaking down disaccharides to monosaccharides. As we shall see, the three most ubiquitous monosaccharides in our diets are fructose (also known as *fruit sugar*), glucose, and galactose (Figure 12.25). These tiny carbohydrates are small enough to be absorbed into the cells that line the intestinal walls.

Fructose, glucose, and galactose are the monomer units that make up most known polysaccharides. Thus, by learning something about the structures of these simple molecules and about how they link to one another, we can learn a lot about larger carbohydrates. Notice in Figure 12.25 that the basic carbohydrate structure contains carbon, hydrogen, and oxygen. In fact, all three of the structures shown in this illustration have the same molecular formula: $C_6H_{12}O_6$. If you rearrange the formula, you can write it as $C_6(H_2O)_6$. Thought of this way, it makes sense that the traditional name for these molecules is *carbo-* + *—hydrate*, which loosely translates as *carbon* and *water*. Today we know that the structure does not actually include water molecules, but the name has stuck nevertheless. (The name was coined before the structures of carbohydrates were known.)

Another thing to notice in Figure 12.25 is that carbohydrates can exist as rings containing different numbers of atoms. While five- and six-membered rings are by far the most common, other rings sizes are known. For glucose and galactose, the ring contains five of the six carbon atoms, and the sixth carbon atom, the only one not part of the ring, is appended to carbon 5. For fructose, the ring contains only four of the six carbon atoms, with carbons 1 and 6 outside the ring.

QUESTION 12.8 In each pair, which carbohydrate is larger? (*a*) oligosaccharide and simple sugar, (*b*) polysaccharide and monosaccharide, (*c*) galactose and maltose.

ANSWER 12.8 (*a*) oligosaccharide, (*b*) polysaccharide, (*c*) maltose.

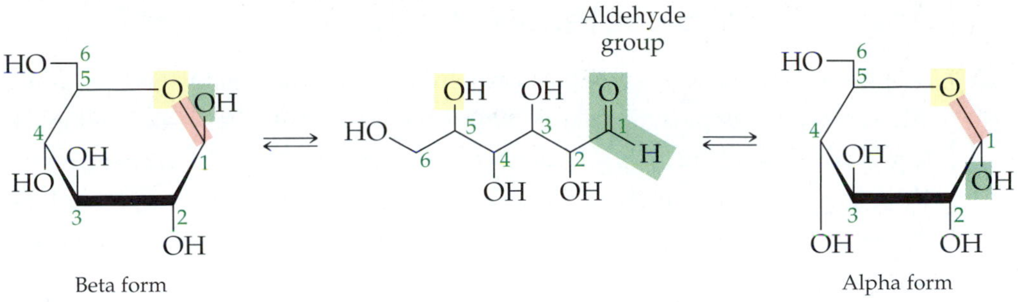

Aldehyde group

Beta form

Alpha form

▲ **FIGURE 12.26** **Three forms for one simple sugar.** Glucose can exist in three interconvertible forms: beta, open chain, and alpha. The red-highlighted bond is the bond that breaks when the ring opens up. The oxygen atom highlighted yellow, the one that is part of the ring in the alpha and beta forms, becomes part of the carbon 5 hydroxyl group in the open-chain form. The green-highlighted oxygen atom, part of an —OH group in the ring forms, becomes part of the aldehyde carbonyl group in the open-chain form.

LIVE ART
Forms of Sugar

QUESTION 12.9 Calculate the molar mass of each monosaccharide in Figure 12.25.

ANSWER 12.9 All have the molecular formula $C_6H_{12}O_6$, so they all have a molar mass of $6(12.01 \text{ g/mol}) + 12(1.01 \text{ g/mol}) + 6(16.00 \text{ g/mol}) = 180.18 \text{ g/mol}$.

Notice in Figure 12.25 that in each case an oxygen atom is part of the ring structure. Also, there are several alcohol groups appended to the carbon atoms in the ring. These alcohol groups impart polarity to the molecule and form hydrogen bonds with water readily. Thus, we expect carbohydrates to be soluble in water, and many are.

In solution, a carbohydrate ring is able to open up into an *open-chain form* and then close up again. Figure 12.26 shows how this happens with glucose, and the process is similar for other monosaccharides. The ring opens at the bond highlighted red in both ring forms, which is the bond between carbon 1 and the ring oxygen atom (highlighted yellow). In the resulting open-chain form, there is an alcohol group at the one end (on carbon 6). At the other end is a functional group we are meeting for the first time. It consists of a carbonyl group (—C=O) with a hydrogen attached to the carbonyl carbon and is known as the *aldehyde functional group*. Molecules containing this functional group are of course called **aldehydes**, and the generic structure of one is shown in Figure 12.27, along with the structures of some notable aldehydes.

Once in the open-chain form, the glucose molecule can do one of two things: it can go back to its original structure, or it can form a slightly different type of ring. These two forms are labeled beta (β) and alpha (α) in Figure 12.26. The only difference between the alpha form and the beta form is the position of the alcohol group attached to carbon 1. This small change might not seem important, but these two forms of glucose, which are referred to as **anomers** of each other, play different roles in the body, especially in the formation of polysaccharides, as we shall soon see. When the alcohol group points downwards (in the direction opposite the direction of carbon 6), we have the *alpha* anomer. When the alcohol group on carbon 1 points upwards (or, more precisely, when it points in the same direction as carbon 6), we have the *beta* anomer of glucose. In any aqueous solution of glucose, most of the glucose (99.98 percent) is in the ring form, but about

▼ FIGURE 12.27 **Aldehydes.** In any aldehyde molecule, the carbonyl carbon is bonded to one hydrogen atom and one hydrocarbon group, represented as usual by R or, as in the case of formaldehyde, another H atom. Aldehydes are known for their distinctively sweet odor and taste. Benzaldehyde is a flavoring molecule that gives almonds their characteristic smell and taste. If you have taken any biology laboratory courses, you may be familiar with the smell of formaldehyde, the primary ingredient in embalming fluid and the solution used to preserve specimens for dissection. Acetaldehyde is used in the manufacture of food flavorings and perfumes.

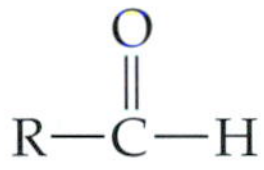

$$R-\overset{\overset{\displaystyle O}{\|}}{C}-H$$

Generic formula

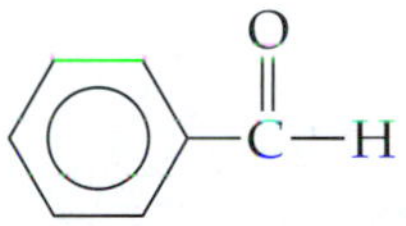

Benzaldehyde

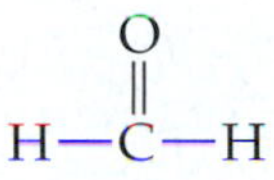

$$H-\overset{\overset{\displaystyle O}{\|}}{C}-H$$

Formaldehyde

$$CH_3\overset{\overset{\displaystyle O}{\|}}{C}-H$$

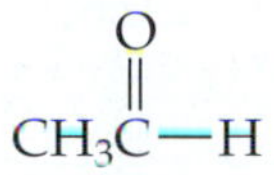

Acetaldehyde

STUDENTS OFTEN ASK

Are Artificial Sweeteners Really Okay to Eat?

Despite the emphasis on complex carbohydrates in the USDA food pyramid, there are many people who believe that carbohydrates are the weak point in the American diet and are in large part responsible for the obesity problems we face. There are weight-loss diets that require dieters to eat little or no carbohydrate and large amounts of protein, and the battle over whether these high-protein diets are healthy rages on. With all the hype about the evils of carbohydrates, products that contain no sugars are now targeting a market that formerly included mainly diabetics. Today, folks on high-protein diets can choose everything from sugar-free candy bars to sugar-free ice cream.

To make sugar-free products palatable, there are several choices of *artificial sweetener* on the market (Figure 12.28). Because many Americans have sworn off carbohydrates altogether, these little packets of sweetness offer a satisfying alternative, even if we do not always know exactly what they are made of or what effect they will have on our bodies. Every year, the average American consumes enough artificial sweetener to equal 20 lb of real sugar. Because we are eating so much of it, and because we are given choices of colored packages to choose from, perhaps we should investigate the choices and see what they have to offer.

The little pink packet has been around for decades. It contains *saccharin*, a molecule discovered at the end of the nineteenth century. When the U.S. Food and Drug Administration (FDA) was created at the beginning of the twentieth century, it could not possibly carry out extensive tests on every kind of food being consumed. Thus, if a food did not cause overt health problems and if it had been used for years, it was deemed *GRAS*, which stands for generally recognized as safe. Saccharin was put into this category. In the 1960s and 1970s, however, saccharin was the subject of many clinical studies that showed a definite link between its use and high incidences of bladder cancer. As a result of these findings, in 1977 the FDA removed the GRAS rating on saccharin. Congress passed the Saccharin Study and Labeling Act, which required that saccharin-containing products carry the warning "May be dangerous to your health. This product has been determined to cause cancer in laboratory animals." Nowadays, the medical community seems less concerned about saccharin, partly because its use has declined.

Looking at the structure of saccharin,

Saccharin

you can easily see that this molecule does not bear any resemblance to a monosaccharide molecule. Clearly, a molecule does not have to look like sugar to be sweet.

The structure of the next artificial sweetener to come along—*aspartame*—also has nothing in common with the structure of a monosaccharide:

Aspartame

▲ FIGURE 12.28   **Artificial sweeteners.**   The familiar pink, blue, and yellow packages help weight-conscious diners choose from among these three sweeteners nestled together in the "sugar" bowl.

In fact, this is a modified dipeptide formed from the amino acids aspartic acid and phenylalanine, with a methyl group added to the carboxyl end of the phenylalanine. This artificial sweetener, which received FDA approval in 1981, was put into little blue packets and dubbed *Nutrasweet*. It is the most common artificial sweetener used in diet sodas today, but has not escaped criticism. Aspartame is broken down in the stomach into its two constituent amino acids. You might think that this breakdown is innocuous because these amino acids are naturally present in the body anyway. However, some individuals have a rare medical condition called *phenylketonurea* (PKU), and these persons cannot metabolize phenylalanine properly. There is a warning to this effect on all aspartame-containing products, and individuals who have PKU should not drink them.

There is a potential for trouble from aspartame even for those of us who do not have PKU. For us, the culprit is the methyl group added to the phenylalanine part of the dipeptide. When the aspartame molecule is broken down, methanol forms. In very small quantities, methanol is not considered harmful, but if large quantities are ingested, methanol poisoning causes blindness and can be fatal. So the trick is to limit your intake of aspartame so that the amount of methanol in your body does not reach harmful levels. The FDA has set the upper limit for aspartame consumption at 50 mg per kilogram of body mass per day. This means that, ironically, the heavier you are, the more diet soda you can drink! One can of the typical diet soda contains about 180 mg of aspartame. Thus, a 125-lb (56.7-kg) woman can drink more than 15 diet sodas before she reaches her limit.

Sucralose

Amanda is taking in more aspartame than is considered safe. You might want to advise her to cut back and explain to her the hazards associated with high levels of methanol in the body.

The yellow packet in Figure 12.28 contains the artificial sweetener *sucralose*, marketed under the brand name *Splenda*®. This product is three times sweeter than aspartame and 600 times sweeter than table sugar, which, as noted earlier, is the disaccharide sucrose. Since its FDA approval in 1998, sucralose has found its way into everything from "lite" juice drinks, to "sugar-free" maple syrup, to "no-sugar" applesauce. Of the artificial sweeteners we have see thus far, this is the only one that resembles a carbohydrate molecule. In fact, sucralose has the same structure as sucrose, except that three of the alcohol groups of sucrose have been replaced by chlorine atoms in sucralose:

Sucrose

QUESTION 12.10 Your friend Amanda takes pride in her health. At 138 lb on a 5-ft 11-in., frame, she is extremely slender. One of her secrets is that she drinks diet sodas rather than carbohydrate-laden juices. In fact, Amanda never goes anywhere without a can of diet soda in her hand. If Amanda drinks 18 cans of aspartame-containing soda every day, what would you say about the state of her health? Use the conversion factor 1 lb = 0.453 kg to answer this question.

ANSWER 12.10 Let us do the math:

How much aspartame she consumes each day:

$$\frac{180 \text{ mg aspartame}}{1 \text{ can}} \times 18 \text{ cans} = 3200 \text{ mg aspartame}$$

The FDA upper limit for her body mass:

$$138 \text{ lb} \times \frac{0.453 \text{ kg}}{1 \text{ lb}} \times \frac{50 \text{ mg aspartame}}{1 \text{ kg}} =$$

$$3100 \text{ mg aspartame}$$

One advertising phrase used for Splenda says, "Made from sugar, so it tastes like sugar." Indeed, sucralose is made by adding chlorine atoms to sucrose molecules in a five-step process. With Splenda, the irony is that the advertisements tout the fact that sucralose has a sugar-based structure, but the products are marketed as "sugar free." Splenda seems to offer consumers the best of both worlds.

So far, negative reports about sucralose have been few. However, there does seem to be an increase in the number of reports of two side effects of sucralose use: mood changes and anxiety attacks. Only time will tell whether this new sweetener can pass muster with the American medical community and with consumers. One thing is certain: many Americans are willing to accept the risks associated with eating artificial sweeteners.

Nature has her own supersweet offerings, such as *stevia*, a plant product used by diabetics around the world who do not have ready access to a diet-soda vending machine (Figure 12.29). Proponents of holistic medicine suggest stevia as an alternative to artificial sweeteners. They also offer up this idea: why can't we simply eat less sugar?

(a)   (b)

◀ **FIGURE 12.29 A natural sweetener much sweeter than table sugar. (a)** The stevia plant (*Stevia rebaudiana*) is the source of a mixture of natural substances that is many times sweeter than table sugar (sucrose). **(b)** "How much sweeter?" you ask. This bottle containing a mere 10 g of the extract is listed as containing 400 servings!

64 percent of that 99.98 percent is in the beta form, because this form is slightly more stable than the alpha form. The ratio of one anomer to the other in solution and the amount of the open-chain form that exists vary from one monosaccharide to another.

When two monosaccharides combine to form a disaccharide, the bond usually forms at carbon 1 of one of the monosaccharides. We refer to this carbon atom in any monosaccharide as the **anomeric carbon**. (Recall from our discussion of Figure 12.26 that the difference between the alpha and beta forms of anomers lies in the position of the —OH group on carbon 1; hence this name for this carbon atom.) We shall look at two disaccharides to demonstrate how monosaccharides link together.

The structure of maltose, the disaccharide we met in our earlier chewing-bread-for-a-long-time experiment, is shown in Figure 12.30*a*. Maltose is made up of two glucose molecules that form a linkage from the anomeric carbon of one glucose to carbon 4 of the other glucose. The anomeric carbon may have its alcohol group pointing either in the same direction as carbon 6, in which case we have the beta form, or in the direction opposite that of carbon 6, in which case we have the alpha form (Figure 12.26). If you look at the two glucoses in Figure 12.30*a*, you can see that the glucose contributing its carbon 1 to the linkage is in the alpha form. For this reason, we call the linkage an *alpha 1 → 4 linkage*.

Maltose is found in beer as well as in bread.

▶ **FIGURE 12.30 Two disaccharides. (a)** Maltose is a disaccharide made up of two glucose molecules held together with an alpha 1 → 4 linkage. **(b)** Lactose is a disaccharide that is also know as *milk sugar*. It is made up of a galactose molecule and a glucose molecule held together with a beta 1 → 4 linkage.

The other disaccharide we shall consider is lactose, obtained in the diet from milk (Figure 12.30b). Lactose is made up of one galactose molecule linked via its anomeric carbon (carbon 1) to carbon 4 of a glucose molecule. As you can see from the figure, the galactose is in its beta form. Thus, we call this a *beta* $1 \rightarrow 4$ *linkage*.

In the small intestine, the cleavage of disaccharides to monosaccharides is done with enzymes specific to different disaccharides. Remarkably, enzymes "know" the difference between an alpha $1 \rightarrow 4$ linkage and a beta $1 \rightarrow 4$ linkage. For example, the enzyme *β-galactosidase* breaks down the disaccharide lactose to the monosaccharides galactose and glucose. Because infants rely on milk for nutrition, they have large amounts of β-galactosidase available for digesting lactose. However, as we get older and start eating solid foods, many of us no longer make this enzyme. The intestinal distress that arises when some adults drink milk is due to *lactose intolerance*, a condition that results from the loss of the body's ability to make β-galactosidase.

QUESTION 12.11 While $1 \rightarrow 4$ linkages are indeed common, $1 \rightarrow 2$ and $1 \rightarrow 6$ linkages are also common. Gentiobiose, a plant disaccharide, has the structure

Gentiobiose

(a) Identify the linkage as alpha or beta, and state the positions of the carbon atoms taking part in the linkage. (b) Which monosaccharide(s) make(s) up gentiobiose?

ANSWER 12.11 (a) A beta linkage between carbon 1 of one monosaccharide and carbon 6 of the other monosaccharide. (b) Glucose and glucose.

A Practical Guide to Dining Out Before Being Murdered

Little did Jane Bock know that her understanding of how complex carbohydrates break down would draw her into the world of homicide investigation. Today a professor emerita at the University of Colorado in Boulder, Bock has studied carbohydrate-laden plant tissues for many years and is considered an authority on the subject. One day early in her academic career, she received a telephone call from the local coroner, who was performing an autopsy on a murder victim and wanted to know more about the victim's last meal. Could she look at the stomach contents and find clues about the homicide that had taken place? Could she identify the foods the victim had eaten? Bock's identification of the foods in the victim's stomach helped to convict the killer by proving that the victim had eaten his last meal with the killer. After that initial success, Bock routinely was asked to investigate stomach contents and to testify in criminal cases. Today, she, along with her husband, David Norris, is recognized as a pioneer in the area of *forensic botany*.

We mentioned that the mouth contains enzymes that begin to break down carbohydrates and that this process then continues in the stomach and in the small intestine. However, not all polysaccharides can be digested by digestive enzymes. Figure 12.31 shows us why. The three polysaccharides shown in this illustration are made up solely of glucose. As you can see, all three contain $1 \rightarrow 4$

linkages. In glycogen and starch, the linkages are alpha $1 \rightarrow 4$. These linkages are easily degraded by human digestive enzymes specific to them. Thus, glycogen, the storage form of glucose in animals, is easy to digest. Likewise, starch, which stores glucose in such plants as potatoes and peas, is also made up of alpha $1 \rightarrow 4$ linkages and so is broken down by these digestive enzymes.

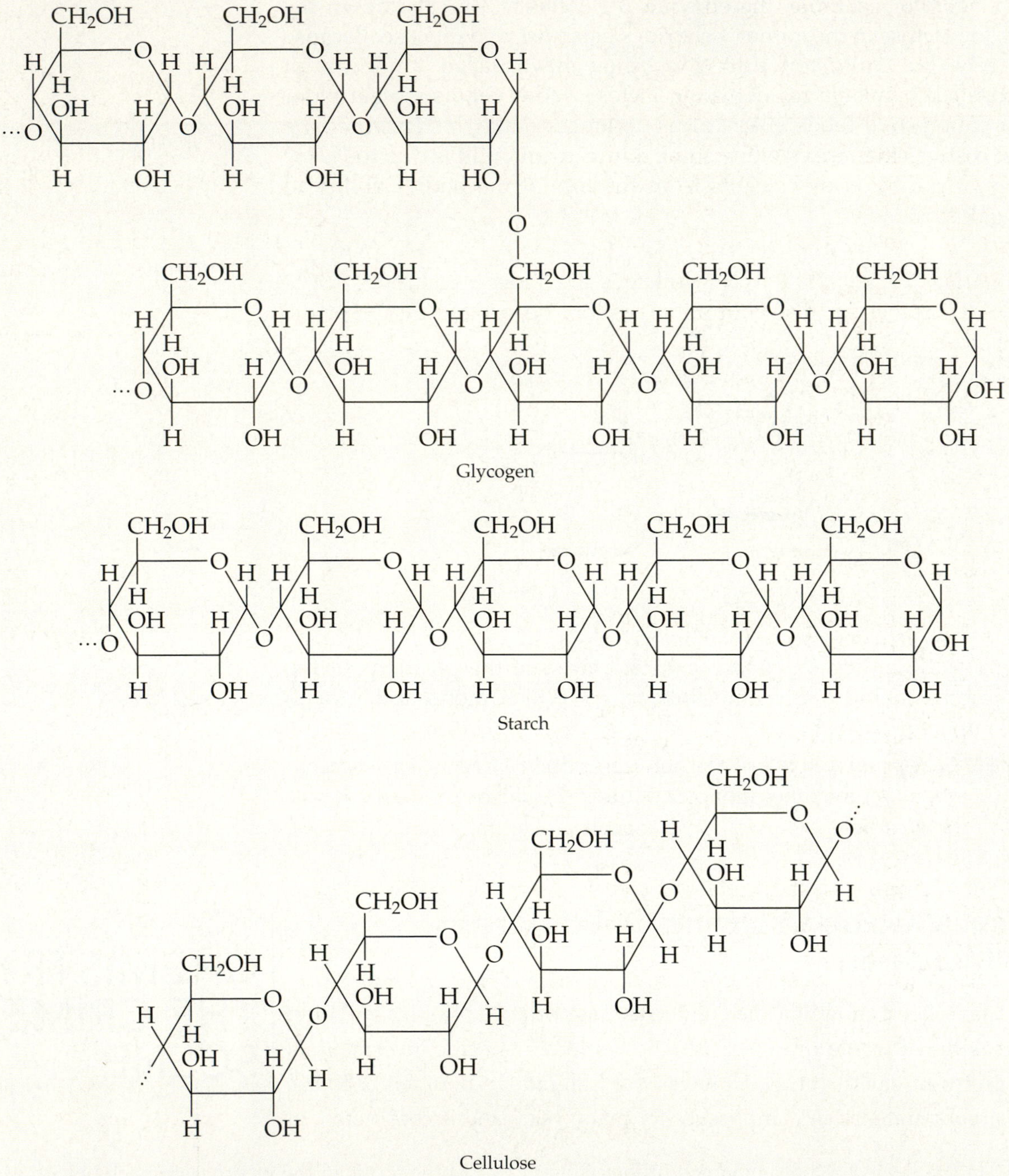

▲ FIGURE 12.31 **The three major polysaccharides in the human body.** Glycogen and starch are digestible by enzymes in the human body; cellulose is not. Note that all three polysaccharides are composed of glucose units. The crucial difference as far as digestion is concerned is in the $1 \rightarrow 4$ linkages, which are alpha in glycogen and starch but beta in cellulose. Note that glycogen and starch are identical molecules, except that glycogen has branches off of the main chain about every 8 to 14 monomer units, while starch has a branch about every 24 to 30 monomer units.

The third polysaccharide in Figure 12.31 is cellulose, a structural polysaccharide that provides rigidity and strength to trees and other plants. In this molecule, the monomer linkages are not alpha, but rather beta. Because humans do not have enzymes capable of breaking beta $1 \rightarrow 4$ linkages, they cannot digest cellulose. (The bacteria that live in the gut of termites *do* possess the enzyme needed to break these linkages in cellulose, which is why termites are able to eat wood.)

We refer to all the solid, indigestible components of food, such as cellulose, as *dietary fiber*. Anyone who watches television has heard about the benefits of eating a high-fiber diet. Because fiber remains intact and is bulky, it acts like a kind of mopping-up agent in the body. Current thinking says that a diet high in fiber not only makes digestion easy and regular, but also helps to absorb unwanted molecules and remove them from the body. If you are thinking about increasing the amount of fiber in your diet, you will want to eat beans, wheat bran, brown rice, and peas.

So why are we talking about the benefits of fiber in a section that is supposed to be about homicide investigations? The answer is that polysaccharides that are not degraded in the digestive tract (most often cellulose) provide useful information about the last meal a person ate before dying. Furthermore, because the valve separating the stomach from the small intestine closes when a person dies, the stomach contents remain in the stomach and can provide clues to the time of death. The degree of digestion of the digestible parts of a meal allow investigators to estimate how long ago the meal was eaten. The indigestible parts can frequently be identified as coming from a specific food, often a cellulose-based tissue. Some cellulose-containing foods are easier to identify than others, however. In fact, Jane Bock will tell you that a few foods can be pinpointed immediately. As it happens, for example, partially digested pineapple is one of the easiest foods to identify. So if you are dining out with a potential murderer, you may want to order the pineapple upside-down cake for dessert.

12.5 | Weight Watching

Humans are remarkable organisms. We have learned to walk on two feet rather than four, developed enormous brains, and found our way into every habitable nook and cranny on the planet. When you think about it, human evolution is astonishing (although many would argue that we are going to evolve ourselves right into extinction). Anthropologists William Leonard and Marcia Robertson think about this every day. They believe that humans have become who they are in large part because of food and that the wanderlust inherent in humans is derived from a primal desire for a variety of food.

Generally, animals with larger brains need a richer diet than do simpler animals, because big brains use calories the way a gas-guzzling SUV uses fuel. Thus, when the brains of humans became larger and larger compared with those of their ancestors, the need for food became more urgent and more specific. In fact, if you have an average-size brain (about 1350 g), about 20 to 25 percent of your resting body's energy needs goes to it. Leonard and Robertson also believe that humans became bipedal because walking on two feet requires fewer calories than walking on four. Of course, early humans were not concerned about burning calories and losing weight; if they had been, we might still be quadrupeds.

Twenty-first-century humans are not as concerned with foraging for food and with conserving energy. In fact, diet-crazed Americans do just the opposite: we try to burn excess calories through exercise to make up for a food supply that requires no more than a trip to the refrigerator. Despite all the books written about dieting, however, people are still perplexed about

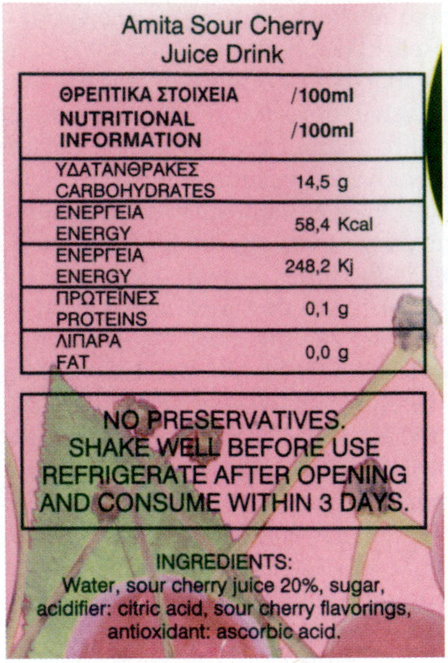

Amita Sour Cherry Juice Drink

ΘΡΕΠΤΙΚΑ ΣΤΟΙΧΕΙΑ NUTRITIONAL INFORMATION	/100ml /100ml
ΥΔΑΤΑΝΘΡΑΚΕΣ CARBOHYDRATES	14,5 g
ΕΝΕΡΓΕΙΑ ENERGY	58,4 Kcal
ΕΝΕΡΓΕΙΑ ENERGY	248,2 Kj
ΠΡΩΤΕΙΝΕΣ PROTEINS	0,1 g
ΛΙΠΑΡΑ FAT	0,0 g

**NO PRESERVATIVES.
SHAKE WELL BEFORE USE
REFRIGERATE AFTER OPENING
AND CONSUME WITHIN 3 DAYS.**

INGREDIENTS:
Water, sour cherry juice 20%, sugar,
acidifier: citric acid, sour cherry flavorings,
antioxidant: ascorbic acid.

▲ **FIGURE 12.32** Keeping track of energy. This label from a Greek juice drink lists energy in kcal units (Cal) and in kilojoules (kJ).

what it takes to lose fat from the body. Is exercising better than dieting? Is there such a thing as a fat-burning diet? The answer to questions like these is really very simple. We take in calories via the food we eat, and we burn calories by being active. If we take in more than we burn, our weight goes up. If we burn more than we take in, our weight goes down. What could be easier? Clearly, part of the confusion associated with weight management comes from the fact that different numbers of calories are associated with different food groups. Before we can look at why different foods have different numbers of calories, however, we must first know something about how calories are counted.

When you are counting calories, you are actually counting **kilocalories** (kcal), a traditional unit for expressing energy. By convention, the kilocalorie is equivalent to a Calorie (Cal, with a capital C), but in everyday usage, it is often written with a small c. So when people are talking about calories in food, it is acceptable to say either *calorie* or *Calorie*, but in both cases it means kilocalorie. As you know from Section 3.3, the SI unit for energy is the *joule*, J, and most scientific work expresses energy in this unit. Because joules and kilocalories are both units of energy, we can interconvert them:

$$1 \text{ kcal} = 1 \text{ Cal} = 1000 \text{ cal} = 4184 \text{ J} = 4.184 \text{ kJ}$$

Nowadays you can even find nutritional labels on food that provide energy values in joules (Figure 12.32).

The average adult man requires about 2500 Cal per day to maintain a constant body weight. Because 1 J is a small amount of energy, human energy requirements are sometimes expressed in kilojoules (kJ), which means that this typical man needs

$$2500 \text{ Cal} \times \frac{4184 \text{ J}}{1 \text{ Cal}} \times \frac{1 \text{ kJ}}{1000 \text{ J}} = 10,460 \text{ kJ}$$

If a man expends 10,460 kJ of energy each day and wants to maintain his weight at a constant level, he must eat food that contains 10,460 kJ (2500 Cal) of energy. Some of that energy is used for mechanical jobs—by muscle tissue, for example. Some is used to fuel the brain. If he consumes fewer calories than he needs, the energy provided by the food molecules is used up, the body must tap its reserves, and the reading on the bathroom scale goes down. Conversely, if he consumes more calories than he needs, the excess energy is stored and reserved for later use, and the reading on the bathroom scale goes up. It would be easy to count calories if a given mass of food could be equated with a specific amount of energy. In fact, though, different food groups provide our bodies with different amounts of energy, because the foods are broken down in different ways.

QUESTION 12.12 (*a*) Express the 425 Cal in a slice of banana cream pie in kilojoules. (*b*) Convert the 636 kJ required to break 1 mol of carbon–carbon double bonds to calories. (*c*) A 250-lb person playing Frisbee for 2 h burns 250 kcal. How many joules is this? (*d*) If a 180-lb person burns 2170 kJ playing Frisbee for 2 h, who uses more energy, the heavier person or the lighter person?

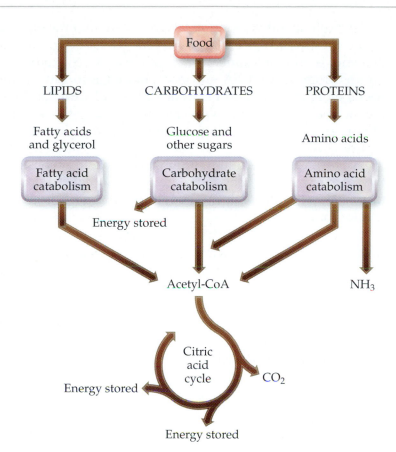

◀ **FIGURE 12.33   A simplified overview of catabolic pathways in the body.**   Each of the three major food groups follows a separate catabolic pathway in the initial stages of catabolism. The three pathways then converge at the citric acid cycle to produce carbon dioxide and stored energy, among other things.

ANSWER 12.12

(a) $425 \; \cancel{\text{Cal}} \times \dfrac{1 \; \cancel{\text{kcal}}}{1 \; \cancel{\text{Cal}}} \times \dfrac{4184 \; \cancel{\text{J}}}{1 \; \cancel{\text{kcal}}} \times \dfrac{1 \; \text{kJ}}{1000 \; \cancel{\text{J}}} = 1780 \; \text{kJ}$

(b) $636 \; \cancel{\text{kJ}} \times \dfrac{1000 \; \cancel{\text{J}}}{1 \; \cancel{\text{kJ}}} \times \dfrac{1 \; \cancel{\text{kcal}}}{4184 \; \cancel{\text{J}}} \times \dfrac{1000 \; \text{cal}}{1 \; \cancel{\text{kcal}}} = 152{,}000 \; \text{cal}$

(c) $250 \; \cancel{\text{kcal}} \times \dfrac{4184 \; \text{J}}{1 \; \cancel{\text{kcal}}} = 1{,}050{,}000 \; \text{J}$

(d) The lighter person, because 2170 kJ equals 2,170,000 J.

The three major food biomolecules—lipid, carbohydrate, and protein—are all sources of energy for our bodies. A highly simplified version of the metabolic pathways involved in the breakdown of lipids, carbohydrates, and proteins in the body is shown in Figure 12.33. At the top of the diagram are the intact food biomolecules we eat. As you move from top to bottom down the chart, these molecules get dismantled into their component parts. Lipids are broken down to fatty acids and glycerol, carbohydrates to glucose and other monosaccharides, and proteins to amino acids. Using separate catabolic pathways, most of these smaller molecules are broken down further, eventually to two-carbon units called *acetyl groups*. These units are attached to a carrier molecule called *coenzyme A*. The combination of this two-carbon unit plus coenzyme A is called *acetyl CoA* (Figure 12.34), and it serves as the primary carbon "currency" in the cells of the body.

▼ **FIGURE 12.34   Acetyl coenzyme A.** An acetyl group (blue) is attached to the carrier molecule known as coenzyme A (red). The acetyl group is the form in which carbon atoms are transported in the body in two-carbon units.

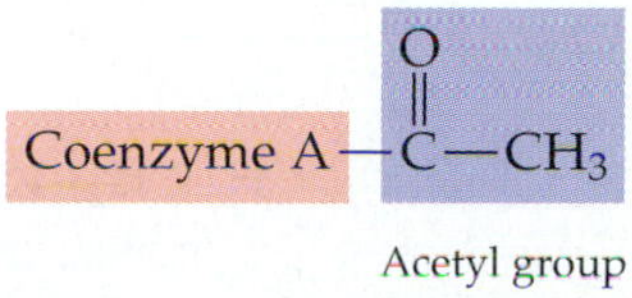

$$\text{Coenzyme A} - \overset{\displaystyle \overset{\text{O}}{\|}}{\text{C}} - \text{CH}_3$$

Acetyl group

The cell operates a kind of triage system that routes carbon currency in the form of acetyl CoA in different ways, depending on where the demand is. If, for example, the cell wants to produce TAGs, acetyl groups from acetyl CoA are incorporated into fatty acids. If the cell needs more energy, then the bonds of acetyl CoA can be broken down into a one-carbon molecule—carbon dioxide—by the *citric acid cycle*. This is an energy-producing, circular pathway shown at the bottom of Figure 12.33.

Now consider the output of these catabolic reactions. For example, in the catabolism of carbohydrates, glucose and oxygen are the reactants of the final steps of the pathway, and the products are carbon dioxide, water, and lots of energy:

$$C_6H_{12}O_6 + 6\,O_2 \longrightarrow 6\,CO_2 + 6\,H_2O + energy$$

Glucose (fuel)   Oxygen (oxidizer)   Usual combustion products

Glucose combustion

You might recognize this as a combustion reaction. In our discussion of explosives in Chapter 9, we saw several examples of combustion reactions, such as the one for the hydrocarbon octane, a component of gasoline:

$$2\,C_8H_{18} + 25\,O_2 \longrightarrow 16\,CO_2 + 18\,H_2O + energy$$

Octane (fuel)   Oxygen (oxidizer)   Usual combustion products

Octane combustion

In your car's engine, octane is present in the fuel, and it mixes forcefully with oxygen under pressure to produce little explosions that move the mechanical parts in the engine. In your body, combustion reactions are also used to produce energy, but they are carried out gradually and carefully so that the energy released is used efficiently by your cells. Like the combustion of octane, the energy-producing reactions taking place in all your cells require oxygen and produce carbon dioxide and water. The fuel used in the primary energy-producing pathways in the living cell is glucose. The glucose comes from the food you eat, and oxygen comes from the air you inhale. You get rid of the carbon dioxide when you exhale, and the water produced in the combustion is either eliminated or used by the body.

QUESTION 12.13 Write a balanced equation for the combustion of (*a*) 2-methylpropene, C_4H_8; (*b*) ethanol, C_2H_6O; (*c*) sucrose, $C_{12}H_{22}O_{11}$.

ANSWER 12.13

(*a*) $C_4H_8 + 6\,O_2 \rightarrow 4\,CO_2 + 4\,H_2O$

(*b*) $C_2H_6O + 3\,O_2 \rightarrow 2\,CO_2 + 3\,H_2O$

(*c*) $C_{12}H_{22}O_{11} + 12\,O_2 \rightarrow 12\,CO_2 + 11\,H_2O$

We just considered the combustion of a carbohydrate, glucose. Now let us look at the combustion of lipids. The combustion of palmitic acid, a fatty acid in the human diet, can be represented as

$$CH_3(CH_2)_{14}COOH + 23\,O_2 \longrightarrow 16\,CO_2 + 16\,H_2O + lots\ of\ energy$$

Palmitic acid (fuel)   Oxygen (oxidizer)   Same products again

Palmitic acid combustion

This reaction is no different from the combustion of glucose or the combustion of octane. However, although both reactions produce a lot of energy, the combustion of palmitic acid produces about 2.5 times more energy per gram than the combustion of glucose. To see why, let us consider how much energy is produced when 1 g of any food biomolecule undergoes combustion, a number referred to as the food's *fuel value*. The fuel values for the three major food groups are

$$\text{Carbohydrate} = 15.6 \text{ kJ/g} \quad \text{or} \quad 3.73 \text{ Cal/g}$$
$$\text{Protein} = 15.6 \text{ kJ/g} \quad \text{or} \quad 3.73 \text{ Cal/g}$$
$$\text{Lipid} = 38.9 \text{ kJ/g} \quad \text{or} \quad 9.30 \text{ Cal/g}$$

Why is the lipid fuel value so different from the other two? The answer lies in the molecular formulas, and the fact that *the reactions that break down food biomolecules in the cell are oxidative in nature*. That is, they rely on the ability of oxygen molecules to add to the carbon atoms in a biomolecule and produce carbon dioxide. The higher the number of oxygen molecules that can be added to a biomolecule, the more oxidation takes place and the more energy is produced.

QUESTION 12.14   How much energy, in kilojoules and in kilocalories, is contained in (*a*) 2.25 kg of table sugar (sucrose), (*b*) a 2.25-kg bottle of protein powder (assume that the powder is 100 percent protein), (*c*) 2.25 kg of olive oil? Use the three food values just given to calculate your answers. (*d*) Compare your answers for parts (*a*), (*b*), and (*c*).

ANSWER 12.14

(*a*) $2.25 \text{ kg} \times \dfrac{1000 \text{ g}}{1 \text{ kg}} \times \dfrac{15.6 \text{ kJ}}{1 \text{ g}} = 35{,}100 \text{ kJ}$

$35{,}100 \text{ kJ} \times \dfrac{1000 \text{ J}}{1 \text{ kJ}} \times \dfrac{1 \text{ kcal}}{4184 \text{ J}} = 8390 \text{ kcal}$

(*b*) Same answer as (*a*).

(*c*) $2.25 \text{ kg} \times \dfrac{1000 \text{ g}}{1 \text{ kg}} \times \dfrac{38.9 \text{ kJ}}{1 \text{ g}} = 87{,}500 \text{ kJ}$

$87{,}500 \text{ kJ} \times \dfrac{1000 \text{ J}}{1 \text{ kJ}} \times \dfrac{1 \text{ kcal}}{4184 \text{ J}} = 20{,}900 \text{ kcal}$

(*d*) The energy in 2.25 kg of oil (20,900 kcal) is much greater than the energy in 2.25 kg of sucrose (8390 kcal) or in 2.25 kg of protein (also 8390 kcal) because lipids store much more energy per unit mass than carbohydrates or proteins do.

The balanced equation for the combustion of glucose shows that six molecules of oxygen must be added to each molecule of glucose to create the products carbon dioxide and water. The formula for palmitic acid, $CH_3(CH_2)_{14}COOH$, tells us that this molecule contains two atoms of oxygen and a long hydrocarbon chain. *Hydrocarbon chains must react with a lot of oxygen to be completely converted to carbon dioxide.* Indeed, our equation for the combustion of palmitic acid shows that the combustion requires *23 molecules of oxygen*! The amount of energy produced from burning palmitic acid—as well as any other lipid—is high because it does

▲ FIGURE 12.35 **Using stored lipids for survival.**   This newborn hooded seal pup is on its own only four days after birth. The huge amount of lipids stored in its plump little body will allow it to survive as it hones its instinctive hunting skills.

not contain much oxygen to begin with. Thus, lipids make good fuels because they contain, in a given mass, much more energy than is contained in an equal mass of carbohydrates or proteins. The role of lipids is to store energy because they can stockpile so much of it and release it when it is needed.

Lipids are the most space-efficient biomolecules for storing energy in our bodies. However, some organisms, such as plants, use carbohydrates to store energy because, for a plant, mobility is not a big priority. For organisms that must move around, though, such as humans, lipids minimize the volume of stored energy that has to be carried around. Because we live in a society that is obsessed with weight loss, we are conditioned to think of storing lipids as a bad thing. However, many animals intentionally build up lipid reserves for energy storage and also for insulation. Consider what happens with hooded seals (Figure 12.35). After nursing their newborn pups for four days, the mother seals take off and leave the pups to fend for themselves. The four-day-old pups are nearly too fat to move at this point, but the stored lipids responsible for all that mass are the key to their survival. Clearly, the baby seal in the figure does not have a favorable BMI, but in this case that is definitely a health benefit.

12.6 | Fake Food

Think about the food you have eaten in the last 24 hours and the money you spent on it. How much of that food was processed, and how much of it was unadulterated? Processed foods include almost anything found in a grocery store, including the fresh produce section. That "100 percent bran" cereal you had for breakfast, for instance, though healthful, is processed, and that apple you had for lunch may have been coated with a layer of a shellaclike substance (made from insects) aimed at making it shiny and tempting. Most everything you drink is processed, including juices and sodas. Your water, too, is processed, unless you are drinking it directly from a stream in your backyard. The average American spends about 90 percent of her or his food budget on processed foods. These products are big business, and food manufacturers compete aggressively to win the approval of our taste buds.

Unfortunately, the methods used to process food so that they do not spoil or overripen before reaching the consumer often kill much of the natural flavor. For this reason, food companies put the flavor back in the form of flavorings, both natural and artificial. If the phrase *natural flavoring* on a food label conjures up in your mind visions of flavorings made from hand-pressed olive oil or fresh-picked berries, think again. The word *natural* when used this way means only that the flavoring was made from natural sources. Those natural sources may include fruits, vegetables, herbs, and spices, but animals are also natural sources, and a flavoring obtained from the by-products of beef or chicken processing is considered to be natural. Thus, do not ever mistakenly equate the term *natural* with *organic* or even *vegetarian*. In fact, many natural flavorings come from sources that would raise the hair on a vegetarian's neck. The distinction between natural and artificial is not always a useful one for consumers.

In 1992, the USDA introduced a label that could be placed on all legitimate organic foods. According to their definition, "Organic food is produced without using most conventional pesticides; fertilizers made with synthetic ingredients or sewage sludge; bioengineering; or ionizing radiation."

Consider the taste of pineapple. The molecule butyl butanoate,

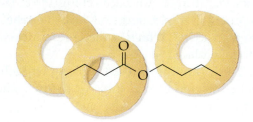

a typical pleasant-smelling ester, has the unmistakable taste of pineapple. If you would like to create a product with a pineapple flavor, you can use a real pineapple. Pineapples are not cheap, however, and their flavor is diminished by food processing. What is more, we are accustomed to flavors that are powerful. In fact, many of the processed foods we eat taste more like a certain flavor than the natural source of that flavor. Thus, to obtain a product with a super-duper pineapple flavor, you could synthesize butyl butanoate in a laboratory and use it in high concentrations in your food product. After all, esters can be synthesized by mixing the appropriate carboxylic acid with the appropriate alcohol:

$$\underset{\text{Butanol}}{\text{OH}} + \underset{\text{Butanoic acid}}{\text{HO}} \longrightarrow \underset{\text{Butyl butanoate}}{} + \text{HOH}$$

Often, a synthetic route will present a practical, cost-saving alternative to using a natural flavoring.

The problem, however, is that most artificial flavorings have been optimized so that they are a mixture of several molecules, not just one. Thus, pineapple flavoring contains a high percentage of butyl butanoate, the primary flavor molecule, but it also contains other flavoring molecules.

In his *Atlantic Monthly* article entitled "Why McDonald's Fries Taste So Good," Eric Schlosser writes about his visits to chemical companies that produce flavorings and perfumes:

> The federal Food and Drug Administration does not require companies to disclose the ingredients of their color or flavor additives so long as all the chemicals in them are considered by the agency to be GRAS (generally recognized as safe). This enables companies to maintain the secrecy of their formulas. It also hides the fact that flavor compounds often contain more ingredients than the food to which they give taste. The phrase *artificial strawberry flavor* gives little hint of the chemical wizardry and manufacturing skill that can make a highly processed food taste like strawberries. A typical artificial strawberry flavor, like the kind found in a . . . strawberry milk shake, contains the following ingredients . . . .

Schlosser then goes on to name the ingredients—all 48 of them!—in one brand of artificial strawberry flavoring.

One artificial flavoring that is difficult to create is the charred flavor we love in fried and grilled foods. In fact, reproducing this flavor is one of the remaining frontiers in the food industry. The secret to the flavor and color obtained from grilling and frying is the high temperature that can be reached in the frying pan or barbecue grill. For example, a potato that is boiled in water can be heated only as high as the boiling point of water, 100°C, and this temperature is not high enough to cause browning. When that potato is put in a skillet with some oil, however, the oil can reach a temperature of more than 150°C, which is the temperature at which browning reactions begin to take place.

Anyone who knows anything about cooking will tell you that the secret to making a great stew is to brown some of the ingredients. In one recipe, for instance, the first step is

> 1. Place flour in a plastic bag. Add meat cubes and shake until meat is coated with flour. In a large skillet, brown half of the meat in 1 tablespoon of the oil, turning to brown evenly. Brown remaining meat in remaining oil. Drain off fat.

The chemistry of browning is an extraordinarily complex and poorly understood process. What we do know is that browning begins when some simple carbohydrates react with an amine group, most frequently one that is part of a protein side chain—for instance, a protein side chain in a chunk of stewing beef. The initial carbohydrate–amine reaction is called the *Maillard reaction.* Exactly what happens after this initial step is largely a mystery. However, it is known that, after several reactions, large, colored polymeric compounds are formed. These polymers, known as *melanoidins,* are dark brown and give food a characteristic charred flavor:

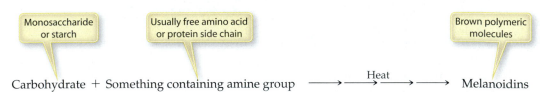

Flavoring chemists working on the development of artificial grilled flavoring often begin with a Maillard reaction between a readily available monosaccharide and an amine. They then turn up the heat and wait for the melanoidins and the "brown flavor" to emerge. The artificial smoky flavor of some maple syrups, chocolates, and coffee are based on this reaction. (Interestingly, melanoidin molecules have a structure similar to the structure of melanin, the molecule that gives human skin its brown pigmentation.)

It is common knowledge in the food industry that the color of food is critically important. One study performed in the 1970s showed that if a food is not the color we expect it to be, most of us will not eat it. Colorants are categorized by the FDA via a method similar to the categorization of flavorings. Natural colorants are derived from natural sources, and artificial colorants are synthesized in the laboratory. Artificial dyes have come under heavy scrutiny because they have been shown to cause certain cancers in laboratory rats. In fact, the Delaney Clause, part of a 1960 law on the uses of colorants for foods, states that the FDA cannot approve any dye for use in food if there is any risk, however small, of carcinogenicity. Thus, certain kinds of red dye, once used in everything from cherries to toaster pastries to jelly, went by the wayside in 1990 when studies showed a very low incidence of cancer in rats fed large amounts.

Nowadays, many food manufacturers use a natural red dye called *carmine* to enhance the color of food. You might expect that this change from artificial to natural would allow healthy eaters to breathe a collective sigh

Honey-Mustard Glazed Irradiated Pork Tenderloin with Grilled Vegetables

The list of ingredients for the recipe for this dish might read, "12 ounces of pork tenderloin, irradiated with gamma rays from radioactive cobalt-60." Recall from Chapter 6 that many nuclear reactions produce gamma rays in addition to alpha or beta particles. One such reaction is the radioactive decay of cobalt-60.

The gamma rays emitted from cobalt-60 are extremely potent, and gamma radiation is the highest-energy radiation in the electromagnetic spectrum. Because it is high in energy, gamma radiation is able to penetrate deep into living tissues, like those of our pork tenderloin. Gamma radiation is so energetic that it is able to break covalent bonds, including those in the molecules that make up bacterial cells. For this reason, gamma radiation is an extremely effective, noninvasive antibacterial agent.

The idea seems sound: shoot rays at food to kill bacteria. Not only is food treated in this way free of dangerous microbes such as botulism; the food also requires no antibacterial additives to prevent spoilage. Yet the irradiation of food has many people up in arms. For some people, the idea that something "nuclear" is involved in the process is enough to scare them, even though radioactive cobalt does not come in direct contact with the food. (Now that you have a solid foundation in chemistry, you are not at all intimidated by the word *nuclear*, of course.)

A more reasonable objection to food irradiation goes like this: if gamma rays break the covalent bonds in bacteria, then they must also break the covalent bonds in the food itself. Unknown and unpredictable reactions may occur when biomolecules (such as the protein found in our pork entrée) break down, including the formation of new, possibly carcinogenic compounds. Studies have shown that the carcinogen benzene is produced in meat when it is irradiated. Fruit juices are also irradiated to kill bacteria and to extend shelf life. Studies have shown that the carbohydrates found in fruit juices form *malondialdehyde* when irradiated. In our bodies, this small organic molecule can attach to proteins and DNA, and these modified biomolecules may affect the body in unpredictable ways.

Despite concerns voiced by consumer advocacy groups, in 1999 the U.S. government approved the irradiation of meat sold to the public. The law that was passed requires that any irradiated product be labeled with the *radura* symbol, shown in Figure 12.36.

The jury is still out on food irradiation. On one hand, irradiation kills microbes very efficiently, and this is comforting to many people, especially in an age of anthrax scares and international terrorism. On the other hand, the risks associated with food irradiation are largely unknown. Only time will tell how these risks and benefits balance each other.

▲ **FIGURE 12.36 Nuclear chemistry at work.** The radura symbol is required on the label of any food that has been irradiated with a radioactive isotope.

of relief. The production of carmine dye begins in Peru, where female insects regularly dine on a variety of cactus. The insects ingest a red pigment from the cactus and then pass this pigment into their eggs, which take on the red hue. To harvest the red dye, insect corpses and eggs are brushed off the cactus plant, crushed, and cooked. The result is a brilliant red pigment that is used today to color everything from ice cream to lipstick to breakfast cereal. So if you are looking for natural ingredients in the grocery store, remember that *natural* does not imply *gentle* and that while some products may not be tested on animals, they may be made from them.

As we wrap up our discussion, let us take one more trip to the grocery store. Our anthropologists William Leonard and Marcia Robertson would applaud the extraordinary range of choices we have as we walk up and down the aisles. According to these experts, one of the most remarkable outcomes of human evolution is the astounding variety of foods we are able to eat. Because humans can incorporate almost any combination of food groups into their diet, Leonard and Robertson marvel at humans who believe that limiting food variety is the answer to their obesity problems. Protein-only diets, low-fat diets, juice fasts—Americans will deprive themselves of anything to lower the number on the bathroom scale. The answer, according to Leonard and Robertson, is to keep the variety you love in your diet, but eat only enough to meet your body's energy needs. These researchers do not address the effect of artificial food additives and colorings that have entered the modern American diet, however. If humans can evolve to eat foods from a wide variety of food groups, can they also adapt to the foreign molecules that make processed foods taste so good? Food for thought, wouldn't you say?

SUMMARY

We started this chapter with fat, a problem many people battle every day. The term *fat* is generally used to describe lipids that are solids at room temperature. Because saturated fats pack efficiently together, they have relatively high melting points and so tend to exist in the solid phase at room temperature. Oils, on the other hand, are usually composed of unsaturated lipids, and the distortion caused by the double bonds in these hydrocarbon chains makes them pack less efficiently. Oils therefore tend to have relatively low melting points and exist in the liquid phase at room temperature. Fatty acids contain long hydrocarbon chains and are insoluble in aqueous solutions. They are joined in groups of three in triacylglycerides (TAGs), the primary storage form of fatty acids, and in groups of two in phospholipids, the primary lipid component of biological membranes.

The cholesterol molecule has a hydrocarbon backbone and a rigid structure that results from an inflexible system of rings. Cholesterol is unpopular because high levels of it are associated with heart disease. However, the human body needs cholesterol and produces it, and it is used by the body to adjust the rigidity of cell membranes in response to environmental changes. Cholesterol is transported through the body in large protein complexes called high-density lipids (HDLs) and low-density lipids (LDLs). The health community encourages high levels of HDLs and low levels of LDLs, and one way to achieve this aim is through the consumption of more polyunsaturated fats and fewer saturated fats.

The food we eat is composed of biomolecules from three groups: lipid, carbohydrate, and protein. When we eat food, these biomolecules begin to break down in the mouth, but most digestion occurs when food reaches the stomach. The conditions of very low pH found in the stomach promote acid-catalyzed hydrolysis of the bonds between monosaccharide units in carbohydrates, of the ester bonds in TAG molecules, and of the peptide bonds between amino acids in proteins. In the stomach and small intestine, digestive enzymes chew away at biomolecules until the molecules are small enough to be absorbed into the bloodstream.

Most enzymes are proteins, and they are the catalysts for chemical reactions that make up metabolic pathways. The vast majority of enzymes are specific to one particular substrate. An enzyme works by binding its substrate(s) in a three-dimensional pocket. In order to bind, the substrate must become distorted into a shape that allows the reaction to reach its transition state. The reaction coordinate for a chemical reaction shows how the energy changes as reactants are converted to products and includes an energy hill that must be traversed in order for reaction to occur. When an enzyme catalyzes a reaction, it reduces the size of the energy hill, and consequently the reaction proceeds at a faster rate.

Carbohydrates can vary in size from monosaccharides (containing one saccharide unit) to polysaccharides (a polymer with many units). Carbohydrates found in foods are mono- and disaccharides (simple sugars) as well as polysaccharides (complex carbohydrates or dietary fiber). The linkages between each unit of a polysaccharide determines whether human digestive enzymes can dismantle the polysaccharide into small units that may be absorbed and metabolized by the body. The most common polysaccharides in nature are glycogen, starch, and cellulose. Of these three, only cellulose cannot be digested by humans.

It is possible to write a balanced equation for the combustion of any organic molecule that contains only oxygen, carbon, and hydrogen. The combustion of lipids releases more energy per unit mass of biomolecule than the combustion of carbohydrates and proteins. The high fuel value of lipids is a result of the paucity of oxygen atoms and the presence of a large number of hydrogen atoms in lipid structures. By comparison, a carbohydrate contains several oxygen atoms, and so the combustion of a carbohydrate will not produce as much energy as the combustion of a lipid. In cells and tissues, anabolic reactions require energy to build up new biomolecules. Catabolic reactions break down biomolecules and release energy when it is needed. Because our bodies require energy to function, the amount of food we eat should provide approximately the same amount of energy that our bodies need to function.

KEY TERMS

aldehyde	catabolic pathway	hydrogenation	monosaccharide	substrate
anabolic pathway	catabolism	induced-fit model	oligosaccharide	transition state
anabolism	catalyst	kilocalorie	phospholipid	triacylglyceride
anomer	disaccharide	lipid	polysaccharide	(TAG)
anomeric carbon	enzyme	lipoprotein	reaction coordinate	
carbohydrate	fatty acid	metabolic pathway	saccharide	

QUESTIONS

The Painless Questions

1. Carbohydrate terminology can be confusing. In your own words, describe what is meant by the following terms: simple sugar, monosaccharide, polysaccharide, disaccharide, complex carbohydrate, dietary fiber.

2. A murder victim was known to eat only potatoes at every meal. Given that potatoes are made mostly from starch molecules, speculate on the usefulness of data collected from the victim's stomach contents.

3. Identify the following fatty acids as unsaturated, monounsaturated, or polyunsaturated. Indicate whether each double bond is in the cis or trans conformation:

(a)

(b)

(c)

(d)

4. Of the four major classes of biomolecules—DNA and RNA, protein, carbohydrate, and lipid—what do the first four have in common that is not shared by lipids?

5. Comment on this statement: Enzymes are a subclass of proteins that carry out chemical reactions, and there is one enzyme for each specific chemical reaction carried out in the cell.

6. How many two-carbon units would be needed to build the framework of a cholesterol molecule?

7. Describe the two major mechanisms by which food is broken down in your stomach. Describe the role of catalysts in each process.

8. What is the fastest enzyme known? How many times does it turn over reactants every microsecond?

9. (a) Sketch a reaction coordinate for two atoms D and G that come together to form a chemical bond. (b) Modify your drawing from part (a) for the same reaction, but one in which a catalyst participates.

10. What is meant by the acronym GRAS? Why was this term employed for the approval of foods in the early years of government control over food products?

11. Anyone who watches professional sports hears stories about drug testing and steroid use. The steroids used by athletes and body builders are anabolic steroids. Why are they referred to as anabolic?

12. If you are writing a balanced equation for a chemical reaction, what must be true of a catalyst that is acting on the reactants?

13. Here are three ways nutritional information can be presented on food products:

Simplified Format

Nutrition Facts		
Serving Size 35 Pieces (40g) Servings Per Package About 3		
Amount Per Serving		
Calories 140		
		%Daily Value*
Total Fat 0g		0%
Sodium 10mg		0%
Total Carbohydrate 37g		12%
Sugars 28g		
Protein 0g		
Not a significant source of calories from fat, saturated fat, cholesterol, dietary fiber, vitamin A, vitamin C, calcium, and iron.		

Simplified Tabular Format

Nutrition Facts	Amount/Serving	% DV*	Amount/Serving	% DV*
Serv. Size 35 Pieces (40g)	Total Fat 0g	0%	Total Carb. 37g	12%
Servings About 6	Saturated Fat 0g	0%	Dietary Fiber 0g	0%
Calories 140	Cholest. 0mg	0%	Sugars 28g	
Calories from Fat 0	Sodium 10mg	0%	Protein 0g	
*Percent Daily Values are based on a 2000 calorie diet	Vitamin A 0%	• Vitamin C 0%	• Calcium 0%	• Iron 0%

Simplified Linear Format

Nutrition Facts Serv Size: 1 Package (40g) Servings: 1 Amount Per Serving: **Calories** 140, **Total Fat** 0g (0% DV), **Sodium** 15mg (1% DV), **Total Carb.** 37g (12% DV), Sugars 28g, **Protein** 0g, Percent Daily Values (DV) are based on a 2,000 calorie diet.

All three formats show the same information. Is this food most likely to be (a) a piece of hard candy, (b) a small bag of potato chips, (c) a carton of eggs, (d) a bottle of canola oil?

14. Describe the benefits and drawbacks of using gamma radiation to kill microorganisms in food.

More Challenging Questions

15. (a) Sketch the structure of cholesterol. (b) Why is cholesterol classified as a lipid even though its structure does not resemble the structure of other lipids? (c) Why does the word *cholesterol* end in the syllable -*ol*? (d) Is cholesterol a saturated lipid or an unsaturated lipid? (e) What feature of cholesterol's structure allows it to add rigidity to cell membranes?

16. Describe the lock-and-key and induced-fit models for enzyme action. How are they alike? How are they different? Why does the induced-fit model provide a more accurate picture of how enzymes behave?

17. In this chapter, we discussed two enzymes that break down carbohydrates: amylase and β-galactosidase. Based on what you know about the jobs of these two enzymes, which of the two is more specific for one reactant? Why?

18. Write a balanced equation showing the acid hydrolysis of a TAG molecule that contains myristic acid, arachidonic acid, and linolenic acid.

19. In separate sketches, draw two glucose molecules connected by (a) an alpha 1 → 4 linkage, (b) a beta 1 → 6 linkage, (c) an alpha 1 → 6 linkage, (d) a beta 1 → 4 linkage.

20. Working from the balanced equation in Figure 12.9 (page 471), determine the number of molecules of water produced when 1.3 mol of glycerol reaches the smoke point if the reaction goes to completion.

21. We discussed three artificial sweeteners in this chapter. (a) One is a small organic molecule that does not resemble a biomolecule. Identify this sweetener and sketch its structure. (b) The other two are modified versions of a biomolecule. Identify these two sweeteners and indicate what the modifications are. (c) What kind of biomolecule does each of the two in part (b) resemble?

22. Describe what happens when you heat a cooking oil. Begin with the structure of the oil molecules at room temperature, and continue to the point at which the oil begins to smoke.

23. Take a look at the structure of acetyl-CoA in Figure 12.34 on page 493. (a) Which organic functional group do you recognize in this structure? (b) How many acetyl groups would be required to make one molecule of the fatty acid myristic acid, shown in Table 12.1 on page 467?

24. If you heat a carbohydrate—glucose, say—to a very high temperature for a long time, you will notice steam escaping and you will be left with a pile of black soot. Based on what you know about the origin of the word *carbohydrate*, account for these observations.

25. One unit of the repeating structure of the polysaccharide chitin is

CH₂OH ... (structure)

N-Acetylglucosamine   N-Acetylglucosamine

Chitin

(a) Name the linkage that connects the two monosaccharides. (b) On what monosaccharide structure is the monomer N-acetylglucosamine based? (c) Would humans be able to break down this polymer?

26. The structure of saccharin is

O

N—H

S

O   O

Saccharin

(a) If you add 0.650 g of saccharin to your morning coffee, how many molecules have you added? (b) Speculate on where this molecule got its name.

27. (a) Describe what is meant by a two-tailed lipid and a three-tailed lipid. (b) What uses for these two types of lipid are discussed in this chapter? (c) How do the structures of these two lipids allow them to do their jobs in the cell?

28. Ramón is on a diet. What is the maximum number of cans of a typical diet soda he can drink in one day if he is 6 ft, 2 in., tall and weighs 235 lb (1.0 lb = 0.453 kg), assuming that he wishes to stay below the recommended maximum intake of aspartame?

29. Describe how an enzyme is able to increase the rate of a chemical reaction and how the enzyme affects the structure of the reaction's transition state.

The Toughest Questions

30. Three of the 48 ingredients listed by Eric Schlosser as being part of artificial strawberry flavor are ethyl heptanoate, ethyl acetate, and butanoic acid. (a) Draw a line structure for each of these molecules. (b) For each of the esters, write a balanced equation showing how the ester was formed from an alcohol and a carboxylic acid.

31. The structure of one specific phospholipid is

O
‖
CH_2—O—C$(CH_2)_{16}CH_3$
| O
| ‖
HC—O—C$(CH_2)_{16}CH_3$
|
O
‖
$(CH_3)_3\overset{\oplus}{N}$—$(CH_2)_2$—O—P—O—$CH_2$
|
$\overset{\ominus}{O}$

Phosphatidylcholine

(a) Write a balanced equation for the formation of this phospholipid from two fatty acids and a derivative of glycerol. Identify each fatty acid. (b) Write the chemical formula for this molecule and calculate its molar mass. (c) What three-dimensional shape would you expect the molecule to have?

32. Figure 12.26 (page 485) shows the two anomers of the simple sugar glucose, along with the open-chain form of this molecule. (a) Using this illustration as a guide, draw the open-chain form of fructose and galactose, the other simple sugars shown in Figure 12.25 (page 484). (b) Describe all the differences you spot in the three open-chain forms. (c) Draw the anomer for the fructose and galactose molecules shown in Figure 12.25.

33. Figure 12.6 (page 469) shows the three-dimensional shapes of several fatty acids. (a) Identify each double bond as cis or trans. (b) Which molecule appears to be the most distorted from a linear shape? Which type of double bonds does this molecule contain? How many of them? (c) Describe the three-dimensional shape of arachidonic acid. (d) Which will pack more efficiently in the solid phase, arachidonic acid or your choice in part (b)?

34. You and a friend are cooking dinner together. You are trying to decide which way the chicken you are planning to sauté will brown faster: soaked in marinade (water, salt, vinegar, herbs, olive oil) before cooking or sautéed dry. Using your knowledge of the browning process, suggest an answer to this question.

35. Comment on this statement: Trans fats are a by-product of the hydrogenation of unsaturated oils, a process intended to produce completely saturated hydrocarbon chains. Although trans fats were not the original desired product of the reaction, they have physical properties that are desirable in lipid products that must have a firm consistency.

36. Proteins undergo hydrolysis reactions that are identical to those of carbohydrates. Knowing this, (a) draw a mechanism for the acid-catalyzed hydrolysis of the dipeptide shown (b) Name the two products of the reaction.

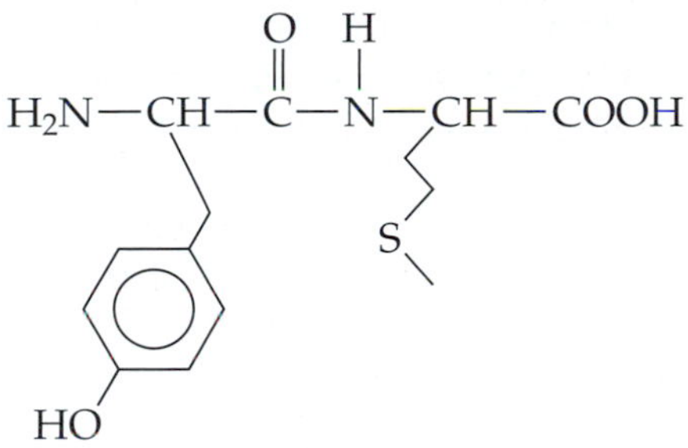

37. A variety of modified polysaccharides exist in the cells of living organisms, and some of these polysaccharides have exotic linkages. The polysaccharide dermatan sulfate, for instance, has a beta $1 \rightarrow 3$ linkage between the monosaccharides iduronate and N-acetylgalactosamine-4-sulfate:

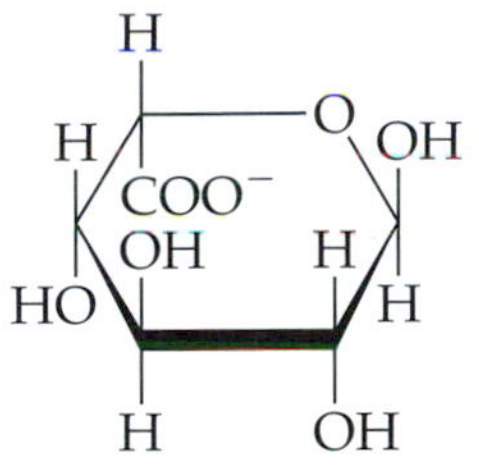

Iduronate 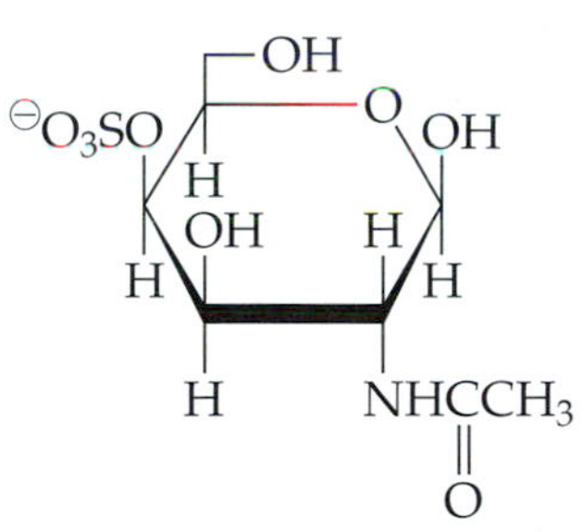   N-Acetylgalactosamine-4-sulfate

Draw these units connected together by the appropriate linkage.

38. The equation for the complete combustion of glucose is given on page 494. (a) Write a balanced chemical equation for the combustion of lactose and one for the combustion of a trisaccharide made of three repeating units of glucose. (b) Write a balanced chemical equation for the combustion of myristic acid and one for the combustion of linoleic acid. You can see the formulas for these acids in Table 12.1 on page 467. (c) Compare your results from parts (a) and (b).

39. If a nutritional label tells you that a food contains 34 g of carbohydrate, 8.2 g of fat, and 13.2 g of protein, how many Calories would this represent? Convert each value to joules.

40. Although not all organic molecules find uses as fuels, we can write an equation for the combustion of any organic molecule composed of carbon, hydrogen, and oxygen. From the group shown here, choose the molecule that has the largest fuel value and the molecules that has the smallest. Briefly explain your choices.

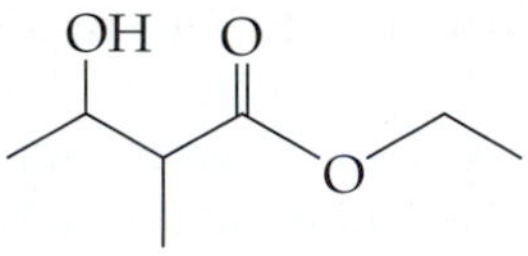

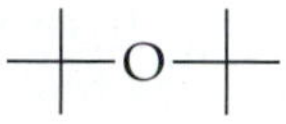

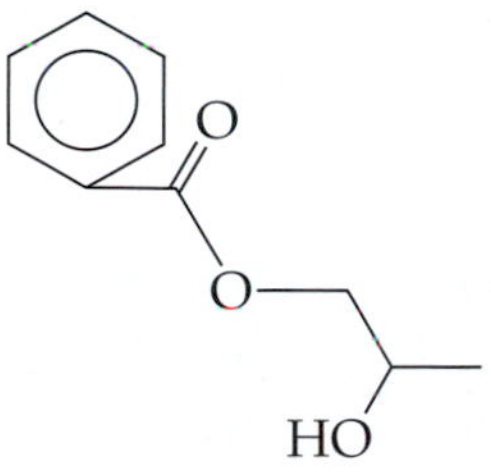

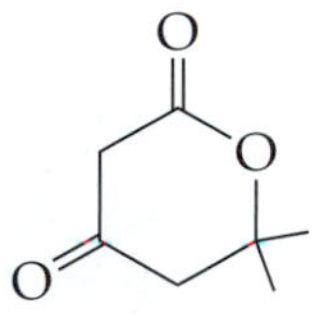

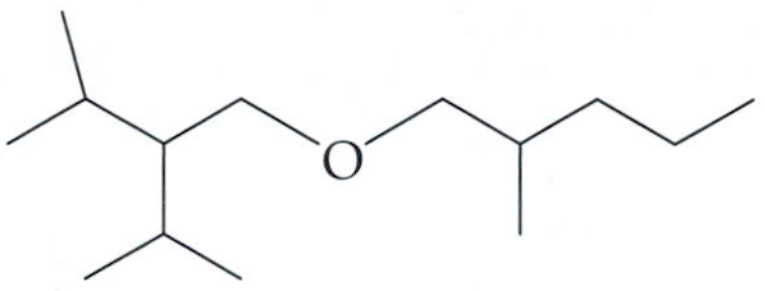

41. In this chapter, we considered some molecules that have color, such as melanoidin and red dyes. (a) Based on what you learned in Sections 5.7 and 5.8 about color and molecules, describe what structural feature these colored molecules, and others, share. (b) In which part of the electromagnetic spectrum do all colored molecules absorb?

42. Predict the products for each of these acid-catalyzed hydrolysis reactions, and for each reactant, identify the organic functional groups it contains:

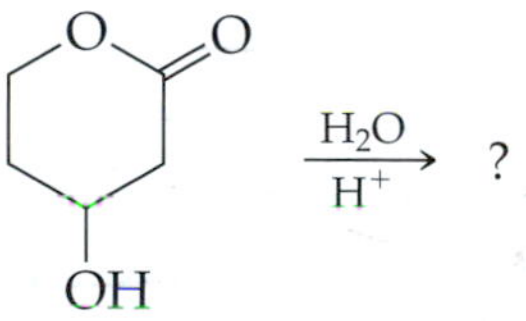

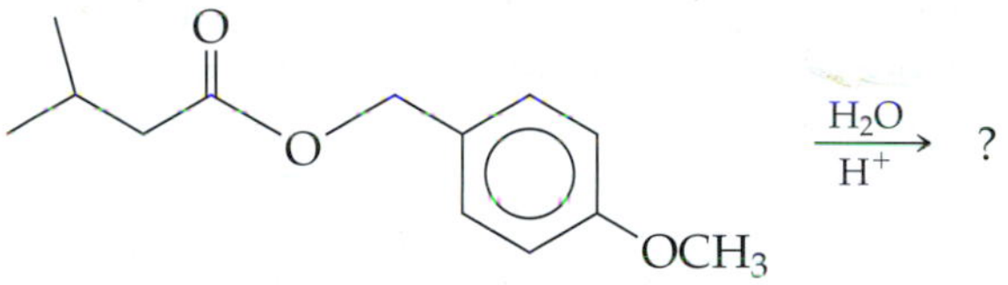

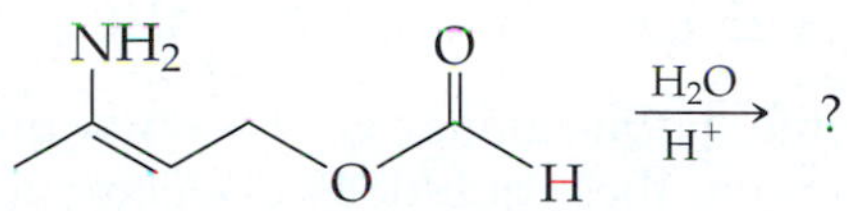

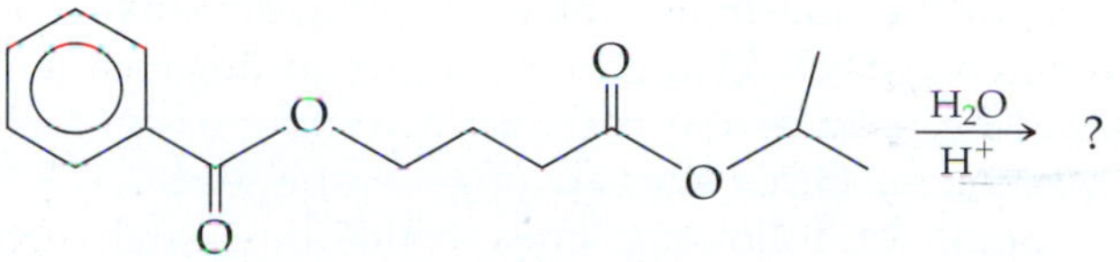

43. (a) Sketch the full structure of the aspartame molecule and write its chemical formula. (b) Amide bonds undergo hydrolysis reactions in acidic solution in the same way that esters do. Knowing this, write the three products of the complete hydrolysis of aspartame.

44. From this nutritional label from a popular brand of soup, determine (a) the total number of calories in one serving and (b) the number of calories from fat in one serving:

Nutrition Facts

Serving Size 1 cup (246g)
Servings Per Container About 2

Amount Per Serving		
Calories	Calories from Fat	
		% Daily Value*
Total Fat 1g		2%
Saturated Fat 0g		0%
Cholesterol 15mg		5%
Sodium 990mg		42%
Total Carbohydrate 21g		7%
Dietary Fiber 2g		8%
Sugars 5g		
Protein 8g		

E-Questions

Go to **www.prenhall.com/waldron** to find these questions in electronic form, complete with hyperlinks directly to the various websites cited in the questions.

45. In our discussion of food additives, we did not discuss one of the most ubiquitous: monosodium glutamate (MSG). Read the short article "What's That Stuff? Monosodium Glutamate?" at **pubs.acs**, and then answer the following questions: (a) How is MSG produced? (b) Draw the chemical reaction (use line structures) that converts glutamic acid and NaOH to MSG. (c) What taste category does MSG fall into? What other substances are in this category? (d) Is it possible for a food to contain a form of MSG, but not list it on its label of ingredient? Explain.

46. Read the article "Irradiated Apple Juice" at **chemistry**. To find the article, search for *irradiated* or for *McCue*, the author's name. After reading the article, answer the following questions: (a) How is the browning of apple juice measured? (b) Does irradiation improve the rate of browning in apple juice? (c) How was the effect of irradiation on antioxidant activity in apple juice measured? What did the researchers find? (d) What one drawback was reported as a result of irradiation?

47. Read about the enzyme OMP decarboxylase at **USC**, and then answer the following questions: (a) Why is OMP decarboxylase a superstar in the world of enzymes despite the fact that the rate at which it catalyzes reactions is unimpressive? (b) Describe the active site of the enzyme, which is the three-dimensional region where the substrate binds. What happens when the substrate binds?

48. Read the article "Are Trans Fats a Health Hazard?" at **benbest**, and then answer the following questions: (a) How is it possible for nonhydrogenated food products to contain trans fats? (b) What do the authors say about the effect of trans fats on cell membranes in neural and cardiac tissues? (c) Do the authors recommend eating saturated fat over trans fat?

49. The article "Food Preservatives" gives an overview of all additives used to preserve food. Find it at **pubs.acs**. After reading it, answer the following questions: (a) What are the three major categories of food preservatives? What is the role of each? (b) What kinds of foods from the grocery store may not require any kind of added preservative? (c) How do weak organic acids act as preservatives? In which of the three categories of preservatives do they belong? (d) Describe how overripening of fresh fruit and vegetables is prevented (or slowed) by adding preservatives. What enzyme is associated with this process?

WORLD WIDE WEB RESOURCES

As with the E-Questions, go to **www.prenhall.com/waldron** to find these questions in electronic form, complete with hyperlinks directly to the various websites cited in the questions. Some of following links contain research articles related to the topics in this chapter and could be used as the basis for a writing assignment in your course. Other sites are of general interest.

Some of following links contain research articles related to the topics in this chapter and could be used as the basis for a writing assignment in your course. Other sites are of general interest.

• Read a story about one company's battle with trans fats at **usatoday**. In this article, the author describes how the company is cutting trans fats and the effect this will have on the products it sells and the people who buy them.

• Visit the website **sandyhershelman**, and read a story about the author's invitation to a bug-laden dinner. In it, she discusses the history of bug eating and how Western countries are in the minority when it comes to this dietary practice. She describes some of the dishes served and recommends a cookbook for people interested in trying some insect cuisine. Most interesting is her discussion of the major food groups, as well as vitamins, that can be found in bugs.

- The article "How Do Food Manufacturers Calculate the Calorie Count of Packaged Foods?" walks you through the calculation that leads from grams of a certain food group to total calories. To find it, go to **scientificamerican**. This article also includes information on counting calories in alcohol and in high-fiber foods. There is an error in the first paragraph of the article. Can you spot it?

- Two articles, one at **niddk.nih** and one at **weight-loss-i**, offer information and statistics on the problems of overweight America. In addition, highly specific nutritional information about foods can be found at the USDA's website, **nal.usda**.

- Olestra is an artificial fat that was first brought to market in the United States in the 1990s. Since its debut, it has been one of the most highly criticized artificial food additives ever. First read an article touting the virtues of Olestra at **acsh**. Then turn to a later article that includes information about the public's response to Olestra: "New Olestra Complaints Bring Close to 20,000—More than All Other Food Additive Complaints in History Combined," at **cspinet**.

CHEMISTRY TOPICS
IN CHAPTER 13

- Structure–activity relationships [13.1]
- Effective dose and toxic dose of drugs, lethal dose of poisons and toxins [13.2]
- Enzyme inhibition [13.3]
- Chirality and handedness [13.4]
- Isomers [13.4]
- Enantiomers [13.4]
- Racemic mixtures [13.5]

"I always keep a supply of stimulant handy in case I see a snake—which I also keep handy."

W. C. FIELDS

Drugs I

HOW DRUGS ARE DESIGNED TO BENEFIT THE HUMAN BODY

The April 12, 2002, issue of *Science* magazine includes a research paper submitted by scientists at the University of Texas Southwestern Medical Center in Dallas. This report, which made the science-highlight reels in 2002, explains the mechanism by which muscles become "exercised." The paper provides a step-by-step account of how the effects of muscle conditioning, such as an increased capacity for using oxygen, are produced—all very interesting stuff if you are an exercise physiologist or a cell biologist, but this research finding was also reported widely in the lay media, and here's why: it shows that the biochemical reactions which take place when a muscle is exercised can be activated by the addition of a biomolecular trigger (in other words, a drug) to the muscle tissue. This means that a muscle can, at least in theory, be exercised simply by being exposed to a drug, gaining all of the properties of a fit muscle without any of the pain.

The authors of the *Science* article suggest that their finding may one day allow individuals who suffer from diseases that restrict activity, such as congestive heart failure or respiratory insufficiency, to get the exercise they need. The media were more cynical, though, reporting that the "exercise pill" will one day be a way for us to get the equivalent of a heart-pounding workout while sitting and watching television (Figure 13.1)!

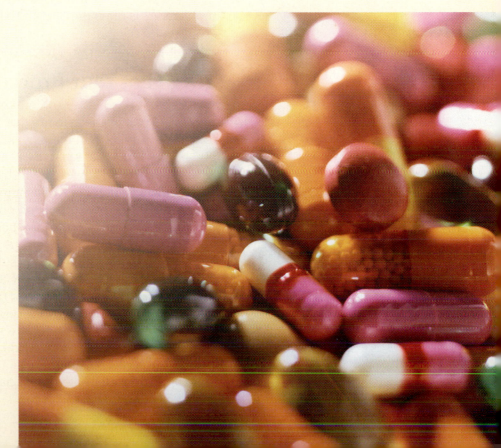

▲ **FIGURE 13.1** **No pain, no gain?** A man watching television may one day be able to get the equivalent of a heart-pounding workout without leaving his chair.

The exercise pill, if it becomes a reality, will join a host of other drugs being marketed to a new group of pharmaceutical consumers: healthy people. One popular television advertisement shows a carefree woman running through a field of wildflowers. All we know about her is that she owes her happiness to the pill being advertised. In some advertisements, the name of a drug is given, but the drug's purpose is not. Instead, you are encouraged to *ask your doctor about* . . . .

These ads leave many healthy people wondering whether they are healthy enough and what they can do to stave off disease. Thus, *prevention* is the new buzzword for today's pharmaceutical industry. Not far down the road, for instance, are chemoprotective drugs like *oltipraz*, otherwise known as "the chemical mop." The designers of oltipraz say it stimulates the production of a cancer-preventing biomolecule already in the body. According to the April 4, 2001, issue of the respected journal *New Scientist*, "Half of all cancers may soon be prevented simply by taking a pill once a week . . . . [These] drugs . . . activate our body's natural defense mechanisms." In the future, we may be popping anticancer drugs as routinely as we swallow multivitamins today.

In this chapter, we shall focus on how drugs do what they do. That is, we shall investigate the mechanisms of drug action and how drugs produce beneficial effects in our bodies. The next chapter, "Drugs II," takes on the sinister side of this topic: the misuse and abuse of drugs. We shall begin our foray into the drug world by learning how a drug is designed to target a particular problem. As is often the case, the best clues come from mother nature, as we are about to see.

13.1 | Eat Your Broccoli!

Broccoli is one of those foods that are supposed to be good for you. Until recently, this piece of folklore was something that our mothers told us simply because that is what *their* mothers told them. Scientific evidence released in 1992, however, vindicated broccoli-pushing mothers everywhere. This study showed that a molecule found in broccoli has the ability to stimulate the production of proteins that flush **carcinogenic** (cancer-causing) compounds out of the body. The molecule, called *sulforaphane*, is present not only in broccoli, but also in Brussels sprouts and cabbage. The most concentrated source of sulforaphane is broccoli sprouts, a product that has been patented by researchers at Johns Hopkins University. In fact, the label on their *Brocco Sprouts*, shown in Figure 13.2, mentions the university's contribution to the development of this product.

Sulforaphane is a known cancer-fighting substance, but how can we get a continuous supply of it in our diets? One way would be to eat either nine stalks of broccoli or about half an ounce of Brocco Sprouts every day. A more appealing alternative, especially for folks who do not like the taste of broccoli, is to get sulforaphane in pill form. Researchers at the University of Illinois at Chicago have been working on extracting sulforaphane from vegetables and putting it into a capsule. In the process, they have run into two problems that give us a glimpse of the challenges associated with drug design. The first problem is that sulforaphane is toxic in large quantities; the second problem is that it is very expensive to produce.

Faced with these problems, the researchers sat down with the structure of the sulforaphane molecule (Figure 13.3) to see how they could make the compound less toxic and less pricey. After considering the organic functional groups in the molecule, the researchers hypothesized that the right end was the part that makes sulforaphane toxic. Once they confirmed this hypothesis, they determined that it was the left end that gives this compound its anticarcinogenic properties. To make synthesizing it easier and less costly, the researchers changed the right end of the molecule to make it less toxic and then made minor alterations to the left end, being careful to preserve its three-dimensional shape.

The new compound they created, named *oxomate*, is also shown in Figure 13.3. This molecule maintains the anticarcinogenic properties of sulforaphane. In fact, the researchers state, "When female rats were exposed to carcinogens but also given oxomate, their incidence of breast cancer was 50 percent lower than that of a control group." In addition, the oxomate molecule is seven times less toxic than sulforaphane and many times less expensive to synthesize.

The sulforaphane research made use of two common principles of medicinal chemistry and drug design: (1) drug designers often begin with one of mother nature's ideas and work from there, and (2) it is the three-dimensional shape of a molecule that determines what it can and will do as a drug.

▲ **FIGURE 13.2 Cancer-fighting foods.** Broccoli sprouts have been marketed under the name "Brocco Sprouts," a product patented by Johns Hopkins University that is known to have anticarcinogenic properties.

> **QUESTION 13.1** When sulforaphane is converted to oxomate (Figure 13.3), what happens at the left end of the molecule? What does *not* happen at the left end of the molecule?
>
> **ANSWER 13.1** A sulfur atom double-bonded to oxygen is replaced by a carbon atom double-bonded to oxygen, but the three-dimensional shape of the molecule does not change.

Now let us consider this question: why is three-dimensional shape so critical in the design of drug molecules? The answer is that most drugs work by binding to natural biomolecules, which are three dimensional. We refer to a biomolecule that binds a drug as a **receptor**. In most cases, the receptors in our bodies are proteins, and as we learned in Chapter 11, most proteins have a rounded, globular shape. More important, proteins have binding sites located in various nooks and crannies on their surface (Figure 13.4*a*), which is not unlike the surface of an English muffin (Figure 13.4*b*).

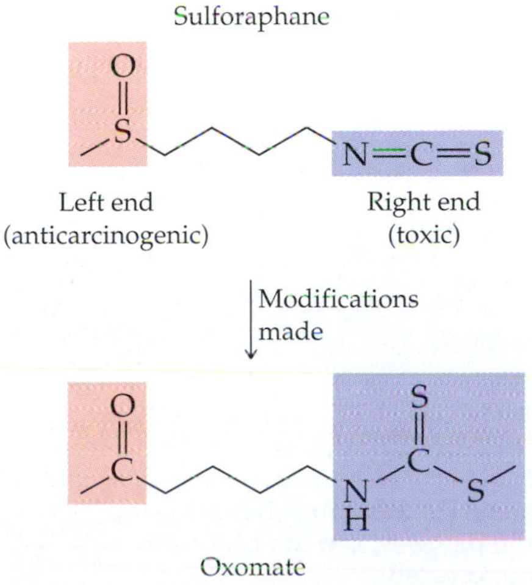

▶ **FIGURE 13.3 Chemical mop.** The left end of the sulforaphane molecule gives this substance its anticarcinogenic properties. The right end of the molecule contributes to the substance's toxicity. The oxomate molecule has a left end that is a bit different from the left end of suforaphane but preserves that molecule's three-dimensional structure. The right end of the oxomate molecule is completely different from the right end of the sulforaphane molecule. This difference makes oxomate less toxic than sulforaphane.

Sulforaphane

Left end
(anticarcinogenic) — Right end (toxic): N=C=S

Modifications made

Oxomate

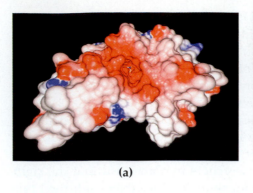
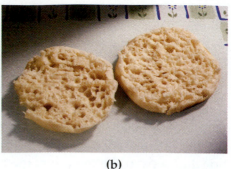

(a) (b)

▶ **FIGURE 13.4 Nooks and crannies.** (a) A space-filling model of a globular protein molecule. The shaded areas represent different regions of positive charge (blue) and negative charge (red) on the surface. (b) The surface of an English muffin is similar to the pitted surface of a globular protein molecule.

Now let us look at how a drug does its job. Figure 13.5a shows a hypothetical four-step sequence leading to a cancerous state somewhere in the body. This sequence can be interrupted by a certain molecule present in the body, which we shall call molecule B. When B interrupts the sequence, as in Figure 13.5b, the cancerous state is never reached, and the body remains healthy. As long as B is available, there is no problem. Now suppose, though, that some protein happens to bind a small organic molecule A (Figure 13.6a). Let's say A binds to the protein at site 1, and when this happens, the protein changes shape slightly so that an indentation that is perfect for binding B is created at site 2 on the protein (Figure 13.6b). By binding B at the newly created site 2 (Figure 13.6c), the protein takes B out of circulation and prevents it from performing its anticarcinogenic function. Thus, the upshot of A's binding to site 1 is to remove B from circulation, leaving the sequence in Figure 13.5a to proceed to the cancerous state.

This type of scenario is typical in the cells of our bodies. In the metabolic pathways (Section 12.3) that exist in every cell, messages are transmitted from one biomolecule to another. If the message says "Initiate uncontrolled cell growth," the metabolic pathway may be responsible for the formation of a cancerous tumor. If the message says "Intercept and stop uncontrolled cell growth," the pathway may inhibit the formation of a cancerous tumor. Once medical researchers understand a metabolic pathway, they can target a new drug that either boosts the pathway or interrupts it, as shown in Figure 13.7. In our hypothetical example, a drug C could be designed to bind at site 2 on the protein surface. This binding would prevent molecule B from binding at site 2 and allow B to continue its work as an anticarcinogen.

Alternatively, a new drug might mimic molecule B's activity and be able to interrupt the pathway of Figure 13.5 exactly as B does. That would also enhance anticarcinogenic activity. Can you see a third alternative approach in this case?

The common denominator in all these interactions is this: *molecules, be they small organic molecules or large biomolecules, interact with one another by contact between specific regions on their three-dimensional surfaces.*

▶ **FIGURE 13.5 A hypothetical metabolic pathway.** (a) A four-step sequence eventually leads to a cancerous state. (b) Substance B blocks step 2 of the pathway, and the cancerous state is not created.

i $\longrightarrow$ ii $\longrightarrow$ iii $\longrightarrow$ iv $\longrightarrow$ cancerous state

(a)

i $\longrightarrow$ ii $\xrightarrow{\text{B}}$ no products

(b)

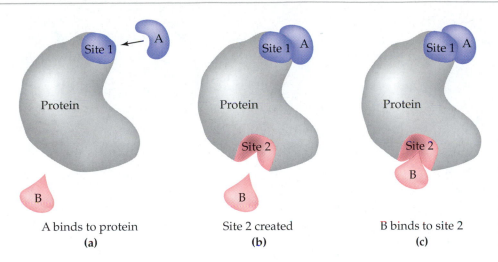

Creating binding sites on proteins. **(a)** Molecule A binds to site 1 on a protein surface. **(b)** When molecule A binds, site 2 is created at another location on the protein surface. **(c)** Once its binding site has been created, B binds to that site and is taken out of circulation.

QUESTION 13.2 The binding of Q to X causes X to bind Z. The binding of Z to X keeps Z from inactivating T, a cancer-causing molecule. Given these facts, which of the following will reduce the activity of T?

(a) Administering a drug that mimics the action of Z.

(b) Administering a drug that prevents Q from binding to X.

(c) Administering a drug that binds to the Z-binding site on X.

ANSWER 13.2 All of the above. (*a*) A drug that mimics Z will increase levels of Z, so that even though some Z binds to X, there is free Z left over and available to inactivate T. (*b*) If Q does not bind to X, then X will not bind to Z, and Z will be free to inactivate T. (*c*) If Z cannot bind to X, then Z is free to inactivate T.

LIVE ART
Binding Sites on Proteins

The contours of two interacting molecules must match in order for binding to occur, as we just saw, but the chemical interactions that take place at those surfaces must also match. Throughout this book, we have discussed various types of interactions between molecules. Recall that the bonds holding atoms together *within* molecules are almost always covalent bonds (Chapter 2). These are strong, long-lasting bonds that can be made and unmade during chemical reactions. However, as we discussed in Chapters 4 and 7, molecules can also interact via shorter-term, less permanent *intermolecular interactions*. Recall from these earlier chapters that intermolecular interactions are those which most often exist *between* molecules and are *noncovalent* in nature. Because intermolecular interactions typically occur when two biomolecules dock together, let's review some of the basic types.

RECURRING THEME IN CHEMISTRY The individual bonds between atoms dictate the properties and characteristics of the substances that contain them.

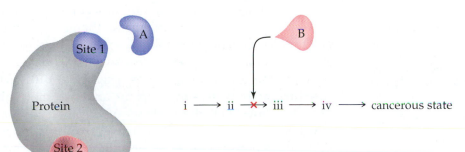

Stopping an undesirable reaction. Molecule C, a drug molecule in our example, also binds to site 2. In the presence of C, molecule B will not bind as often to site 2 and will be available to stop the reactions following step ii in Figure 13.5.

i ⟶ ii ⟶✗⟶ iii ⟶ iv ⟶ cancerous state

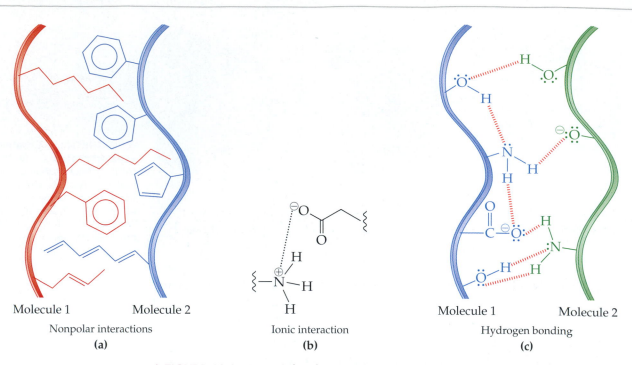

Molecule 1   Molecule 2

Nonpolar interactions

(a)

Ionic interaction

(b)

Molecule 1   Molecule 2

Hydrogen bonding

(c)

▲ **FIGURE 13.8   Intermolecular interactions.   (a)** The surfaces of two molecules are held together by nonpolar interactions between hydrophobic groups on each surface. **(b)** A positively charged group on one molecule is attracted to a negatively charged group on another molecule. The result is an ionic interaction. **(c)** The surfaces of two molecules are held together by hydrogen bonds.

One type of intermolecular interaction is the *nonpolar interaction,* in which groups with little or no polarity are weakly attracted to one another. Figure 13.8*a* depicts two molecules docking together via nonpolar regions on their surfaces. Molecules also interact via attraction between a positive and negative charge, and this *ionic bond* represents another type of intermolecular interaction (Figure 13.8*b*). These two types of intermolecular interactions represent the extremes, with nonpolar interactions weakest and ionic interactions strongest. There are several types of intermediate intermolecular interactions between these extremes. In Chapter 7, we discussed one of these intermediate-strength interactions, the type known as *hydrogen bonding.* Recall that a hydrogen bond can form between any group X and any group —XH (where X is oxygen, fluorine, or nitrogen), as shown in Figure 13.8*c*.

All varieties of intermolecular interaction play a role in the docking of one biomolecule to another. The relative strengths and polarities of the various types of interaction are summarized in Figure 13.9.

Two molecules that have matching contours can interact if the distribution of chemical groups on one molecule complements the groups on the other molecule. The two molecules may be held together by only one type of intermolecular interaction or by a combination of several types (Figure 13.10). The **structure–activity relationship** for any two molecules describes how well the surfaces of the molecules are matched to each other. The stronger the structure–activity relationship for any pair of molecules, the more tightly they bind together. This relationship is especially important to chemists designing drug molecules, of course, for whom the binding of interest is that between the drug molecule and its receptor. Most modern drug

▼ **FIGURE 13.9   The noncovalent interactions.**   There is a continuum of intermolecular interactions that ranges from ionic bonds (very polar and very strong) to nonpolar interactions (very nonpolar and very weak).

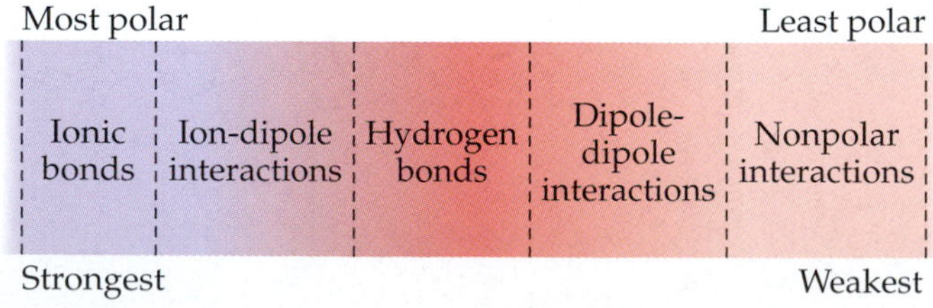

Most polar				Least polar
Ionic bonds	Ion-dipole interactions	Hydrogen bonds	Dipole-dipole interactions	Nonpolar interactions
Strongest				Weakest

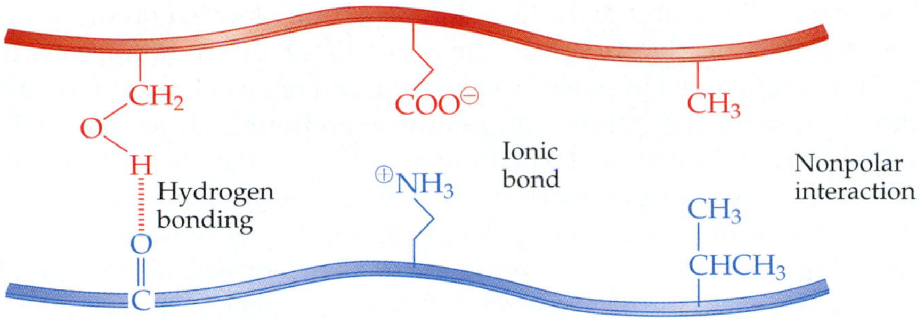

◀ **FIGURE 13.10 Modes of interaction.** All varieties of intermolecular interaction play a role in the binding of one molecular surface to another. In this figure, hydrogen bonding, ionic bonding, and nonpolar interaction are shown.

design efforts are based on optimizing the structure–activity relationship, because *optimal binding between drug and receptor means optimal drug activity.*

QUESTION 13.3 In the two binding surfaces depicted here, which type of intermolecular interaction will allow the left side to bind to the right?

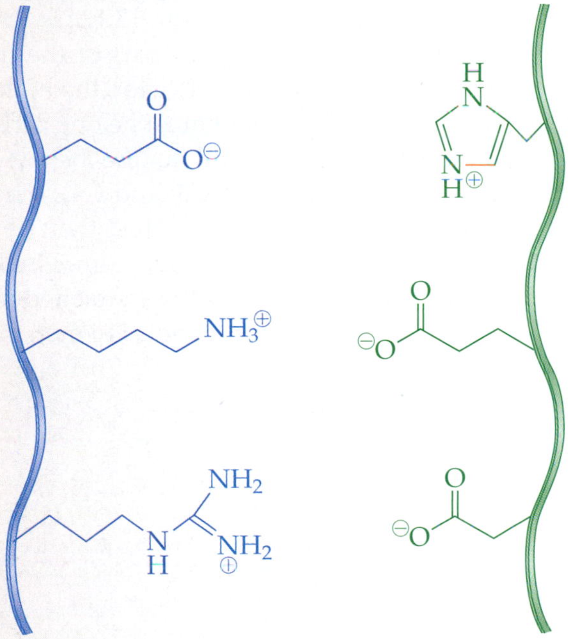

ANSWER 13.3 The charges on the left surface are opposite the charges on the right surface. The two surfaces will therefore line up to form three ionic bonds.

Let us consider a real drug design example and how structure–activity relationships are used to enhance the effectiveness of a drug. Anyone who has seen the film *The Wizard of Oz* will remember the scene in which Dorothy falls asleep in a field of poppies and is then rescued by the Good Witch of the North (Figure 13.11). In real life, you cannot feel the effects of poppies by sleeping in a poppy field, although the poppy plant has been lauded (and cursed) since 4000 B.C. because of its medicinal properties. Poppy pods contain several medicinally useful compounds, and of these, *morphine* is present in highest quantity.

The morphine molecule, shown in Figure 13.12a, acts by binding to a protein in the brain called an *opiate receptor*. We refer to a drug that either binds to an opiate receptor or affects the binding of other molecules to an

▲ **FIGURE 13.11 No longer in Kansas.** Dorothy sleeps in a field of wild poppies, plants containing high concentrations of morphine.

MOLECULE
Heroin

▶ **FIGURE 13.12 Molecules in the morphine family.** **(a)** The structure of the morphine molecule. The parts of the molecule required for opiate receptor binding are highlighted in red. **(b)** In the heroin molecule, the —OH at carbons C8 and C9 of morphine are replaced by —OOCCH₃ groups. **(c)** In the codeine molecule, the —OH group at C9 of morphine is replaced by a —OCOCH₃ group. **(d)** In the powerful tranquilizer etorphine, the group at C8 is —OCH₃ instead of —OH, and a bridging group (highlighted blue) connecting C8 and C2 gives the molecule additional hydrophobic character.

opiate receptor as an **opiate**. Opiate receptors are located on the surfaces of selected neurons that make up the nerve tissue of the brain. In drug-free brain tissue, a nerve impulse travels to the end of one neuron, and its arrival there prompts the release of **neurotransmitters**. These molecules are released from the end of one neuron, travel across the *synaptic cleft* that separates one neuron from another, and bind to receptors at the beginning of the next neuron. This binding event creates a new nerve impulse in the receiving neuron, and the process continues from neuron to neuron. The neurotransmitter in the synaptic cleft is then broken down by an enzyme that is specific to that molecule. Opiates and other drugs that affect the brain cause an imbalance in the natural neurotransmitter levels in the synaptic cleft. This imbalance produces the analgesic (painkilling) and euphoric (exaggerated happiness) effects for which morphine is famous.

Morphine has been used as a painkiller for centuries, but it is also an extraordinarily addictive substance. Thus, drug designers have tried to mimic the beneficial analgesic properties of morphine in a new drug molecule that is not addictive. Studies have revealed that only the portion of the morphine structure highlighted in red in Figure 13.12*a* is responsible for its opiatelike properties. Let's consider how this part of the molecule is configured. First, in order to bind to an opiate receptor, the molecule must contain a benzene ring. Attached to that, there must be one carbon atom (C1 in the figure) which, besides being attached to the benzene ring, is also attached to three other carbon atoms (C2, C3, and C4). One of those three carbon atoms (C2 in our numbering scheme) must then be connected to a carbon atom (C5) that is connected to a nitrogen atom. Finally, the nitrogen atom, part of an amine functional group, must be attached to two other carbon atoms (C6 and C7). When these requirements are met, the molecule will bind to an opiate receptor.

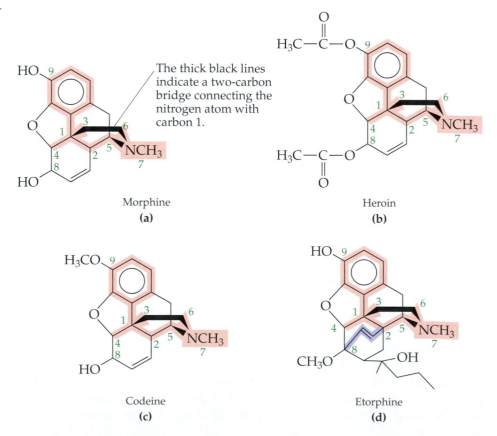

QUESTION 13.4   Which of these molecules are likely to bind to opiate receptors in the brain?

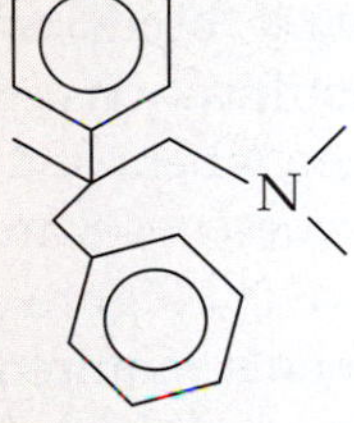

(a)

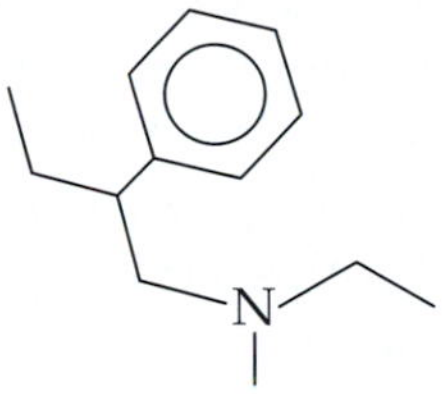

(b)

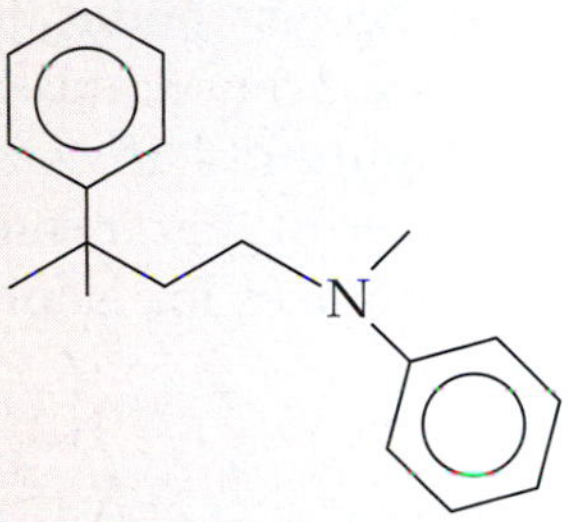

(c)

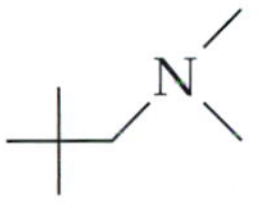

(d)

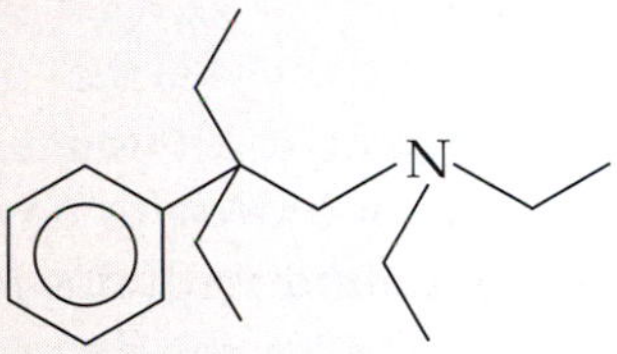

(e)

ANSWER 13.4   Only (*c*) includes the required structural elements, which are highlighted in red:

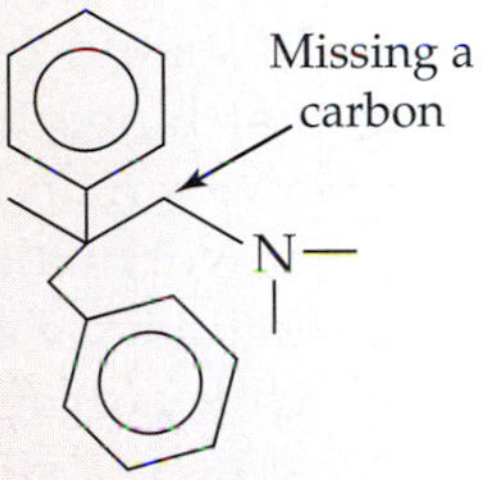

Missing a carbon

(a)

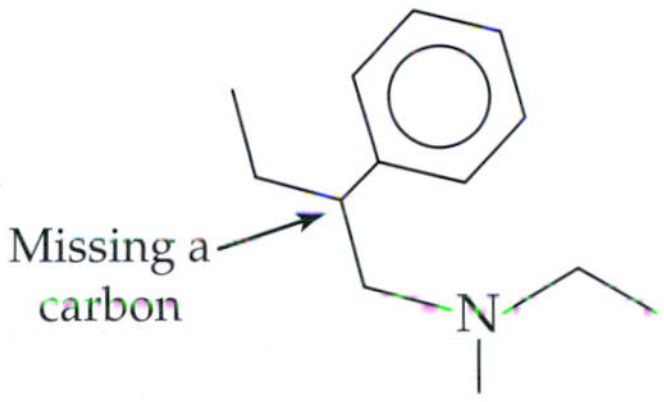

Missing a carbon

(b)

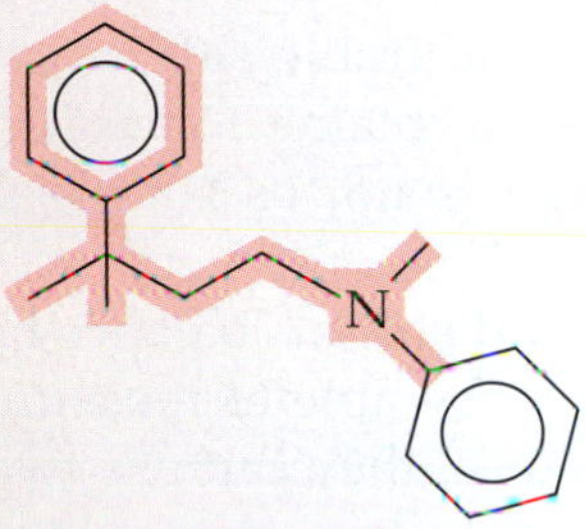

(c)

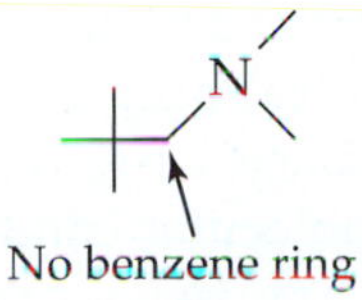

No benzene ring

(d)

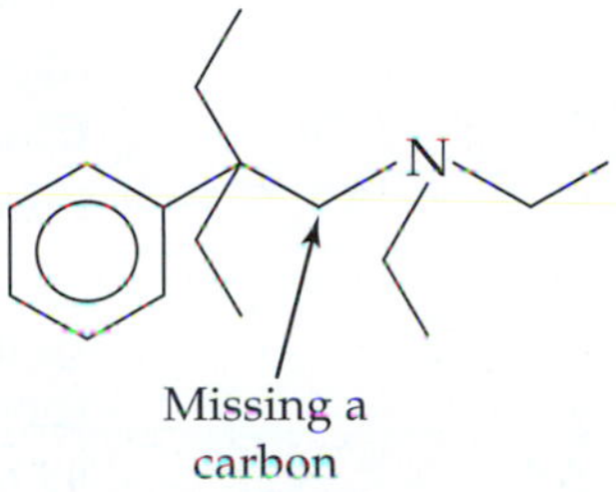

Missing a carbon

(e)

The rules for morphine binding were elucidated very early in the course of modern medicinal chemistry, and the structure–activity relationship of the morphine molecule was one of the first elucidated by scientists. You can see that morphine is more complex than the rules require. The extraneous parts of the molecule, which are the regions *not* highlighted in Figure 13.12*a*, affect how tightly the molecule binds to the receptor, but those parts are not required for binding. Modifications of these extraneous parts produce similar molecules that are all members of the opiate family of drugs. For example, replacing the —OH groups at C8 and C9 in morphine with acetyl groups (H_3CCOO—) gives *heroin*. Figure 13.12*b* shows the structure of heroin, with the structural elements required for opiate receptor binding highlighted. To convert morphine to *codeine*, an opiate found in over-the-counter cough medicines, you must modify the group attached to C9 by replacing the H with a methyl group, as shown in Figure 13.12*c*. *Etorphine*, a morphine derivative with a modification at C8 plus an added hydrophobic group bridging C2 and C8 (Figure 13.12*d*), is 2000 times more potent than morphine. Etorphine is so powerful that it cannot be used on humans. Instead, its main use is as a tranquilizer for elephants and other very large animals.

13.2 | The Ideal Drug

The structure–activity relationship for morphine says (1) that only very specific parts of the molecule are required for binding to an opiate receptor and (2) that those parts must be present in order for a drug to have morphine-like properties. It is often true that one compound with desirable (but not ideal) drug properties will act as a starting point for the design of a superior drug, as in the case of morphine. These original molecules are known as **lead compounds**. Drug designers seek to maintain a lead compound's desirable properties and eliminate its bad ones, as when the lead compound sulforaphane was modified to create the more useful oxomate (Figure 13.3, page 511). Ideally, the new drug will be less addictive, be more effective, and have fewer side effects.

One lead compound may lead to thousands of drug candidates. This being so, how do we know which molecule should be chosen for the marketplace? It may seem that the drug that binds its receptor most tightly would be the best candidate. However, as in the case of etorphine, it is possible to design a drug that works too well and is too powerful for humans. Thus, the best drug should have **efficacy**: the ability to produce some desirable effect. Also, while a drug may bind readily to its receptor, it may also bind to other things, a situation that usually leads to undesirable side effects. Thus, a good drug candidate should also have **specificity** for its receptor and should not bind to other biomolecules in the body. These undesirable side reactions, referred to as *nonspecific interactions*, are often to blame for the toxicity of drugs. Clearly, the process of drug design is not as simple as matching one binding surface to another.

If the ideal drug candidate should strike a balance between specificity and efficacy, drug designers must be able to measure these variables for each molecular candidate. To do so, they can use two indicators. The TD_{50} for any drug candidate is the dose that elicits some specific, measurable toxic effect, such as a severe headache, in 50 percent of test subjects. (Thus, TD stands for "toxic dose.") Because toxicity often arises as a result of nonspecific interactions, this measurement tells us something about the drug's specificity or lack thereof. The **ED$_{50}$** for any drug candidate is the dose that produces a therapeutic effect, such as the disappearance of a rash, in 50 percent of test subjects. (Thus, ED stands for "effective dose.") This measurement tells us about the efficacy of the drug.

Figure 13.13 gives an example of an ED$_{50}$ test run. The horizontal axis of the graph shows the dose, in milligrams, of a drug given to 100 test subjects. Those subjects are monitored for signs of the drug's therapeutic action. For an anti-inflammatory drug, for example, the test subjects are monitored for a reduction of inflammation. The vertical axis shows the number of subjects experiencing the desired effect. At the point indicated with the black arrows, 50 out of 100, or 50 percent, of the test subjects demonstrated the therapeutic effect. Therefore, the ED$_{50}$ for this drug is 100 mg. The TD$_{50}$ would be measured in a similar way, except that the vertical axis would measure a toxic effect.

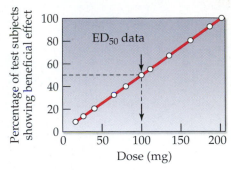

▲ **FIGURE 13.13** **Quantifying the effects of drugs.** A graph showing ED$_{50}$ data for a group of 100 test subjects. The vertical axis is the percentage of the group showing a benefit from the drug. The dose at which 50 percent of the test subjects show a benefit is 100 mg.

QUESTION 13.5 A new drug for treating depression is administered to 100 subjects to test for toxicity. The data collected yield the following graph:

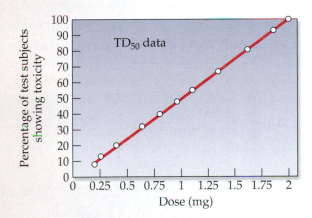

(*a*) What is the TD$_{50}$ for this drug? (*b*) Which is more desirable, a low value for TD$_{50}$ or a high value?

ANSWER 13.5 (*a*) If you locate the 50 percent mark on the vertical axis and run your finger to the right until you intersect the graph line, you see that the corresponding value on the horizontal axis is about 1 mg. (*b*) A high TD$_{50}$ is more desirable because it means that more drug is required to produce a toxic effect.

QUESTION 13.6 Look again at the data in Question 13.5. (*a*) What is the minimum dosage that will produce a toxic effect in all individuals in the test sample? (*b*) If you were to extend the graph line to the left, would you expect the line to cross the point where both axes have a value of zero? Explain.

ANSWER 13.6 (*a*) Any amount greater than 2 mg. (*b*) Yes, because if no drug is administered, there should be no measurable toxic effect.

Both TD_{50} data and ED_{50} data are required to understand the toxicity of a drug. Toxicity is usually blamed on poor specificity, which means that molecules of the drug bind to sites other than their target receptors. The ratio of these values, TD_{50}/ED_{50}, provides an overall indicator of drug performance and is referred to as the **therapeutic index**:

$$\text{therapeutic index} = \frac{TD_{50}}{ED_{50}}$$

An ideal drug will have a high TD_{50}, which means that a large quantity of the drug is required to attain a toxic effect, as just noted in the answer to Question 13.5. It should also have a low ED_{50}, which means that little of it is required to produce the desired effect. Thus, a drug with a large therapeutic index is likely to have low toxicity and high efficacy. In the pharmaceutical industry, drugs that meet these criteria are rare.

STUDENTS OFTEN ASK

Why Do Patients Undergoing Chemotherapy for Cancer So Often Get Sick from the Medicine?

The efficacy of many anticancer drugs is impressive, but one look at an individual who has undergone extensive chemotherapeutic treatment tells a different story about specificity, which is notoriously bad for chemotherapeutic agents. There are dozens of varieties of cancer, and different types of cancer cells react differently to drugs. One thing all cancer cells have in common, however, is uncontrolled growth of new cancer cells. Thus, one way to battle cancer to is to interrupt the reproduction cycle of cancer cells.

When a cell—any cell, cancerous or healthy—divides, it must produce one copy of its DNA for each of its progeny. Therefore, drugs that are effective at interrupting DNA synthesis are often used to treat cancer today. One drawback of these chemotherapeutic drugs, however, is that they affect *all* cells that are dividing rapidly. This poses a big problem because, in the human body, not only cancerous cells, but also cells in the bone marrow, hair follicles, and lining of the stomach, are targeted by the drug. Because all these types of cells are destroyed by anticarcinogenic drugs, many chemotherapy patients have weakened immune systems as a result of the deleterious effects the drugs have on the bone marrow, where blood cells are manufactured. Figure 13.14 shows a chemotherapy patient who is in a sterile room because his immune system is susceptible to opportunistic infections. This patient has lost his hair, a side effect of the drug's action on hair follicles. Chemotherapy patients also suffer from chronic nausea, a side effect of the destruction of cells in the stomach.

Having read all this, you will not be surprised to hear that much of today's research on chemotherapeutic agents is focused on drugs that target only cancerous

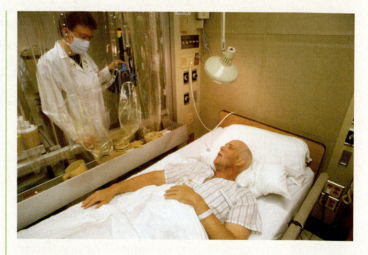

▲ **FIGURE 13.14** **Targeting cancer.** Patients undergoing a chemotherapy regimen are often susceptible to opportunistic infections and must be isolated. Hair loss is also a common side effect of chemotherapy, which targets rapidly dividing cells, including hair follicles.

cells and leave other rapidly dividing cells alone. Efforts are also aimed at the blood supply that keeps tumors growing. New drugs that limit the growth of blood vessels to tumors, called *antiangiogenesis* drugs, show promise for recovery from advanced cancers.

TABLE 13.1	Reported Side Effects of a Popular Antianxiety Drug

Abdominal discomfort	Insomnia
Agitation	Leucopenia[b]
Anterograde amnesia[a]	Memory impairment
Blurred vision	Nausea
Clumsiness	Sedation
Depression	Sleepiness
Disorientation	Tachycardia[c]
Dizziness	Unsteadiness
Hallucinations	Visual problems
Headache	Weakness
Inappropriately grandiose behavior	

[a] Inability to recall events during period of drug use.
[b] Abnormal decline in number of white blood cells.
[c] Rapid or irregular heart rhythm.

If you have ever seen or heard an advertisement for a prescription drug, you know that the list of side effects can be daunting. For example, the deleterious side effects of a popular antianxiety drug are listed in Table 13.1. The good news is this: any worries one might have about the list of side effects should seem unimportant if the antianxiety drug is doing its job.

The terms *toxin* and *poison* both refer to substances that are highly toxic in small quantities. For these substances, the term LD_{50} is defined as the amount of substance that kills 50 percent of subjects exposed to it (and so you can think of LD as standing for "lethal dose.") The lower the LD_{50}, the smaller the amount of toxin or poison needed to cause death and the higher the toxicity of the substance. Table 13.2 lists the LD_{50} values for several poisons, including the popular insecticide parathion. Each molecule listed in the table acts by inactivating an enzyme required for normal cellular function. In each case, the highlighted part of the poisonous molecule is displaced when the molecule binds to the active site of an enzyme.

Technically speaking, we use the term *toxin* when the harmful substance comes from a natural source, such as a bacterium, while the word *poison* is used when the harmful substance is artificial, such as a drain cleaner or pesticide.

Let us look at an example of how LD_{50} data are used. We shall consider *sarin*, the extremely poisonous nerve gas described in Section 8.1. Table 13.2 shows that the LD_{50} value for sarin is 24 mg/kg of body mass. Using this value, we can estimate the mass of sarin to which an 85-kg man could be exposed and have a 50 percent chance of survival:

$$85 \text{ kg body mass} \times \frac{24 \text{ mg sarin}}{1 \text{ kg body mass}} = 2.0 \times 10^3 \text{ mg sarin} = 2.0 \text{ g sarin}$$

Death following exposure to nerve gas is usually rapid, with the victim often succumbing in less than 5 minutes.

QUESTION 13.7 What mass of sarin, in milligrams, could a 50-kg boy be exposed to and have a 50 percent chance of survival?

ANSWER 13.7

$$50 \text{ kg body mass} \times \frac{24 \text{ mg sarin}}{1 \text{ kg body mass}} = 1.2 \times 10^3 \text{ mg sarin}$$

TABLE 13.2 LD$_{50}$ Data for Selected Poisons

Poison	Chemical structure[a]	Use	LD$_{50}$ (mg/kg of body mass)	How substance enters body
Parathion		Insecticide	2	Inhalation
Malathion		Insecticide	2800	Oral
Diisopropyl fluorophosphate		Chemical nerve gas	0.5	Inhalation
Sarin		Chemical nerve gas	24	Through skin
Tabun		Chemical nerve gas	9.3	Through skin

[a] The highlighted groups are eliminated from the molecule when it binds to its target.

13.3 | Gumming Up the Works

You now know that proteins are often the target of drug design. However, *enzymes*—those proteins which catalyze chemical reactions (Section 12.3)—are much better targets than other types of proteins because their active sites are designed to bind other molecules that then undergo some kind of chemical reaction. Before we continue, let us clarify the difference between the terms *binding site* and *active site*. A protein that is not an enzyme binds other molecules at *binding sites* on the protein surface. This is the type of interaction taking place at sites 1 and 2 in Figure 13.6 (page 513). A protein that is an enzyme has *a special type of binding site* called an *active site* on its surface. Your average enzyme will typically have one active site, and very often a metabolic pathway will include a series of reactions catalyzed at the active sites of a series of enzymes.

Enzymes that are part of a metabolic pathway are good targets for drug design because, as catalysts, these enzymes must bind tightly to the molecule they react with (called the *substrate*, as we learned in Chapter 12). Drugs can be designed to mimic a natural substrate in the body so that the drug molecule rather than the natural substrate molecule binds at the enzyme's active site. We refer to drugs that interfere with enzymatic reactions as *inhibitors*, and the interactions involved can be either covalent or noncovalent. When a drug binds *covalently* (and therefore tightly) to an enzyme's active site, we refer to the drug as an **irreversible inhibitor**, because it removes that enzyme molecule from the "workforce" of available enzyme molecules. In this section, we shall look at an example of a drug that acts as a **reversible inhibitor**—a molecule that binds *noncovalently* to an enzyme's active site. We shall first look at how the enzyme functions with its natural substrate and then see how an inhibitor drug prevents the substrate from doing its job.

QUESTION 13.8 In the metabolic pathway molecule D is a known carcinogen, and molecule F, which binds to enzyme 2 and inhibits it, has broad-ranging anticancer properties. You are designing a drug to reduce the amount of D produced by the body. (*a*) Would a drug that mimics molecule F produce the desired effect? (*b*) Is there an approach to reducing the concentration of D that might be more beneficial to a patient?

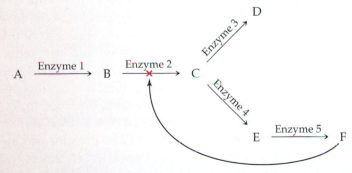

ANSWER 13.8 (*a*) Yes, a drug that mimics the action of F would reduce the conversion of B to C; reduced levels of C would mean reduced levels of D. (*b*) Yes. The pathway described in (*a*) will also reduce the amount of F in the body by blocking the pathway B→C→E→F. Because F has broad-ranging anticancer benefits, this is not desirable. A more beneficial approach would be to create a drug that inhibits enzyme 3 rather than enzyme 2, so that F can continue to be produced.

This lesson in drug design will focus on reversible inhibitors designed to fight the *human immunodeficiency virus* (HIV), the cause of *autoimmune deficiency syndrome* (AIDS). This virus is the only pathogen discovered in the twentieth century to cause a global epidemic. Figure 13.15*a* shows World Health Organization (WHO) statistics on the number of people living with the virus as of the end of 2003. Clearly, Africa has borne the brunt of the outbreak. Figure 13.15*b* shows four epidemiological studies charting how the virus has spread in Africa since 1988. The virus began its slow march through Africa in the early 1980s, striking the worst blow in Uganda first (indicated on the 1988 map). Since then, Uganda has managed to significantly reduce its HIV-infected population, something no other African country has been able to do. Despite this isolated success, the death toll in Africa continues to rise, especially in the southern portion of the continent. Tiny Swaziland (labeled in the 2003 map of Figure 13.15*b*) has been hit hardest, with 38.8 percent of its adult population infected.

As the virus has spread across Africa and around the globe, never before has one disease or one virus been the focus of so much research in the medical and scientific communities. Because of an enormous international research effort undertaken to understand and combat HIV and AIDS, our understanding of this virus surpasses our understanding of other pathogens that have been known for many decades. However, many aspects of HIV remain enigmatic because it does not follow the "rules" we learned from studying simpler viruses. HIV attacks cells of the immune system and kills them. (Most other viruses do not kill their host cells.) The death of infected immune-system cells causes an immune deficiency characterized by a greater-than-normal susceptibility to opportunistic infections and produces the symptoms associated with AIDS. Thus, because their immune systems are compromised, individuals with AIDS cannot fight off innocuous infections that do not cause the average person to miss a day of work.

Viruses are broadly defined as infectious pathogens that carry their own genetic material. They are, in fact, little more than tiny pouches made of protein containing either DNA or RNA (Section 11.4). HIV belongs to a class of viruses known as *retroviruses*. Their mode of action is illustrated in Figure 13.16 for an HIV particle invading a human white blood cell. The genetic material of retroviruses is single-stranded RNA. When a retrovirus is taken up by a healthy cell, the viral RNA is transcribed into viral DNA by the enzyme *reverse transcriptase* (steps 1 and 2). The process is the reverse of the normal process of DNA transcription. The newly encoded viral DNA then enters the nucleus of the host cell and is integrated into its hereditary material, which is also composed of double-stranded DNA (step 3). Recall from our discussion of protein synthesis in Chapter 11 that, in a healthy cell, DNA is transcribed into RNA, and RNA is then translated into protein. In a cell that has been infected by a retrovirus, the viral DNA is then transcribed and translated along with the cell's normal DNA. The proteins produced by the viral DNA make up the coat of new viral particles. In this way, new viral particles are produced by the infected cell (steps 4, 5, and 6).

A newly produced viral particle that buds off from the infected cell is initially inactive. However, once two connected proteins—*gag* and *pol*—inside

▶ **FIGURE 13.15** **HIV statistics.** **(a)** The number of adults and children estimated to be living with the human immunodeficiency virus at the end of 2003. **(b)** Epidemiological data for Africa for 1988, 1993, 1998, and 2003. The growing area coded dark blue (indicating 20 to 39 percent of population infected) indicates how the disease has spread over the 15-year period shown.

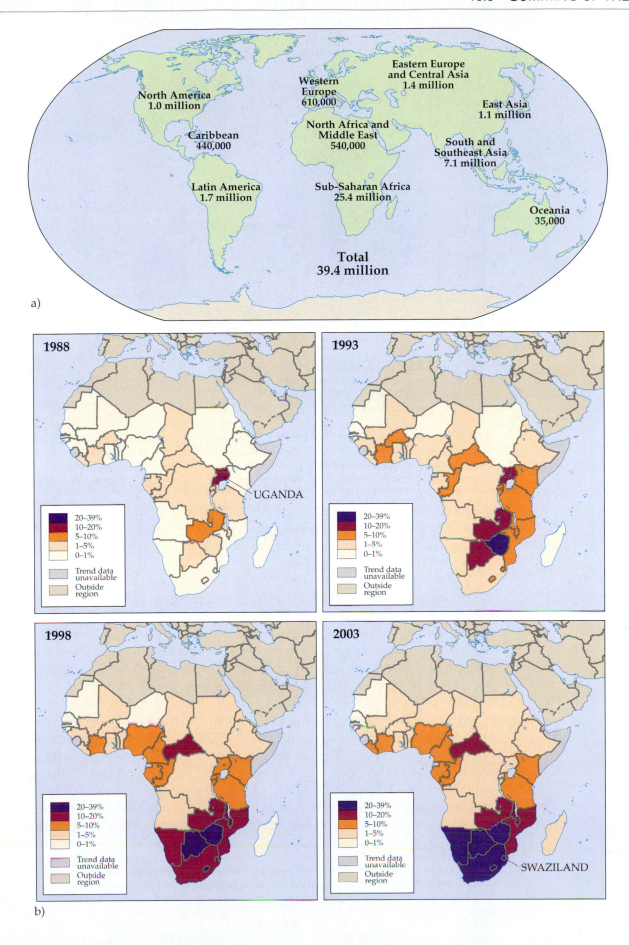

a)

b)

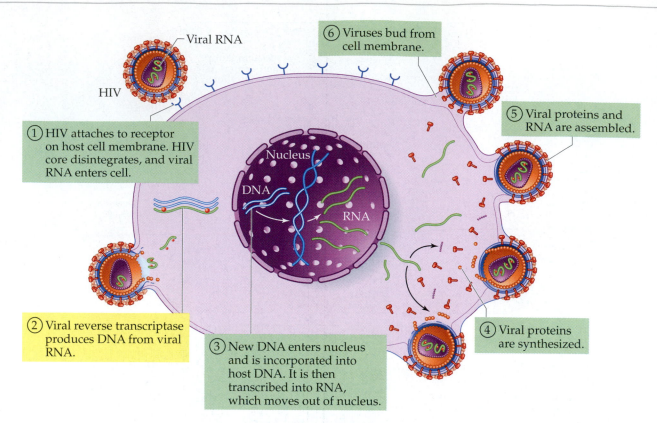

Viral RNA

HIV

① HIV attaches to receptor on host cell membrane. HIV core disintegrates, and viral RNA enters cell.

Nucleus

DNA

RNA

② Viral reverse transcriptase produces DNA from viral RNA.

③ New DNA enters nucleus and is incorporated into host DNA. It is then transcribed into RNA, which moves out of nucleus.

⑥ Viruses bud from cell membrane.

⑤ Viral proteins and RNA are assembled.

④ Viral proteins are synthesized.

▲ **FIGURE 13.16** **How retroviruses work.** Six steps in the infection of a healthy white blood cell by the retrovirus HIV. The result is the production of new viral particles that bud off of the cell and that can infect other cells. Step 2, which involves the enzyme reverse transcriptase, is one of the two most frequently targeted steps in the design of HIV drugs.

MOLECULE
Azidodeoxythymidine (AZT)

the particle are cleaved from each other, as represented in Figure 13.17, the virus becomes active. (We do not need to concern ourselves here with the formal names and molecular structures of *gag* and *pol*.) This reaction is catalyzed by the enzyme *HIV protease*.

There have been two major drug-design strategies used in the fight against AIDS, one targeting reverse transcriptase (step 2 in Figure 13.16) and the other targeting HIV protease (Figure 13.17). Researchers were able to design several drugs that inhibit reverse transcriptase, the most well-known being *azidodeoxythymidine (AZT)*. Unfortunately, HIV is able to mutate quickly so that it is no longer susceptible to the drug. Thus, AZT can stave off AIDS for a while, but it is not an effective long-term medication for the treatment of the disease.

HIV protease, the second target of drugs designed to combat AIDS, is called a protease because it cleaves other proteins, in this case the proteins *gag* and *pol*. If you are designing a drug to inhibit HIV protease, what type of molecule might you choose? Because the substrate for this enzyme is a protein, it would be best to choose an inhibitor molecule that is also a protein, so that it will bind at the HIV protease active site and thereby prevent gag-pol from binding instead. Because large proteins are difficult to produce, your inhibitor should be a short peptide, and you should use the least number of amino acids

▼ **FIGURE 13.17** **Cleaving gag-pol.** In a newly formed viral particle, the protein *gag-pol* is cleaved into the two proteins *gag* and *pol* by the enzyme HIV protease. The viral particle is active and infectious only after this cleaving occurs. Inhibition of HIV protease is the second of the two most frequently targeted steps in the design of HIV drugs.

$$\text{gag—pol} \xrightarrow{\text{HIV protease}} \text{gag} + \text{pol}$$

possible to simplify the synthesis of the drug. This is a fine strategy, but it is fraught with difficulties. For example, a drug that is a peptide will be broken down by natural proteases in the body before it can reach a target cell. Furthermore, peptides do not enter cells easily, and so there is a problem with the delivery of the drug.

The most promising HIV drugs available at the time of this writing are all HIV protease inhibitors that *resemble* proteins, but are modified to allow them to be delivered to infected cells and to prevent attack by natural proteases. Two typical examples are shown in Figure 13.18, with each molecule's peptide-mimicking backbone highlighted in blue. The peptidelike backbone allows the inhibitor molecule to fit into the active site on an HIV protease molecule, which is shaped to fit peptides. The interactions between HIV protease molecules and inhibitor molecules are most often noncovalent: ionic bonds, nonpolar interactions, and hydrogen bonds. The combination of all three types results in effective binding of the inhibitor at the active site. In the design of drugs, the binding of the inhibitor should be as strong as possible, because an inhibitor that remains bound at the active site will prevent binding of the natural substrate.

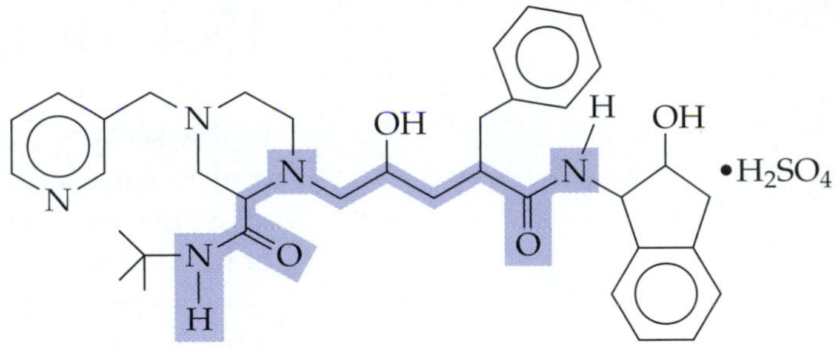

So why aren't all drugs designed to bind permanently and irreversibly to an enzyme active site? The answer is that irreversible inhibitors are usually highly reactive molecules that interact with other molecules in the body. As a result, they have deleterious side effects. For this reason, many drugs on the market are reversible inhibitors. The newest drugs used to treat AIDS, such as the reversible inhibitors shown in Figure 13.18, are examples of how creativity is required in the design of effective drugs.

Indinavir

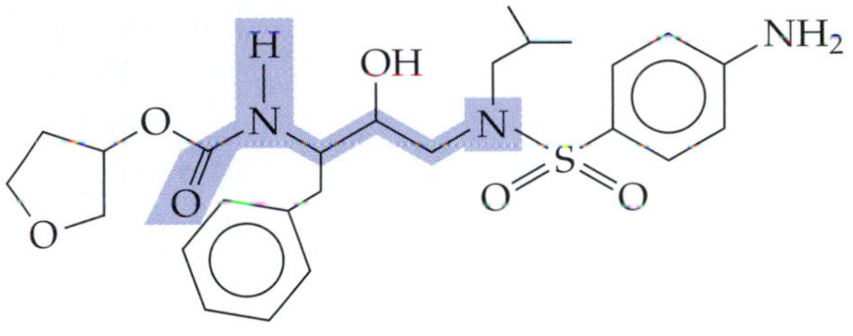

Amprenavir

▲ **FIGURE 13.18 HIV drugs.** Structures of two reversible HIV protease inhibitors.

QUESTION 13.9 Amide bonds make up a peptide's backbone. In Figure 13.18, identify all the amide bonds in each molecule.

ANSWER 13.9 The amide bonds are highlighted in red:

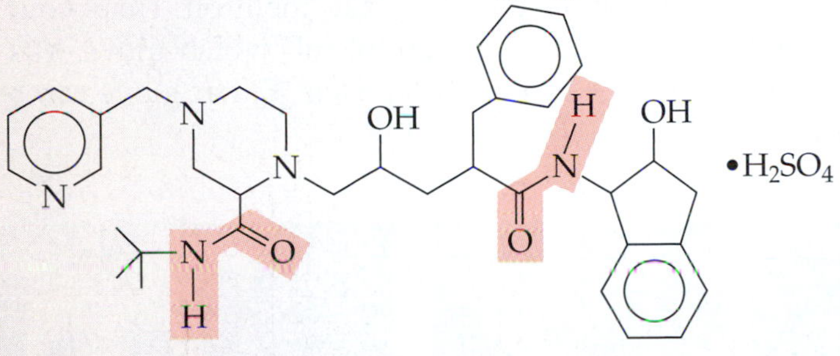

Indinavir

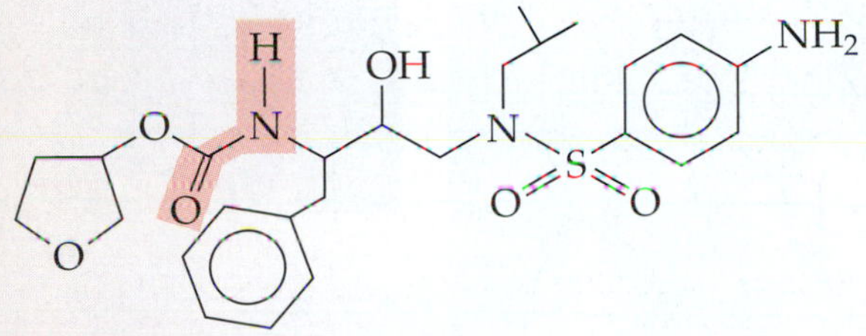

Amprenavir

13.4 | On the Other Hand . . .

In this section, we introduce the notion of handedness as it applies to molecules, especially drug molecules. Because this topic is often difficult to visualize, we begin with a thought experiment to demonstrate the meaning of handedness.

Imagine that you are buying gifts for various people in your family. Your list includes a baseball glove for your sister and a pair of winter gloves for your grandfather. When you begin to shop, you realize that you do not remember whether your sister and grandfather are right handed or left handed. Luckily, you know that this will not be an issue in one of the two cases. Which gift will fit either a right-handed or a left-handed person?

Figure 13.19 shows the two gifts. Because baseball gloves do not come in pairs, you must know whether your sister is right handed or left handed before you buy one. The blue winter gloves will fit your grandfather no matter whether he is right handed or left handed, because the gloves are designed to allow either glove to fit either hand. Wait a minute, though: you also are considering another pair of gloves for your grandfather, the black gloves shown in Figure 13.20, which have a different design. Could you buy these gloves for your grandfather without remembering whether he is right handed or left handed? The answer is yes, of course, because right-handedness and left-handedness are irrelevant when the right-hand and left-hand gloves are bought as a pair.

Clearly, the blue and black gloves are both suitable gifts for your grandfather, but these two pairs of gloves differ in an important way: unlike the blue gloves, the black gloves are not interchangeable. That is, you cannot put the glove shown on the left in Figure 13.20*a* on your left hand or the glove shown on the right on your right hand. Because each black glove is designated for a particular hand, we say that each glove has **handedness**.

The blue gloves of Figure 13.19 do not have the property of handedness (even though they are shaped like a hand), and this lack of handedness means that the two gloves are *identical to each other*. Figure 13.21 shows that the front of each blue glove looks the same as the back. In fact, if you put these gloves on your hands and then take them off, close your eyes, and toss them around in the air, you cannot tell which glove was on which hand, because the gloves *are exactly the same*. Either glove can be worn on either hand.

► **FIGURE 13.19   Happy holidays.** Two items on your gift list: **(a)** a baseball glove for your sister and **(b)** a pair of gloves for your grandfather.

(a)

(b)

Mirror

(a) **(b)**

▲ **FIGURE 13.20** **Gloves with handedness.** **(a)** Each of these black gloves has a "palm" side that is different from the "knuckle" side of the glove. For this reason, these gloves are designed for specific hands: one must go on a left hand, and one must go on a right hand. These gloves have handedness. **(b)** The gloves are nonsuperimposable mirror images of each other: no matter how you twist and turn the glove shown to the left of the mirror, you cannot position it to duplicate the image shown to the right of the mirror.

(a)

The two black gloves are *not identical* to each other, because they have the property of handedness. You know this because you cannot pick up one black glove and superimpose it exactly on the other. To visualize the situation, use your mind's eye to move one of the black gloves in Figure 13.20a up in the air over the other black glove. Move the upper glove this way and that as much as you like, and then decide whether they are identical or not. Examining the two gloves, one hovering in the air over the other, you can see that they are not identical to each other. You can also think about it this way: one black glove is the mirror image of the other black glove. That is, if you put one of the gloves next to a mirror, the image you will see in the mirror is that of the other glove. Thus, we say that the black gloves are *mirror images* of each other. Because the two gloves are not identical, we say that the black gloves are *nonsuperimposable mirror images* of each other, as shown in Figure 13.20b.

When one object is a nonsuperimposable mirror image of a second object, the two objects are *similar* in many ways—texture, color, and size, for example—because they are perfect "reflections" of each other. Because they are not superimposable mirror images of each other, however, they are not *identical*. In other words,

(b)

> Related objects that have handedness, like the black gloves, come in pairs. Such objects are nonsuperimposable mirror images of each other and are *not identical* to each other.

▶ **FIGURE 13.21** **Gloves without handedness.** **(a)** Each of these blue gloves does not have a "palm" side that is distinguishable from the "knuckle" side. The two gloves are indistinguishable from each other, and if you own such a pair, you probably wear each glove randomly on either hand without thinking about it. **(b)** In this image, the glove on the right in part (a) has been flipped over. Note that there is no way to distinguish it either from its mate to its left or from the glove on the left in part (a). **(c)** Sliding the glove on the left in part (b) until it rests on top of its mate shows that the two gloves are identical to each other.

(c)

If two objects are mirror images of each other and *can* be superimposed on each other, like the blue gloves, then they are *identical* to each other. Thus, we can say that the two blue gloves, which are both without handedness, are *identical*. There is only one version of the blue glove, and when you buy the blue gloves, *you are buying two copies of the exact same glove.* Because they are identical, each blue glove fits either hand. We can also say that the two black gloves, which both have handedness, are *not identical*. One black glove fits only the left hand, and the other fits only the right hand. Thus, when you purchase the black gloves, you are buying two different gloves, one for the left hand and one for the right hand.

QUESTION 13.10 Try this mental exercise with your hands rather than with a pair of gloves. Imagine that your hands are not attached to your arms, and determine whether they are superimposable mirror images of each other (like the blue gloves) or nonsuperimposable mirror images of each other (like the black gloves). Do your hands have handedness?

ANSWER 13.10 Barring scars and imperfections, your hands probably look the same; each one has five digits and a front and a back. They are only *similar*, however, rather than identical, to each other. If you place one hand on top of the other, you will see that they are not identical because one cannot be superimposed on the other. Thus, your hands are nonsuperimposable mirror images of each other, and they do indeed have handedness.

QUESTION 13.11 Which of these everyday items have handedness: a straightback wooden chair, a pencil, and a screw?

ANSWER 13.11 Only the screw has handedness. (The threads twist in one direction along the screw's shaft.)

Now consider again the baseball glove for your sister. Does it possess handedness? That is, does it come in right-handed and left-handed versions? The answer is that a baseball glove fits only one hand or the other, and you must know the dominant hand of the ballplayer when buying a glove. If your sister is left handed, you must buy a mitt that fits her right hand (because a baseball glove always goes on the nondominant hand). Since a baseball glove fits only one hand, we say that it has handedness. We also know that a right-handed baseball glove is the mirror image of a left-handed baseball glove, but that one cannot be superimposed on the other (Figure 13.22). Thus, a left-hander's glove and a right-hander's glove are not identical.

Many molecules possess handedness, but many others do not. The criterion for distinguishing them from one another parallels the thought experiment we just completed, as we are about to see. Before we begin, however, here is a warning: it is extremely difficult for some people to visualize molecules in three dimensions. For these people, picking up a virtual molecule and mentally twisting it around is nearly impossible. If you are one of these people, *don't panic.* The importance of this topic can be understood even if you cannot visualize every molecular manipulation, and we shall teach you a trick for identifying organic molecules that have handedness.

▼ FIGURE 13.22 **Batter up!** Baseball gloves are designed to be worn on the player's nondominant hand. Therefore, if you wish to buy a baseball glove as a gift, you must know which hand is dominant for the person who will be wearing it.

Mirror

Before we take on the subject of handedness in drug molecules, let us begin with a very simple biomolecule. Recall from Section 11.2 that amino acids all have the same fundamental structure, which is shown in Figure 13.23. There is a central carbon attached to an amine group, a hydrogen atom, a carboxylic acid group, and a side chain (represented by R in the figure). Each of the 20 amino acids found in the human body has a different side chain, and these 20 amino acids are linked together to make protein molecules in the body. Let us begin our look at handedness with the amino acid glycine. Figure 13.24 shows the three-dimensional structure of this molecule, along with its mirror image. (If you need to brush up on what the solid wedge and dashed line represent, go back and reread Question 3.2 and its answer, page 84) Is the glycine molecule identical to its mirror image (that is, are the two superimposable), or is the molecule not identical to its mirror image (that is, are the two nonsuperimposable)?

By mentally picking up the glycine molecule in Figure 13.24 and rotating it as you move it rightward until it hovers over the mirror image, you should be able to compare the two to see if they are identical. This process is illustrated in Figure 13.25 and shows that the glycine molecule is indeed superimposable on its mirror image. Therefore, the two drawings in Figure 13.24 are different representations of the same molecule. The two molecules are identical to each other. Thus, there is only one version of the glycine molecule, in the same way that there is only one version of the blue glove.

Now try comparing the amino acid alanine and its mirror image (Figure 13.26). Are the two superimposable? If you follow the steps in Figure 13.25, you should conclude that you cannot

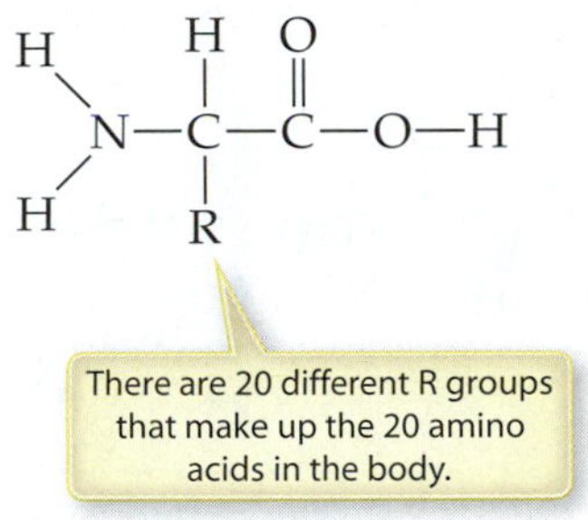

There are 20 different R groups that make up the 20 amino acids in the body.

▲ **FIGURE 13.23**   **Structure of amino acids.**

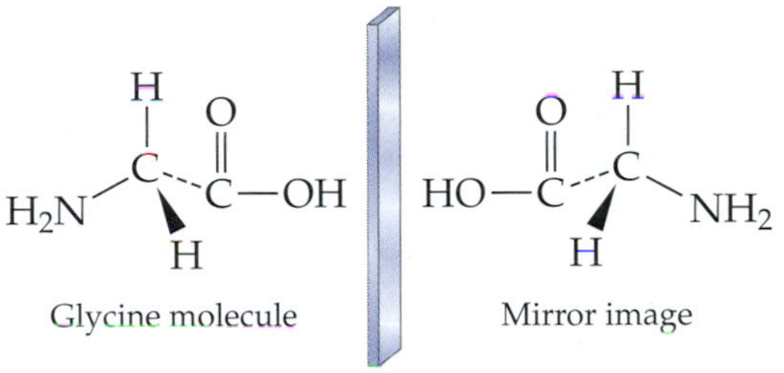

Glycine molecule   Mirror image

▲ **FIGURE 13.24**   **The amino acid glycine and its mirror image.**   Either of these structures is superimposable on the other, telling you that the two structures are identical to each other. What we have here are two ways of drawing the same molecule.

MOLECULE
Glycine

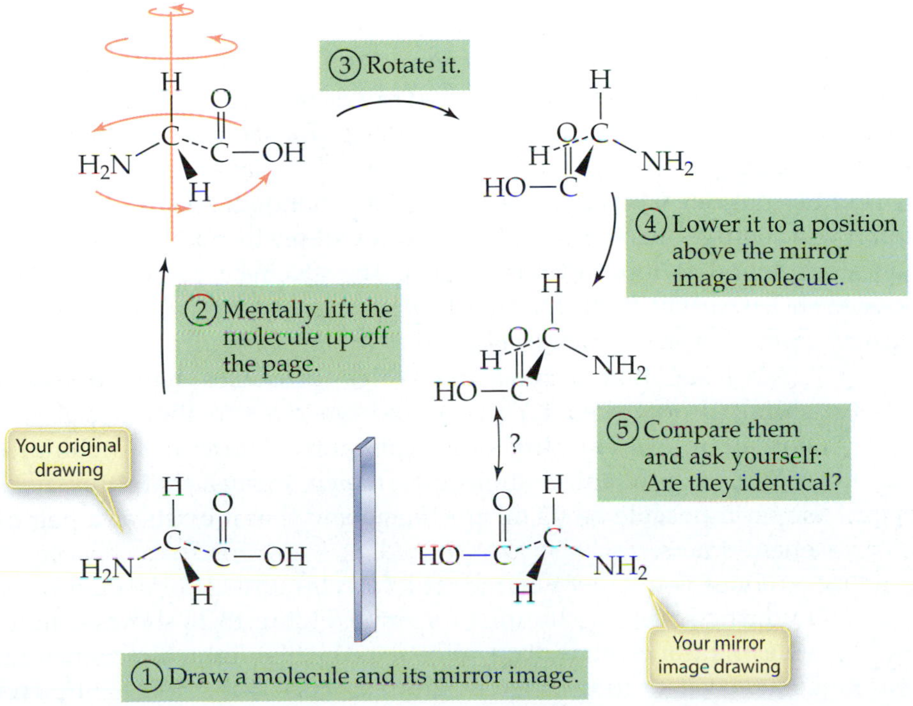

◀ **FIGURE 13.25**   **Determining whether two molecules are identical.** This process allows you to figure out whether two molecules that look similar are superimposable mirror images or nonsuperimposable mirror images. If the two are superimposable mirror images, they are identical to each other and you are looking at two different renderings of the same molecule. If the two are not superimposable, you have two different molecules.
① Draw the two molecules side by side, showing a three-dimensional rendering around a central carbon atom in the case of glycine.
② and ③ Mentally pick up the first molecule and rotate it slowly about a vertical axis through the central carbon atom.
④ Hold the rotated first molecule just above the mirror-image molecule.
⑤ Can the first molecule be superimposed on the mirror-image molecule? If the answer is yes, the two are identical. In the case of glycine, the molecule used in this example, the two are superimposable and therefore identical.

③ Rotate it.

④ Lower it to a position above the mirror image molecule.

② Mentally lift the molecule up off the page.

Your original drawing

⑤ Compare them and ask yourself: Are they identical?

Your mirror image drawing

① Draw a molecule and its mirror image.

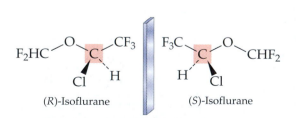

CH3, O
| ||
H2N—C—C—OH
|
H

H3C O
| ||
HO—C—C
| NH2
H

Alanine molecule   Mirror image

▲ **FIGURE 13.26** **The amino acid alanine and its mirror image.**   Are we looking at two different molecules or two versions of the same molecule? The process described in Figure 13.25 will help you answer this question.

MOLECULE
Alanine

① O
    ||
    C
H3C   CH2—CH2—OH
Molecular formula: $C_4H_8O_2$

② OH
    |
H2C=CH—CH—CH2OH
Molecular formula: $C_4H_8O_2$

③ O
    ||
H3C—CH—C—CH3
    |
    OH
Molecular formula: $C_4H_8O_2$

④ H3C—O—CH=CH—CH2OH
Molecular formula: $C_4H_8O_2$

▲ **FIGURE 13.27** **Four isomers.** All of the molecules shown here have the same molecular formulas, even though they contain different organic functional groups. These molecules are all constitutional isomers of one another.

F2HC   O   CF3
    \ | /
      C
    / \
  Cl   H

F3C   O   CHF2
   \ | /
     C
   / \
  H   Cl

(R)-Isoflurane   (S)-Isoflurane

▲ **FIGURE 13.28** **A pair of enantiomers.**   The R and S forms of the anesthetic isoflurane.

superimpose the alanine molecule on its mirror image. Therefore, the two drawings of alanine in Figure 13.26 represent two different, distinct molecules that are nonsuperimposable mirror images of each other. Thus, glycine does not possess handedness, but alanine does. Glycine is just glycine and, like the blue gloves, does not exist in two forms. Alanine, because it has handedness, exists in a right-handed form and a left-handed form, just as the black gloves do.

We call two molecules that have different structures, but the same molecular formula, **isomers** of each other. This is a general term, and there are several categories of isomers. For example, consider the molecular formula $C_4H_8O_2$. Figure 13.27 shows some of the molecules having this molecular formula. Each structure has a different name, and they all differ in the way the atoms are connected. Note that the molecules may contain different functional groups. For example, molecules 1 and 3 both contain a ketone group and an alcohol group, but the way the groups are connected is different in the two molecules. Molecule 2 has two alcohol groups and one alkene bond, and molecule 4 contains an ether group, an alcohol group, and an alkene bond. All of these molecules are **constitutional isomers** of one another: they all contain the same atoms, but the atoms are connected differently in each case.

Two molecules that are nonsuperimposable mirror images of each other are isomers also, but they are not constitutional isomers, because all the atoms are connected in the same way in the two molecules. For example, the alanine molecule in Figure 13.26 and its isomer on the other side of the mirror each have the molecular formula $C_3H_7O_2N$. For both the left and right molecules, a central carbon atom is attached to one —COOH group, one methyl group, one amine group, and one hydrogen atom. We say that two molecules such as these—nonsuperimposable mirror images of each other—are **enantiomers**, a special kind of isomer. Two molecules must be *non*superimposable in order to be enantiomers. The two glycine molecules of Figure 13.24, for example, are not enantiomers of each other, because the molecule on the left is identical to, and superimposable on, the mirror-image molecule. Thus glycine is just glycine; it does not exist in right and left-handed forms.

Let us consider why some molecules have handedness and others do not. If you compare alanine and glycine, you will see that alanine has a carbon atom bound to four *different* groups. The glycine molecule does not possess such a carbon atom. *The presence of a carbon atom that is bound to four different groups is a predictor for handedness in organic molecules.*

We say that a molecule which possesses handedness is **chiral** (pronounced KYE-rul), and we refer to the special carbon atom as a **chirality center**. Because alanine has a chirality center, it is a chiral molecule, which means that it is nonsuperimposable on its mirror image and that it exists as a pair of enantiomers.

Once you know what to look for, chirality centers are easy to find, and so let's go hunting for some. Figure 13.28 shows a molecule of the anesthetic drug *isoflurane*. The highlighted carbon atom is attached to four different groups: —OCHF2, —Cl, —H, and —CF3. Therefore, this carbon is a chirality center. Because the molecule has a chirality center, we know that it has a nonsuperimposable mirror image.

There is a complicated exception to the rule that every molecule containing a chirality center must have a nonsuperimposable mirror image, but we shall not discuss that exception here.

QUESTION 13.12 Consider the full structure of bromobenzene that follows. (*a*) Is the highlighted carbon atom a chirality center? Explain your answer. (*b*) Is it ever possible for a carbon atom in a benzene ring to be a chirality center?

ANSWER 13.12 (*a*) No. The carbon is bound to only three different things, not four, and so the requirement for chirality is not met. To be a chirality center, the carbon atom must have bonds to four different groups. (*b*) No.

It would be convenient to have names that distinguish two enantiomers so that we do not have to call them "the one on the left" and "the one on the right." Enantiomers can be described with the notation R (from *rectus*, Latin for "right") and S (*sinister*, Latin for "left"). The assignment of R and S to enantiomer molecules is an esoteric and complex process that we shall not describe here. Instead, when necessary, we shall merely indicate the R or S designator for each molecule of an enantiomeric pair, as shown for isoflurane in Figure 13.28.

If you go back and look at the structures of the 20 amino acids shown in Table 11.1 (page 433), you will see that all of them are chiral except one. As we already know, the one **achiral** amino acid is glycine. The central carbon atom of glycine is not a chirality center because there are not four different groups attached to it. The other 19 amino acids each contain at least one chirality center and therefore exist as enantiomeric pairs. As we are about to see, nature almost always uses just one enantiomer of a pair. Natural amino acids, for example, are overwhelmingly S enantiomers.

When amino acids are connected by peptide bonds into a protein chain, each amino acid in the chain (except glycine) possesses a chirality center. Thus, all protein molecules are chiral. But if all proteins are chiral, then enzymes must also be chiral. If you could explore an enzyme's active site, you would find the enzyme's amino acid side chains protruding into the space where chemical reactions take place. All of those amino acids except glycine are chiral. Each side chain is attached to a chirality center (except in the case of glycine), and so the entire three-dimensional interior contour of an enzyme active site is chiral. It should come as no surprise, therefore, that enzyme active sites usually bind only one enantiomer of a chiral substrate molecule. It follows that many drugs—especially those which target enzyme active sites—are chiral. In the next section, we will see how the prevalence of chiral drugs has mushroomed in recent years.

QUESTION 13.13 Locate any chirality centers in each molecule:

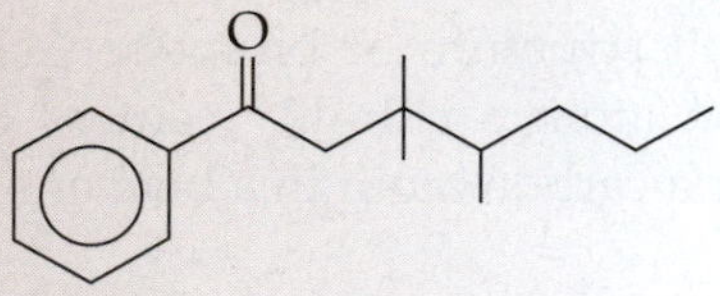

(a)

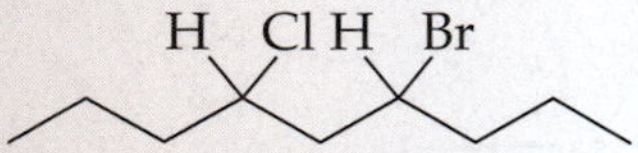

(b)

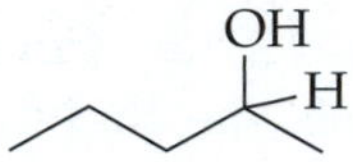

(c)

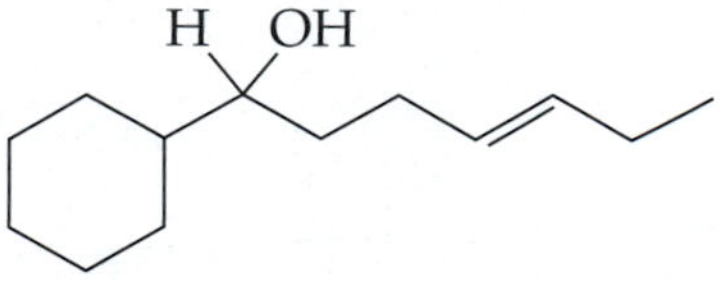

(d)

ANSWER 13.13

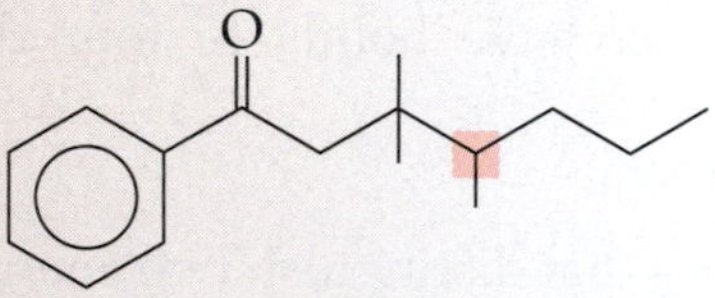

(a)

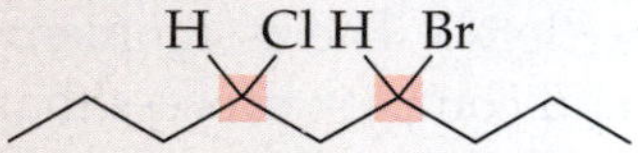

(b)

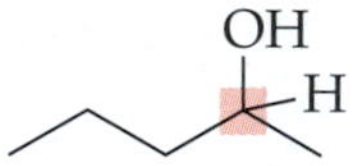

(c)

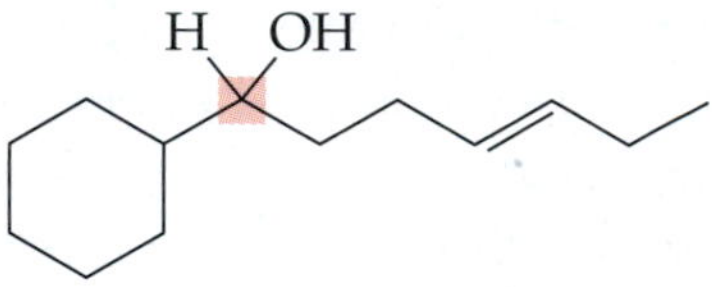

(d)

Note that you must consider everything attached to a carbon atom when determining chirality. In (a), for instance, you might have thought that the highlighted carbon was not chiral because it is attached to three carbon atoms (as well as one H). Each of those three attached carbon atoms is part of a different group, however, so the chiral carbon is indeed attached to four different entities.

QUESTION 13.14 Locate any chirality centers in each molecule:

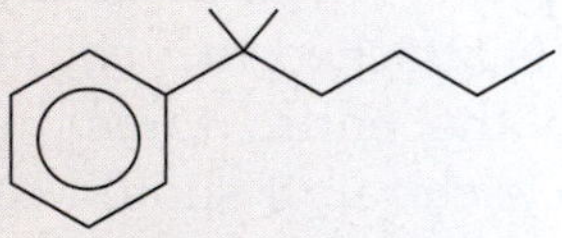

(a)

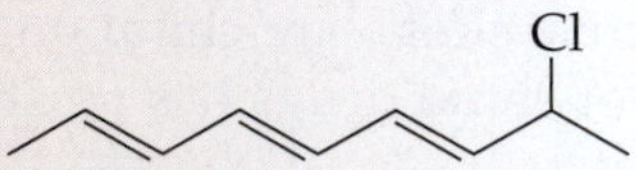

(b)

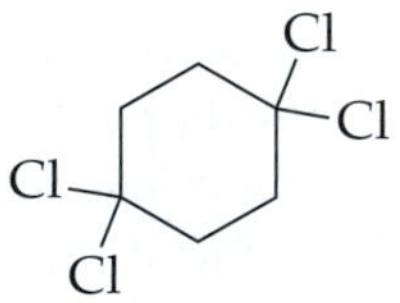

(c)

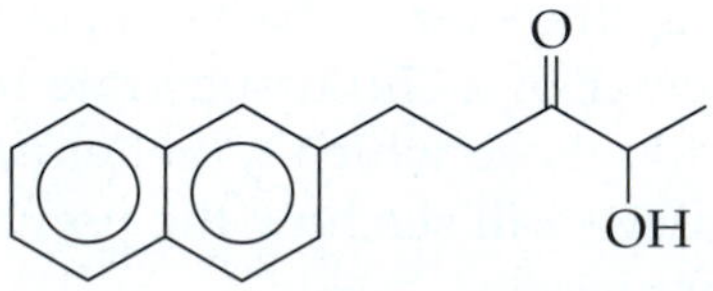

(d)

ANSWER 13.14 Only (b) and (d) have any chirality centers:

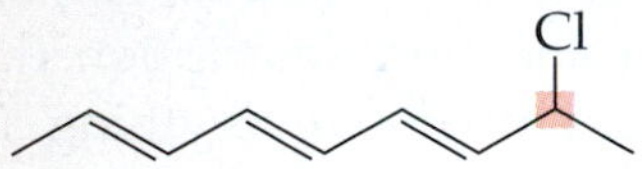

(b)

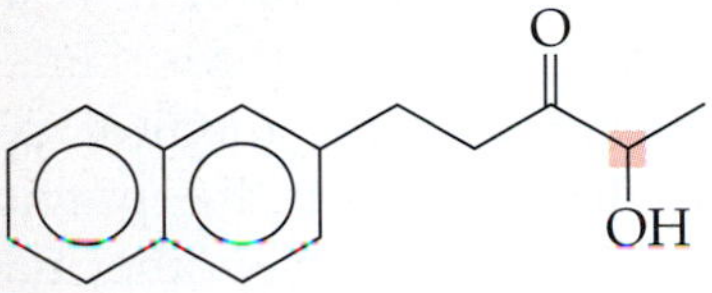

(d)

QUESTION 13.15 Draw the mirror image of each molecule, and identify any chirality centers:

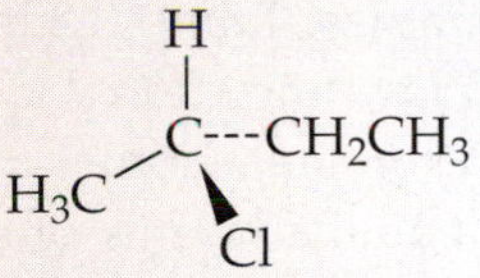

(a)

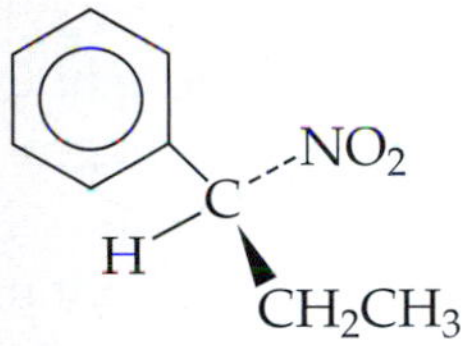

(b)

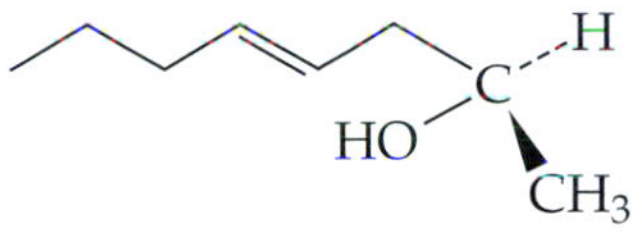

(c)

ANSWER 13.15

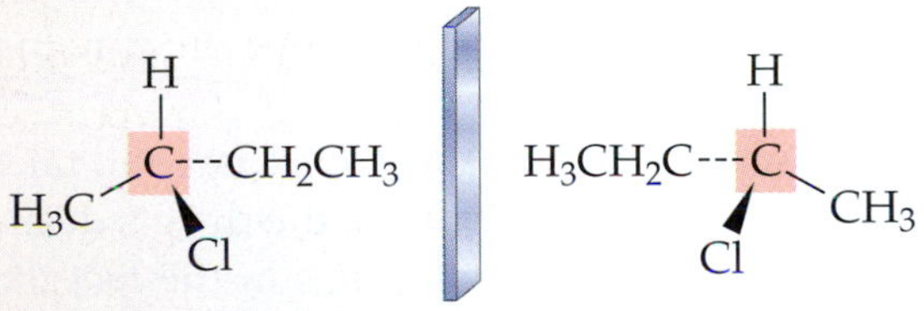

(a)

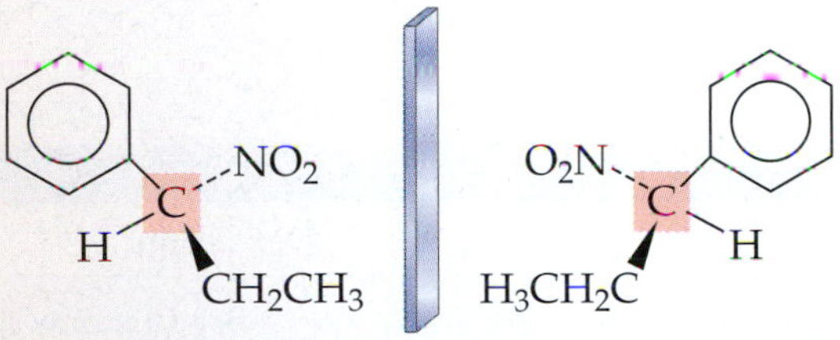

(b)

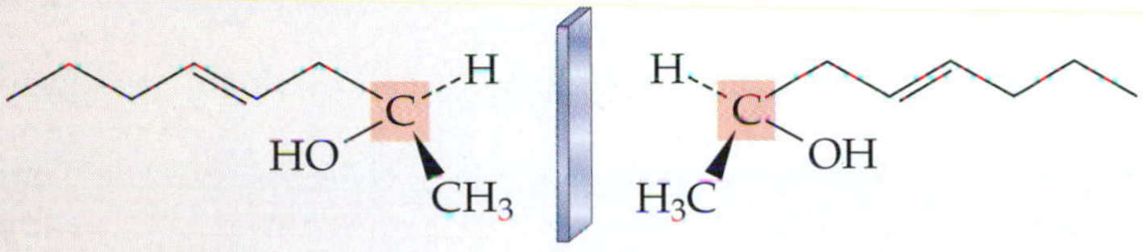

(c)

13.5 | Left or Right, Right or Wrong?

The *United States Pharmacopoeia and National Formulary* is a comprehensive listing of all prescription and nonprescription drugs legally sold in the United States. In it, you will find everything from acetaminophen to Zyrtec®, including many drugs with names designed to appeal to pill-popping consumers: Celebrex®, Claritin®, Ambien®, Singulair®, Allegra®, and Avandia®. A given drug can have up to three names to describe it: a *brand name*, used to entice consumers and medical professionals (examples include Tylenol® and the drugs listed in the previous sentence); a *generic name* (such as acetaminophen, which is another name for Tylenol), and a *formal chemical name*, which is not generally used, except by chemists who synthesize drug molecules. The formal chemical name for acetaminophen is N-4-(hydroxyphenyl)acetamide, but if you ask your pharmacist for this, you are more likely to get a puzzled look than a bottle of Tylenol.

If you look through this pharmacopoeia, you might notice that many generic names for drugs begin with *lev-* (Greek *levo,* "left") or *dex-* (Greek *dexios,* "right"). These two prefixes represent another way to distinguish left-handed molecules from right-handed ones. Because they have handedness, chiral drugs, such as *dex*amethasone, *lev*oxyl, *dex*ferrum, and *lev*aquin, often show off their chirality at the beginning of their names. Table 13.3 lists some popular drugs whose generic names begin with lev- and dex-.

Until recently, many drugs on the market were **racemates** (sometimes called *racemic mixtures*), meaning that they contained equal amounts of the *R* and *S* enantiomers. One classic example is the antidepressant Prozac® (the generic name is fluoxetine). This drug is the most widely sold drug for depression ever marketed, and in its heyday it earned the pharmaceutical company Eli Lilly more than $2.5 billion annually. If you swallow a Prozac tablet, you are taking a 50:50 racemic mixture of the *R* and *S* enantiomers (Figure 13.29). As we already know, most of the biomolecules in our bodies are chiral and interact with only one molecule of an enantiomer pair. Thus, it is typical for only one enantiomer of a racemic mixture to produce a desirable drug effect.

This being true, why should a Prozac user have to take a racemic mixture when only the *R* enantiomer produces the drug's desirable antidepressant effects? The answer to this question lies in the fact that when a chiral drug is synthesized, the product is usually a racemate, and separating one

TABLE 13.3   Examples of Popular Chiral Drugs

Lev drugs			Dex Drugs		
Generic name	Brand name	Used as . . .	Generic name	Brand name	Used as . . .
*Lev*othyroxine	*Lev*o-T	Treatment for thyroid deficiency	*Dex*fenfluramine	Redux	Antiobesity agent
*Lev*onorgestrel	Norplant	Contraceptive	*Dex*troamphetamine	*Dex*edrine	Respiratory stimulant
Penbutolol sulfate	*Lev*atol	Treatment for heart ailments	*Dex*tromethorphan	N/A	Cough suppressant
			*Dex*brompheniramine	Drixoral	Antihistamine

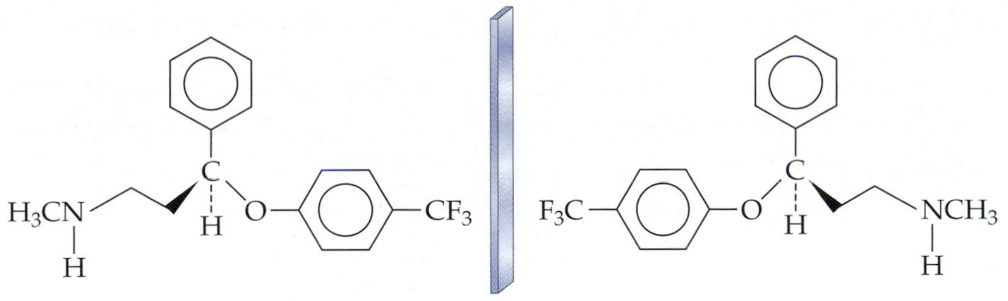

◀ **FIGURE 13.29 Another pair of enantiomers.** The R and S enantiomers of fluoxetine. The racemate is marketed under the brand name Prozac.

MOLECULE
Prozac (fluoxetine hydrochloride)

of the isomers from the other can be an arduous and expensive task. To see why this is so, just look at Table 13.4, which lists some physical properties of the fluoxetine enantiomers. The data clearly show that the two enantiomers have identical physical properties, as all pairs of enantiomers do. Because they have identical physical properties, separating enantiomers from one another is no simple task. Therefore, to save time and money, many drugs have been manufactured as racemates. As we are about to see, though, racemates can present legal problems that are not easily solved.

The physical properties of two enantiomers, such as boiling point and melting point, are always the same because each enantiomer molecule has the same atoms connected in the same order. However, we should expect the LD_{50} values of the two enantiomers to be different because toxicity involves the interaction between a drug and biomolecules in the body, which, as we know, are usually chiral. For example, the S enantiomer of fluoxetine is retained in the body much longer than the R enantiomer, which is readily eliminated. Further, only the R enantiomer has been shown to have antidepressant properties. This being true, why not market (R)-fluoxetine as an antidepressant in place of the racemic mixture? The answer is that Eli Lilly has identified another chiral drug, Cymbalta® (duloxetine hydrochloride), that works better than Prozac (or Prozac's R enantiomer). Meanwhile, the patent on the racemate Prozac expired in 2001, and Eli Lilly's profits from the drug dropped. The company decided to put its resources into the single-enantiomer drug Cymbalta rather than toward a single enantiomer of Prozac. Cymbalta was given FDA approval in 2004.

QUESTION 13.16 Which of the following properties of the two enantiomers of a chiral drug would you expect to be identical: (a) melting point, (b) solubility in water, (c) ED_{50}, (d) toxicity?

ANSWER 13.16 (a) and (b) would be identical. Properties that involve interactions with biomolecules, such as (c) and (d), would not be the same for the two enantiomers.

TABLE 13.4 Selected Physical Properties of Fluoxetine Enantiomers

Property	(R)-Fluoxetine	(S)-Fluoxetine
Melting point (°C)	158	158
Density (g/mL)	1.2	1.2
Solubility in H_2O (mg/mL)	50	50
Molar mass (g/mol)	309.35	309.35

STUDENTS OFTEN ASK

Why Are There So Many Names for the Same Drug?

The answer to this question has to do with patent law. For example, the compound *N,N*,6-trimethyl-2-*p*-tolyl-imidazo [1,2-α] pyridine-3-acetamide, designed and developed by G.D. Searle, is an effective sleep-inducing drug. The company was granted a patent that gave it the exclusive right to sell this drug for a specific number of years, and the name they chose was *Ambien*. Once

that patent expired, however, the drug was up for grabs, which meant that any other company could manufacture *N,N*,6-trimethyl-2-*p*-tolyl-imidazo [1,2-α] pyridine-3-acetamide and sell it as a sleep aid. Of course, this is exactly what one manufacturer did, naming its product *zolpidem*, after the generic name of the drug, *zolpidem tartrate*. Both names—Ambien and zolpidem—refer to the same molecule, shown in Figure 13.30. This tactic, whereby a generic drug is made once the patent on a brand-name drug expires, is known as a *generic switch*.

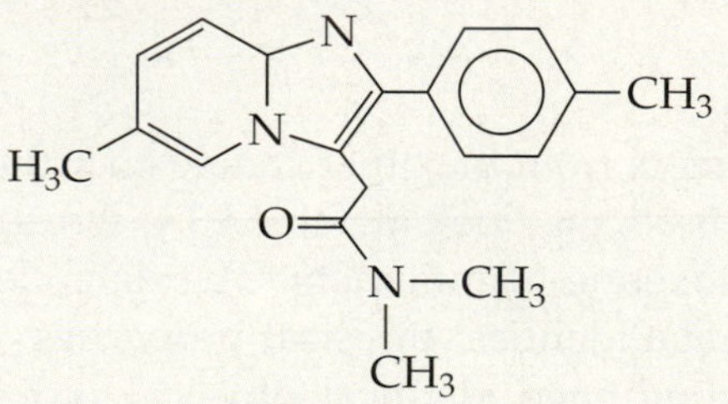

▲ **FIGURE 13.30** **Drugs often have many names.** Ambien, a popular sleep aid, is also known by the generic name zolpidem.

In some cases, a drug formulated as a racemic mixture is not as safe as a drug made of only the active enantiomer. One example is perhexiline, used for many years to treat patients with angina, a type of heart-associated chest pain. Perhexiline is another example of a racemic mixture in which only one of the enantiomers is active. The active enantiomer is readily metabolized in the body, but the opposite enantiomer is not and builds up gradually. As a result, several individuals who took the drug in the 1980s were killed by the toxic effects of the unused enantiomer. The drug, marketed under the brand name Pexsig™, was taken off the market.

As noted in the "Students Often Ask . . . " box, when a pharmaceutical company produces a drug, the company receives a patent that provides a certain number of years of exclusive rights. Drug patents typically last for 20 years, but can be for longer or shorter periods, depending on the class of drug and the time spent waiting for FDA approval. The income the company makes during the years of exclusivity allows the company to earn back money spent on researching and developing the drug. When a chiral drug is patented, the patent protects the racemic mixture. However, if an enantiomer is found to have some activity that is not the same as that of the racemate, any company may patent that enantiomer for the new use. One example is the racemic drug sold under the brand name Celexa™, a popular antidepressant that was approved by the FDA and given 5.5 years of marketing exclusivity. The structure of Celexa (generic name citalopram) is shown in Figure 13.31. In its peak selling years, this drug earned more than $1 billion dollars annually for its maker, Forest Laboratories. Further studies on the drug and its enantiomers showed that one of the two enantiomers was only responsible for its antidepressant action. The company submitted a patent for the active enantiomer and was awarded a new patent on it that included 3.5 additional years of marketing exclusivity.

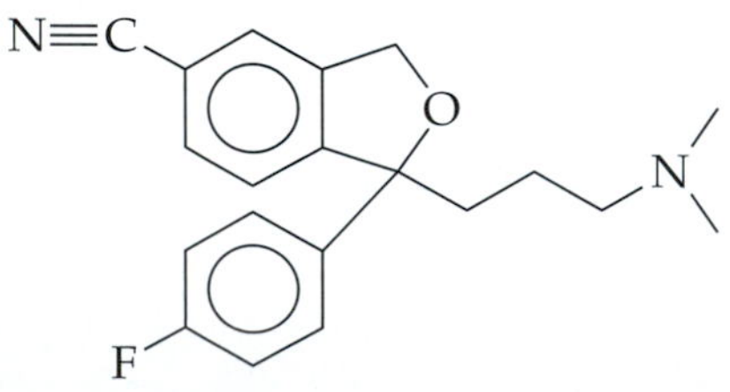

A new trend in the pharmaceutical world is the submission of separate patents for a racemate and both enantiomers at the initial time of discovery. This protects the designer of a racemic drug from having another company

▼ **FIGURE 13.31** **Just one enantiomer needed.** The chiral antidepressant drug citalopram is marketed under the brand name Celexa.

submit a patent for one enantiomer, a situation that has occurred in recent years. For example, U.S. Patents 5 114 714 and 5 114 715 both cover the anesthetic properties of isoflurane enantiomers. In the first patent, every mention of the chemical name is preceded by the letter R. In the second, every mention of the chemical name is preceded by the letter S. Both claim that the enantiomer is superior in action to the racemate. However, it is clear that one enantiomer has to be more effective than the other. This sort of legal shenanigan has become more common as pharmaceutical companies grapple with the difficulties associated with chiral drugs. In fact, because one enantiomer is almost always superior to the racemate, the number of racemic drugs taken to market has declined dramatically over recent years. The percentage of racemic drugs going to market decreased from 32 percent in 1990 to 0 percent in 2001. The percentage of chiral drugs introduced increased from 33 percent to 72 percent over the same period. Thus, in the future, you can expect to see more drugs with generic names beginning with lev- and dex-.

13.6 | Brand Name or Generic?

Imagine what this would be like: your heart suddenly begins to beat very quickly, above 200 beats per minute, and this racing continues 24 hours a day. You feel exhausted more often than rested, and you are not able to work or exercise. This state continues for more than one month, and during that time, you are constantly aware of your beating heart, something that most of us rarely notice. This situation is a reality for thousands of people who suffer from atrial fibrillation (AF). Drug designers have responded to the need for something to bring the heartbeat of AF patients into check, and there are now several drugs on the market that mitigate some of the unpleasantness associated with the condition. Unfortunately, many patients cannot tolerate some of these medications, and some fail to respond to them.

The drug sold under the brand name Cordarone®, with molecular structure shown in Figure 13.32, has been beneficial to many AF patients who do not respond to other, milder drugs. This drug works by slowing the electrical impulses in the heart and, hopefully, converting the heart back to its normal rhythm. Patients who respond well to Cordarone find that their episodes of AF are less frequent and shorter in duration. Early in 2003, however, the patent protection for Cordarone, held by the pharmaceutical company Wyeth Pharma, expired. At that point, a generic version of the drug could be put on the market by any drug company. The generic version of the drug, called amiodarone, was made available immediately. Eager to save money by switching to the generic version, many faithful Cordarone patients changed loyalties. It did not take long for feedback from patients taking amiodarone to start pouring in. A conspicuous number of patients claimed that the generic drug caused breathing difficulties in addition to other new side effects. What is more, blood levels of the generic drug were found to be quite different from the blood levels obtained from the same dosage of Cordarone. Upon switching back to Cordarone, patients found their adverse reactions subsiding immediately.

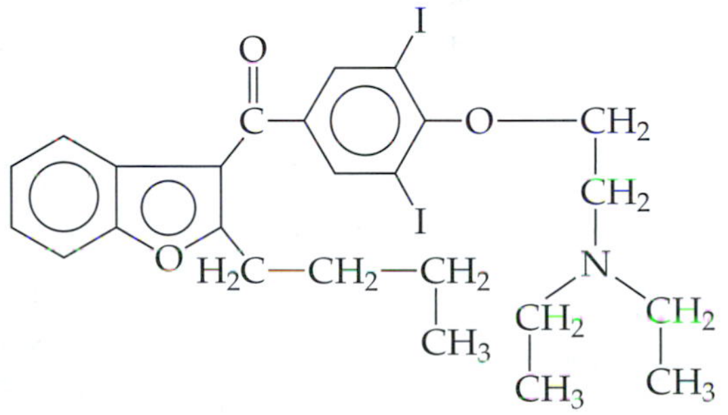

▲ **FIGURE 13.32 Keeping the heart under control.** The drug amiodarone is used to treat atrial fibrillation. It is marketed under the brand name Cordarone.

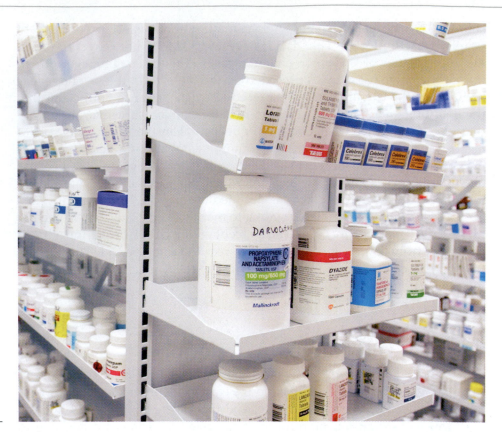

▶ **FIGURE 13.33 Generics replacing brand-name drugs.** Thousands of prescription drugs are available today to treat our ills. When a patent on a brand-name medication expires, drug companies rush to bring generic versions to the market. Most of the time, the generics work as well as the brand-name patent drugs, but sometimes some patients need to stick with the brand-name drug.

QUESTION 13.17 Look at the structure of Cordarone in Figure 13.32. (*a*) Does this molecule possess a chirality center? If so, identify it. (*b*) No indication of left-handedness or right-handedness is provided in the brand, generic, or formal chemical name of the drug. Is this because the drug is manufactured as a racemic mixture?

ANSWER 13.17 (*a*) No. (*b*) No; a molecule that is achiral is not a racemate.

For the vast majority of generic switches, the generic drug behaves exactly like the brand-name drug. The Cordarone–amiodarone story demonstrates, however, that there are rare instances in which a generic drug does not successfully reproduce the beneficial effects of the brand-name drug (Figure 13.33). How can this happen when both the brand-name drug and the generic drug contain exactly the same drug molecule? The answer is that, although a generic must contain the same *active* ingredient, it is not required to have the same **excipients**—the *inactive* ingredients used to make medications hold together, taste better, easier to swallow, and look prettier. There are almost as many excipients as there are drugs because it is usually not possible to swallow the active ingredient in its pure form. Thus, excipients are available to help in all of the processes associated with the manufacture and delivery of the active ingredient.

Table 13.5 lists the excipients in the popular cholesterol-lowering drug sold under the brand name Lipitor®. The list includes substances that are binders, which aid in compressing the drug mixture into tablets (calcium carbonate, lactose, microcrystalline cellulose, simethicone emulsion). Lipitor tablets also contain a colorant (opadry white) and a lubricant (magnesium stearate), which prevent tablets from sticking to the machinery that

TABLE 13.5 Excipients in Lipitor	
Excipient	**Purpose**
Calcium carbonate	Tablet binder (salt)
Microcrystalline cellulose	Tablet binder (polysaccharide)
Lactose	Tablet binder (sugar)
Croscarmellose sodium	Dissolving agent (polysaccharide)
Polysorbate 80	Dissolving agent (polysaccharide)
Hydroxypropylcellulose	Coating agent (polysaccharide)
Magnesium stearate	Lubricant (fat)
Opadry white YS-1-7040	Colorant (titanium dioxide)
Simethicone emulsion	Tablet binder (silicon polymer)
Candelia wax	Coating agent (hydrocarbon)

makes them. The list also includes dissolving agents (polysorbate 80 and croscarmellose sodium), which help the solid tablet mass to dissolve immediately when it reaches its destination. Finally, coating agents (hydroxypropylcellulose and candelia wax) seal the drug mixture within the tablet.

QUESTION 13.18   Cellulose, the primary structural component of trees and plants, is the most ubiquitous polysaccharide on the planet. Which excipients in Table 13.5 contain cellulose?

ANSWER 13.18   Microcrystalline cellulose and hydroxypropylcellulose.

Excipients are like the people who retrieve balls at a tennis match: they should go unnoticed. In a chemical sense, excipients must be inert and should neither have druglike properties nor interfere with the active ingredient. As we have seen with amiodarone, however, excipients can cause problems. In some cases, problems arise because a patient has an allergy to some excipient in the generic drug. In other cases, all patients who make the switch to the generic drug experience new side effects. This can occur, for example, when the binders or dissolving agents in the excipients change the rate at which the drug enters the bloodstream. In fact, there are many instances in which a generic drug is not marketed when patents on the brand-name drug expires. This often happens when the timing of the release and the dosage must be precisely regulated. For these drugs, which are said to have a *narrow therapeutic index*, the manufacture of a generic version is often overwhelmingly difficult.

This chapter has focused on the drugs we take and how they benefit us. We have looked at the design process, issues surrounding drug patents, how drugs are delivered in the body, and what they do once they arrive at their designated targets. Although we have covered a lot of ground, it is impossible to survey the entire realm of drugs in the limited space we have here. Consequently, there is drug-related terminology that we have not touched on and do not have the space to mention. We feel that an educated consumer, such as yourself, would benefit from a digest of common terminology used to describe over-the-counter and prescription drugs, and such

STUDENTS OFTEN ASK

How Do Time-release Medications Work?

If you create a tablet with the ingredients compressed inside a simple thin coating, the ingredients will all be available as soon as the coating dissolves. This is advantageous if you want your drug released all at once so that it can relieve a problem quickly. In many cases, however, it is more desirable to deliver a drug slowly over a longer period, so that the drug is discharged at a constant rate, up until the time the next dose is taken.

In one design used in time-release medications, the drug is dissolved in a biodegradable polymer matrix that breaks down slowly after ingestion (Figure 13.34). This polymer matrix is surrounded by a thick layer of a different biodegradable polymer that breaks down in the stomach. The outer layer is designed to survive the initial onslaught of salivary and digestive enzymes that every tablet encounters on its trip down the gullet. By the time the pill has reached the stomach or small intestines, the outer layer has dissolved, and the interior polymer matrix is exposed and breaks down over time, slowly releasing the drug.

Time-release technology is making way for new drug-delivery systems that are more precise and more adaptable. Transdermal (*trans-*, "across"; *dermal*, "skin") patches that deliver drugs to the bloodstream through the skin present an easy-to-use alternative to popping pills. The technology is limited, however, to small molecules, such as nicotine (Figure 13.35).

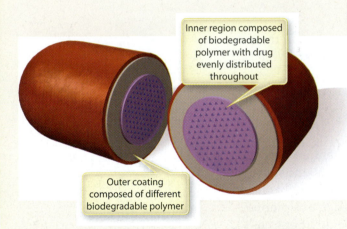

Inner region composed of biodegradable polymer with drug evenly distributed throughout

Outer coating composed of different biodegradable polymer

▲ **FIGURE 13.34** **Slow drug delivery.** In this design for a time-release medication, the outer biodegradable coating breaks down, leaving an inner matrix that contains drug molecules embedded in a second biodegradable polymer. This latter polymer degrades slowly, and as a result, the drug is released into the body gradually.

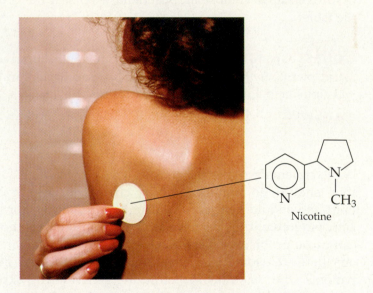

Nicotine

▲ **FIGURE 13.35** **The transdermal patch.** Small molecules such as nicotine can be administered through patches that deliver a drug through the skin.

a digest is presented in Table 13.6. Some of these terms are mentioned in this chapter, and others are not. You may notice that psychoactive drugs— narcotics, barbiturates, hallucinogens, and the like—are missing from this table. These drugs are the ones most heavily abused and most commonly

The newest drug-delivery frontier mixes medicine with nanotechnology. One example is the silicon microchip, on which hundreds or thousands of tiny, tiny, *tiny* reservoirs can be filled with any combination of drugs (Figure 13.36). When stimulated with a weak elec- trical pulse, the reservoirs can be selectively opened to release their contents. Once the potential of nanotechnology is realized, it should be possible to release a drug on cue. In about 20 years from now, in fact, the pills we eagerly swallow today may be obsolete.

(a) (b)

▲ **FIGURE 13.36 Precision drug delivery.** **(a)** Silicon microchips can be used as miniature implants that release a drug on cue. **(b)** One compartment in such a device before and after its drug has been released.

found on the street. Because drugs of abuse are a painful, but real, part of our society, Chapter 14 is devoted solely to them. There you will find ter- minology related to psychoactive drugs and other drugs of abuse, as well as a summary table similar to Table 13.6 at the end of that chapter.

TABLE 13.6	Categories of Over-the-Counter and Prescription Drugs		
Drug action	**Drug category**	**What drug does**	**Examples**
Reduces pain, inflammation, or fever	Analgesic	Raises pain tolerance	Acetaminophen, aspirin, ibuprofen, naproxen
	Anesthetic	Kills pain	Novocaine, benzocaine
	Anti-inflammatory	Reduces tissue inflammation	Acetaminophen, aspirin, ibuprofen, naproxen
	Antipyretic	Reduces fever	Acetaminophen, aspirin, ibuprofen, naproxen
Treats flu, cold, or allergy	Antihistamine	Reduces allergy symptoms	Diphenhydramine
	Antitussive	Cough suppressant	Dextromethorphan, codeine
	Decongestant	Reduces sinus and chest congestion	Ephedrine
	Expectorant	Brings mucus up out of bronchial tubes	Guaifenesin
Fights invading pathogens	Antibiotic	Fights bacterial pathogens	Cephalexin, penicillin, tetracycline
	Antiviral	Fights viral pathogens	Acyclovir, AZT
Treats heart ailments	ACE inhibitor[a]	Lowers blood pressure, reduces risk of heart failure	Captopril, enalapril
	Anticoagulant	Lowers blood pressure, reduces blood clotting	Aspirin, heparin, warfarin
	Beta blocker	Slows heart rate, treats abnormal heart rhythms	Bisoprolol, propranolol
	Diuretic	Reduces blood volume and pressure	Diamox, benzthiazide
Targets cancerous tissues	Antimetabolite	Inhibits DNA–RNA synthesis	Cisplatin, mercaptopurine
Acts on endocrine system	Hormone drugs	Mimic, enhance, or block hormone action	Prednisone, cortisone (reduces inflammation); estrogen, progesterone (birth control pills)

[a] ACE stands for *angiotensin converting enzyme*. ACE inhibitors block the action of this enzyme.

Chapter 13

SUMMARY

In this chapter, we considered the way drugs are supposed to work. Drug designers often begin their work with a lead compound, a molecule found in nature. In many cases, it is possible to improve on Mother Nature by making small modifications to the molecule. Drugs typically are designed to interrupt a biochemical pathway that leads to some deleterious condition. The series of steps that convert one molecule to another in the cell is referred to as a metabolic pathway. Because most biochemical reactions are catalyzed by enzymes, metabolic pathways are controlled by enzymes at each step. Thus, the active sites of enzymes are frequently the target of new drugs.

The three-dimensional structure of a drug determines what the drug will do once it is in the body. Because biomolecules are also three dimensional, the structure of drug molecules must be finely tuned to maximize binding to a receptor on the biomolecule's surface. The way the activity of a molecule changes as its three-dimensional structure is manipulated is referred to as its structure–activity relationship, a variable that is optimized in the design of new drugs.

The activity of drug molecules can be quantified with the use of three terms: the ED_{50}, TD_{50}, and LD_{50} tell us how much drug will cause 50 percent of test subjects taking the drug to elicit a beneficial effect, a toxic effect, and death, respectively. The ratio of TD_{50} to ED_{50} is the therapeutic index, an all-inclusive value that combines the specificity and efficacy of a drug candidate.

Drugs that target enzymes can bind permanently to the active site via covalent bonds. We refer to these drugs as irreversible inhibitors. Reversible inhibitors bind to their targets via intermolecular interactions, such as hydrogen bonding and ionic interactions. Anti-AIDS drugs that target either reverse transcriptase or HIV protease are examples of reversible inhibitors. We considered examples of drugs that bind these two enzymes. Because irreversible inhibitors can bind to other biomolecules and cause severe side effects, many new drugs are reversible inhibitors of specific enzymes.

Organic molecules that are isomers of one another have the same molecular formula, but different structures. Two molecules that are mirror images of each other represent a special kind of isomerism, and we refer to them as enantiomers of each other. An organic molecule is chiral and exists as a pair of enantiomers when at least one of its carbon atoms is bound to four different groups. We refer to this carbon as a chirality center. A molecule that possesses a chirality center is not superimposable on a mirror image of itself, and a molecule that does not contain a chirality center is identical to its mirror image. Many drugs are chiral, and in most cases the activity of a chiral drug can be attributed to only one of the two enantiomers. In past years, numerous drugs were marketed as racemic mixtures of the two enantiomers. However, because only one enantiomer produces the beneficial effect, there has been a move toward the manufacture of single-enantiomer drugs. Several stories about specific chiral drugs were discussed in the text.

KEY TERMS

achiral	ED_{50}	irreversible inhibitor	opiate	structure–activity
carcinogenic	efficacy	isomer	racemate	relationship
chiral	enantiomer	LD_{50}	receptor	TD_{50}
chirality center	excipient	lead compound	reversible inhibitor	therapeutic index
constitutional isomer	handedness	neurotransmitter	specificity	virus

QUESTIONS

The Painless Questions

1. Comment on this statement: Time-release medications often are designed with an outer coating around the drug molecules, which are released all together in the stomach when the coating dissolves.

2. The structure of the ibuprofen molecule is

(a) Does this molecule possess a chirality center?
(b) Name one organic functional group in the molecule.

3. Describe the differences between reversible and irreversible inhibitors. Summarize the benefits and drawbacks of each in the design of new drugs.

4. Which of these items possesses handedness: (a) a CD disc, (b) an ice cream cone, (c) a rope, (d) the silver part that makes up the base of a light bulb?

5. Write three plausible structures that have the molecular formula C_3H_6O. Name the organic functional groups in each molecule you draw.

6. In the text, we provide examples of how a drug can be designed, starting with a naturally occurring lead compound. Provide one example from the text of the design of a new drug from a natural one, and describe how the new drug performs.

7. Why is it necessary to use both ED_{50} and TD_{50} to describe the overall success of a drug, rather than just one value or the other? What is the name given to the ratio of these two values?

8. Which of these organic molecules has a chirality center?

(a)

(b)

(c)

(d)

9. Name four roles that excipients play in a drug formulation. What kinds of chemical compounds are used as excipients?

10. Which of these carbon atoms *could be* a chirality center: (a) a carbon atom attached to an alcohol, (b) the carbon atom of a ketone group, (c) a carbon atom that is part of a benzene ring, (d) a carbon atom attached to an ether group?

11. The term *suicide inhibitor* is sometimes used in place of *irreversible inhibitor*. Why would either name be appropriate for this type of drug?

12. Comment on this statement: The molecule amiodarone, shown in Figure 13.32 on page 539, was marketed briefly in generic form after the patent for Cordarone® expired. The *R* enantiomer of the drug is used in the treatment of atrial fibrillation.

13. Explain what a virus is. What retrovirus was discussed in detail in this chapter? What disease does it cause in humans?

14. Briefly describe the different kinds of names a drug molecule can have. Why is it uncommon to hear the full chemical name for a drug molecule?

15. Name one example of an achiral drug discussed in this chapter.

More Challenging Questions

16. Drug designers look for maximized *efficacy* and *specificity* in new drug candidates. Describe what these terms mean. How is each term related to a drug's side effects and beneficial drug interactions?

17. In this chapter, we discussed a nitrogen-containing organic functional group found in biomolecules that is susceptible to cleavage. Which functional group is it? In what type of biomolecule would you find it? What organic functional groups arise when this organic functional group is cleaved?

18. Create your own hypothetical three-step metabolic pathway using Figure 13.5 (page 512) as a guide, arbitrarily classifying each substance in the pathway as desirable or undesirable. Which parts of your pathway would serve as targets for new drugs, and what effect would you expect an efficacious drug to have on these parts?

19. Which of these molecules should show opiate activity? Explain your answer.

(a)

(b)

(c)

(d)

20. (a) What do proteases do? (b) Which step in the production of new retroviruses requires a protease enzyme? (c) What job does this protease perform for the virus?

21. When we use the term *chemical mop*, what do we mean? What types of chemicals are being mopped in the example given in the text?

22. An enzyme called H has two sites where molecules bind. Site 1 binds the substrate C, and site 2 binds the substrate E. When C is bound to site 1, the rate at which E binds at site 2 is 25 times greater than the rate at which E binds when site 1 is empty. E is known to be carcinogenic, and your job is to prevent it from participating in a series of reactions unrelated to C and H. (*a*) Sketch a schematic that depicts the action of enzyme H. (*b*) Suggest two approaches to limiting the effect of E.

23. What is the difference, if any, between the terms *receptor* and *active site*? Is one a subset of the other? Is one broad and one specific? In what context would you use each word?

24. The structure of oxomate is

(*a*) You should recognize two of the organic functional groups in this molecule. What are they? (*b*) The left and right ends of the molecule have different functions for this drug. What are these functions? (*c*) What are the molecular formula and molar mass of this molecule?

25. Which of these drugs has the largest therapeutic index?

Drug	TD$_{50}$ (mg)	ED$_{50}$ (mg)
P	4.5	6.5
Q	14.8	3.4
V	34.9	2.87
W	12.2	14.3
Z	14.1	0.35

26. In the evolution of drugs used to fight AIDS, there have been two primary targets for drug design. (*a*) The drug AZT is an example of a drug that targets which part of the HIV life cycle? (*b*) Why has a new target been chosen for the design of AIDS drugs? (*c*) What is the target of the newest class of anti-AIDS drugs? Sketch the structure of one of these new drugs.

27. Draw the mirror images of each of these organic molecules. Indicate whether each molecule is identical to the mirror image you have drawn.

28. What four types of rapidly growing cells are targeted by chemotherapeutic agents? Which of these cell types give rise to the side effects associated with chemotherapy? What side effects are attributable to the death of each cell type?

The Toughest Questions

29. (*a*) Identify each of the molecules shown as either an ester or an amide. (*b*) Each molecule undergoes a cleavage reaction. Show the products of each reaction. (HINT: To answer this question, you may want to review Section 10.6.)

30. Describe how retroviruses replicate by using healthy cells. In particular, point out the type of nucleic acid

(RNA or DNA, single stranded or double stranded) that participates in each step of replication. Include a description of how a newly formed virus particle becomes an active, infectious virus particle.

31. Refer back to Figure 13.30 (page 538), which shows the structure of the drug zolpidem. (*a*) What are the molecular formula and molar mass of this compound? (*b*) If you take a pill that contains 10 mg of the active ingredient, how many molecules of zolpidem are you ingesting? (*c*) Does zolpidem possess any chirality centers? If so, identify them.

32. Use Figure 13.10 (page 515) and your knowledge of intermolecular interactions to sketch an interaction between the surfaces of two biomolecules. Your sketch should include an ionic interaction, a nonpolar interaction, and a hydrogen bond. Rank the different intermolecular interactions in your sketch according to relative strengths.

33. Use Figure 13.3 (page 511) to determine (*a*) the molecular formulas of sulforaphane and oxomate and (*b*) the molar mass of each. (*c*) If the LD_{50} values are 2400 mg/kg for sulforaphane and 4600 mg/kg for oxomate, what mass, in milligrams, would be required to kill 50 percent of 75-kg patients who consume each molecule? (*d*) If 12.5 percent of subjects who take oxomate die, how many milligrams did each person ingest?

34. There are many constitutional isomers that have the molecular formula $C_{10}H_{10}O_3$. Sketch ten of them. In each structure, draw a circle around any chirality centers.

35. Drug molecules can be designed to be highly polar, highly nonpolar, or anything in between, depending on their targets and the way they are administered. (*a*) Using your knowledge of molecular polarity, rank etorphine (Figure 13.12*d*), codeine (Figure 13.12*c*), and amprenavir (Figure 13.18) according to polarity. Provide a reason for your ranking. (*b*) Which of these three compounds is most soluble in water? Least soluble? (Figure 13.12 is on page 516, and Figure 13.18 is on page 527.)

36. The molecules methadone, a drug used to treat heroin addicts, and Demerol, a painkiller, are shown below. (*a*) Can either of these molecules be classified as an opiate? How do you know? (*b*) Go back to the structure of morphine (Figure 13.12*a*, page 516). How does the structure of methadone differ from the structure of morphine? How does the structure of Demerol® differ from that of morphine? (*c*) List the organic functional groups present in each of the three molecules.

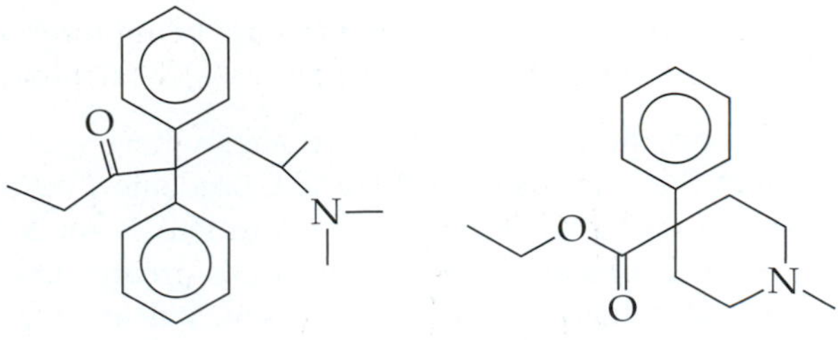

Methadone   Demerol®

37. Compare the structure of morphine with the structures of codeine, heroin, and etorphine, all shown in Figure 13.12 on page 516. Given that the extraneous parts of each molecule dictate the way it bonds to the opiate receptor, do you think that polar or nonpolar groups enhance the binding of these molecules? How do you know?

38. Two types of intermolecular interactions we discussed earlier, but not here in Chapter 13, are dipole–dipole interactions (Section 7.2) and ion–dipole interactions (Section 4.3). Sketch two protein molecules that interact via these two types of intermolecular interactions.

39. Which of the amino acids shown in Table 11.1 (page 433) have two chirality centers? Sketch their three-dimensional structures and their mirror images.

E-Questions

Go to **www.prenhall.com/waldron** to find these questions in electronic form, complete with hyperlinks directly to the various websites cited in the questions.

40. Antiangiogenesis drugs promise to be the next great treatment for cancerous tumors. David Cheresh, a scientist at Scripps in San Diego, has found a way to deliver potent drugs specifically to young blood vessels that are destined to provide blood to a growing tumor. His method is described in the article titled "Nanoparticle Guided Missiles" in **CENews**. Read this one-page story, and then answer the following questions: (*a*) What specifically are Cheresh's "guided missiles" designed to target? (*b*) What are these guided missiles made of? Are they charged or neutral? (*c*) What do they deliver to the target they seek? How does their weapon attack newly forming blood vessels that supply tumors? (*d*) Why is his method superior to a method that uses artificial viruses to deliver drugs?

41. The article "Chemical Mop" at **NewScientist** describes the new drug oltipraz, introduced in the opening of this chapter. Read this one-page report, and then answer the following questions: (*a*) What biomolecule does oltipraz stimulate in the body? (*b*) What beneficial effects does this biomolecule have?

42. Do a Web search for the molecule *diisopropylfluorophosphate*, or DFP for short. As part of your search, locate information on what biomolecule this toxin affects in the body. Sketch the structure of the molecule, and calculate its molar mass. Look for toxicity data, and report the LD_{50} value. To what class of poisons discussed in this chapter does this molecule belong?

43. The Hudson Institute, a think tank for political policy, publishes the journal *American Outlook Today*. One 2002 article titled "Hidden Costs of Generic Drugs" can be found at **Outlook**. This article presents a point of view on the generic issue that has not been considered in this text. Read the short, two-page article, and then answer the following questions: (*a*) Whose side is the author defending in the article, brand-name drug manufacturers or generic drug manufacturers? (*b*) Briefly summarize the reasons the author uses to defend this side of the issue. (*c*) Do you feel that this report is a fair and balanced look at the issue, or did you perceive a bias in the author? Explain your choice.

44. The article "Promiscuity Doesn't Pay" offers an explanation for the prevalence of false positive results that drug designers observe. The article can be found at **CENews**. Read this brief "news of the week" article, and then answer the following questions: (*a*) What do they mean by the terms *compound library* and *false positive*? (*b*) What false impression has led scientists to call certain drug molecules "promiscuous"? What do they mean by this label? (*c*) To what do they attribute the frequency of false positive results?

WORLD WIDE WEB RESOURCES

As with the E-Questions, go to **www.prenhall.com/waldron** to find these questions in electronic form, complete with hyperlinks directly to the various websites cited.

   Some of the links that follow contain research articles related to the topics in this chapter and could be used as the basis for a writing assignment in your course. Other sites are of general interest.

- The article "Placebo Produces Surprise Biological Effect" can be found at **NewScientist**. In it, you will learn about an interesting study that measures the physiological effects of placebo ingestion.

- The NIH maintains statistics on hundreds of topics that are important to the medical community. At **NIH**, you will find detailed statistical data on HIV and AIDS worldwide. The end of this NIH fact sheet provides references that will lead you to additional information.

- Learn more about how drugs are designed to be taken up readily in the body in the article "New Ways to Assess Drug Candidates" at **CENews**. In this article, the authors discuss new research on specific aspects of molecular structure that give rise to high oral availability.

- Learn more about the benefits of vitamins in fruits and vegetables in the **Vitamin Guide**. This site includes dozens of links to specific foods and discusses the beneficial biomolecules they contain.

ABC NEWS VIDEO
Drug Company Investigation

CHEMISTRY TOPICS IN CHAPTER 14

- Primary, secondary, and tertiary amines [14.1]

- Illicit drugs, neurotransmitters, and the human brain [14.2]

- Qualitative tests for amines [14.3]

- Quantitative versus qualitative analysis [14.4]

- Antibodies and antigens [14.4]

- Steroids [14.4]

- Quaternary ammonium salts [14.5]

- Free bases and hydrochloride salts [14.5]

- Electron-withdrawing and electron-donating groups [14.6]

"Each fingernail a crimson petal, seen
Through a pale garnishing of nicotine"

A LETTER TO THE EDITOR, ROBERT HILLYER

Drugs II

THE DARK SIDE OF DRUG USE

In the world of sports, the British are notoriously gifted soccer players, experts at cricket, and accomplished equestrians. In road cycling, though, England has never made much of a splash, with the exception of a young Brit by the name of Tommy Simpson. In the early 1960s Simpson was the first English-speaking cyclist to make a showing in Europe, the undisputed hub of international cycling, and in 1965 he won the World Championship. He immediately became a central figure in the British sporting world, taking on veteran cyclists on the Continent and beating them. His warmth and charm made him a darling of the British press, displacing prominent athletes in England's more popular sports.

Following a respectable showing in the 1966 Tour de France, Simpson made his way toward the front of the pack during the 1967 Tour. After the 12th stage of the race, he sat in an impressive seventh place. The 13th stage, through the

French countryside from Marseille to Carpentras, included a climb up the unforgiving Mount Ventoux (Figure 14.1). According to reports, the temperature reached 40 °C (104 °F) that day on a barren mountain pass that afforded no shade to the riders. Simpson had joined a knot of cyclists out in front and was within 2 km of the summit when he began to list and wobble, losing his bike from beneath him.

 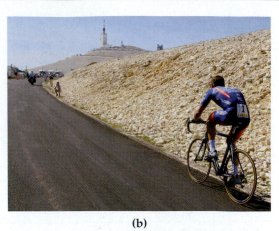

(a)   (b)

He asked the spectators to put him back into the saddle and give him a push. About 200 m later, Simpson collapsed and never regained consciousness. An autopsy showed measurable quantities of amphetamine and methamphetamine in his bloodstream that day, both organic molecules used by athletes to enhance their performance.

This was the first highly publicized case of an athlete's use of performance-enhancement drugs, and it spurred the international sporting community to consider testing athletes for these drugs, especially those which present a health risk. Since that day in 1967, though, the situation has not improved much. We now routinely test athletes, but the creation of new drugs has outpaced the creation of tests that can detect them in the body. More than almost any other sport, professional cycling continues to be fraught with drug scandals. The problem came to a head at the 1998 Tour de France, when dozens of riders tested positive for performance enhancers and were expelled from the race. Other cyclists refused to finish, claiming that the rights of riders were being violated. In all, only 96 of the 198 starters finished the 1998 Tour, the lowest percentage in the race's long history.

Clearly, illicit-drug use is not limited to shadowy alleyways and squalid back rooms. In this chapter, we shall examine the role these drugs play on our streets, in our sports arenas, and in our courtrooms. We shall see how drugs can have deleterious effects on our bodies and why we can tolerate them only in limited quantities. Because most illicit drugs contain nitrogen, we shall examine the amine functional group and look at the properties it confers on organic and biological molecules. We shall look at "uppers," "downers," hallucinogens, and narcotics to see what gives each its specific properties and how we can distinguish one of these drugs from another.

We begin, as we usually do, with the structure of the molecules in question and then, with that information in hand, survey the drugs some humans mistakenly abuse.

14.1 | That Ubiquitous Nitrogen Atom

The presence of nitrogen atoms confers special properties to molecules, and most illicit-drug molecules contain at least one nitrogen atom. Carbon and hydrogen atoms form the scaffolding of organic molecules, but atoms such as nitrogen, oxygen, and sulfur are sites of activity and change, places where electron density either accumulates or is drawn away.

The nitrogen atom is part of several organic functional groups, and we are already familiar with some of them. Figure 14.2 shows the generic

▼ **FIGURE 14.2 Three nitrogen-containing functional groups.** Nitrogen atoms are found in the amide functional group, the amine functional group, and the imine functional group.

$$
\begin{array}{c}
\quad\ \ \overset{\displaystyle O}{\underset{}{\parallel}} \\
R-C-\ddot{N}-R'' \\
\quad\ \ \ \ \ \ \ | \\
\quad\ \ \ \ \ \ R'
\end{array}
$$

Amide functional group

$$
\begin{array}{c}
R-\ddot{N}-R' \\
\ \ \ \ \ | \\
\ \ \ \ R''
\end{array}
$$

Amine functional group

$$
\begin{array}{c}
\ \ \ \ R' \\
\ \ \ \ | \\
R-C=\ddot{N}-R''
\end{array}
$$

Imine functional group

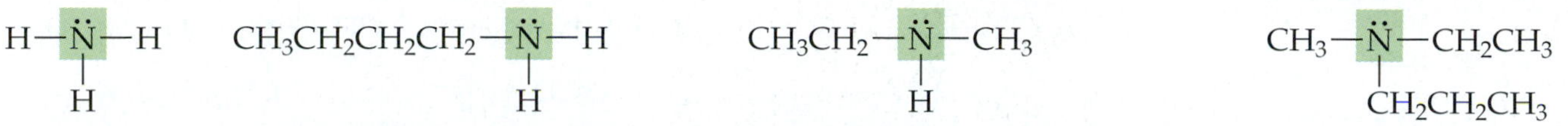

$$H\!-\!\overset{\cdot\cdot}{N}\!-\!H$$
$$\underset{\displaystyle H}{\vert}$$

Ammonia

(a)

$$CH_3CH_2CH_2CH_2\!-\!\overset{\cdot\cdot}{N}\!-\!H$$
$$\underset{\displaystyle H}{\vert}$$

Butylamine, a primary amine

(b)

$$CH_3CH_2\!-\!\overset{\cdot\cdot}{N}\!-\!CH_3$$
$$\underset{\displaystyle H}{\vert}$$

Ethylmethylamine, a secondary amine

(c)

$$CH_3\!-\!\overset{\cdot\cdot}{N}\!-\!CH_2CH_3$$
$$\underset{\displaystyle CH_2CH_2CH_3}{\vert}$$

Ethylmethylpropylamine, a tertiary amine

(d)

▲ **FIGURE 14.3 Ammonia plus primary, secondary, and tertiary amines.** The nitrogen atom of each amine is highlighted in green. **(a)** Ammonia is the smallest amine, with three hydrogen atoms attached to a central nitrogen atom. The nitrogen also has a nonbonding pair of electrons. **(b)** Primary amines have two hydrogen atoms and one alkyl group attached to the central nitrogen atom, in addition to the nonbonding pair of electrons. **(c)** Secondary amines have one hydrogen atom and two alkyl groups attached to the central nitrogen atom, in addition to the nonbonding pair of electrons. **(d)** Tertiary amines have no hydrogen atoms and three alkyl groups attached to the central nitrogen atom, in addition to the nonbonding pair of electrons.

structures of the three most common nitrogen-containing functional groups. Notice a few things about them. First, each carbon atom in each structure makes four bonds to other atoms, and each nitrogen atom makes three. As we know from previous chapters, this will always be the case in any electrically neutral organic molecule. Second, each nitrogen atom has one nonbonding pair of electrons. Recall from Section 3.6 that nitrogen has five valence electrons available for bonding and forms three bonds in any electrically neutral molecule. Because each bond it makes uses one electron, two must be left over, and they are indicated by a pair of dots, as shown in the figure.

Now let us look at the amine functional group. In Figure 14.2, the R groups (shown as R, R′, and R″) can be a hydrogen atom or any kind of hydrocarbon group. For example, the amine *ammonia* has R = R′ = R″ = H. That is, all of the R groups are hydrogen atoms (Figure 14.3*a*). For the amine *butylamine*, one of the R groups is a butyl group and the other two are hydrogen atoms (Figure 14.3*b*). Butylamine is an example of a **primary amine**, defined as an amine in which one of the three hydrogen atoms of ammonia has been replaced by a hydrocarbon group.

Now consider the molecule in Figure 14.3*c*. Can you predict its name? If you guessed *ethylmethylamine*, you are right. This molecule, in which two of the three H atoms of ammonia have been replaced by hydrocarbon groups, is referred to as a **secondary amine**. In naming secondary amines, the hydrocarbon groups are alphabetized. Thus *ethyl* precedes *methyl* in the name (even though the ethyl group contains more carbon atoms). **Tertiary amines**—those in which all three hydrogen atoms of ammonia have been replaced by hydrocarbon groups—are named the same way. For example, *ethylmethylpropylamine*, shown in Figures 14.3*d* and 14.4, has hydrocarbon groups containing two, one, and three carbon atoms, respectively. Note from Figure 14.4 that you can draw a tertiary amine with the alkyl groups and the nonbonding pair in any order around the central nitrogen atom. The same is true for primary and secondary amines, which means that you

RECURRING THEME IN CHEMISTRY The structure of carbon-containing substances is predictable because each carbon atom almost always makes four bonds to other atoms.

▼ **FIGURE 14.4 Three versions of one molecule.** All positions around a central nitrogen atom are equivalent in amines.

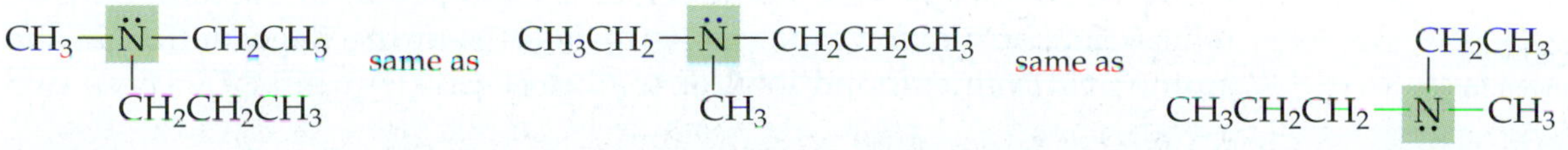

$$CH_3\!-\!\overset{\cdot\cdot}{N}\!-\!CH_2CH_3$$
$$\underset{\displaystyle CH_2CH_2CH_3}{\vert}$$

same as

$$CH_3CH_2\!-\!\overset{\cdot\cdot}{N}\!-\!CH_2CH_2CH_3$$
$$\underset{\displaystyle CH_3}{\vert}$$

same as

$$\overset{\displaystyle CH_2CH_3}{\vert}$$
$$CH_3CH_2CH_2\!-\!\underset{\cdot\cdot}{N}\!-\!CH_3$$

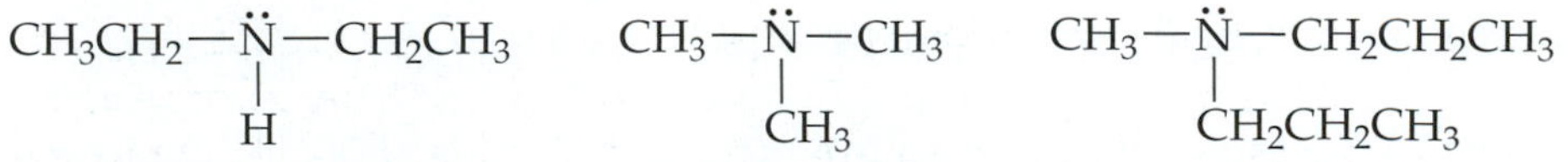

$$CH_3CH_2-\overset{..}{N}-CH_2CH_3 \qquad CH_3-\overset{..}{N}-CH_3 \qquad CH_3-\overset{..}{N}-CH_2CH_2CH_3$$

$$\overset{|}{H} \qquad\qquad \overset{|}{CH_3} \qquad\qquad \overset{|}{CH_2CH_2CH_3}$$

Diethylamine   Trimethylamine   Methyldipropylamine

▲ FIGURE 14.5   Naming amines.   The prefixes *di-* and *tri-* tell you when two or three of the R groups in an amine are identical. Note from the molecule on the right that the prefix does not count when you are placing the R groups in alphabetical order in the amine name.

are free to switch around the H atoms, alkyl groups, and nonbonding electrons in Figure 14.3 any way you please.

In the amines we just looked at, each R group is different from every other R group in the amine, but this is not always the case. How do we name an amine that contains either two or three identical hydrocarbon groups? As we have done with the names of other organic molecules, we use the Greek prefixes *di-* and *tri-* to indicate two and three like alkyl groups, respectively. Now try drawing the structures for *diethylamine, trimethylamine,* and *methyldipropylamine.* The structures of these three molecules are shown in Figure 14.5. Notice from the name *methyldipropylamine* that the *di-* is not used in alphabetization. Thus, the name *dipropylmethylamine,* which is what you would get if your thinking is "d comes before m," is incorrect.

QUESTION 14.1   Indicate whether each amine in Figure 14.5 is primary, secondary, or tertiary.

ANSWER 14.1   Diethylamine is a secondary amine, and trimethylamine and methyldipropylamine are tertiary amines.

QUESTION 14.2   Draw a line structure for each amine in Figure 14.5.

ANSWER 14.2

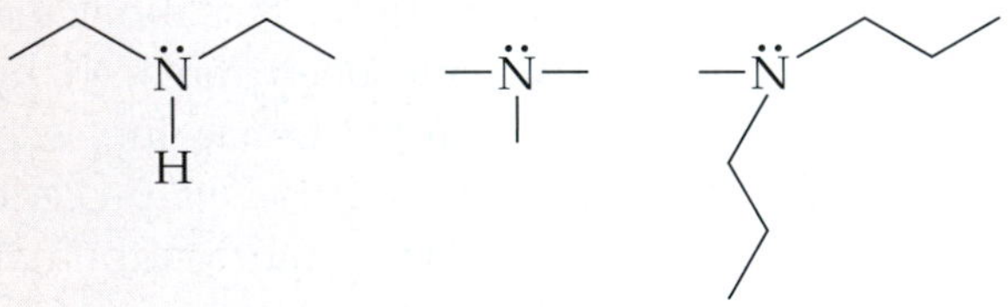

Notice that all groups attached to the nitrogen atom in an amine should be indicated. Thus, the hydrogen atom attached to the nitrogen atom in diethylamine is always drawn.

▼ FIGURE 14.6   **Malodorous molecules.**   Putrescine and cadaverine are both examples of amines that have an unpleasant odor. The boiling point range of cadaverine is higher than that of putrescine because cadaverine has the greater molar mass.

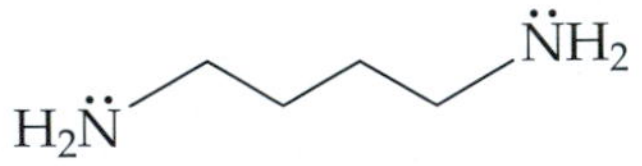

Putrescine
Boiling point range 158–160 °C

Cadaverine
Boiling point range 178–180 °C

So that we can better understand how amines can be illicit drugs, let us examine the properties of some other interesting amines. The structures of putrescine and cadaverine, two notoriously stinky amines, are shown in Figure 14.6. The smell of rotting meat comes from putrescine, and the smell of decaying human cadavers comes from cadaverine. These two molecules are both formed by the action of bacteria on proteins. Cadaverine, for example, is formed from the degradation of the amino acid lysine, found in almost all proteins (Figure 14.7). These two amines are identical except for one —CH$_2$— group, but Figure 14.6 tells us that the boiling points differ by about 20 C°. This should come as no surprise because we know that larger molecules boil at higher temperatures.

Both putrescine and cadaverine are primary amines. Now let us compare the boiling points of one primary, one secondary, and one tertiary amine, shown in Figure 14.8a. All three compounds have the same molecular formula (C$_3$H$_9$N) and therefore the same molar mass (59 g/mol). However, the boiling points of these three amines are dramatically dissimilar. The highest boiling point of the three is that of propylamine, a primary amine, followed by that of ethylmethylamine, a secondary amine, and finally that of trimethylamine, a tertiary amine. We can understand this trend by recalling that molecules containing an —OH or —NH group can form hydrogen bonds with an O or N atom on a different molecule. Figure 14.8b shows propylamine molecules forming hydrogen bonds to one another.

The formation of hydrogen bonds between molecules makes those molecules less likely to leave the liquid phase to become a gas. Because it is a primary amine, a propylamine molecule is able to make hydrogen bonds to other propylamine molecules through its two hydrogen atoms and its nitrogen atom. This makes its boiling point comparatively high. Ethylmethylamine, a secondary amine, can also make hydrogen bonds with other amine molecules, as shown in Figure 14.8c. In this case, however, the ethylmethylamine molecule has only one hydrogen and its nitrogen available for hydrogen bonding, which means that the intermolecular interactions are weaker than in propylamine and the boiling point is lower. The tertiary amine in our example here, trimethylamine, does not meet the criterion for hydrogen bonding, because it does not possess a nitrogen atom bound to a hydrogen atom. Thus, since there is no hydrogen bonding between molecules of a tertiary amine, their boiling points are comparatively low. This trend exists for any series of primary, secondary, and tertiary amine having the same molecular formula.

Protein

↓ Breakdown

Amino acids, including

Lysine

↓ Breakdown

$H_2\ddot{N}$ ⌂ $\ddot{N}H_2$ + CO_2

Cadaverine

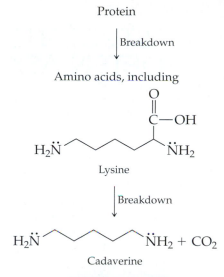

▲ FIGURE 14.7   **Formation of cadaverine.**   When proteins break down into their component amino acids, the amino acid lysine can break down further to form the very smelly molecule cadaverine.

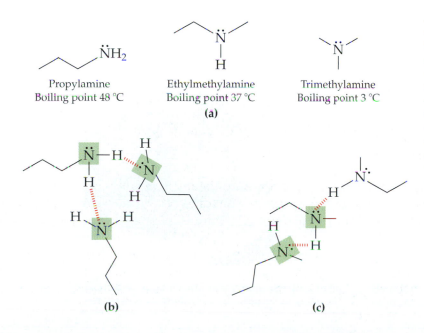

Propylamine
Boiling point 48 °C

Ethylmethylamine
Boiling point 37 °C

Trimethylamine
Boiling point 3 °C

(a)

(b)

(c)

◄ FIGURE 14.8   **Boiling-point trends in amines.**   **(a)** Primary amines have a greater capacity for hydrogen bonding than secondary amines, and this makes their boiling points higher. Tertiary amines cannot make any hydrogen bonds and have low boiling points as a result. **(b)** and **(c)** Primary and secondary amines make hydrogen bonds both with other amine molecules. The hydrogen bonds are shown in red.

STUDENTS OFTEN ASK

What Causes Bad Breath?

Bad breath is caused by the presence of anaerobic bacteria in your mouth. The adjective *anaerobic* means that these microbes must live in an environment devoid of oxygen. In the mouth, an oxygen-deficient environment can be created by the buildup of plaque on the teeth, and whenever a layer of plaque reaches a thickness of about 1 to 2 mm, it is possible for anaerobic bacteria to grow within it. This is why your dentist urges you to keep your mouth free of plaque and to use antibacterial mouthwash.

Anaerobic bacteria produce offensive-smelling molecules as waste products, and these are the odors that we associate with bad breath. Two of these waste products are the amines putrescine and cadaverine, the stinky molecules in Figure 14.6. Amines are not the only culprits in bad breath, however, as Figure 14.9 shows. The four molecules whose structures are depicted here are also produced by anaerobic bacteria in the mouth, in addition to being produced by the sources indicated in the figure.

CH_3-S-CH_3

Dimethyl sulfide, also associated with the odor of boiled cabbage

Isovaleric acid, also associated with the odor of feet at the end of an active day

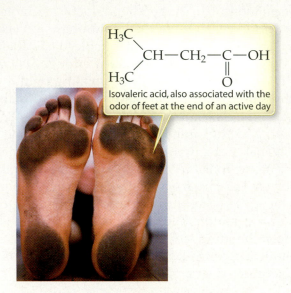

H_2S

Hydrogen sulfide, also associated with the odor of rotten eggs

Skatole, also associated with the odor of cooked beets

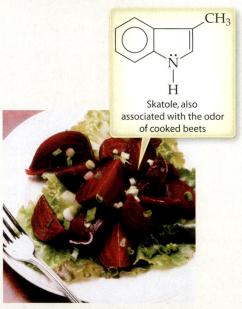

▲ **FIGURE 14.9 Bad-breath molecules.**   Boiled cabbage, smelly feet, rotten eggs, and cooked beets all have distinctive odors that arise from the same molecules that contribute to bad breath.

QUESTION 14.3 For each pair of amines, predict which of the two has the lower boiling point:

Amine 1
Amine 2

(a)

Amine 3
Amine 4

(b)

ANSWER 14.3 (*a*) The secondary amine, amine 2, has a lower boiling point than the primary amine, amine 1. (*b*) The tertiary amine, amine 4, has a lower boiling point than the secondary amine, amine 3.

QUESTION 14.4 Provide a name and molecular formula for each molecule shown in Question 14.3.

ANSWER 14.4 Amine 1: hexylamine, $C_6H_{15}N$; amine 2: dipropylamine, $C_6H_{15}N$; amine 3: butylmethylamine, $C_5H_{13}N$; amine 4: diethylmethylamine, $C_5H_{13}N$.

14.2 | "Speed"

To see how mind-altering drugs can have a striking effect on the brain, we must return to the chemistry that takes place in the spaces between neurons, the nerve cells of our bodies. We saw in Section 13.1 how opiates can bind to opiate receptors in the brain and produce euphoria. In the unaddicted brain, a nerve impulse travels to the end of one neuron, and its arrival there prompts the release of neurotransmitter molecules into the synaptic cleft that separates one neuron from another (Figure 14.10). The neurotransmitter

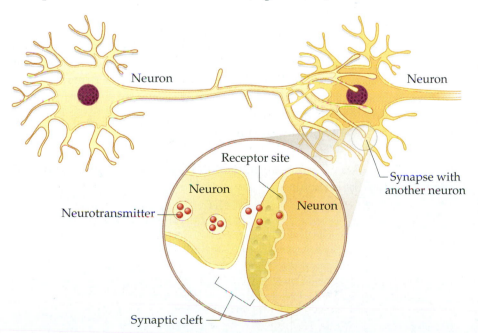

◄ **FIGURE 14.10 Neurotransmission in the undrugged brain.**
Neurotransmitter molecules are released from one neuron, travel across the synaptic cleft, and bind to a receptor on another neuron. Drugs that affect behavior often disrupt the natural levels of neurotransmitters in the brain.

MOLECULE
Serotonin

MOLECULE
Dopamine

TABLE 14.1   Three Neurotransmitters Found in Humans

Neurotransmitter	Structure	Normal physiological effects
Serotonin		Involved with learning, sleep, mood control
Norepinephrine		Increases heart rate and blood pressure in response to stress
Dopamine		Involved with emotional response and ability to experience pleasure and pain

diffuses across the cleft and binds to a receptor on the next neuron in the sequence. This binding creates a new nerve impulse in that neuron, and the process continues from neuron to neuron. The neurotransmitter that was released into the synaptic cleft is then broken down by an enzyme that is specific for that molecule. Opiates and other drugs that affect the brain cause an imbalance in the natural neurotransmitter levels in the synaptic cleft.

Now we are prepared to understand why illicit drugs often contain amine functional groups. All neurotransmitters are small organic molecules, and all the major human neurotransmitters are amines. Each neurotransmitter is associated with a specific receptor and binds only to that receptor. Each neurotransmitter is also associated with certain striking physiological effects when it is functioning normally and when that normal function is disrupted. Table 14.1 shows the structures of the three main neurotransmitters in the human body and lists the physiological effects each causes in a brain that is functioning normally. Look closely, and you will see that the three structures are similar. All three have a benzene ring on one end and an amine group on the other. All three contain a primary amine. All three contain at least one alcohol group—one in serotonin, two in dopamine, and three in norepinephrine. In fact, the structures of norepinephrine and dopamine differ by only one —OH group.

QUESTION 14.5 The neurotransmitter *epinephrine* (also known as *adrenaline*) has the structure

Epinephrine (adrenaline)

Both epinephrine and norepinephrine allow us to react to short-term stress by increasing our heart rate and blood pressure. (*a*) How does the structure of epinephrine differ from that of norepinephrine? (*b*) Which of these two molecules do you suppose has the stronger intermolecular interactions with other like molecules? Why?

ANSWER 14.5   (*a*) Epinephrine has a methyl group attached to what is the primary amine group in norepinephrine. (*b*) Norepinephrine, because it is a primary amine; the intermolecular interactions between primary amines are stronger than those between secondary amines.

Given the similarity in the three neurotransmitters shown in Table 14.1, how is it possible for receptors for one neurotransmitter to distinguish among molecules having nearly identical structures? As we have seen time and time again throughout this text,

the three-dimensional contour of a molecule often determines whether or not it will bind to another molecule.

In this instance, one molecule of the binding pair is either a neurotransmitter or a mind-altering drug and the other is a receptor, usually a protein, located on a neuron. The amine groups of drugs and neurotransmitters help to make these molecules recognizable to the receptor molecule.

Figure 14.11 shows some fragments of organic molecules. You can see from these drawings that the molecular geometry of a molecule depends on the atoms it contains and on how those atoms are bonded to their neighbors. (We first talked about molecular geometry in Section 3.8, and you may want to review the terminology used in Table 3.4, p. 115). An uncharged nitrogen atom always has one nonbonding electron pair and always make three bonds, and when those bonds are all single bonds, the molecular geometry is pyramidal (Figure 14.11*a*). The basic skeleton of most organic molecules is a carbon chain with each carbon atom attached to four other atoms, and so each carbon atom creates a local tetrahedral molecular geometry (Figure 14.11*b*). When a primary amine is inserted into a carbon chain, as in Figure 14.11*c*, there is a very subtle new feature on the surface—this section of the molecule now has a pyramidal geometry around the nitrogen atom.

In addition to determining molecular geometry, the types of functional groups in a molecule also determine how that molecule binds to another molecule. For example, the presence of any group that can participate in

---

▼ **FIGURE 14.11   Molecular geometry is dictated by individual atoms.**   The three-dimensional shape that exists around a nitrogen or carbon atom depends on the number of bonds the atom makes to neighboring atoms. **(a)** When an uncharged nitrogen atom bonds with three other atoms, the molecular geometry is pyramidal. **(b)** The molecular geometry of a saturated hydrocarbon chain is tetrahedral because each carbon atom bonds with four other atoms. **(c)** Inserting an amine group into the chain changes the local geometry from tetrahedral to pyramidal.

Pyramidal molecular geometry

(a)

Tetrahedral molecular geometry

(b)

Local pyramidal molecular geometry
because of inserted amine group

(c)

hydrogen bonding, such as —OH or —NH (Section 7.2), in a molecule allows the molecule to interact strongly with other molecules. Because uncharged primary and secondary amines have the capacity to hydrogen bond, this affords another means to an exact fit with another molecule. Thus, the presence of amine groups gives organic molecules a characteristic shape and provides the molecules with hydrogen bonding contact sites.

For this reason, we find amine groups in molecules, such as neurotransmitters, that can bind with remarkable specificity to their mates. As we have noted, dopamine and norepinephrine differ by only one —OH group. Like amines, alcohol groups have the capacity for hydrogen bonding. The extra —OH in norepinephrine means that the hydrogen bonding pattern in this molecule differs from that in dopamine. This difference offers a sufficient basis for discriminating between the two molecules.

QUESTION 14.6 The drug ephedrine was, until recently, marketed as Ephedra, an appetite suppressant used for weight loss.

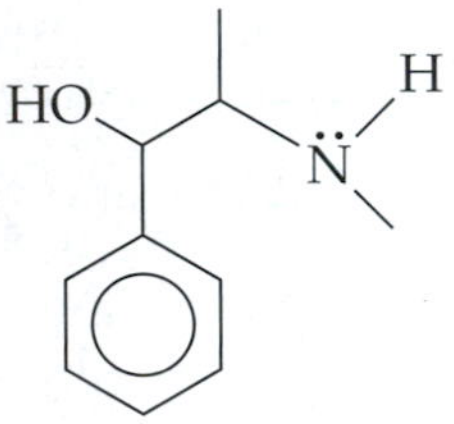

Ephedrine

A form of ephedrine can also be found in over-the-counter cold and flu medications. (*a*) How does the structure of ephedrine compare with that of epinephrine? (The structure of epinephrine is shown in Question 14.5, page 558) (*b*) Is ephedrine a primary, secondary, or tertiary amine?

ANSWER 14.6 (*a*) The two structures differ in two ways: (1) ephedrine has an additional methyl group on the carbon atom adjacent to the nitrogen; (2) epinephrine has two —OH groups attached to its benzene ring, and ephedrine has none. (*b*) Secondary.

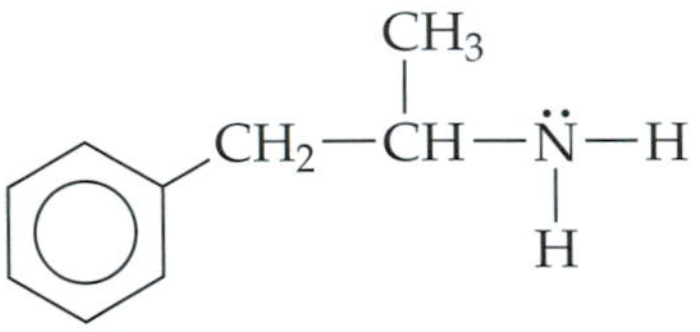

Amphetamine, a primary amine

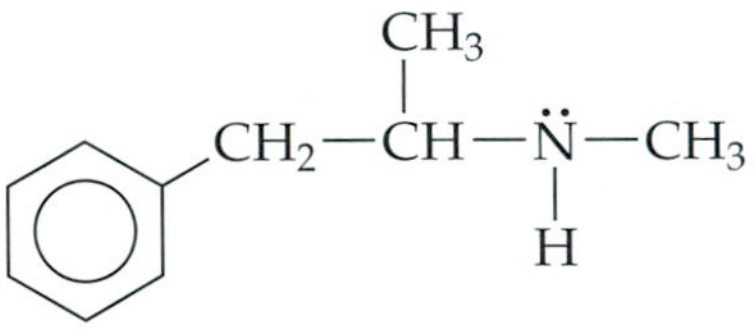

Methamphetamine, a secondary amine

▲ **FIGURE 14.12 Two amphetamines.** The molecules amphetamine and methamphetamine differ by one methyl group on the N atom. Both these molecules are members of the class of drugs known as amphetamines.

MOLECULE
Amphetamine

MOLECULE
Methamphetamine

We already know a formal naming system for amines. Many molecules, however—especially drug molecules—also have common names, and we shall use these common names in this text. *Dopamine* is an example of a common name for an amine.

Now let us return to Tommy Simpson and take a look at the drugs he took on that summer day in 1967: amphetamine and methamphetamine, the latter commonly referred to as "speed." Amphetamine is composed of a benzene ring attached to a chain of three carbon atoms, with the middle carbon attached to a primary amine group (Figure 14.12). Methamphetamine, as its name suggests, is a methylated version of the amphetamine molecule. The methyl group is attached to the nitrogen atom. Thus, methamphetamine is a secondary amine.

Dopamine (Table 14.1) is the neurotransmitter associated with a variety of emotional responses, and it controls our ability to experience pain and pleasure. Amphetamines and cocaine both cause an increase in the amount of dopamine in the synaptic cleft, but they do this via different mechanisms. Amphetamines increase the amount of dopamine released

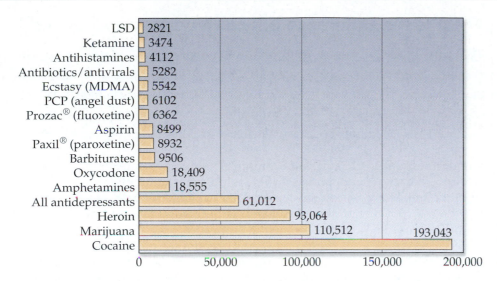

◀FIGURE 14.13 **Illicit-drug use that results in ER visits.** These data for emergency room visits are for 2001.

into the synaptic cleft, whereas cocaine blocks the uptake of dopamine from the cleft. Because both drugs result in higher-than-normal levels of dopamine in the synaptic cleft, both cause an overstimulation of the neuron. The result is a feeling of exhilaration, increased energy, and confidence.

Why wouldn't everyone want to feel exhilarated, more energetic, and confident? The problem, of course, is that there is a serious tradeoff involved. Amphetamines and cocaine also cause extreme nervousness, an accelerated heartbeat, hyperexcitability, and insomnia. Consequently, there are serious long-term consequences to amphetamine and cocaine use. Prolonged use of these drugs puts enormous stress on the heart muscle, making cardiac arrest possible. Fatal seizures caused by respiratory arrest also result from excessive cocaine use. Figure 14.13 shows data on drug-related visits to emergency rooms across the United States for 2001. According to these data, cocaine resulted in nearly 200,000 trips to the ER that year, accounting for almost twice the number of trips attributed to the second most troublesome drug, marijuana.

Amphetamines are less likely to send you to the emergency room, according to the data in Figure 14.13. However, as we saw in the case of Tommy Simpson, amphetamine use can have serious consequences. Furthermore, more than any other group of drugs, amphetamines are associated with antisocial and violent behavior. Adolf Hitler was a rabid amphetamine junkie, receiving injections daily during the last three years of his life. People who are able to tolerate prolonged amphetamine use often end up in amphetamine psychosis, an advanced condition marked by paranoid delusions and auditory hallucinations.

Another problem with cocaine and amphetamine use, and with drugs such as heroin and morphine, is that they are physiologically addictive. But how can a drug cause the body to crave its continued use? This situation arises when the natural balance of a neurotransmitter is knocked out of kilter. For example, cocaine and amphetamines both increase the levels of dopamine in the synaptic cleft, as noted earlier. When this happens repeatedly, the body adjusts to increased dopamine concentrations in an effort to counteract the increased activity of these neurons. It does this by *decreasing* the number of available dopamine

The term *amphetamine* refers both to the specific molecule shown in Figure 14.12 and to an entire class of drugs. Thus, you may hear someone refer to methamphetamine as *an amphetamine*.

What Is the Difference between Ecstasy and Speed?

STUDENTS OFTEN ASK

The street drug *ecstasy* is formally known as 3,4-methylenedioxy-N-methylamphetamine, abbreviated MDMA. As its name implies, ecstasy is an amphetamine. Its structure is identical to that of methamphetamine, with the addition of a —O—CH₂—O— group attached to the benzene ring, as shown in Figure 14.14a. Although ecstasy has many of the same physiological effects as other amphetamines, such as increased alertness and heart rate, the addition of this new group gives the drug many effects that are not typical of other amphetamines.

Ecstasy is classified both as a stimulant (because it is an amphetamine) and as a hallucinogen. The drug causes feelings of profound empathy, enhances sound and color sensations, and is an "ego softener," qualities sometimes associated with hallucinogens such as lysergic acid diethylamide (LSD, Figure 14.14b). A look at the structure of LSD shows that it bears some resemblance to serotonin (Figure 14.14c), a neurotransmitter associated with mood, among other things. LSD works by binding directly to serotonin receptors in the brain, and studies show that ecstasy also affects the natural balance of serotonin. Thus, ecstasy alters not only brain dopamine levels (as do other amphetamines) but also serotonin levels. This explains its dual effects. Speed, on the other hand, is pure stimulant and exhibits only amphetaminelike properties.

Ecstasy does not account for many drug-related visits to the emergency room (Figure 14.13). However, its long-term effects may be more severe than those of other drugs. For example, it has been shown that long-term ecstasy use causes permanent damage to dopamine receptors, producing symptoms like those associated with Parkinson's disease. Prolonged use has also been linked to permanent memory loss as well as liver and kidney disorders.

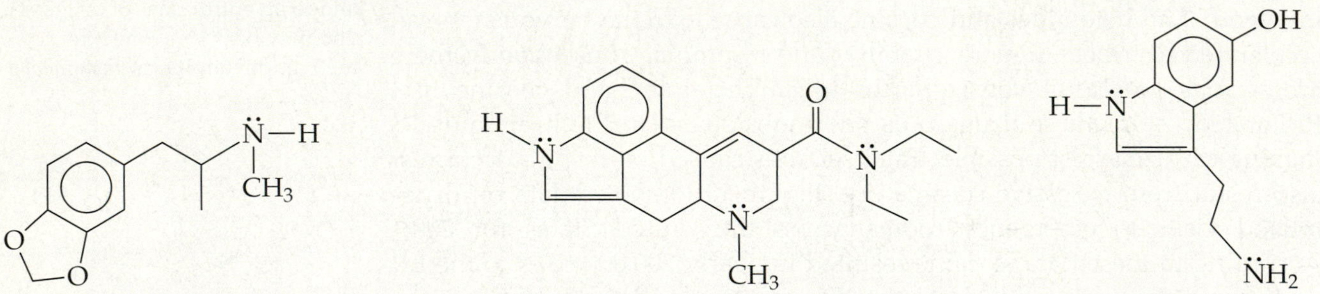

Ecstasy (MDMA), a hallucinogenic amphetamine
(a)

Lysergic acid diethylamide (LSD), a hallucinogen
(b)

Serotonin, a neurotransmitter
(c)

▲ FIGURE 14.14 Hallucinogens. **(a)** The structure of the drug ecstasy, which is classified as both an amphetamine and a hallucinogen. **(b)** The structure of the drug LSD, a hallucinogen. **(c)** In the body, the levels of the neurotransmitter serotonin are changed in the presence of hallucinogenic drugs.

receptors. As the body reduces the number of dopamine receptors, more and more of the drug is required to produce the high associated with elevated levels of dopamine. If the supply of the drug is cut off, the reduced number of dopamine receptors causes the addict to "crash," and only more drug will bring back the feelings of euphoria. It can take weeks or months of unpleasant rehabilitation to return dopamine receptors to their normal levels, and there are new data to suggest that, for some addicts, the number of receptors never returns to normal, even after extensive rehabilitation.

QUESTION 14.7 Point out all the organic functional groups in the three molecules in Figure 14.14, and indicate whether each amine you identify is primary, secondary, or tertiary.

ANSWER 14.7

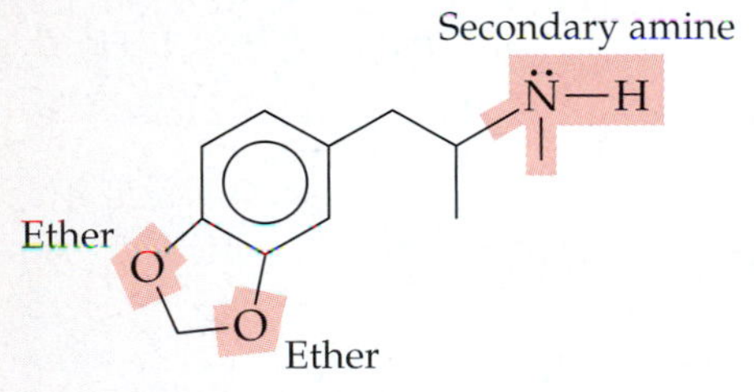

Secondary amine
N—H
Ether
O
Ether
Ecstasy

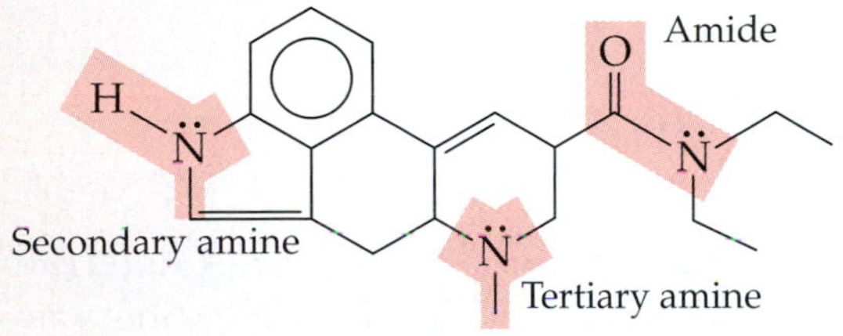

Amide
H
N
Secondary amine
N
Tertiary amine
LSD

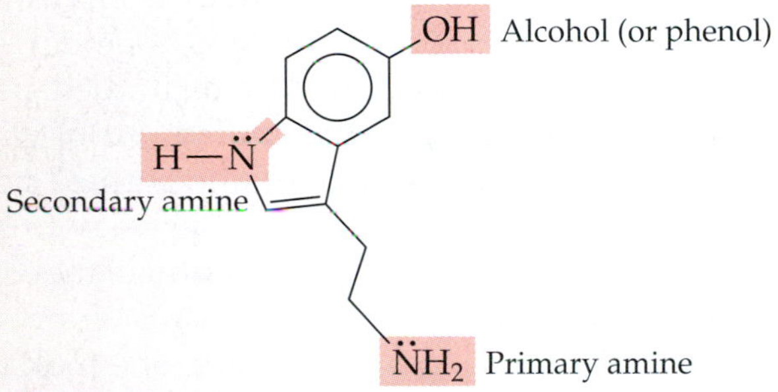

OH Alcohol (or phenol)
H—N
Secondary amine
NH₂ Primary amine
Serotonin

QUESTION 14.8 Nicotine, shown here, is a highly addictive molecule that disrupts normal levels of dopamine. (*a*) Redraw nicotine with a full structure, showing each atom and bond. (*b*) Write the molecular formula, and determine the molar mass, of nicotine. (*c*) If a piece of nicotine gum contains 0.0020 g of the drug, how many moles of nicotine does it contain?

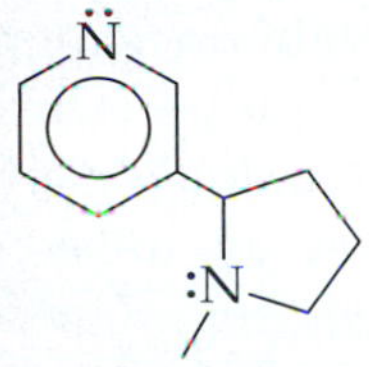

Nicotine

ANSWER 14.8 (*a*)

(*b*) $C_{10}H_{14}N_2$; 162.23 g/mol.

(*c*) $0.0020 \text{ g} \times \dfrac{1 \text{ mol}}{162.23 \text{ g}} = 0.000012$ mol nicotine

14.3 | Who's on Drugs?

Lindsay Earls of Tecumseh, Oklahoma, was determined to make the most of her high school years (Figure 14.15). She wanted to go to college—specifically, Dartmouth—and she loved to sing. So Lindsay worked hard to earn good grades, joined the Academic Team, sang in two choirs at her high school, and stayed away from drugs. Never having been in any trouble, Lindsay was surprised one day to learn that a mandatory drug test was required for her continued participation in some extracurricular activities. She was pulled out of class and tested three times by teachers. To Lindsay, something seemed horribly wrong. After all, the Fourth Amendment guarantees privacy to all Americans, even high school students, right? Having studied the Constitution in school, she knew that there must be probable cause for a search, including a search of her urine for illegal substances. She took the case to the American Civil Liberties Union, whose lawyer took it from Oklahoma to the U.S. Supreme Court. On June 27, 2002, the Court handed down its decision: random drug testing for students wanting to take part in extracurricular activities does not violate the Fourth Amendment because the testing is a reasonable step in schools' efforts to stem drug abuse. Even though she lost her case, Lindsay's stand against her school administration brought the issue of student privacy to the fore. She graduated from Dartmouth in 2005.

As Lindsay's case shows, the analysis of human body fluids for drugs is a complex issue. Beyond the issue of privacy, there are questions of the accuracy and specificity of the tests performed. How do we know that a test performed on our urine was done properly? Can we be sure that an innocuous substance did not cause a positive result? Can eating a poppy-seed muffin, which could contain trace quantities of opiates, cause you to test positive for heroin or morphine? During a time when drug testing is routinely required to get a job or play a professional sport, the tests must be accurate because the answers they give can change lives and end careers.

These issues make the design of tests for illicit drugs tricky. There are tests on the market designed to be used by untrained, nonmedical personnel,

▲ FIGURE 14.15 **Lindsay Earls.** While in high school, she took her school to the U.S. Supreme Court for performing random drug testing.

such as police officers and parents. These tests must be easy to use and cannot contain hazardous chemicals that need to be disposed of by special means. They must also be reasonably accurate. The tests used by police officers for the roadside evaluation of drugs must be accurate enough to create probable cause for arrest. In these cases, the tests usually tell you whether some illicit drug is present, but not how much is present or even which drug is involved.

Let us look at a few examples to see how drug tests work. The Marquis test is a classic, easy-to-use, portable test for a variety of drugs. It is based on a simple chemical reaction that produces a color change, and the color produced suggests which drug might be present. The test requires that a small sample of the drug be placed in a test tube and a solution containing formaldehyde added. Because so many illicit drugs contain amine functional groups, the Marquis test was designed to show a positive test in the presence of amines, as illustrated generically in Figure 14.16a. Figure 14.16b shows two examples of a Marquis test for specific drugs. In the top reaction, mescaline, a hallucinogen derived from the peyote cactus (*Lophophora williamsii*), reacts with formaldehyde to create a bright orange product. In the lower reaction, amphetamine reacts with formaldehyde to produce a yellow-orange product that then changes to brown. Other illicit drugs produce other colors—purple for morphine, for instance, and violet black for ecstasy.

▼ **FIGURE 14.16** The Marquis test for the qualitative detection of amines.
(a) The test involves a reaction between an amine and formaldehyde. The color of the product depends on the amine present. **(b)** When the amine being tested is the hallucinogen mescaline, the product is orange. When the amine is amphetamine, the product is initially yellow-orange and later brown.

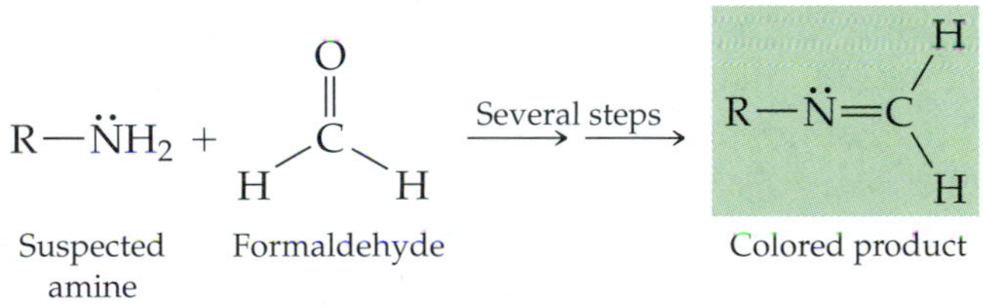

(a)

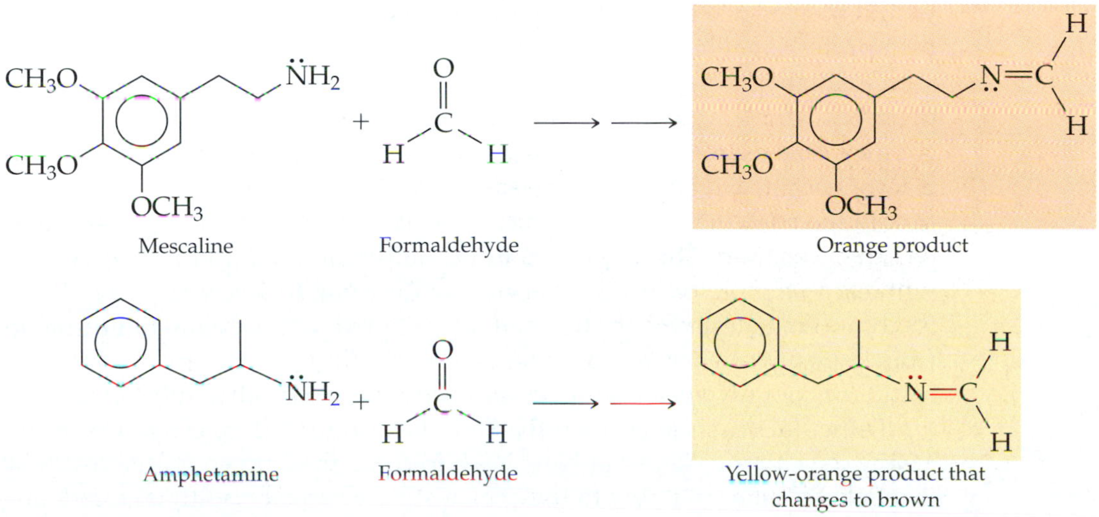

(b)

QUESTION 14.9 Phentermine is the "phen" part of the now-illegal appetite suppressant *phen-fen*.

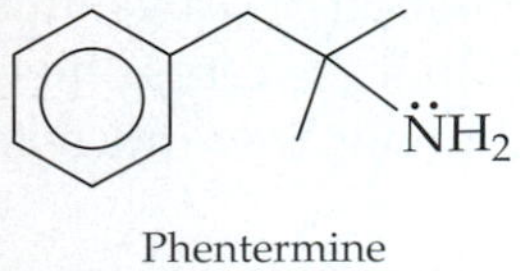

Phentermine

Like mescaline, phentermine gives an orange color with the Marquis test. Suggest a structure for the orange-colored product in the phentermine Marquis test.

ANSWER 14.9 Looking at the reactions shown in Figure 14.16 shows you how the formaldehyde adds to the amine portion of the drug molecule:

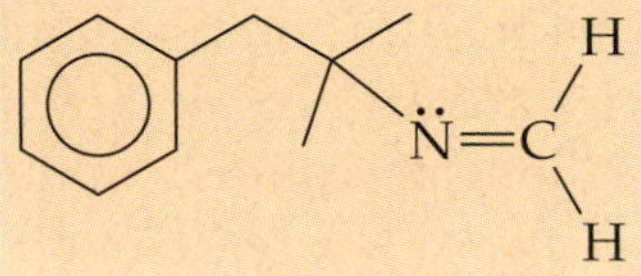

We refer to a test that *identifies* a compound as a **qualitative test** (because it tells us the "quality," or nature, of the compound). It may seem that the Marquis test is a good qualitative test for illicit drugs because it is easy to use, gives straightforward results, and can occasionally tell us which drug is present. However, since most amines give a positive result in the Marquis test, it is almost impossible to identify a drug by using it. For example, while the oft-abused drug morphine gives a purple color, so does codeine, a prescription drug most often used for legitimate medical purposes. Amphetamine and methamphetamine both give the same color result (yellow-orange changing to brown), and psilocybin, the major active component in hallucinogenic mushrooms, gives a product that is the same color (orange) as the product yielded when the test is run on mescaline. Thus, the Marquis test is nonspecific, because it cannot distinguish one drug from another, and it is only loosely qualitative, because certain colors suggest certain *classes* of amines rather than specific ones. Despite the test's limitations, though, it is still considered valid enough to raise probable cause in a roadside drug test. A suspect who tests positive in such a test would then be taken to a place where a more specific test can be performed by a professional laboratory.

Qualitative tests for illicit drugs have become more sophisticated as the drug problem has caused them to be more and more widely used in our country. One example is a response to the need for a more reliable portable test for cocaine, an amine that does not give a clear result with the Marquis test. In the past, the Scott test, which gives a color change in the presence of cocaine, was used. This three-step test is notoriously complicated, however, with each step providing an opportunity for error. In a recent study, out of 55 compounds tested with the Scott test, 22 gave a result indistinguishable from a positive result for cocaine. Moreover, the test can give a negative result for cocaine when that drug is mixed with some other substances.

Difficulties associated with the Scott test prompted investigators at the Marcy Psychiatric Center in New York to come up with a cocaine test that is quick, specific, and easy to use. The test they came up with takes advantage of a feature in the structure of the cocaine molecule: a benzene ring that

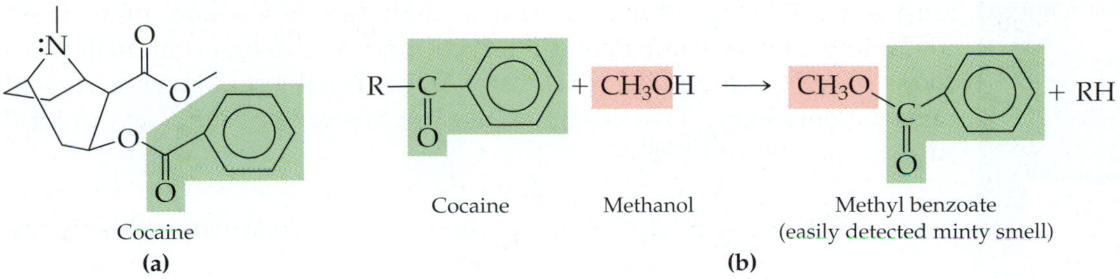

Cocaine

(a)

Cocaine   Methanol   Methyl benzoate
                                       (easily detected minty smell)

(b)

▲ **FIGURE 14.17   The Marcy test for cocaine.   (a)** The cocaine molecule contains a benzene ring attached to an ester group. The Marcy test for cocaine is based on this feature of the molecule. **(b)** In the Marcy reaction, cocaine and methanol react to produce methyl benzoate, which has a distinctively minty smell.

is adjacent to an ester group. Cocaine is the only street drug that has this feature, highlighted green in Figure 14.17a, and it is this part of the molecule that takes part in the reaction that is the basis of the Marcy test.

QUESTION 14.10   Test your structure-drawing mettle by drawing the full structure of the cocaine molecule. Use Figure 14.17a as a guide.

ANSWER 14.10   In this rendering, the multiple-ring region is highlighted for clarity:

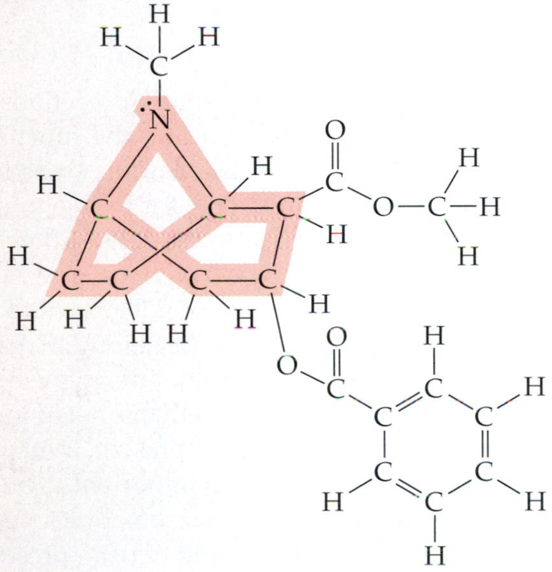

The Marcy test is based on a very simple reaction between the green-highlighted portion of the cocaine molecule in Figure 14.17a and methanol. The reaction is shown in Figure 14.17b, in which the unhighlighted portion of the cocaine molecule is represented by R for simplicity. In the presence of methanol (in addition to a few other reagents that need not concern us here), the cocaine molecule undergoes a reaction in which the red-highlighted group of methanol displaces the R group of cocaine. The product of this reaction is the ester *methyl benzoate*. This test works well because methyl benzoate has a very distinctive, minty smell. Of more than 100 street drugs tested, only cocaine produces a minty smell with this test, because it is the only street drug that produces the methyl benzoate molecule. Thus, a police officer who has stopped someone and found some powder

▲ **FIGURE 14.18 Testing for drugs at home.** Drug tests for home use are available at your local drugstore.

suspected of being cocaine can take a small sample and add a few drops of a methanol-containing solution. If cocaine is present, an unmistakable odor of mint is produced. This test is now available to law enforcement agencies and in the form of a home-use test that can be found at your local drugstore (Figure 14.18).

QUESTION 14.11 Using the reaction that takes place in the Marcy test as a guide, predict the products of this reaction:

$$CH_3CH_2CH_2OC-\bigcirc + CH_3OH \longrightarrow ?$$

Would any of the products of the reaction have a distinctive odor?

ANSWER 14.11 One product is methyl benzoate, which produces a minty odor:

$$CH_3CH_2CH_2OC-\bigcirc + CH_3OH \longrightarrow CH_3OC-\bigcirc + CH_3CH_2CH_2OH$$

Methyl benzoate

14.4 | The State of the Art

Qualitative tests for drugs are based on specific reactions between small organic molecules, and so the tests depend on the functional groups the drug molecules contain. We saw that the Marquis test identifies amines, which are present in most illicit drugs. The test is able to distinguish only the broadest classes of amine, though, and the color differences that distinguish different compounds are not always obvious. The Marcy test is specific for cocaine because it relies on a structural feature that is unique to the cocaine molecule. Although these tests may be helpful in demonstrating the presence of some illegal substance, they cannot distinguish between highly similar drugs, such as amphetamine and methamphetamine, and they cannot tell us *how much* of the compound is present. With the importance placed on drug tests in today's society, it is important to be able to identify a drug beyond a reasonable doubt. It is also sometimes necessary to use a **quantitative test**, one that tells us the *amount* of drug present.

State-of-the-art methods are available to address this need. Let us go back to a class of drugs we first learned about in Chapter 13: opiates. The structures of three common opiates are shown in Figure 14.19. The way this illustration is drawn—with one molecule containing R and R' groups

MOLECULE
Morphine

▶ **FIGURE 14.19 The opiate family of molecules.** The drugs in the opiate family all have the molecular structure shown here. Modifications indicated for the groups labeled R and R' give the molecules heroin, morphine, and codeine.

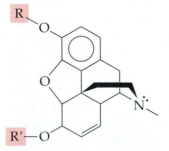

Called HEROIN when R and R' are $-\overset{\overset{\displaystyle O}{\|}}{C}-CH_3$

Called MORPHINE when R and R' are $-H$

Called CODEINE when R' is $-H$ and R is $-CH_3$

accompanied by a list telling us what R and R' represent in each molecule—emphasizes the similarities in the three. All three have a tertiary amine group, an ether group, a benzene ring, and a fused hexane ring containing one double bond.

These molecules differ from one another only in the identity of the groups R and R'. In fact, codeine and morphine differ only by one methyl group, which means that their molecular masses are 299.36 g/mol and 285.33 g/mol, respectively, a difference of less than 5 percent. It can be especially difficult to distinguish between big organic molecules like these, because their differences are dwarfed by the size of the identical parts of the molecules. For instance, when a very sensitive detection method is required, the level of discrimination must be extraordinarily high. Clearly, tests based on simple chemical reactions, like the Marquis test for cocaine, will not pass muster. The Marcy test is specific, but it works well only because there are no other compounds similar to cocaine that must be distinguished from it. What is needed to identify most drugs is a versatile, highly specific method for identifying *one single molecule*. As we often do, we can turn to Mother Nature for a clue on how to proceed.

QUESTION 14.12 *Oxycodone* (also known as oxycontin and Percocet®) is a highly effective pain reliever that is used by people suffering from the intense pain of a disease or an injury and abused by people seeking the high of an opiate in a prescription drug. How does the structure of oxycodone differ from that of codeine?

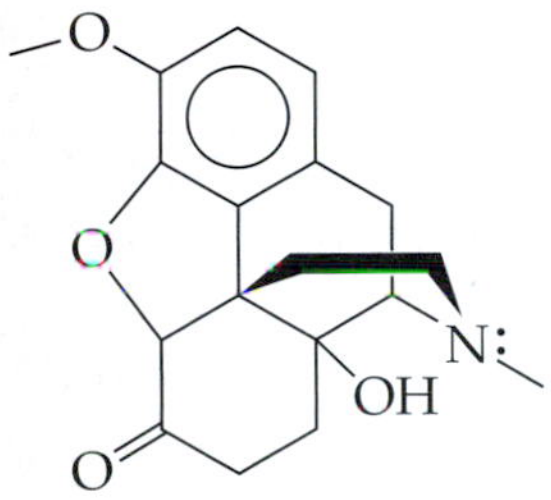

Oxycodone (Percoset®)

ANSWER 14.12 In the bottom six-membered ring, oxycodone has no double bond and has a carbonyl group in place of the alcohol found in codeine. Oxycodone also has an alcohol functional group attached to this ring where the ring is joined to the six-membered ring just above.

We already know that the interactions between biomolecules, such as enzymes and their substrates, are amazingly specific. The same level of discrimination is found with **antibodies** (also called *immunoglobulins*)—proteins created in the body to defend against specific invading molecules called **antigens**. Most antibodies are specific for only one antigen and are able to single out that antigen over other, nearly identical, ones.

A typical antibody is Y shaped, as shown in Figure 14.20, and is composed of two short protein chains (labeled 2 and 3 in the figure) and two long protein chains (1 and 4). Each antibody has two sites for antigen binding, as indicated in the figure. One binding site is located between chains 1 and 2, the other between chains 3 and 4. Our bodies create antibodies in response to the presence of an antigen, and the two binding sites

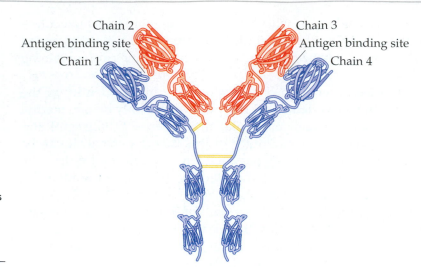

Chain 2
Antigen binding site
Chain 1

Chain 3
Antigen binding site
Chain 4

▶ **FIGURE 14.20 A typical antibody.**
The four protein chains of an antibody
molecule are arranged as two long chains
sandwiching two short chains, to form an
overall Y shape. The chains are linked
together with disulfide bonds (shown in
gold).

We introduced protein structure in
Section 11.3. You may wish to go back
and review that material.

on a given antibody are identical to each other and bind the same antigen
molecule.

The interaction of an antibody with its antigen is much like that of an
enzyme with its substrate: two contoured molecular surfaces come
together by virtue of their specific geometries and the chemical groups
on their surfaces.

The degree of specificity is extraordinarily high, and most antibodies will
recognize only one antigen.

You may be wondering why we are talking about antibodies in a discussion of illicit drugs. The answer is this: because it is possible for the body
to create antibodies to a specific antigen, it may also be possible to create
antibodies to a specific drug molecule. The way this is done is shown in
Figure 14.21. An antigen in the form of some drug, such as heroin, morphine,
or amphetamine, is injected into a mouse, which then produces an antibody
to the drug. Cells from the mouse's spleen, which contain instructions for
making the antibody, are removed from the animal and, in a test tube, fused
with cancerous cells to make *hybrid cells*. These hybrid cells have instructions
from both types of cell: the spleen cells give the hybrid cells instructions for
making antibodies to the desired antigen, and the cancerous cells give the
hybrid cells instructions for uncontrolled cell growth. When grown in mass
culture, the cells produce buckets of the antibody that binds the drug antigen originally injected into the mouse. As the advertisements in Figure 14.22
illustrate, it is possible to get antibodies custom made for many different
antigens (although some are trickier to make than others).

If you could obtain antibodies to our three opiates—morphine, heroin,
and codeine—you could use the antibodies to design a separate test for
each drug. One way these tests are configured is to attach the antibodies to
the walls of a series of wells in a *microtiter plate*—a disposable plastic block
with wells imbedded in it (Figure 14.23). Say you want to test for heroin in
urine samples taken from the members of a chess team. With heroin antibody molecules firmly attached to the sides of the wells, you simply add a
few drops of a urine sample to each well. If any sample contains heroin, the
heroin will bind to the antibodies on the well walls. A subsequent series of
reactions creates a color in the well indicating the presence of the drug in
that sample.

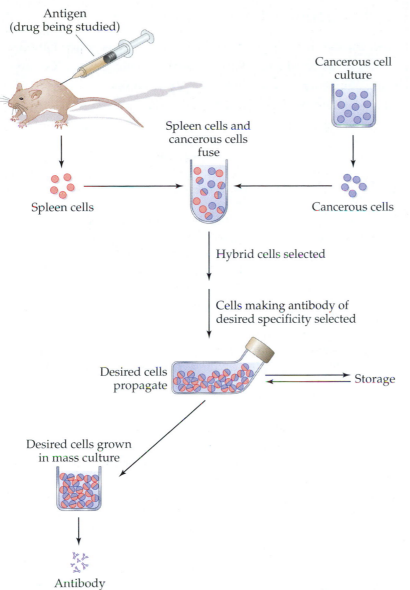

Antigen
(drug being studied)

Cancerous cell
culture

Spleen cells and
cancerous cells
fuse

Spleen cells

Cancerous cells

Hybrid cells selected

Cells making antibody of
desired specificity selected

Desired cells
propagate

Storage

Desired cells grown
in mass culture

Antibody

► FIGURE 14.21 **Creating antibodies
to a specific drug molecule.**
Antibodies can be produced for a specific
antigen by injecting the antigen into a
mouse, which produces antibodies
specific to that antigen. Cells from the
mouse are fused with cancerous cells to
produce hybrid cells that are then grown
in culture. These cells produce an
antibody to the antigen originally injected
into the mouse.

▼ FIGURE 14.22 **Designer
antibodies.** Biotechnology companies
can create antibodies to specific antigens.

groovybluegenes
try us on for size!

▼ FIGURE 14.23 **Quantitative analysis
of drug molecules.** A test for a specific
drug can be performed with a container
called a microtiter plate. An antibody to
the drug being tested for is attached to
the wall of each well of the plate. Then
test specimens are added to the wells.
Any specimen containing the drug causes
a telltale color change in the well.

CHEMISTRY AT THE CRIME SCENE

Drug Testing at Sporting Events

Since the day Tommy Simpson died in 1967, it has gotten harder and harder to get away with using performance-enhancement drugs. Certainly, no athlete using amphetamines would be allowed to compete today. Yet, in cycling, scandals related to drug use persist. Part of the problem is that drugs have always been a mainstay for professional cyclists in the grueling Tour de France. They rely on anti-inflammatory drugs to reduce joint swelling, and pain medications make three consecutive weeks on a bike possible.

How can a line be drawn separating drugs that relieve pain and drugs that enhance performance? Should *all* drugs be banned, so that riders must endure swollen joints and pain throughout a bicycle race? Or, as some people have suggested, should all drugs be allowed and the athlete who learns to use them to the greatest benefit wins?

According to many experts on professional cycling, many athletes are managing to stay one step ahead of the chemists developing the tests for different drugs. One example of a drug that has been hard to pin down is *erythropoietin,* known as *EPO* and originally created for the treatment of kidney disease. EPO increases the percentage of red blood cells (RBCs) in the body and allows an athlete to bind more oxygen molecules, which provides more energy to the muscles. A typical man's blood is between 40 and 50 percent RBCs by volume (36 to 44 percent for women). EPO can increase this value by 3 to 4 percent by stimulating the bone marrow to produce more RBCs.

In response to EPO use among professional cyclists, the cycling authorities decided to allow EPO, but to test athletes' RBC counts: any cyclist whose RBC level is greater than 50 percent is banned from competition. When the RBC level exceeds this limit, there is significant risk of blood clot formation, organ failure, and death. In extreme cases, EPO-enhanced blood can take on a thick, gelatinous consistency, and when this happens, it becomes impossible for the heart to pump the blood through the body. It is estimated that EPO can enhance an athlete's performance by as much as 15 percent. Since it arrived on the scene in the late 1980s, however, the drug has killed more than 26 cyclists and many athletes in other endurance sports.

In a hard stance against performance-enhancement drugs, the International Olympic Committee (IOC) added EPO to its "banned" list in 2000 for the Sydney games. Don Caitlin, head of the IOC-certified laboratory at UCLA, believes, "There are really two sports: with and without EPO....It makes a big difference...and if athletes find a drug and they think we can't detect it, they'll take it." At the 2002 games in Salt Lake City, athletes were screened for more than 400 substances on the IOC's banned-substances list.

On average, about 1 to 2 percent of athletes test positive for drugs at the Winter Games, and more than double that percentage test positive at the Summer Games, which have a greater number of endurance sports. As each Olympic season approaches, we are promised that more and more of these abuses will be uncovered. At the Sydney games, authorities announced that we would be watching games "cleaner than a ping-pong match in a convent." At the 2004 Summer Games in Athens, more than 3000 drug tests were performed, a 25 percent increase over the Sydney games. Also in Athens, for the first time athlete specimens collected during the games were frozen for possible future testing when more advanced tests are available.

The 2005 congressional hearings on performance-enhancement drugs in the world of professional baseball focused on abuse of **anabolic steroids**—muscle-building drugs popular with strength athletes. For many of us, the mention of steroids these days brings to mind outfielders with tree-trunk arms. Steroid drug molecules are more accurately known as *anabolic androgenic steroids*—anabolic because they build up muscle, androgenic because they induce male characteristics, and steroids because they mimic and are similar in structure to testosterone, the male sex hormone. (It is the *androgenic* and *steroid* parts of the name that were responsible, in Olympics past, for women swimmers with facial hair and deep voices.)

The newest steroid to hit the athletic black market is *tetrahydrogestrinone* (THG). Figure 14.24 compares the structure of this molecule with that of another anabolic steroid *trenbolone.* The structure of THG is clearly similar to that of trenbolone, a drug injected into cattle to increase muscle mass. THG is difficult to defect because it tends to decompose during the testing procedure. Despite problems like this, the inexorable IOC had several surprise drug tests lying in wait for athletes who went to Athens in 2004, including a new one for THG.

Steroids are used not only by professional athletes, but also by everyday folks who want to lose weight, build muscle, and look great. Ironically, though, men who use steroids for prolonged periods can develop breasts, lose their hair, and become infertile. Now, that's sexy! Some of the long-term effects of steroid use are extremely serious for both men and women, as Figure 14.25 shows.

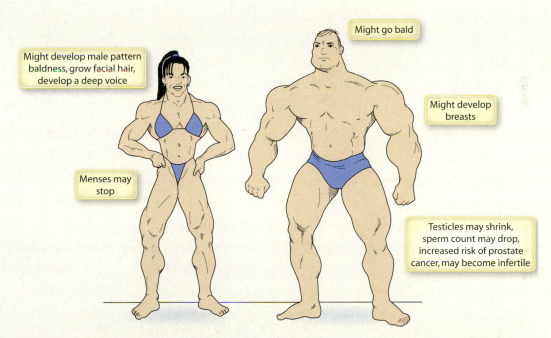

Trenbolone

Testosterone

Tetrahydrogestrinone (THG)

◄ **FIGURE 14.24   New drugs, new tests.**   The illegal steroid tetrahydrogestrinone, favored by athletes for building body mass, has a structure similar to that of trenbolone, a drug used to increase muscle mass in cattle. Both of these molecules have structures similar to the steroid testosterone. Sports-monitoring organizations constantly have to come up with new tests for policing steroid abuse in the sports world.

Might develop male pattern baldness, grow facial hair, develop a deep voice

Menses may stop

Might go bald

Might develop breasts

Testicles may shrink, sperm count may drop, increased risk of prostate cancer, may become infertile

▲ **FIGURE 14.25   Potential drawbacks of steroid use.**   The physiological problems associated with anabolic androgenic steroid use are many.

A standard microtiter plate has 96 wells, and so 96 samples can be run simultaneously. In addition, the test can be quantitative if a *microtiter plate reader* is used. This instrument measures the intensity of color in each well, and that information can be used to calculate the amount of drug the sample contains. Because they are so sensitive and so specific, antibody tests are used widely in drug testing when accuracy is paramount.

QUESTION 14.13 Which of these molecules most resembles testosterone?

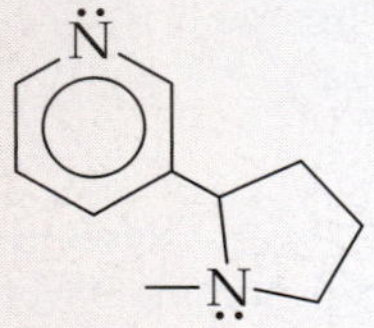

Nicotine

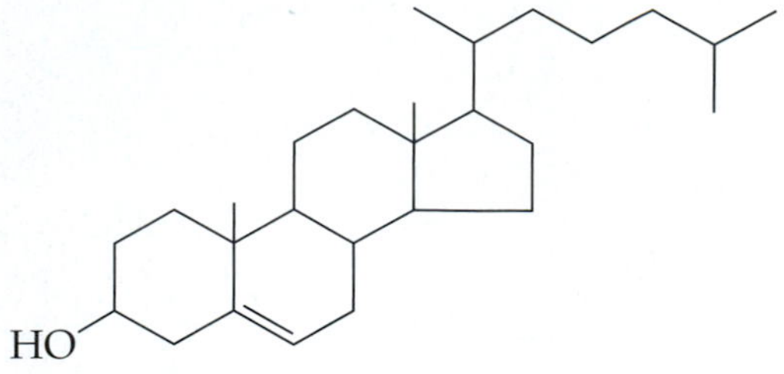

Cholesterol

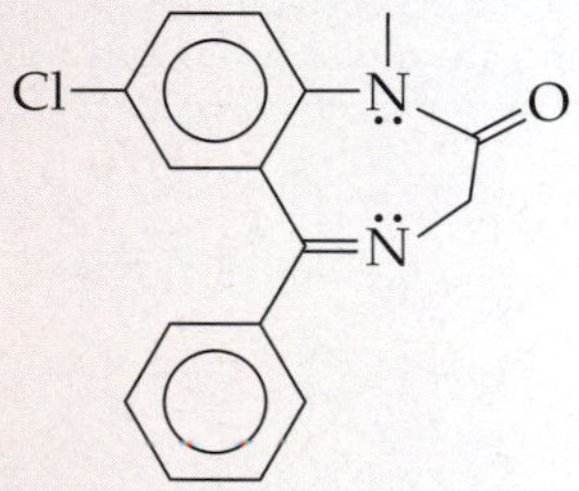

Valium®

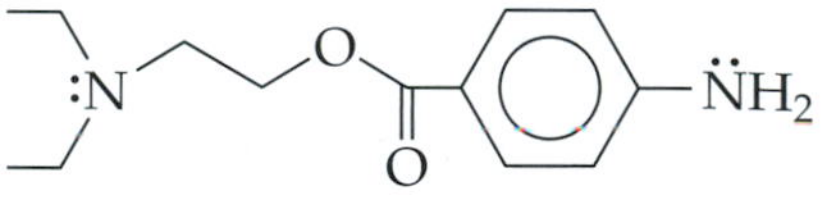

Novocaine

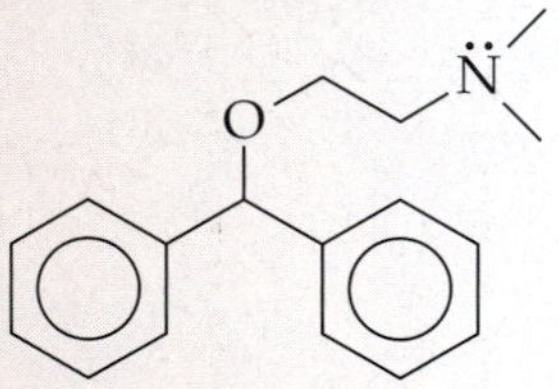

Diphenhydramine (Benadryl®)

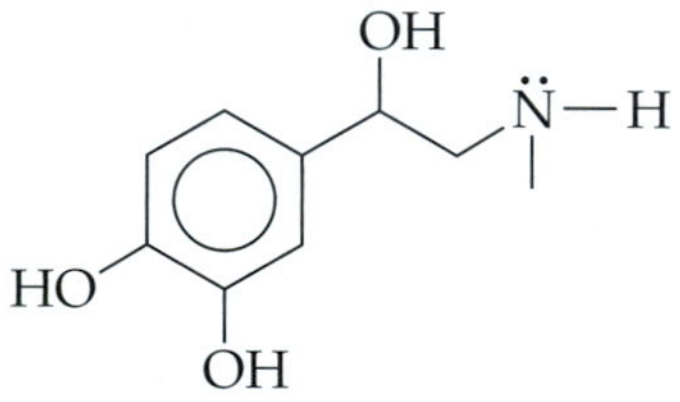

Adrenaline

ANSWER 14.13 The cholesterol molecule has the same framework as testosterone and is a steroid molecule.

QUESTION 14.14 Point out all organic functional groups in each molecule in Question 14.13. Indicate whether each amine group is primary, secondary, or tertiary.

ANSWER 14.14 Nitrogen-containing functional groups are indicated in green and all others are indicated in red.

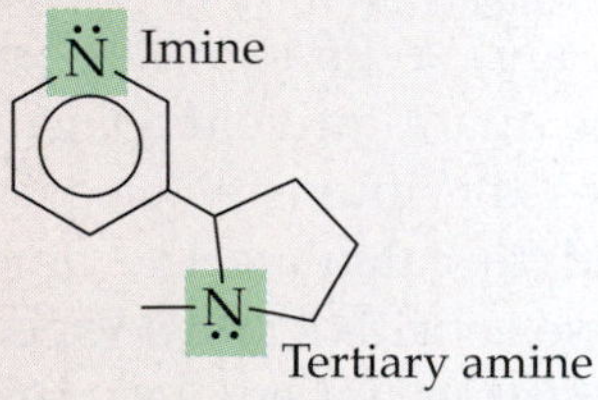

Imine

Tertiary amine

Nicotine

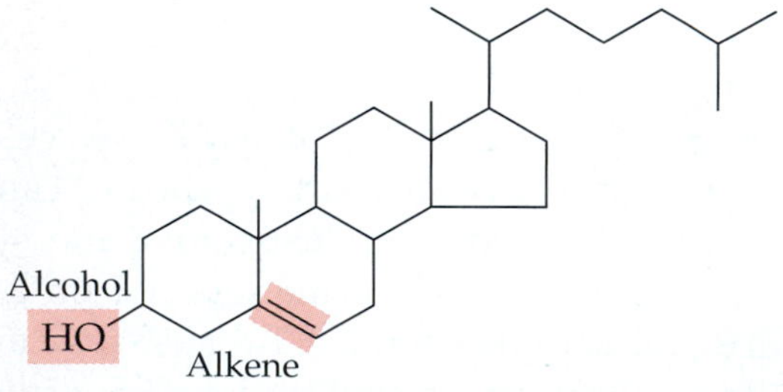

Alcohol

Alkene

Cholesterol

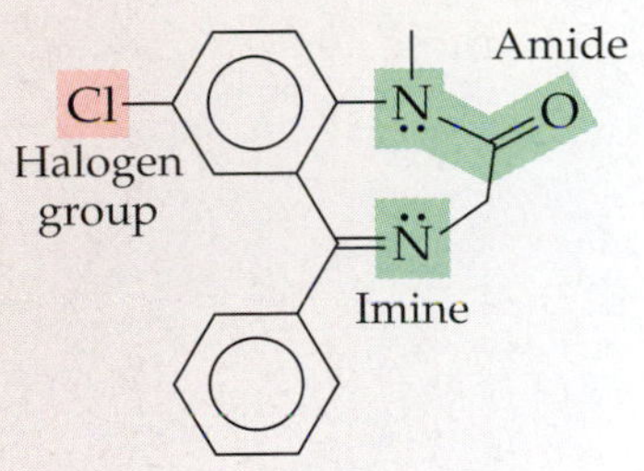

Valium®

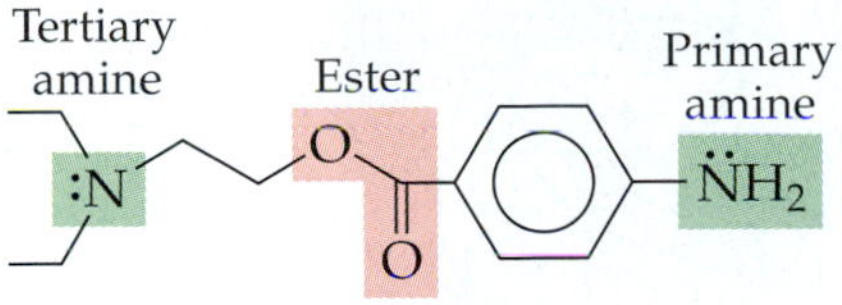

Novocaine

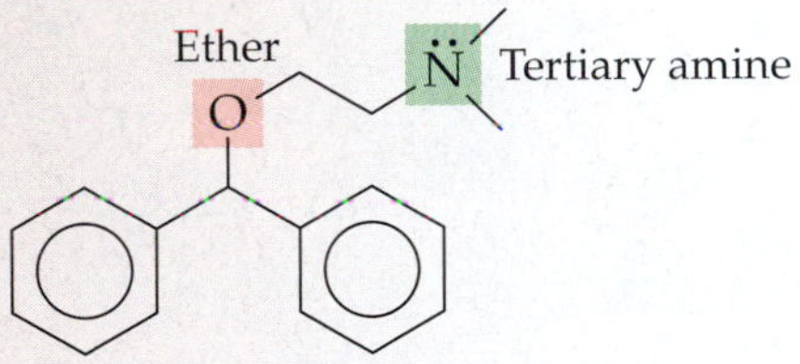

Diphenhydramine

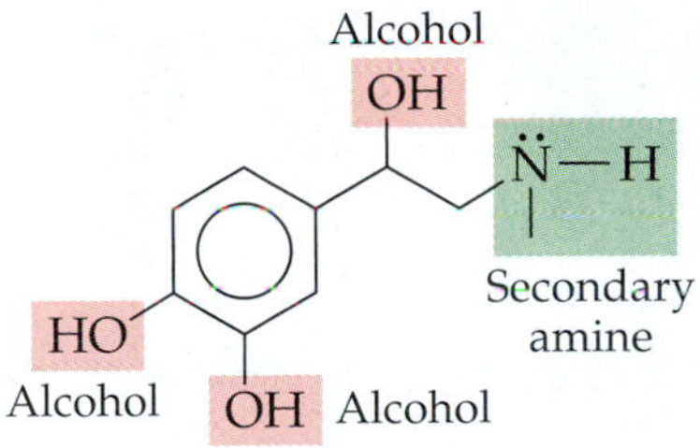

Adrenaline

14.5 | Cocaine, Coca-Cola, Crack

A day in the life of a person addicted to crack is devoted, for the most part, to seeking out the drug, smoking it, feeling high, and then planning for the next hit. One rock of crack, the amount it takes to get high, costs anywhere from $10 to $20, and the intense high lasts for only about 5 to 10 min. It does not take long for the brain to signal another craving, and many crack addicts must smoke more than 10 rocks per day to fulfill that craving. Unlike the powdered form of cocaine, crack gets to the brain immediately, and the result is a high that is more direct and more dramatic, but shorter lived, than that experienced with cocaine. Although both crack and cocaine are incredibly addictive substances, crack can hook a person after only a handful of experiments with it (Figure 14.26).

The lingo associated with cocaine-based drugs can be confusing. What is the difference between crack and cocaine? What is free base cocaine, and how does it differ from the other two? Why are two forms of cocaine smoked (crack and the free base form), while the third is snorted into the nose? These questions are all answered with some basic chemistry you already know from Chapter 4, where we looked at acids and bases. This section will provide you with a brief refresher on acid and bases, and we shall use cocaine as an example of their importance. With this knowledge in hand, you will be able to understand both the fundamental differences among the various forms of cocaine and the reactions that amines undergo.

Recall from Chapter 4 that acids are proton (H^+) donors and bases are proton acceptors. Given this distinction, consider what happens when we put ammonia into water:

$$NH_3 + HOH \rightleftharpoons NH_4^+ + OH^-$$

For this acid–base reaction, water is written as HOH to demonstrate that it is the molecule that donates a proton and thus is behaving as an acid. Ammonia, NH_3, accepts that proton, and so it is behaving as a base. The

▲ FIGURE 14.26 **These photos illustrate the trappings of and culture of crack addiction**

reaction is reversible, which means we can write it in the reverse direction also:

$$NH_4^+ + OH^- \rightleftharpoons NH_3 + HOH$$

In this case, the proton is donated by NH_4^+ (now acting as the acid in the reaction) and accepted by OH^- (now acting as the base).

We call NH_4^+ the *ammonium ion*, which is a fancy name for an ammonia molecule, NH_3, with a proton added to it. The addition of a proton to an atom in a molecule is referred to as **protonation**, and such a group is said to be *protonated*. All other amines can be protonated, not just ammonia, as Figure 14.27 shows for a primary, secondary, and tertiary amine. Notice from these reactions that in each case an H^+ ion is added to the amine, which means that the nonbonding pair of electrons on the amine is replaced by a bond to an H and one positive charge. In general, we can say that

the nitrogen atom in a protonated amine always has four bonds and no unbonded electrons and carries a positive electrical charge, whereas the nitrogen atom in a nonprotonated amine always has three bonds and one nonbonding pair of electrons and carries no electrical charge.

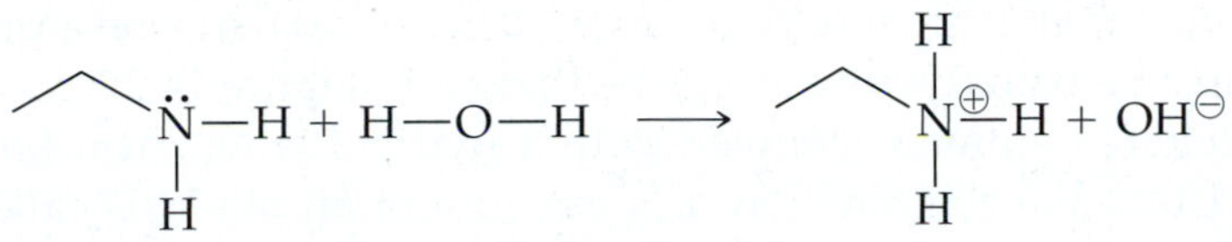

Primary amine

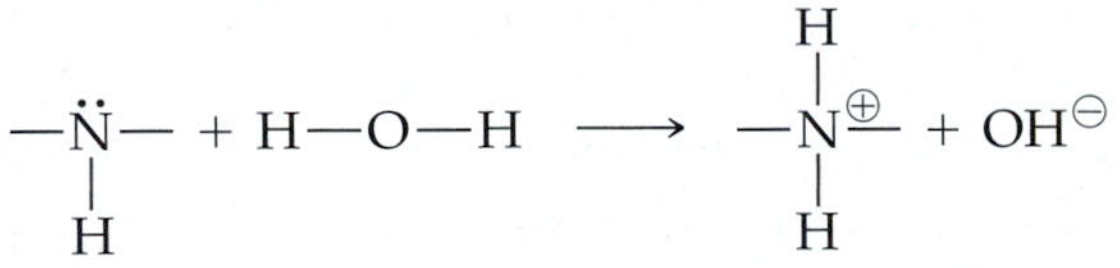

Secondary amine

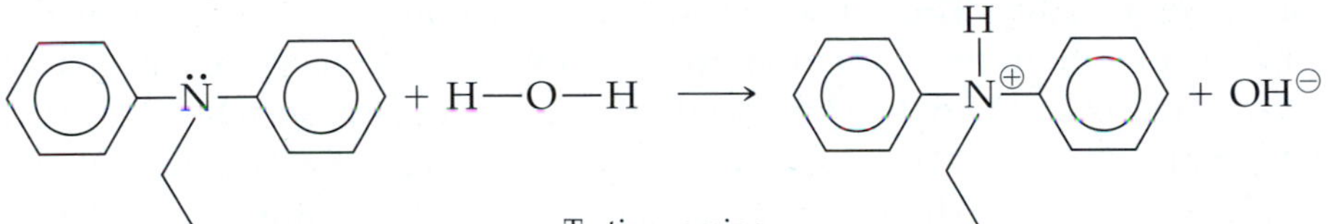

Tertiary amine

◀**FIGURE 14.27** **Amines can be salts.** Amines can form salts by accepting a proton from a water molecule.

QUESTION 14.15 Complete the following acid–base reactions, and name each reactant amine:

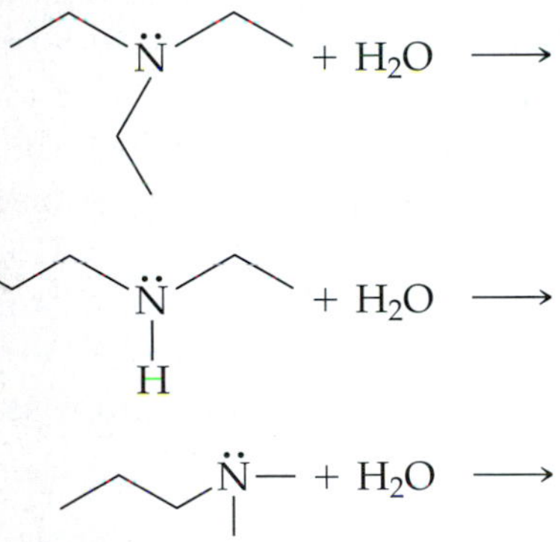

ANSWER 14.15

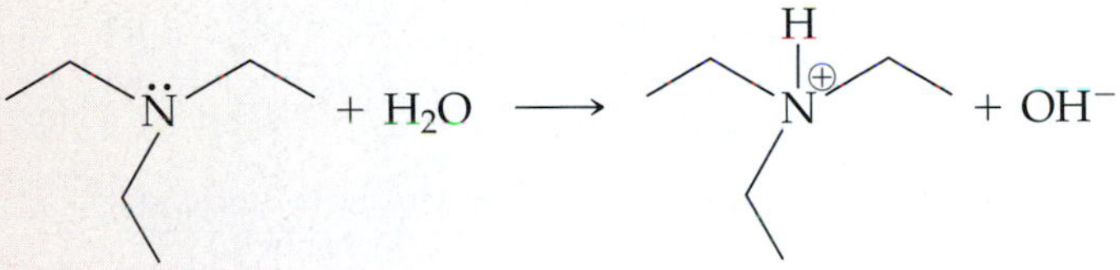

Triethylamine

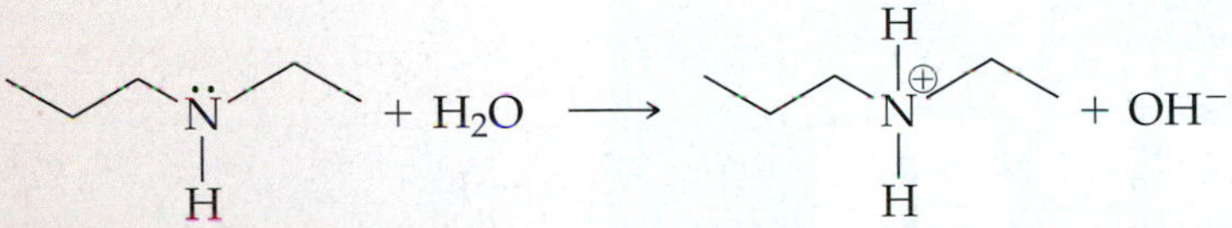

Ethylpropylamine

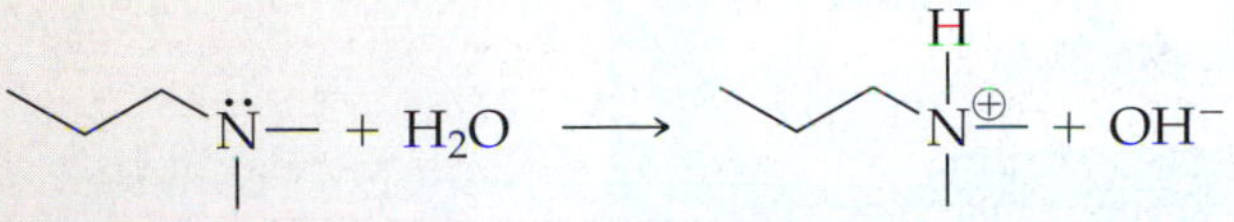

Dimethylpropylamine

We know that every positive charge, such as the charge on a protonated amine, must be paired with a negative charge. In Figure 14.27, for instance, each protonated amine is coupled with a hydroxide ion. You know from Chapter 4 that the combination of a cation and an anion is called a salt, and in the case of protonated amines, chemists use the term **quaternary ammonium salt**. The word *salt* implies that these compounds are ionic and contain equal numbers of positive and negative charges. Quaternary ammonium salts have all of the properties of salts—crystallinity, high melting point, solubility in water, and so forth—that are attributable to the strong ionic bonds between positive and negative ions.

As is true with all other salts, how much saltlike character a given quaternary ammonium compound has depends on how much the charge dominates the structure. Thus, on one hand, an enormous organic molecule that contains one quaternary ammonium salt might retain more of the characteristics we associate with uncharged organic molecules, such as a low boiling point, weak intermolecular interactions, and low water solubility. On the other hand, a small organic molecule containing a quaternary ammonium salt as part of its structure, such as one of the three molecules shown in Figure 14.27, will tend to behave more like an inorganic salt because the charges make a more significant contribution to the molecular structure.

If you read labels on food or medicine packaging, you may have noticed that the words *hydrochloride* or *hydrochloride salt* appear on occasion. A **hydrochloride salt** is what we call a quaternary ammonium salt in which the anion is a chloride ion. Processed foods and drugs contain molecules in their hydrochloride salt forms because the salt form of an amine is more soluble in stomach fluid than the uncharged form. Thus, the salt form dissolves more quickly in the stomach and is sooner able to pass from the stomach to the blood, which is mostly water. Consequently, many molecules that contain amines exist in two forms: as a **free base** (the uncharged amine with a nonbonding pair of electrons on the nitrogen) and as a hydrochloride salt (the quaternary ammonium salt). Figure 14.28a shows

▶ **FIGURE 14.28** **Nitrogen-containing compounds can take two forms.** **(a)** The structure of the free base form and the hydrochloride salt of pyridoxine. **(b)** Pyridoxine hydrochloride is vitamin B$_6$ and is a nutritional additive in many processed foods.

Pyridoxine, free base form

Pyridoxine hydrochloride (vitamin B$_6$)

(a)

(b)

the molecule pyridoxine both in the free base form and as a hydrochloride salt. Another name for this substance is vitamin B_6, and it is used as a nutritive additive in many foods. Look for its name on labels the next time you are in the grocery store (Figure 14.28b).

QUESTION 14.16 Draw the structure of the hydrochloride salt of each of these free bases:

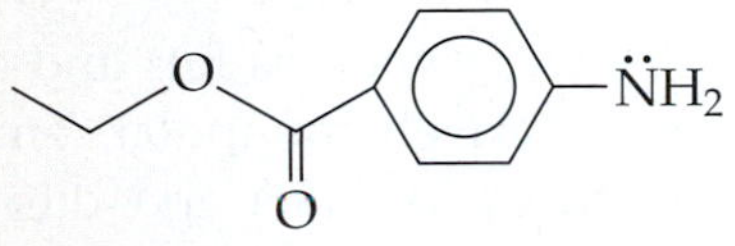

Benzocaine (local anesthetic)

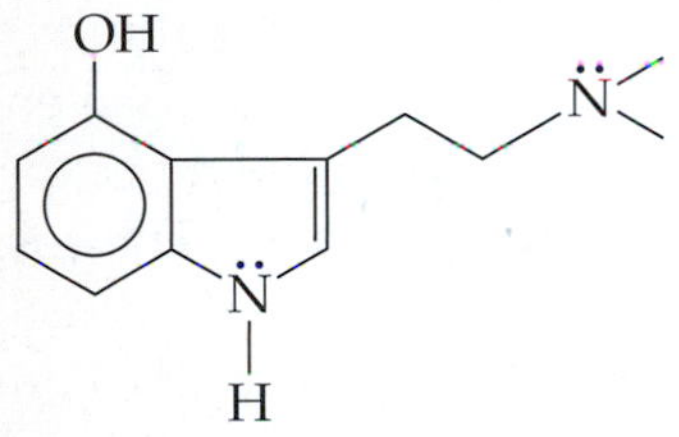

Psilocin (active ingredient in hallucinogenic mushrooms)

ANSWER 14.16

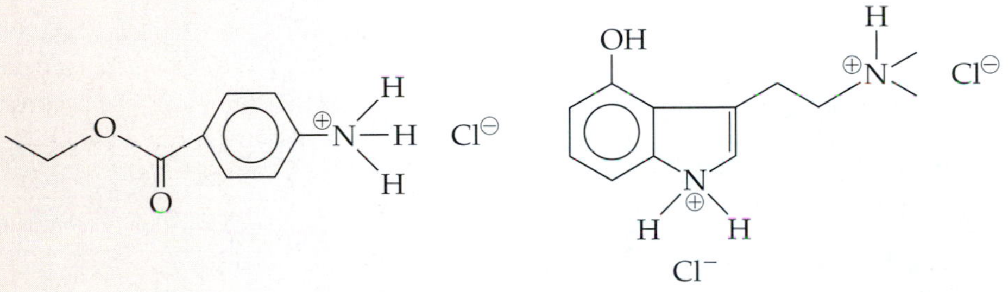

Now we are prepared to tackle the complexities of the cocaine molecule. Cocaine is an example of an **alkaloid**, which is any biologically active nitrogen-containing molecule derived from a plant source. Some other familiar alkaloids are the opiates we looked at in Chapter 13, as well as psilocybin, mescaline, and nicotine. The name *alkaloid* hints at the fundamental nature of these molecules: they are bases, and they create solutions that are basic, or *alkaline*.

Most alkaloids can exist in two or more forms that are distinguished only by the protonation state of their amine group(s), which can be either neutral or positively charged as a result of protonation.

Cocaine is a typical amine. Thus, it is found on the street in two forms: the hydrochloride salt and the free base. Because one form of cocaine is a salt and the other—the free base—is an uncharged molecule, we would expect these two forms of cocaine to have very different physical properties. As Table 14.2 (p. 580) shows, they do. The melting point of the free base is nearly 100 C° lower than that of the hydrochloride salt. This makes sense because the intermolecular interactions between molecules of the free base are much weaker than the ionic interactions holding the ions together in the hydrochloride salt. The hydrochloride salt is soluble in water and highly

TABLE 14.2 Properties of Cocaine

Property	Free base	Hydrochloride salt
Melting point (°C)	98	195
Solubility in water	1 g in 600 mL water	1 g in 0.4 mL of water
Solubility in oils and fats	High	None
Volatility	High	Very low

insoluble in nonpolar substances such as fats and oils, just as salts should be. The free base is comparatively nonpolar, and so it dissolves more readily in nonpolar substances like fats and oils and does not dissolve easily in water.

The last property of the two forms of cocaine covered in the table, volatility, tells us something about how the two forms of this drug are used. Being a volatile molecule, the free base enters readily into the gas phase. Therefore, the free base is smoked because the heat quickly takes the cocaine molecules into the gas phase. The gaseous cocaine then enters the lungs and the bloodstream and travels directly to the brain, resulting in an immediate sensation of euphoria. By contrast, the hydrochloride salt is not amenable to smoking, because it has low volatility. This form is familiar to us as the white powder that is inhaled into the nasal cavity (Figure 14.29), where it is absorbed through the cavity walls, enters the bloodstream, and travels to the brain, a process that takes much longer

▶ FIGURE 14.29 **Cocaine.** (a) Cocaine in the form of a white powder is inhaled into the nose. (b) The powder form is the hydrochloride salt. (c) The hydrochloride salt of any amine is sometimes written with *HCl* next to the amine functional group.

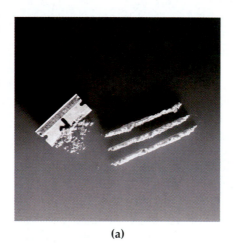

(a)

Cocaine hydrochloride
(b)

Less accurate way of writing cocaine hydrochloride molecule
(c)

Does Coca-Cola Really Contain Cocaine?

The answer is no. Coca-Cola does not contain cocaine, but it used to, and that is where it gets part of its name. Cocaine is extracted from the leaves of the coca plant, and this is the origin of the *coca* part of the name. A nut called the kola nut is used to flavor the drink and add caffeine, and this is where the *cola* part of the name comes from. The original formulation included both cocaine and

caffeine and provided a powerful pick-me-up for people who drank it. Since 1900, though, the coca leaves and kola nuts are both treated by a method similar to the method used in the decaffeination of coffee beans, so that the cocaine is removed from the leaves and the caffeine from the nuts.

Today's Coca-Cola gets its flavor from the extracts of these two plants, in addition to vanilla, orange oil, lemon oil, and sweetener. If you ever try a Coke in another country, you may find its taste to be different from Coca-Cola in the United States. This is because the U.S. formulation uses corn syrup, but the formulations in most other countries use sugar, which is more faithful to the original recipe. The exact recipe for Coca-Cola is not patented, and it is considered to be one of the most closely guarded trade secrets there is.

---

than the route through the lungs. On average, the hydrochloride salt form of cocaine available on the black market in the United States is only 56 percent pure.

There are two kinds of free base cocaine, *freebase* and *crack*, and their structures are identical. They differ only in the way they are prepared from the hydrochloride salt, which is the most stable and most easily transported form of the drug. Both freebase and crack are prepared by treating cocaine hydrochloride with alkaline reagents that deprotonate the salt. The identity of these reagents and the purity of the product differ for the two forms of free base cocaine. For example, the preparation of freebase requires the use of the highly flammable solvent diethyl ether. The preparation of freebase therefore poses a significant risk of fire, and the community of freebase users includes a subset of burn victims.

The term *free base* (two words) is a chemical term used when discussing amines in their uncharged form. The term *freebase* (one word) refers specifically to the free base form of the cocaine molecule that is available as a street drug.

Freebase and crack contain the same form of the cocaine molecule, and both are extraordinarily addictive substances. According to a 2003 survey by the White House Office of National Drug Control Policy, "Approximately 5.9 million [Americans age 12 and older] used cocaine in the past year, and 2.3 million used cocaine within the past month."

14.6 | Smoke and Mirrors

Although we tend to think of cocaine when we hear the words *free base*, the term is also commonly used to describe the basic form of many other organic substances, including the alkaloid nicotine. Nicotine is the most addictive substance known, much more addictive than cocaine or heroin. The hydrochloride salt of nicotine, shown on the left in Figure 14.30, is by far the most prevalent form in the tobacco plant, which uses the compound as a natural and exceptionally toxic insecticide. From what we know about the relative physical properties of the free base form of cocaine and cocaine hydrochloride, it should come as no surprise that the free base form of nicotine, shown on the right in Figure 14.30, is more toxic and many more times addictive than the hydrochloride salt. In fact, the free base form of the drug is often referred to as "crack nicotine."

MOLECULE
Nicotine

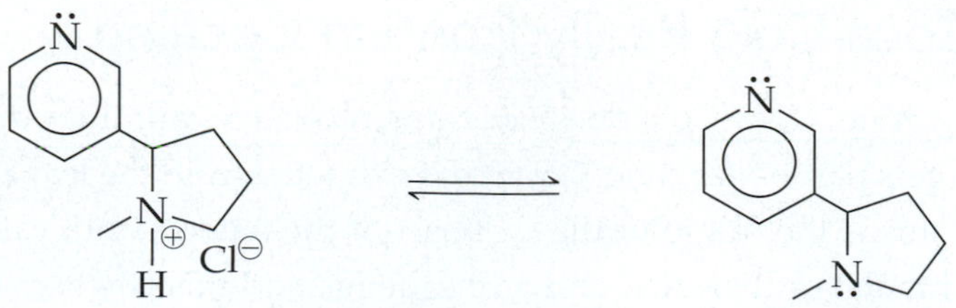

Nicotine hydrochloride, predominates at low pH   Nicotine free base, predominates at high pH

▲ **FIGURE 14.30** **Two forms of nicotine.** Which form you have depends on the pH of the surroundings.

QUESTION 14.17 In addition to forming the hydrochloride salt shown in Figure 14.30, nicotine can form a hydrochloride salt in which both nitrogen atoms are protonated. Draw the structure of this doubly protonated form.

ANSWER 14.17

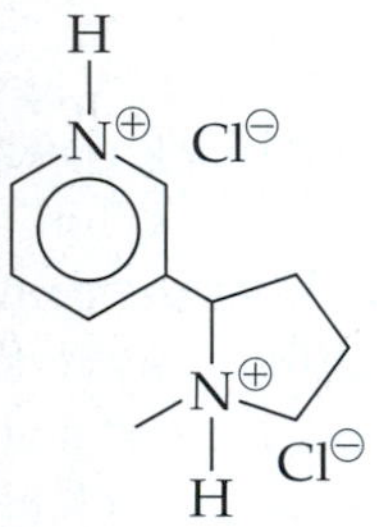

Astute readers probably are wondering why, when there are two amine groups in the nicotine molecule, only one is protonated in Figure 14.30. As Question 14.17 demonstrates, it is possible to protonate both amines and make them both into quaternary ammonium salts. However, the form of nicotine that exists in highest concentration in cigarettes is the hydrochloride salt with only one amine protonated. Why is this the case? The answer lies in the fact that although both amine groups in nicotine have a non-bonding pair of electrons and can act as a base by accepting a proton, they do this to differing extents. Note in Figure 14.30 that one amine group in nicotine is part of a conjugated ring and the second one is part of an unconjugated ring.

Consider for a moment the nonbonding pair of electrons on any amine, which allow the amine to accept a proton. The presence of these nonbonding electrons makes all amines bases. In all our drawings, the pair is always represented by two dots. Depending on the environment around the nitrogen atom, however, those two dots can represent not exactly two electrons, but rather a bit more than two electrons or a bit less. If there are groups near the nitrogen atom that pull electron density away from it, the two dots represent, say, about 1.6 or 1.8 electrons—not the full two. Likewise, if there are groups that push electron density onto the nitrogen atom, the two dots might represent 2.2 or 2.5 electrons. Of course, there is no such thing as part of an electron, but this idea of adding or taking away electron density helps us to understand why one amine might be protonated when another is not.

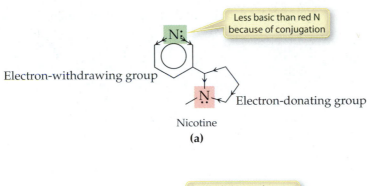

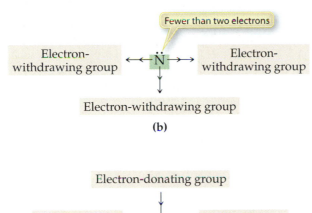

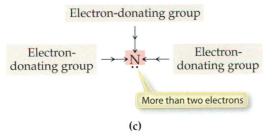

Less basic than red N because of conjugation

◀ **FIGURE 14.31 Relative basicity of amines.** (a) The two amine groups in nicotine are not equally strong bases. Base strength depends on the type of group attached to the amine. **(b)** Because it is part of a conjugated system, the green nitrogen has electron density pulled away from it. This is the weaker of the two amine bases in this molecule, and the two dots represent a bit less than two nonbonding electrons. **(c)** Not being part of a conjugated system, the red nitrogen has electron density pushed onto it. This is the stronger of the two bases, and the dots represent a bit more than two nonbonding electrons. The nonbonding electrons here attract H^+ ions more strongly than the electrons in **(b)** do.

What do we mean when we talk about the groups that can add or subtract electron density from the N atom in an amine? We classify some organic functional groups as being either **electron-withdrawing groups** or **electron-donating groups** (Figure 14.31). Conjugated groups offer a way for a molecule to disperse electron density over several atoms, and so we consider them to be electron withdrawing. Thus, when an amine nitrogen, such as the one highlighted in green in Figure 14.31a, is part of a conjugated system (or directly attached to one), we expect the two dots to represent fewer than two electrons (Figure 14.31b). This means that the nitrogen has less electron density in its two dots and is therefore less able to attract hydrogen atoms to itself. We say that this amine is *less basic* than an amine that is not part of a conjugated system.

An amine attached to unconjugated alkyl groups, such as the one highlighted in red in Figure 14.31a, will have electron density added to it, because these groups are generally electron donating. In this case, the two dots over the nitrogen represent more than two electrons (Figure 14.31c). These amines are *more basic* than amines that are part of a conjugated system and readily attract protons. Thus, when there are only enough protons around to protonate one of the two amines, the red amine is protonated first because it is the stronger base. The green amine can also be protonated, as shown in Answer 14.17, but this second protonation requires a much higher concentration of protons.

QUESTION 14.18   For each pair of molecules, indicate which is the stronger base:

Molecule 1   Molecule 2
(a)

Molecule 3   Molecule 4
(b)

Molecule 5   Molecule 6
(c)

ANSWER 14.18   (*a*) molecule 2, (*b*) molecule 3, (*c*) molecule 6.

What all this means is that nicotine can exist in three forms (Figure 14.32): as the free base, which bears no electrical charges because both amines are unprotonated; as the hydrochloride salt, in which only the amine highlighted in red is protonated; and as the *di*hydrochloride salt, in which both amines are protonated. Which one exists at any given time depends on the pH of the nicotine in an aqueous solution. If the nicotine is in an environment with few protons (basic, high pH), the free base is dominant. If the nicotine is in an environment with a lot of protons present (acidic, low pH), the dihydrochloride salt dominates. At intermediate pH values, the hydrochloride salt is dominant. In fact, at all pH values, there is a mixture of the different forms of nicotine present in solution, but the pH dictates which form is in the highest concentration.

At the pH of a tobacco leaf, the hydrochloride salt is overwhelmingly predominant, but the distribution of the three forms changes as acid or base is added. Nicotine usually exists either as the hydrochloride salt or as the free base and rarely as the dihydrochloride salt. It is the equilibrium between the hydrochloride salt and the free base, shown in Figure 14.30, that is manipulated in the real-world applications we are considering here.

The free base form of nicotine (aka crack nicotine) is made the same way the free base form of cocaine is made: ammonia or some other basic substance is added to the hydrochloride salt to remove the proton of the quaternary ammonium salt. This shifts the equilibrium shown in Figure 14.30 toward the free base. Like the free base form of cocaine, the free base form of nicotine is highly volatile and so gets to the brain much more quickly than the hydrochloride salt. Because nicotine is so exceptionally toxic, though, the consequences of smoking the free base form of nicotine can be more serious than the consequences of smoking the free base form of cocaine (either in the form of freebase or in the form of crack). It is estimated that a person

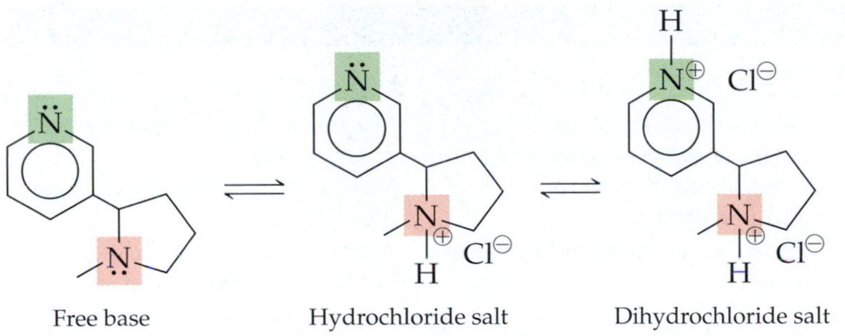

Free base   Hydrochloride salt   Dihydrochloride salt

▲ **FIGURE 14.32 Three forms of nicotine.** Which form exists in highest concentration at any given time depends on the pH of the solution. In a very basic solution, with few H^+ ions around, most of the nicotine molecules in the solution are in the free base form. In a solution of intermediate pH, the more basic of the two nitrogen atoms accepts protons, and the hydrochloride salt prevails. In a very acidic solution, both nitrogen atoms are protonated, and we find mainly the dihydrochloride salt.

smoking 40 mg of pure free base nicotine would have about a 50 percent chance of surviving the experience.

QUESTION 14.19 Imagine that a hypothetical cigarette contains 0.20 mg of the free base form of nicotine and that the lethal dose of this substance in rats is 3.6 mg/kg. If a particular rat has a mass of 620 g, how many cigarettes would it have to smoke before it would die of nicotine poisoning? .

ANSWER 14.19

$$\frac{3.6 \text{ mg}}{1 \text{ kg}} \times \frac{1 \text{ kg}}{1000 \text{ g}} \times 620 \text{ g} \times \frac{1 \text{ cigarette}}{0.20 \text{ mg}} = 11 \text{ cigarettes}$$

The existence of free base nicotine and the dangers it poses to those who smoke it were revealed in a 1997 study performed by environmental scientists at Oregon Health and Science University (*Environmental Science & Technology*, volume 31, issue 8 [1997], p. 2428). Given these findings, one might expect tobacco producers to minimize the levels of free base nicotine in their products. However, the 1997 study and subsequent research (*Chemical Research in Toxicology*, volume 16 [2003], p. 1014) showed that all popular brands of cigarette contain some free base, and links have been made between the addictiveness of certain brands and their high levels of the free base. The cigarette brands surveyed contained anywhere from 1 percent free base to a whopping 36 percent free base!

Here is a question, though: If most of the nicotine in tobacco leaves is in the form of the hydrochloride salt, how does the free base make its way into cigarettes? The answer is that when the tobacco is processed, basic substances are added to change the hydrochloride salt to the free base. As more and more of these alkaline compounds are added, the amount of free base in the cigarette rises. The FDA has yet to make any recommendations on what maximum level of the free base form of nicotine should be allowed in cigarettes sold in the United States.

This chapter has focused on the chemistry of abused drugs. We realize that there are a myriad of illegal drugs, and we have discussed many of them here. However, a chapter surveying every abused drug would be very

TABLE 14.3   Major Classes of Illegal Drugs

Class	Effect	Examples
Narcotics	Alleviate pain and may produce sense of well-being	Morphine, heroin, codeine, other opiates
Hallucinogens	Alter mood, thought, and perception	Marijuana, LSD, phencyclidine (PCP), ecstasy
Stimulants	Enhance alertness and physical activity	Amphetamines (including methamphetamine and ecstasy), cocaine, nicotine
Depressants	Reduce physical activity and mental processes	Alcohol, barbiturates, diazepam (Valium®), methaqualone (Quaaludes™)

long indeed. As a concerned citizen, you will see that the various classes of illegal drugs are mentioned regularly in the newspaper. For this reason, we have provided a list of the four major classes of illegal drugs in Table 14.3.

In the United States, drugs have been classified differently, with a system that puts all illegal drugs into five categories according to the risk they pose to the public. The Controlled Substances Act lists these five categories (called *schedules*), the drugs they include, and the penalties associated with their possession and sale. Schedule I drugs supposedly have the highest risk of abuse and pose the greatest risk to human life. Therefore, the abuse of drugs in this class, which includes ecstasy, marijuana, and LSD, carry the strictest penalties. Schedule V drugs are the least harmful, are the least addictive, and have the lightest penalties. An official list of drug schedules is maintained by the Drug Enforcement Agency, part of the U.S. Department of Justice.

You can visit the DEA online at www.usdoj.gov/dea. To view the list of drug schedules, look for "Drug Schedules" in the "Drug Policy" section.

Postscript

The chemistry of everything is an overwhelming subject to tackle. How can we wrap our minds around the diversity of everything in the physical world? We humans make attempts at simplification: we sort, separate, organize, classify, and inventory. And so we can separate all matter into inorganic and organic; or gas, liquid, and solid; or red things, blue things, green things, and yellow things; or animate and inanimate. For each of these categorizations, though, we know there will be a list of exceptions. For example, how do we categorize the minerals necessary for human life? Are they still inorganic when we ingest them and they undergo reactions with biomolecules, which are decidedly organic? What about substances that are a combination of gas and solid (like the catalyst in your catalytic converter) or liquid and solid (like glass)? What do we do with things that do not fit into the categories we create?

It boils down to this: the natural world defies categorization. Our attempts to pigeonhole everything inevitably wind up incomplete. For thousands of years now, scientists have been frustrated with and awed by the remarkable variety of everything. Here in the twenty-first century, scientists are still grappling with the intricacies of the natural world, and we are nowhere close to a full understanding. We have made big strides, certainly. In 2003, for example, the human genome was completely sequenced. In and of itself, the sequence does not help us much, however. The result of this project is an enormous new assignment for modern scientists: figuring out what each of the genes in the sequence does and what the pieces in between might mean.

In this book, we have returned again and again to one lesson of chemistry that helps us come to grips with the diversity of everything in the physical world: when we look at the structures of things at the atomic and molecular level, we are often able to see why things behave the way they do. When we know what holds something together—is it ionic (a salt) or is it covalent (a molecule)?—we can begin to understand its physical properties—those we can see, taste, and smell. What is more, when we look at the structures of biologically important molecules, we can see how they interact and how they might dock together. Only an understanding of a biomolecule at the level of molecular contours and bonds allows us to understand how that molecule can make us healthy or make us sick.

Scientists of the twentieth century took us beyond the boundaries of the natural world for the first time. We figured out how to synthesize substances in the laboratory, such as polymers. We have the ability now to manipulate the atom and to tap the energy of the nucleus. Through human innovation, we can also tamper with the genomes of living things. However, these new capabilities bring with them a level of responsibility that did not weigh on us before. Now that we have moved outside nature's boundaries, how can we keep our new-found knowledge in check? Will we use or abuse these new technologies? This is a question that you will see debated over and over again by our country's leaders, intellectuals, and scientists.

Watch for it in the media. You will see questions about how we should distribute the drugs we create, how we should regulate the wastes we produce, how we should monitor the genes we modify, and how we should make use of energy sources available to us. Many of these are scientific issues that we humans are struggling with for the first time. You can help to guide the legislation of these issues, and we hope that the principles you have learned in this course give you a better understanding of new scientific dilemmas and how you feel about them. We also hope you will keep this book on hand, on an easy-to-reach shelf in your home, close by when you are reading the Sunday paper or deciding how to vote. Ultimately, we hope that, in some small way, this book will allow you to make informed decisions about the world around you.

KAW

Chapter 14

SUMMARY

We began this chapter with the functional group most frequently found in illicit drugs: the amine. There are different types of amines, depending on the number of hydrogen atoms attached to the nitrogen atom. Those with one alkyl group are primary, those with two alkyl groups are secondary, and those with three alkyl groups are tertiary. With all uncharged amines, the nitrogen atom is attached to three groups and there is a nonbonding pair of electrons on it. Primary and secondary amines are capable of hydrogen bonding—primary amines more so because they have two hydrogen atoms available for use. Because primary and secondary amines can hydrogen bond, their intermolecular interactions are stronger that those in tertiary amines, and their boiling points are higher. For a series of primary, secondary, and tertiary amines having the same molecular formula, the tertiary amine will typically have the lowest boiling point because of its inability to form hydrogen bonds.

All major neurotransmitters are amines. The addition of amine groups changes the surface contour of a molecule, and it is their shape, in part, that allows amine-containing molecules to be identified by the biomolecules to which they bind. Each neurotransmitter has specific receptors in the brain, and most drugs that affect the brain disrupt the natural flow of neurotransmitters in the synaptic cleft. Sometimes the drug causes neurotransmitters to be released at an accelerated rate; other times the drug binds to receptors, blocking them. Many of these drugs mimic the structures of neurotransmitters, and most contain amine functional groups.

Analytical tests are broadly divided into two categories: qualitative (those which tell you what a drug is) and quantitative (those which tell you how much of the drug is present). We looked at two examples of qualitative tests for illicit drugs. Each is relatively simple to perform and can be used in the field—for example, as a roadside test. Qualitative tests often involve a color change to tell the analyst whether or not a compound is present. However, many of these tests are neither specific enough nor accurate enough to be used in a court of law.

Antibody tests are based on the fact that antibodies have a very high level of selectivity for one antigen molecule. Using a procedure outlined in this chapter, chemists can produce antibodies to a specific target molecule and use those antibodies to create an accurate qualitative and quantitative test for that specific molecule.

Amines are bases and take part in acid–base reactions. When an amine becomes protonated, there are four atoms attached to the nitrogen atom, which now carries a positive charge. The combination of the protonated form of an amine and its corresponding anion is called a quaternary ammonium salt. By accepting a proton, the amine changes from a uncharged form (often referred to as a free base) to a charged form. When the positively charged amine is balanced with a chloride anion, we refer to it as a hydrochloride salt. Many foods and drugs contain hydrochloride salts because this form is more soluble in water than free bases. The free base is used if volatility is desired—for example, to create the smokable form of cocaine or nicotine.

KEY TERMS

alkaloid	electron-donating group	free base	qualitative test	secondary amine
anabolic steroid	electron-withdrawing group	hydrochloride salt	quantitative test	tertiary amine
antibody		primary amine	quaternary ammonium salt	
antigen		protonation		

QUESTIONS

The Painless Questions

1. When the Marcy test is performed, what molecule is the source of the minty smell that is observed in the presence of cocaine? Draw the full structure of this molecule, and write its molecular formula.

2. Describe how signals are transmitted between neurons in the human body. Draw full structures for the three major neurotransmitters in the body.

3. Indicate whether each drug test is qualitative or quantitative:

   (a) The test gives you a purple color if a specific drug is present.

   (b) The test tells you how many milligrams of a drug were in the original sample.

   (c) The test tells you that 23 percent of a sample is morphine.

   (d) The test tells you that the drug in question was an amphetamine.

4. Lidocaine (also known as xylocaine) is an anesthetic used in medicine. Identify each organic functional

group in the lidocaine molecule and state whether each amine is primary, secondary, or tertiary.

Lidocaine

5. What is the most addictive substance known? Draw the full structure of this molecule in its free base form.

6. Which of these two amines has the lower boiling point? Write the molecular formula for each.

7. The Marquis test not only provides a colored product in the presence of amines; it also gives different colors when different amines are tested—purple with heroin, for example, and orange changing to brown with amphetamines. Given this specificity, why is the Marquis test considered inadequate as a quantitative test for illicit drugs?

8. Referring to Figure 14.13 (page 561), rank the following drugs according to the number of emergency room visits they caused in 2001: barbiturates, ecstasy, heroin, aspirin, oxycodone, antihistamines, antidepressants, and marijuana.

9. Draw line structures for the following amines, and indicate for each whether it is primary, secondary, or tertiary: (*a*) tributylamine, (*b*) dimethylamine, (*c*) ethylmethylamine, (*d*) diethylamine.

10. What is the difference between a qualitative analytical test and a quantitative one? List two situations mentioned in this chapter in which quantitative tests were described.

11. Draw full structures for the neurotransmitters epinephrine and norepinephrine. Would it be possible for a biological molecule to discriminate between these two similar molecules? Why or why not?

12. We know from the "Students Often Ask" box on page 556 that the amines putrescine and cadaverine are two of the molecules responsible for bad breath. Which atom other than the N of amines is sometimes found in organic molecules associated with unpleasant odors? (Hint: See Figure 14.9, page 556.)

13. Describe how a drug can become physiologically addictive. Specifically, indicate how neurotransmitter levels in the brain can lead to cravings for a drug.

14. Is it possible to categorize the ammonia molecule as a primary, secondary, or tertiary amine? Why or why not?

15. Rank these amines in order of increasing boiling point, and indicate whether each is primary, secondary, or tertiary:

More Challenging Questions

16. Make a sketch showing hydrogen-bonding interactions that might exist among two molecules of diethylamine and two molecules of water.

17. Describe why crack and cocaine hydrochloride are administered by users in different ways. Use your understanding of the acid–base chemistry of amines in your explanation.

18. Write line structures for the products formed when the Marquis test is used on each of these amines:

Formaldehyde

19. Parkinson's disease is associated with damage to dopamine receptors in the brain. Given this association, why can the use of some drugs elicit symptoms like those of Parkinson's disease? Name two drugs that fall into this category.

20. Draw line structures for (a) butyldimethylamine, (b) octyldipentylamine, (c) triphenylamine (a phenyl group is the same as a benzene ring), (d) heptylhexylnonylamine. In your drawings, label each amine as primary, secondary, or tertiary.

21. Caffeine is a dear friend to many of us who are addicted to coffee. (a) Identify all the organic functional groups in caffeine, and indicate whether each amine is primary, secondary, or tertiary. (b) Write the molecular formula for caffeine, and determine its molar mass. (c) Draw line structures for caffeine hydrochloride and for caffeine dihydrochloride. Which salt predominates in a solution of very low pH?

Caffeine

22. Rank these amines in order of increasing boiling point:

23. Describe how the three-dimensional shape of an amine in an organic molecule allows the molecule to be recognized by a binding site in a biomolecule. What is the molecular geometry of the amine, and how does that geometry differ from the molecular geometry of a long alkyl chain?

24. Imagine that a new designer drug has just hit the black market. It is causing problems for law enforcement agents because it produces a minty smell when the Marcy test for cocaine is performed on it. (a) Is it theoretically possible for a substance other than cocaine to create a product that has a minty smell when the Marcy test is performed? (b) In order for this new drug to give a positive result with the Marcy test, what must be true about the molecular structure of the drug?

25. The street drug ecstasy is classified both as a stimulant and as a hallucinogen. Explain the reason for this dual classification, and describe how the ecstasy molecule affects specific nerve receptors in the brain.

26. None of the molecules whose structures are shown here can interfere with a test for cocaine. However, some of them give a positive Marcy test. Which ones? Explain your choices.

27. Amphetamine is a chiral drug. Identify its chirality center (Section 13.4), and then draw the full structure of both amphetamine enantiomers.

28. There is at least one mistake in each of the amine names that follow. Offer a corrected alternative, and draw the line structure of each: (a) diphenylbutyl-amine (a phenyl group is the same as a benzene ring), (b) diethyldimethylamine, (c) triethylpentylamine.

29. Describe how biological molecules are able to bind small organic molecules with an extraordinarily high degree of specificity. In your explanation, include a discussion of shape recognition and of functional-group interactions.

30. Given what you know about the colors produced by different drugs in the Marquis test, predict the color produced when oxycodone (Percocet®) is tested. The structure of oxycodone is shown in Question 14.12, page 569.

The Toughest Questions

31. Novocaine (also known as procaine) is an anesthetic commonly used in dental work. (a) Identify each of its organic functional groups, and indicate whether each amine is primary, secondary, or tertiary. (b) Write the molecular formula for novocaine, and determine its molar mass. (c) Identify the amine functional group that is most basic in the novocaine structure, and provide a rationale for your choice. (d) Draw line structures for the hydrochloride salt and dihydrochloride salt of novocaine. Which form predominates in a solution that has a very low pH?

Novocaine

32. In Chapter 13, we talked at length about chirality in organic molecules, especially drug molecules. Identify all chirality centers in these illicit-drug molecules:

Ecstasy   Nicotine

Fenfluramine

33. Recall from Chapter 12 that esters are hydrolyzed in water (in the presence of either an acid or a base) to produce a carboxylic acid and an alcohol. The cocaine molecule is broken down into three pieces when its esters are hydrolyzed. Write the chemical equation representing this hydrolysis.

34. The *naphthyl* group is a two-ringed hydrocarbon group. If you look in a chemical catalog, it is possible to find 1-naphthylamine and 2-naphthylamine. However, you will not find 3-naphthylamine or 4-naphthylamine. Why will you not find these molecule names listed?

Naphthyl group   1-Naphthylamine

35. Psilocybin is the active ingredient in hallucinogenic mushrooms. (a) The lethal dose of psilocybin for rats is 275 mg per kilogram of body mass. If a rat is given 135 mg of the drug, what minimum mass must the rat have to avoid certain death? (b) One of the two amines in psilocybin is protonated much more readily than the other. Which is it? Briefly explain your answer.

Psilocybin

36. A certain cigarette contains 2.0 mg of nicotine, and 6.5 percent of that is the free base. (*a*) If an 80-mg dose of the free base causes certain death in humans, how many cigarettes would a person have to smoke consecutively to accumulate this level of the free base? (*b*) How many molecules of free base nicotine are contained in an 80-mg pure sample?

37. Extracts of the West African plant *Cryptolepis sanguinolenta* are widely used as an herbal antimalarial drug by native populations. Very often, ancient herbal remedies, such as this one, can be a source of interesting drug molecules. In this case, the plant extract has been found to contain molecules that act as anticancer agents. Another molecule found in the mixture is the alkaloid *quindolinocryptotackieine*:

(*a*) Locate all the amines in this drug molecule. (*b*) Write the molecular formula for this molecule, and calculate its molar mass to two decimal places. (*c*) Draw the full structure for the molecule, showing each amine in the form of its hydrochloride salt. (*d*) At which of the following pH values would you be most likely to find the version of the molecule you drew in part (*c*): 2.0, 4.0, 7.0, 9.0 or 11.0? (*e*) Look for chirality centers in this molecule, and circle any that you find.

38. Stanozolol is an anabolic steroid that has been used by some athletes:

Stanozolol

(*a*) Identify all chirality centers in this molecule. (*b*) Does this structure bear any resemblance to the structure of other anabolic steroids? (*c*) Draw the full structure of the drug, and determine its molecular formula and molar mass.

E-Questions

Go to **www.prenhall.com/waldron** to find these questions in electronic form, complete with hyperlinks directly to the various websites cited in the questions.

39. Go to the website for the United Nations Office on Drugs and Crime at **UNODC** to find out more about the cocaine test developed at the Marcy Psychiatric Center. Read the four-page article titled "A Simple, Sensitive, Specific Field Test for Cocaine Based on the Recognition of the Odour of Methyl Benzoate as a Test Product." Then answer these questions: (*a*) Outline the drawbacks of the Scott test described in this article. (*b*) What four reasons do the authors give for the efficacy and validity of the new test?

40. Psilocybin was the molecule of the month at the University of Bristol's School of Chemistry in October 1999. In celebration, the chemists there posted detailed information on the drug at **Bristol**. Read this short article, and then answer these questions: (*a*) What two species of plant contain the highest concentrations of psilocybin? (*b*) How are the structures of psilocybin and mescaline related? (*c*) Which neurotransmitter does psilocin, a derivative of psilocybin, most closely resemble? (*d*) What are indole-amines, and what physical properties and physiological effects do they exhibit?

41. To learn more about the chemistry of ecstasy (and related drugs), go to the article "Ecstasy: Science and Speculation" at **RSC**. Read this five-page article, and then answer the following questions: (*a*) What is the structure of ephedrine, and where is its chirality center? (*b*) What molecule was used in the synthesis of ecstasy in 1912? Show its structure and the structure of any intermediates of the chemical reaction. (*c*) What is the serotonin syndrome? What physiological effects does it cause? (*d*) How does the number of deaths from ecstasy use compare with the number of deaths attributed to alcohol poisoning?

42. Read about "Drug Testing Basics" at **Erowid**. After reading this four-page article, answer the following questions: (*a*) What is a barbiturate? Draw the structure of one. (*b*) What are the pros and cons related to the testing of hair for illicit drugs? (*c*) What is considered the most obtrusive method of drug testing? (*d*) How many nanograms of cannabis can an antibody test (immunoassay) detect in a 1-mL sample?

43. Find out more about cocaine use and abuse at **FreeDictionary**. Read this six-page article, and then answer the following questions: (*a*) How did cocaine first make its way to Europe, and what were its advertised benefits at the time? (*b*) Outline the mechanism of action of cocaine. Specifically, what kinds of receptors in the brain are affected by it? (*c*) What compounds are most commonly used to "cut" cocaine?

WORLD WIDE WEB RESOURCES

As with the E-Questions, go to **www.prenhall.com/waldron** to find these questions in electronic form, complete with hyperlinks directly to the various websites cited.

Some of the links that follow contain research articles related to the topics in this chapter and could be used as the basis for a writing assignment in your course. Other sites are of general interest.

- For more information on the use of steroids and steroid substitutes in sporting events, especially the Olympics, check out the following articles: "Steroid Substitutes: No-Win Situation for Athletes" at **OpenSeason** and "THG: The Hidden Steroid" at **Steroids**.

- Read more about the history of Coca-Cola, including changes made to its formulation at **Cola**. The company's website (**CocaCola**) also has an abundance of information, and **Snopes** has additional information on the folklore surrounding Coca-Cola, including a story about how an advertisement in the 1930s produced the man we now know as Santa Claus.

- Learn more about the issues of student privacy, the Fourth Amendment, and drug testing at **ACLU**. The article titled "Ask Sybil Liberty About Your Right to Privacy" outlines how drug and alcohol tests and metal detectors in schools are viewed from the perspective of law enforcement agencies. Several legal precedents in the area of student privacy are discussed.

- Choose any well-known prescription drug or drug of abuse, and find out more about it. The Partnership for a Drug-Free America at **DrugFreeAmerica** provides free information on prescription drugs and drugs of abuse. For the drug you have chosen, find out what it looks like, the dosages used, drug nicknames, drug class, and the various forms it can take. Research its short-term and long-term effects.

ABC NEWS VIDEO
Getting an Edge

Working with Measured Numbers

PART 1: SIGNIFICANT FIGURES

Scientific measurements can be performed with very primitive or very sophisticated equipment. For this reason, it is important to be able to distinguish one type of measurement from the other by looking at the value measured. For example, let's say you use a kitchen scale that you bought at your local department store for $14.00 to determine the mass of a jelly doughnut. The balance tells you that it has a mass of 86 grams (g). By convention, the last number in a measurement is the number that is least reliable. For this measurement, the number 6 is least reliable. That is, this number could vary by one unit in either direction. Thus, this doughnut could have a mass of 87 g or 86 g or 85 g. Another way to say this is "The jelly doughnut has a mass of 86 g, plus or minus 1 g (±1 g)."

If, instead, you go to people from the National Institute of Standards and Technology and ask them to determine the mass of your jelly doughnut, they may be able to use a very fancy balance that can determine mass to ±0.00001 g. Suppose that after a very careful measurement, the mass of your jelly doughnut is officially reported to be 85.32055 g. If you report the mass of your doughnut to others, you now have the right to report it with decimal places, whereas your kitchen balance would only allow you to report your mass with zero decimal places. You would not, for example, round off your very exact measurement to zero decimal places, because you earned those five decimal places by using an extremely sophisticated balance. Thus, *the way we express a measured value tells us something about the equipment or method that was used to make that measurement.*

Because the last number of any measurement is estimated, we refer to it as *uncertain* and the other numbers preceding it as *certain*. Scientists usually keep track of certain and uncertain numbers by using *significant figures*, the total number of both in a measurement. To determine the number of significant figures in a measured value with a decimal point, you begin at the left end of the number and move to the right. When you reach the first nonzero digit (that is, the first digit that is not a zero), you stop. That nonzero digit and every digit to its right are considered to be significant. Consider this example:

$$0.0000\textcolor{red}{9}800600 \text{ second}$$

Beginning at the left end of the number and moving right, you reach the first nonzero digit at the red number 9. This is the first significant figure. Each of the remaining numbers to its right is also significant. Therefore, this number has seven significant figures. We also know that this measured value was obtained by using a more exact timing device than, for example, the value 0.00009800 second, which has only four significant figures.

Let's consider another set of measured values:

1.45 g has three significant figures.
1.450 g has four significant figures.
1.4500 g has five significant figures.
1.45000 g has six significant figures.

Clearly, the topmost mass measurement is the least exact, and the measurements become more exact down the list. From this example, you can see that the zeroes at the end of these measurements are an indicator of exactness. But what about numbers without a decimal point? What do we do with the zeroes in such numbers? The answer is that this is an ambiguous situation, and different numbers of significant figures can be counted. For example,

45,000 mg has either two, three, four, or five significant figures.

300 mg has either one, two, or three significant figures.

BUT

300.00 mg definitely has five significant figures.

We have seen that significant figures are an indicator of exactness. The following example gives another reason why they are important (in this equation, measured values are multiplied and divided):

$$\frac{281.00 \text{ g} \times 0.00066 \text{ g}}{351 \text{ g}} = ?$$

If you punch this equation into your calculator, you will get 0.0005283761 g. (Your calculator may show more or fewer digits, depending on how it is programmed.) How do we know how many of these numbers we can include? The answer comes from the number of significant figures in each value in our calculation:

281.00 g has five significant figures.

0.00066 g has two significant figures.

351 g has three significant figures.

Whenever we are performing a calculation in which measured values are multiplied and divided, as in this example, *we limit the number of significant figures in the answer by the fewest found in the measured values in the calculation.* The smallest number of significant figures in this case is two (in 0.00066 g), so our answer should have two significant figures as well: 0.00053 g.

QUESTION A.1 For the following measured values, what is the number of significant figures contained in each? *(a)* 0.0900009; *(b)* 800; *(c)* 760.000; *(d)* 30,004.

ANSWER A.1 *(a)* 6; *(b)* 1, 2, or 3; *(c)* 6; *(d)* 5.

QUESTION A.2 What is the correct answer to each of these calculations?

(a) $\dfrac{5.600 \text{ kg} \times 0.0008 \text{ kg}}{7.9223 \text{ kg}} =$

(b) $\dfrac{4.555 \text{ L} \times 0.000551 \text{ L} \times 102.8 \text{ L}}{0.9009 \text{ L} \times 300.0 \text{ L}} =$

ANSWER A.2 *(a)* 0.0006 kg; *(b)* 0.000955 L.

QUESTION A.3 In the 1940s, both steel pennies and copper pennies were minted in the United States. They can be distinguished by their masses: the steel penny has a mass of 2.70 g, and the copper penny has a mass of 3.11 g. If you are going to buy a balance that you will use to distinguish one type of penny from the other, what is the minimum number of decimal places that your balance should display?

ANSWER A.3 If you round off both masses to one number only, both masses will be 3 g. If you round off to one decimal point, the masses will be 2.7 g and 3.1 g, respectively. Thus, a balance that determines mass to one decimal place should be adequate to distinguish one type of penny from another.

In cases where measured values are added or subtracted, there is a different rule for determining the number of significant figures in the answer. Consider this example:

$$2.335 \text{ s} + 33.20 \text{ s} + 6.1 \text{ s} = ?$$

When adding and subtracting, we rely on the number of decimal places that each number has, rather than the number of significant figures. In this calculation, the measured values have three, two, and one decimal place, respectively. Therefore, the answer will be limited to the smallest of these: 1. If you punch this equation into your calculator, you will get the number 41.635 s. We must round this off so that it has only one decimal place. Therefore, the correct answer is 41.6 s.

QUESTION A.4 Perform the following calculations:

    (a) 3206.56 s − 289.6554 s =

    (b) 32.78900 g + 25.544 g + 6.32200 g =

ANSWER A.4 (a) Two decimal places are permitted in this answer: 2916.90 s; (b) Three decimal places are permitted in this answer: 64.655 g.

When we perform calculations in chemistry, we often must use numbers that are not measured values. For example, let's say you measured the length of your car *very* precisely and found it to be 4670.15 millimeters. If you would like to express this length in meters, you must convert millimeters to meters with the following conversion factor:

$$1000 \text{ mm} = 1 \text{ m}$$

If you set up a dimensional analysis, it will look like this:

$$(4{,}670.15 \text{ mm})\left(\frac{1 \text{ m}}{1{,}000 \text{ mm}}\right) = 4.67015 \text{ m}$$

It might seem like we violated one of the rules we just learned. If the value 1 m has only one significant figure, then why is our answer not limited to one significant figure? The answer is that the values 1 m and 1000 mm are not *measured* numbers. Rather, they are *exact numbers*, which are values that are defined by convention or that are counted. In this case, the S.I. system of measurement defines 1000 mm as equal to exactly 1 m. Exact numbers such as these are considered to have an infinite number of significant figures. Therefore, in the preceding calculation, this conversion factor does not limit the number of significant figures, but the measured value 4670.15 mm does.

The other instance in which we see exact numbers is when things are counted. For example, if you use the conversion factor 12 eggs = 1 dozen, this is an exact number because the number 12 is counted. Because counted values increase in whole-number increments, they are also considered to have an infinite number of significant figures and will not limit the number of significant figures in a calculation.

QUESTION A.5 Which of the following conversion factors would limit the number of significant figures in a calculation? (a) 1,000,000,000 μm = 1 km; (b) 58.93 g = the mass of 1 mole of Fe atoms; (c) 24,500 seats = the capacity of your local football stadium; and (d) 0.08350 nm = the average diameter of a human hair.

ANSWER A.5 Choices (b) and (d) must be measured values. Therefore, the number of significant figures in these values must be considered when they are used in a calculation. Choice (a) is a metric unit definition and choice (c) is a counted number.

PART 2: ACCURACY AND PRECISION

In Part 1, we discussed individual measured values and their significance. However, we should also understand how to evaluate a collection of measured values. How can we tell whether such data are valuable to us or not? When we talk about "confidence" in an experiment, what exactly do we mean? Certainly, we mean that we have been meticulous in our measurements, and scientists use two terms to describe the worth and quality of data. The first term, *precision*, tells you the closeness of measured values to one another.

As an example, let us consider a series of actual experiments that were performed to determine the average distance that a cookie is dunked into a cup of tea in Great Britain. In this experiment, a cup with gradations on the inside was used to determine how much liquid tea was displaced when the cookie was dunked. Take a look at the data for successive measurements on the height of tea displaced, in centimeters, in one experiment:

	Trial A (centimeters)	Trial B (centimeters)	Trial C (centimeters)
measurement 1	3.4	4.0	5.0
measurement 2	3.3	2.9	2.4
measurement 3	3.4	2.7	3.5
measurement 4	3.5	3.7	2.2
measurement 5	3.2	3.5	3.7
average	3.4	3.4	3.4
range	3.2–3.5	2.7–4.0	2.2–5.0

Which trial is most precise? All three give the same average value, but Trial A is most precise because the measured values are closest to one another. As you go from A to B to C, the values become less precise. That is, they are more widely scattered around the average value.

What could cause three measurements of the same thing to have different levels of precision? In this case, it is possible that different measuring cups were used in the trials and one cup may have finer lines than the others. It is also possible that three different people performed these measurements. Maybe the person who measured the values in Trial C needs a new pair of glasses. This brings us to an important lesson in experimental design: it is important to keep the conditions as uniform as possible from one measurement to the next. One person using the same measuring device will provide the most consistent results.

The second term, *accuracy*, describes the closeness of a measurement to a known value. For the preceding experiment, you cannot evaluate the accuracy. Why? Because there is no known standard to which you can compare the measured values. So let's look at a slightly different experiment, one that has the benefit of a known standard. If you have a 5.0000-g mass that has been calibrated by a team of internationally recognized scientists at the National Institute of Standards and Technology, you can use that mass to determine the accuracy of three analytical balances in your laboratory. Suppose you collect the following data for your three balances, using your 5.0000-g standard:

	Mass from balance I (grams)	Mass from balance II (grams)	Mass from balance III (grams)
measurement 1	5.4403	5.0009	4.9988
measurement 2	5.3344	4.8992	4.9907
measurement 3	4.6677	5.1200	5.0120
measurement 4	4.8009	5.0130	5.0066
measurement 5	5.2230	4.8895	5.0004
average	5.0933	4.9845	5.0017
range	4.6677–5.4403	4.8895–5.1200	4.9907–5.0120

Because we have a known standard, we can compare the averaged measured values to it. Of your three balances, balance III is most accurate because its average value, 5.0017 g, is closest to the standard. Balance I ranks second in accuracy, and balance II ranks last. As for precision, the values from balance III are tightly grouped together (in other words, the range is very narrow), and all of the values average to a number close to the standard. Thus, we can say that balance III is both accurate and precise. Balance I is less accurate than balance II, and its precision is lower because the balance I values span a larger range than the values for balance II. Thus, we say that balance I is less accurate and less precise than balance II. Both I and II need a tune-up to reach the levels of accuracy and precision found for balance III.

> **QUESTION A.6** Is it possible for a balance to be *(a)* accurate and precise; *(b)* inaccurate and imprecise; *(c)* precise but not accurate; *(d)* inaccurate and precise?
>
> **ANSWER A.6** *(a)* and *(b)* Yes. *(c)* Yes. A balance can give a wide range of values, making the balance imprecise, but if the values average to a value close to the known value, the balance is accurate (although you may still be skeptical about the value of the masses that it provides). *(d)* Yes. A balance can have very closely grouped readings, but those readings may not be close to the known value.

You know by now that the number of decimal places you are justified in reporting for a given measurement depends on the measuring device you use. Very often, however, you must manipulate data—put it through some mathematical hoops, so to speak—to get your desired answer. So here are a few words about handling numbers and about how we will perform calculations in this book.

There are different ways to do a calculation if you're using a handheld calculator. Try the following using yours:

$$\frac{5.60 \times 4.51 \times 7.80 \times 1.11}{3.40} = ?$$

Depending on what you did when you punched the numbers in, and depending on how your calculator is set, you get different results, most likely either 64.3136611765 or 64.4117647059. You get the former value if you don't round off after every individual operation and the latter if you round off as you go. Try it both ways and see. Because each value in the equation contains three significant figures, we must have three significant figures in the answer, so we can round these two answers off to 64.3 and 64.4, respectively. Thus, your answer changes by 0.1, depending on when (and how often) you round off. In this book, we will round off only at the end of a calculation. Therefore, you should get the same numbers that we do if you plug everything in and *then* think about how much rounding off should be done.

When taking measurements, scientists make every effort to be consistent and to follow some simple rules. They keep careful records, perform multiple trials in every experiment, use instruments that are properly calibrated, and agree upon internationally acceptable units of measure. Despite these efforts, though, blunders do occur, often with disastrous results. In the 1970s, a report was published by a group of nutritionists who measured the nutritional content in spinach and found it to be a remarkable source of iron. They contended that every 100 g of spinach contained 34 mg of iron, a value *much* higher than in most other foods. As a result, millions of children were forced to eat piles of spinach. In fact, the super strength of one of our dearest cartoon characters, Popeye, was attributed to his love of the green, soggy stuff. As it turns out, the people who transcribed the iron content misplaced the decimal point. Indeed, 100 g of spinach contains only 3.4 mg of iron!

Another equally disastrous measuring blunder occurred when the *Mars Climate Orbiter*, sent into space in 1999, simply disappeared. The calculations needed to determine how to track the orbiter were done by combining measurements from navigation teams in Colorado and California. As it turns out, one team was using English units and the other was using metric. So the lesson here is this: whether you're measuring spacecraft orbits at NASA or the parameters for dunking biscuits in Great Britain, be scrupulous in your experimental design, keep track of units, and don't forget to eat your spinach!

Basic Algebra Review

This text must use mathematics minimally, because sometimes math is required to understand a fundamental chemical concept. Most often, basic chemistry problems use algebra much more often than, say, geometry or trigonometry. For this reason, we have provided an essential review of algebra for those of you who may be rusty.

A few simple rules will allow us to solve most algebraic problems in chemistry. For each rule presented, examples will accompany it, and you will given the opportunity to try working problems on your own.

RULE 1 *To solve an equation for a variable, you can perform identical operations on both sides of the equation in order to isolate the variable.* It is common to see algebraic problems in which there is a variable, which we might call x, as well as numbers in an equation. To find the value of x, we must move everything but x to one side of the equation. We will demonstrate how this is done with a straightforward example. Consider the equation

$$4.56 + 6.34 + x = 98.65$$

If we are asked to solve this equation for x, we must rearrange the equation and simplify it so that x is equal to one number. In other words, our answer should be in the form "$x = \underline{\hspace{1cm}}$," where the numerical value of x is in the blank space. To isolate x on the left side of the equation, we can perform identical mathematical operations on each side. For example, because we want to move 6.34 from the left side to the right side, we perform an operation that will make it disappear. In this case, we subtract 6.34 from both sides:

$$4.56 + 6.34 - 6.34 + x = 98.65 - 6.34$$

gives

$$4.56 + x = 98.65 - 6.34$$

Now we can do the same thing for 4.56. We will remove it from the left side by subtracting it from both sides:

$$4.56 - 4.56 + x = 98.65 - 6.34 - 4.56$$

gives

$$x = 98.65 - 6.34 - 4.56$$

Now that we have successfully isolated x on one side, we need only perform the subtraction on the right with a calculator, bearing in mind that our answer should retain two decimal places (for a review of significant figures and calculations, see Appendix A):

$$x = 87.75$$

QUESTION B.1 Solve the following equations for x:

(a) $23.655 - 356.320 + x = 258.567$

(b) $5.2 + 8.9 - 4.0 + 6.2 = x + 2.2 - 3.5$

ANSWER B.2 (a) 591.232; (b) 17.6.

The isolation of x on one side of an equation may require multiplication or division on both sides, as the following example demonstrates. Consider this equation:

$$(86.23)(23.10) = (8.500)\left(\frac{x}{3.222}\right)$$

To solve for x, we can begin by multiplying both sides of the equation by 3.222. This removes 3.222 from the right side and moves it to the left side:

$$(3.222)(86.23)(23.10) = (8.500)\left(\frac{x}{3.222}\right)(3.222)$$

gives

$$(3.222)(86.23)(23.10) = (8.500)(x)$$

To remove 8.500 from the right side of the equation, we can divide both sides of the equation by 8.500:

$$\frac{(3.222)(86.23)(23.10)}{8.500} = \frac{(8.500)(x)}{8.500}$$

gives

$$\frac{(3.222)(86.23)(23.10)}{8.500} = x$$

Now that we have successfully isolated x on one side, we need only perform the multiplication and division on the left with a calculator, bearing in mind that our answer should retain four significant figures:

$$x = 755.0$$

If you plugged in everything before rounding off, then your calculator should read 755.0521984. . . . Because we want to round this number off to four significant figures, we must leave only the first decimal place. In this case, the second decimal place is a 5, so rounding off is not as easy as if the number had been a 6 or more (round up) or a 4 or less (round down). There are different ways to round off numbers in this situation. In this book, we will do the following: when the number rounded is a 5, you adjust the previous number to the *nearest even number*. In this example, the previous number is a zero. Therefore, we leave it as zero. If, instead, we were rounding off 755.15 to one decimal place, we would change the 1 to a 2, to give 755.2.

QUESTION B.2 Solve the following equations for x:

(a) $\dfrac{56.3}{x} = \dfrac{32.6}{(4.20)(9.36)}$

(b) $(0.0300)(5.62)(8.00) = \dfrac{(x)(8.50)(3.22)}{0.00222}$

ANSWER B.2 (a) $x = 67.9$; (b) $x = 0.000109$.

RULE 2 *When doing algebra with measured units, you can only add and subtract values with the same measured unit.* Consider the following equation:

$$0.0465 \text{ kg} + 45.9 \text{ g} + 766 \text{ mg} = x$$

You cannot simply add these values, because they are all given in different metric mass units. The easiest way to add the values is to convert them all to the same unit. In this case, we will convert everything to grams. (Metric conversions are covered in the main text in Sections 1.4 and 1.5.) Once this is done, the values can be added, because the units are all the same:

$$46.5 \text{ g} + 45.9 \text{ g} + 0.766 \text{ g} = x$$

gives

$$x = 93.2$$

> **QUESTION B.3** Find the value of x in units of liters (L):
>
> $$x = 4.56 \text{ L} + 650.0 \text{ mL} + 1{,}203.3 \text{ mL}$$
>
> **ANSWER B.3** $x = 4.56 \text{ L} + 0.6500 \text{ L} + 1.2033 \text{ L} = 6.41 \text{ L}.$

RULE 3   *When doing multiplication and division with measured units, units can-cel each other just as numbers do for these mathematical operations.* This rule comes in handy if you have a poor memory, because, very often, the way that units cancel will give you a clue to how the numbers should be manipulated. We shall use a typical chemistry example to demonstrate this technique:

> *The density of gasoline is 0.692 g/mL. If your gas tank holds exactly 50,000.0 mL of gasoline, what is the difference in the mass of your car when it is full of gas, compared with when it is completely empty?*

The question being posed is this: what is the mass of the gasoline in your full gas tank? You are given two numbers to work with. The density has units of mass (g) in the numerator and volume (mL) in the denominator. The volume of your tank (and, therefore, of the gas in it) has units of milliliters (mL). The question asks for mass, for example, in grams.

In the text, we shall teach you this equation:

$$\text{density} = \frac{\text{mass}}{\text{volume}}$$

You can rearrange the equation to give mass, your desired value, by multiplying both sides of the equation by volume, as we learned earlier using numbers:

$$(\text{volume})(\text{density}) = \frac{\text{mass}}{\cancel{\text{volume}}}(\cancel{\text{volume}})$$

gives

$$(\text{volume})(\text{density}) = \text{mass}$$

In this way, you can create an equation that will give you your desired value, mass. But what if you can't remember this equation? What can you do? How would you know that you should simply multiply volume and density to get mass? The an-swer is that the units for each value can guide you. In many, many cases, if your units cancel to give the correct unit, you have done your calculation correctly.

In this case, the original question gives you these measured values:

$$\frac{0.692 \text{ g}}{1 \text{ mL}} \quad \text{and} \quad 50{,}000.0 \text{ mL}$$

What could you do with these values to get a value with mass only? What would happen if you divided them? Divided them the opposite way? Multiplied them?

You could divide them to get

$$\frac{\dfrac{0.692 \text{ g}}{1 \text{ mL}}}{50{,}000.0 \text{ mL}} = 0.0000138 \text{ g/mL}^2$$

You could divide the other way to get

$$\frac{50{,}000.0 \text{ mL}}{\dfrac{0.692 \text{ g}}{1 \text{ mL}}} = 72{,}200 \text{ mL}^2/\text{g}$$

Clearly, these are not the units you want! What happens if you multiply the two values?

$$\frac{0.692 \text{ g}}{1 \text{ mL}} \times 50{,}000.0 \text{ mL} = 34{,}600 \text{ g}$$

Eureka! This is the correct unit, and we've also gotten the correct answer. In this case, and in many others, you will find that you don't need to memorize equations such as the one for density. This method of canceling units can often guide you to the correct answer.

QUESTION B.4 You may know how velocity and distance are related, but see if you can figure out the answer to this question without a memorized equation:

At 4:45 PM, Little Red Riding Hood leaves for her grandmother's house in her new sports car at a velocity of 175 km/h. If grandmother's house is 525 km away, what time will she arrive?

ANSWER B.4 The desired unit is hours. We have a value with units of km/h (the velocity) and a value in km units (the distance). If we could somehow multiply or divide these to get rid of the km, then we would be left with hours. The solution is as follows:

$$\frac{525 \text{ km}}{\dfrac{175 \text{ km}}{1 \text{ h}}} = 3.00 \text{ h}$$

If she does not get a speeding ticket, Little Red Riding Hood should arrive at 7:45 P.M. sharp.

This mathematics review should give you the basics you need to work through problems in this text. However, it is only a brief overview of what you need to know, and you will be asked to apply these math skills in different ways as you learn fundamental chemistry concepts. The trick is to be able to apply your math to new and different situations. After all, mathematics, and chemistry problems that use it, can be mastered only by practice.

Glossary

achiral   Not chiral; not possessing handedness or a chirality center.

acid   A source of H^+ ions; a substance that, when added to pure water, decreases the pH of the solution by increasing the concentration of H^+.

acid rain   Rain that has an abnormally low pH, often produced as a result of dissolved pollutant gases.

addition polymer   A macromolecule formed by the addition of monomers, one by one, to a chain.

addition reaction   A chemical reaction in which the number of bonds between atoms is reduced as a molecule is added across a bond.

alcohol   An organic molecule containing an —OH group; tends to form hydrogen bonds with water.

aldehyde   A molecule containing an aldehyde functional group.

aldehyde functional group   A functional group that consists of a carbonyl group with a hydrogen attached to the carbonyl carbon.

alkali metal   An element (except for hydrogen) in column 1 of the periodic table.

alkaline earth metal   An element in column 2 of the periodic table.

alkaloid   Any biologically active nitrogen-containing molecule derived from a plant source; typically, a basic molecule that creates an alkaline solution.

alkane   A member of a class of organic molecules that have only single bonds.

alkene   A member of a class of organic molecules that contain at least one double bond.

alkyne   A member of a class of organic molecules that contain at least one triple bond.

allotrope   Any one of two or more different forms of an element.

alpha decay   A type of radioactivity in which an alpha particle is emitted.

alpha helix   The type of secondary structure in globular proteins that consists of a right-handed coil with three to four amino acids making up each turn; depicted as a coiled ribbon or a cylinder.

alpha particle   A helium nucleus, which is a helium-4 atom stripped of its electrons and carrying a charge of +2; symbolized by α.

amide   A molecule containing an amide functional group.

amide functional group   A functional group that contains a carbonyl group, a nitrogen attached to one side of the carbonyl group, and a hydrocarbon attached to the other side.

amine   A molecule containing an amine functional group.

amine functional group   The functional group —NH_2.

amino acid   A molecule that contains an amine functional group and a carboxylic functional group.

Amontons' law   A principle about the behavior of gases that relates temperature and pressure: If the volume of a gas is fixed, an increase in gas temperature results in an increase in gas pressure; $P \propto T$.

amorphous region   In a polymer, a region that lacks specific shape and order.

amphipathic   Containing both a hydrophobic portion and a hydrophilic portion. The solubility of an amphipathic molecule is dictated by both hydrophobic and hydrophilic portions of the molecule.

anabolic pathway   A metabolic pathway in which new biomolecules are built up from smaller molecules.

anabolic steroid   Muscle-building drug popular with strength athletes.

anabolism   Any of the metabolic processes that include anabolic pathways.

anion   A negatively charged ion.

anode   In a battery, the electrode where oxidation takes place; also called a negative terminal.

anomer   A cyclic sugar, such as glucose, that can form another, slightly different type of ring. Pairs of anomers are distinguished as alpha or beta forms.

anomeric carbon   Carbon 1 of a monosaccharide that combines to form a disaccharide.

antibody   A protein created in the body to defend against antigens; also called an immunoglobulin; typically Y shaped and composed of two short protein chains and two long protein chains. Most antibodies are specific for only one antigen.

antigen   An invading molecule that may elicit a response—the production of antigen-specific antibodies—from the body's immune system.

antioxidant   A molecule that can donate an electron to a free radical.

aqueous solution   A watery solution.

atmospheric pressure   The pressure exerted on Earth's surface by the air in the environment.

atom   The smallest possible unit of matter that cannot be broken down by chemical or physical means.

atomic mass   The mass in amu of 1 atom of an element.

atomic number   The number of protons contained in a given element; the defining characteristic of each element.

autoionization   The reaction by which pure water (H_2O) produces the ions OH^- and H^+.

Avogadro's law   A description of how different numbers of atoms and molecules contribute to the properties of gases: As the number of moles of gas increases, so does the volume; $V \propto n$.

Avogadro's number   The number 6.02×10^{23}. One mole of anything contains Avogadro's number of that thing.

balanced equation   A chemical equation in which the numbers of each kind of atom on the left side of the equation match the numbers of each kind of atom on the right side.

base   A source of OH^- ions; a substance that, when added to pure water, increases the pH of the solution by increasing the concentration of OH^- ions.

base pair   A pair of nitrogenous bases that are complementary to each other; includes adenine and thymine (in DNA), adenine and uracil (in RNA), and cytosine and guanine (in both DNA and RNA).

battery A device in which an electric current exists as a result of electrochemical half reactions; a sophisticated version of an electrochemical cell in which different ions have different affinities for electrons, resulting in redox reactions that create a useful current.

belt of stability The region where stable isotopes fall on a plot of number of neutrons versus number of protons for a given nucleus; a graph showing relative stabilities of radioisotopes.

beta decay A type of radioactivity in which beta particles (electrons) are emitted.

beta particle An electron emitted from an atom's nucleus during beta decay; symbolized by β or $_{-1}^{0}\beta$.

beta pleated sheet The type of secondary structure in globular proteins that consists of an amino acid chain configured in a pleated shape that zigzags back and forth; depicted as a fat arrow.

biodegradable Describes molecules that can be broken down into smaller pieces by something in the environment, such as microorganisms or sunlight.

boiling point The temperature at which a liquid becomes a gas; the temperature at which the pressure inside a submerged bubble equals the atmospheric pressure pushing down from above.

bond energy The strength of a bond, measured in energy units (joules in the SI system). Double bonds are stronger than single bonds but weaker than triple bonds.

bond length The distance between atoms in a bond; shorter in triple bonds than in double bonds, and shorter in double bonds than in single bonds.

Boyle's law A principle about the behavior of gases that relates volume and pressure: Volume and pressure are inversely proportional to each other; $P \propto 1/V$.

branched hydrocarbon A hydrocarbon in which the main chain (the longest carbon chain in the molecule) has branches consisting of shorter carbon chains.

carbohydrate Any saccharide, large or small.

carbon dating A technique used to determine the age of organic substances; measures ratios of carbon isotopes, which remain constant in living organisms but change when the organisms die.

carbonyl group The $C=O$ bond found in esters, carboxylic acids, amides, and several other organic functional groups.

carboxylic acid An organic molecule containing a carboxylic acid functional group.

carboxylic acid functional group A functional group made up of a carbonyl group, $-C=O$, and a hydroxyl group, $-OH$, attached to the carbon atom of the carbonyl group.

carcinogenic Cancer causing.

catabolic pathway A metabolic pathway in which the large biomolecules in food are broken down into smaller molecules (and ultimately to carbon dioxide and water in human cells).

catabolism Any of the metabolic processes that include catabolic pathways.

catalyst A molecule that facilitates, and therefore speeds up, a chemical reaction.

cathode In a battery, the electrode where reduction takes place; also called a positive terminal.

cation A positively charged ion.

Celsius temperature scale The temperature scale, using the unit °C, that is based on assigning 0 °C as the melting point of water and 100 °C as its boiling point.

chain reaction A series of fissions that occurs when an initial collision between a neutron and certain radioactive atoms (such as uranium-235) produces multiple neutrons, which impinge on other radioactive atoms and amplify the reaction.

Charles's law A principle about the behavior of gases that relates volume and temperature: When the temperature of a gas is increased, the volume also increases; $V \propto T$.

chemical bond The interaction between joined atoms.

chemical change A change that results in the breakdown of a pure substance into one or more other pure substances.

chemical equation A notation chemists use to represent a chemical reaction; consists of symbols for a reactant species and product species separated by an arrow.

chemical formula A combination of elemental symbols that describes the composition of a chemical compound; consists of letters and, where applicable, subscript numbers (e.g., carbon dioxide: CO_2).

chemical reaction An interaction between substances such that atoms in the reacting substances can give up electrons to other atoms, take away electrons from other atoms, or share electrons with other atoms.

chemically neutral Having a pH of 7.00 in water; neither acidic nor basic.

chiral Referring to a molecule that possesses handedness and has a chirality center.

chirality center A carbon atom bound to four different groups. The presence of a chirality center is a predictor of handedness.

chlorofluorocarbons (CFCs) Small, simple organic molecules that contain carbon, chlorine, fluorine, and sometimes hydrogen; used as refrigerants (like Freon); degrade the ozone layer.

cis–trans isomerism When a double bond exists between two carbon atoms, the orientation of groups around the double bond may be designated as cis or trans; a kind of isomer that is unique to alkenes.

closed system A defined space where nothing is exchanged with the surroundings.

coefficient A number before a reactant or product in a chemical equation, indicating the number of molecules present before and after the reaction. When no number is given, the coefficient is understood to be 1.

combustion A heat-producing chemical reaction that involves the interaction between a fuel source and oxygen.

complementary base For a given nitrogenous base in RNA or DNA, the base that pairs with it in the complementary strand. For DNA, A pairs with T and C pairs with G. For RNA, C pairs with G and A pairs with U.

compound Any substance made up of atoms of two or more elements.

concentration gradient A difference in concentration on the two sides of a membrane; changed by the movement of substances across the membrane (which may require an input of energy).

condensation Phase change from gas to liquid.

condensation polymer A polymer formed from small polymers that condense into larger ones.

condensed structure A timesaving abbreviation for a full organic structure in which methyl groups are represented as $-CH_3$ and methylene groups as $-CH_2-$.

conjugated molecule   A molecule in which the electrons do not remain between the atoms of one bond, but instead spread out and become delocalized over all the areas of the molecule in which alternating double and single bonds exist; often brightly colored.

constitutional isomer   An isomer that contains the same atoms as another molecule, but the atoms are connected differently.

control experiment   An experiment in which control samples are compared with samples of interest; because control samples are constant and predictable, they are useful for comparisons with unknown samples.

conversion factor   A fraction in which the numerator is equal to the denominator; equal to 1.

copolymer   A polymer in which the repeating unit is made up of two or more different monomers.

core electron   In an atom, an electron that is not a valence electron.

covalent bond   An interaction between two atoms that share electrons.

critical mass   The minimum amount of radioactive material required to sustain a nuclear chain reaction.

cross-link   A connector of one or more atoms that joins two polymer chains together.

crystallite   Any highly ordered, crystalline region in a polymer. The more crystallites a polymer contains, the harder it is and the more resistant to external influences.

delocalization   The spreading out and diffusing of electrons in a region of alternating double and single bonds, providing extra stability (resonance stabilization) for the molecule.

density   The mass of any substance squeezed into a given volume (mass divided by volume).

deoxyribonucleic acid (DNA)   The hereditary material present in the nuclei of living cells; a type of nucleic acid that contains the sugar deoxyribose (instead of ribose, as in RNA).

detergent   An amphipathic molecule that contains a polar head group and a hydrocarbon tail; can form organized structures in water; often associated with cleaning.

diatomic   Describes a molecule with two like atoms.

diffusion   The process by which molecules of a gas quickly become integrated into the fast-moving molecules of another gas.

dimensional analysis   A system of cancellation and conversion factors that simplifies unit conversions within the metric system and between metric units and other types of units.

dipole   The imbalance of charge within a polar molecule.

dipole–dipole interaction   An interaction, such as that between two water molecules, in which there is no full charge involved; includes hydrogen bonding; a common type of noncovalent interaction.

disaccharide   A carbohydrate made up of two monomers; also called a sugar or simple sugar.

distillation   The process by which a mixture of molecules can be separated into components. The mixture is heated to a high temperature and then allowed to travel, in the gas phase, up a fractionating column, where different mixture components eventually cool and condense. Smaller molecules travel farther up the column before condensing.

disulfide bond   The link connecting cysteine amino acids in proteins; a sulfur-containing bond which acts as a reinforcement that helps to hold the overall protein structure together.

doping   The addition of an impurity to enhance a substance's ability to conduct electricity under specific conditions.

double bond   A multiple bond composed of a pair of bonds, each containing two electrons, for a total of four shared electrons; represented by two lines between atoms.

dry cell   A type of battery that contains an electrolyte paste rather than aqueous solutions, preventing the battery from leaking.

dynamite   An explosive, more stable than pure nitroglycerin, that consists of nitroglycerin mixed with an inert, solid substance such as charcoal or porous silica.

ED$_{50}$ The dose of a drug candidate that produces a therapeutic effect in 50 percent of test subjects; ED stands for "effective dose."

efficacy   The ability of a drug to produce some desirable effect appropriate for humans.

electrochemical cell   A device that uses chemical reactions to push electrons from one electrode to another as a result of differences in reduction potential between the two electrode half reactions.

electrode   A solid piece of metal in an electrochemical cell which can be immersed in a solution that conducts an electric current; connected to a second electrode by means of a metal wire.

electrolyte   An ionic compound (a salt) that, when dissolved in water, is capable of conducting electricity.

electromagnetic radiation   Light, a form of energy; represented on the electromagnetic spectrum according to wavelength.

electromagnetic spectrum   A continuum of light energies, including all types of electromagnetic energy; a representation of electromagnetic radiation, arranged by wavelength.

electron   A particle, located outside the nucleus of an atom, that carries a negative electrical charge.

electron configuration   A notation for multielectron atoms that lists sublevels of increasing energy in a horizontal series. The sublevel letter has a superscript number to its right, indicating the number of electrons in the sublevel.

electron density   A "smearing" of the average positions of all electrons in an atom.

electron dot structure   A sketch indicating the position of all valence electrons in a molecule; also called a Lewis structure.

electron-donating group   An organic functional group that adds electron density to other parts of the molecule that contains it.

electronegativity   The tendency of an atom to draw electrons toward itself; the difference in electronegativity between two atoms is a gauge of a bond's polarity.

electron-pair geometry   A sketch depicting all the atoms surrounding a central atom in a molecule as far away from one another as possible; not a true three-dimensional representation of a molecule, because it includes nonbonding pairs at the structure's vertices. (See also: molecular geometry and VSEPR theory.)

electron-withdrawing group   An organic functional group that pulls electron density away from other parts of the molecule that contain it.

electrophile   A region of positive charge that is attracted to regions of negative charge. (See also: nucleophile.)

element   Any substance composed of only one type of atom.

enantiomer   A molecule that has a nonsuperimposable mirror image; a molecule that is chiral, exhibiting handedness. (See also: handedness.)

endothermic reaction   Any reaction that requires heat.

energy level   A discrete energy value that an electron can have. Electrons are arranged in a series of energy levels that get progressively farther from the nucleus, with the level closest to the nucleus defined as the level of lowest energy.

energy sublevel   A subdivision of an energy level in a multi-electron atom as a result of electron–electron repulsions; designated with a number and a letter (e.g., 3s or 4p).

entropy   Disorder.

enzyme   A biomolecule, usually a protein, that acts as a catalyst; holds the substrate in a particular position in the active site, and distorts the substrate into a shape more like that of the transition state.

equilibrium   A condition in which there is no net change in a system as reactions proceed in both directions at equal rates.

ester   A molecule containing an ester functional group.

ester functional group   A functional group consisting of a carbon atom plus two oxygen atoms.

ether   A class of volatile organic molecules that undergoes combustion reactions and is similar to hydrocarbons, except that ethers have an oxygen inserted into the middle of the hydrocarbon chain; an oxygen link between two hydrocarbon chains; the generic structure R—O—R'.

evaporation   Phase change from liquid to gas.

excipient   An inactive ingredient used to hold together the active ingredients in a pill tablet or capsule; also used to make pills taste better, easier to swallow, and look prettier.

excited state   The state of an atom when its electrons are not located in the lowest available energy level(s); can be formed when energy is added to the atom.

exothermic reaction   Any reaction that gives off heat.

explosion   A violent expansion of gases and release of energy produced by a fast chemical reaction.

family   A column in the periodic table of elements; also called a group.

fatty acid   A simple type of lipid composed of a long hydrocarbon chain with a carboxylic acid functional group attached at one end; classified as nonpolar.

first law of thermodynamics   Energy can neither be created nor destroyed. (See also: second law of thermodynamics.)

fixation   The process by which an organism takes some or all of the atoms in a gaseous molecule and fixes them by incorporating them into a larger, nongaseous molecule.

formula unit   The smallest repeating unit in a salt.

fractionating column   A tall column, used in distillation, that separates a mixture into fractions containing molecules having similar masses and physical properties.

free base   An uncharged amine with a nonbonding pair of electrons on the nitrogen.

free radical   A molecule containing unpaired electrons; can be formed through homolytic cleavage.

freezing point   The temperature at which a liquid becomes a solid.

full structure   A sketch indicating every single atom in a molecule; more time consuming than other, abbreviated ways of drawing organic molecules.

functional group   Within an organic molecule, a group that is a modification of a simple hydrocarbon structure.

gamma radiation   A type of radioactivity in which very high energy, highly penetrating radiation in the form of gamma rays (symbolized by γ) is produced.

gas chromatography   One of the most well-known and reliable methods for the analysis of complex mixtures, by which different molecules are separated according to size and bulk. A printout of such an analysis is a gas chromatogram.

gene   A small segment of DNA in which the nucleotide sequence provides instructions for making a specific protein.

general gas law   A principle about the behavior of gases that relates pressure, volume, number of moles of gas, and temperature: PV/nT = constant.

genetic code   The system by which the four types of nitrogenous base in RNA are translated into one of the 20 types of amino acid during protein synthesis; mRNA triplets are equated with specific amino acids.

genetic engineering   Technology that allows us to manipulate the hereditary material contained in the cells of living things.

global warming   The unnatural increase in the temperature of Earth's atmosphere as a result of human activities.

globular protein   A roughly spherical biomolecule that results when a protein chain systematically folds upon itself. The bulk of known proteins are globular proteins.

greenhouse effect   The phenomenon by which lower-energy radiation reflected from Earth cannot pass through the atmosphere, remains where it is, and causes things to heat up; the primary cause of global warming.

greenhouse gas   A heat-absorbing gas that contributes to global warming.

ground state   The state of an atom in which its electrons are located in the lowest available energy level(s).

group   A column in the periodic table of elements; also called a family.

half reaction   An oxidation reaction or a reduction reaction; when combined, an oxidation reaction and a reduction reaction make up a redox reaction.

half-life   The time it takes for one-half of a radioactive sample to decay. The longer the half-life of a given radioactive element, the more slowly the element decays.

halogen   An element in column 17 of the periodic table.

handedness   The property by which two molecules are chiral and are enantiomers of each other; they have the same molecular formula but are nonsuperimposable mirror images of each other; having a right-handed form and a left-handed form. (See also: enantiomer.)

hard ion   An ion such as Ca^{2+} or Mg^{2+}, that makes water hard. The smaller the volume available to contain the charges on an ion, the harder the ion.

heterogeneous mixture   A mixture that has visible boundaries where one component stops and another begins because the composition changes from one region of a sample to another.

heterolytic cleavage   The breaking of a single covalent bond whereby both electrons go with one atom of the bond and the other atom takes no electrons.

homogeneous mixture   A mixture that has a smooth texture, making it impossible to see boundaries between the various components; also called a solution.

homologous series   A list in which the only difference from one organic molecule on the list to the next is that each molecule contains one carbon and two hydrogen atoms more than the molecule immediately before it.

homolytic cleavage The breaking of a single covalent bond whereby each atom of the bond takes one electron away with it, resulting in free radicals.

hydrated ion An ion that is surrounded by water.

hydration The process by which an ion is surrounded by water; a result of ion–dipole interactions.

hydrochloride salt A quaternary ammonium salt in which the anion is a chloride ion. (See also: quaternary ammonium salt.)

hydrogen bond An especially strong type of dipole–dipole interaction that has a substantial positive charge on a hydrogen atom because it is attached to a highly electronegative atom (such as oxygen, nitrogen, or fluorine). The hydrogen bond forms between the positively charged hydrogen atom and a nearby electronegative atom.

hydrogenation An addition reaction in which hydrogen atoms are added across a multiple bond.

hydrolysis A reaction in which the agent that breaks a chemical bond is a water molecule.

hydrophilic Dissolving readily in water; refers to molecules that possess charge, a dipole, the capacity for hydrogen bonding, or a combination of these properties.

hydrophobic Not dissolving readily in water; avoiding water; refers to highly nonpolar molecules.

hypothesis An initial best guess about some question concerning nature that may then be tested by experimentation. (See also: theory.)

ideal gas A gas that is infinitely dilute and in which there are no chemical interactions between the individual atoms or molecules. (See also: real gas.)

immiscible Unable to mix; refers to two liquids that do not mix.

indicator Typically a brightly colored organic molecule that changes color abruptly as the pH of a solution is changed.

induced-fit model The model of enzyme–substrate interaction whereby the substrate is distorted as it binds to the enzyme's active site in a way that forces the substrate's shape to be more like the transition state's shape.

infrared light A low-energy form of light; associated with heat.

inorganic Nonorganic; not carbon–based.

ion An electrically charged atom; if positively charged, a cation; if negatively charged, an anion.

ion exchanger A device that can transform hard water to soft water by removing hard ions.

ion–dipole interaction The interaction between an ion and a dipole, such as water; includes hydration of a cation or anion; the reason salts dissolve readily in water.

ionic bond The strong electrical attraction between a cation (positively charged) and an anion (negatively charged); the strongest type of noncovalent interaction. (See also: salt.)

ionic interaction An interaction between positive and negative charges.

ionizing radiation Any radiation that causes water molecules to ionize; radiation with an energy greater than 1200 kJ/mol.

irreversible inhibitor A drug that binds covalently (and therefore tightly) to an enzyme's active site, making the enzyme unavailable; usually a highly reactive molecule that interacts with other molecules of the body and therefore has deleterious side effects.

isomer A molecule that has a different structure, but the same molecular formula, as another molecule; includes constitutional isomers and enantiomers.

isotope One of two or more atoms that have the same number of electrons and the same number of protons, but different numbers of neutrons.

joule The SI unit for energy; abbreviated J.

kelvin The base unit of the Kelvin temperature scale.

Kelvin temperature scale The temperature scale, using the base unit kelvin, that is based on a fundamental reference point of 0 K (zero kelvin); the temperature at which all motion stops; used in scientific work.

ketone An organic functional group that has a carbonyl group attached to two hydrocarbon groups.

kilocalorie A traditional unit for expressing energy; equivalent to a Calorie, but sometimes written as "calorie" in everyday usage; equal to 1 kcal = 4.184 kJ.

LD_{50} The amount of toxin or poison that kills 50 percent of subjects exposed to it; "LD" stands for "lethal dose."

lead compound A molecule that acts as a starting point for the design of a superior drug.

line spectrum The pattern that results when light emitted from excited atoms of one element passes through a prism and separates into specific wavelengths (colors).

line structure A timesaving abbreviation of a full structure in which the chemical symbols for H atoms and C atoms are not shown.

lipid A biological molecule that dissolves readily in nonpolar solvents; includes fats, which are solids at room temperature, and oils, which are liquids at room temperature.

lipoprotein An enormous composite molecule that carries cholesterol through the bloodstream. There are two varieties of lipoproteins: low-density lipoproteins (LDLs), which have a low protein-to-lipid ratio and tend to form artery-clogging plaque, and high-density lipoproteins (HDLs), which have a high protein-to-lipid ratio and tend to reduce the buildup of plaque.

magnetism A property of materials that contain unpaired electrons.

mass number The total number of neutrons plus protons in the nucleus of an atom; sometimes shown as a superscript before the symbol of the element (e.g., ^{17}O) or as a number following the name of the element (e.g., oxygen-17).

matter The "stuff" of which all physical material is composed. Matter is made up of atoms.

mean free path The average distance a gas molecule travels between collisions.

melting point The temperature at which a solid becomes a liquid.

messenger RNA (mRNA) The RNA that is made in the nucleus by transcription and that travels to a ribosome—a protein-making center, located outside the nucleus, where mRNA is used to produce proteins by translation.

metabolic pathway Any series of reactions that take place in a living organism.

metal An element to the left of the thick, stepped line in the periodic table (except for hydrogen); typically a shiny, lustrous solid.

metallic bond An interaction in which atoms are ionized and electrons are shared in a fluid sea of electrons; allows solid metals to be malleable and to conduct heat and electricity readily.

methyl group One carbon and three hydrogen atoms; typically attached to one end of an organic molecule; written in condensed form as $—CH_3$.

methylene group   A carbon in the middle of a chain in an organic molecule; written in condensed form as —CH_2—.

micelle   A spherical structure, composed of surfactants with hydrocarbon tails, with a highly nonpolar interior (made of hydrocarbon tails) and a polar exterior (made of polar head groups) when exposed to an aqueous solvent.

millimeters of mercury (mmHg)   The distance, in millimeters, that the mercury in a barometer tube travels as a result of applied gas pressure; atmospheric pressure at sea level is typically about 760 mmHg.

millimolar   A unit of concentration giving the number of millimoles of some substance per liter of solution; abbreviated mM. There are 1000 millimoles in 1 mole.

mineral   A naturally occurring inorganic substance that is a crystalline solid.

mixture   Two or more pure substances combined together; may vary in composition from one part of the sample to another.

model   A design, such as the periodic table or the structure of a molecule, used to classify and organize information; a means for understanding nature.

molar mass   The mass of 1 mole of a substance.

molar volume   The volume occupied by 1 mol of an ideal gas at standard temperature and pressure (0 °C and 760 mmHg): exactly 22.414 L.

molarity   A unit of concentration that indicates the number of moles of some substance per liter of solution; also called molar concentration; abbreviated M.

mole   A counting device equal to 6.02×10^{23} (Avogadro's number) of anything; abbreviated mol. One mole is formally defined as the number of carbon atoms in 12.000 g of carbon-12.

molecular geometry   A molecule's three-dimensional shape, which does not include nonbonding pairs at the structures vertices. A sketch of molecular geometry shows the three-dimensional arrangement of only the atoms in the molecule and the bonds between them. (See also: electron-pair geometry and VSEPR theory.)

molecule   The smallest unit that makes up a substance in which the atoms share electrons.

monomer   A small organic molecule; one of the building blocks of a polymer.

monosaccharide   A carbohydrate made up of only one monomer; also called a sugar or a simple sugar.

multiple bond   An interaction in which more than two electrons are shared by two atoms; commonly found between period 2 elements because they are very small and can get very close to one another. (See also: uniqueness principle.)

nanotube   A hollow carbon cage roughly one ten-thousandth the width of a human hair; under development for use in many practical applications.

network solid   A solid that has an extended system of repeating covalent bonds; the system tends to make network solids, including diamond, very robust.

neurotransmitter   A molecule that is released from the end of one neuron, travels across the synaptic cleft which separates that neuron from another one, and binds to receptors on the end of the next neuron.

neutron   A particle, located in the nucleus of an atom, that carries no electrical charge.

nitrogenous base   The nitrogen-containing part of a nucleotide.

noble gas   An element in column 18 of the periodic table. Each noble gas has an ideal number of electrons and does not interact with anything else under ordinary conditions.

noble gas configuration   The electron distribution of a noble gas; an ideal number of electrons, making the element stable, so it will not react with anything else.

nomenclature   A standard naming system—e.g., for hydrocarbons.

nonbonding pair   A pair of valence electrons that does not take part in bonding.

noncovalent interaction   An interaction in which no electrons are shared; includes ion–dipole, dipole–dipole, and ionic interactions; also includes hydrogen bonding.

nonmetal   An element to the right of the thick, stepped line in the periodic table; typically a gas or a brittle, dull solid.

nuclear fission   A nuclear reaction whereby an unstable nucleus splits into roughly equal parts to give two nuclei of similar size; produces enormous amounts of energy; used to produce power in nuclear power plants.

nuclear fusion   A nuclear reaction whereby nuclei combine to form a larger nucleus. All the energy we receive from the Sun is created by solar fusion reactions.

nuclear reaction   A reaction in which the number of protons or neutrons (or both) in the reacting atoms changes in some way, such that atoms of one element become atoms of another element; includes nuclear fusion and radioactive decay and, in general, gives off large amounts of energy compared with chemical reactions.

nucleation center   The first few hydrogen-bonded water molecules that form the basis of the crystalline lattice of ice during freezing.

nucleic acid   One class of biopolymers, existing primarily in a cell's nucleus, with nucleotides as the monomers; includes DNA and RNA.

nucleophile   A region of negative charge that seeks out and attacks regions of accumulated positive charge on other atoms. (See also: electrophile.)

nucleotide   A monomer that consists of a phosphate group, a sugar, and a nitrogenous base; the building block of a nucleic acid.

nucleus   The region of an atom containing protons (particles that are positively charged) and neutrons (particles that carry no electrical charge).

octane number   A number that is assigned to a hydrocarbon fuel to indicate the relative amount of knocking it causes when it undergoes combustion in an engine. A fuel with an octane number of 100 causes no knocking; a fuel with an octane number of 0 causes constant knocking.

octet rule   The principle whereby the atoms in most organic molecules have a total of eight valence electrons, including the electrons they share with other atoms; useful in drawing electron dot structures. H is an exception; it must have two valence electrons.

oligosaccharide   A carbohydrate made up of a few to about 100 monomers.

open system   A defined space that is always exchanging matter and/or energy freely with the environment surrounding it; can be heated or cooled by the surrounding environment.

opiate   A drug that either binds to an opiate receptor or affects the binding of other molecules to an opiate receptor; affects the brain and causes an imbalance in the natural neurotransmitter levels in the synaptic cleft, producing analgesic and euphoric effects.

organic compound   A compound with a structure based on a carbon framework.

osmosis   The movement of water molecules across a semipermeable membrane from a region of low ion concentration to a region of higher ion concentration.

oxidation reaction   A reaction characterized by a loss of electrons; a half reaction. (See also: reduction reaction.)

oxidizing agent   A species that takes electrons from another species undergoing an oxidation reaction. (See also: reducing agent.)

oxygenate   A source of oxygen that promotes the combustion of gasoline.

ozone layer   The thin, fragile shield that protects Earth from harmful ultraviolet rays; composed, in part, of ozone molecules (O_3).

partial charge   Part of a charge in one portion of a polar molecule, balanced by a part of a charge of opposite sign in another portion of the molecule; less than a full charge.

parts per billion   A unit of concentration that is used to express a very small amount of one thing mixed with something else; abbreviated ppb; 1 ppb = 1000 ppm; one part of one substance mixed with 1 billion parts of another.

parts per million   A unit of concentration that is used to express a very small amount of one thing mixed with something else; abbreviated ppm; 1 ppm = 0.001 ppb; one part of one substance mixed with 1 million parts of another.

pascal   The SI unit for pressure.

peptide   A short protein; also called a polypeptide.

peptide bond   The amide bond between an amine functional group and a carboxylic acid functional group in a protein backbone.

period   A horizontal row in the periodic table of elements.

periodic table   An arrangement of elements into rows (called periods) and columns (called groups or families).

pH   The negative logarithm of the hydrogen ion molar concentration: pH = $-\log[H^+]$; a measure of how acidic (pH < 7) or basic (pH > 7) a solution is.

phase   The state of matter—solid, liquid, or gas.

phenol   An alcohol in which the —OH group is attached to a benzene ring.

phospholipid   A molecule composed of a glycerol backbone with two fatty acid molecules attached and the third position on the glycerol backbone occupied by a phosphate group connected to some small organic group; the primary lipid component of biological membranes.

physical change   A process that alters a substance without changing it into some other substance; often a change of phase, as from solid to liquid.

plastic   An item, made from polymers, that can often be molded, stretched, or flattened into a variety of conceivable shapes.

pOH   The negative logarithm of the hydroxide ion molar concentration: pOH = $-\log[OH^-]$.

polar bond   A bond in which the electrons are shared unequally, making one side of the molecule electron rich and the other side electron poor.

polar molecule   A molecule with one end that is relatively more positive and another end that is relatively more negative, forming an imbalance called a dipole.

polyatomic ion   An ion that contains more than one atom.

polymer   A long chain composed of monomers.

polypeptide   A short protein; also called a peptide.

polysaccharide   A carbohydrate made up of hundreds or even thousands of monomers; also called a complex carbohydrate.

precipitation   The return of an ion in solution to a solid phase crystalline lattice; a substance "falling out" of solution.

pressure   The phenomenon of molecules exerting a force against a surface. More collisions equal higher pressure; fewer collisions equal lower pressure.

primary amine   An amine in which one of the three hydrogen atoms of ammonia has been replaced by a hydrocarbon group.

primary structure   The sequence of amino acids in a protein; the backbone of a protein structure.

product   One of the substances present after a chemical reaction has been carried out.

protein   A natural polyamide; a polymer chain composed of amino acids linked together.

protein folding   The process whereby a protein chain forms contacts and coalesces upon itself to create the tertiary structure of a protein.

proton   A particle, located in the nucleus of an atom, that is positively charged.

protonation   The addition of a proton (H^+) to an atom; an atom to which a proton has been added is said to be protonated.

pure substance   Matter that has a definite composition; either an element in its pure form or a compound with a fixed ratio of one atom to another.

qualitative test   An analytical test that identifies the compound being tested.

quantitative test   An analytical test that tells us the amount of compound present.

quaternary ammonium salt   A protonated amine coupled with an anion; a type of ionic species that can exist in an organic molecule. (See also: hydrochloride salt.)

racemate   A drug that contains equal amounts of *R* and *S* enantiomers; also called a racemic mixture.

radiation   The products of a nuclear reaction in the form of alpha decay, beta decay, or gamma radiation.

radioactive decay   A process whereby an atom gives off some energetic product, in the form of either a particle or radiation.

radioactive decay series   A sequence of nuclear reactions that continues until a stable isotope is reached.

radioactivity   The emission of radiation.

random coil   In a globular protein, a short segment of the amino acid chain that often connects segments of alpha helix and beta pleated sheet.

reactant   One of the starting materials in any chemical reaction.

reaction coordinate   A diagram that plots energy (on the vertical axis) versus the progress of a chemical reaction (on the horizontal axis).

real gas   A gas in which interactions between the individual atoms or molecules are not negligible. (See also: ideal gas.)

receptor   A biomolecule that binds a drug. Most receptors are proteins. (See also: structure–activity relationship.)

recombinant DNA   DNA that has been genetically altered (broken and then recombined).

redox reaction   The combination of a reduction reaction and an oxidation reaction (two half reactions).

reducing agent   A species that provides electrons to another species undergoing a reduction reaction. (See also: oxidizing agent.)

reduction potential   A measure of the potential for reduction of metal ions.

reduction reaction   A reaction characterized by a gain of electrons; a half reaction. (See also: oxidation reaction.)

rem   A *roentgen equivalent for man*; the commonly used unit for effective radiation dose; 1 rem = 0.01 Sv.

resonance stabilization   The extra stability that the delocalization of electrons provides to a molecule that contains alternating double and single bonds.

restriction enzyme   An enzyme that cuts DNA, typically at a specific location in a DNA sequence; makes genetic engineering possible.

reversible chemical reaction   A reaction that can proceed in two directions: from reactants to products and from products to reactants.

reversible inhibitor   A molecule that binds noncovalently to an enzyme's active site and tends to have fewer deleterious side effects than an irreversible inhibitor does.

ribonucleic acid (RNA)   A type of nucleic acid that contains the sugar ribose (instead of deoxyribose, as in DNA).

saccharide   The monomer in a polysaccharide; a single sugar unit.

salt   A neutral compound formed from the complete transfer of one or more electrons from one atom to another to create a cation and an anion; any ionic solid or ionic liquid. (See also: ionic bond.)

saturated   Containing only single bonds. A saturated molecule will not accept any more H atoms.

saturated solution   A solution that contains the maximum amount of a dissolved substance, often a salt.

scientific method   The general approach to experimentation that includes verifying the reproducibility of results and peer review.

scientific notation   A convenient way of dealing with exceedingly small or exceedingly large numbers; involves moving a number's decimal point to result in a number between 1 and 10 that is multiplied by 10 raised to a power equal to the number of places the decimal point was moved.

second law of thermodynamics   A fundamental law of nature which states that the total amount of disorder in the universe is always increasing; total disorder (entropy) must increase whenever any event takes place. (See also: first law of thermodynamics.)

secondary amine   An amine in which two of the three hydrogen atoms of ammonia have been replaced by hydrocarbon groups.

secondary structure   The local folding of an amino acid chain, dictated by the amino acid sequence (primary structure) of the protein. Common secondary structures are alpha helices and beta pleated sheets.

semiconductor   A substance, such as silicon, that is conductive only under specific conditions, such as when a minimum amount of energy is available in the form of heat or high-energy light, allowing valence electrons to jump to a higher energy level.

semipermeable membrane   A barrier that allows only certain molecules to pass through it.

shortened condensed structure   A timesaving abbreviation of a full structure, simplified further than a condensed structure, in which all methylene groups are represented by one methylene unit plus a subscript indicating how many of these units there are in the molecule (e.g., $CH_3(CH_2)_4CH_3$).

side chain   The R group of an amino acid.

sievert   The SI unit for the amount of radiation to which a human being is exposed; abbreviated Sv; an effective unit, because it takes into account the differences in penetrating power of the three types of radiation; 1 Sv = 100 rem.

single bond   An interaction in which two electrons are shared by two atoms in a molecule.

solubility in water   The degree to which a substance will dissolve in water; often expressed in molar units or in mass/volume units.

solution   A mixture that has a smooth texture, making it impossible to see boundaries between the various components; also called a homogeneous mixture.

specific heat   The amount of heat absorbed by 1 g of a substance when its temperature increases by 1 °C.

specificity   The ability of a drug to bind to its receptor only, not to other biomolecules in the body.

standard temperature and pressure (STP)   The combination of the temperature 0 °C and the pressure 760 mmHg; a convenient standard set of conditions for gas-phase experiments.

steric hindrance   A phenomenon in which the size and bulk of a molecule hinder its range of motion or the approach of other groups to it.

sticky end   The uneven end of a double-stranded DNA fragment cut by a restriction enzyme.

strong nuclear force   The very strong attractive force responsible for the attraction between protons and protons, between protons and neutrons, and between neutrons and neutrons; released in nuclear reactions, the strong nuclear force is what accounts for the large amount of energy given off.

structural isomer   A molecule that has the same molecular formula as, but a structure different from, that of one or more other molecules.

structural protein   A nonglobular (typically fibrous) protein that has a protective role, does mechanical work, or is stretchy and flexible.

structure–activity relationship   For any two molecules, a description of how well matched the surfaces of the molecules are to each other. The stronger the structure–activity relationship for any pair of molecules, the more tightly they bind together; usually refers to the interaction between a drug and its target molecule. (See also: receptor.)

substituent   A group appended to the main chain of a hydrocarbon.

substrate   A reactant in an enzymatic reaction.

sugar–phosphate backbone   The alternating pattern of phosphate groups and sugars to which the nitrogenous bases attach in a nucleic acid.

surface tension   The forces between water molecules on the water surface; the result of water molecules at the surface being pulled only sideways and downward, not upward.

surfactant   An amphipathic molecule that keeps its hydrophobic tail(s) away from water; typically depicted with a circle representing the polar head group and a wavy tail representing each hydrocarbon chain; often forms structures on surfaces, such as the interface between lung tissue and air in the lungs.

sustainability   Relating to the method of harvesting a resource so that the resource is not depleted or permanently damaged.

TD$_{50}$ The dose of a drug candidate that elicits some specific, measurable toxic effect in 50 percent of test subjects; TD stands for "toxic dose."

tertiary amine   An amine in which all three of the hydrogen atoms of ammonia have been replaced by hydrocarbon groups.

tertiary structure   The fully folded, compact, three-dimensional shape of a protein, stabilized by disulfide bonds, hydrophobic interactions between amino acid side chains, and an extensive network of hydrogen bonds.

tetrahedron   A pyramid made up of four faces, each an equilateral triangle; a very common structural motif in carbon-containing molecules.

theory   An explanation of observed phenomena based on collected data. Evidence can support or disprove a theory, but can never prove a theory, which provides the latest and best interpretation of experimental results. (See also: hypothesis.)

therapeutic index   The ratio TD$_{50}$/ED$_{50}$, which provides an overall indicator of drug performance.

thermoplastic polymer   A polymer that, when heated, can re-form crystallites and therefore retain its shape; a recyclable polymer.

thermosetting polymer   A polymer that, when heated, does not change shape; a nonrecyclable polymer.

transcription   The process in which an mRNA sequence is made from a DNA template; a step in protein synthesis. (See also: translation.)

transgenic   Refers to an organism that has been modified, through genetic engineering, to contain hereditary material from another organism.

transition metal   An element in the center block of the periodic table—in the transition between the left and right sides of the table; a d-block element.

transition state   A shape into which a substrate must become distorted in order for a chemical reaction to proceed.

translation   The process in which an mRNA sequence is translated into a sequence of amino acids; a step in protein synthesis. (See also: transcription.)

transmutation   The conversion of one element into another.

triacylglyceride (TAG)   A molecule that contains three fatty acids connected by glycerol; also called a triglyceride.

triple bond   A multiple bond composed of a triplet of bonds, each containing two electrons, for a total of six shared electrons; represented by three lines between atoms; the C≡C bond makes up the alkyne organic functional group.

triplet   A group of three mRNA nitrogenous bases; in protein synthesis, the group that is equated with a specific amino acid according to the genetic code.

ultraviolet light   A high-energy form of light; associated with sunburns; abbreviated UV.

uniqueness principle   A property of period 2 elements whereby an element other than the top one in the group defines the size trend of the atoms in that group. The paucity of electrons and their proximity to the nucleus make atoms of period 2 elements very small.

unsaturated   Containing one or more multiple bonds. An unsaturated molecule can accept additional H atoms.

vacuum   A space in which no matter is present.

valence electron   One of the outermost electrons of an atom. Only valence electrons can be added to or taken away from an atom.

vapor pressure   The pressure exerted by a liquid's vapor in the air above the liquid when the two phases are in equilibrium; determined at a specific temperature.

variable   Quantity that can change in an experiment.

virus   An infecting pathogen that carries its own genetic material; a tiny pouch made of protein and containing either DNA or RNA.

visible light   A midenergy form of light; the only light observable with human eyes.

volatile   Referring to molecules that escape easily into the air.

VSEPR (valence shell electron pair repulsion) theory   The theory stating that all the atoms surrounding a central atom in a molecule tend to get as far away from one another as possible. (See also: electron pair geometry and molecular geometry.)

vulcanization   A process in which sulfur is added to rubber, forming disulfide cross-links.

water vapor   The gaseous form of water.

Answers to Odd-numbered Questions

Chapter 1

1.1 These discoveries were revolutionary because they changed the belief that matter was inert and rigid by demonstrating that matter could absorb and emit energy. Discoveries include medical x-rays, nuclear weapons, and instrumentation to determine the structure of DNA and other chemical molecules.

1.3 **a.** 12 atoms, six carbon atoms and six hydrogen atoms
b. Two atoms, both bromine
c. Three atoms, two nitrogen atoms and one oxygen atom

1.5 **a.** compound   **b.** pure substance
c. compound   **d.** pure substance

1.7 compound, pure substance

1.9 **a.** compound   **b.** element
c. element   **d.** compound

1.11 5 seconds

1.13 4.5 g

1.15 $C_3H_8 + 5\,O_2 \longrightarrow 3\,CO_2 + 4\,H_2O$

1.17 **a.** micromole   **b.** femtometer
c. megamole   **d.** decisecond

1.19 An element consists of only one type of atom, existing individually or connected in three-dimensional arrangements. Atoms of an element have the same number of protons and electrons.

1.21 **a.** the coefficient in front of HCl is incorrect.
b. six   **c.** All atoms are balanced.

1.23 Samples are injected into instruments that are computer controlled.

1.25 **a.** 20,000 seconds, too long to wait
b. 333 minutes
c. 5.6 hours
d. 0.000634 years

1.27 34,500,000 mcd

1.29 **a.** 67   **b.** 5   **c.** 58

1.31 **a.** mass units (g or kg)
b. volume units (mL or L)
c. the volume

1.33 **a.** electron: 0.00000000000000000000000000091 g;
proton: 0.0000000000000000000000017 g
b. proton
c. 1868

1.35 3.5 atoms

1.37 The samples from Quebec, Morocco, and Norway are impure.

1.39 No. A chemical reaction or change would be required.

1.41 17 K

1.43 45 minutes

Chapter 2

2.1 iron > sodium > hydrogen > phosphorus > barium > vanadium. Iron, sodium, barium, and vanadium are metals; phosphorus and hydrogen are nonmetals.

2.3 **a.** 1×10^{-10} m   **b.** 0.1 nm

2.5 5.6×10^9 nm or 5.6 m, 4.76×10^{-5} g or 47.6 μg, 4.5×10^3 mg or 4.5 g

2.7 brown asbestos

2.9 The statement is correct.

2.11 **a.** The copper atoms are mostly Cu^{2+} surrounded by a sea of electrons.
b. The charge on the atom is balanced by electrons rather than by anions.

2.13 naturally occurring and inorganic

2.15 **a.** $Cu^{2+} + 2$ electrons $\longrightarrow Cu$
b. $Cu \longrightarrow Cu^{2+} + 2$ electrons; oxidation

2.17 **a.** BaO **b.** Rb_3N **c.** MgF_2

2.19 125.40 g/mol

2.21 239.3 g/mol

2.23 Crystal, mineral, and salt are related terms. A crystal is a broader term and can include things that are not minerals. Salts are minerals. Salt and formula unit are also related. Salt is any ionic compound and formula unit is the smallest repeating unit in a salt.

2.25 **a.** ionic
b. polar covalent (slightly)
c. nonpolar covalent
d. ionic
e. polar covalent
polarity a > d > e > b > c

2.27 **a.** A mole is a convenient unit of quantity that is independent of the size of the atom or molecule.
b. Avogadro's number is large because atoms and molecules are so small.

2.29 **a.** Unnaturally elevated levels of trace minerals in dirt indicate that the object or person has been to an area containing dirt composed of trace minerals at the levels found.
b. Taking samples from various locations and measuring the amounts of different trace minerals makes it possible to determine the geographical source of the dirt. Soil samples are deposited onto a surface in the order in which each location was visited.

2.31 Ohgreatium-232 has 116 neutrons; ohgreatium-234 has 118 neutrons; ohgreatium-235 has 119 neutrons.

2.33 The magnetite crystals in the Mars meteorite have the same shape and size as the magnetite crystals formed by bacteria on Earth. Geologically formed magnetite crystals have a different shape.

2.35 Elements in the same column have the same number of valence electrons. Elements in the same row have their valence electrons in the same shell.

2.37 879.29 g/mol

2.39 Dirt samples around the bank and from Mississippi should be tested in an identical way.

2.41 **a.** Magnetite is magnetic. It is present if a magnet is attracted to the meteorite.

b. Meteorites are heterogeneous mixtures containing substances in various proportions.

c. The magnetic crystals made by bacteria are small, oblong, and defect-free.

2.43 a. redox—Fe^{2+} is the reducing agent, Cu^+ is the oxidizing agent.

b. reduction, oxidizing agent

c. oxidation, reducing agent

Chapter 3

3.1 diamond (clear, colorless, hard solid with three-dimensional network of C—C bonds), graphite (shiny, gray, slippery solid with layers of carbon rings that slide across one another), and buckminsterfullerene (powdery, dark solid with large geodesic balls consisting of 60 carbon atoms)

3.3 Both are inorganic because both lack carbon and hydrogen.

3.5 b, d, e, and f

3.7

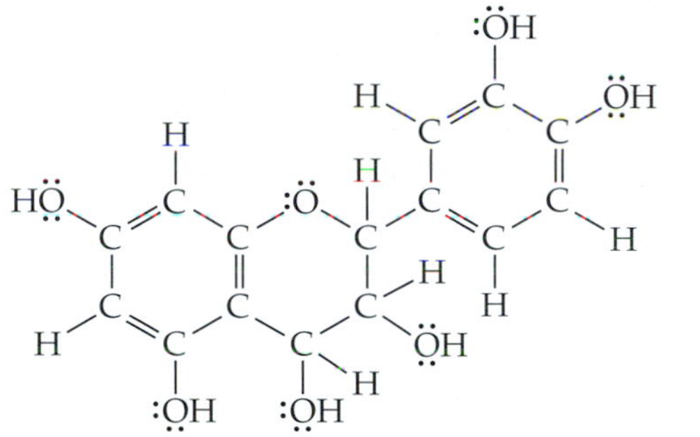

3.9 a. 1.39 Å **b.** 0.139 nm **c.** 0.000139 μm

3.11 $2 H_2(g) + O_2(g) \longrightarrow 2 H_2O(l)$; 7 moles H_2, 3.5 moles O_2

3.13 The two nonbonding electron pairs take up slightly more space than the electrons in the O—H bonds, thereby bending the two O—H bonds closer together.

3.15 a. $\cdot\dot{P}\cdot$ **b.** 3 bonds **c.** nitrogen

3.17 3138 kJ

3.19 a. 3.66 L **b.** 42.7 g/mile

3.21 P-type and n-type silicon crystals are placed side by side. When electrons from the n-type crystal travel to fill holes in the p-type crystal, the system releases energy in the form of light.

3.23 one

3.25 Buckyballs are small geodesic-shaped balls that can tumble over each other.

3.27 a.

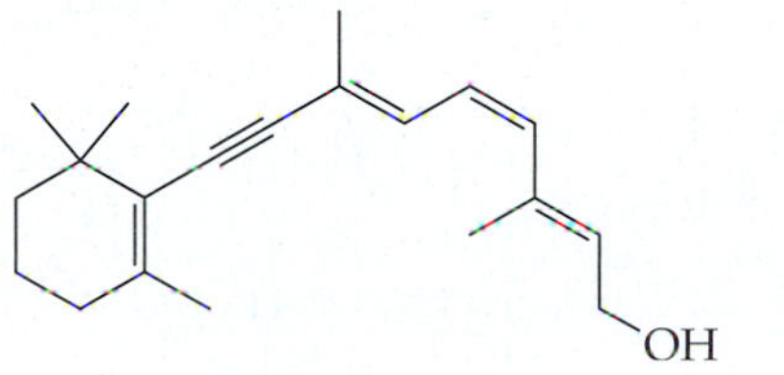

b.

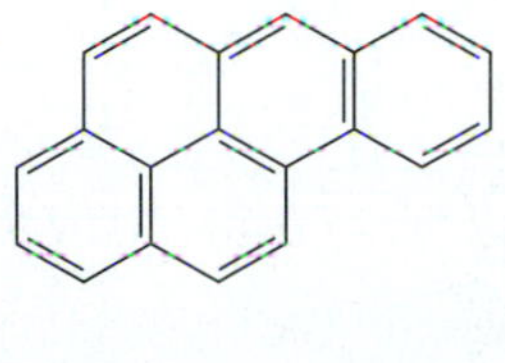

3.29

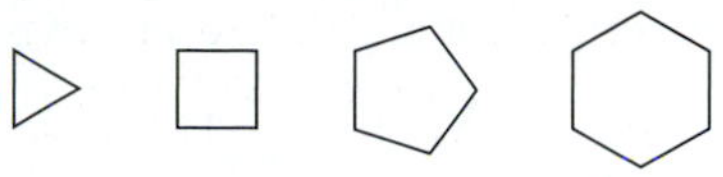

3.31 Prozac
a. Prozac: $C_{17}H_{18}F_3NO$; Cocaine: $C_{17}H_{21}NO_4$
b. Prozac: 309.3 g/mol; Cocaine: 303.4 g/mol
c.

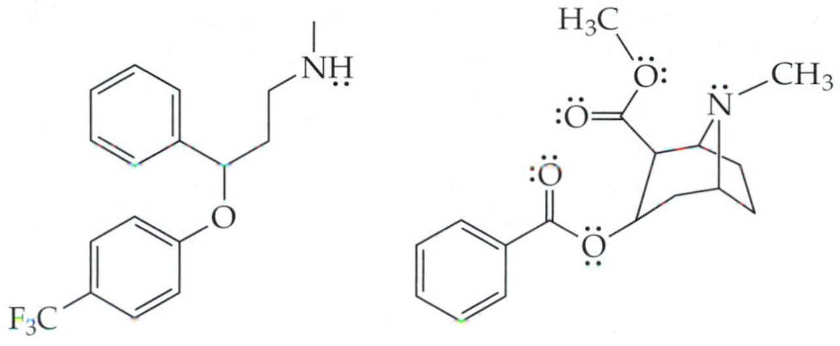

Prozac®   Cocaine

3.33

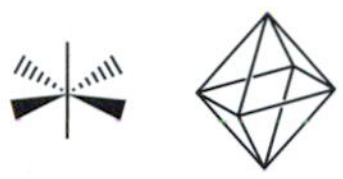

3.35 a.

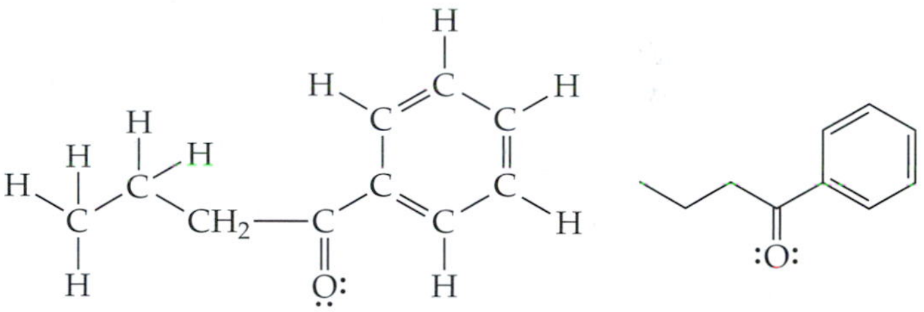

b.

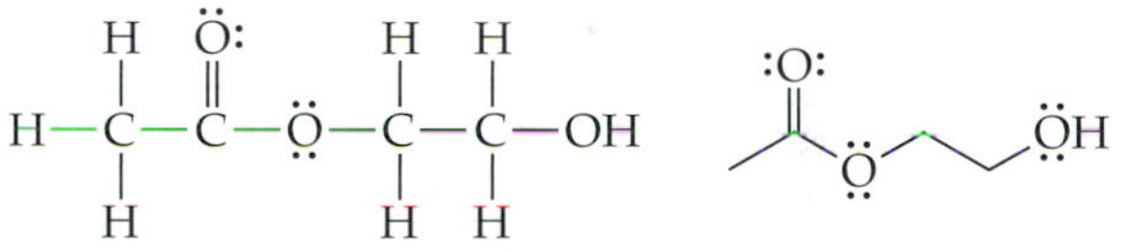

3.37 a.

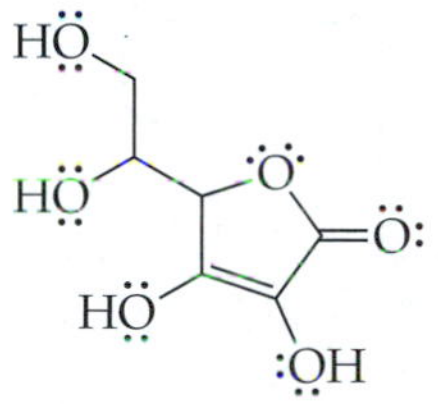

b. $C_6H_8O_6$; 176.14 g/mol

3.39 a.

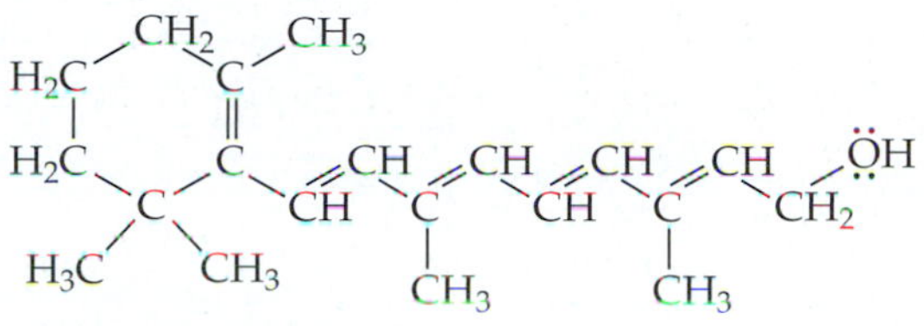

b. $C_{20}H_{30}O$, 286.5 g/mol
c. The alternating single and double carbon bonds result in resonance stabilization.

3.41 Several possible structures (but not all) are shown below.

a.

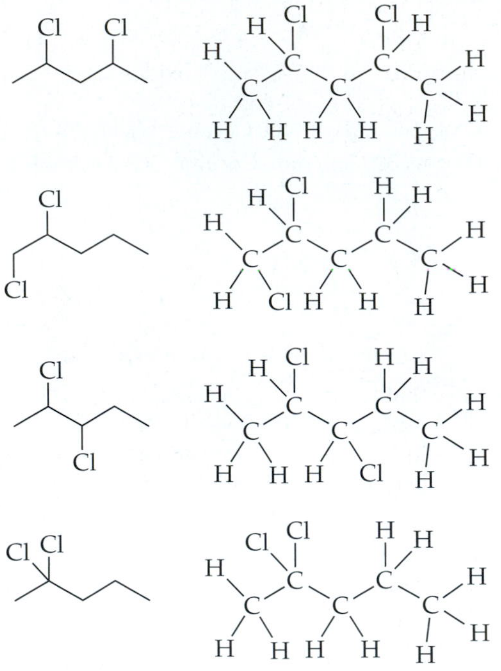

b.

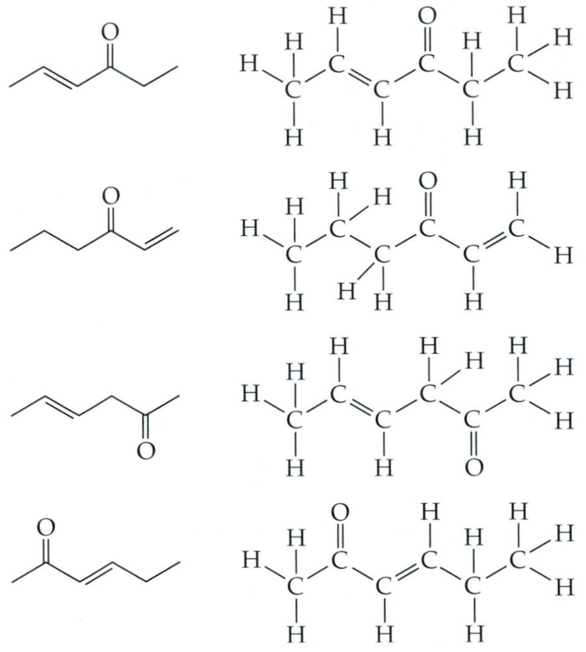

c.

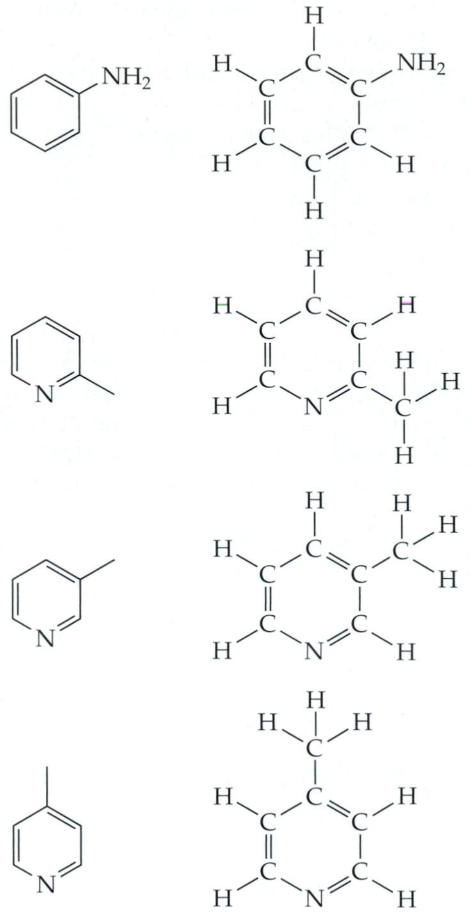

d.

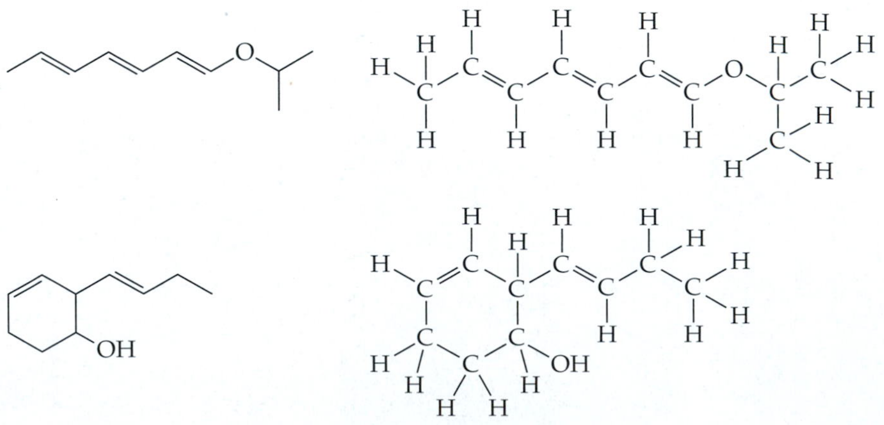

Chapter 4

4.1   **a.** Na^+: 1.16 Å; Cl^-: 1.76 Å

   b. Na^+: 0.116 nm; Cl^-: 0.176 nm

4.3   **a.**

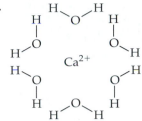

   b.

4.5   **a.** Li_2S, lithium sulfide

   b. Na_2O, sodium oxide

   c. $MgBr_2$, magnesium bromide

   d. CaF_2, calcium fluoride

   e. SrI_2, strontium iodide

4.7   yellow is KOH; red is $HClO_4$

4.9   **a.** acidic   **b.** basic   **c.** basic

   d. acidic   **e.** slightly basic

4.11   Hard water contains Mg^{2+} and Ca^{2+} ions. The inexpensive ion exchanger cannot regenerate after it has given all its Na^+ ions to the water in exchange for Mg^{2+} and Ca^{2+}. To remedy the problem, repeatedly wash the resin with a highly concentrated NaCl solution.

4.13   **a.** Tomato juice would not be too bad. The K^+ level (56 mM) is higher than needed, but the Na^+ level (4.4 mM) is lower than needed. The sugar level is low enough that stomach cramps should not be a problem. The tomato juice without salt would be better because the Na^+ concentration in salted tomato juice is higher than that found in human blood. Tomato juice with salt will likely cause dehydration.

   b. Tomato juice is a better source of Mg^{2+} and K^+ than Gatorade, but tomato juice is not as high in Na^+.

   c. Tomato juice with no salt is best because the sodium, magnesium, and potassium concentrations are, as a whole, closest to those found in human sweat.

4.15   3680 ppm

4.17   **a.** salt; $KClO_4 \xrightarrow{H_2O} K^+(aq) + ClO_4^-(aq)$

   b. base; $RbOH \xrightarrow{H_2O} Rb^+(aq) + OH^-(aq)$

   c. acid; $H_2CrO_4 \xrightarrow{H_2O} 2 H^+(aq) + CrO_4^{2-}(aq)$

   d. salt; $Ca(CN)_2 \xrightarrow{H_2O} Ca^{2+}(aq) + 2 CN^-(aq)$

   e. acid; $HNO_3 \xrightarrow{H_2O} H^+(aq) + NO_3^-(aq)$

   f. salt: $MgBr_2 \xrightarrow{H_2O} Mg^{2+}(aq) + 2 Br^-(aq)$

4.19   **a.** sulfur oxides and nitrogen oxides

   b. sulfuric acid and nitric acid

   c. $H_2SO_4 \xrightarrow{H_2O} 2 H^+(aq) + SO_4^{2-}(aq)$

   $HNO_3 \xrightarrow{H_2O} H^+(aq) + NO_3^-(aq)$

4.21

   Electron pair geometry   Molecular geometry

4.23   To effectively replenish species lost due to sweating one would need to ingest both salt and water. The salt tablet alone will actually dehydrate you as it attracts water to itself while it begins to dissolve.

4.25   **a.** greater   **b.** less   **c.** equal

   d. greater   **e.** less

4.27   **a.** 70,000 ppb   **b.** 70 ppm   **c.** ppm

4.29 $[Mg^{2+}] = 0.18$ M; $[Cl^-] = 0.36$ M

4.31

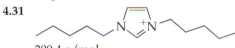

   209.4 g/mol

4.33   **a.** 3.5×10^{-5} moles

   b. $Mg(OH)_2 \xrightarrow{H_2O} Mg^{2+}(aq) + 2 OH^-(aq)$

   c. 10.45

   d. most will remain as a solid in equilibrium with the solution.

4.35   **a.** 9.84×10^{-3} mole

   b. 0.0104 Molar

4.37   **a.** gastric juice   **b.** urine

   c. pancreatic juice   **d.** gastric juice

4.39 $HNO_3 \xrightarrow{H_2O} H^+(aq) + NO_3^-(aq)$; pH = 3.35

4.41   4.0

4.43   **a.** true, strong acids fully dissociate

   b. false, ions of any kind will work

   c. true, the intensity of the light is related to the number of ions present

Chapter 5

5.1   **a.** reduction   **b.** oxidation

   c. reduction   **d.** oxidation

5.3   6.0 V

5.5   reaction c is wrong; $Br + 1$ electron $\longrightarrow Br^-$

5.7   **a.** $F\ 1s^2 2s^2 2p^5$

   b. $Si\ 1s^2 2s^2 2p^6 3s^2 3p^2$

   c. $Be\ 1s^2 2s^2$

   d. $P\ 1s^2 2s^2 2p^6 3s^2 3p^3$

5.9   Niépce discovered that light could produce an image on an asphalt coated plate. Daguerre discovered that silver coating on a copper plate could work better than asphalt and that the image could be fixed by treatment with mercury vapor.

5.11 Given the proper amount of energy any electron can be moved to a higher energy level.

5.13 **a.** All colors but red are absorbed.

b. All colors but green are absorbed.

c. No visible light is absorbed.

d. All colors of visible light are absorbed.

e. All colors but blue are absorbed.

5.15 B and Al

5.17 **a.** $Mg \longrightarrow Mg^{2+}(aq) + 2$ electrons (oxidation)

$O + 2$ electrons $\longrightarrow O^{2-}(aq)$ (reduction)

b. $Mg + O \longrightarrow Mg^{2+}(aq) + O^{2-}(aq)$

c. magnesium oxide

d. MgO

5.19 Fruits that do not turn brown contain ascorbic acid.

5.21 In modern devices such as cameras, pacemakers, cell phones, and hearing aids, it is necessary to produce a large amount of power in a small, lightweight battery. The challenge has been to produce a battery that contains all of the essential components while still fitting into a tiny space.

5.23 **a.** $O^{2-} 1s^22s^22p^6$

b. $S^{2-} 1s^22s^22p^63s^23p^6$

c. $Na^+ 1s^22s^22p^6$

d. $Be^{2+} 1s^2$

5.25 The anode is the location of the oxidation reaction. Meanwhile the cathode is the site of the reduction. Batteries typically contain an electrolyte paste suitable for each electrode separated by an inert spacer. A light or small motor connected by wire to the two terminals could be used to check for current.

5.27 In order for this electrochemical cell to work, the electrode and the ionic solution into which it is immersed must be composed of the same metal. The copper electrode should be immersed in the Cu^{2+} and the zinc electrode should be immersed in the Zn^{2+} solution.

5.29 Natural defense mechanisms include vitamins, such as vitamin E, and antioxidant enzymes, such as superoxide dismutase and glutathione peroxidase. Two naturally occurring free radicals are the superoxide radical and the class of lipid free radicals.

5.31 **a.** $Br^-(aq) \longrightarrow Br + 1$ electron

b. $Mg \longrightarrow Mg^{2+}(aq) + 2$ electrons

c. $K \longrightarrow K^+(aq) + 1$ electron

5.33 $Cu^{++} + 2$ electrons $\longrightarrow Cu$ (reduction)

$Fe \longrightarrow Fe^{++} + 2$ electrons (oxidation)

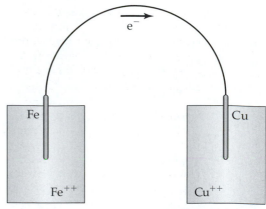

5.35 Lithium batteries use lithium as the anode (oxidation) and titanium as the cathode (reduction). The casing contains both the anode and the cathode, and the interior of the battery is composed of two types of electrolyte paste separated by an inert spacer. This allows the battery to be smaller and lighter. Electrons flow from the anode to the cathode.

5.37 c

5.39 Tocopherols have a benzene ring and an —OH group that allow for the elimination of lipid free radicals, thereby keeping the breakfast cereal fresh and preserving its flavor.

5.41 Silver iodide would be more like silver bromide because of the relative size of the anions.

5.43 The color of light given off corresponds to the energy gap between the electron in the excited state and the electron in the ground state. The gap is different for each element.

5.45 Molecule A will absorb visible light, Molecule B will absorb UVA light, and Molecule C will absorb UVB light.

5.47 Pigments in leaves, sunscreen ingredients, and antioxidants are all highly conjugated molecules, allowing them to absorb light and stabilize free electrons.

5.49 blue light

Chapter 6

6.1 **a.** $^4_2\alpha$ **b.** $^0_{-1}\beta$ **c.** ^1_0n

6.3 **a.** nuclear **b.** chemical **c.** chemical

6.5 **a.** 82% **b.** 10% **c.** 61%

6.7 Alchemists did mix science with magic and religion. They wanted to create gold from other elements, but they also sought elixirs that would prolong life.

6.9 Beta particles are small and lighter and would penetrate farther into tissue.

6.11 1.50×10^{11} kJ

6.13 Montana, South Dakota

6.15 **a.** 50% **b.** 98% **c.** 24% **d.** 57%

6.17 **a.** $^{254}_{102}No$

b. $^{82}_{36}Kr$

6.19 **a.** 8 **b.** 6

6.21 **a.** $^0_{-1}\beta$ **b.** $^0_{-1}\beta$

6.23 **a.** chemical

b. Nuclear reactions produce more energy per mass than do chemical reactions.

c. Nuclear fuel has the potential to cause very serious accidents, and the United States is not sure what to do with the waste.

6.25 The Big Bang theory states that an explosion of an enormous amount of energy created neutrons, protons, and electrons. This began a series of fusion reactions that produced very hot stars. Debris from an exploding star formed our solar nebula. Fusion was responsible for the heavy elements.

6.27 Elements with atomic numbers above 92 are created in the laboratory rather than naturally occurring and exist for only a short period of time. Several scientists may be working on the discovery of a new element at the same time, and it is difficult to prove that something new was created. Thus, there may be debate about who discovered it first or made the most important contributions.

6.29 Cobalt-60 with its half-life of greater than five years is used to produce gamma radiation in a medical device called a "gun." This radiation is directed to the site of a tumor and is used to destroy these cancer cells. Technetium-99 is taken internally and is used as an imaging agent.

6.31 a. In both cases the energy produced is used to heat water into steam.

   b. The steam turns turbines to produce electricity.

   c. The main difference is the waste produced. Traditional power plants produce carbon dioxide, while nuclear power plants produce hazardous nuclear waste.

6.33 a. The strong nuclear force holds the nucleus of atoms together.

   b. Neutrons are necessary to moderate the repulsive force between protons. In large nuclei, more neutrons are needed to make the nuclear force more powerful than the proton-proton repulsive force. As a result, the average number of neutrons per proton increases with increasing atomic size.

6.35 a. $^{235}_{92}U + ^{1}_{0}n \longrightarrow 2\ ^{106}_{46}Pd + 24\ ^{1}_{0}n$

   b. Yes. The reaction produces many neutrons that can be used in subsequent fission reactions.

6.37 a. Extreme heat is required to produce the needed energy for nuclear fusion. Fission is also easier to control and initiate.

   b. Current technology is not available to produce nuclear power reactors that can handle the high heat required.

6.39 a. No. $4\ C_7H_5N_3O_6 + 21\ O_2 \longrightarrow 28\ CO_2 + 10\ H_2O + 6\ N_2$

   b. 2.72×10^{11} kJ

6.41 76 rem/year; temporary decrease in white blood cell count

6.43 Uranium deposits on Earth are plentiful and have been decaying for millions of years. These isotopes are radioactive and decay at a measurable rate. Additionally, the steps between radon and lead are all relatively fast, meaning that if a person breathes in radon gas there will be a large number of radioactive events in a short period of time. The lead will then remain in the lungs a long time, as its half-life is 22 years.

6.45 a. $^{238}_{92}U \longrightarrow ^{234}_{90}Th + ^{4}_{2}\alpha$

   b. Even though the half-life of uranium-238 is 4.5 billion years, it is still producing thorium-234 at a slow rate. Thus, small amounts of radon are being produced. In a well ventilated home, the risk would be minimal.

   c. There would likely be lead in the glaze. Because lead should not be consumed, these plates should not be used to serve food.

Chapter 7

7.1 Water is polar and can form hydrogen bonds. With a variety of possible interactions, water can dissolve many different substances.

7.3 Different regions have different distributions of hydrogen isotopes in the water. The bird eats insects that have eaten plants containing water with a distinctive isotopic ratio. The feathers can be collected, the isotope distribution of the water contained in the feathers determined, and this ratio compared to water found at various places along the migration route.

7.5

7.7 a. ion-dipole

   b. hydrogen bonding

   c. neither ion-dipole nor dipole-dipole

7.9

7.11 The acetone nail polish remover container is empty. Acetone evaporates faster due to a lack of hydrogen bonding between the molecules.

7.13

7.15 2.41×10^4 T atoms

7.17 Yes

7.19 Oatmeal will be undercooked.

7.21 Cooking will take longer and require more energy. Ready-to-eat meals would be a good alternative.

7.23 A freezer works by removing heat from the compartment.

7.25 The variation in tritium amounts will likely be too small to measure accurately. She would be better off measuring the ratio of the hydrogen isotopes.

7.27 The oxygen, nitrogen, and fluorine use a lone pair to participate in hydrogen bonding. Also, oxygen, nitrogen, and fluorine are very electronegative and draw electron density away from the hydrogen, making it partially positive.

7.29 False. Because our bodies are mostly water, we are able to maintain a stable body temperature.

7.31 $CH_3CH_2CH_2\ddot{O}H$

7.33 **a.**

b. 537.1 g/mol

c. 3.19×10^{21} molecules

d. For molecules of similar polarity, the extent of intermolecular interactions increases as the molecular weight of the substance increases. The greater the intermolecular interactions, the higher the melting point and boiling point.

7.35 As long as the volume is constant, when you increase the pressure, you increase the temperature and when you decrease the pressure, you decrease the temperature.

7.37 The pressure at extreme depths in the ocean is very high compared to atmospheric pressure.

7.39 The hydrogen atoms are bonded to the carbon atom and not the oxygen atom. Acetone does not have an —OH group.

7.41 The interaction between two water molecules will have a lower value (ion-dipole interactions are stronger). The interaction between a lithium ion and a chloride ion will have a higher value (ionic bonding is stronger).

Chapter 8

8.1 phosphorus bonded to oxygen

8.3 Air is compressible, liquids and solids are not.

8.5 Get as far away from the agent as possible and seek a large area with as much ventilation as possible.

8.7 Gas molecules do not interact with each other, as they are too far apart.

8.9 No. The average speed means some molecules are moving faster and some are moving slower.

8.11 Low boiling point (highly volatile) and inert; CFC and HFC deplete the ozone layer.

8.13 smaller

8.15 Y is the most dense; Z is the least dense.

8.17 **a.** The two structures are resonance forms.

b. No

8.19 A bee's wings get coated with pollutants as they fly around. When returning to the hive, they beat their wings furiously to cool off the temperature of the hive and, in doing so, release the pollutants. Bees have been trained to sniff out various compounds, such as the TNT in land mines.

8.21 1.86 kPa

8.23 In clean air, the concentration of carbon dioxide is very small relative to nitrogen and oxygen. As a result, it mixes completely with the other components in clean air.

8.25 a, b, and c are true

8.27 The ratio of oxygen to nitrogen does not change, but the pressure is decreased, making less oxygen available.

8.29 **a.** 7.81×10^4 molecules

b. 4.70×10^{23} molecules

8.31 **a.** yes

b. yes

c. yes

d. A similar amount of gas B and gas C would likely form two layers due to dramatically different molar masses.

8.33 When the atmosphere contains high quantities of gas molecules, the amount of radiated heat from the Earth that is absorbed and retained by the atmosphere increases. This ultimately increases the temperature of the atmosphere. Like a car sitting in the sun on a hot day, the radiated heat from the car's interior is trapped inside.

8.35 **a.** inversely proportional

b. As the temperature decreases, more moles of gas can occupy a container at constant pressure and volume.

8.37 764 mm Hg; 30.1 inches Hg

8.39 Planes can take off more easily when the air is more dense. Humid air is actually less dense than dry air, because regardless of the composition of the air, the molecules are about the same distance apart. Dry air is comprised of nitrogen (28 grams/mole) and oxygen (32 grams/mole). The water in humid air is only 18 grams/mole, resulting in a lower average density of the air. Therefore the plane should take off more easily when the air is dry.

8.41 **a.** In order to determine the number of molecules in the neon sign it would be necessary to know the volume and temperature of the glass tube.

b. If the tube is punctured and the neon gas escapes, the pressure in the tube will become equal to the atmospheric pressure surrounding the neon sign.

8.43 **a.** H_2 **b.** H_2

8.45 A thermos bottle does not contain a complete vacuum. There are a few atoms and molecules of gas remaining in the space between the two containers. Thus, over time, there is some convection through the evacuated space.

8.47 625 kPa

8.49

Breaking O_2 requires more energy. The double bond of O_2 is stronger, so a shorter wavelength of light is required.

Chapter 9

9.1 **a.** alkene **b.** alkyne **c.** alkene

d. alkane **e.** alkane **f.** alkane

9.3   **a.** $C_{18}H_{38}$

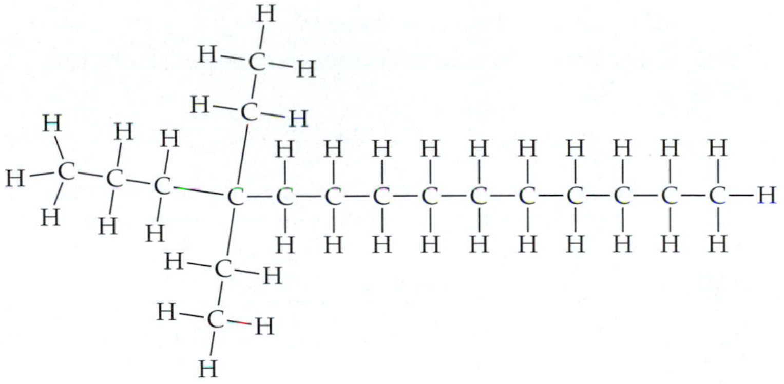

b. C_9H_{18}

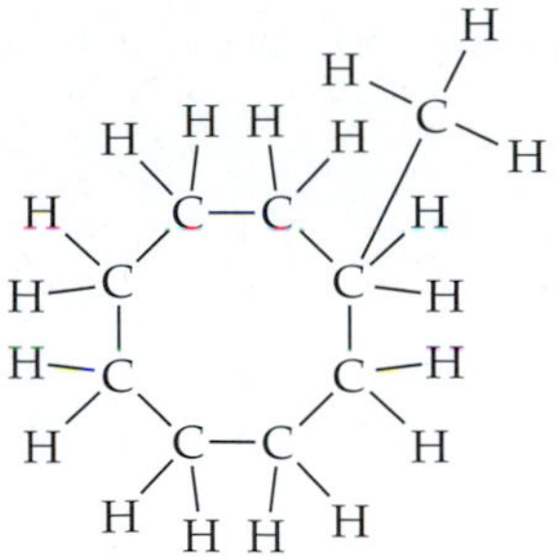

c. $C_{24}H_{50}$

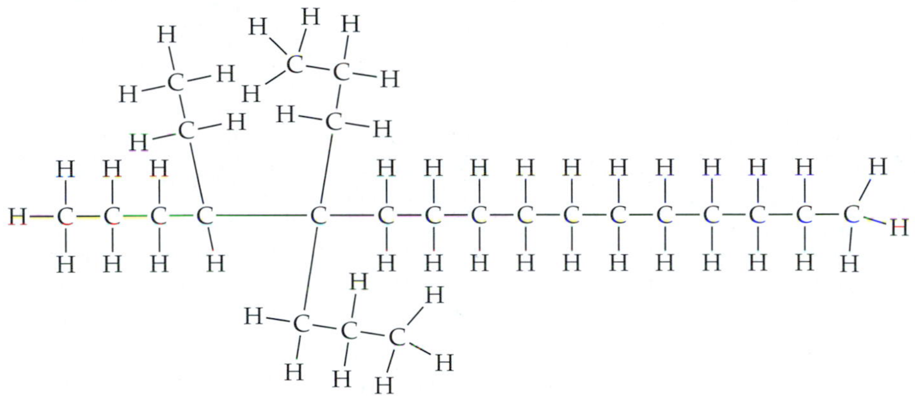

d. $C_{13}H_{28}$

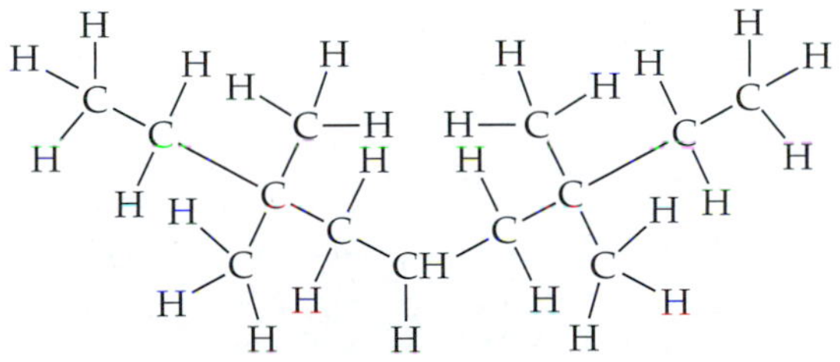

9.5 $CH_4 + 2\,O_2 \longrightarrow CO_2 + 2\,H_2O$

9.7   **a.** increasing   **b.** increasing
         c. decreasing   **d.** increasing

9.9   True

9.11 Complete combustion produces a maximum amount of carbon dioxide. Incomplete combustion produces carbon monoxide as well as carbon dioxide.

9.13 A homologous series has an increasing number of carbons, not an increasing number of bonds.

9.15 **a.** alcohol   **b.** ether
         c. ether   **d.** alcohol

9.17 **a.** *trans*-3-nonene
         b. 3,4,7,8-tetramethylcyclooctene
         c. *cis*-3,6-diethyl-4-decene
         d. 2,3-dipropylcyclopentene

9.19

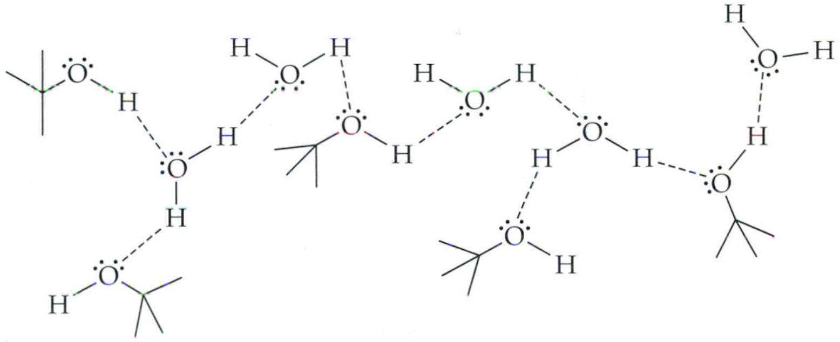

9.21 **a.** 16 carbons   **b.** alkane   **c.** hexadecane

9.23 **a.**

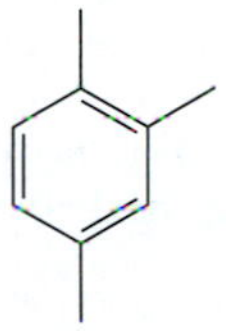

b.

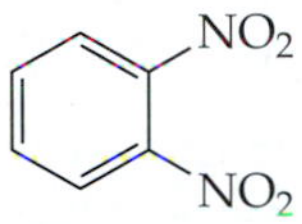

c.

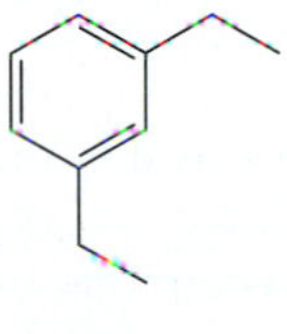

d.

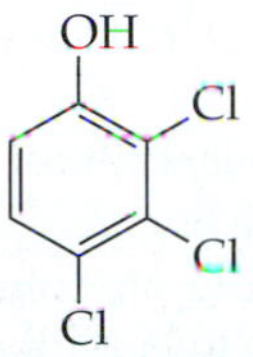

9.25 3-phenylhexane

9.27 **a.** Yes   **b.** Yes

    c. Too much solid residue was formed.

9.29 **a.** 2,6-dipropylphenol

    b. 2,3-dinitrotoluene

    c. 3,4-diethylcyclohexene

    d. 3,4-dinitrocyclohexene

9.31 **a.** Reaction always goes forward.

    b. Reaction may go forward, depending upon the temperature.

    c. Reaction may go forward, depending upon the temperature.

    d. Reaction will never go forward.

9.33 False; entropy must always increase somewhere in the system.

9.35 **a.**

    b. Because of it shape, *trans*-2-pentene molecules can pack more closely together and form weak intermolecular bonds.

9.37 **a.** 0.227 kg   **b.** 1 mole C   **c.** 0.012 kg

9.39 **a.**

$$2\ C_4H_{10} + 13\ O_2 \longrightarrow 8\ CO_2 + 10\ H_2O$$

$$C_4H_8 + 6\ O_2 \longrightarrow 4\ CO_2 + 4\ H_2O \qquad C_4H_8 + 6\ O_2 \longrightarrow 4\ CO_2 + 4\ H_2O \qquad 2\ C_4H_6 + 11\ O_2 \longrightarrow 8\ CO_2 + 6\ H_2O$$

    b. *n*-Butane

    c. No. These molecules have the same molecular formula and the same number and type of bonds.

9.41 One must be branched and therefore have a different shape.

9.43 Crude oil is a mixture of many compounds that can be separated by distillation. The different compounds are separated by boiling point in a fractionating column that is several feet tall. Fractions with low boiling points, such as natural gas, propane, and butane, climb to the top of the column, whereas fractions with high boiling points, such as gasoline, jet fuel, and diesel, climb less far.

9.45 **a.** Toluene has a methyl on position one already.

2, 3-dinitrotoluene

    b. Not alphabetical; , methyl propyl ether

    c. The number one is not needed; methylcyclobutane

    d. Trans alkenes do not exist is small rings;

cis-cyclohexene

Chapter 10

10.1 true

10.3 It is not biodegradable.

10.5 Polylactic acid hydrolyzes easily, and the fabric would break down in water. Polyesters do not hydrolyze easily in water.

10.7 The water evaporates.

10.9 **a.** The starting materials for synthetic plastics originate from crude oil.

    b. They sit in the landfill for decades as they are broken down very slowly.

    c. Despite recycling programs, an enormous amount of plastic ends up in landfills.

10.11 They have been sorted. They still need to be cleaned, shredded, and remolded.

10.13 **a.** Thermoplastic polymers are composed of strands of polymer that can stack or align themselves into a well-ordered matrix. These highly ordered crystalline regions are called crystallites. Heat changes crystallites into amorphous regions, and the polymer becomes more pliable. When cooled, the crystallites reform, and the plastic retains its shape. Thermosetting polymers do not have this structural property.

    b. Thermoplastic polymers are more easily recycled because they can be heated and remolded.

    c. Bakelite

    d. no

10.15 In order to maintain the integrity of the sample, examine it under a microscope first, determine its melting, and then check solubility by dissolving it in various solvents.

10.17 **a.**

    b.

Polyethylene is more flexible.

10.19 **a.** methyl butanoate

    b. ethyl heptanoate

    c. hexyl acetate

10.21 Newspapers, dozens of years old, not decomposing; pessimistic

10.23 Recycling is a mechanical, physical process, whereas biodegradability is a chemical breakdown into smaller pieces.

10.25 an electron rich species, electrophile, negative

10.27 addition

10.29 a. Crosslinking with sulfur interconnects the chains.

   b. There is an absence of individual chains in both molecules. In vulcanized rubber, chains are crosslinked. Thermosetting polymers exist as one enormous molecule.

10.31 a.

b.

c.

10.33 It reacts with water. The tube contains monomer and becomes polymer after leaving tube.

10.35 a. splitting by water

   b. a and b

   c.

10.37 a. amide, alkyne, benzene, alkene, ester, ether, alcohol, ketone

   b. $C_{21}H_{25}O_6N$

   c. 387.4 g/mol

10.39 a. The negative charge on A will attack the partial positive charge on B.

   b. no

   c. yes

10.41 5.40×10^{26} g/mol

10.43 a. ester, alcohol   **b.** CH_3OH

Chapter 11

11.1 a.

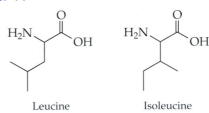

Leucine   Isoleucine

   b. both $C_6H_{13}NO_2$

   c. different R group

   d. inside, away from water

11.3 a. The blue ribbon or strand is representative of the phosphate-sugar backbone (in this case phosphate and deoxyribose), while the tube-shaped objects are the nitrogenous bases.

   b. Yes, there are four different colors. Additionally, dark blue is always paired with red, and light blue is always paired with yellow.

   c. the phosphate-sugar backbone

11.5 DNA. Deoxyribose is the sugar.

11.7 a. CUAAUGU   **b.** Leu-Met

11.9 a.

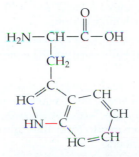

b. $C_{11}H_{12}N_2O_2$

c. 204.23 g/mol

11.11 a. Ether   **b.** Carboxylic acid

     c. Ester   **d.** Alcohol

11.13 a. Peptides are short amino acid chains, generally having 100 or fewer amino acids, and proteins are much longer.

     b. Peptide is a subset of protein.

     c. Protein

11.15 a. GCAUUAAGCC

     b. GCCAUAAAUGCC

     c. GGAUUAGCAAUU

11.17 Usually where proline and glycine are adjacent to each other. Proline causes the chain to bend and glycine is small enough to allow adjacent amino acids to get close together around the bend.

11.19 they contain sulfur

11.21 A C–G pair is held together by three hydrogen bonds, while an A–T pair is held together by two hydrogen bonds.

11.23 a. The H on the OH is acidic.

     b. The oxygen atoms contribute to polarity and the acid is sometimes found in its negatively charged carboxylate form.

11.25 a. primary   **b.** tertiary   **c.** secondary

11.27 Hydroxyproline is responsible for the rigidity in the collagen structure because it allows each collagen chain to make hydrogen bonds with other collagen chains. Vitamin C is required for the conversion of proline to hydroxyproline. Vitamin C deficiency, scurvy, causes weakening of the collagen.

11.29 False. Nonessential amino acids can be synthesized by the body, while essential amino acids must be eaten in the diet.

11.31 a. GTTTTTTTAGGAATAGCCTTG

     b. GUUUUUUUAGGAAUAGCCUUG

     c. Val-Phe-Leu-Gly-Ile-Ala-Leu

     d. Interior

11.33 a. Molecule A $C_8H_{10}O$; Molecule B $C_8H_{16}O_2$; Molecule C $C_{12}H_{10}O_2$; Molecule D $C_8H_{18}O_2$

     b. Molecule A 122.1 g/mol; Molecule B 144.1 g/mol; Molecule C 186.1 g/mol; Molecule D 146.1 g/mol

     c. Molecules B and D can contribute hydrogen atoms for hydrogen bonding.

     d. All four can contribute electronegative atoms for hydrogen bonding.

11.35 Both polymers contain amide functional groups and a hydrocarbon group separates any two adjacent amide groups. Unlike nylon, the separating hydrocarbon group in proteins is a methylene group with one of 20 possible R group side chains. Because proteins are composed of 20 different monomers, they are much more diverse.

11.37 a. This opposite-twists pattern creates a very tight weave. This strategy is frequently used to make very strong ropes.

     b. The non-polar amino acids allow the formation of a tough, inflexible left-handed helix, while the polar hydroxyproline amino acids form hydrogen bonds between the collagen chains to form the right-handed helix of the collagen fibers.

11.39 a. Benzene ring, 2 amides, carboxylic acid, sulfide (R-S-R').

     b. yes

     c.

$$\underset{R}{\overset{O}{\|}}\text{—OH} \longrightarrow \underset{R}{\overset{O}{\|}}\text{—O}^{\ominus}$$

11.41 a. Many facets of our physical characteristics are determined by DNA. In fact, our DNA serves as our biochemical fingerprint.

     b. DNA evidence is often used to identify criminals.

     c. No

11.43 It arranges so that polar groups can be on one side near to water, and the nonpolar groups can be on the other side facing the interior of the protein.

11.45 Corn has been modified to resist insect pests and to produce a natural polymer. Recombinant techniques were employed to insert the hereditary material from one organism into another.

Chapter 12

12.1 Simple sugars are one-unit or two-unit carbohydrates. Monosaccharides are single unit carbohydrates. Polysaccharides are carbohydrate polymers consisting of more than 100 monomer units. Disaccharides are two-unit carbohydrates. Complex carbohydrates are polysaccharides consisting of hundreds or thousands of individual units. Dietary fiber is the term given to all solid indigestible components of food, such as the complex carbohydrate cellulose.

12.3 a. saturated

     b. monounsaturated, trans

     c. polyunsaturated, both trans

     d. monounsaturated, cis

12.5 The statement is incorrect. Not all enzymes are proteins and some enzymes can catalyze a class of reactions.

12.7 Acids nonspecifically catalyze some reactions, while enzymes specifically catalyze others.

12.9

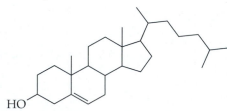

Dashed line is without catalyst, solid line is with catalyst.

12.11 Anabolic steroids regulate processes that build new biomolecules from smaller ones.

12.13 hard candy

12.15 a.

     b. fairly nonpolar

     c. alcohol

d. monounsaturated

e. Cholesterol is rather flat and rigid because most of the carbon atoms are part of closed rings.

12.17 Amylase is general and can catalyze a variety of carbohydrate cleaving reactions. Beta-galactosidase is more specific, because it only breaks down lactose into galactose and glucose.

12.19 a.

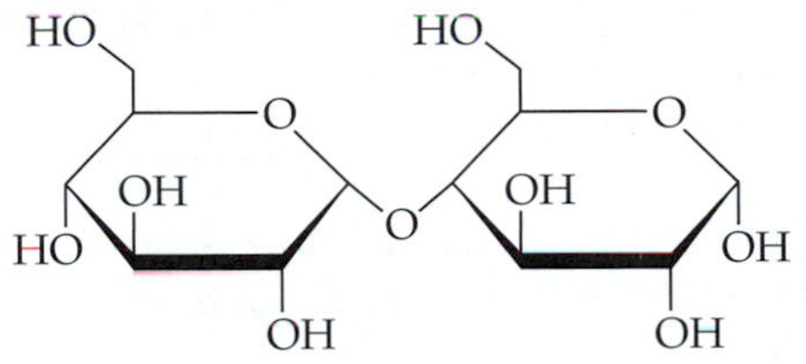

b.

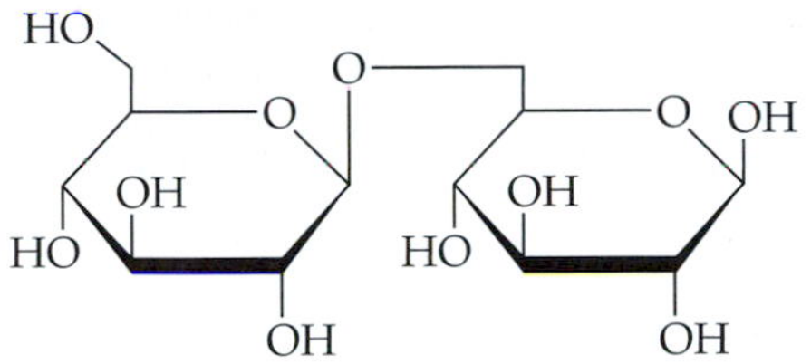

c.

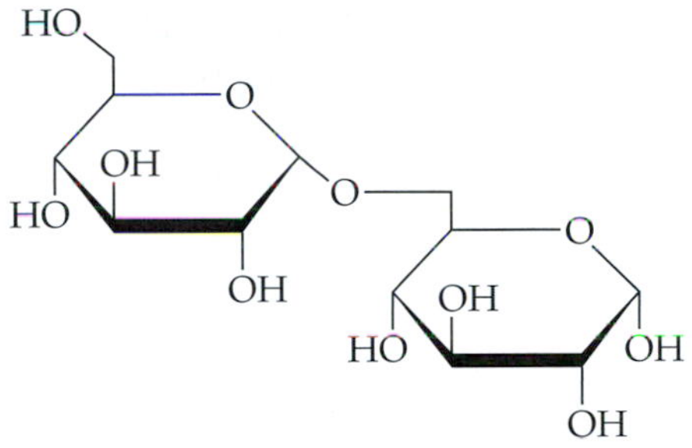

d.

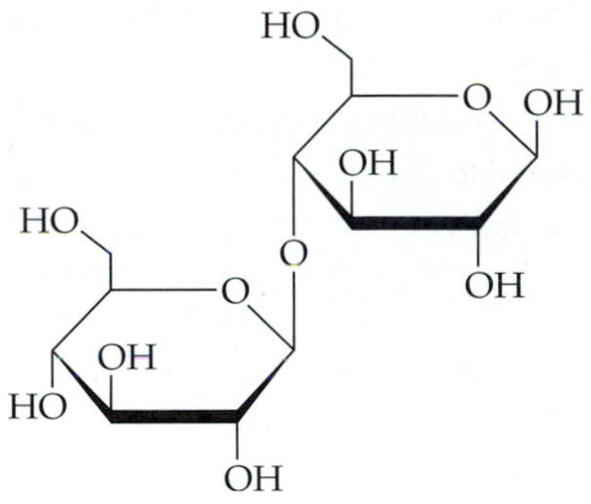

12.21 a. saccharin

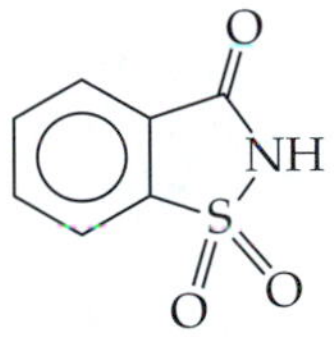

b. aspartame and sucralose

c. Aspartame is a modified dipeptide and sucralose is a modified carbohydrate.

12.23 a. ketone, methyl group   **b.** seven

12.25 a. β 1-4   **b.** β-glucose   **c.** no

12.27 a. TAGs have three fatty acids attached, while phospholipids have two.

b. TAGs are stored energy, while phospholipids build the structure of cell membranes.

c. TAGs are nonpolar, while phospholipids are both nonpolar and polar.

12.29 Lowers activation energy; stabilizes transition state.

12.31 a.

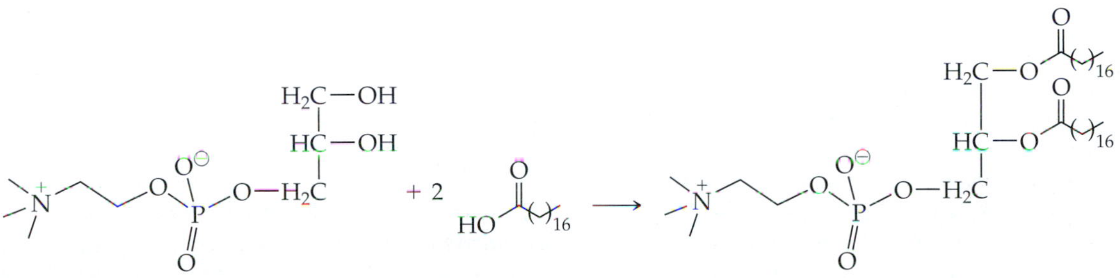

b. $C_{44}H_{88}PNO_8$, molar mass 790.30 g/mol

c. relatively straight and rigid

12.33 a. All are cis.

b. Oleic acid, cis, one

c. Flat, s-shaped

d. Arachidonic

12.35 The statement is true except that the hydrogenation reaction is generally intended to partially hydrogenate the oil.

12.37

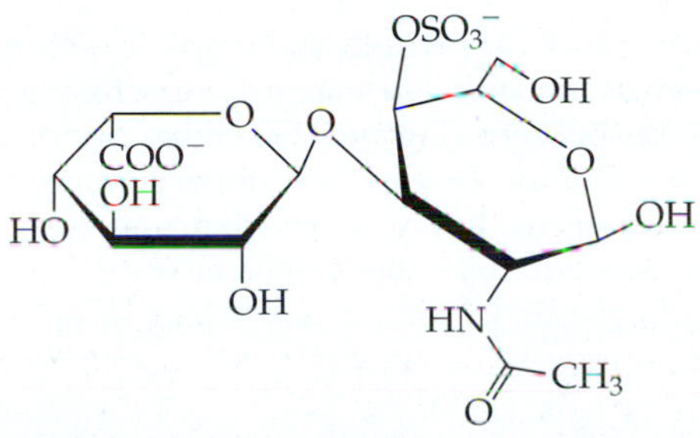

12.39 252 Calories; 531,400 J carbohydrates; 318,000 J fat; 205,000 J protein; 1,054,400 joules total

12.41 a. an extensive network of conjugated double bonds

   b. visible region

12.43 a.

, $C_{14}H_{18}N_2O_5$

   b.

   CH_3OH

Chapter 13

13.1 Not true. The outer coating begins to dissolve slowly in the stomach and a polymer matrix that holds the drug dissolves slowly as well.

13.3 Irreversible inhibitors covalently bind to an enzyme's active site, whereas reversible inhibitors bind noncovalently to the active site. Irreversible inhibitors bind permanently to the active site, but are highly reactive molecules with deleterious side effects. Reversible inhibitors have fewer side effects but are typically not as effective or effective only for relatively short periods of time.

13.5

   Ketone   Aldehyde   Alkene, alcohol

13.7 ED_{50} describes the effective dose, and TD_{50} describes the toxic dose. The ratio is called the therapeutic index.

13.9 dissolving agents (polysaccharides), binders (salts, polysaccharides), coating (polysaccharides, hydrocarbon), lubricants (fat), colorant (titanium dioxide).

13.11 Irreversible, or suicide, inhibitors bind permanently to the enzyme, thus rendering both the enzyme and the inhibitor inactive.

13.13 Viruses are tiny protein pouches containing either DNA or RNA. The retrovirus is HIV, and the disease is AIDS.

13.15 amiodarone

13.17 Amide, found in proteins, produces a carboxylic acid and an amine.

13.19 None of the molecules would show opiate activity. The molecule must contain a benzene ring, must have a carbon bonded to the benzene ring and three other carbon atoms, must have one of the three carbon atoms bonded to a carbon that is connected to a nitrogen atom, and the nitrogen must be attached to two other carbon atoms.

13.21 Substances that can remove disease-causing chemicals from the body. Carcinogenic compounds.

13.23 An active site is an area on an enzyme where a reaction occurs. Receptors are biomolecules that simply bind to other molecules.

13.25 Z

13.27 a. identical, achiral

   b.

   , nonidentical chiral

   c. identical, achiral

   d. identical, achiral

13.29 a. amide

   b. ester

   c. ester

   d. amide

13.31 a. $C_{19}H_{21}N_3O$, 307.4 g/mol

   b. 1.96×10^{19} molecules

   c. no

13.33 a. sulforaphane $C_6H_{11}NOS_2$; oxomate $C_8H_{15}NOS_2$

   b. sulforaphane 177.29 g/mol; oxomate 205.38 g/mol

   c. sulforaphane 1.8×10^5 mg; oxomate 3.5×10^5 mg

   d. assuming a linear decline, 8.63×10^4 mg

13.35 a. amprenavir (several polar functional groups) > codeine (one OH group) > etorphine (OH groups, but a large hydrophobic bridging group)

   b. The more polar a compound, the greater its solubility in water. Amprenavir should be the most soluble, and codeine should be the least soluble.

13.37 Nonpolar groups enhance the binding of these molecules. The more lipophilic the molecule, the more potent it is.

13.39   Isoleucine

NH₂

OH

O

NH₂

OH

O

NH₂

OH

O

NH₂

OH

O

Threonine

NH₂

OH

OH   O

NH₂

OH

OH   O

NH₂

OH

OH   O

NH₂

OH

OH   O

Chapter 14

14.1

Methyl benzoate
$C_8H_8O_2$

14.3   **a.** qualitative   **b.** quantitative
        c. quantitative   **d.** qualitative

14.5

Nicotine

14.7   Since most amines give a positive result in the Marquis test, it is almost impossible to positively identify the presence of an illicit drug using it.

14.9   **a.** tertiary

        b. secondary

        NH

c. secondary

NH

d. secondary

NH

14.11   HO

HO

NH₂

OH

Norepinephrine

HO

HO

NH

OH

Epinephrine

Yes. Monoclonal antibodies are able to distinguish between the two structures. Because it is a primary amine, norepinephrine will exhibit stronger intermolecular interactions.

14.13   When neurotransmitter levels increase, the body adjusts by decreasing the number of neurotransmitter receptors. As a result, it takes more of the drug to have the same effect.

14.15   All are primary amines, so the boiling point increases with the molecular weight.

14.17   Crack is neutral and would easily vaporize so it is smoked. Cocaine hydrochloride is a salt and is water soluble so it is snorted.

14.19   Many drugs cause an increase in dopamine in the synapse. This causes a decrease in the number of dopamine receptors. Amphetamine and cocaine are two drugs that cause an increase in dopamine.

14.21   **a.** Caffeine has two amide groups, one amine (tertiary), one imine, and one alkene.

        b. The molecular formula is $C_8H_{10}N_4O_2$, and its molar mass is 194.2 g/mol.

        c. The complete structures for caffeine monochloride and caffeine dichloride are:

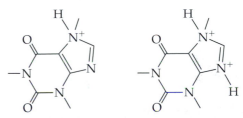

The dihydrochloride predominates at very low pH.

14.23   The molecular geometry of an amine is pyramidal, whereas the geometry of a carbon atom in an alkyl chain is tetrahedral.

14.25   Ecstasy is both an amphetamine (increases the amount of dopamine) and a hallucinogen (affects the natural balance of serotonin).

14.27

NH₂

NH₂

14.29   The binding site works mostly by noncovalent interactions, so the location of groups in the molecule is critical. The binding site is also a very specific shape, so the molecular shape is very important.

14.31 a. a tertiary amine, ester, benzene ring, primary amine

b. $C_{13}H_{20}N_2O_2$, molar mass 236.33 g/mol

c. The tertiary amine is more basic, as the other amine is next to a conjugated ring, which reduces its basicity.

d.

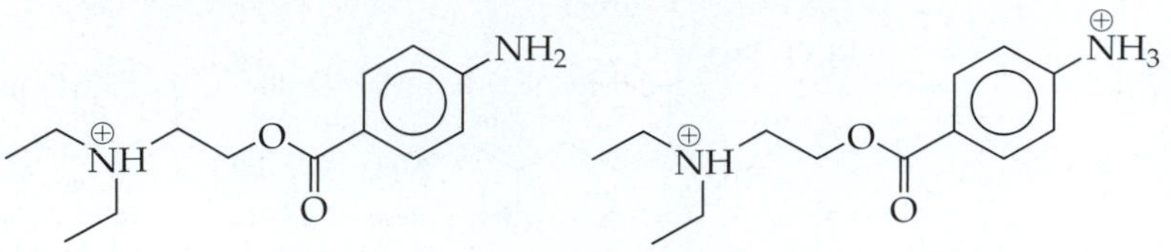

The dihydrochloride will predominate at very low pH.

14.33

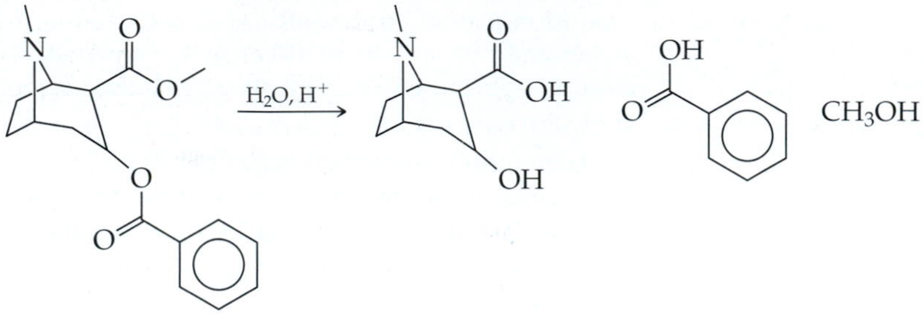

14.35 a. 0.49 kg

b. The tertiary amine, because carbon groups are donating. The secondary amine is next to a conjugated system which decreases its basicity.

14.37 a. There are three "amines" and one imine in this molecule.

b. $C_{31}H_{20}N_4$, molar mass 448.31 g/mol

c.

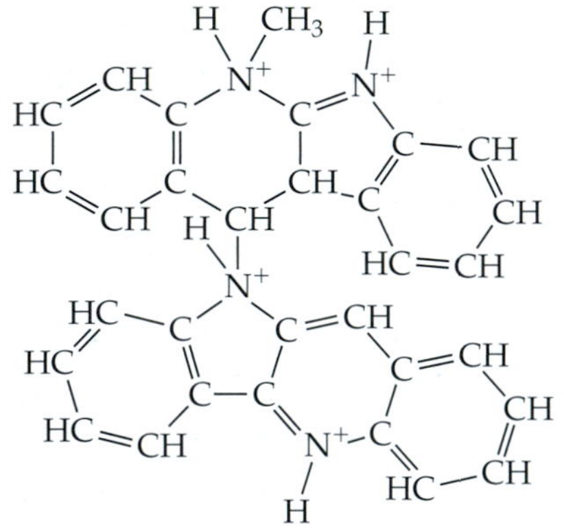

d. pH 2

e. No chirality centers.

Credits

Frontmatter

p. v: Jeff Navarro. p. viii: Steve Cole/Photodisc. p. ix: Nick Koudis/Photodisc. p. x (top): StockTrek/Photodisc. (bottom): Don Farrall/Photodisc. p. xi: Hisham F. Ibrahim/Photodisc. p. xii: PhotoLink/Photodisc. p. xiii: Janis Christie/Photodisc. p. xx: Fundamental Photographs, NYC.

Chapter 1

Chapter Opener (left): Peter Steiner/Corbis/Bettmann. (right): Mark Kemp/Digital Vision. Figure 1.1: Corbis/Bettmann. Figure 1.2: Hulton Archive/Getty Images. Figure 1.3: Edwin Levick/Hulton Archive / Getty Images. Figure 1.4: Kiryat-Tivon, Israel. Figure 1.5 (bottom): British Crystallographic Association, www.argonet.co.uk. (top): *The New York Times*. Figure 1.6: Getty Images Inc. - Hulton Archive Photos. Figure 1.7a: SPL/Photo Researchers, Inc. Figure 1.7b: Vittorio Luzzati/Eurelios Photographic Press Agency. Figure 1.9: Getty Images/Time Life Pictures. Figure 1.10: Tom Pantages. Figure 1.15a (middle): Historical Museum of Kiev. (left): Henry Diltz/Getty Images, Inc. (right): E.R. Degginger/Color-Pic, Inc. Figure 1.15b (left): Matt Meadows/Peter Arnold, Inc. (middle): Jack Andersen/PictureArtsCorbis/Bettmann. (right): Tom Pantages. Figure 1.16: www.CartoonStock.com. Page 14: Ewing Galloway/Index Stock Imagery, Inc. Figure 1.18: Tom Pantages. Figure 1.19: Tom Pantages. Figure 1.21a: IBM Communications Media Relations. Figure 1.21b: IBM Research, Almaden Research Center. Figure 1.22: Color-Pic, Inc. Figure 1.25: Evan Sklar/FoodPix/PictureArts Corporation. Page 24: CREASTOCK/Getty Images, Inc. - Taxi. Figure 1.28: Reprinted with permission of THE ONION. Copyright 2001, by ONION, INC. http://www.theonion.com/Getty Images, Inc. Figure 1.29: Richard Megna/Fundamental Photographs, NYC. Figure 1.30: Courtesy of Steven Larson and Alex McPherson. Larson, S.B., Day, J., Greenwood, A. and McPherson, A. (1998) Refined structure of satellite tobacco mosaic virus at 1.8A resolution. *J. Molecular Biology* 277, 37–59. Figure 1.31: Paul Axelsen, M.D./Phototake NYC. Figure 1.32: Science Museum/Science & Society Picture Library/The Image Works. Page 33: ©Bettmann/CORBIS. Page 34: Photodisc/Getty Images. Page 35: FoodPix/PictureArts Corporation. Page 36 (top): © Photographers Consortium/ eStock Photo. (bottom): Picture Desk, Inc./Kobal Collection.

Chapter 2

Chapter Opener (left, top): Walter Bibikow/Taxi/Getty Images. (left, second from top): Corbis Royalty Free. (left, second from bottom): Theo Allofs/Zefa/Corbis/Bettmann. (left, bottom): Owen Franken/Corbis/Bettmann. (right): Steve Cole/Photodisc. Quotation: *Dirt: The Ecstatic Skin of the Earth* by W.B. Logan is printed with permission from The Aaron Priest Literary Agency. Figure 2.8b: Andrew Syred/SPL/Photo Researchers, Inc. Page 53 (bottom): Dave Bunk Collection/Scovil Photography. Figure 2.12a: The Parthenon, Athens, Greece. Blaine Harrington III/Stock Market. Figure 2.12b: Nick Nicholls © The British Museum. Figure 2.13 (bottom): Scott Pike. Figure 2.15: © Dorling Kindersley. Page 61 (bottom): Kimberley Waldron. Figure 2.16: Richard Megna/Fundamental Photographs, NYC. Figure 2.17: Roger RessmeyerCorbis/Bettmann. Figure 2.18a: Jeffrey A. Scovil. Figure 2.18b: Ian Steele & Ian Hutcheon/Photo Researchers, Inc. Figure 2.19: Carol Cohen/Corbis/Bettmann. Figure 2.22: Reuters/Corbis/Bettmann. Figure 2.23a: Raymond Gehman/Corbis/Bettmann. Figure 2.23b: © Dr. Richard Busch/Courtesy American Geological Institute, AGI. Figure 2.24a: Corbis/Bettmann. Figure 2.24b: Grant Heilman Photography, Inc. Figure 2.25: AP Wide World Photos. Figure 2.26a: Barry Runk/Grant Heilman Photography, Inc. Figure 2.26b: Laguna Photo/Getty Images, Inc - Liaison. Figure 2.27 (top): "Adolph Coors III Disappears; FBI Enters Search", in the February 11, 1960 edition of *The Denver Post*. (bottom left): February 1966/ Reprinted with permission of the *Rocky Mountain News*. (bottom right): AP Wide World Photos. Page 78 (top left): Chip Clark. (bottom left): Breck P. Kent/Animals Animals/Earth Scenes. (top right): George D. Lepp/Corbis/ Bettmann. (bottom right): Jeffrey A. Scovil. Page 79 (top): Jeffrey A. Scovil. (bottom): Jeffrey A. Scovil. Page 80: Chip Clark.

Chapter 3

Chapter Opener (left): Steve Cole/Photodisc. (right): Paramount//Picture Desk, Inc./Kobal Collection. Figure 3.5: Reprinted from Diamond and Related Materials, Vol. 10, Issue 11, A.R. Krauss et al. pp.1952–1961 Fig. 10(middle) (2001) Elsevier. Figure 3.6: Prof. Stacey Bent. Figure 3.14: NaturalHandyman.com. Figure 3.17a: Thinkstock. Figure 3.17b: Dave G. Houser/Corbis/Bettmann. Figure 3.21: Courtesy of Lawrence Berkeley National Laboratory. Figure 3.22: CSIRO Molecular Science. Figure 3.24: Scanning electron micrographs by Dr. Robert S. Preston and Prof. Joseph E. Hawkins, Kresge Hearing Research Institute, University of Michigan Medical School. Figure 3.36: Photo courtesy of Nite Ize, Inc. Page 120: Phil Degginger/Color-Pic, Inc.

Chapter 4

Chapter Opener (left): Nick Koudis/Photodisc. (right): Courtesy of Morton International, Inc. All rights reserved. Figure 4.4: AP Wide World Photos. Figure 4.9: Phil Degginger/Color-Pic, Inc. Figure 4.10b (right): Richard Hamilton Smith/Corbis/Bettmann. Figure 4.10b (left): Phil Degginger/Color-Pic, Inc. Figure 4.11: Adam G. Sylvester/Photo Researchers, Inc. Figure 4.12: Fundamental Photographs, NYC. Page 140: Tom Pantages. Figure 4.15 (top): © Dr. David Phillips/Visuals Unlimited. Figure 4.16: Jackson Vereen/Food Pix/PictureArts Corporation. Page 145 (left): David Roth/Getty Images Inc. - Stone Allstock. (right): © Dave Blazek - 2002 Tribune Media Services. Reprinted with permission, 2007. Figure 4.23: Richard Megna/Fundamental Photographs, NYC. Figure 4.24a: Tom Pantages. Figure 4.24b: Richard Megna/Fundamental Photographs, NYC. Figure 4.25: University of Wisconsin Environmental Remote Sensing Center. Page 160: Tom Pantages. Page 161 (left): NYC Parks Photo Archive/Fundamental Photographs, NYC. (right): Kristen Brochman/Fundamental Photographs, NYC.

Chapter 5

Chapter Opener 5 (left): VCL/Spencer Rowell/Taxi/Getty Images. (right): Steve Cole/Photodisc. Figure 5.1a: Thomas A. Vincent, Santa Rosa, CA. (inset): Camera Obscura of San Francisco (Giant Camera), www.Giantcamera.com. Figure 5.1b: NMPFT/SSPL/The Image Works. Figure 5.2: Topham/The Image Works. Figure 5.6: Getty Images, Inc. Figure 5.8a: Courtesy of © Eastman Kodak Company. Figure 5.8b: M. Rutz/Satchen/Niven/Corbis/Bettmann. Figure 5.9: Color-Pic, Inc. Page 177: Serengeti Eyewear(R). Figure 5.10b: Courtesy of © Eastman Kodak Company. Figure 5.11a: Everett Collection. Figure 5.11b: DANJAQ/EON/UA/Picture Desk, Inc./Kobal Collection. Figure 5.12: James King-Holmes/Photo Researchers, Inc. Figure 5.15 (left): Tony Freeman/PhotoEdit. Figure 5.18: Richard T. Nowitz/Photo Researchers, Inc. Figure 5.19: Tom Pantages. Figure 5.22: Courtesy of SE-IR Corporation, Goleta, California. Figure 5.26: Richard Megna/Fundamental Photographs, NYC. Figure 5.27 (left): Robert K. MacDowell, Waterford, Virginia, USA. Member International Dark Sky Association, www.darksky.org. (right): Phil Degginger/Color-Pic, Inc. Figure 5.30: Henry Diltz/Corbis/Bettmann. Figure 5.31: Herman Agopian/Getty Images Inc. - Stone Allstock. Figure 5.34: E.R. Degginger/Color-Pic, Inc. Page 202: Color-Pic, Inc.

Chapter 6

Chapter opener (left): StockTrek/Photodisc. (right): Getty Images, Inc. Figure 6.1: J-L Charmet/Photo Researchers, Inc. Figure 6.4: UKAEA Culham Division. Figure 6.5: Scala/Art Resource, N.Y. Figure 6.7: Fermilab Visual Media Services. Figure 6.8: Browne, Malcolm W., "US-Russian Team May Have Created Ultra-Heavy Element" (© 1999 by The New York Times Company. Reprinted with permission.) Figure 6.9 (top): Browne, Malcolm W., "Team Adds Two Elusive Elements to the Periodic Table" (© 1999 by The New York Times Company. Reprinted with permission.) (bottom): Johnson, George, "At Lawrence Berkeley, Physicists Say a Colleague Took Them for a Ride" (© 2002 by The New York Times Company. Reprinted with permission.) Figure 6.10: Lawrence Berkeley Nat'l Lab. Figure 6.14: AIP/Photo Researchers, Inc. Figure 6.15: AFP/Getty Images, Inc. Figure 6.16: Photo by Heka Davis, courtesy AIP Emilio Segre Visual Archives, Physics Today Collection. Figure 6.17: Achiv zur Geschichte der Max-Planck-Gesellschaft, Berlin-Dahlem. Page 228: © Dave Blazek- 2002 Tribune Media Services. Reprinted with permission, 2007. Figure 6.22: Environmental Instruments Canada Inc. Figure 6.23: Anne Purkiss. Figure 6.26 (top): David Martyn Hughes, images-of-france/Alamy Images Royalty Free. Figure 6.27b: © Royalty-Free/CORBIS. Figure 6.31: Getty Images, Inc. Figure 6.32: Getty Images, Inc. Figure 6.33a: SPL/Photo Researchers, Inc. Figure 6.36: Martin Dohrn/Photo Researchers, Inc. Figure 6.38: Simon Fraser/RVI RMPD/Photo Researchers, Inc. Figure 6.39: Visuals Unlimited. Page 247 (left): Neil Borden/Photo Researchers, Inc. (right): Jack Vearey/Getty Images Inc. - Stone Allstock. Page 248: Photo by Erik Hunter.

Chapter 7

Chapter Opener (left): Randy Wells/Corbis/Bettmann. (right): Don Farrall/Photodisc. Figure 7.1: Ernst Haas/Getty Images Inc. - Hulton Archive Photos. Figure 7.3: Bruce James/Foodpix/PictureArts Corporation. Figure 7.4 (left): James Zipp/Photo Researchers, Inc. (right): Photo D.R. Rubenstein. Figure 7.10: Herman Eisenbeiss/Photo Researchers, Inc. Figure 7.15a: Inmagine/Alamy Images Royalty Free. Figure 7.16: Jim Brandenburg/Minden Pictures. Figure 7.17a: National Hurricane Center, National Oceanic and Atmospheric Administration. Figure 7.17b: Harold V. Thurman. Figure 7.21: Carla Breeze/Fundamental Photographs, NYC. Figure 7.22: Stone/Getty Images. Figure 7.23: Corbis/Bettmann. Figure 7.24: P. Rona/National Oceanic and Atmospheric Administration. Figure 7.25a: Debbie

Kelley, University of Washington. Figure 7.25b: Adam Jones/Photo Researchers, Inc. Figure 7.25c: Reprinted with permission from Kelley, D. et al *Science* 307:1428–34 Fig.4A, March 4, 2005. © 2005 AAAS. Figure 7.26: Pressure Canner image provided by National Presto Industries, Inc. Figure 7.34: Photofest. Page 288: John Burke/Index Stock Imagery, Inc.

Chapter 8

Chapter Opener (left): StockTrek/Photodisc. (right): Laurence Dutton/The Image Bank/Getty Images. Figure 8.1: NASA/Johnson Space Center. Figure 8.2a: Corbis/Bettmann. Figure 8.2b: Robert Essel NYC/Corbis/Bettmann. Figure 8.2c: © 2002 Thinkstock LLC/Punchstock. Figure 8.4: Fundamental Photographs, NYC. Figure 8.6 (photo): Getty Images - Photodisc-. Figure 8.8: Pam Francis Photography. Figure 8.9 (top): Sandia National Laboratories. (bottom): Todd Goodrich/The University of Montana. Figure 8.11a (left): Max Alexander © Dorling Kindersley, Courtesy of South Street Seaport Museum, New York. (right): Douglas Whyte/Corbis/Bettmann. Figure 8.11b (left): David Young-Wolff/PhotoEdit. (right): Color-Pic, Inc. Figure 8.12a (left): Harold M. Lambert/Getty Images Inc. - Hulton Archive Photos. Figure 8.12b (left): Hulton Archive/Getty Images. Figure 8.15a: Hulton-Deutsch Collection/Corbis. Figure 8.15b: Corbis/Bettmann. Figure 8.15c: Hulton-Deutsch Collection/Corbis/Bettmann. Figure 8.16: Tom Pantages. Figure 8.19: Hulton-Deutsch Collection/Corbis. Figure 8.23: © Greg Von Doersten. Figure 8.27: Richard Megna/Fundamental Photographs, NYC. Figure 8.29: Photography courtesy of Cessna Aircraft Company. Figure 8.31a: Walter Bibikow/Index Stock Imagery, Inc. Figure 8.31b: Courtesy of Sierra Club. Figure 8.33: Comstock/PunchStock. Figure 8.35: Richard T. Wright. Figure 8.36b: Philippe Hays/AFP/Getty Images, Inc. - Agence France Presse. Page 330: CLAVER CARROLL/Photolibrary.Com. Figure 8.39: Photograph courtesy of University Communications, University of California, Irvine, Copyright 1975. Figure 8.42: NASA Earth Observing System. Page 331: RubberBall/Superstock Royalty Free. Page 332: Nicholas Rigg, Taxi/Getty Images.

Chapter 9

Chapter Opener (left): Photo by Rico Torres © 2003 Miramax/Columbia Pictures, All Rights Reserved, Kobal Collection - The Picture Desk. (right): Hisham F. Ibrahim/Photodisc. Figure 9.1: Getty Images Inc. - Hulton Archive Photos. Figure 9.2: Rysher Entertainment/Picture Desk, Inc./Kobal Collection. Figure 9.4: William Weddle, Leslie Garland Picture Agency/Alamy Images. Figure 9.5: Jeff Hunter/Image Bank/Getty Images. Figure 9.9: Erik Freeland/Corbis/SABA Press Photos, Inc. Figure 9.11 (top left): Tom Pantages. (top center): Tom Pantages. (top right): Tom Pantages. (middle left): E.R. Degginger/Color-Pic, Inc. (bottom left): Richard Megna/Fundamental Photographs, NYC. (bottom right): Photo courtesy NeoPaws. Figure 9.16a: Corbis/Bettmann. Figure 9.20: Time Life Pictures/Mansell/Time Life Pictures/Getty Images. Page 365: www.CartoonStock.com. Figure 9.22: Richard Megna/Fundamental Photographs, NYC. Figure 9.23 (photo): Ewing Galloway/Index Stock Imagery, Inc. Figure 9.24: James Shaffer/PhotoEdit. Page 375 (top): ©The New Yorker Collection 2001 Leo Cullum from cartoonbank.com. All Rights Reserved. Figure 9.29: Greg Keller, Portland Fire & Rescue. Page 377: Mark & Audrey Gibson/The Stock Connection.

Chapter 10

Chapter Opener (left): Richard Hamilton Smith/Corbis/Bettmann. (right): PhotoLink/Photodisc. Figure 10.1 (top left): Thayer, Mark/Index Stock Imagery, Inc. (top center): Image Source/Photolibrary.Com. (top right): Richard Pasley/The Stock Connection. (bottom left): Ryan McVay/Photodisc Green/Getty Images - Photodisc-. (bottom center): Tom Pantages. (bottom right): Deshakalyan Chowdhury/AFP/Getty Images, Inc. - Agence France Presse. Figure 10.2a: Tootsie Roll Industries. Figure 10.2b: © Bettmann/CORBIS All Rights Reserved. Figure 10.4a: Richard T. NowitzCorbis/Bettmann. Figure 10.4b: © Richard Nowitz/www.novitz.com. Figure 10.8b: Sergio Piumatti. Figure 10.10b: Tom Pantages. Figure 10.11 (bottom left): John Fairfax Publications Pty Limited/Photolibrary.Com. (bottom center): Pascal Parrot/Corbis/Sygma. (bottom right): Photo courtesy of Bestop Acrylic Ware Product Ltd, www.acrylical.com. Figure 10.12a: Lynn Goldsmith/Corbis/Bettmann. Figure 10.12b: Douglas Whyte/Corbis/Bettmann. Figure 10.20a: Danny Lehman/Corbis/Bettmann. Figure 10.20b: Otis Imboden/ NGS Image Collection. Figure 10.21: Peter Arnold, Inc. Figure 10.26: Gabe Palmer/Corbis/Bettmann. Figure 10.27 (bottom): Corbis Royalty Free. Figure 10.29a: Photo courtesy of DuPont. Figure 10.33a: Roger Ressmeyer/Corbis/Bettmann. Figure 10.33b: Manchan/Photodisc Red/Getty

Images - Photodisc-. Figure 10.34: Pearson Education/PH College. Figure 10.35a: Andrew Syred/Photo Researchers, Inc. Figure 10.35b: Susumu Nishinaga/Photo Researchers, Inc. Figure 10.35c: Dennis Kunkel Microscopy, Inc. Figure 10.35d: Tom Pantages. Figure 10.37b: © Alen MacWeeney/CORBIS. Figure 10.38 (top left): Bob Krist/Corbis/Bettmann. (top right): Alan Towse; Ecoscene/Corbis/Bettmann. (bottom left): Alan Towse; Ecoscene/Corbis/Bettmann. (bottom right): Richard T. Nowitz/Corbis/Bettmann. Page 413: Reuters/Corbis/Bettmann. Page 415 (left): Comstock/PunchStock. (right): EyeWire Collection/Getty Images - Photodisc-. Page 417 (left): Corbis Royalty Free. (right): Tom Pantages.

Chapter 11

Chapter Opener (left): Richard T. Nowitz/Corbis/Bettmann. (right): Nick Koudis/Photodisc. Figure 11.1: Tom Pantages. Figure 11.6: Hulton-Deutsch Collection/Corbis/Bettmann. Figure 11.7a (right): Julie Newmar U.S Patent 4,003,094 Jan. 18, 1977 Fig. 2 drawing/http://www.uspto.gov/patft/index.html. Figure 11.7a (left): Getty Images Inc. - Hulton Archive Photos. Figure 11.7 (bottom): © Stacy Morrison/zefa/Corbis. Figure 11.15 (right): © Judith Miller/Dorling Kindersley / Cristobal. Figure 11.26a: Michael Lustbader/Photo Researchers, Inc. Figure 11.26b: H. Turvey/Photo Researchers, Inc. Figure 11.26c: Janet Foster/Wonderfile Royalty Free. Figure 11.26d: Reportage/Getty Images. Figure 11.28b: Tom Pantages. Figure 11.30: PRNewsFoto/NewsCom. Figure 11.35 (right): Dr. A. Lesk, Laboratory of Molecular Biology/Science Photo Library/Photo Researchers, Inc. Page 457: Digital Art/Corbis/Bettmann. Page 458: America 24-7/Getty Images. Page 460: Images.com/Corbis/Bettmann. Page 461: Vaughan Stephen/Corbis/Bettmann.

Chapter 12

Chapter Opener (left): John Martin/Images.com/Corbis/Bettmann. (right): PhotoLink/Photodisc. Figure 12.5a: David Raymer/Corbis/Bettmann. Figure 12.5b: Michelle Garrett/Corbis/Bettmann. Figure 12.7: AP Wide World Photos. Figure 12.13: Jacques Langevin/Corbis/Sygma. Figure 12.15: Tom Pantages. Figure 12.16: Color-Pic, Inc. Figure 12.20 (left): Ariel Skelley/Corbis/Bettmann. (right): Getty Images/Digital Vision. Figure 12.28: Tom Pantages. Figure 12.29a: Color-Pic, Inc. Figure 12.29b: Color-Pic, Inc. Figure 12.32: Tom Pantages. Figure 12.35: Roy Tanami/Ursus/SeaPics.com. Page 502 (left): Phil Degginger/Color-Pic, Inc. (right): Pearson Education/PH College. Page 503: Thinkstock. Page 504: Corbis Royalty Free.

Chapter 13

Chapter Opener (left): Eyewire (Photodisc)/PunchStock. (right): Janis Christie/Photodisc. Figure 13.1: Getty Images/Digital Vision. Figure 13.2: Phil Degginger/Color-Pic, Inc. Figure 13.4a: Petrey, D. and Honig, B. (2003) GRASP2: Visualization, Surface Properties, and Electrostatics of Macromolecular Structures and Sequences. *Methods in Enzymology*. 374, 492–509. (http://trantor.bioc.columbia.edu/). Figure 13.4b: © 2005 Michael Dalton - Fundamental Photographs. Figure 13.11: Everett Collection. Figure 13.14: Leif Skoogfors/Corbis/Bettmann. Figure 13.19a: Stockbyte. Figure 13.19b: Kimberley Waldron. Figure 13.20: Kimberley Waldron. Figure 13.21: Kimberley Waldron. Figure 13.33: AP Wide World Photos. Figure 13.35: Amet Jean Pierre/Corbis/Sygma. Figure 13.36a: AP Wide World Photos. Figure 13.36b: SEM micrographs by Dr. Zouhair Sbiaa, provided courtesy of MicroCHIPS, Inc.

Chapter 14

Chapter Opener (left): © Alfred Saerchinger/Zefa/Corbis. (right): PhotoLink/Photodisc. Quotation: Hillyer, Robert, "A Letter to the Editor," *The Collected Verse of Robert Hillyer* (© 1934, Knopf). Figure 14.1a: Archives/AFP/Getty Images, Inc. - Agence France Presse. Figure 14.1b: Thierry Roge/Reuters/Corbis/Bettmann. Figure 14.7: Stone/Getty Images. Figure 14.9 (top left): Martin Jacobs/FoodPix/PictureArts Corporation. (top right): oote boe/Alamy Images. (bottom left): Corbis Royalty Free. (bottom right): Thom DeSanto Photography Inc./StockFood America. Figure 14.15: AP Wide World Photos. Figure 14.18: Tom Pantages. Figure 14.22: Logo Designed and developed by: Tymic Designs Inc. 6537 Cedar Rapids Crescent Mississauga, Ontario Canada L5N 7P5, www.tymic.com. Figure 14.23: Pete Saloutos/Corbis/Bettmann. Figure 14.26 (top): CORBIS- NY. (bottom): Viviane Moos/Corbis/Bettmann. Figure 14.28b: Tom Pantages. Figure 14.29a: Lester Lefkowitz/Corbis/Bettmann. Page 589 (left): Anne Domdey/Corbis/Bettmann. (right): Michael DeYoung/Corbis/Bettmann. Page 590: Images.com/Corbis/Bettmann. Page 592: Michael Stroud/Express/Getty Images.

Index

A

Absolute temperature scale, 25
Absolute zero, 25, 173
ACE inhibitor, 544
Acetaldehyde, 485
Acetate fiber, 396
Acetic acid, 104–105
Acetone, 15
Acetyl CoA, 493–494
Acetylene, 116, 340
Acetyl groups, 493
Achiral amino acid, 533
Acid rain, 156, 157–158
Acid(s), 149, 575–576
   strong, 149, 154
   weak, 149
Acrolein, 471
Acrylic, 396, 406
Active ingredients, in drugs, 540
Active site, 480
   binding site *vs.*, 523
Addiction, physiological, 561–562
   drug design and, 516
   to nicotine, 581, 585
Addition polymers, 388, 389, 394, 395
Addition reaction, 89–90, 474
Adenine, 446, 447
Adrenaline, 574, 575
Aerosols, 278
Africa, AIDS statistics in, 524, 525
Agrobacterium tumefaciens, 454–455
AIDS, 524, 525
Air
   composition of clean, 292, 293
   density of, 305–306
   evidence of, 291–292
   pollutants in, 155–157, 292, 313–314
Aircraft
   detecting corrosion using pH indicator,
      152–153
   relation between thrust and altitude, 312
Airport security, gas chromatography as aid
      for, 374–375
Air pressure, tires and, 301–302, 308
ALA (alpha-linolenic acid), 475
Alanine, 433, 443, 531–532
Alcaligenes eutrophus, 424, 426
Alcer's Miracle Sports Water, 140
Alchemist, The, 214
Alchemy, 214–215
Alcohols, 189, 392, 428, 429
   ethers *vs.*, 354–355
   hydrogen bonding in, 560
   miscibility of, 281–282
   in neurotransmitters, 558
   as nucleophiles, 395
Aldehydes, 485
Alkali metals, 41, 42
Alkaline, 579
Alkaline earth metals, 41, 42
Alkaloid, 579

Alkanes, 344, 345–346, 348, 353
   naming, 341–342
   physical properties, 351–352
Alkenes, 344, 346, 348, 392
   naming, 344–345
Alkynes, 92, 346, 392
   naming, 344–345
Allene, 92
Allotrope, 90
Allyl mercaptan, 299
Alpha anomer, 485
Alpha decay, 219–220
Alpha emission, 222
Alpha helix, 439
Alpha 1→ 4 linkage, 488, 490
Alpha particles, 215, 219, 230
Alpha radiation, 237, 239–240
Alpha symbol, 219
Altitude
   aircraft thrust and, 312
   air density and, 305
   boiling point and, 272
Aluminum, 117, 118
Ambien (zolpidem tartrate), 538
American Civil Liberties Union, 564
American Journal of Epidemiology, 72
Americium-241, 231
Amethyst, 70
Amide bonds, 527
Amides, 401–402
   nitrogen atoms in, 552
Amines, 100, 428
   attraction to carboxylic acid functional
      group and, 431
   basicity of, 583
   binding specificity and, 559–560
   boiling points, 555
   free base, 578–581
   as hydrochloride salt, 578–581
   hydrogen bonding in, 559–560
   naming, 554
   neurotransmitters and, 558
   nitrogen atoms in, 552–553
   odorous, 554–555, 556
   primary, 553, 555, 577
   protonation of, 575–577, 583
   as salts, 577, 578–581
   secondary, 553, 577
   tertiary, 553, 577
   testing for, 565–566
Amino acids, 427–437
   achiral, 533
   catabolism of, 493
   chiral, 532–533
   essential, 452
   handedness and, 531
   nonessential, 451–452
   20 protein-building, 433–423
   side chain polarity classification of, 436
Amiodarone (Cordarone), 539–540
Ammonia, 10

ball-and-stick molecule of, 112
bonding rules, 102
hydrogen bonds in, 260
molar mass of, 63
R groups, 553
single bond of, 93
three-dimensional structure of, 111, 112
Ammonia gas, 315–316
Ammonium carbonate, 131
Ammonium ion, 130
Amontons, Guillaume, 312
Amontons' law, 311, 312, 317, 321
Amorphous regions, 403
Amphetamine psychosis, 561
Amphetamines, 552, 560–561
   Ecstasy *vs.*, 562
   Marquis test result, 565, 566
   side effects, 561
Amphipathic substances, 282–283
Amprenavir, 527
Amylase, 483–484
Anabolic pathways, 478
Anabolic steroids (anabolic androgenic
      steroids), 572–573
Anabolism, 478
Anaerobic bacteria, 556
Analgesic, 544
Androstenone, 322
-ane, 341
Anesthetics, 544
   ether, 354, 356
Angina
   nitroglycerin for, 362
   perhexiline for, 538
Anion, 49
Anode, 183, 184
Anomeric carbon, 488
Anomers, 485, 488
Antiangiogenesis drugs, 520
Antianxiety drug, side effects, 521
Antibiotic, 544
Antibodies, 441, 569–570
   creating for specific drugs, 570–571
Anticoagulant, 544
Antigens, 569–570
Antihistamine, 544
Anti-inflammatory, 544
Antimetabolite, 544
Antioxidant enzymes, 190
Antioxidants, 188–191
   content of some foods, 191–192
   recommended daily allowances of, 190
Antipyretic, 544
Antitussive, 544
Antiviral drug, 544
Apalachicola Bay (Florida), metal levels in,
      43–45
Aqueous solution, 139
Arachidonic acid, 469
Arginine, 434
Argon, 47, 171, 173, 293

Argonne National Laboratory, 87
Arsenic, 117–118
    electron configuration, 169–171
Arson, 370–375
Artificial flavoring, 497–498
Artificial sweeteners, 486–487
Asbestos, 71–72
Ascorbic acid (vitamin C), 191
Asparagine, 433
Aspartame, 486–487
Aspartic acid, 434, 486
Asphalt, 165
Asteroids, 64
Atlantic Monthly, 466, 497
Atmospheric pressure, 273–275, 306–307
Atomic bomb, 4, 5, 214, 236–237
Atomic number, 8–9
Atom(s), 7–9
    counting, 61
    defined, 7
    energy absorption by, 193–194
    energy emission by, 193–194
    fusion reaction, 210–214
    modern view of, 9
    probing, 3
    radii, 20–22
    radioactive decay and, 219–225
    size of, 8, 18, 84
    splitting, 4
Atrial fibrillation, drugs for, 539
Aum Shinrikyo terrorist attack, 293–294
Autoimmune deficiency syndrome (AIDS),
        524, 525
Autoionization, of water, 147–153
Avogadro's law, 302–304, 305, 317, 321
Avogadro's number, 61
Azidodeoxythymidine (AZT), 526

B
Bacteria
    anaerobic, 556
    as source of polymers, 424–425, 426
Bakelite, 407
Balanced equation, 13
Ball-and-stick model, 112
Barium, 225, 235
Barium hydroxide, 154
Barometer, 306–307
Base pair, 447
Base(s), 149–150, 575–576
    nitrogenous, 446–447, 449
    strong, 154
Base unit, 17
Basicity, of amines, 583
Battery(ies), 179, 183–186
    lithium, 184, 186
    miniaturized, 184
    in series, 185
Becquerel, Antoine Henri, 2–3
Beef fat, 472
Beer, effects of sunlight on, 199–200
Bees, detection of land mines using, 298
Beeswax, 280
Belt of stability, 223
Bends, 279
Benedryl (diphenhydramine), 574, 575
Bent, Stacey, 87
Bent molecular geometry, 113, 114, 115

of water, 251, 252
Benzaldehyde, 485
Benzene, 95–96
    irradiated food and, 499
Benzene ring, 189
    in explosives, 363–365
Benzocaine, 579
Benzophenone, 199
Beryl, 70
Beta anomer, 485
Beta blockers, 544
Beta-carotene, 198
Beta decay, 220–221
Beta emission, 222
Beta emitter, 220
Beta 1→ 4 linkage, 489, 491
Beta particles, 220, 221
Beta pleated sheet, 439–440, 443
Beta radiation, 237, 239–240
"Better Killing Through Chemistry," 297
BHA, 188, 189
BHT, 188, 189
Bicycle tire, air pressure and, 301–302, 308
Big Bang theory, 215
Binding site, 513
    active site *vs.*, 523
Biodegradable, 411–412
Biodegradable plastic, 422
Biodegradable polymers, 423–425
Biological agents, 296–299
Biomineralization, 68
Bird migration patterns, distribution of
        isotopes in water and, 253–255
Bismuth, 45
Bismuth-210, 221–222, 223
Black powder, 336
Black smokers, 274
Black-throated warbler, migration patterns,
        253–255
BMI. *See* Body mass index
bmim$^+$, 127–128
Bock, Jane, 489, 491
Body mass index (BMI), 464
Boiling points, 256–257
    of amines, 555
    bmim$^+$, 128
    of fats and oils, 470
    of hydrocarbons, 351–352
    intermolecular interactions and, 274
    in structural isomers, 354–355
    of water, 256–257, 259, 272–276, 276
Bombs, atomic, 4, 5, 214, 236–237
Bond, James, 179
Bond energy, 91
    for nitroglycerin reaction, 360–362
Bond lengths, 91, 116
Boron, 102
Bovine chymotrypsinogen, 440–441
Bow tying analogy for enzyme action, 480
Boyle, Robert, 308
Boyle's law, 308, 317–318, 321
Branched hydrocarbon, 343
Brand name drugs, 536, 539–544
Breath, bad, 556
Brief History of Time, A, 215
Broccoli, sulforaphane in, 510
Brocco Sprouts, 510–511
Bromenshenk, Jerry, 298

Bromine, 48
Bromobenzene, 533
Browning, chemistry of, 498
Bryce Canyon (Utah), 274, 275
Bubble gum, invention of, 384–385
Buckminsterfullerene (buckyballs), 94–97
Bush, George W., 238
Butane, 100, 277, 280, 294, 342
Butanol, 281
n-Butanol, 354–355, 356
2-Butanone, 401
Butter, 472
Butylamine, 553
Butyl butanoate, 497
5-Butyldecane, 347
tert-Butyl group, 347–348
tert-Butyl mercaptan, 299–300
Butyl propyl ether, 353, 354

C
C$_{60}$, 94–95
Cadaverine, 100, 554–555, 556
Cade, Robert, 138
Caffeine, 581
Caitlin, Don, 572
Calcite, 57
Calcium
    creation of ion, 48–49
    electron configuration, 173
    ionization of, 173–174
    isotopes, 55
Calcium carbonate, 106–107
Calcium hydroxide, 154
Calories, 91
    daily requirements, 491–492
Camera obscura, 165
Campbell, Bob, 358
Cancer
    chemotherapy, 520
    sequence leading to, 512
Carbohydrates, 478, 483–491
    complex, 483
    digestion of, 483–484, 493
    forensic study of breakdown of, 489–491
    metabolic pathway, 493, 494
    structure of, 484
    types of, 483
Carbon. *See also* Diamonds
    anomeric, 488
    atomic size, 84
    bonding patterns in molecules
            containing, 91–92
    C$_{60}$, 94–95
    in dirt, 40
    four bonds of, 114–115
    graphite, 90–94
    in gunpowder, 336
    isotopes, 55
    molar mass, 62
    nanotubes, 97–100
    uniqueness of, 84–86
    valence electrons, 102
    VSEPR rules and, 116
Carbon-14, 221
Carbonate ions, 131
Carbon-carbon bond length, 116
Carbon dating, 54
Carbon dioxide

in clean air, 293
as combustion product, 337, 339
as propellant, 278–279
release of at Lake Nyos, 314, 315
Carbonic anhydrase, 481
Carbon monoxide
in air, 155
as byproduct of gasoline combustion, 357
as combustion product, 337, 339
Carbon tetrachloride, 104, 115
Carbonyl group, 395
ester vs., 396
in ketones, 400–401
Carboxylic acids, 428–429
attraction to amine functional group, 431
in fatty acids, 467
hydrogen bonds in, 430
naming, 429
proton loss in, 430
Carcinogenic compounds, 510
Carmine dye, 498–499
Catabolic pathways, 478, 493–494
Catabolism, 478
Catalysts, 480
Catalytic converters, 339
Categorization, of natural world, 586–587
Cathode, 183–184
Cation, 49
Celexa (citalopram), 538
Cell membrane
ion concentrations and, 143
phospholipids in, 472
role of cholesterol in, 476
Cells, osmosis in living, 143, 144
Cellulose, 406, 490, 491, 541
Celsius temperature scale, 17, 19, 24–26
Center for the Advancement of Public
Health, 465
Centimeter (cm), 17
Cesium, 52
Chain reaction, 235–236
Chalcopyrite, 53
Chaos, 367–370
Charcoal, in gunpowder, 336
Charged amino acids, 436
Charles, Jacques, 311
Charles's law, 311, 317, 321
Chemical bonds, 11. See also Covalent
bonds; Hydrogen bonds
breaking, 93–94
carbon, 91–92
disulfide, 438
double, 87, 88, 91–93
ionic, 49–52, 257, 261, 514, 515
metallic bond, 53
multiple, 87, 88, 90–94, 115–116
single, 87, 88, 91–92
triple, 87, 88, 92–93
Chemical change, 16
Chemical equation, 11
components, 12
nuclear equation vs., 211, 212
Chemical formulas, 12
Chemically neutral, 148
Chemical reactions, 9–14
addition reactions, 89–90, 474
diamonds in, 87, 88
reaction coordinate, 482

reversible, 177
sources of energy for, 166
Chemotherapy, 520
Chernobyl nuclear disaster, 243
Chewing gum, 385–386
Chicken feathers, recycling, 423, 426
China, ideal-diet graphic, 466
Chiral, 532
Chiral drugs, 536
Chirality center, 532–535
Chloride ions, 49
Chlorine, 48
diatomic, 63
isotopes, 55–56
molar mass, 62–63
Chlorofluorocarbons (CFCs), 323–325
Chloroform, 104
Chlorophyll, 201
Cholesterol
conversion of squalene to, 476
effects of, 476–477
line structure, 109
miscibility of, 284
as steroid, 574
structure, 4
Chromatographic column, 371–372, 373
Chromium, 75
Chymotrypsinogen, 440–441
Cigarettes, addictiveness of, 585
Cis, 348, 349
cis-trans isomerism, 349
Citalopram (Celexa), 538
Citric acid cycle, 493–494
Citrine, 70
Clamshells, recycling, 423, 426
Clean Air Act (1970), 157
Clean Air Act (1991), 356, 357, 358
Clinton, Bill, 238
Cloning, 5, 28
Closed systems, 299–305
Clouds, 265–266
C–N bond, 402
Coating agents, in drugs, 541
Cobalt-60, 237, 240, 499
Coca-Cola, cocaine and, 581
Cocaine, 560–561, 579–581
Coca-Cola and, 581
properties of, 580
side effects, 561
testing for, 566–568, 569
Codeine, 516, 518
antibodies to, 570
Marquis test result, 566
testing for, 566, 569
Coefficient, 13
Coenzyme A, 493
Collagen, 443–445
Color, of food, 498–499
Colorado potato beetle, pheromones from,
322
Color film, 200–202
Combustion engine, 337–338, 349, 350
Combustion reaction, 336–340
car engine, 337–338, 349, 350
catabolism and, 494
incomplete, 339
increase in entropy and, 368–369
Complementary base-pairing, 447, 448

Compound(s), 9, 17
elements vs., 10
inorganic, 40, 131
organic, 40
Computerized tomography (CT), 241
Concentration gradient, 143
Condensation, 269, 276–277
Condensation polymers, 388, 394–395
Condensed structure, 107
polymers, 385
Conduction, 309–310
Conjugated double bonds, 197–198
Constitutional isomers, 532
Control experiment, 75
Controlled Substances Act, 586
Control rods (nuclear), 234
Control samples, 75
Convection, 309–310
Conversion factors, 19–22, 23
Cooling, evaporative, 323
Coors, Adolph, III, 73–75
Copolymer, 394
Copper
as antioxidant, 190, 192
in electrochemical cells, 179–183
Cordarone (amiodarone), 539–540
Core electrons, 47
Corn, transgenic, 436
Corundum, 70
Cotton, 406
transgenic, 436
Covalent bonds, 46, 47–48
electronegativity and, 51–52
in network solids, 85–86
nonpolar, 48
polar, 48
Crack cocaine, 575–576, 579–581
Crack nicotine, 581, 582, 584–585
Crick, Francis, 4
Critical mass, to sustain chain
reaction, 235
Crocidolite asbestos, 71–72
Cross-links, 398–399, 402–403
Crude oil, 352–353
Crutzen, Paul, 325
Crysotile asbestos, 71–72
Crystal, mineral vs., 68
Crystalline solids, 53
Crystallite, 403, 404
Crystal shapes, in photographic film,
177–178
C-terminus, 435
Cubane, 366
Cubic centimeter, 23
Curie, Irene, 224–225
Curie, Marie Sklodowska, 3, 224
Curie, Pierre, 3
Curl, Robert, 94, 95
Cyanide ion, 130
Cycling, drug abuse in professional,
551–552, 572
Cyclo-, 345
Cyclohexanone, 401
Cyclohexene, 105–106
Cyclopentanol, 356
Cyclopentene, 349
Cylinder, 349
Cymbalta (duloxetine hydrochloride), 537

Cysteine, 434
Cytosine, 446, 447

D

Dacron, 412
  synthesis of, 394–395
Daguerre, Louis, 166
Daguerreotypes, 166
Darmstadtium, 218
Davis, James, 129
Death, determining time of, 148
Decane, 342
Decompression sickness, 279
Decongestant, 544
Degree symbol, 25
Delaney Clause (1960), 498
Delocalization, 95
Density
  of air, 305–306
  of gas, 305–306
  of ice, 263, 264
  of water, 263, 264
Deoxyribose, 446
Depressants, 586
Desiccant, 132
Detergents, 284–285
Deuterium, 211, 252–253, 254–255, 256
dex-, 536
Dex drugs, 536
DHA (docosahexenoic acid), 475
di-, 554
Dialcohol, 395
Diamonds, 53, 83
  addition reaction on surface of, 89–90
  in chemical reaction, 87, 88
  covalent bonds in, 85–86
  double and single bonds in, 99
  endurance of, 86–90
  structure of, 85, 92
  as thermal conductor, 86–87
Diatomic molecule, 47
Diemer, Walter, 384–385
Diester, 395
Dietary fiber, 463, 491
Diethyl ether, 354–356, 581
3,3-Diethyloctane, 347
2,3-Diethylphenol, 364
Diffusion, 296
Digestion, 477–482
  of carbohydrates, 483–484
  of disaccharides, 489
  of polysaccharides, 490–491
Dihydrochloride salt, nicotine as, 584, 585
Diisopropyl fluorophosphate, 522
Dimensional analysis, 21–22
Dimethylamine, 260
2,2-Dimethylbutane, 358
Dimethyl ether, 354, 356
3,3-Dimethylnonane, 347
Dimethyl sulfide, 556
Dinitrobenzene, 365
Dinitrotoluene (TNT), 12, 364-365
  detecting, 374, 375
Diode, 118
Diphenhydramine (Benedryl), 574, 575
Dipole, 133
Dipole-dipole interaction, 257–259, 261

Dirt
  atmospheric, 64–67
  composition of, 39–46
  as forensic evidence, 72–76
  salt concentration in, 141, 143–144
Disaccharides, 483, 488–489
Dissolving agents, in drugs, 541
Distillation, 353
Disulfide bond, 438
Disulfide cross-links, 399
Diuretic, 544
DNA (deoxyribonucleic acid), 445–452
  discovery of structure of, 4–5
  forensics and analysis of, 455
  genetic engineering, 452–455
  recombinant, 453–454
Dodecane, 280, 352
Dodecapeptide, 435
Dogs, searching for arson evidence, 375
Dopamine
  cocaine and, 560–561
  effects of, 558
  physiological drug addiction and, 561–562
  structure, 558
Doping/dopants, 117–118
Double bonds, 87, 88, 91–93
  combustion and, 340
  conjugated, 197–198
  fatty acid melting point and, 468, 469
  in lipids, 474
Double helix, DNA, 448, 449
Double-stranded DNA, 447–448
Drop, 23, 24
Drowning, evidence of, 282–283
Drug Enforcement Agency, 586
Drugs
  brand name vs. generic, 536, 539–544
  categories of over-the-counter and prescription, 544
  chemotherapy, 520
  chiral, 536
  design of, 511, 512–518
    for AIDS drugs, 526
    enzymes as targets of, 523–524
    proteins as targets of, 510–518
  handedness, 528–535
  ideal, 518–522
  illicit, 551–587
    classes of, 586
    cocaine, 579–581
    crack cocaine, 575–576, 579–581
    mind-altering, 557–564
    nitrogen atoms in, 552–557
  names, 536, 538
  nicotine, 581–585
  performance-enhancing, 551–552, 572–573
  racemates, 536–539
  testing, 564–568
    qualitative, 564–568
    quantitative, 568–575
    at sporting events, 572–573
  three-dimensional shape of molecules, 511–512
  time-release, 542–543
Dry cell, 183, 184
Dry ice, 10

Dubble Bubble, 385
Duloxetine hydrochloride (Cymbalta), 537
Dyes
  artificial food, 498
  fabric, 198
Dynamite, 11, 359

E

Earls, Lindsay, 564
Earth
  creation of, 215
  elements, 41, 216
  gold in crust of, 46
  increase in average temperature, 327
  sunlight reaching, 213
Eaton, Phil, 366
EcoRI, 452–453
Ecstasy (MDMA), 562
  Marquis test result, 565
ED$_{50}$ (effective dose), 519–520
Effective unit, 241
Efficacy, of drug, 518
Einstein, Albert, 224
Elastin, 445
Electrically neutral, 148
Electrical power, from nuclear fission, 234–235
Electric current, 180
Electric eels, 185
Electrochemical cells, 179–183
  details in, 182
  diagram, 180
Electrocytes, 185
Electrodes, 179–180
Electrolysis, 16
Electrolyte balance, 138
Electrolytes, 137–141
  concentration of in some liquids, 139
Electromagnetic radiation, 192, 193
Electromagnetic spectrum, 192
Electron configuration, 169–172, 173–174
Electron density, 9
Electron-donating groups, 583
Electron dot structure, 100–106
  drawing, 130
  polyatomic ions, 130–131
  tetrahedral molecule, 113
Electronegativity, 51–52, 133
Electron-pair geometry, 111–112
  tetrahedral, 114, 251, 252
  of water, 251, 252
Electron(s), 9
  color film and, 200–202
  core, 47
  electrochemical cell, 179–183
  electron configuration, 169–172
  electron dot structure, 100–106
  energy levels, 167–169
  light-driven reactions and, 192–198
  movement of, 172–174
    in organic molecules, 186–192
    through film, 174–178
  sun protection and, 198–200
  symbol for, 219
  valence, 47–48, 102, 172
Electron transfer, 48–49
Electron-withdrawing groups, 583
Electrophile, 389

Electrophorus electricus (electric eel), 185
Element 114, 217
Element 116, 218
Element 118, 218
Element(s), 7, 17
    bonding rules for, in uncharged organic
        molecules, 102
    compounds *vs.*, 10
    confirmation of new, 218
    creation of, 217
    distribution in dirt, 40
    in Earth's crust, 41
    group names for, 41–42
    in human body, 41
    identifying with flame test, 194, 196
    naming, 218
    number of known, 217
    origins of, 216
Elgin marbles, 56–58
Eli Lilly, 536, 537
Emerald, 70
Enantiomers, 532–533
    fluoxetine, 537
    isoflurane, 539
Endothermic reactions, 339–340, 370
-ene, 344
Energy
    absorption by atoms, 193–194
    conservation of, 209–210
    emission by atoms, 193–194
    production in catabolism, 494–495
    as reactant and product, 362
    Sun as source of, 210
Energy levels, 167–169
Energy sublevels, 168–169, 171, 193–194
Entropy, 367–369
Environmental Protection Agency, 155
Enzymes, 441
    antioxidant, 190
    bow tying analogy, 480
    in digestion, 480–482
    induced-fit model, 482
    lock-and-key model, 481–482
    restriction, 452–453
    as targets of drug design, 523–524
Ephedrine, 560
Epinephrine, 558–559
EPO (erythropoietin), 572
Epsom salt (magnesium sulfate
    heptahydrate), 14, 15
Equilibrium
    evaporation and condensation at,
        276–277
    osmosis and, 142
    systems in, 136–137
Erythropoietin (EPO), 572
Essential amino acids, 452
Ester hydrolysis, 411, 412
Esters, 392–396, 428, 429
    carbonyl *vs.*, 396
    in fruits, 393
    naming, 393
Ethane, 280, 342
    combustion of, 339–340
Ethanol, 356–357
    hydrogen bonds in, 259–260
    miscibility of, 281
Ethene, combustion of, 340

Ethers, 353–356, 428, 429
    alcohols *vs.*, 354–355
    MTBE, 357
Ethyl formate, 394
Ethyl group, 343
Ethylmethylamine, 553, 555
Ethyl methyl ether, 353, 354
Ethylmethylpropyline, 553
Ethyne, combustion of, 340
Etorphine, 516, 518
Evaporation, 269, 276–277
Evaporative cooling, 323
Evidence, theory and, 31
Excipients, 540–541
Excited state, 167–168
Exercise pill, 509–510
Exothermic reactions, 339–340, 340
    entropy and, 368, 370
    of nitroglycerin, 359–361
Expectorant, 544
Experiment, control, 75
Explosion, 337
    of nitroglycerin, 359
Explosives, 335–336
    arson and analysis of, 370–375
    combustion reaction, 336–340
    gunpowder, 335, 336–337
    high and low, 363–366
    instability of, 360
    Molotov cocktail, 350–351
    most explosive compound, 366
    nitroglycerin, 359–362

F

Fabre, Jean-Henri, 322
Fabrics, synthetic, 396. See also Polyesters
Fahrenheit temperature scale, 25–26
Family, of elements, 7
Fats, 465–471
    digestion of, 478–479
    double bonds in, 474
    as energy stores, 496
    fatty acid content, 470
    free radicals and, 187–188
    metabolism of, 493, 494–496
    obesity and, 465–466
    phospholipids, 472–473
    saturated, 467, 469, 470, 474, 476
    trans, 473–475
    unsaturated, 467, 469, 474
Fatty acids, 467–471
    biologically important, 467
    catabolism of, 493
    in cooking lipids, 472
    melting points, 468–469
    omega-3, 475
    omega-6, 475
    saturated, 467, 469, 470, 474
    unsaturated, 467, 469
Fawkes, Guy, 335, 336
FDA. *See* Food and Drug Administration
Femtosecond, 27
Fermi, Enrico, 224
Fermi National Accelerator Laboratory, 217
Ferrosoferric oxide, 66
Fiber
    dietary, 463, 491
    synthetic, 396, 405–406

Film
    color, 200–202
    composition of, 174–178
    creation of, 165–166
    crystal shapes in, 177–178
    Polaroid, 178
Finland, Molotov cocktail, 350–351
Fireworks, 194
First law of thermodynamics, 209
Fixation, 425
Flame test, 194, 195, 196
Flat triangular molecular geometry,
    115
Flavoring
    artificial, 497–498
    esters and, 393
    natural, 496
Fleer Chewing Gum Company, 384–385
Fluoride, in toothpaste, 106
Fluorine, 47, 48, 52, 102
Fluorine gas, 10, 11
Fluorite, 53
Fluoxetine (Prozac), 536–537
Food
    antioxidants in, 191–192
    browning, 498
    carbohydrates, 483–491
    color of, 498–499
    desire for variety of, 491–492
    digestion of, 477–482
    fats, 465–477
    irradiated, 499
    nutritional information, 463–464
    processed, 496–500
    weight maintenance and, 491–496
Food and Drug Administration (FDA), 474,
    486, 498
Food pyramid, 464, 466
Forces
    pressure *vs.*, 307
    strong nuclear, 221
Forces of attraction, for hydrocarbon
    molecules, 351
Forensics
    arson and, 370–375
    determining time of death, 148
    dirt and, 72–76
    DNA analysis and, 455
    evidence of drowning, 282–283
    forensic botany, 489–491
    synthetic fibers and, 405–406
Formaldehyde, 485
    Marquis test and, 565
Formic acid, 103–104
Formula units, 50, 61
Fortified water, 140
Fossil fuels, 210, 352–353
Fractionating column, 353
Fractionation, 353
France
    nuclear fission reactors in, 232–234
    recycled radioactive waste in, 238
Francium, 52
Frankincense, 387
Franklin, Rosalind, 4
Fraud, creation of new elements and,
    218
Freebase, 581

Free base, 578-581
  freebase *vs.*, 581
  nicotine, 582, 584, 585
Free radicals, 186–192
  breakdown of Freon and, 324–325
  creating, 240–241
  oxidation and, 187
  ultraviolet light and creation of, 199–200
Freezing points
  of alkanes, 351
  defined, 264
Freon, 323
Freon-11, 324–325
Friable, 71
Frisch, Otto, 225
Fructose, 484
Fruit
  esters in, 393
  oxidation of, 191
Fuel cells, 98
Fuel rods (nuclear), 234
Fuels
  arson and analysis of, 370–375
  fossil, 210, 352–353
Fuel source for combustion reaction, 337
Fuel value, 495
Fuller, Buckminster, 94
Full structures, 107
Function, structure and, 587
Functional groups, 354
  alcohols. *See* Alcohols
  aldehydes, 485
  alkenes, 344–345, 346, 348, 392
  alkynes, 92, 344–345, 346, 392
  amides, 401–402
  amines. *See* Amines
  benzene rings, 392
  carboxylic acid, 428–429, 431, 467
  esters, 392–396, 428, 429
  ethers, 353–356, 428, 429
  ketones, 400–401
Fusion reaction, 210–214
  first artificial, 217

G
gag-pol, cleaving, 525, 526
Galactose, 484
β-Galactosidase, 489
Gallon, 23
Gamma radiation, 213, 221, 226
  effects on living organisms, 237, 239–240
  irradiated food, 499
Gamma symbol, 219
Garbage Project, 413
Gas chromatography, fuel analysis and, 371–375
Gas(es)
  closed systems and, 299–305
  differences with liquids and solids, 293–295
  diffusion of molecules in, 296
  dissolution in other gases, 294, 296
  global warming and, 323–328
  greenhouse, 326–328
  homogeneity of, 293
  ideal, 294
  laws governing
    Amontons' law, 311, 312, 317, 321

Avogadro's law, 302–304, 305, 317, 321
    Boyle's law, 308, 317–318, 321
    Charles's law, 311, 317, 321
    general, 316–318
  mixing of, 314–316
  molecular speed in, 295
  moles of gas and, 302, 317–319
  motion of molecules in, 306
  noble, 41, 42, 46–47, 50
  open systems and, 299, 300–301
  pheromones, 322
  pressure and, 302, 305–308, 312–313, 317–319
  pressurized, 278–279
  real, 294
  symbol, 12
  temperature and, 302, 308–316, 317
  volume and, 302–304, 308, 317
Gasoline
  arson and, 370
  combustion of, 337–338
  ethanol as additive to, 356–357
  gas chromatogram of, 372–374
  MTBE in, 356–358
  octane number, 349–350
Gatorade, 138–141
G.D. Searle, 538
Geiger counter, 229, 230
Gelatin, film, 176–177
Gemstones, 70
Gene, 448
General gas law, 316–318
Generic drugs, 536, 539–544
Generic formal chemical name, of drug, 536
Generic switch, 538
Genetic code, 450
Genetic engineering, 425–426, 436–437, 452–455
Geodesic domes, 94–95
Germanium-germanium bonds, 116–117
Gernsheim Collection (Austin, TX), 165
GLACÉAU smartwater, 140
GLACÉAU vitaminwater, 140
Glass electrode, 152
Global warming, 326
Globular proteins, 439–440, 441, 443, 511–512
Glucose, 484
  anomers of, 485
  breakdown of, 368, 480, 494, 495
Glue
  super, 390, 391
  white, 389
Glutamic acid, 434
Glutamine, 433
Glutathione peroxidase, 190
Glycerol, 471, 472
Glycine, 433, 531, 532, 533
Glycogen, 490
Glyphosate (Roundup), 436–437
Goettlich, Paul, 437
Gold, 10, 11
  alchemy and attempts to create, 214–215
  in Earth's crust, 46
Goodyear, Charles, 398
Grains, 176
Grams, 19, 26–27

Graphite, 90–94
  structure of, 90–91, 92
GRAS (generally recognized as safe), 486
Gravity, pressure and, 305
Greenhouse effect, 326
Greenhouse gases, 326–328
Ground state, 167–168
Group, of elements, 7
Guanine, 446, 447
Gum base, 385
Gum Base Company, 386
Gunpowder, 335, 336
Gutta-percha, 387, 398, 399

H
Hahn, Otto, 224–225
Hairston, Jack, 138
Half-life, 228–230
Half reactions, 65–67, 175, 181–182
  in electrochemical cell, 180–182
Halide, 177
Hallucinogens, 562, 586
Halogens, 41, 42, 84
Hancox, Paul, 64
Handedness, 528–535
Hard ions, 145
Hassium, 219
Hawking, Stephen W., 215
HDLs. *See* High-density lipoproteins
HDPE (high-density polyethylene), 409
Head group, 282–283
Headspace analysis, 371
Heat. *See also* Temperature
  electron movement and, 172–173
  from explosive combustion, 366
  infrared light and, 193
  from nitroglycerin reaction, 361
  specific, 270, 272
Heliograph, 165, 166
Helium, 47
Hemoglobin, 441
Hepburn, Audrey, 83
Heptane, 280
Heptanol, 281
Heroin, 516, 518, 561
  antibodies to, 570
Heterogeneous mixture, 15, 17
Heterolytic cleavage, 186
Hexadecane, 352
Hexane, 280
Hexanol, 281
Hexapeptide, 435
Hexokinase reaction, 480, 481
High-density lipoproteins (HDLs), 477
High explosives, 365–366
Hiroshima, 4, 237
Histidine, 434
Hitler, Adolf, 561
HIV (human immunodeficiency virus), 524, 525
HIV protease, 526–527
HIV protease inhibitors, 527
Hodgkin, Dorothy, 3–4
Holmes, Richard, 253, 254
Homogeneous mixture, 15, 17
Homologous series, 341, 342
Homolytic cleavage, 186
Hooded seals, stored lipids in, 496

Hormone drugs, 544
Hormones, 441
Human immunodeficiency virus (HIV), 524, 525
Humans
  catabolic pathways, 493
  effect of specific heat of water on, 270, 272
  effects of lead on, 45
  elements in, 41
  health effects of radiation on, 242–244
  major polysaccharides in, 490
  neurotransmitters in, 558
  pheromones in, 322
  radiation exposure and, 241–244
  use of polymers by, 384–387
Hurricane Andrew, 267
Hurricanes, ice-crystal formation and, 266–267
Hybrid cells, 570
Hydration, 134, 149
Hydriodic acid, 154
Hydrobromic acid, 154
Hydrocarbons, 100
  as amino acid side chain, 435
  branched, 343
  combustion of, 337–340
  in different phases, 351–352
  engine knocking and, 349–350
  as explosives, 365
  immiscibility of, 280–282
  nomenclature, 341–350
Hydrochloric acid, 154
Hydrochloride salt, 578–581
  of nicotine, 581, 582, 584
Hydrofluorocarbons, 325–326
Hydrogen, 48
  energy levels in atom of, 167–168
  ions in water, 147–148
  isotopes, 211, 252–256
  line spectrum of, 194, 195
Hydrogenation, 474, 475
Hydrogen bomb, 4, 5, 214, 236–237
Hydrogen bond acceptor, 355
Hydrogen bond donor, 355
Hydrogen bonds, 259–261, 514, 515
  in alcohols, 560
  in amines, 559–560
  in beta pleated sheet, 443
  in carboxylic acids, 430
  in ice, 264
  in liquid water, 263
  phase changes and, 555
Hydrogen chloride, 257–258
Hydrogen isotope ratios, use of, 253–255
Hydrogen peroxide, 15
Hydrogen sulfide, 114, 115, 556
Hydrolysis, 479
  ester, 411, 412
  polylactic acid, 411–412
  triacylglycerol, 478–479
Hydronium ion, 241
Hydrophilic molecules, 280–281
Hydrophobic molecules, 281
Hydrothermal vents, 274–275
Hydroxide ions, 130
  in water, 147–148

Hydroxyalkanoate, 424, 425
Hydroxyl radical, 241
Hypothesis, 29

I

Ianthanum, 224
Ice, 263–268, 294
  density of, 263, 264
  hurricanes and, 266–267
  hydrogen bonds in, 264
  melting, 266, 268–269
Iceman, carbon dating of, 54
Ideal-diet graphics, 464, 466
Ideal gas, 294
Identical molecules, 530, 531–532
I.E. Dupont de Nemours, 401
Imine functional group, nitrogen atoms in, 552
Immiscible, 280
Immunoglobulins, 569–570
Inactive ingredients, in drugs, 540
Incomplete combustion, 339
Indicators, 152–153
Indinavir, 527
Induced-fit model of enzyme action, 482
Infrared light, 192, 193
Inhibitors
  irreversible, 523
  reversible, 523–524
Inorganic compounds, 40, 131
Insects, walking on water, 262
Insulin, 435, 437–438
Intermolecular interactions, 274
  biomolecule docking and, 513–514
  liquids and solids and, 293–294
  solubility and, 280
Intramolecular interactions, 274
Inverse logarithm, 151
Iodine, 48
  tincture of, 15
Ion concentrations
  determining time of death and, 148
  inside and outside a cell, 143
Ion-dipole interactions, 134, 261
Ion exchanger, 145–146
Ion-exchange resin, 145–146
Ion hydration, 149
Ionic bonds, 49–50, 257, 261, 514, 515
  electronegativity and, 51–52
Ionic interaction, 126
Ionic liquids, 126–129
Ionic solid, 126–127
Ionizing radiation, 240–241
Ion pumps, 144
Ion replacement drink, 141
Ions
  creation of, 48–49
  electric current and, 138
  fortified water and replacement of, 140
  from groups 1, 2, 16, and 17, 50
  hard, 145
  hydrated, 134
  polyatomic, 129–132
  removal from water, 145–146
Iron
  in meteorites, 64
  oxidation of, 64–65, 66
Iron-56, 232

Iron-59, 228
Irradiation, 499
Irreversible inhibitor, 523
Isoflurane, 532–533
Isoflurane enantiomers, 539
Isoleucine, 433
Isomerism, *cis-trans*, 349
Isomers, 532
  constitutional, 532
  enantiomers, 532–533, 537
  structural, 347–348
Isooctane, 100, 349–350
Isopentyl acetate, 100
cis-Isoprene, 397–398
Isopropyl group, 348
Isotopes, 54
  hydrogen, 252–256
  natural occurrences of, 55–58
  oxygen, 252
  push toward more stable, 232
  radioactive, 221–223
Isovaleric acid, 556

J

Jet Propulsion Laboratory, 99
Joule (J), 91, 492
Judd, Alan, 341
Juneau (Alaska), molar volume and, 308, 310

K

Kaysing, Bill, 291
Keller, Greg, 375
Kelvin temperature scale, 17, 19, 24, 25–26
Ketones, 400–401
Kevlar, 401–403
Kilocalories (kcal), 492
Kilometer (km), 17
Kilopascal (kPa), 307
Knocking, engine, 349–350
Kroto, Harry, 94, 95
Krypton, 47, 225, 235
Kwolek, Stephanie, 401
Kyoto Protocol (1997), 326–328

L

Lactic acid, 423, 424, 425
Lactose, 488, 489
Lactose intolerance, 489
Lake Nyos (Cameroon), release of carbon dioxide in, 314, 315
Land mines, detection of, 298
Latex, 397
Laue, Max von, 3
Lawrence Berkeley Laboratory, 218
LD_{50} (lethal dose), 521–522
LDLs. *See* Low-density lipoproteins
LDPE (low-density polyethylene), 409
Lead
  in dirt, 42–45, 45
  effects on human body, 45
  levels in river and reservoir sediments, 43–45
Lead-206, 222, 227
Lead compounds, 518
Leaf color, 200, 201
Leonard, William, 491, 500
Leucine, 433

Leukemia, radiation exposure and, 244
lev-, 536
Lev drugs, 536
Lewis structures, 100
Lieberman, Gloria, 407
Light
    classification of, 193
    electron movement and, 172
    infrared, 192, 193
    ultraviolet, 192
    visible, 192, 193
    white, 195–196, 197
Light bulbs, 196
Light-driven reactions, 192–198
Light-emitting diode (LED), 118
Lightweighting, 413
Limelight, 32
Limestone, 73, 74
Lindbergh, Charles, 73
Linear molecular geometry, 115
Line spectrum, 194, 195
Line structure, 108–109
Linoleic acid, 467, 468, 469, 471
Linolenic acid, 469
Lipids. *See* Fats
Lipitor, excipients in, 540–541
Lipoproteins, 477
Liquid(s)
    differences with solids and gases,
        293–295
    fatty acid, 468
    hydrocarbons as, 351–352
    ionic, 126–129
    symbol, 12
Lise Meitner: A Life in Physics, 225
Liter(s), 17, 19, 23
Lithium, 171
Lithium-6, 232
Lithium batteries, 184, 186
Lithium hydroxide, 154
Local pyramidal molecular geometry, 559
Locard, Edmond, 73
Locard's principle, 405
Lock-and-key model of enzyme action,
        481–482
Logarithm, inverse, 151
Lost City, 274, 275
Low-density lipoproteins (LDLs), 477
Low explosives, 365–366
LSD (lysergic acid diethylamide), 562
Lysine, 434

M
Macromolecules, 384
Magnesium, 47, 53
Magnesium chloride, 131
Magnesium sulfate heptahydrate (Epsom
        salt), 14, 15
Magnetism, 67–68
Magnetite, 64, 66, 67–72
Magnetotactic bacteria, 68, 69
Maillard reaction, 498
Malathion, 522
Malondaldehyde, 499
Maltose, 484, 488
Manganese, 40, 75, 131
Marble, 56–58
Marcy test, for cocaine, 566–568

Margarine, 473–474, 475
Marine animals, effects of plastic waste on,
        410–411
Marquis test, 565–566
Mass number, 54
Mathews, Robert, 30
Matsumoto, Masakazu, 263
Matter
    classification scheme for, 17
    defined, 7
    division of, 14–15
    phases of, 294–295
MDMA (Ecstasy), 562, 565
Mean free path, 306, 311
Measurement
    conversion factors, 19–22, 23
    metric system, 18–19, 23–24, 26–27
    SI system, 17–19
    temperature, 24–26
Medicine, radioactivity in, 237, 240–241
Meitner, Lise, 224–225
Meitnerium, 225
Melanoidins, 498
Melting points
    defined, 264
    of fatty acids, 468–469
    of free base *vs.* hydrochloride salt, 579,
        580
    of polymers, 405–406
    of salts, 126–128
    of thermoplastic polymers, 407–408
Membrane, 141
    cell, 143, 472, 476
    semipermeable, 141–142
Mercury, in shellfish, 75–76
Mercury vapor lamps, 194, 196
Mescaline, Marquis test result, 565, 566
Messenger RNA (mRNA), 449
Meta-, 365
Metabolic pathways, 478
    drug design and understanding of, 512
    hypothetical, 512
Metallic bond, 53
Metals, 41, 42
    alkali, 41, 42
    alkaline earth, 41, 42
    chemical bonding in, 53
    as electrical conductors, 118
    ionic bonds and, 52–53
    in soil, 41–42
    transition, 41, 42
Meteorites, 64
    magnetite in, 67, 69
Meteroids, 64
Meters, 17, 19–20
Methamphetamine, 552, 560
    Marquis test result, 566
Methane, 351
    combustion of, 213–214
    disappearing ships and pockets of, 341
    electron dot structure, 100–101
    forming layer of, 316
    fractionating column and, 353
    molecular geometry, 115
    single bond of, 93
Methanol, in Marcy test, 566–567
Methionine, 433
Methylbenzene (toluene), 364

Methyl benzoate, 566–567
Methylcyanoacrylate, 390
Methylene chloride, 104, 113
Methylene group, 107–108
Methyl group, 107, 343
Methyl methacrylate, 390–391
Methyl pentyl ether, 353, 354
3-Methylphenol, 364
Metric drop, 24
Metric system, 18
    brownie recipe, 23
    prefixes, 18
    SI system *vs.*, 19
Micelles, 285
Microsecond, 27
Microtiter plate, 570, 571, 573
Microtiter plate reader, 573
Midgely, Thomas, 323
Milliliter (mL), 23
Millimeter (mm), 17
Millimeters of mercury (mmHg), 307
Millimolar (mM), 139
Millisecond (ms), 27
Minerals, 40, 46–53
    colors of, 70
    crystal *vs.*, 68
    harmful, 71–72
Mirror image, 529, 532
Miscible, 281
Mixture, 15–16, 17
    defined, 14
    heterogeneous, 15, 17
    homogeneous, 15, 17
Model, 31–32
Molarity (molar concentration), 139
Molar mass, 62–63
Molar volume, 310
Mole, 19, 61–62, 308
Molecular geometry, 111–115, 559
    bent, 113, 114, 115, 251, 252
    flat triangular, 115
    linear, 113, 115
    pyramidal, 112, 115
    tetrahedral, 113, 114, 115, 559
    of water, 251, 252
Molecules, 11
    blueprint of, 4
    breaking bonds in, 93–94
    models of, 31–32
    three-dimensional shape of, 110–116
Moles of gas, volume of gas and, 302–304,
        317
Molina, Mario, 324
Molotov cocktail, 350–351
Monomers, 383
    building polymer proteins, 432, 434
    creation of polymers from, 388–391
Monosaccharides, 483
Montreal Protocol (1987), 325
Moon walk, refutation of *Apollo 11*, 291, 292
Morphine, 515–518, 516, 561
    antibodies to, 570
    Marquis test and, 565, 566
    testing for, 565, 566, 569
Mount Pendelikon (Greece), 57–58
mRNA (messenger RNA), 449
Mt. Everest, air density and, 306
MTBE, in gasoline, 356–358

Multiple bonds, 87, 88, 90–94
 graphite and, 90–94
 molecular geometry and, 115–116
Mummification project, 129–130, 137
Muscle conditioning, 509
Musser, George, 297, 299
Myoglobin, 441
Myrrh, 387

N

n-, 342
Namath, Joe, 427, 428
Nanoears, 99
Nanosecond, 27
Nanotubes, 99
Nanotubules, 97–100
Narcotics, 586
Narrow therapeutic index, 541
National Air Quality Standards (NAAQS), 155
National Institutes of Health, 464
National Renewable Energy Laboratories (NREL), 98
National Toxicology Program, 72
Natron, 129–130, 132
Natural flavoring, 496
Natural polymers, 422–427
 synthetic *vs.*, 432
Navigation, magnets and, 68
Negative (photographic), 176
Negative terminal, battery, 184
Neon, 10, 11, 46–47, 102
Neptunium, 219, 220–221, 224
Nerve agents, 296–299
 obtaining chemicals for, 297, 299
Network solid, 85
 silicon, 116–117
Neurotransmitters, 296, 297
 drug action and, 516, 557–558
Neutrons, 7–8
 importance of, 54–58
 symbol for, 219
New car smell, 110
Newmar, Julie, 427, 428
New Scientist, 410, 510
New Yorker magazine, 72
Nicotine, 563–564, 574, 581–585
 crack, 581, 582, 584–585
 as dihydrochloride salt, 584, 585
 free base, 582, 584, 585
 transdermal patch, 542
Niepce, Nicophore, 200
Niépce, Nicophore, 165, 166
Nitric acid, 154, 157
Nitrogen
 in air, 292, 293
 as combustion product, 337
 in diatomic molecule, 48
 illicit-drug molecules containing, 552–557
 in organic molecules, 100
 valence electrons, 102
 VSEPR rules and, 116
Nitrogen dioxide, 155, 157
Nitrogenous base, 446–447, 449
Nitroglycerin
 chemical reaction, 12–13
 detecting, 374, 375

 as explosive, 9, 11, 12–13, 359–362, 368
 as heart drug, 362
Nobel, Alfred, 11, 359, 362
Noble gas configuration, 47, 173
Noble gases, 41, 42, 46–47, 50
Noddack, Ida, 224
Noddack, Walter, 224
Nomenclature, for organic molecules, 341–350
Nonanol, 281
Nonbonding pairs, 102
Noncovalent interactions, 134, 514
 water and, 257–261
Nonessential amino acids, 451–452
Nonmetals, 41, 42
 ionic and covalent bonds and, 52–53
Nonpolar amino acids, 436
Nonpolar covalent bonds, 48
Nonpolar interaction, 514, 515
Nonspecific interactions, 518
Nonsuperimposable mirror images, 529
Norepinephrine, 558, 559, 560
Norris, David, 489
Novocaine, 574, 575
N-terminus, 435
n-type dopants, 118
Nuclear energy, 27–28
 percentage of energy from nuclear power in various countries, 233
 use among countries, 234
Nuclear equations, 211– 213
 symbols used in, 219
Nuclear fission, 225, 231–237
Nuclear fusion, 211–214
Nuclear reaction, 211–214
Nuclear reactors, 232–234, 234–235
Nuclear Test Ban Treaty (1962), 256
Nucleation center, 264–267
Nucleic acids, 445
Nucleophiles, 388, 389, 395
Nucleotides, 445–446
Nucleus, atomic, 7–8
Numbers in science
 conversion factors, 19–22, 23
 scientific notation, 58–60
 SI units, 17–19
 very large and very small, 58–63
Nutrasweet, 487
Nutritional information, 463–464, 466, 470
Nutritional labeling, 475
Nylon
 creating polymers, 430–431
 fiber, 396, 427, 428
 properties of, 405–406
 structure of, 406
Nylon 6, 431, 432

O

-oate, 393
Obesity, 464–466, 500
Octane, 342
 combustion of, 338
Octane number, 349–350
Octanitrocubane, 366
Octanol, 281
1-Octene, 349
3-Octene, 349
trans-2-Octene, 349

Octet rule, 101
Octyl acetate, 393
Octyl methoxycinnamate, 199
Odor, 295, 299
 amines and, 554–555
Ohmine, Iwao, 263
-oic acid, 429
Oil of wintergreen, 109
Oils, 470
Oleic acid, 89, 468–469, 471
Oligopeptide, 435
Oligosaccharides, 483
Olive oil, 472
Oltipraz, 510
Olympic Games, drug testing and, 572
Omega-3 fatty acids, 475
Omega-6 fatty acids, 475
OMP decarboxylase, 481
-one, 401
Onion, The, 29
Open-chair form, of carbohydrate ring, 485
Open system, 299, 300–301
Opiate receptor, 515–517
Opiates, 516
 testing for, 568–569
Oregon Recycling Act (1991), 410
Organic, 496
Organic compounds, 40
 volatile, 110
Organic molecules
 bonding rules for elements in uncharged, 102
 electron dot structures for, 100–106
 electron movement in, 186–192
 nomenclature, 341–350
Organic pigments, 200, 201
Orr, John, 370
Ortho-, 365
Osmosis, 142
 in living cells, 143, 144
Oxidation reaction, 64–65
 digestion and, 495
 free radicals and, 187
 of fruit, 191
 of iron, 64–65, 66
Oxidizer, 338
Oxidizing agent, 64
Oxomate, 511
Oxycodone (oxycontin; Percocet), 569
Oxygen
 in air, 292, 293
 combustion reaction and, 337
 as diatomic molecule, 48
 in dirt, 40, 45
 formation of free radicals and, 189–190
 isotopes, 55, 252
 metabolism and, 495–496
 in organic molecules, 100
 valence electrons, 102
 VSEPR rules and, 116
 in water, 256–262
Oxygen-16, 252
Oxygen-17, 252, 253
Oxygen-18, 252, 253
Oxygenate, 357
Ozone, 155, 324–325
Ozone hole, 325
Ozone layer, 213, 324

P

Packaging
plastic, 421–422
super glue, 391
Palmitic acid, 469, 471
combustion of, 494–496
Pantyhose, 427, 428
Para-, 365
Parathion, 522
Parthenon, identifying origin of marble in, 56–58
Partial charges, 133
Partially hydrogenated, 475
Particle accelerator, 217
Parts per billion (ppb), 155
Parts per million (ppm), 155
Pascal (Pa), 307
Passive solar energy, 272
PCR. *See* Polymerase chain reaction
Peer review, 28
Penicillin, structure of, 4
Pentafluoroethane, 325–326
Pentane, 352
Pentanol, 281
Pentyne, 345
Peptide, 435
protein *vs.*, 435
Peptide bonds, 432
Pepto-Bismol, 15, 45
Perchloric acid, 154
Percocet (oxycodone), 569
Performance-enhancing drugs, 551–552
Perhexiline, 538
Period, 7
Periodic table of elements, 7
determining electron configurations using, 169–171
group names for elements, 42
metals and nonmetals, 41, 42
updating, 31
Peristalsis, 478
Permangate ion, 131
Peroxide ions, 131
Perpetual-motion machines, 209, 210
PET (polyethylene terephthalate), 409
Petroleum, 352–353
Pexsig (perhexiline), 538
Peyote cactus (*Lophophora williamsii*), 565
pH, 148–153
determining time of death and, 148
indicators, 152–153
meter, 152
physiological, 436
scale, 151
PHAs. *See* Polyhydroxyalkanoates
Phase, of matter, 252
Phen-fen, 566
Phenol, 189, 363
Phenolphthalein, 152–153
Phentermine, 566
Phenylalanine, 433, 486
Phenylketonuria (PKU), 487
Phospholipid bilayer, 472–473
Phospholipids, 472–473
Phosphor, 230
Photogray sunglasses, 177
Photosynthesis, 210
Phycocyanin, 201

Physical change, 16
Physiological pH, 436
Picometers, conversion to meters, 20
Picric acid (trinitrophenol), 363–364
Pike, Scott, 56
Pineapple, taste of, 497
PKU. *See* Phenylketonuria
PLA. *See* Polylactic acid
Plants
as source of polymers, 425–426
transgenic, 426, 436–437, 452, 454–455
Plasmid, 453–454
Plastics, 384, 397–403
from bacteria, 424–425, 426
biodegradable, 422
disposal of, 407–414
in landfills, 413
natural, 422–427
packaging, 421–422
recycled, 421
recycling, 408, 410
world consumption of, 387
Plexiglas, 390, 391
Plutonium, 219, 220–221
p-n junction, 118
pOH, 148, 150
Poison, 521–522
Polar amino acids, 436
Polar bonds, 51
Polar covalent bonds, 48
Polar molecules, 133
Polaroid film, 178
Polonium-210, 221–222, 223
Poly-, 383
Polyamides, 430
Polyatomic ions, 129–132
Polyesters, 387, 392–396, 412
Polyethylene, 386, 387, 403, 404, 422
Polyethylene terephthalate (PET), 409, 412
Polyhydroxyalkanoates (PHAs), 424–425
cis-Polyisoprene, 397–398
trans-Polyisoprene, 398
Polylactic acid (PLA), 411, 412, 423–425, 426
Polymerase chain reaction (PCR), 455
Polymers, 383–384
addition, 388, 389, 394, 395
biodegradable, 411–412, 423–425
condensation, 388, 394–395
condensed structure, 385
human uses of, 384–387
melting points, 405–406
natural, 385, 422–427, 432
recycling, 408, 410
relation of structure to function, 390–391
see-through, 400
steric hindrance and, 403–404
synthetic, 385–386, 432
thermoplastic, 403, 407–408, 410
thermosetting, 407
in time-release medications, 542
Polymethylcyanoacrylate, 390–391, 403, 404
Polypeptide, 435
Polypropylene, 409, 411
Polysaccharides, 383, 484
digestion of, 490–491
in human body, 490
Polystyrene, 387, 409
Polyunsaturated, 470

Polyvinylacetate (PVA), 386, 389
Polyvinylchloride (PVC), 387, 409
Portugal, ideal-diet graphic, 466
Positive (photographic), 176
Positive terminal, battery, 184
Potassium carbonate, 337
Potassium hydroxide, 154
Potassium nitrate, 336
Potassium sulfate, 337
Potassium sulfide, 337
Potassium superoxide, 63
Precipitation, 136
Prefixes
metric, 18
for multiple substituent groups in organic molecules, 346
Pressure, 301–302
atmospheric, 273–275, 306–307
effect of temperature on gas, 312–313, 317
force *vs.*, 307
gases and, 305–308, 317
relation to volume of gas, 308, 317
vapor, 277
Pressure cookers, 274–275
Pressurized gas, 278–279
Prevention, health and, 510
Primary amines, 553, 555, 577
Primary structure, protein, 437–438
Primer (explosive), 365
Prism, 196
Processed foods, 496–500
Products, 11, 12
Proline, 433, 444–445
Propane, 280
Propellant, 278
Propene, 345
Propyline, 555
Prostaglandin A$_2$, 430
Protactinium-234, 228
Protein backbone, 432
Protein folding, 440
Protein(s), 431, 437–445
creating binding sites on, 513
digestion of, 478
diverse capabilities of, 441
DNA and synthesis of, 449–451
globular, 439–440, 441, 443, 511–512
metabolic pathway, 493
peptide *vs.*, 435
primary structure, 437–438
ropes, 444
secondary structure, 439–440
structural, 441–445
as targets of drug design, 510–518
tertiary structure, 440–441
three-dimensional shape of, 440–441
Protonation, 575–577
of amines, 583
Protons, 7–8, 156
addition to atom in molecule, 575–577
loss in carboxylic acid functional group, 430
symbol for, 219
Prozac (fluoxetine), 536–537
Psilocin, 579
Psilocybin, 566
p-type dopants, 117
Public debate on scientific research, 28

Pure substance, 14–15, 17
Purity, 14–15
Putrescine, 554–555, 556
PVC (polyvinyl chloride), 409
Pyramidal molecular geometry, 112, 115, 559
Pyridoxine, 578–579
Pyridoxine hydrochloride (vitamin B$_6$), 578–579

Q

Qualitative drug tests, 566
Quantitative drug tests, 568–575
Quart, 23
Quartz, 70
Quaternary ammonium salt, 578

R

Racemates (racemic mixtures), 536–539
Radiation, 219
  alpha, 237, 239–240
  alpha decay, 219–220
  beta, 237, 239–240
  beta decay, 220–221
  discovery of, 2–3
  electromagnetic, 192, 193
  exposure to, 228–230
  gamma, 221, 226, 237, 239–240
  half-life, 228–230
  health effects of on humans, 242–244
  human exposure to, 241–244
  ionizing, 240–241
  living organisms and, 237–244
  relative penetrating powers of, 237, 239
  X-ray, 237
Radioactive decay, 219–225
Radioactive decay series, 223
Radioactive tracer, 241
Radioactive waste, 238–239
Radioactivity, 219–225
  discovery of, 3
  identifying, 230
  in medicine, 237, 240–241
Radio-labeled sodium iodide, 241
Radon, 47, 226–228
  exposure to, 228–230
  levels in United States, 227
Radon-222, 226–228
Radura symbol, 499
Rainbow, 196
Random coil, 440
Rauch, Jonathan, 466
Rawalt, Ronald, 73
Reactants, 11, 12
Reaction coordinate, 482
Reactor core, 234
Reading Prong, 229
Real gas, 294
Receptor, 511
Recombinant DNA, 453–454
Recycling
  of natural materials, 423, 426
  of plastics, 408, 410, 421
Redox reactions, 65–67
  in electrochemical cells, 180–181
  film and, 175
Reducing agent, 65
Reduction potential, 181

electric eels and, 185
Reduction reaction, 65
Refrigerant, 323–325
Reindeer, lipid composition in leg of, 473
rem (roentgen equivalent for man), 242
R enantiomers, 533
Reproducibility, 28
Resilin, 442
Resin ID code, 408, 409
Resonance stabilization, 95–97
Restriction enzyme, 452–453
Retroviruses, 524, 526
Reverse transcriptase, 524, 526
Reversible chemical reactions, 177
Reversible inhibitor, 523–524
Rhenium, 224
Ribose, 446
Ribosome, 449
Rigor mortis, 148
Ritter, Bill, 322
RNA (ribonucleic acid), 446, 449–451
Robertson, Marcia, 491, 500
Roentgen, Wilhelm, 2
Roundup (glyphosate), 436–437
Rowland, F. Sherwood, 324
Rubber, 387, 397–399
  structure of, 386
  synthesis of, 398, 399
Rubenstein, Dustin, 253, 254
Ruby, 70
Rust, 66
Rutherford, Ernest, 217, 230

S

Saccharides, 483
Saccharin, 486
Saccharin Study and Labeling Act (1977), 486
Saito, Shinji, 263
Saline solution, 16
Salt Lake City (Utah), smog and, 313, 314
Saltpeter, 336
Salts, 49, 50. See also Table salt
  amines as, 577
  defined, 126
  as desiccants, 132–137
  destructive property of, 137
  dissolution in water, 134–135
  electrolytes, 137–141
  hydrochloride, 578–581
  ionic liquids, 126–129
  melting point, 126
  polar bonds in, 51
  polyatomic ions and, 129–132
  quaternary ammonium, 578
  useful properties of, 137
Samples, control, 75
Sapphire, 70
Sarin, 293–294, 296–297, 521, 522
Satellite tobacco mosaic virus, 32
Saturated fats, 476
Saturated fatty acids, 467, 469, 470, 474, 476
Saturated molecules, 89
Saturated solution, 135, 136
Savitch, Pavel, 224
Scanning tunneling microscope (STM), 17–18
Schedules, drug, 586

Schlosser, Eric, 497
Schmidt, Walter, 423
Science, 509
Science, impact on society, 27, 28
Scientific American, 297
Scientific calculators
  determining pH value using, 148
  inverse logarithm operation, 151
  scientific notation on, 60
Scientific discovery, 1–5
Scientific hypothesis, 29
Scientific knowledge, responsibility for use of, 5–6
Scientific method, 28–31
Scientific notation, 58–60
Scientific research, selection of topics, 28
Scintillation, 230
Scott test, for cocaine, 566
Scuba diving, 279
Scurvy, 445
Seaborg, Glenn, 218, 231
Seaborgium, 218
Sea level, global warming and increases in, 327
Secondary amines, 553, 577
Secondary structure, protein, 439–440
Second law of thermodynamics, 367–368
Second (time), 19, 27
Seddon, Kenneth, 128
Sediments, metal levels in, 43–45
Selenium, as antioxidant, 190, 192
Semiconductors, 117–118
Semipermeable membrane, 141–142
S enantiomers, 533
Serine, 433, 443
Serotonin, 558, 562
Shanklin, Jonathan, 325
Shell lacca, 386–387
Ships, methane and disappearing, 341
Shires, Dana, 138
Shortened condensed structure, 108
Side chain, 432, 435
  polarity of, 436
Side effects, drug, 521
  amphetamine, 561
  cocaine, 561
  HIV drugs, 527
Sievert (Sv), 241–242
Significant figures, rules for, 63
Silicon, 86, 116–118
  in dirt, 40, 45
  possibility of life-form based on, 116
Silicon microchip, drug delivery via, 543
Silicon-silicon bonds, 116, 117
Silk protein, 443–444
Siloxanes, 110
Silver bromide, 174–176
Silver chloride, 174, 177
Silver halide, 177
Silver iodide, 174
  cloud seeding and, 265–266
Silver salts, film and, 166
Sime, Ruth, 225
Simple sugar, 483, 485
Simpson, Tommy, 551–552, 560
Single bonds, 87, 88, 91–92
  combustion and, 340

Skatole, 556
Skeletal structure, of molecule, 101
Smalley, Richard, 94, 95
Smog, 313, 314
Smoke
    as byproduct of combustion, 337
    from overheated fat or oil, 470–471
Smoke detectors, 231
Snapple Rain, 140
Snow, 268
Sobrero, Ascanio, 362
Society, impact of science on, 27, 28
Soda cans, 279
Sodium bicarbonate, 130
Sodium carbonate, 130
Sodium chloride, 50, 51, 125, 126, 127–128,
        130, 266
Sodium hydroxide, 154
Sodium lauryl sulfate, 107–109
Sodium sulfate, 130
Sodium vapor lamps, 194, 196
Soft contact lens, 400
Soil, 40
Solid(s)
    crystalline, 53
    differences with liquids and gases, 293–295
    fatty acid, 468
    hydrocarbons as, 351–352
    ionic, 126–127
    network, 85
    symbol for, 12
    water as, 263–268
Solid support, in chromatographic column,
        371
Solubility
    of fibers, 405
    of free base *vs.* hydrochloride salt,
        579–580
    of MTBE, 357–358
    in water, 135–136
Solutions, 15, 17
    aqueous, 139
    saturated, 135, 136
Solvent(s)
    ketones as, 401
    universal, 251, 279–285
Sommerville, Richard, 327
Soybeans, transgenic, 436
Space-filling models
    of globular protein molecule, 512
    hexokinase reaction, 481
    of protein surface, 440–441
Specific heat, 270, 272
Specificity, of drug, 518
Splenda, 487
Sports drinks, 138–141
Squalene, 476
Standard temperature and pressure (STP),
        310
Starch, 490
Stearic acid, 89, 468–469
Stephenson, Richard, 137
Steric hindrance, 403–404
Steroids, anabolic, 572–573
Stevia (*Stevia rebaudiana*), 487–488
Sticky end, 453
Stiction, 86
Stimulants, 562, 586

STP. *See* Standard temperature and pressure
Strands, DNA, 447
Strassman, Fritz, 224–225
Strong acids, 149, 154
Strong bases, 154
Strong nuclear force, 221
Structural isomers, 347–348
Structural proteins, 441–445
Structure
    condensed, 107
    full, 107
    function and, 587
    line, 108–109
    shortened condensed, 108
Structure-activity relationship, 514–515
    for morphine, 518
Styrene butadiene, 386
Subscript, 211
Substituents, 343, 346
Substrates, 480, 523
Sucralose, 487
Sucrose, 483, 487
Sugar-phosphate backbone, 446, 447
Sugar(s), 483
    artificial, 486–487
    simple, 483, 485
Sulfate polyatomic ion, 130
Sulforaphane, 510–511
Sulfur, 47
    energy levels in, 169
    in gunpowder, 336
    ionization of, 173–174
    in organic molecules, 100
    rubber synthesis and, 399
    VSEPR rules and, 116
Sulfur dioxide, 155, 156, 157
Sulfuric acid, 154, 156–157
Sun, fusion reaction in, 210–214
Sunlight
    damaging effects of, 198–200
    in electromagnetic spectrum, 195–196
    reaching Earth, 213
Sunscreen, 198–199, 199
Supercritical mass, 236
Super glue, 390, 391
Superoxide dismutase, 190
Superoxide ion, 189
Superscript, 211
Surface tension, 262
Surfactants, 282–285
Sustainability, 426–427
Synthetic fibers, 405–406
Synthetic polymers, natural *vs.*, 432
Systéme International (SI), 17–19
Szent-Györgyi, Albert, 191

T

Table salt (sodium chloride), 50, 51, 125,
        126, 127–128. *See also* Salts
    boiling point of water and, 276
    melting ice and, 266
Tabun, 522
TAGs. *See* Triacylglycerides
Talc, 72
TD_{50} (toxic dose), 518–520
Technetium-99, 241
Telluride Airport (Colorado), 312

Temperature. *See also* Heat
    boiling points and, 272–273
    effect of specific heat of water on
        fluctuations in, 270–272
    effect on aircraft thrust, 312
    effect on gas pressure, 312–313, 317
    effect on volume of gas, 310–311, 317
    gas, 308–316, 317
    increase in Earth's average, 327
    maintenance of, 309–310
    phases of fatty acids and, 468
    phases of hydrocarbons and, 351–352
    phases of water and, 268, 269–270
    snow and, 268
Template, 448
Temple of Luxor, 137
Teniers, David, the Younger, 214
Tertiary amines, 553, 577
Tertiary structure, protein, 440–441
Testosterone, 573
Tetrahedral electron-pair geometry, 114
    of water, 251, 252
Tetrahedral molecular geometry, 113, 114,
        115, 559
Tetrahedron, 84, 85
Tetrahydrogestrinone (THG), 572, 573
Theory, 30–31
Therapeutic index, 520
Thermal conduction, diamonds and, 86–87
Thermal inversion, 313–316
Thermodynamics
    first law of, 209
    second law of, 367–369
Thermonuclear bomb, 4, 5, 214, 236–237
Thermoplastic polymers, 403, 407–408, 410
Thermos, 309–310
Thermosetting polymers, 407
THG. *See* Tetrahydrogestrinone
Thiol, 100
Thorium-234, 228
Threonine, 433
Thymine, 446, 447, 450
Time, units of, 27
Time magazine, 5
Time-release medications, 542–543
Tincture of iodine, 15
Tin isotopes, 55
Tires, air pressure and, 301–302, 308
Titanium isotopes, 54, 55
TNT (2,4,6-trinitrotoluene), 12, 298, 364–365
    detecting, 374, 375
Toast drop experiment, 29–31
Tocopherols, 188
Toluene (methylbenzene), 364
Tomatoes, soil requirements, 141, 143–144
Toothpaste, 15, 106–110
Tour, James, 297
Tour de France, drug abuse and, 551–552,
        572
Toxin, 521
Trans, 348, 349
Transcription, 449
Transdermal patches, 542
Trans fats, 473–475
Transgenic plants, 426, 436–437, 452,
        454–455
Transition metals, 41, 42
Transition state, 482

Translation, 450–451
Transmutation, 214, 222
Trenbolone, 572, 573
tri-, 554
Triacontanyl hexadecanoate, 108
Triacylglycerides (TAGs, triglycerols), 471
   hydrolysis of, 478–479
Trimethylamine, 555
2,3,5-Trimethylphenol, 364
Trinitrophenol (picric acid), 363–364
Trinitrotoluence (TNT), 12, 298, 364–365,
      374, 375
TRINITY Natural Mineral Dietary
      Supplement, 140
Tripeptide, 435
Triple bonds, 87, 88, 92–93
   combustion and, 340
Triplet, 450
Tritium, 252, 253, 256
Trout Lake (Wisconsin), acid rain study,
      157–158
Tryptophan, 434
Tumors, radiation treatment of, 237, 240
Tuvalu, sea level increase and, 327
Tyrosine, 434

U
Ultraviolet light (UV), 192
   damaging effects of, 198–200
Ultraviolet radiation, ozone layer and, 324
Uniqueness principle, 84
   water and, 256–257
United States
   average temperature fluctuation in, 271
   obesity in, 464–466
U.S. Geological Survey (USGS), 43
*United States Pharmacopoeia and National
      Formulary*, 536
U.S. Supreme Court, on drug testing, 564
Universal indicator, 153
Universal solvent, 251, 279–285
University of Arizona, 413
Unsaturated fats, 474
Unsaturated fatty acids, 467, 469
Unsaturated molecules, 89
Uracil, 449, 450
Uranium, 216–217
   experiments on, 224–225
   half-life, 228
   weapons-grade, 236
Uranium-234, 236
Uranium-235, 235
Uranium-238, 236
   radon and decay of, 226–227
UVA light, 198–199
UVB light, 198–199

V
Vacuoles, 144
Vacuum, 310
Valence electrons, 47–48, 102, 172
   electron dot structure and, 100–101
Valence shell electron pair repulsion
      (VSEPR) theory, 111–115
Valine, 433
Valium, 574, 575
Vapor pressure, 277
Variables, 303
VELCRO, 98
Vernier caliper, 87
Video pill, 179
Vinyl acetate, 389
Viruses, 524, 526
Visible light, 192, 193
Vitamin A, 190–191
Vitamin B_6, 578–579
Vitamin C, 190, 191
   role in collagen, 444–445
Vitamin E, 188, 189, 190
Volatile molecules, 277
Volatile organic compounds, 110
Volatility, 580
Voltage, 185
Volume
   amount of gas and gas, 302–304
   effect of temperature on gas, 310–311,
      317
   gas pressure and gas, 308, 317
   molar, 310
Vulcanization, 398–399

W
Water, 10, 294. *See also* Ice; Water vapor
   acid rain, 156, 157–158
   autoionization of, 147–153
   boiling point of, 256–257, 259, 272–276
   cleaning with ionic-liquid application,
      128–129
   density of, 263, 264
   dipole-dipole interactions and, 258
   dissolution of salt in, 134–135
   distribution of isotopes, 252–255
   electrolysis of, 16
   fortified, 140
   home purifiers, 145–146
   hydrogen bonds in, 258–259
   interaction with radiation, 240
   molecular geometry of, 132–133, 251,
      252
   phase changes in, 276
   pH of, 154–158
   as purifier, 251, 252
   shifting phases in, 268–272
   as solid (ice), 263–268
   solubility and, 135–136, 355
   specific heat of, 270, 272
   surface tension, 262
   tetrahedral electron-pair geometry, 251,
      252
   three-dimensional structure of, 112–114
   as universal solvent, 251, 279–285
   with water, 256–262
Water vapor, 255, 269, 276–277, 294
   formation of in clouds, 265
   pressure, 277
Watras, Stanley, 226, 229–230
Watson, James, 4
Wavelength, 193
Wax, 281
Weak acids, 149
Weapons-grade uranium, 236
Weight, units of, 26–27
We Never Went to the Moon, 291
White glue, 389
White House Office of National Drug
      Control Policy, 581
White light, 195–196, 197
Wilkins, Maurice, 4
Williams, Wayen, 405, 406
Winchester drop, 24
World Health Organization, 524
World Trade Center collapse, asbestos and,
      72
Wyeth Pharma, 539

X
Xenon, 22, 47
X-rays, 237
   to discover molecule blueprint, 3–4
   discovery of, 2

Y
Yellow topaz, 70
-yl, 343
-yne, 344
Yucca Mountain (Nevada), 238, 239

Z
Zero, absolute, 25, 173
Zettl, Alex, 97
Zinc, 45
   as antioxidant, 190, 192
   in dirt, 42–45
   in electrochemical cells, 179–183
   levels in river and reservoir sediments,
      43–45
Zinc oxide cream, 198
Zinc sulfide, 230
Zolpidem (zolpidem tartrate), 538

Waldron
The Chemistry of Everything
Student Accelerator CD
ISBN 0-13-187973-1
CD License Agreement

Pearson Prentice Hall
Pearson Education, Inc.
Upper Saddle River, NJ 07458

YOU SHOULD CAREFULLY READ THE TERMS AND CONDITIONS BEFORE USING THE CD-ROM PACKAGE. USING THIS CD-ROM PACKAGE INDICATES YOUR ACCEPTANCE OF THESE TERMS AND CONDITIONS.
Pearson Education, Inc. provides this program and licenses its use. You assume responsibility for the selection of the program to achieve your intended results, and for the installation, use, and results obtained from the program. This license extends only to use of the program in the United States or countries in which the program is marketed by authorized distributors.

LICENSE GRANT
You hereby accept a nonexclusive, nontransferable, permanent license to install and use the program ON A SINGLE COMPUTER at any given time. You may copy the program solely for backup or archival purposes in support of your use of the program on the single computer. You may not modify, translate, disassemble, decompile, or reverse engineer the program, in whole or in part.

TERM
The License is effective until terminated. Pearson Education, Inc. reserves the right to terminate this License automatically if any provision of the License is violated. You may terminate the License at any time. To terminate this License, you must return the program, including documentation, along with a written warranty stating that all copies in your possession have been returned or destroyed.

LIMITED WARRANTY
THE PROGRAM IS PROVIDED "AS IS" WITHOUT WARRANTY OF ANY KIND, EITHER EXPRESSED OR IMPLIED, INCLUDING, BUT NOT LIMITED TO, THE IMPLIED WARRANTIES OR MERCHANTABILITY AND FITNESS FOR A PARTICULAR PURPOSE. THE ENTIRE RISK AS TO THE QUALITY AND PERFORMANCE OF THE PROGRAM IS WITH YOU. SHOULD THE PROGRAM PROVE DEFECTIVE, YOU (AND NOT PEARSON EDUCATION, INC. OR ANY AUTHORIZED DEALER) ASSUME THE ENTIRE COST OF ALL NECESSARY SERVICING, REPAIR, OR CORRECTION. NO ORAL OR WRITTEN INFORMATION OR ADVICE GIVEN BY PEARSON EDUCATION, INC., ITS DEALERS, DISTRIBUTORS, OR AGENTS SHALL CREATE A WARRANTY OR INCREASE THE SCOPE OF THIS WARRANTY. SOME STATES DO NOT ALLOW THE EXCLUSION OF IMPLIED WARRANTIES, SO THE ABOVE EXCLUSION MAY NOT APPLY TO YOU. THIS WARRANTY GIVES YOU SPECIFIC LEGAL RIGHTS AND YOU MAY ALSO HAVE OTHER LEGAL RIGHTS THAT VARY FROM STATE TO STATE.
Pearson Education, Inc. does not warrant that the functions contained in the program will meet your requirements or that the operation of the program will be uninterrupted or error-free. However, Pearson Education, Inc. warrants the CD-ROM(s) on which the program is furnished to be free from defects in material and workmanship under normal use for a period of ninety (90) days from the date of delivery to you as evidenced by a copy of your receipt. The program should not be relied on as the sole basis to solve a problem whose incorrect solution could result in injury to person or property. If the program is employed in such a manner, it is at the user's own risk and Pearson Education, Inc. explicitly disclaims all liability for such misuse.

LIMITATION OF REMEDIES
Pearson Education, Inc.'s entire liability and your exclusive remedy shall be:
1. the replacement of any CD-ROM not meeting Pearson Education, Inc.'s "LIMITED WARRANTY" and that is returned to Pearson Education, or
2. if Pearson Education is unable to deliver a replacement CD-ROM that is free of defects in materials or workmanship, you may terminate this agreement by returning the program.

IN NO EVENT WILL PEARSON EDUCATION, INC. BE LIABLE TO YOU FOR ANY DAMAGES, INCLUDING ANY LOST PROFITS, LOST SAVINGS, OR OTHER INCIDENTAL OR CONSEQUENTIAL DAMAGES ARISING OUT OF THE USE OR INABILITY TO USE SUCH PROGRAM EVEN IF PEARSON EDUCATION, INC. OR AN AUTHORIZED DISTRIBUTOR HAS BEEN ADVISED OF THE POSSIBILITY OF SUCH DAMAGES, OR FOR ANY CLAIM BY ANY OTHER PARTY. SOME STATES DO NOT ALLOW FOR THE LIMITATION OR EXCLUSION OF LIABILITY FOR INCIDENTAL OR CONSEQUENTIAL DAMAGES, SO THE ABOVE LIMITATION OR EXCLUSION MAY NOT APPLY TO YOU.

GENERAL
You may not sublicense, assign, or transfer the license of the program. Any attempt to sublicense, assign or transfer any of the rights, duties, or obligations hereunder is void. This Agreement will be governed by the laws of the State of New York.

Should you have any questions concerning this Agreement, you may contact Pearson Education, Inc. by writing to:
ESM Media Development
Higher Education Division
Pearson Education, Inc.
1 Lake Street
Upper Saddle River, NJ 07458
Should you have any questions concerning technical support, you may write to:
New Media Production
Higher Education Division
Pearson Education, Inc.
1 Lake Street
Upper Saddle River, NJ 07458
or contact:
Pearson Education Product Support Group at (800) 677-6337 Monday through Friday, 8 AM to 8 PM, Eastern time, and Sunday, 5 PM to 12 AM, Eastern time, or anytime at http://247.prenhall.com.
YOU ACKNOWLEDGE THAT YOU HAVE READ THIS AGREEMENT, UNDERSTAND IT, AND AGREE TO BE BOUND BY ITS TERMS AND CONDITIONS. YOU FURTHER AGREE THAT IT IS THE COMPLETE AND EXCLUSIVE STATEMENT OF THE AGREEMENT BETWEEN US THAT SUPERSEDES ANY PROPOSAL OR PRIOR AGREEMENT, ORAL OR WRITTEN, AND ANY OTHER COMMUNICATIONS BETWEEN US RELATING TO THE SUBJECT MATTER OF THIS AGREEMENT.

SYSTEM REQUIREMENTS:
Windows:
Pentium II 300 MHz processor
Windows 98, NT, 2000, ME, or XP
64 MB RAM (128 MB RAM required for Windows XP)
4.3 MB available hard drive space
(optional—for minimum QuickTime installation)
800 x 600 resolution
8x or faster CD-ROM drive
QuickTime 6.x
Sound Card

Macintosh:
Power PC G3 233 MHz or better
Mac OS 9.x or 10.x
64 MB RAM
10 MB available hard drive space for Mac OS 9, 19 MB on OS X
(optional—if QuickTime installation is needed)
800 x 600 resolution
8x or faster CD-ROM drive
QuickTime 6.x

Some Metric Units of Length, Mass, and Volume

Length

1 kilometer (km) = 1000 meters (m)

1 meter (m) = 100 centimeters (cm)

1 centimeter (cm) = 10 millimeters (mm)

1 millimeter (mm) = 1000 micrometers (μm)

Mass

1 kilogram (kg) = 1000 grams (g)

1 gram (g) = 1000 millligrams (mg)

1 milligram (mg) = 1000 micrograms (μg)

Volume

1 liter (L) = 1000 milliliters (mL)

1 milliliter (mL) = 1000 microliters (μL)

1 milliliter (mL) = 1 cubic centimeter (cm^3)

Some Conversions Between Common and Metric Units

Length

1 mile (mi) = 1.61 kilometers (km)

1 yard (yd) = 0.914 meter (m)

1 inch (in.) = 2.54 centimeters (cm)

Mass

1 pound (lb) = 454 grams (g)

1 ounce (oz) = 28.4 grams (g)

1 pound (lb) = 0.454 kilogram (kg)

Volume

1 U.S. quart (qt) = 0.946 liter (L)

1 U.S. pint (pt) = 0.473 liter (L)

1 fluid ounce (fl oz) = 29.6 milliliters (mL)

1 gallon (gal) = 3.78 liters (L)

Some Conversion Units for Energy

1 calorie (cal) = 4.184 joules (J)

1 British thermal unit (Btu) = 1055 joules (J) = 252 calories (cal)

1 food Calorie = 1 kilocalorie (kcal) = 1000 calories (cal) = 4184 joules (J)

List of Recurring Themes in *The Chemistry of Everything*

1. **Recurring Theme in Chemistry** Chemical reactions are *nothing more than exchanges or rearrangements of electrons*, which is why we pay particular attention to electrons throughout this book. Chapter 1, p. 9; Chapter 2, pp. 54, 64; Chapter 3, p. 100; Chapter 4, p. 147; Chapter 5, pp. 166, 172; Chapter 6, p. 211.

2. **Recurring Theme in Chemistry** Chemical reactions do not create or destroy matter. Therefore, the types and numbers of atoms must be the same on the two sides of a chemical equation. Chapter 1, p. 13; Chapter 2, p. 66; Chapter 5, pp. 181, 190; Chapter 9, p. 360.

3. **Recurring Theme in Chemistry** The individual bonds between atoms dictate the properties and characteristics of the substances that contain them. Chapter 2, p. 46; Chapter 4, p. 128; Chapter 7, p. 259; Chapter 10, pp. 391, 399; Chapter 12, p. 470; Chapter 13, p. 513.

4. **Recurring Theme in Chemistry** Control experiments are included in every respectable scientific analysis. For example, chemists studying a molecule that is dissolved in a watery medium must run the same tests on the watery medium that they run on the molecule. Doing so establishes that the medium is not playing a role in the results obtained on the molecule of interest. Chapter 2, p. 75; Chapter 4, p. 157.

5. **Recurring Theme in Chemistry** The structure of carbon-containing substances is predictable because each carbon atom almost always makes four bonds to other atoms. Chapter 3, p. 84; Chapter 7, pp. 277, 285; Chapter 9, p. 337; Chapter 13, p. 553; Chapter 14, p. 553.

6. **Recurring Theme in Chemistry** Like dissolves like. Chapter 4, p. 134; Chapter 5, pp. 180, 199; Chapter 7, p. 281; Chapter 9, p. 355; Chapter 10, p. 399; Chapter 11, p. 441; Chapter 12, p. 467.

7. **Recurring Theme in Chemistry** Systems naturally seek out equilibrium, and so energy is required to move a system away from equilibrium. Chapter 4, p. 143; Chapter 8, p. 301.

8. **Recurring Theme in Chemistry** *Intra*molecular bonds, those *within* a molecule, are covalent. *Inter*molecular *interactions*, those either *between* two molecules or between two parts of a large molecule, are noncovalent. Chapter 7, p. 274; Chapter 8, p. 295; Chapter 9, p. 352; Chapter 11, pp. 440, 448.

9. **Recurring Theme in Chemistry** Making bonds releases energy; breaking them requires an input of energy. Chapter 9, p. 361.

10. **Recurring Theme in Chemistry** Many chemical reactions, especially organic reactions, are driven by the attraction between positive and negative parts of molecules. Chapter 10, p. 395; Chapter 11, p. 431.